T0181709

Lecture Notes in Computer Science   12266

Anne L. Martel · Purang Abolmaesumi ·
Danail Stoyanov · Diana Mateus ·
Maria A. Zuluaga · S. Kevin Zhou ·
Daniel Racoceanu · Leo Joskowicz (Eds.)

Medical Image Computing and Computer Assisted Intervention – MICCAI 2020

23rd International Conference
Lima, Peru, October 4–8, 2020
Proceedings, Part VI

 Springer

Editors
Anne L. Martel 🆔
University of Toronto
Toronto, ON, Canada

Purang Abolmaesumi 🆔
The University of British Columbia
Vancouver, BC, Canada

Danail Stoyanov 🆔
University College London
London, UK

Diana Mateus 🆔
École Centrale de Nantes
Nantes, France

Maria A. Zuluaga 🆔
EURECOM
Biot, France

S. Kevin Zhou 🆔
Chinese Academy of Sciences
Beijing, China

Daniel Racoceanu 🆔
Sorbonne University
Paris, France

Leo Joskowicz 🆔
The Hebrew University of Jerusalem
Jerusalem, Israel

ISSN 0302-9743   ISSN 1611-3349   (electronic)
Lecture Notes in Computer Science
ISBN 978-3-030-59724-5   ISBN 978-3-030-59725-2   (eBook)
https://doi.org/10.1007/978-3-030-59725-2

LNCS Sublibrary: SL6 – Image Processing, Computer Vision, Pattern Recognition, and Graphics

This Springer imprint is published by the registered company Springer Nature Switzerland AG
The registered company address is: Gewerbestrasse 11, 6330 Cham, Switzerland

Preface

The 23rd International Conference on Medical Image Computing and Computer-Assisted Intervention (MICCAI 2020) was held this year under the most unusual circumstances, due to the COVID-19 pandemic disrupting our lives in ways that were unimaginable at the start of the new decade. MICCAI 2020 was scheduled to be held in Lima, Peru, and would have been the first MICCAI meeting in Latin America. However, with the pandemic, the conference and its program had to be redesigned to deal with realities of the "new normal", where virtual presence rather than physical interactions among attendees, was necessary to comply with global transmission control measures. The conference was held through a virtual conference management platform, consisting of the main scientific program in addition to featuring 25 workshops, 8 tutorials, and 24 challenges during October 4–8, 2020. In order to keep a part of the original spirit of MICCAI 2020, SIPAIM 2020 was held as an adjacent LatAm conference dedicated to medical information management and imaging, held during October 3–4, 2020.

The proceedings of MICCAI 2020 showcase papers contributed by the authors to the main conference, which are organized in seven volumes of *Lecture Notes in Computer Science* (LNCS) books. These papers were selected after a thorough double-blind peer-review process. We followed the example set by past MICCAI meetings, using Microsoft's Conference Managing Toolkit (CMT) for paper submission and peer reviews, with support from the Toronto Paper Matching System (TPMS) to partially automate paper assignment to area chairs and reviewers.

The conference submission deadline had to be extended by two weeks to account for the disruption COVID-19 caused on the worldwide scientific community. From 2,953 original intentions to submit, 1,876 full submissions were received, which were reduced to 1,809 submissions following an initial quality check by the program chairs. Of those, 61% were self-declared by authors as Medical Image Computing (MIC), 6% as Computer Assisted Intervention (CAI), and 32% as both MIC and CAI. Following a broad call to the community for self-nomination of volunteers and a thorough review by the program chairs, considering criteria such as balance across research areas, geographical distribution, and gender, the MICCAI 2020 Program Committee comprised 82 area chairs, with 46% from North America, 28% from Europe, 19% from Asia/Pacific/Middle East, 4% from Latin America, and 1% from Australia. We invested significant effort in recruiting more women to the Program Committee, following the conference's emphasis on equity, inclusion, and diversity. This resulted in 26% female area chairs. Each area chair was assigned about 23 manuscripts, with suggested potential reviewers using TPMS scoring and self-declared research areas, while domain conflicts were automatically considered by CMT. Following a final revision and prioritization of reviewers by area chairs in terms of their expertise related to each paper, over 1,426 invited reviewers were asked to bid for the papers for which they had been suggested. Final reviewer allocations via CMT took account of reviewer bidding, prioritization of area chairs, and TPMS scores, leading to allocating about 4 papers per reviewer. Following an initial double-blind review phase by reviewers, area chairs provided a meta-review summarizing key points of reviews and a recommendation for each paper. The program chairs then evaluated the reviews and their scores, along with the recommendation from the area chairs, to directly accept 241 papers (13%) and reject 828 papers (46%); the remainder of the papers were sent for rebuttal by the authors. During the rebuttal phase, two additional area chairs were assigned to each paper using the CMT and TPMS scores while accounting for domain conflicts. The three area chairs then independently scored each paper to accept or reject, based on the reviews, rebuttal, and manuscript, resulting in clear paper decisions using majority voting. This process resulted in the acceptance of a further 301 papers for an overall acceptance rate of 30%. A virtual Program Committee meeting was held on July 10, 2020, to confirm the final results and collect feedback of the peer-review process.

For the MICCAI 2020 proceedings, 542 accepted papers have been organized into seven volumes as follows:

- Part I, LNCS Volume 12261: Machine Learning Methodologies
- Part II, LNCS Volume 12262: Image Reconstruction and Machine Learning
- Part III, LNCS Volume 12263: Computer Aided Intervention, Ultrasound and Image Registration
- Part IV, LNCS Volume 12264: Segmentation and Shape Analysis
- Part V, LNCS Volume 12265: Biological, Optical and Microscopic Image Analysis
- Part VI, LNCS Volume 12266: Clinical Applications
- Part VII, LNCS Volume 12267: Neurological Imaging and PET

For the main conference, the traditional emphasis on poster presentations was maintained; each author uploaded a brief pre-recorded presentation and a graphical abstract onto a web platform and was allocated a personal virtual live session in which they talked directly to the attendees. It was also possible to post questions online allowing asynchronous conversations – essential to overcome the challenges of a global conference spanning many time zones. The traditional oral sessions, which typically included a small proportion of the papers, were replaced with 90 "mini" sessions where all of the authors were clustered into groups of 5 to 7 related papers; a live virtual session allowed the authors and attendees to discuss the papers in a panel format.

We would like to sincerely thank everyone who contributed to the success of MICCAI 2020 and the quality of its proceedings under the most unusual circumstances of a global pandemic. First and foremost, we thank all authors for submitting and presenting their high-quality work that made MICCAI 2020 a greatly enjoyable and successful scientific meeting. We are also especially grateful to all members of the Program Committee and reviewers for their dedicated effort and insightful feedback throughout the entire paper selection process. We would like to particularly thank the MICCAI society for support, insightful comments, and continuous engagement with organizing the conference. Special thanks go to Kitty Wong, who oversaw the entire process of paper submission, reviews, and preparation of conference proceedings. Without her, we would have not functioned effectively. Given the "new normal", none of the workshops, tutorials, and challenges would have been feasible without the true leadership of the satellite events organizing team led by Mauricio Reyes: Erik Meijering (workshops), Carlos Alberola-López (tutorials), and Lena Maier-Hein (challenges). Behind the scenes, MICCAI secretarial personnel, Janette Wallace and Johanne Langford, kept a close eye on logistics and budgets, while Mehmet Eldegez and his team at Dekon Congress and Tourism led the professional conference organization, working tightly with the virtual platform team. We also thank our sponsors for financial support and engagement with conference attendees through the virtual platform. Special thanks goes to Veronika Cheplygina for continuous engagement with various social media platforms before and throughout the conference to publicize the conference. We would also like to express our gratitude to Shelley Wallace for helping us in Marketing MICCAI 2020, especially during the last phase of the virtual conference organization.

The selection process for Young Investigator Awards was managed by a team of senior MICCAI investigators, led by Julia Schnabel. In addition, MICCAI 2020 offered free registration to the top 50 ranked papers at the conference whose primary authors were students. Priority was given to low-income regions and Latin American students. Further support was provided by the National Institutes of Health (support granted for MICCAI 2020) and the National Science Foundation (support granted to MICCAI 2019 and continued for MICCAI 2020) which sponsored another 52 awards for USA-based students to attend the conference. We would like to thank Marius Linguraru and Antonion Porras, for their leadership in regards to the NIH sponsorship for 2020, and Dinggang Shen and Tianming Liu, MICCAI 2019 general chairs, for keeping an active bridge and engagement with MICCAI 2020.

Marius Linguraru and Antonion Porras were also leading the young investigators early career development program, including a very active mentorship which we do hope, will significantly catalize young and briliant careers of future leaders of our scientific community. In link with SIPAIM (thanks to Jorge Brieva, Marius Linguraru, and Natasha Lepore for their support), we also initiated a Startup Village initiative, which, we hope, will be able to bring in promising private initiatives in the areas of MICCAI. As a part of SIPAIM 2020, we note also the presence of a workshop for Peruvian clinicians. We would like to thank Benjaming Castañeda and Renato Gandolfi for this initiative.

MICCAI 2020 invested significant efforts to tightly engage the industry stakeholders in our field throughout its planning and organization. These efforts were led by Parvin Mousavi, and ensured that all sponsoring industry partners could connect with the conference attendees through the conference's virtual platform before and during the meeting. We would like to thank the sponsorship team and the contributions of Gustavo Carneiro, Benjamín Castañeda, Ignacio Larrabide, Marius Linguraru, Yanwu Xu, and Kevin Zhou.

We look forward to seeing you at MICCAI 2021.

October 2020

Anne L. Martel
Purang Abolmaesumi
Danail Stoyanov
Diana Mateus
Maria A. Zuluaga
S. Kevin Zhou
Daniel Racoceanu
Leo Joskowicz

Organization

General Chairs

Daniel Racoceanu — Sorbonne Université, Brain Institute, France
Leo Joskowicz — The Hebrew University of Jerusalem, Israel

Program Committee Chairs

Anne L. Martel — University of Toronto, Canada
Purang Abolmaesumi — The University of British Columbia, Canada
Danail Stoyanov — University College London, UK
Diana Mateus — Ecole Centrale de Nantes, LS2N, France
Maria A. Zuluaga — Eurecom, France
S. Kevin Zhou — Chinese Academy of Sciences, China

Keynote Speaker Chair

Rene Vidal — The John Hopkins University, USA

Satellite Events Chair

Mauricio Reyes — University of Bern, Switzerland

Workshop Team

Erik Meijering (Chair) — The University of New South Wales, Australia
Li Cheng — University of Alberta, Canada
Pamela Guevara — University of Concepción, Chile
Bennett Landman — Vanderbilt University, USA
Tammy Riklin Raviv — Ben-Gurion University of the Negev, Israel
Virginie Uhlmann — EMBL, European Bioinformatics Institute, UK

Tutorial Team

Carlos Alberola-López (Chair) — Universidad de Valladolid, Spain
Clarisa Sánchez — Radboud University Medical Center, The Netherlands
Demian Wassermann — Inria Saclay Île-de-France, France

Challenges Team

Lena Maier-Hein (Chair)	German Cancer Research Center, Germany
Annette Kopp-Schneider	German Cancer Research Center, Germany
Michal Kozubek	Masaryk University, Czech Republic
Annika Reinke	German Cancer Research Center, Germany

Sponsorship Team

Parvin Mousavi (Chair)	Queen's University, Canada
Marius Linguraru	Children's National Institute, USA
Gustavo Carneiro	The University of Adelaide, Australia
Yanwu Xu	Baidu Inc., China
Ignacio Larrabide	National Scientific and Technical Research Council, Argentina
S. Kevin Zhou	Chinese Academy of Sciences, China
Benjamín Castañeda	Pontifical Catholic University of Peru, Peru

Local and Regional Chairs

Benjamín Castañeda	Pontifical Catholic University of Peru, Peru
Natasha Lepore	University of Southern California, USA

Social Media Chair

Veronika Cheplygina	Eindhoven University of Technology, The Netherlands

Young Investigators Early Career Development Program Chairs

Marius Linguraru	Children's National Institute, USA
Antonio Porras	Children's National Institute, USA

Student Board Liaison Chair

Gabriel Jimenez	Pontifical Catholic University of Peru, Peru

Submission Platform Manager

Kitty Wong	The MICCAI Society, Canada

Conference Management

DEKON Group
Pathable Inc.

Program Committee

Ehsan Adeli	Stanford University, USA
Shadi Albarqouni	ETH Zurich, Switzerland
Pablo Arbelaez	Universidad de los Andes, Colombia
Ulas Bagci	University of Central Florida, USA
Adrien Bartoli	Université Clermont Auvergne, France
Hrvoje Bogunovic	Medical University of Vienna, Austria
Weidong Cai	The University of Sydney, Australia
Chao Chen	Stony Brook University, USA
Elvis Chen	Robarts Research Institute, Canada
Stanley Durrleman	Inria, France
Boris Escalante-Ramírez	National Autonomous University of Mexico, Mexico
Pascal Fallavollita	University of Ottawa, Canada
Enzo Ferrante	CONICET, Universidad Nacional del Litoral, Argentina
Stamatia Giannarou	Imperial College London, UK
Orcun Goksel	ETH Zurich, Switzerland
Alberto Gomez	King's College London, UK
Miguel Angel González Ballester	Universitat Pompeu Fabra, Spain
Ilker Hacihaliloglu	Rutgers University, USA
Yi Hong	University of Georgia, USA
Yipeng Hu	University College London, UK
Heng Huang	University of Pittsburgh and JD Finance America Corporation, USA
Juan Eugenio Iglesias	University College London, UK
Madhura Ingalhalikar	Symbiosis Center for Medical Image Analysis, India
Pierre Jannin	Université de Rennes, France
Samuel Kadoury	Ecole Polytechnique de Montreal, Canada
Bernhard Kainz	Imperial College London, UK
Marta Kersten-Oertel	Concordia University, Canada
Andrew King	King's College London, UK
Ignacio Larrabide	CONICET, Argentina
Gang Li	University of North Carolina at Chapel Hill, USA
Jianming Liang	Arizona State University, USA
Hongen Liao	Tsinghua University, China
Rui Liao	Siemens Healthineers, USA
Feng Lin	Nanyang Technological University, China
Mingxia Liu	University of North Carolina at Chapel Hill, USA
Jiebo Luo	University of Rochester, USA
Xiongbiao Luo	Xiamen University, China
Andreas Maier	FAU Erlangen-Nuremberg, Germany
Stephen McKenna	University of Dundee, UK
Bjoern Menze	Technische Universität München, Germany
Mehdi Moradi	IBM Research, USA

Dong Ni	Shenzhen University, China
Marc Niethammer	University of North Carolina at Chapel Hill, USA
Jack Noble	Vanderbilt University, USA
Ipek Oguz	Vanderbilt University, USA
Gemma Piella	Pompeu Fabra University, Spain
Hedyeh Rafii-Tari	Auris Health Inc., USA
Islem Rekik	Istanbul Technical University, Turkey
Nicola Rieke	NVIDIA Corporation, USA
Tammy Riklin Raviv	Ben-Gurion University of the Negev, Israel
Hassan Rivaz	Concordia University, Canada
Holger Roth	NVIDIA Corporation, USA
Sharmishtaa Seshamani	Allen Institute, USA
Li Shen	University of Pennsylvania, USA
Feng Shi	Shanghai United Imaging Intelligence Co., China
Yonggang Shi	University of Southern California, USA
Michal Sofka	Hyperfine Research, USA
Stefanie Speidel	National Center for Tumor Diseases (NCT), Germany
Marius Staring	Leiden University Medical Center, The Netherlands
Heung-Il Suk	Korea University, South Korea
Kenji Suzuki	Tokyo Institute of Technology, Japan
Tanveer Syeda-Mahmood	IBM Research, USA
Amir Tahmasebi	CodaMetrix, USA
Xiaoying Tang	Southern University of Science and Technology, China
Tolga Tasdizen	The University of Utah, USA
Pallavi Tiwari	Case Western Reserve University, USA
Sotirios Tsaftaris	The University of Edinburgh, UK
Archana Venkataraman	Johns Hopkins University, USA
Satish Viswanath	Case Western Reserve University, USA
Hongzhi Wang	IBM Almaden Research Center, USA
Linwei Wang	Rochester Institute of Technology, USA
Qian Wang	Shanghai Jiao Tong University, China
Guorong Wu	University of North Carolina at Chapel Hill, USA
Daguang Xu	NVIDIA Corporation, USA
Ziyue Xu	NVIDIA Corporation, USA
Pingkun Yan	Rensselaer Polytechnic Institute, USA
Xin Yang	Huazhong University of Science and Technology, China
Zhaozheng Yin	Stony Brook University, USA
Tuo Zhang	Northwestern Polytechnical University, China
Guoyan Zheng	Shanghai Jiao Tong University, China
Yefeng Zheng	Tencent, China
Luping Zhou	The University of Sydney, Australia

Mentorship Program (Mentors)

Ehsan Adeli	Stanford University, USA
Stephen Aylward	Kitware, USA
Hrvoje Bogunovic	Medical University of Vienna, Austria
Li Cheng	University of Alberta, Canada
Marleen de Bruijne	University of Copenhagen, Denmark
Caroline Essert	University of Strasbourg, France
Gabor Fichtinger	Queen's University, Canada
Stamatia Giannarou	Imperial College London, UK
Juan Eugenio Iglesias Gonzalez	University College London, UK
Bernhard Kainz	Imperial College London, UK
Shuo Li	Western University, Canada
Jianming Liang	Arizona State University, USA
Rui Liao	Siemens Healthineers, USA
Feng Lin	Nanyang Technological University, China
Marius George Linguraru	Children's National Hospital, George Washington University, USA
Tianming Liu	University of Georgia, USA
Xiongbiao Luo	Xiamen University, China
Dong Ni	Shenzhen University, China
Wiro Niessen	Erasmus MC - University Medical Center Rotterdam, The Netherlands
Terry Peters	Western University, Canada
Antonio R. Porras	University of Colorado, USA
Daniel Racoceanu	Sorbonne University, France
Islem Rekik	Istanbul Technical University, Turkey
Nicola Rieke	NVIDIA, USA
Julia Schnabel	King's College London, UK
Ruby Shamir	Novocure, Switzerland
Stefanie Speidel	National Center for Tumor Diseases Dresden, Germany
Martin Styner	University of North Carolina at Chapel Hill, USA
Xiaoying Tang	Southern University of Science and Technology, China
Pallavi Tiwari	Case Western Reserve University, USA
Jocelyne Troccaz	CNRS, Grenoble Alpes University, France
Pierre Jannin	INSERM, Université de Rennes, France
Archana Venkataraman	Johns Hopkins University, USA
Linwei Wang	Rochester Institute of Technology, USA
Guorong Wu	University of North Carolina at Chapel Hill, USA
Li Xiao	Chinese Academy of Science, China
Ziyue Xu	NVIDIA, USA
Bochuan Zheng	China West Normal University, China
Guoyan Zheng	Shanghai Jiao Tong University, China
S. Kevin Zhou	Chinese Academy of Sciences, China
Maria A. Zuluaga	EURECOM, France

Additional Reviewers

Alaa Eldin Abdelaal
Ahmed Abdulkadir
Clement Abi Nader
Mazdak Abulnaga
Ganesh Adluru
Iman Aganj
Priya Aggarwal
Sahar Ahmad
Seyed-Ahmad Ahmadi
Euijoon Ahn
Alireza Akhondi-asl
Mohamed Akrout
Dawood Al Chanti
Ibraheem Al-Dhamari
Navid Alemi Koohbanani
Hanan Alghamdi
Hassan Alhajj
Hazrat Ali
Sharib Ali
Omar Al-Kadi
Maximilian Allan
Felix Ambellan
Mina Amiri
Sameer Antani
Luigi Antelmi
Michela Antonelli
Jacob Antunes
Saeed Anwar
Fernando Arambula
Ignacio Arganda-Carreras
Mohammad Ali Armin
John Ashburner
Md Ashikuzzaman
Shahab Aslani
Mehdi Astaraki
Angélica Atehortúa
Gowtham Atluri
Kamran Avanaki
Angelica Aviles-Rivero
Suyash Awate
Dogu Baran Aydogan
Qinle Ba
Morteza Babaie

Hyeon-Min Bae
Woong Bae
Wenjia Bai
Ujjwal Baid
Spyridon Bakas
Yaël Balbastre
Marcin Balicki
Fabian Balsiger
Abhirup Banerjee
Sreya Banerjee
Sophia Bano
Shunxing Bao
Adrian Barbu
Cher Bass
John S. H. Baxter
Amirhossein Bayat
Sharareh Bayat
Neslihan Bayramoglu
Bahareh Behboodi
Delaram Behnami
Mikhail Belyaev
Oualid Benkarim
Aicha BenTaieb
Camilo Bermudez
Giulia Bertò
Hadrien Bertrand
Julián Betancur
Michael Beyeler
Parmeet Bhatia
Chetan Bhole
Suvrat Bhooshan
Chitresh Bhushan
Lei Bi
Cheng Bian
Gui-Bin Bian
Sangeeta Biswas
Stefano B. Blumberg
Janusz Bobulski
Sebastian Bodenstedt
Ester Bonmati
Bhushan Borotikar
Jiri Borovec
Ilaria Boscolo Galazzo

Alexandre Bousse
Nicolas Boutry
Behzad Bozorgtabar
Nadia Brancati
Christopher Bridge
Esther Bron
Rupert Brooks
Qirong Bu
Tim-Oliver Buchholz
Duc Toan Bui
Qasim Bukhari
Ninon Burgos
Nikolay Burlutskiy
Russell Butler
Michał Byra
Hongmin Cai
Yunliang Cai
Sema Candemir
Bing Cao
Qing Cao
Shilei Cao
Tian Cao
Weiguo Cao
Yankun Cao
Aaron Carass
Heike Carolus
Adrià Casamitjana
Suheyla Cetin Karayumak
Ahmad Chaddad
Krishna Chaitanya
Jayasree Chakraborty
Tapabrata Chakraborty
Sylvie Chambon
Ming-Ching Chang
Violeta Chang
Simon Chatelin
Sudhanya Chatterjee
Christos Chatzichristos
Rizwan Chaudhry
Antong Chen
Cameron Po-Hsuan Chen
Chang Chen
Chao Chen
Chen Chen
Cheng Chen
Dongdong Chen

Fang Chen
Geng Chen
Hao Chen
Jianan Chen
Jianxu Chen
Jia-Wei Chen
Jie Chen
Junxiang Chen
Li Chen
Liang Chen
Pingjun Chen
Qiang Chen
Shuai Chen
Tianhua Chen
Tingting Chen
Xi Chen
Xiaoran Chen
Xin Chen
Yuanyuan Chen
Yuhua Chen
Yukun Chen
Zhineng Chen
Zhixiang Chen
Erkang Cheng
Jun Cheng
Li Cheng
Xuelian Cheng
Yuan Cheng
Veronika Cheplygina
Hyungjoo Cho
Jaegul Choo
Aritra Chowdhury
Stergios Christodoulidis
Ai Wern Chung
Pietro Antonio Cicalese
Özgün Çiçek
Robert Cierniak
Matthew Clarkson
Dana Cobzas
Jaume Coll-Font
Alessia Colonna
Marc Combalia
Olivier Commowick
Sonia Contreras Ortiz
Pierre-Henri Conze
Timothy Cootes

Luca Corinzia
Teresa Correia
Pierrick Coupé
Jeffrey Craley
Arun C. S. Kumar
Hui Cui
Jianan Cui
Zhiming Cui
Kathleen Curran
Haixing Dai
Xiaoliang Dai
Ker Dai Fei Elmer
Adrian Dalca
Abhijit Das
Neda Davoudi
Laura Daza
Sandro De Zanet
Charles Delahunt
Herve Delingette
Beatrice Demiray
Yang Deng
Hrishikesh Deshpande
Christian Desrosiers
Neel Dey
Xinghao Ding
Zhipeng Ding
Konstantin Dmitriev
Jose Dolz
Ines Domingues
Juan Pedro Dominguez-Morales
Hao Dong
Mengjin Dong
Nanqing Dong
Qinglin Dong
Suyu Dong
Sven Dorkenwald
Qi Dou
P. K. Douglas
Simon Drouin
Karen Drukker
Niharika D'Souza
Lei Du
Shaoyi Du
Xuefeng Du
Dingna Duan
Nicolas Duchateau

James Duncan
Jared Dunnmon
Luc Duong
Nicha Dvornek
Dmitry V. Dylov
Oleh Dzyubachyk
Mehran Ebrahimi
Philip Edwards
Alexander Effland
Jan Egger
Alma Eguizabal
Gudmundur Einarsson
Ahmed Elazab
Mohammed S. M. Elbaz
Shireen Elhabian
Ahmed Eltanboly
Sandy Engelhardt
Ertunc Erdil
Marius Erdt
Floris Ernst
Mohammad Eslami
Nazila Esmaeili
Marco Esposito
Oscar Esteban
Jingfan Fan
Xin Fan
Yonghui Fan
Chaowei Fang
Xi Fang
Mohsen Farzi
Johannes Fauser
Andrey Fedorov
Hamid Fehri
Lina Felsner
Jun Feng
Ruibin Feng
Xinyang Feng
Yifan Feng
Yuan Feng
Henrique Fernandes
Ricardo Ferrari
Jean Feydy
Lucas Fidon
Lukas Fischer
Antonio Foncubierta-Rodríguez
Germain Forestier

Reza Forghani
Nils Daniel Forkert
Jean-Rassaire Fouefack
Tatiana Fountoukidou
Aina Frau-Pascual
Moti Freiman
Sarah Frisken
Huazhu Fu
Xueyang Fu
Wolfgang Fuhl
Isabel Funke
Philipp Fürnstahl
Pedro Furtado
Ryo Furukawa
Elies Fuster-Garcia
Youssef Gahi
Jin Kyu Gahm
Laurent Gajny
Rohan Gala
Harshala Gammulle
Yu Gan
Cong Gao
Dongxu Gao
Fei Gao
Feng Gao
Linlin Gao
Mingchen Gao
Siyuan Gao
Xin Gao
Xinpei Gao
Yixin Gao
Yue Gao
Zhifan Gao
Sara Garbarino
Alfonso Gastelum-Strozzi
Romane Gauriau
Srishti Gautam
Bao Ge
Rongjun Ge
Zongyuan Ge
Sairam Geethanath
Yasmeen George
Samuel Gerber
Guido Gerig
Nils Gessert
Olivier Gevaert

Muhammad Usman Ghani
Sandesh Ghimire
Sayan Ghosal
Gabriel Girard
Ben Glocker
Evgin Goceri
Michael Goetz
Arnold Gomez
Kuang Gong
Mingming Gong
Yuanhao Gong
German Gonzalez
Sharath Gopal
Karthik Gopinath
Pietro Gori
Maged Goubran
Sobhan Goudarzi
Baran Gözcü
Benedikt Graf
Mark Graham
Bertrand Granado
Alejandro Granados
Robert Grupp
Christina Gsaxner
Lin Gu
Shi Gu
Yun Gu
Ricardo Guerrero
Houssem-Eddine Gueziri
Dazhou Guo
Hengtao Guo
Jixiang Guo
Pengfei Guo
Yanrong Guo
Yi Guo
Yong Guo
Yulan Guo
Yuyu Guo
Krati Gupta
Vikash Gupta
Praveen Gurunath Bharathi
Prashnna Gyawali
Stathis Hadjidemetriou
Omid Haji Maghsoudi
Justin Haldar
Mohammad Hamghalam

Bing Han
Hu Han
Liang Han
Xiaoguang Han
Xu Han
Zhi Han
Zhongyi Han
Jonny Hancox
Christian Hansen
Xiaoke Hao
Rabia Haq
Michael Hardisty
Stefan Harrer
Adam Harrison
S. M. Kamrul Hasan
Hoda Sadat Hashemi
Nobuhiko Hata
Andreas Hauptmann
Mohammad Havaei
Huiguang He
Junjun He
Kelei He
Tiancheng He
Xuming He
Yuting He
Mattias Heinrich
Stefan Heldmann
Nicholas Heller
Alessa Hering
Monica Hernandez
Estefania Hernandez-Martin
Carlos Hernandez-Matas
Javier Herrera-Vega
Kilian Hett
Tsung-Ying Ho
Nico Hoffmann
Matthew Holden
Song Hong
Sungmin Hong
Yoonmi Hong
Corné Hoogendoorn
Antal Horváth
Belayat Hossain
Le Hou
Ai-Ling Hsu
Po-Ya Hsu

Tai-Chiu Hsung
Pengwei Hu
Shunbo Hu
Xiaoling Hu
Xiaowei Hu
Yan Hu
Zhenhong Hu
Jia-Hong Huang
Junzhou Huang
Kevin Huang
Qiaoying Huang
Weilin Huang
Xiaolei Huang
Yawen Huang
Yongxiang Huang
Yue Huang
Yufang Huang
Zhi Huang
Arnaud Huaulmé
Henkjan Huisman
Xing Huo
Yuankai Huo
Sarfaraz Hussein
Jana Hutter
Khoi Huynh
Seong Jae Hwang
Emmanuel Iarussi
Ilknur Icke
Kay Igwe
Alfredo Illanes
Abdullah-Al-Zubaer Imran
Ismail Irmakci
Samra Irshad
Benjamin Irving
Mobarakol Islam
Mohammad Shafkat Islam
Vamsi Ithapu
Koichi Ito
Hayato Itoh
Oleksandra Ivashchenko
Yuji Iwahori
Shruti Jadon
Mohammad Jafari
Mostafa Jahanifar
Andras Jakab
Amir Jamaludin

Won-Dong Jang
Vincent Jaouen
Uditha Jarayathne
Ronnachai Jaroensri
Golara Javadi
Rohit Jena
Todd Jensen
Won-Ki Jeong
Zexuan Ji
Haozhe Jia
Jue Jiang
Tingting Jiang
Weixiong Jiang
Xi Jiang
Xiang Jiang
Jianbo Jiao
Zhicheng Jiao
Amelia Jiménez-Sánchez
Dakai Jin
Taisong Jin
Yueming Jin
Ze Jin
Bin Jing
Yaqub Jonmohamadi
Anand Joshi
Shantanu Joshi
Christoph Jud
Florian Jug
Yohan Jun
Alain Jungo
Abdolrahim Kadkhodamohammadi
Ali Kafaei Zad Tehrani
Dagmar Kainmueller
Siva Teja Kakileti
John Kalafut
Konstantinos Kamnitsas
Michael C. Kampffmeyer
Qingbo Kang
Neerav Karani
Davood Karimi
Satyananda Kashyap
Alexander Katzmann
Prabhjot Kaur
Anees Kazi
Erwan Kerrien
Hoel Kervadec

Ashkan Khakzar
Fahmi Khalifa
Nadieh Khalili
Siavash Khallaghi
Farzad Khalvati
Hassan Khan
Bishesh Khanal
Pulkit Khandelwal
Maksym Kholiavchenko
Meenakshi Khosla
Naji Khosravan
Seyed Mostafa Kia
Ron Kikinis
Daeseung Kim
Geena Kim
Hak Gu Kim
Heejong Kim
Hosung Kim
Hyo-Eun Kim
Jinman Kim
Jinyoung Kim
Mansu Kim
Minjeong Kim
Seong Tae Kim
Won Hwa Kim
Young-Ho Kim
Atilla Kiraly
Yoshiro Kitamura
Takayuki Kitasaka
Sabrina Kletz
Tobias Klinder
Kranthi Kolli
Satoshi Kondo
Bin Kong
Jun Kong
Tomasz Konopczynski
Ender Konukoglu
Bongjin Koo
Kivanc Kose
Anna Kreshuk
AnithaPriya Krishnan
Pavitra Krishnaswamy
Frithjof Kruggel
Alexander Krull
Elizabeth Krupinski
Hulin Kuang

Serife Kucur
David Kügler
Arjan Kuijper
Jan Kukacka
Nilima Kulkarni
Abhay Kumar
Ashnil Kumar
Kuldeep Kumar
Neeraj Kumar
Nitin Kumar
Manuela Kunz
Holger Kunze
Tahsin Kurc
Thomas Kurmann
Yoshihiro Kuroda
Jin Tae Kwak
Yongchan Kwon
Aymen Laadhari
Dmitrii Lachinov
Alexander Ladikos
Alain Lalande
Rodney Lalonde
Tryphon Lambrou
Hengrong Lan
Catherine Laporte
Carole Lartizien
Bianca Lassen-Schmidt
Andras Lasso
Ngan Le
Leo Lebrat
Changhwan Lee
Eung-Joo Lee
Hyekyoung Lee
Jong-Hwan Lee
Jungbeom Lee
Matthew Lee
Sangmin Lee
Soochahn Lee
Stefan Leger
Étienne Léger
Baiying Lei
Andreas Leibetseder
Rogers Jeffrey Leo John
Juan Leon
Wee Kheng Leow
Annan Li

Bo Li
Chongyi Li
Haohan Li
Hongming Li
Hongwei Li
Huiqi Li
Jian Li
Jianning Li
Jiayun Li
Junhua Li
Lincan Li
Mengzhang Li
Ming Li
Qing Li
Quanzheng Li
Shulong Li
Shuyu Li
Weikai Li
Wenyuan Li
Xiang Li
Xiaomeng Li
Xiaoxiao Li
Xin Li
Xiuli Li
Yang Li (Beihang University)
Yang Li (Northeast Electric Power
    University)
Yi Li
Yuexiang Li
Zeju Li
Zhang Li
Zhen Li
Zhiyuan Li
Zhjin Li
Zhongyu Li
Chunfeng Lian
Gongbo Liang
Libin Liang
Shanshan Liang
Yudong Liang
Haofu Liao
Ruizhi Liao
Gilbert Lim
Baihan Lin
Hongxiang Lin
Huei-Yung Lin

Jianyu Lin
C. Lindner
Geert Litjens
Bin Liu
Chang Liu
Dongnan Liu
Feng Liu
Hangfan Liu
Jianfei Liu
Jin Liu
Jingya Liu
Jingyu Liu
Kai Liu
Kefei Liu
Lihao Liu
Luyan Liu
Mengting Liu
Na Liu
Peng Liu
Ping Liu
Quande Liu
Qun Liu
Shengfeng Liu
Shuangjun Liu
Sidong Liu
Siqi Liu
Siyuan Liu
Tianrui Liu
Xianglong Liu
Xinyang Liu
Yan Liu
Yuan Liu
Yuhang Liu
Andrea Loddo
Herve Lombaert
Marco Lorenzi
Jian Lou
Nicolas Loy Rodas
Allen Lu
Donghuan Lu
Huanxiang Lu
Jiwen Lu
Le Lu
Weijia Lu
Xiankai Lu
Yao Lu

Yongyi Lu
Yueh-Hsun Lu
Christian Lucas
Oeslle Lucena
Imanol Luengo
Ronald Lui
Gongning Luo
Jie Luo
Ma Luo
Marcel Luthi
Khoa Luu
Bin Lv
Jinglei Lv
Ilwoo Lyu
Qing Lyu
Sharath M. S.
Andy J. Ma
Chunwei Ma
Da Ma
Hua Ma
Jingting Ma
Kai Ma
Lei Ma
Wenao Ma
Yuexin Ma
Amirreza Mahbod
Sara Mahdavi
Mohammed Mahmoud
Gabriel Maicas
Klaus H. Maier-Hein
Sokratis Makrogiannis
Bilal Malik
Anand Malpani
Ilja Manakov
Matteo Mancini
Efthymios Maneas
Tommaso Mansi
Brett Marinelli
Razvan Marinescu
Pablo Márquez Neila
Carsten Marr
Yassine Marrakchi
Fabio Martinez
Antonio Martinez-Torteya
Andre Mastmeyer
Dimitrios Mavroeidis

Jamie McClelland
Verónica Medina Bañuelos
Raghav Mehta
Sachin Mehta
Liye Mei
Raphael Meier
Qier Meng
Qingjie Meng
Yu Meng
Martin Menten
Odyssée Merveille
Pablo Mesejo
Liang Mi
Shun Miao
Stijn Michielse
Mikhail Milchenko
Hyun-Seok Min
Zhe Min
Tadashi Miyamoto
Aryan Mobiny
Irina Mocanu
Sara Moccia
Omid Mohareri
Hassan Mohy-ud-Din
Muthu Rama Krishnan Mookiah
Rodrigo Moreno
Lia Morra
Agata Mosinska
Saman Motamed
Mohammad Hamed Mozaffari
Anirban Mukhopadhyay
Henning Müller
Balamurali Murugesan
Cosmas Mwikirize
Andriy Myronenko
Saad Nadeem
Ahmed Naglah
Vivek Natarajan
Vishwesh Nath
Rodrigo Nava
Fernando Navarro
Lydia Neary-Zajiczek
Peter Neher
Dominik Neumann
Gia Ngo
Hannes Nickisch

Dong Nie
Jingxin Nie
Weizhi Nie
Aditya Nigam
Xia Ning
Zhenyuan Ning
Sijie Niu
Tianye Niu
Alexey Novikov
Jorge Novo
Chinedu Nwoye
Mohammad Obeid
Masahiro Oda
Thomas O'Donnell
Benjamin Odry
Steffen Oeltze-Jafra
Ayşe Oktay
Hugo Oliveira
Marcelo Oliveira
Sara Oliveira
Arnau Oliver
Sahin Olut
Jimena Olveres
John Onofrey
Eliza Orasanu
Felipe Orihuela-Espina
José Orlando
Marcos Ortega
Sarah Ostadabbas
Yoshito Otake
Sebastian Otalora
Cheng Ouyang
Jiahong Ouyang
Cristina Oyarzun Laura
Michal Ozery-Flato
Krittin Pachtrachai
Johannes Paetzold
Jin Pan
Yongsheng Pan
Prashant Pandey
Joao Papa
Giorgos Papanastasiou
Constantin Pape
Nripesh Parajuli
Hyunjin Park
Sanghyun Park

Seyoun Park
Angshuman Paul
Christian Payer
Chengtao Peng
Jialin Peng
Liying Peng
Tingying Peng
Yifan Peng
Tobias Penzkofer
Antonio Pepe
Oscar Perdomo
Jose-Antonio Pérez-Carrasco
Fernando Pérez-García
Jorge Perez-Gonzalez
Skand Peri
Loic Peter
Jorg Peters
Jens Petersen
Caroline Petitjean
Micha Pfeiffer
Dzung Pham
Renzo Phellan
Ashish Phophalia
Mark Pickering
Kilian Pohl
Iulia Popescu
Karteek Popuri
Tiziano Portenier
Alison Pouch
Arash Pourtaherian
Prateek Prasanna
Alexander Preuhs
Raphael Prevost
Juan Prieto
Viswanath P. S.
Sergi Pujades
Kumaradevan Punithakumar
Elodie Puybareau
Haikun Qi
Huan Qi
Xin Qi
Buyue Qian
Zhen Qian
Yan Qiang
Yuchuan Qiao
Zhi Qiao

Chen Qin
Wenjian Qin
Yanguo Qin
Wu Qiu
Hui Qu
Kha Gia Quach
Prashanth R.
Pradeep Reddy Raamana
Jagath Rajapakse
Kashif Rajpoot
Jhonata Ramos
Andrik Rampun
Parnesh Raniga
Nagulan Ratnarajah
Richard Rau
Mehul Raval
Keerthi Sravan Ravi
Daniele Ravi
Harish RaviPrakash
Rohith Reddy
Markus Rempfler
Xuhua Ren
Yinhao Ren
Yudan Ren
Anne-Marie Rickmann
Brandalyn Riedel
Leticia Rittner
Robert Robinson
Jessica Rodgers
Robert Rohling
Lukasz Roszkowiak
Karsten Roth
José Rouco
Su Ruan
Daniel Rueckert
Mirabela Rusu
Erica Rutter
Jaime S. Cardoso
Mohammad Sabokrou
Monjoy Saha
Pramit Saha
Dushyant Sahoo
Pranjal Sahu
Wojciech Samek
Juan A. Sánchez-Margallo
Robin Sandkuehler

Rodrigo Santa Cruz
Gianmarco Santini
Anil Kumar Sao
Mhd Hasan Sarhan
Duygu Sarikaya
Imari Sato
Olivier Saut
Mattia Savardi
Ramasamy Savitha
Fabien Scalzo
Nico Scherf
Alexander Schlaefer
Philipp Schleer
Leopold Schmetterer
Julia Schnabel
Klaus Schoeffmann
Peter Schueffler
Andreas Schuh
Thomas Schultz
Michael Schwier
Michael Sdika
Suman Sedai
Raghavendra Selvan
Sourya Sengupta
Youngho Seo
Lama Seoud
Ana Sequeira
Saeed Seyyedi
Giorgos Sfikas
Sobhan Shafiei
Reuben Shamir
Shayan Shams
Hongming Shan
Yeqin Shao
Harshita Sharma
Gregory Sharp
Mohamed Shehata
Haocheng Shen
Mali Shen
Yiqiu Shen
Zhengyang Shen
Luyao Shi
Xiaoshuang Shi
Yemin Shi
Yonghong Shi
Saurabh Shigwan

Hoo-Chang Shin
Suprosanna Shit
Yucheng Shu
Nadya Shusharina
Alberto Signoroni
Carlos A. Silva
Wilson Silva
Praveer Singh
Ramandeep Singh
Rohit Singla
Sumedha Singla
Ayushi Sinha
Rajath Soans
Hessam Sokooti
Jaemin Son
Ming Song
Tianyu Song
Yang Song
Youyi Song
Aristeidis Sotiras
Arcot Sowmya
Rachel Sparks
Bella Specktor
William Speier
Ziga Spiclin
Dominik Spinczyk
Chetan Srinidhi
Vinkle Srivastav
Lawrence Staib
Peter Steinbach
Darko Stern
Joshua Stough
Justin Strait
Robin Strand
Martin Styner
Hai Su
Pan Su
Yun-Hsuan Su
Vaishnavi Subramanian
Gérard Subsol
Carole Sudre
Yao Sui
Avan Suinesiaputra
Jeremias Sulam
Shipra Suman
Jian Sun

Liang Sun
Tao Sun
Kyung Sung
Chiranjib Sur
Yannick Suter
Raphael Sznitman
Solale Tabarestani
Fatemeh Taheri Dezaki
Roger Tam
José Tamez-Peña
Chaowei Tan
Jiaxing Tan
Hao Tang
Sheng Tang
Thomas Tang
Xiongfeng Tang
Zhenyu Tang
Mickael Tardy
Eu Wern Teh
Antonio Tejero-de-Pablos
Paul Thienphrapa
Stephen Thompson
Felix Thomsen
Jiang Tian
Yun Tian
Aleksei Tiulpin
Hamid Tizhoosh
Matthew Toews
Oguzhan Topsakal
Jordina Torrents
Sylvie Treuillet
Jocelyne Troccaz
Emanuele Trucco
Vinh Truong Hoang
Chialing Tsai
Andru Putra Twinanda
Norimichi Ukita
Eranga Ukwatta
Mathias Unberath
Tamas Ungi
Martin Urschler
Verena Uslar
Fatmatulzehra Uslu
Régis Vaillant
Jeya Maria Jose Valanarasu
Marta Vallejo

Fons van der Sommen
Gijs van Tulder
Kimberlin van Wijnen
Yogatheesan Varatharajah
Marta Varela
Thomas Varsavsky
Francisco Vasconcelos
S. Swaroop Vedula
Sanketh Vedula
Harini Veeraraghavan
Gonzalo Vegas Sanchez-Ferrero
Anant Vemuri
Gopalkrishna Veni
Ruchika Verma
Ujjwal Verma
Pedro Vieira
Juan Pedro Vigueras Guillen
Pierre-Frederic Villard
Athanasios Vlontzos
Wolf-Dieter Vogl
Ingmar Voigt
Eugene Vorontsov
Bo Wang
Cheng Wang
Chengjia Wang
Chunliang Wang
Dadong Wang
Guotai Wang
Haifeng Wang
Hongkai Wang
Hongyu Wang
Hua Wang
Huan Wang
Jun Wang
Kuanquan Wang
Kun Wang
Lei Wang
Li Wang
Liansheng Wang
Manning Wang
Ruixuan Wang
Shanshan Wang
Shujun Wang
Shuo Wang
Tianchen Wang
Tongxin Wang

Wenzhe Wang
Xi Wang
Xiangxue Wang
Yalin Wang
Yan Wang (Sichuan University)
Yan Wang (Johns Hopkins University)
Yaping Wang
Yi Wang
Yirui Wang
Yuanjun Wang
Yun Wang
Zeyi Wang
Zhangyang Wang
Simon Warfield
Jonathan Weber
Jürgen Weese
Donglai Wei
Dongming Wei
Zhen Wei
Martin Weigert
Michael Wels
Junhao Wen
Matthias Wilms
Stefan Winzeck
Adam Wittek
Marek Wodzinski
Jelmer Wolterink
Ken C. L. Wong
Jonghye Woo
Chongruo Wu
Dijia Wu
Ji Wu
Jian Wu (Tsinghua University)
Jian Wu (Zhejiang University)
Jie Ying Wu
Junyan Wu
Minjie Wu
Pengxiang Wu
Xi Wu
Xia Wu
Xiyin Wu
Ye Wu
Yicheng Wu
Yifan Wu
Zhengwang Wu
Tobias Wuerfl

Pengcheng Xi
James Xia
Siyu Xia
Yingda Xia
Yong Xia
Lei Xiang
Deqiang Xiao
Li Xiao (Tulane University)
Li Xiao (Chinese Academy of Science)
Yuting Xiao
Hongtao Xie
Jianyang Xie
Lingxi Xie
Long Xie
Xueqian Xie
Yiting Xie
Yuan Xie
Yutong Xie
Fangxu Xing
Fuyong Xing
Tao Xiong
Chenchu Xu
Hongming Xu
Jiaofeng Xu
Kele Xu
Lisheng Xu
Min Xu
Rui Xu
Xiaowei Xu
Yanwu Xu
Yongchao Xu
Zhenghua Xu
Cheng Xue
Jie Xue
Wufeng Xue
Yuan Xue
Faridah Yahya
Chenggang Yan
Ke Yan
Weizheng Yan
Yu Yan
Yuguang Yan
Zhennan Yan
Changchun Yang
Chao-Han Huck Yang
Dong Yang

Fan Yang (IIAI)
Fan Yang (Temple University)
Feng Yang
Ge Yang
Guang Yang
Heran Yang
Hongxu Yang
Huijuan Yang
Jiancheng Yang
Jie Yang
Junlin Yang
Lin Yang
Xiao Yang
Xiaohui Yang
Xin Yang
Yan Yang
Yujiu Yang
Dongren Yao
Jianhua Yao
Jiawen Yao
Li Yao
Chuyang Ye
Huihui Ye
Menglong Ye
Xujiong Ye
Andy W. K. Yeung
Jingru Yi
Jirong Yi
Xin Yi
Yi Yin
Shihui Ying
Youngjin Yoo
Chenyu You
Sahar Yousefi
Hanchao Yu
Jinhua Yu
Kai Yu
Lequan Yu
Qi Yu
Yang Yu
Zhen Yu
Pengyu Yuan
Yixuan Yuan
Paul Yushkevich
Ghada Zamzmi
Dong Zeng

Guodong Zeng
Oliver Zettinig
Zhiwei Zhai
Kun Zhan
Baochang Zhang
Chaoyi Zhang
Daoqiang Zhang
Dongqing Zhang
Fan Zhang (Yale University)
Fan Zhang (Harvard Medical School)
Guangming Zhang
Han Zhang
Hang Zhang
Haopeng Zhang
Heye Zhang
Huahong Zhang
Jianpeng Zhang
Jinao Zhang
Jingqing Zhang
Jinwei Zhang
Jiong Zhang
Jun Zhang
Le Zhang
Lei Zhang
Lichi Zhang
Lin Zhang
Ling Zhang
Lu Zhang
Miaomiao Zhang
Ning Zhang
Pengfei Zhang
Pengyue Zhang
Qiang Zhang
Rongzhao Zhang
Ru-Yuan Zhang
Shanzhuo Zhang
Shu Zhang
Tong Zhang
Wei Zhang
Weiwei Zhang
Wenlu Zhang
Xiaoyun Zhang
Xin Zhang
Ya Zhang
Yanbo Zhang
Yanfu Zhang

Yi Zhang
Yifan Zhang
Yizhe Zhang
Yongqin Zhang
You Zhang
Youshan Zhang
Yu Zhang
Yue Zhang
Yulun Zhang
Yunyan Zhang
Yuyao Zhang
Zijing Zhang
Can Zhao
Changchen Zhao
Fenqiang Zhao
Gangming Zhao
Haifeng Zhao
He Zhao
Jun Zhao
Li Zhao
Qingyu Zhao
Rongchang Zhao
Shen Zhao
Tengda Zhao
Tianyi Zhao
Wei Zhao
Xuandong Zhao
Yitian Zhao
Yiyuan Zhao
Yu Zhao
Yuan-Xing Zhao
Yue Zhao
Zixu Zhao
Ziyuan Zhao
Xingjian Zhen
Hao Zheng
Jiannan Zheng
Kang Zheng

Yalin Zheng
Yushan Zheng
Jia-Xing Zhong
Zichun Zhong
Haoyin Zhou
Kang Zhou
Sanping Zhou
Tao Zhou
Wenjin Zhou
Xiao-Hu Zhou
Xiao-Yun Zhou
Yanning Zhou
Yi Zhou (IIAI)
Yi Zhou (University of Utah)
Yuyin Zhou
Zhen Zhou
Zongwei Zhou
Dajiang Zhu
Dongxiao Zhu
Hancan Zhu
Lei Zhu
Qikui Zhu
Weifang Zhu
Wentao Zhu
Xiaofeng Zhu
Xinliang Zhu
Yingying Zhu
Yuemin Zhu
Zhe Zhu
Zhuotun Zhu
Xiahai Zhuang
Aneeq Zia
Veronika Zimmer
David Zimmerer
Lilla Zöllei
Yukai Zou
Gerald Zwettler
Reyer Zwiggelaa

Contents – Part VI

Breast Imaging

Heart and Lung Imaging

Musculoskeletal Imaging

Angiography and Vessel Analysis

Lightweight Double Attention-Fused Networks for Intraoperative Stent Segmentation

Yan-Jie Zhou[1,3]([✉]), Xiao-Liang Xie[1,3], Zeng-Guang Hou[1,2,3], Xiao-Hu Zhou[1], Gui-Bin Bian[1], and Shi-Qi Liu[1]

[1] State Key Laboratory of Management and Control for Complex Systems, Institute of Automation, Chinese Academy of Sciences, Beijing 100190, China
`zhouyanjie2017@ia.ac.cn`
[2] CAS Center for Excellence in Brain Science and Intelligence Technology, Beijing 100190, China
[3] School of Artificial Intelligence, University of Chinese Academy of Sciences, Beijing 100049, China

Abstract. In endovascular interventional therapy, the fusion of preoperative data with intraoperative X-ray fluoroscopy has demonstrated the potential to reduce radiation dose, contrast agent and processing time. Real-time intraoperative stent segmentation is an important pre-requisite for accurate fusion. Nevertheless, this task often comes with the challenge of the thin stent wires with low contrast in noisy X-ray fluoroscopy. In this paper, a novel and efficient network, termed Lightweight Double Attention-fused Network (LDA-Net), is proposed for end-to-end stent segmentation in intraoperative X-ray fluoroscopy. The proposed LDA-Net consists of three major components, namely feature attention module, relevance attention module and pre-trained MobileNetV2 encoder. Besides, a hybrid loss function of both reinforced focal loss and dice loss is designed to better address the issues of class imbalance and misclassified examples. Quantitative and qualitative evaluations on 175 intraoperative X-ray sequences demonstrate that the proposed LDA-Net significantly outperforms simpler baselines as well as the best previously-published result for this task, achieving the state-of-the-art performance.

Keywords: Stent segmentation · Intraoperative X-ray fluoroscopy · Convolution neural networks

1 Introduction

Abdominal aortic aneurysm (AAA) has been the most common aneurysm, which is usually asymptomatic until it ruptures, with an ensuring mortality 85% to 90% [1]. Clinical evidence-based research shows a lower perioperative morbidity and mortality, and similar long-term survival, for endovascular aortic repair (EVAR) compared with open repair of suitable AAAs. Meanwhile, recent technological

© Springer Nature Switzerland AG 2020
A. L. Martel et al. (Eds.): MICCAI 2020, LNCS 12266, pp. 3–13, 2020.
https://doi.org/10.1007/978-3-030-59725-2_1

advances in EVAR make it the treatment of choice for most AAA patients [2]. However, due to the complexity of the EVAR, long-term radiation and large doses of contrast agents are usually required during the intervention, which will lead to common complications for patients, such as renal insufficiency. It is therefore of special concern to reduce the procedure time of EVAR.

Fusion of preoperative data with intraoperative X-ray fluoroscopy to guide the intervention has been proved to reduce contrast agent and radiation dose [3]. The preoperative data is obtained by 3D computed tomography (CT). However, the fusion may become inaccurate due to patient motion and deformation of the vessels caused by interventional instruments. To avoid repeated use of contrast agent, comparing the stent segmentation with preoperative information can assess and monitor the quality of the current fusion throughout the intervention [4]. Hence, real-time and accurate intraoperative stent segmentation is imperative. Nevertheless, fully automatic stent segmentation is not straightforward for the following reasons: (1) The morphological variation of stents in different interventions affects visual features such as shape and size. (2) The low ratio of stent wire pixels to background pixels results in class imbalance. (3) The contrast agent, artifacts from the spine and wire-like structures such as guidewire interfere with the classification accuracy of edge pixels of stents.

Although the guidewire segmentation [5] and catheter segmentation [6] in X-ray fluoroscopy have received widespread interest, less attention has been spent on stent segmentation. Previously, Demirci *et al.* [7] proposed a model-based method that relies on Hessian-based filtering for preprocessing. Although this method can directly recover the shape of the stent in 3D, it needs to define the model of the stent in advance and is limited to a certain stent shape. Recently, deep learning has achieved promising results in medical image segmentation [8,9] and provide a data-driven approach to address stent segmentation. Breininger *et al.* [4] presented a fully convolutional network with a contraction and expansion path to segment aortic stents. However, due to the utilization of residual units as its backbone, the real-time requirements were not met.

To address above-mentioned concerns, the Lightweight Double Attention-fused Networks (LDA-Net) is proposed for real-time stent segmentation in intraoperative X-ray fluoroscopy. Firstly, aggregation for multi-scale features is conducive to capturing the shape and size features of stents at different scales. Hence, the feature attention module is employed to fuse different scale dense features. Secondly, the relevance attention module is designed in gating to disambiguate irrelevant and noisy responses in skip connections. Thirdly, the pretrained MobileNetV2 encoder can reduce network parameters and improve model processing speed while ensuring performance. Additionally, the designed hybrid loss function with dice loss to address extreme class imbalance and reinforced focal loss to force model to focus on the pixels easily misclassified.

Our main contributions can be summarized as follows: (1) To the best of our knowledge, this is the first real-time approach that achieves fully automatic stent segmentation at the inference rate of 12.6 FPS in intraoperative X-ray fluoroscopy. (2) The designed double attention modules and hybrid loss improve model sensitivity to stent wire pixels without requiring complicated heuristics. (3) The proposed LDA-Net achieves the state-of-the-art segmentation performance on three different datasets, namely SeTaX, PUGSeg and NLM Chest.

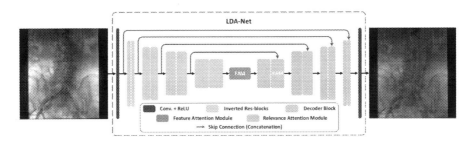

Fig. 1. An overview of the Lightweight Double Attention-fused Networks (LDA-Net). It contains double attention modules, namely feature attention module (FAM) and relevance attention module (RAM).

2 Method

In this section, we first present a general architecture of our proposed network and then introduce the designed double attention modules, namely feature attention module (FAM) and relevance attention module (RAM). Finally, we describe the hybrid loss function of both reinforced focal loss and dice loss.

2.1 Lightweight Double Attention-Fused Networks

The architecture of the proposed LDA-Net is shown in Fig. 1. The proposed network takes the original intraoperative X-ray images as input and outputs the predicted mask for stent without any post-processing. The network is a novel encoder-decoder structure, where the pre-trained MobileNetV2 [10] is employed as the backbone in the encoder stage. The depth-wise separable convolutions in the MobileNetV2 replace the standard convolutional layers, thereby reducing considerable computational burden. The FAM is utilized to gather dense pixel-level feature from the output of MobileNetV2.

Each decoder block in decoder consists of transposed convolution and batch normalization, aims to recover the resolution of the feature map from 16×16 to 512×512. In order to highlight salient features useful for the stent wire and disambiguate irrelevant and noisy responses in skip connections, the RAMs are designed and employed in the decoder stage.

Feature Attention Module. To gather precise dense pixel-level features, the FAM is integrated into the network, as shown in Fig. 2(a). The FAM combines features from three different scales by U-shape architecture. In order to better extract the context from different level scales, the 3×3, 5×5 and 7×7 convolutions are utilized in the structure of FAM, respectively. Because of the low resolution of the high-level feature maps, using large kernel size does not increase computational complexity by much [11]. This structure gradually integrates the information of different scales through up and down sampling, which can integrate the adjacent scales of context features more accurately. Then, after passing through a 1×1 convolution of the original features from the encoder part, multiply the pixel-wisely by the different level attention feature. Specifically, the adaptive average pooling is used to improve model performance further.

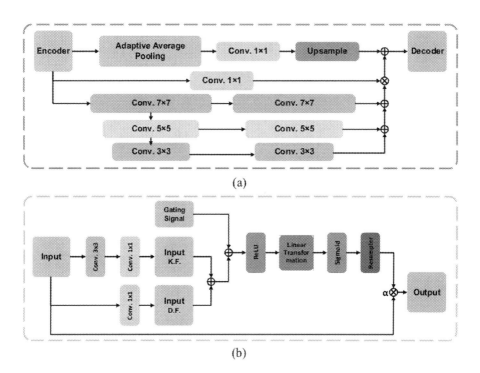

Fig. 2. (a) Schematic of the Feature Attention Module (FAM). (b) Schematic of the Relevance Attention Module (RAM).

Relevance Attention Module. In order to capture a sufficiently large receptive domain to obtain semantic context information, our designed RAMs are integrated into the LDA-Net. Compared with two-stage networks, RAM gradually suppresses the feature responses of irrelevant background regions without the necessity of region of interest (ROI). As shown in Fig. 2(b), the input of RAMs can be divided into two parts. The first part is to obtain the key feature map (K.F.) by a series of convolution 3×3, BN and ReLU. Another part is to directly adjust the feature map (D.F.) to the universal. Then making the summation of two parts to enhance the nonlinearity. The output of RAMs is the element-wise multiplication of input feature maps and attention coefficients: $\widehat{x}_{i,c}^l = x_{i,c}^l \cdot \alpha_i^l$. The attention coefficient $\alpha_i \in [0,1]$ identifies image salient regions to preserve the activation relevant to the stent wire. In the default setting, a single scalar attention value is calculated for each pixel vector $x_{i,c}^l$, where F_l corresponds to the number of feature maps in layer l. The gating vector g_i is used for each pixel i to determine the focus regions. The gating vector consists of contextual information and removes lower-level feature responses as recommended in [12]. Additive attention is employed to obtain the gating coefficient [13]. The RAM is represented as follows:

$$\alpha_i^l = \sigma_2(\psi^T(\sigma_1(W_x^T x_i^l + W_g^T g_i + b_g)) + b_\psi) \tag{1}$$

where σ_1 and σ_2 represent the ReLU activation and sigmoid activation respectively. The W_x and W_g correspond to the weights of linear transformation, and b_g and b_ψ are the bias. In order to reduce trainable parameters and computational complexity of RAMs, linear transformation ($1 \times 1 \times 1$ convolution) is performed without any spatial support, and the input feature map is down-sampled to the resolution of the gating signal. Grid re-sampling of the attention coefficients is employed by trilinear interpolation. The designed RAMs are merged into our network to highlight salient features useful for the stent wire. The information extracted from coarse scale is utilized in gating to disambiguate irrelevant and noisy responses in skip connections, thereby improving the accuracy and sensitivity of the model for edge misclassified pixels prediction.

2.2 Hybrid Loss

In the task of stent segmentation, the thin stent wire results in class imbalance. Meanwhile, due to the contrast agent, artifacts from the spine and wire-like structures, the edge pixels of the stent turn into the misclassified samples. The huge number of easy and background samples tend to overwhelm the training. Dice loss performs relatively better than cross entropy loss when the training samples are highly imbalanced [9]. However, dice loss fails to capture the pixels on the border which are difficult to classify. The modulating factor in focal loss can automatically reduce the weight of easy examples in the training process and quickly focus the model on misclassified examples [14]. To this end, we design a hybrid loss function of both reinforced focal loss and dice loss to better address the issues of class imbalance and misclassified examples. The hybrid loss function is formulated as follows:

$$L = L_{R-Focal} + \lambda L_{Dice} \tag{2}$$

$$L_{R-Focal} = \begin{cases} -\alpha(1-p_i)^\gamma \log p_i \\ -p_i^\gamma \log(1-p_i) \end{cases} \quad y_i = 1 \; y_i = 0 \tag{3}$$

where L_{Dice} is the dice loss function. y_i is the label of the i_{th} pixel, 1 for stent wire, 0 for background and p_i is the prediction probability of the i_{th} pixel. The weighting factor α and the modulating factor γ are tunable within the range of $\alpha, \gamma \geq 0$. And we have strengthened the role of weighting factor α to increase the weight contribution of the stent wire, thus solving the extreme class imbalance more efficiently. Besides, λ is also a hyper-parameter coordinating the balance between reinforced focal loss and dice loss, which is set to 0.75 in this work.

3   Experiments

In this section, quantitative and qualitative evaluations for the proposed LDA-Net are carried out on three different datasets, namely SeTaX, PUGSeg and NLM Chest X-ray Database.

3.1   Datasets

SeTaX is a intraoperative stent dataset based on 2D X-ray fluoroscopy, which is provided by Peking Union Medical College Hospital. This dataset consists of 1269 images (20 patients) in training set, 381 images (6 patients) in testing set and 254 images (4 patients) in validation set. Each image has 512×512 pixels. **PUGSeg** is an interventional tool dataset containing various stiff guidewires, which are provided by Shanghai Huadong Hospital and Peking Union Medical College Hospital. It consists of 1585 images for training, 476 images for testing and 317 images for validation. Each image has a resolution of 512×512 pixels. **NLM Chest X-ray Database** is the standard digital image database for tuberculosis [15]. The chest X-rays are from out-patient clinics, and were captured as part of the daily routine using Philips DR Digital Diagnose systems. This dataset contains 336 cases with tuberculosis and 326 normal cases.

3.2   Implementation Details

The proposed framework was implemented on PyTorch library (version 0.4.1) with one NVIDIA TITAN Xp (12 GB). To ensure the validity of the experimental evaluation, the patient data of the training set, validation set and testing set are independent of each other. Stochastic gradient descent (SGD) was used as optimizer with an initial learning rate of 0.001, weight decay of 0.0005 and momentum of 0.9. To find the optimal performance, the poly learning rate policy is employed, the learning rate is multiplied by the factor of 0.9 when the validation accuracy was saturated. Moreover, we set the batch size of 8, and 180 epochs was used for each model training.

We report mean precision, sensitivity and F_1-Score to evaluate the segmentation performance. The mean processing time is calculated to verify the real-time performance. To obtain the processing time, we load the sequence into the proposed framework and compute each frame parallelly offline. The total processing time T can be computed after getting the results of all the frames (N). Therefore, we obtain the inference rate N/T frames per second (FPS) and processing time $1000 \times T/N$ ms.

Table 1. Ablation study on SeTaX. BaseNet is the regular U-Net. BCE represents Binary Cross Entropy Loss. DL and FL represent Dice Loss and Focal Loss respectively. DRF represents the proposed hybrid loss with both reinforced Focal Loss and Dice Loss.

Method	Backbone	Loss	F_1-Score	Time (ms)
BaseNet	MobileNetV2	DRF	0.898 ± 0.009	62.6 ± 1.5
BaseNet+FAM	MobileNetV2	DRF	0.946 ± 0.023	73.5 ± 1.9
BaseNet+RAM	MobileNetV2	DRF	0.925 ± 0.011	68.5 ± 0.9
LDA-Net	MobileNetV2	DRF	0.969 ± 0.015	79.6 ± 1.3
LDA-Net	ResNet-50	DRF	0.970 ± 0.018	143.6 ± 1.7
LDA-Net	ResNet-101	DRF	0.973 ± 0.016	174.2 ± 2.3
LDA-Net	VGG-11	DRF	0.955 ± 0.008	142.4 ± 2.5
LDA-Net	VGG-16	DRF	0.964 ± 0.014	158.9 ± 1.9
LDA-Net	MobileNetV2	BCE	0.835 ± 0.009	–
LDA-Net	MobileNetV2	DL	0.916 ± 0.018	–
LDA-Net	MobileNetV2	FL	0.932 ± 0.012	–

3.3   Results on SeTaX

Ablation Study. To evaluate the contribution of different modules on our approach, we conduct experiments with different settings. As shown in Table 1, the double attention modules improve the model performance significantly. In details, we first conduct BaseNet with FAM, which improves the performance from 0.898 to 0.946. Then, we implement BaseNet with RAM, which yields 0.227 improvement in mean F_1-Score. When both of FAM and RAM are integrated into BaseNet, the mean F_1-Score reaches 0.969 and improves by 7.91 % over baseline. Specifically, it can be seen from the processing time that double attention modules do not bring much computational burden.

To verify the performance of backbone and loss function, we first replace the backbone of the original network with wildly-used backbones ResNet and VGGNet. As shown in Table 1, it clearly demonstrates the promotion in processing speed brought by the pre-trained MobileNetV2, reducing mean processing time from 174.2 ms for ResNet-101 to 79.6 ms. Then, we employ our model on three different loss function, which are Binary Cross Entropy Loss, Dice Loss and Focal Loss respectively. Every baseline loss function is set with the best hyper-parameters. The hyper-parameter settings of our proposed hybrid loss function are as follows: $\lambda = 0.75$, $\alpha = 100$ and $\gamma = 2.5$. As shown in Table 1, our proposed hybrid loss outperforms the other three baseline loss functions remarkably.

Comparing with the State-of-the-art. To demonstrate the advantage of our proposed approach, we compare it with three widely-used networks (U-Net, LinkNet and TernausNet), two attention-based networks (Attention U-Net and CS-Net) and a previously-proposed approach on SeTaX. It is worth noting that

Table 2. Quantitative comparison with state-of-the-art approaches on SeTaX.

Method	Precision	Sensitivity	F_1-Score	Time (ms)
U-Net [8]	0.890	0.903	0.896	104.5
LinkNet [16]	0.914	0.932	0.924	179.3
TernausNet [17]	0.939	0.923	0.932	142.8
Attention U-Net [18]	0.951	0.940	0.945	125.4
CS-Net [19]	0.942	0.955	0.948	125.8
KBS [4]	0.960	0.934	0.945	750
LDA-Net	0.962	0.978	0.969	79.6

we implement other approaches with best parameters. As shown in Table 2, it clearly demonstrates that our approach achieves better accuracy than other existing approaches in terms of mean F_1-Score and processing time. As can be seen in Fig. 3, the proposed approach is robust to all kinds of intraoperative stents in different interventions, and the segmentation results are accurate without any post-processing. Besides, mean processing time per image of our proposed network is about 79.6 ms (12.6 FPS), which meets real-time requirements [20].

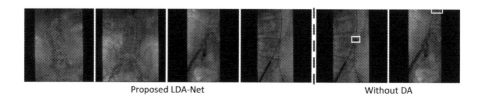

Proposed LDA-Net   Without DA

Fig. 3. Visualization results on SeTaX. DA represents double attention modules.

Table 3. Quantitative comparison on PUGSeg and NLM Chest X-ray Database.

Method	PUGSeg				NLM Chest X-ray Database			
	Seq. 1	Seq. 2	Seq. 3	Mean F_1	Seq. 1	Seq. 2	Seq. 3	Mean F_1
U-Net [8]	0.884	0.909	0.911	0.901	0.899	0.907	0.864	0.890
LinkNet [16]	0.889	0.918	0.917	0.908	0.902	0.915	0.872	0.896
TernausNet [17]	0.916	0.933	0.923	0.924	0.910	0.922	0.898	0.910
Att. U-Net [18]	0.928	0.941	0.945	0.938	0.931	0.945	0.916	0.931
CS-Net [19]	0.928	0.946	0.946	0.940	0.943	0.951	0.929	0.941
LDA-Net	0.938	0.955	0.961	0.951	0.956	0.968	0.934	0.953

3.4   Results on PUGSeg and NLM Chest X-Ray Database

To further verify the effectiveness of our proposed LDA-Net, we conduct experiments on PUGSeg and NLM Chest X-ray Database. As shown in Table 3, it clearly demonstrates that our proposed LDA-Net is superior to other wildly-used networks and attention-based networks in terms of F_1-Score. Visualization results of different approaches are shown in Fig. 4. Compared with U-Net and Attention U-Net, our proposed LDA-Net can capture better contours which are usually considered as hard samples, obtaining more accurate and smooth segmentation masks. The qualitative comparison on PUGSeg and NLM Chest X-ray Database also indicates the success of our proposed approach.

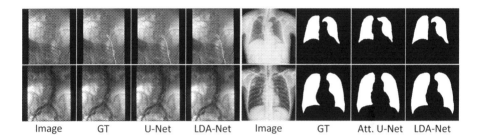

Fig. 4. Visualization results on PUGSeg and NLM Chest X-ray Database.

4   Conclusion

In this paper, we have proposed Lightweight Double Attention-fused Networks (LDA-Net) to address the challenging task of real-time stent segmentation in intraoperative X-ray fluoroscopy. Quantitative and qualitative evaluations on SeTaX, PUGSeg and NLM Chest X-ray database demonstrate that our approach achieves significant improvement in terms of both accuracy and robustness. The ablation experiments prove the effectiveness of double attention modules (FAM and RAM) and hybrid loss. By integrating these components into the network, our proposed LDA-Net effectually addresses the issues of class imbalance and misclassified examples, achieving the state-of-the-art performance. Specifically, the inference rate of our approach is approximately 12.6 FPS, which enables for real-time computer-assisted interventions.

Acknowledgments. This work was supported in part by the National Key Research and Development Plan of China (2019YFB1311700), the National Natural Science Foundation of China (U1913210, U1613210, 61533016), the CAMS Innovation Fund for Medical Sciences (2018-I2M-AI-004), and the Youth Innovation Promotion Association of CAS (2020140).

References

1. Kent, K.C.: Abdominal aortic aneurysms. N. Engl. J. Med. **371**(22), 2101–2108 (2014)
2. Buck, D.B., Van Herwaarden, J.A., Schermerhorn, M.L., Moll, F.L.: Endovascular treatment of abdominal aortic aneurysms. Nat. Rev. Cardiol. **11**(2), 112 (2014)
3. Schulz, C.J., Schmitt, M., Böckler, D., Geisbüsch, P.: Fusion imaging to support endovascular aneurysm repair using 3D–3D registration. J. Endovasc. Ther. **23**(5), 791–799 (2016)
4. Breininger, K., Albarqouni, S., Kurzendorfer, T., Pfister, M., Kowarschik, M., Maier, A.: Intraoperative stent segmentation in X-ray fluoroscopy for endovascular aortic repair. Int. J. Comput. Assist. Radiol. Surg. **13**(8), 1221–1231 (2018). https://doi.org/10.1007/s11548-018-1779-6
5. Zhou, Y.-J., et al.: Real-time guidewire segmentation and tracking in endovascular aneurysm repair. In: Gedeon, T., Wong, K.W., Lee, M. (eds.) ICONIP 2019. LNCS, vol. 11953, pp. 491–500. Springer, Cham (2019). https://doi.org/10.1007/978-3-030-36708-4_40
6. Ambrosini, P., Ruijters, D., Niessen, W.J., Moelker, A., van Walsum, T.: Fully automatic and real-time catheter segmentation in X-ray fluoroscopy. In: Descoteaux, M., Maier-Hein, L., Franz, A., Jannin, P., Collins, D.L., Duchesne, S. (eds.) MICCAI 2017. LNCS, vol. 10434, pp. 577–585. Springer, Cham (2017). https://doi.org/10.1007/978-3-319-66185-8_65
7. Demirci, S., et al.: 3D stent recovery from one X-ray projection. In: Fichtinger, G., Martel, A., Peters, T. (eds.) MICCAI 2011. LNCS, vol. 6891, pp. 178–185. Springer, Heidelberg (2011). https://doi.org/10.1007/978-3-642-23623-5_23
8. Ronneberger, O., Fischer, P., Brox, T.: U-Net: convolutional networks for biomedical image segmentation. In: Navab, N., Hornegger, J., Wells, W.M., Frangi, A.F. (eds.) MICCAI 2015. LNCS, vol. 9351, pp. 234–241. Springer, Cham (2015). https://doi.org/10.1007/978-3-319-24574-4_28
9. Milletari, F., Navab, N., Ahmadi, S.A.: V-net: fully convolutional neural networks for volumetric medical image segmentation. In: 3DV, pp. 565–571. IEEE (2016)
10. Sandler, M., Howard, A., Zhu, M., Zhmoginov, A., Chen, L.C.: MobileNetV2: inverted residuals and linear bottlenecks. In: CVPR, pp. 4510–4520. IEEE (2018)
11. Li, H., Xiong, P., An, J., Wang, L.: Pyramid attention network for semantic segmentation (2018). arXiv preprint arXiv:1805.10180
12. Wang, F., Jiang, M., Qian, C., Yang, S., Li, C., Zhang, H.: Residual attention network for image classification. In: CVPR, pp. 3156–3164. IEEE (2017)
13. Luong, M.T., Pham, H., Manning, C.D.: Effective approaches to attention-based neural machine translation (2015). arXiv preprint arXiv:1508.04025
14. Lin, T.Y., Goyal, P., Girshick, R., He, K., Dollár, P.: Focal loss for dense object detection. In: ICCV, pp. 2980–2988. IEEE (2017)
15. Candemir, S., Jaeger, S., Palaniappan, K., Musco, J.P.: Lung segmentation in chest radiographs using anatomical atlases with nonrigid registration. IEEE Trans. Med. Imag. **33**(2), 577–590 (2013)
16. Chaurasia, A., Culurciello, E.: Linknet: Exploiting encoder representations for efficient semantic segmentation. In: VCIP, pp. 1–4. IEEE (2017)
17. Iglovikov, V., Shvets, A.: Ternausnet: U-Net with VGG11 encoder pre-trained on ImageNet for image segmentation (2018). arXiv preprint arXiv:1801.05746
18. Oktay, O., Schlemper, J., Folgoc, L.L., Lee, M., Heinrich, M.: Attention U-Net: learning where to look for the pancreas (2018). arXiv preprint arXiv:1804.03999

19. Mou, L., et al.: CS-Net: channel and spatial attention network for curvilinear structure segmentation. In: Shen, D., et al. (eds.) MICCAI 2019. LNCS, vol. 11764, pp. 721–730. Springer, Cham (2019). https://doi.org/10.1007/978-3-030-32239-7_80
20. Heidbuchel, F., Wittkampf, F.H., Vano, E., Ernst, S., Schilling, R.: Practical ways to reduce radiation dose for patients and staff during device implantations and electrophysiological procedures. Europace **16**(7), 946–964 (2014)

TopNet: Topology Preserving Metric Learning for Vessel Tree Reconstruction and Labelling

Deepak Keshwani[1(✉)], Yoshiro Kitamura[1], Satoshi Ihara[1], Satoshi Iizuka[2], and Edgar Simo-Serra[3]

[1] Imaging Technology Center, Fujifilm Corporation, Minato, Japan
deepak.keshwani@fujifilm.com
[2] Center for Artificial Intelligence Research, University of Tsukuba, Tsukuba, Japan
[3] Department of Computer Science and Engineering, Waseda University, Shinjuku, Japan

Abstract. Reconstructing Portal Vein and Hepatic Vein trees from contrast enhanced abdominal CT scans is a prerequisite for preoperative liver surgery simulation. Existing deep learning based methods treat vascular tree reconstruction as a semantic segmentation problem. However, vessels such as hepatic and portal vein look very similar locally and need to be traced to their source for robust label assignment. Therefore, semantic segmentation by looking at local 3D patch results in noisy misclassifications. To tackle this, we propose a novel multi-task deep learning architecture for vessel tree reconstruction. The network architecture simultaneously solves the task of detecting voxels on vascular centerlines (i.e. nodes) and estimates connectivity between center-voxels (edges) in the tree structure to be reconstructed. Further, we propose a novel connectivity metric which considers both inter-class distance and intra-class topological distance between center-voxel pairs. Vascular trees are reconstructed starting from the vessel source using the learned connectivity metric using the shortest path tree algorithm. A thorough evaluation on public IRCAD dataset shows that the proposed method considerably outperforms existing semantic segmentation based methods. To the best of our knowledge, this is the first deep learning based approach which learns multi-label tree structure connectivity from images.

Keywords: Deep learning · Vessel tree reconstruction · Vessel segmentation · Liver vessel · Centerline detection · Computed tomography

1 Introduction

Primary liver cancer is the third most common cause of cancer mortality [2]. Liver resection is currently the most prevalent treatment method with 5-year

Electronic supplementary material The online version of this chapter (https://doi.org/10.1007/978-3-030-59725-2_2) contains supplementary material, which is available to authorized users.

A. L. Martel et al. (Eds.): MICCAI 2020, LNCS 12266, pp. 14–23, 2020.
https://doi.org/10.1007/978-3-030-59725-2_2

survival rates of up to 40% [11]. To perform a safe liver resection, preoperative surgical simulations have been found to be very useful [17,19]. In these simulations, a 3D model of the liver which shows the anatomical structures of portal and hepatic veins together with tumour location is reconstructed. Using vascular reconstructions, blood flow patterns around the tumour is analysed to compute the resection region. The analysis of blood flow requires not just segmentation of vascular structures but their representation as a tree structure.

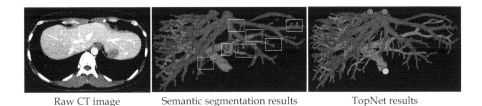

Raw CT image   Semantic segmentation results   TopNet results

Fig. 1. Portal and hepatic vein structures obtained with semantic segmentation and proposed TopNet. The red rectangles indicate vessel misclassifications. (Color figure online)

Rule based methods for hepatic and portal vein reconstruction have been studied and a comprehensive literature survey can be found in [6]. Most popular rule based methods are based on global optimization techniques like graph cuts [1,20]. Graph cut based techniques use a handcrafted energy term between vascular nodes. A downside of such methods is that they work well only in conditions for which the rules were handcrafted. For instance when both portal and hepatic veins are high in contrast [20]. Recently, deep learning-based semantic segmentation methods have been proposed for liver vessel recognition tasks [9,13,14]. In these methods, semantic segmentation is used to classify a voxel by extracting information from a local 3D patch (also known as the network's receptive field) centered around it. However, since portal and hepatic veins look very similar locally, these methods result in noisy vessel misclassifications as shown in Fig. 1. Such noisy misclassifications magnify the tree reconstruction error. Additionally, within semantic segmentation literature, simultaneous segmentation of portal and hepatic vein from CT volumes has not been tackled. Either only hepatic [13,14] or only portal vein [9] segmentation has been targeted.

In this work, we propose TopNet, a deep metric learning method for vascular tree reconstruction. The main idea is to learn the connectivity metric between vascular voxel pairs rather than to assign absolute labels. Conceptually, TopNet is a two stage network. The first stage computes center-voxels along the centerlines of all the vessels. The second stage learns the connectivity metric between center-voxel pairs in the form of topological distance between the voxels along the vascular tree. To generate a global tree, center-voxels are connected consecutively starting from the vessel source using the learned topology metric. We summarize the main contributions of this work as follows:

1. Proposed a novel multi task architecture designed for tree reconstruction, which detects nodes (voxels on vascular center lines) and estimates edges (connectivity between center-voxels) in the tree to be reconstructed.
2. Proposed a novel topology metric which learns both inter-class distance and intra-class topological distance between vascular voxel pairs in multiple trees.
3. Verified that the proposed method achieves higher accuracy than existing methods with a large margin for portal and hepatic vein reconstruction.

2   Related Work

TopNet uses deep metric learning to connect detected center-voxels. In this regard, our approach is closer to deep metric learning approaches for image instance segmentation tasks [4,5,15,18]. These methods map pixels from image space to features space such that pixels belonging to the same instance are close to each other and separated by a margin if they belong to different instances. The methods differ in how they measure similarity in the feature space. Among the similarity metrics, cosine similarity has been used extensively [15,18]. Another way in which these methods differ is the way pixels are clustered in the learned feature space to generate instances. In this work, the task is not just instance segmentation (vessel segmentation) but to learn connectivity within the instance (vessel trees). For this purpose, we propose a novel deep metric learning approach.

3   TopNet

The proposed TopNet is a multi-task 3D Fully Convolutional Neural Network (3D-FCN). All the tasks share a common base encoder and task specific decoders as shown in Fig. 2. The proposed architecture is similar to 3D-UNet [3], with the difference that our approach has three decoders for three different tasks instead of one. Such Multi-task architectures with shared encoder and task specific decoders have also been proposed earlier [12]. For detailed architecture, please refer to the supplementary material. The first task is extraction of all the vessels from the image. The second task is assigning a centerness score to all the voxels within the vessel mask such that the centermost voxel is assigned a minimum value. Using this centerness score, center-voxels are computed using non-maximum suppression. The final task is learning connectivity between center-voxels, which is proportional to topological distance between the center-voxels. Figure 2 shows the conceptual representation of topological distance with solid red lines. The network is trained by minimizing the sum of loss terms from each of the three tasks. The formulation of loss for each task is described in the following sections.

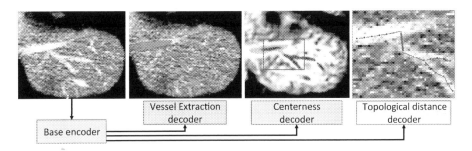

Fig. 2. TopNet methodology with base encoder and three task specific decoders.

3.1   Vessel Extraction Decoder

The task of separating all the vessels from the background is formulated as a semantic segmentation problem. The loss function used for this task is the dice loss function which has shown to improve the accuracy in segmentation tasks with severe class imbalance [16]. We compute vessel extraction loss V_{loss} as

$$V_{loss} = 1 - \frac{2\sum_{i=1}^{N} v_i v'_i}{\sum_{i=1}^{N} v_i + \sum_{i=1}^{N} v'_i}, \tag{1}$$

where the sums run over the N voxels, of the ground truth binary vessel volume $v_i \in V$, and predicted vessel volume $v'_i \in V'$.

3.2   Centerness Decoder

The task of generating a centerness score map is formulated as a regression problem. The ground truth centerness score map is the distance transform of binary volume $m_i \in M$ representing the vessel center voxels. We consider the centerness score map only within the ground truth binary vessel volume V. The voxels outside the vessel region are masked out when computing the training loss. The training loss for the regression task is formulated as weighted $smooth_{L1}$ loss between the ground truth centerness score S_i and predicted centerness score S'_i summed over the voxels belonging to vessel mask V. We use $smooth_{L1}$ function as defined in [8] and the weight is inverse square of true centerness score i.e. S_i^2. Mathematically, the centerness loss C_{loss} is given as

$$C_{loss} = \frac{1}{\sum_{i=1}^{N} v_i} \sum_{i|v_i=1} \frac{1}{S_i^2} (smooth_{L1}(S_i - S'_i)). \tag{2}$$

Weighting the loss function is essential because the loss would otherwise be highly biased towards the voxels away from the center. During the inference, the centerness score is masked using the predicted vessel mask. The vessel center-points are extracted by first thresholding the centerness score map followed by applying non-maximum suppression (NMS) on negative centerness score map. In this work, we use a threshold of 1.5 and NMS window of $5 \times 5 \times 5$.

3.3   Topology Distance Decoder

Topological distance decoder outputs an 8-dimensional feature vector for each voxel in CT volume. The network is trained to map vessel center-voxels into the feature space such that the L_2 norm between two center-voxels in the feature space is proportional to the topological distance between along the vascular trees. For voxel pairs belonging to different vascular trees, the L_2 norm is greater than a set margin. The pairwise loss between the two voxels i and j with associated vessel labels l_i and l_j is defined as

$$Top(x_i, x_j) = \begin{cases} smooth_{L1}\left(\| x_i - x_j \| - \alpha D_{ij}\right) & \forall i, j \mid l_i = l_j, i \neq j, \\ \gamma * \max(0, [K - \| x_i - x_j \|]) & otherwise. \end{cases} \quad (3)$$

Here, $\| x_i - x_j \|$ is the L_2 norm between the feature representations of the two voxels, D_{ij} is the topological distance between the voxels along the vascular tree. α and γ are constants whose values are explained subsequently. To compute the topological loss, all such center voxel pairs which are in the local neighborhood of each other are considered. The total loss then becomes

$$T_{loss} = \frac{1}{n} \sum_{i \mid m_i = 1} \sum_{j \in N_i} Top(x_i, x_j). \quad (4)$$

Here, N_i is the set of voxels in the local neighborhood of voxel i and n is the total number of such voxel pairs for which the loss is computed. In this work, we use all voxel pairs which lie inside a 3D sphere of radius 15 voxels in the image space. Thus, the topological distance is normalized from 0 to 1 using the proportionality factor $\alpha = 1/15$. This roughly means that the voxels belonging to the same vascular tree will follow $0 \leq \| x_i - x_j \| \leq 1$ and if they belong to different vascular trees, then $\| x_i - x_j \| \geq 3$. To balance out the two loss terms, we use $\gamma = 1/3$.

3.4   Vascular Tree Reconstruction

Vascular tree is constructed by aggregating the center-voxels starting from the vessel sources using the topological distance metric learned by the TopNet. We train a separate network which outputs the location of portal and hepatic vein sources (see supplementary material). Typical portal (pink disc) and hepatic vein (blue disc) sources are shown in Fig. 1. For computed vessel sources, the vascular trees are constructed using Dijkstra's multi-source shortest path tree algorithm. To that end, the undirected graph is formulated as follows. The vertices of the graph are center-voxels. The weighted edges of the graph are given by the set $\left\{(w_{ij}/\alpha)^2\right\}_{w_{ij} \leq 2}$. Here, $w_{ij} = \| x_i - x_j \|$, $i \mid m_i = 1$ and $j \mid m_j = 1$ represents two center-voxels in a local neighborhood defined by a 3D sphere of radius 15. The term $(w_{ij}/\alpha)^2$ is essentially the square of topological distance between the two center-voxels i and j. All the pairs for which $w_{ij \geq 2}$ are not considered as feasible edges, since such edges most likely belong to different vascular trees.

Table 1. Characteristics of INTERNAL and public dataset IRCAD.

Dataset	Training	Test	Contrast phase	Slice Thickness
INTERNAL	115 CT volumes	20 CT volumes	Late portal	0.5–1.0 mm
IRCAD [10]	None	20 CT volumes	Late portal	1.0–4.0 mm

4   Experiments

We divide the experiments into comparison and ablation studies. The comparison of TopNet is made against single task 3D-UNet [3] with the dice loss function [16] baseline. This is a standard and competitive baseline for 3D image segmentation and an upgrade over existing 2D-FCN based existing liver vessel segmentation methods [9,13,14]. In the ablation study, we investigate the performance with different learning metrics (dice loss based classification, cosine, and topology) using the same multi-task architecture. For the Multi-task dice loss based classification approach, the last two layers of topology distance decoder are replaced to output a two-channel probability map for each hepatic and portal vein. Here, we use a 2 class dice loss summed over all the center-voxels. Multi-task cosine metric learning method uses the exact same architecture as TopNet but uses cosine metric following from similar works [15,18]. The network uses cosine metric loss between voxel pairs i and j associated with feature vectors x_i and x_j and vessel labels l_i and l_j is defined as

$$L\left(x_i, x_j\right) = \begin{cases} 1 - 0.5(1 + S_{ij}) & \forall i, j \mid l_i = l_j, i \neq j, \\ 0.5(1 + S_{ij}) & otherwise. \end{cases} \tag{5}$$

where, $S_{ij} = \frac{\|x_i^T x_j\|}{\|x_i\|\|x_j\|}$ is the cosine similarity metric. To reconstruct vascular tree using the learned metric as mentioned in Sect. 3.4, the weighted edges of the graph are given by the set $\{E_{ij}(1 - S_{ij})\}_{S_{ij} \leq 0.5}$ Here, E_{ij} is the Euclidean distance between the center-voxels.

4.1   Dataset and Preprocessing

We used an internal dataset for training and both the internal as well as public IRCAD dataset for evaluation [10]. The characteristics of both the datasets is shown in Table 1. For training, the CT values were normalized from 0 to 1, voxel spacing in all three dimensions was normalized to 0.7 mm. We also crop the CT volume using the liver region before setting it as input to the network.

4.2   Evaluation Metrics and Results

We evaluate the results of portal and hepatic vein reconstruction by comparing respective ground truth and predicted centerlines. When comparing the center-lines, an exact overlap is not possible. For this, we introduce a variable tolerance

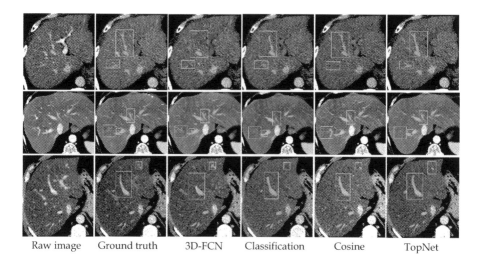

Raw image   Ground truth   3D-FCN   Classification   Cosine   TopNet

Fig. 3. Comparison between 3D-UNet and proposed TopNet for hepatic and portal vein extraction. Hepatic vein is shown in green while portal vein is shown in blue. Red boxes show regions of interest for comparing misclassifications. (Color figure online)

term δ which is equal to the radius defined at a ground truth centerpoint following from [7]. Based on this, an illustration of True Positives (TP), False Positives (FP), False Negatives (FN), and True Negatives (TN) of portal vessel class is shown in the supplementary material. Using these metrics, we compute the dice coefficient, sensitivity and specificity for hepatic and portal centerlines. Predicted centerpoints for which there are no associated ground truth centerpoints within the tolerance are ignored while computing the metrics. Usually, these points would be counted as false positives but we ignore them because in the IRCAD dataset, small vessels are often unlabelled.

The proposed TopNet considerably outperforms the baseline i.e. single task 3D-UNet method with over 7–8% improvement in all metrics as shown in Table 2. When Multi-task dice loss classification is compared to single-task 3D-UNet which also uses the dice loss, a performance improvement of 3–4% is seen. By this, we can say that multi-stage vessel segmentation works better than single-stage method. Within the Multi-task network architecture, TopNet gives the best results. We observe that if not for topology metric learning, the problem of abrupt change in vessel labels along the vascular tree (large and medium red boxes) persists as shown in Fig. 3. Such misclassifications also occur when portal and hepatic veins cross each other in close vicinity (small red boxes in second and third row). We reason that such misclassifications arise because conventional margin based techniques like semantic segmentation or even cosine metric learning classify or seperate the center-voxels based on frequently seen vessel patterns in different liver regions within the training dataset. If unique vessel patterns are observed in test data, the classification fails. With Topology metric learning on the other hand, we explicitly constrain the network to learn to trace the vessels. To verify this direction of reasoning, we show the similarity metric field from a reference center-voxel to it's neighborhood voxels in a misclassified region in Fig. 3. It can be seen that the Topology metric shows a smooth transitioning of the similarity field as compared to cosine metric which shows an abrupt change. Since the cosine metric learning method does not learn absolute labels, such misclassifications propagate as shown with the yellow box in Fig. 4. This explains why the cosine metric learning works poorly as compared to the multi-task classification approach as shown in Table 2.

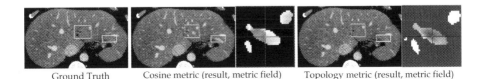

| Ground Truth | Cosine metric (result, metric field) | Topology metric (result, metric field) |

Fig. 4. Comparison between Topology and cosine metric learning. Hepatic vein is shown in green while portal vein is shown in blue. Boxes show regions of interest for comparing misclassifications. (Color figure online)

Table 2. Comparison between TopNet and existing methods.

Method	Vessel	IRCAD			INTERNAL		
		Dice	Specificity	Sensitivity	Dice	Specificity	Sensitivity
Single Task							
3D-UNet	Portal	0.84	0.85	0.82	0.88	0.86	0.87
[3]	Hepatic	0.84	0.82	0.85	0.86	0.87	0.86
Multi Task							
Classification	Portal	0.87	0.85	0.88	0.92	0.93	0.90
	Hepatic	0.86	0.88	0.85	0.91	0.90	0.93
Cosine metric	Portal	0.85	0.85	0.81	0.88	0.84	0.9
	Hepatic	0.80	0.82	0.86	0.85	0.90	0.84
TopNet	Portal	**0.91**	**0.96**	**0.91**	**0.95**	**0.96**	**0.94**
	Hepatic	**0.92**	**0.91**	**0.96**	**0.94**	**0.94**	**0.96**

5 Conclusion

We proposed a deep metric learning method which learns multi-label tree structure connectivity from images. We converted a hard problem of assigning absolute labels to vessels to a simpler one. Topological metric learning is simpler because it just learns local connectivity and does not require global image context. For global connectivity, we use the shortest path tree algorithm which uses the learned metric to connect vascular voxels. The results show that using topological metric learning, the issue of noisy misclassifications is resolved. We believe that the approach is general enough to be applicable to vessel segmentation tasks in other organs. This is because the network is constrained to learn connectivity rather than other absolute labels. There are a few downsides to this approach as well. First, the method has multiple manually fine tuned parameters involved both in detecting the center-voxels and creating vascular trees using the learned topology metric. Second, the shortest path tree algorithm which is used to create global connectivity is sensitive to misdetection of vascular center voxels. Due to this, in the future it would be interesting to explore a method which can embed tree generation algorithms like the shortest path into the training process.

Acknowledgements. This research was done in cooperation with the Hepato-Biliary-Pancreatic Surgery Division, the University of Tokyo Hospital. We would like to thank Prof. Kiyoshi Hasegawa, Dr. Junichi Kaneko, Dr. Ryugen Takahashi, and Dr. Yusuke Kazami for their valuable advice and curating a liver vessel dataset.

References

1. Bauer, C., Pock, T., Sorantin, E., Bischof, H., Beichel, R.: Segmentation of interwoven 3D tubular tree structures utilizing shape priors and graph cuts. Med. Image Anal. **14**(2), 172–184 (2010)
2. Center, M.M., Jemal, A.: International trends in liver cancer incidence rates. Cancer Epidemiol. Prev. Biomark. **20**(11), 2362–2368 (2011)
3. Çiçek, Ö., Abdulkadir, A., Lienkamp, S.S., Brox, T., Ronneberger, O.: 3D U-Net: learning dense volumetric segmentation from sparse annotation. In: Ourselin, S., Joskowicz, L., Sabuncu, M.R., Unal, G., Wells, W. (eds.) MICCAI 2016. LNCS, vol. 9901, pp. 424–432. Springer, Cham (2016). https://doi.org/10.1007/978-3-319-46723-8_49
4. De Brabandere, B., Neven, D., Van Gool, L.: Semantic instance segmentation with a discriminative loss function (2017). arXiv preprint arXiv:1708.02551
5. Fathi, A., et al.: Semantic instance segmentation via deep metric learning (2017). arXiv preprint arXiv:1703.10277
6. Fraz, M.M., et al.: Blood vessel segmentation methodologies in retinal images-a survey. Comput. Methods Programs Biomed. **108**(1), 407–433 (2012)
7. Gegúndez-Arias, M.E., Aquino, A., Bravo, J.M., Marín, D.: A function for quality evaluation of retinal vessel segmentations. IEEE Trans. Med. Imaging **31**(2), 231–239 (2011)
8. Girshick, R.: Fast R-CNN. In: Proceedings of the IEEE International Conference on Computer Vision, pp. 1440–1448 (2015)
9. Ibragimov, B., Toesca, D., Chang, D., Koong, A., Xing, L.: Combining deep learning with anatomical analysis for segmentation of the portal vein for liver sbrt planning. Phys. Med. Biol. **62**(23), 8943 (2017)
10. IRCAD: Ircad dataset for liver vessel segmentation, March 2020. https://www.ircad.fr/research/3d-ircadb-01/

11. Kazaryan, A.M., et al.: Laparoscopic resection of colorectal liver metastases: surgical and long-term oncologic outcome. Ann. Surg. **252**(6), 1005–1012 (2010)

12. Keshwani, D., Kitamura, Y., Li, Y.: Computation of total kidney volume from CT images in autosomal dominant polycystic Kidney disease using multi-task 3D convolutional neural networks. In: Shi, Y., Suk, H.-I., Liu, M. (eds.) MLMI 2018. LNCS, vol. 11046, pp. 380–388. Springer, Cham (2018). https://doi.org/10.1007/978-3-030-00919-9_44

13. Kitrungrotsakul, T., Han, X.H., Iwamoto, Y., Foruzan, A.H., Lin, L., Chen, Y.W.: Robust hepatic vessel segmentation using multi deep convolution network. In: Medical Imaging 2017: Biomedical Applications in Molecular, Structural, and Functional Imaging, vol. 10137, p. 1013711. International Society for Optics and Photonics (2017)

14. Kitrungrotsakul, T., et al.: Vesselnet: a deep convolutional neural network with multi pathways for robust hepatic vessel segmentation. Comput. Med. Imaging Graph. **75**, 74–83 (2019)

15. Kong, S., Fowlkes, C.C.: Recurrent pixel embedding for instance grouping. In: Proceedings of the IEEE Conference on Computer Vision and Pattern Recognition. pp. 9018–9028 (2018)

16. Milletari, F., Navab, N., Ahmadi, S.A.: V-Net: fully convolutional neural networks for volumetric medical image segmentation. In: 2016 Fourth International Conference on 3D Vision (3DV), pp. 565–571. IEEE (2016)

17. Mise, Y., et al.: How has virtual hepatectomy changed the practice of liver surgery?: experience of 1194 virtual hepatectomy before liver resection and living donor liver transplantation. Ann. Surg. **268**(1), 127–133 (2018)

18. Payer, C., Štern, D., Neff, T., Bischof, H., Urschler, M.: Instance segmentation and tracking with cosine embeddings and recurrent hourglass networks. In: Frangi, A.F., Schnabel, J.A., Davatzikos, C., Alberola-López, C., Fichtinger, G. (eds.) MICCAI 2018. LNCS, vol. 11071, pp. 3–11. Springer, Cham (2018). https://doi.org/10.1007/978-3-030-00934-2_1

19. Wakabayashi, G., et al.: Recommendations for laparoscopic liver resection: a report from the second international consensus conference held in morioka. Ann. Surg. **261**(4), 619–629 (2015)

20. Zeng, Y.Z., et al.: Liver vessel segmentation and identification based on oriented flux symmetry and graph cuts. Comput. Methods Programs Biomed. **150**, 31–39 (2017)

Learning Hybrid Representations for Automatic 3D Vessel Centerline Extraction

Jiafa He[1], Chengwei Pan[2], Can Yang[1], Ming Zhang[2], Yang Wang[1], Xiaowei Zhou[3(✉)], and Yizhou Yu[4(✉)]

[1] Department of Mathematics,
The Hong Kong University of Science and Technology, Hong Kong, China
[2] Department of Computer Science, School of EECS, Peking University, Beijing, China
[3] The State Key Lab of CAD&CG, Zhejiang University, Hangzhou, China
xzhou@cad.zju.edu.cn
[4] Deepwise AI Lab, Beijing, China
yizhouy@acm.org

Abstract. Automatic blood vessel extraction from 3D medical images is crucial for vascular disease diagnoses. Existing methods based on convolutional neural networks (CNNs) may suffer from discontinuities of extracted vessels when segmenting such thin tubular structures from 3D images. We argue that preserving the continuity of extracted vessels requires to take into account the global geometry. However, 3D convolutions are computationally inefficient, which prohibits the 3D CNNs from sufficiently large receptive fields to capture the global cues in the entire image. In this work, we propose a hybrid representation learning approach to address this challenge. The main idea is to use CNNs to learn local appearances of vessels in image crops while using another point-cloud network to learn the global geometry of vessels in the entire image. In inference, the proposed approach extracts local segments of vessels using CNNs, classifies each segment based on global geometry using the point-cloud network, and finally connects all the segments that belong to the same vessel using the shortest-path algorithm. This combination results in an efficient, fully-automatic and template-free approach to centerline extraction from 3D images. We validate the proposed approach on CTA datasets and demonstrate its superior performance compared to both traditional and CNN-based baselines.

Keywords: Centerline extraction · Vessel segmentation · Hybrid representations

---

J. He and C. Pan—These authors contributed equally to this work.

Electronic supplementary material The online version of this chapter (https://doi.org/10.1007/978-3-030-59725-2_3) contains supplementary material, which is available to authorized users.

A. L. Martel et al. (Eds.): MICCAI 2020, LNCS 12266, pp. 24–34, 2020.
https://doi.org/10.1007/978-3-030-59725-2_3

1   Introduction

Extracting tubular objects, e.g., blood vessels, has become a crucial task in computer-assisted diagnosis (CAD) of many diseases. For example, vessel lumen segmentation and centerline extraction are prerequisites for vessel curved-planar reconstruction (CPR) [6] from computed tomography angiography (CTA) images, which further facilitates stenosis detection and plaque identification in clinical diagnosis. However, it is usually time-consuming to segment vessel and extract centerline from various medical images. Instead, automatic vessel segmentation and centerline detection play a more and more important role for quantitative analysis of vascular diseases. [1, 7, 26].

Recently, convolutional neural networks (CNNs) have been widely applied in 3D medical image segmentation. However, segmenting vessels in 3D medical images is still very challenging. The blood vessels have delicate tubular structures with a large variety in long-range topology, which cannot be captured by slice-wise or patch-wise convolutional operations in most deep learning based segmentation methods [5, 9, 24]. Moreover, at the presence of imaging artifacts which often exist in medical images, CNN-based segmentation algorithms are prone to missing some segments of vessels, resulting discontinuities in the extracted vessel centerline [18, 23, 28]. To preserve the correct topology of the extracted vessels, previous methods may rely on user input that annotates start and end points of each vessel [17, 22]. Then, a more complete vessel centerline can be found by a minimal-cost path-based algorithm [2, 4, 8, 10]. To avoid manual input, another approach to ensure correct topology is to build an atlas or template of the target vessels from training samples and register the template to the test image [27, 28]. However, this approach is not very generalizable as the vascular structure of the test sample might be very different from the template.

In this paper, we present a novel approach for vessel centerline extraction, which is able to ensure the connectivity of extracted vessels without the need of any manual input or vessel template. The key idea is to use a patch-wise 3D CNNs to segment vessel mask and regress vessel centerline heatmap from the input image, and meanwhile use another point-cloud network to label the extracted vessel segments, such that segments belonging to the same vessel can be connected in a post-processing step. This hybrid approach makes the best use of both worlds: patch-wise CNNs for local appearance learning and point-cloud networks for global geometry learning, resulting in a robust and efficient algorithm for automatic centerline extraction. We also propose a geometry-aware grouping strategy to improve the performance of point-cloud network for vessel labeling. The effectiveness of the proposed framework is validated on two datasets: a public dataset of coronary artery CTA scans and an in-house dataset of head and neck artery CTA scans. Experimental results show that our approach outperforms existing baseline methods in terms of both accuracy and completeness of extracted centerlines.

In summary, we make the following contributions: (1) A novel hybrid representation learning approach for fully-automatic and template-free vessel centerline extraction; (2) A geometry-aware grouping method that utilizes the skele-

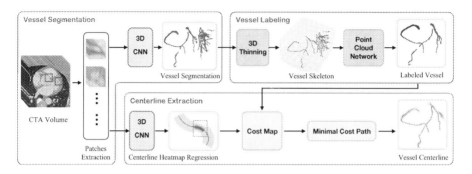

Fig. 1. Overview of the proposed approach. The framework consists of three components: **Vessel segmentation (blue block)**: the input CTA images are divided into 3D patches with overlap and then a 3D CNNs are utilized to efficiently learn local features of vascular objects. **Vessel labeling (red block)**: vascular skeletons produced by thinning the segmentation results are fed into a point-cloud network to learn the global geometry of vessels and realize vascular branch labeling. **Centerline extraction (green block)**: based on a cost map, which is constructed from the centerline heatmap and the labeled skeleton, a minimal cost path algorithm is finally utilized to extract the complete vessel centerline. (Color figure online)

ton's connection property to improve the performance of vessel labeling; (3) The state-of-the-art performance on the public benchmark.

2   Methods

Given a 3D CTA image consisted of a sequence of 2D slices, the objective is to segment the arteries and delineate their centerlines. The state-of-the-art segmentation methods are mostly based on CNNs. Due to the heavy computation of 3D convolutions, an input 3D image needs to be divided into overlapped patches and fed into the segmentation network separately. This leads to restricted local receptive fields, which may not provide sufficient information to distinguish between arteries and veins, resulting false detection. Moreover, there is no guarantee on the connectivity of extracted vessel segments from patch-based CNNs. A solution is to connect the segments that belong to the same vascular branch in post-processing. But to achieve this we need to label all the segments, which is also difficult for a patch-based CNN as vessel labeling requires considering the global geometry of the vessels.

To address these issues, we propose a hybrid approach that consists of a patch-based CNNs for local **vessel segmentation**, a point-cloud network for global **vessel labeling** and a path-finding algorithm for final **centerline extraction**. Figure 1 provides an overview to our approach.

2.1   Vessel Segmentation

At first, we use a 3D CNNs to learn vascular local appearance features from 3D patches of the original CTA images and produce a coarse vessel segmentation. An

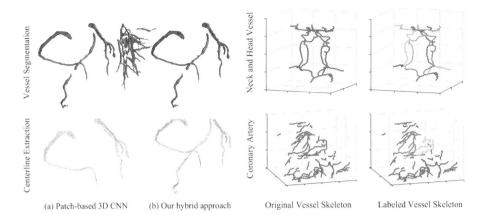

(a) Patch-based 3D CNN   (b) Our hybrid approach   Original Vessel Skeleton   Labeled Vessel Skeleton

Fig. 2. Comparison. Patch-based 3D CNNs extract extra non-vascular tissues and miss some true segments (left). Our hybrid approach is able to remove non-vascular tissues and connect the disjointed segments (right).

Fig. 3. Generated vessel skeletons. Head and neck arteries (the top row) are divided into 17 categories distinguished by color. Coronary artery (the bottom row) are divided into 3 categories including a category for false positives.

UNet [15] backbone architecture with an encoding-decoding module is selected to transform the input image to the segmentation mask. Moreover, to explore long-range contextual information inside 3D patches, we embed a dual attention module [3] on top of the UNet backbone. Finally, a combination of a binary cross-entropy loss and a Dice loss is used as the total segmentation loss. Please refer to the supplementary for more details.

2.2   Vessel Labeling

Due to the lack of the global information in the patch-wise 3D CNNs, the vessel segments obtained from the vessel segmentation procedure tend to contain some false-positive results like veins and also miss some parts of tiny or tortuous vessels. We propose to perform a vessel labeling procedure that classifies the segmented vessels into different branches. Then such semantic labels can be used to remove non-vascular segments and group discontinuous vessel segments, as shown in Fig. 2. The vessel labeling procedure is implemented by first generating a set of points which represent vascular skeleton from the vessel segmentation results and then using a point-cloud network to predict labels of these skeleton points.

Point-Cloud Generation. As it is inefficient to directly label vessel segments in 3D volumes using CNNs, we propose to perform the labeling on vascular skeletons represented by a set of points, which can reduce complexity and preserve original geometric information of vessels. To generate vascular skeletons from

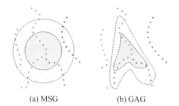

(a) MSG   (b) GAG

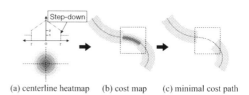

(a) centerline heatmap   (b) cost map   (c) minimal cost path

Fig. 4. Comparison of points grouping methods. (a) **Multi-scale grouping (MSG) in the PointNet++:** the grouping areas are multi-scale ϵ-sphere and grouped points may come from different skeleton components. (b) **Geometry-aware grouping (GAG):** the grouping areas trend to stretch along the lines and group points from the same islands. (Color figure online)

Fig. 5. Centerline Extraction. (a) **Centerline heatmap:** the heatmap has larger value in point closer to the centerline and the points laying outside the vascular radius are set to a step-down low value. (b) **Cost map:** the cost map is built by combining the heatmap and the skeleton lines. (c) **Minimal cost path:** based on the cost map, a minimal cost path can be extracted to connect the broken lines as the complete vessel centerline. (Color figure online)

the vessel segments, a 3D thinning algorithm [11] is applied to erode the vessel segments and finally obtain single-voxel-width skeleton points, as shown in Fig. 3. The generated skeletons are composed of discrete and unordered points lying on the center of the vascular lumen, which represent vascular geometry.

Vessel Labeling Network. Given a set of vascular skeleton points generated from vessel segments $\{Q_i \in \mathbb{R}^3, i \in 1, 2, ..., M\}$, the target of the vessel labeling network is to predict the label of each point, $f(Q_i) \rightarrow Y_j$ $(Y_j \in \mathbb{Z}, j \in 1, 2, ..., K)$, as shown in Fig. 3. It is similar to point-cloud semantic segmentation tasks in 3D computer vision. Any point-cloud network can be adopted, such as the state-of-the-art PointNet++ [14] and dynamic graph CNNs (DGCNN) [21]. We will provide a comparison between them in experiments.

Geometry-Aware Grouping. A particular property of the skeleton points compared to a general point cloud is the given connectivity among adjacent skeleton points. Specifically, skeleton points can be divided into separated components using a connected-component labeling (CCL) algorithm [12]. However, both PointNet++ and DGCNN are realized to group points by k-nearest neighbor (k-NN) algorithm based on L2 distance, which ignores the given geometry of the vessel skeletons. As demonstrated in Fig. 4(a), the L2-based methods are more likely to group points belonging to different components. To address this issue and leverage the skeletons' connection property, we propose a geometry-aware grouping method (GAG) to modify the distance based on the connection relationship. The modified distance between two points is computed by

$$\mathbf{D}_{i,j} = \begin{cases} \lambda_n \left\| \mathbf{x}_i - \mathbf{x}_j \right\|_2 & \text{if } \mathbf{x}_i, \mathbf{x}_j \in C_n. \\ (1 - \lambda_n) \left\| \mathbf{x}_i - \mathbf{x}_j \right\|_2 & \text{otherwise.} \end{cases} \tag{1}$$

where λ_n ($\lambda_n < 0.5$) is a weight, C_n is the n-th component. As shown in Fig. 4(b), the grouping area in GAG is prone to stretching along the skeleton lines and the points belonging to the same connected component are more likely to be grouped. This design facilitates local feature consistency in the same component and consequently improves the accuracy of vessel labeling as evaluated in the experiments.

2.3 Centerline Extraction

After vessel labeling, each vessel segment is assigned a semantic label. The semantic label can be used to remove non-vascular tissues and provide the guidance to connect disjointed vessel segments. Specifically, we first determine which segments should be connected and then use a minimal cost path method to connect them and output final centerlines based on a cost map. The cost map is constructed by a centerline heatmap regressed from another 3D CNNs and the labeled vessel skeletons.

Centerline Heatmap Regression. The centerline heatmap is defined as the opposite of a distance map, where points closer to the vessel centerline have larger values and points outside the vascular radius have a step-down low value, as shown in Fig. 5(a). Considering that the probability map obtained from the segmentation network has only learned the difference between background and the vessel (e.g. edge features) rather than the centerline feature within the vessel lumen, we adopt another 3D CNN similar to the network used in vessel segmentation to regress the centerline heatmap. The mean square error (MSE) loss is used to train the network. Based on the centerline heatmap, we construct a cost map to guide the minimal cost path search algorithm as shown in Fig. 5(b). In areas where vessel segments exist, the cost map directly assigns a large value to the skeleton points (red lines in Fig. 5(b)) and a small value to the points elsewhere (gray areas in Fig. 5(b)).

Minimal Cost Path. Given vessel skeletons and their labels, we successively merge the disjointed segments with the same label. Paired boundary points with the same label are put in a priority queue according to the distance between the two points in the pair. We take a pair successively from the queue and then the minimal cost path is found by Dijkstra algorithm [20] to connect the two boundary points, as shown in Fig. 5(c). If the two segments represented by the two points are connected in the previous step, we skip the pair and go on until the queue is empty.

3 Experiments

We evaluate the proposed method on two datasets: a public coronary artery dataset and a private head and neck artery dataset. The first dataset is mainly

Table 1. Performance comparison of automatic coronary artery centerline extraction methods. OV represents the completeness of extracted vessel centerline and is similar to Dice coefficient. OF determines the ratio of a coronary artery that has been extracted before making an error. OT gives an indication of how much the extracted centerline overlaps with the clinically relevant centerline reference (radius ≥ 0.75 mm).

Methods	OV (%)	OF (%)	OT (%)
ModelDrivenCenterline [28]	92.4	80.6	93.4
SupervisedExtraction [7]	90.6	70.9	92.5
GFVCoronaryExtractor [23]	93.7	74.2	95.9
DepthFirstModelFit [25]	84.7	65.3	87.0
AutoCoronaryTree [19]	84.7	59.5	86.2
Our approach	**95.8**	72.3	**96.3**

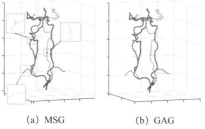

(a) MSG   (b) GAG

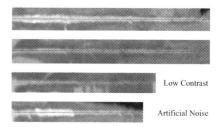

Low Contrast

Artificial Noise

Fig. 6. Comparison of different grouping methods used in vessel labeling networks. It can be seen that the proposed GAG method achieves better local consistency of the predicted vessel labels.

Fig. 7. Vessel curved-planar reconstruction (CPR) of the coronary artery based on the extracted centerline. Our approach can extract complete centerlines under serious imaging artifacts.

used to compare our method with existing baseline methods in literature. The second dataset is mainly used for ablative study to verify our system designs. Following [16], extracted centerlines are evaluated based on three metrics, namely total overlap (OV), overlap until first error (OF), and overlap with the clinically relevant part of the vessel (OT). The stage-wise results for the proposed framework are demonstrated in Fig. 8.

Implementation Details. In vessel segmentation, we randomly crop 3D patches with the size of $256 \times 256 \times 32$ for the head and neck dataset and $32 \times 32 \times 32$ for the coronary artery dataset. The ResNet34 is used as the encoder in the UNet architecture which starts with 32 feature channels that are doubled in each scale, and the max-pooling layer is removed from the original residual network. All convolutions are specified as $3 \times 3 \times 3$ kernels, except the last two ResNet blocks, which are $3 \times 3 \times 1$ to reduce the parameter count. We employ the Adam optimizer with a polynomial learning rate which equals to $0.001 \times (1 - \frac{iter}{total\,iter})^{0.9}$. In vessel labeling, the inputs are skeleton points,

Table 2. Performance comparison of different variants of vessel labeling networks on the private head and neck artery dataset. Four metrics are reported: the accuracy of vessel labeling and OV, OF and OT of four main vessel centerlines including left and right common carotid artery (L/RCCA), left and right vertebral artery (L/RVA).

Vessel Labeling	Accuracy (%)	OV (%)	OF (%)	OT (%)
PointNet [13]	92.2	96.1	84.5	96.6
PointNet++ [14]	95.1	96.5	**85.8**	97.0
PointNet++ (GAG)	97.0	**97.2**	85.7	**97.6**
DGCNN [21]	96.5	96.6	84.4	97.1
DGCNN (GAG)	**97.2**	97.0	84.9	97.4

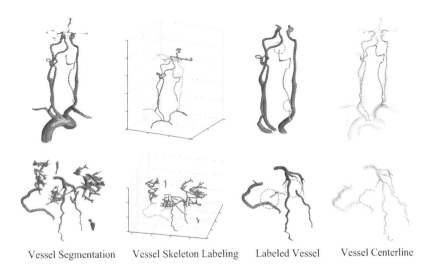

Vessel Segmentation   Vessel Skeleton Labeling   Labeled Vessel   Vessel Centerline

Fig. 8. Vessel segmentation, labeling and centerline extraction. Top: head and neck arteries. Note that the discontinuous RVA is connected in the centerline extraction procedure. **Bottom:** coronary arteries. Note that the non-vascular tissues are removed in the vessel labeling procedure and the discontinuous vessel segments are connected.

which are generated from vessel segments and resampled to 3000 points for each sample. The inital learning rate is set to 0.001, which is reduced by half every 30 epochs. The weight λ_n in GAG is set to be 0.3.

Experiments on the Coronary Artery Dataset. This public dataset contains 100 cardiac CT angiography (CCTA) scans collected from the clinic for training and 32 CCTA scans from [16] for evaluation. We train the vessel segmentation network and vessel labeling network on the annotated coronary artery CTA images and the vessel skeletons are labeled as three categories including right arteries, left arteries and false-positive venous vessels. According to the ablation study in the head and neck artery dataset, we use the Pointnet++ with the GAG module as our vessel labeling network. The quantitative comparison is listed in Table 1 and the visualization of results is showed in Fig. 7. Our hybrid approach achieves the highest performance in terms of OV and OT, respectively, indicating that the centerlines extracted by the proposed method are more complete than those produced by other methods.

Experiments on the Head and Neck Artery Dataset. This private dataset collected from the clinic contains 450 CTA scans, each of which has a manually annotated vessel mask. The dataset is split into 380 scans for training, 20 for validation and 50 for testing. In the dataset, vessel skeletons are labeled as 17 categories including left and right common carotid artery (L/RCCA), left and right vertebral artery (L/RVA), etc. Table 2 shows the evaluation results of several variants of our system with different point-cloud network designs. It can be seen that, with the geometry-aware grouping (GAG) method, the vessel labeling accuracy of both PointNet++ and DGCNN can be improved. Figure 6 shows the GAG can facilitate local consistency of the skeleton components.

4   Conclusions

We propose an automatic and template-free approach to 3D vessel centerline extraction based on hybrid representations, which ensures the connectivity of extracted centerlines. We show that the hybridization between learning local appearance with patch-based CNNs and learning global geometry with point-cloud networks results in an efficient and robust framework to extract geometric objects from 3D data. We demonstrate superior performance on artery centerline extraction from CTA images and believe that the proposed approach can also be applied in other centerline or skeleton extraction tasks.

Acknowledgements. This work is funded by National Key Research and Development Program of China (No. 2019YFC0118101), and is partially supported by National Key Research and Development Program of China with Grant No. 2018AAA0101900/2018AAA0101902, Beijing Municipal Commission of Science and Technology under Grant No. Z181100008918005, the National Natural Science Foundation of China (NSFC Grant No. 61772039 and No. 91646202), and Hong Kong Research Grant Council [12301417, 16307818, 16301419]; Hong Kong University of Science and Technology [R9405, IGN17SC02, Z0428].

References

1. Cetin, S., Unal, G.: A higher-order tensor vessel tractography for segmentation of vascular structures. IEEE Trans. Med. Imaging **34**(10), 2172–2185 (2015)
2. Deschamps, T., Cohen, L.D.: Fast extraction of minimal paths in 3D images and applications to virtual endoscopy. Med. Image Anal. **5**(4), 281–299 (2001)
3. Fu, J., et al.: Dual attention network for scene segmentation. In: CVPR (2019)
4. Gülsün, M.A., Funka-Lea, G., Sharma, P., Rapaka, S., Zheng, Y.: Coronary centerline extraction via optimal flow paths and CNN path pruning. In: Ourselin, S., Joskowicz, L., Sabuncu, M.R., Unal, G., Wells, W. (eds.) MICCAI 2016. LNCS, vol. 9902, pp. 317–325. Springer, Cham (2016). https://doi.org/10.1007/978-3-319-46726-9_37

5. Kamnitsas, K., Ledig, C., Newcombe, V.F., Simpson, J.P., Kane, A.D., Menon, D.K., Rueckert, D., Glocker, B.: Efficient multi-scale 3D CNN with fully connected CRF for accurate brain lesion segmentation. Medical image analysis **36**, 61–78 (2017)
6. Kanitsar, A., Fleischmann, D., Wegenkittl, R., Felkel, P., Gröller, M.E.: CPR: curved planar reformation. In: VIS. IEEE Computer Society (2002)
7. Kitamura, Y., Li, Y., Ito, W.: Automatic coronary extraction by supervised detection and shape matching. In: ISBI. IEEE (2012)
8. Li, H., Yezzi, A.: Vessels as 4-D curves: global minimal 4-D paths to extract 3-D tubular surfaces and centerlines. IEEE Trans. Med. Imaging **26**(9), 1213–1223 (2007)
9. Litjens, G., et al.: A survey on deep learning in medical image analysis. Med. Image Anal. **42**, 60–88 (2017)
10. Moriconi, S., Zuluaga, M.A., Jäger, H.R., Nachev, P., Ourselin, S., Cardoso, M.J.: VTrails: inferring vessels with geodesic connectivity trees. In: Niethammer, M., et al. (eds.) IPMI 2017. LNCS, vol. 10265, pp. 672–684. Springer, Cham (2017). https://doi.org/10.1007/978-3-319-59050-9_53
11. Palágyi, K., et al.: A sequential 3D thinning algorithm and its medical applications. In: Insana, M.F., Leahy, R.M. (eds.) IPMI 2001. LNCS, vol. 2082, pp. 409–415. Springer, Heidelberg (2001). https://doi.org/10.1007/3-540-45729-1_42
12. Park, J.M., Looney, C.G., Chen, H.C.: Fast connected component labeling algorithm using a divide and conquer technique. Comput. Their Appl. **4**, 4–7 (2000)
13. Qi, C.R., Su, H., Mo, K., Guibas, L.J.: Pointnet: deep learning on point sets for 3D classification and segmentation. In: CVPR (2017)
14. Qi, C.R., Yi, L., Su, H., Guibas, L.J.: Pointnet++: deep hierarchical feature learning on point sets in a metric space. In: NIPS (2017)
15. Ronneberger, O., Fischer, P., Brox, T.: U-Net: convolutional networks for biomedical image segmentation. In: Navab, N., Hornegger, J., Wells, W.M., Frangi, A.F. (eds.) MICCAI 2015. LNCS, vol. 9351, pp. 234–241. Springer, Cham (2015). https://doi.org/10.1007/978-3-319-24574-4_28
16. Schaap, M., et al.: Standardized evaluation methodology and reference database for evaluating coronary artery centerline extraction algorithms. Med. Image Anal. **13**(5), 701–714 (2009)
17. Selvan, R., Petersen, J., Pedersen, J.H., de Bruijne, M.: Extracting tree-structures in CT data by tracking multiple statistically ranked hypotheses (2018). arXiv preprint arXiv:1806.08981
18. Stefancik, R.M., Sonka, M.: Highly automated segmentation of arterial and venous trees from three-dimensional magnetic resonance angiography (MRA). Int. J. Cardiovasc. Imaging **17**(1), 37–47 (2001)
19. Tek, H., Gulsun, M.A., Laguitton, S., Grady, L., Lesage, D., Funka-Lea, G.: Automatic coronary tree modeling. Insight J., 1–8 (2008)
20. Verscheure, L., Peyrodie, L., Makni, N., Betrouni, N., Maouche, S., Vermandel, M.: Dijkstra's algorithm applied to 3D skeletonization of the brain vascular tree: evaluation and application to symbolic. In: EMBC. IEEE (2010)
21. Wang, Y., Sun, Y., Liu, Z., Sarma, S.E., Bronstein, M.M., Solomon, J.M.: Dynamic graph cnn for learning on point clouds (2018). arXiv preprint arXiv:1801.07829
22. Wolterink, J.M., van Hamersvelt, R.W., Viergever, M.A., Leiner, T., Išgum, I.: Coronary artery centerline extraction in cardiac CT angiography using a CNN-based orientation classifier. Med. Image Anal. **51**, 46–60 (2019)

23. Yang, G., Kitslaar, P., Frenay, M., Broersen, A., Boogers, M.J., Bax, J.J., Reiber, J.H., Dijkstra, J.: Automatic centerline extraction of coronary arteries in coronary computed tomographic angiography. Int. J. Cardiovasc. Imaging **28**(4), 921–933 (2012)

24. Yu, Q., Xie, L., Wang, Y., Zhou, Y., Fishman, E.K., Yuille, A.L.: Recurrent saliency transformation network: incorporating multi-stage visual cues for small organ segmentation. In: CVPR (2018)

25. Zambal, S., Hladuvka, J., Kanitsar, A., Bühler, K.: Shape and appearance models for automatic coronary artery tracking. The MIDAS Journal. In: MICCAI Workshop-Grand Challenge Coronary Artery Tracking (2008). http://hdl.handle.net/10380/1420

26. Zhai, Z., et al.: Automatic quantitative analysis of pulmonary vascular morphology in CT images. Med. Phys. **46**(9), 3985–3997 (2019)

27. Zhang, D.P., et al.: Coronary motion estimation from CTA using probability atlas and diffeomorphic registration. In: Liao, H., Edwards, P.J.E., Pan, X., Fan, Y., Yang, G.-Z. (eds.) MIAR 2010. LNCS, vol. 6326, pp. 78–87. Springer, Heidelberg (2010). https://doi.org/10.1007/978-3-642-15699-1_9

28. Zheng, Y., Tek, H., Funka-Lea, G.: Robust and accurate coronary artery centerline extraction in CTA by combining model-driven and data-driven approaches. In: Mori, K., Sakuma, I., Sato, Y., Barillot, C., Navab, N. (eds.) MICCAI 2013. LNCS, vol. 8151, pp. 74–81. Springer, Heidelberg (2013). https://doi.org/10.1007/978-3-642-40760-4_10

Branch-Aware Double DQN for Centerline Extraction in Coronary CT Angiography

Yuyang Zhang, Gongning Luo$^{(\boxtimes)}$, Wei Wang, and Kuanquan Wang$^{(\boxtimes)}$

Harbin Institute of Technology, Harbin, China
{luogongning,wangkq}@hit.edu.cn

Abstract. Accurate coronary artery centerline is essential for coronary stenosis analysis and atherosclerotic plaque analysis. However, the existence of many branches makes accurate centerline extraction a challenging task in coronary CT angiography (CCTA). In this paper, we proposed a branch-aware coronary centerline extraction approach (BACCE) based on Double Deep Q-Network (DDQN) and 3D dilated CNN. It consists of two parts: a DDQN based tracker and a branch-aware detector. The tracker can predict the next action of an agent accurately which can trace the centerline. The detector can detect the branch points and radius of coronary artery, and it makes our BACCE able to trace the branches automatically. Benefiting from the detector, our BACCE only needs one seed point to extract the entire coronary tree. Moreover, we proposed a new reward calculation based on dot product of two vectors and a new agent movement strategy based on twenty-six adjacent voxels, which were proved to improve tracing speed and accuracy. We evaluated the BACCE model on the public dataset in CAT08 challenge and experiment results demonstrated that our method achieved state-of-the-art results in terms of time-cost(7 s), OV(96.2%), OF(88.3%), OT(96.5%) and AI(0.21 mm) metrics. Moreover, we also demonstrated results on qualitative evaluation at the end. Source code and pre-trained models are publicly available: https://github.com/514sz/Branch-aware-centerline-extraction.

Keywords: Coronary centerline extraction · Branch-aware · Double Deep Q-network · Coronary CT angiography

1 Introduction

Coronary artery disease (CAD) is one of the leading causes of death worldwide [1]. Coronary CT angiography (CCTA) is the primary non-invasive imaging modality to diagnose CAD for suspicious patients [2,3]. Because of numerous branches, it is difficult to observe the entire coronary artery in CCTA directly. It is the first step to extract coronary centerline from CCTA, which is used to reconstruct coronary artery. During (semi) automatic centerline extraction procedure, tracing branches and passing through stenosis are one of the hardest problems as shown in Fig. 1.

© Springer Nature Switzerland AG 2020
A. L. Martel et al. (Eds.): MICCAI 2020, LNCS 12266, pp. 35–44, 2020.
https://doi.org/10.1007/978-3-030-59725-2_4

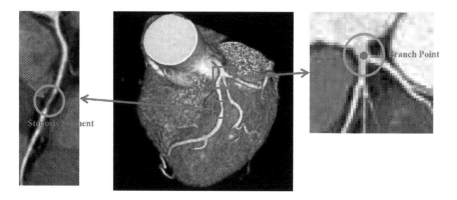

Fig. 1. The accurate centerline tracing in CCTA volumes is a very challenging task, especially in the branch and stenosis region.

The state-of-the-art methods can be mainly divided into three categories, the minimal cost path based method, the segmentation based method, and the iterative tracking based method. The **minimal cost path based** methods compute a minimal cost path between two points by a cost function, which is higher on other locations than on the centerline [4,5]. However, these methods need many interactions to avoid shortcuts off the centerline. The **segmentation based** methods firstly obtain a segmentation of the coronary artery and then extract the centerline [6–8]. Such methods require processing the entire 3D CCTA and are time consuming. For example, Yang et al. [7] used a vesselness filter to enhance coronary artery, which needed exploring the entire CCTA.

The **tracking based** methods depend on iterative tracing [9–14]. Recently, Zhang et al. [15] proposed a deep reinforcement learning based method to extract vessel centerline. This method learns an agent to trace centerline by collecting rewards from interaction with environment. But it cannot solve the problem of branch point detection, so this method cannot trace the entire coronary tree in the case of branches. It also cannot predict the vessel radius that is essential for vessel stenosis and plaque analysis. Wolterink et al. [16] and Yang et al. [17] proposed a CNNTracker and a DCAT, respectively. They are all based on 3D CNN and achieve state-of-the-art performance. However, they need some seeds on branches to extract the entire coronary tree. Each of these seeds extracts a single centerline connecting the ostium and the endpoint of a branch. It causes repeated extraction of the centerline on coronary trunk, which is time consuming.

In order to solve these problems, especially tracing branches and passing through stenoses as showed in Fig. 1, we proposed a branch-aware coronary centerline extraction method (BACCE), which depends on Double Deep Q-Network (DDQN) and 3D dilated CNN. DDQN and 3D dilated CNN implement a tracker and a detector, respectively. When a 3D patch centered at current point is fed to the tracker, it will predict the next action of an agent, which is used to trace coronary centerline. Simultaneously, the same patch is fed to the detector, it can identify proper radius at current location and whether current point is a branch point. Moreover, due to the detector, the BACCE model only needs one seed to extract the entire coronary tree. Benefiting from the learning-based nature, our BACCE can correctly pass through plaque segments or severe stenosis segments.

The contribution of this work is four-fold: (1) We proposed the first DDQN based coronary artery tracing method in CCTA. The tracker learned an agent that interacted with the local 3D patch and traced centerline by collecting rewards from the interaction. (2) We proposed a twenty-six adjacent voxels based movement strategy. It was proved to make the agent trace the centerline faster. (3) We proposed a new reward calculation method based on dot product of two vectors during training. By the dot product based method, the step-wise reward can be calculated more easily and efficiently. (4) We proposed a detector based on 3D dilated CNN to detect the coronary branches and identify lumen radius directly from image data. As a result, our BACCE only needs one seed to trace the entire coronary tree.

2   Methodology

The main architecture of the BACCE is shown in Fig. 2. The BACCE mainly contains a tracker and a detector. Firstly, the tracker predicts the next action of an agent that traces the centerline. To improve tracing efficiency, we proposed a dot product based reward calculation and a twenty-six adjacent voxels based movement strategy. Secondly, when the tracker encounters a branch point, we proposed the detector to detect coronary branches. Then the tracker goes into branches and continues to trace centerline. Moreover, the proposed detector can accurately identify the radius at current location. The detailed information about the BACCE model is described as follows.

2.1   Coronary Centerline Tracing

Given a ground truth centerline $G = [g_0, g_1, ..., g_n]$, the **tracker** learned to predict proper action for an agent to trace the centerline. We represented the optimal tracing path as $P = [p_0, p_1, ..., p_m]$. At step t, the agent received a 3D patch I_t centered at point p_t. Then, according to policy π, the agent chose a_t from action space A. An optimal policy π was defined as $Q^*(I_t, a_t)$. In this paper, the $Q^*(I_t, a_t)$ was approximated as

$$Q^*(I_t, a_t) = r_t + \gamma Q'(I_{t+1}, \arg\max_a Q(I_{t+1}, a)) \qquad (1)$$

where Q is the current network, Q' is the network from some previous iteration and γ is a reward decay coefficient. r_t is a reward measuring the transition from patch I_t to I_{t+1} through action a_t. To learn a better tracker, we used DDQN instead of DQN used in [15]. Compared with DQN, the accumulated reward in DDQN is higher than that in DQN.

Moreover, to improve tracing efficiency, we proposed a twenty-six adjacent voxels action space $A \{left, top-left, bottom-left, right, top-right ... left-$

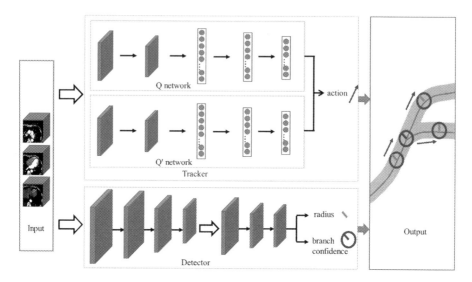

Fig. 2. The framework of the proposed BACCE. The accurate centerline tracing and coronary artery reconstruction can be achieved based on the effective tracker module and branch detection module in an end-to-end deep reinforcement learning framework.

$rear, right - rear\}$. As shown in Fig. 3(a), in the movement strategy used in [15], the agent needs three steps from p_1 to p_4 through p_2 and p_3. While in our movement strategy, the agent only needed one step from p_1 to p_4 directly, which made the tracing process more efficient.

We used the scalar reward $r_t = r_{I_t, a_t}^{I_{t+1}}$ measuring the transition from patch I_t to I_{t+1} through action a_t. Compared with [15], to make the reward calculation easier and more efficient , we proposed a new reward calculation based on dot product of two vectors as

$$r_t = (p_{t+1} - p_t) \cdot (g_{t+1} - g_t) \tag{2}$$

where p_t and p_{t+1} are the current location and the next location of the agent, respectively. g_t and g_{t+1} are closest points on the ground truth centerline to p_t and p_{t+1}, respectively.

2.2 Radius Identification and Branch Detection

To accurately identify radius and detect branches, we proposed the **detector** which worked with the tracker simultaneously. When the tracker reached a point p_t, the detector would identify the radius r of the coronary artery at p_t and detect whether p_t was a branch point. Once the output of the detector was higher than a specified threshold K, we would use ray-burst sampling [18] at p_t to detect branches as illustrated in Fig. 3(b). Compared with [18], we used ray-sampling only at branch points, not the entire coronary tree. It can greatly

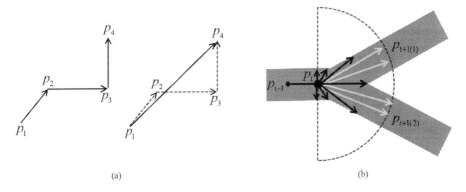

Fig. 3. a) Left: six adjacent voxels based movement strategy; Right: our high-efficiency movement strategy. b) A ray-burst sampling strategy used to detect branches.

reduce the number of ray-burst sampling execution, thereby saving the overall time of coronary artery tracing.

As shown in Fig. 3(b), sampling was restricted to a hemisphere, whose origin located at current point p_t , and the half angle ray represented direction $\overrightarrow{p_{t-1}p_t}$, where p_{t-1} was the last point. We extended all of these rays from current point p_t until their length exceeding a specified threshold (4.0 times the radius r that was identified by the detector), rays that were far away from the direction $\overrightarrow{p_{t-1}p_t}$ would reach the boundary of coronary artery, and their samplings would be terminated. According to the directions, the remaining rays that were still inside the vessel would be divided into different groups, which represented coronary branches. As shown in Fig. 3(b), some new points could be identified in branches. The agent would reach one of these points and trace along the corresponding branch.

Benefiting from the detector, the BACCE only needed one seed to extract the entire coronary tree. In [16] and [17], they used some seeds on branches, each of these seeds extracted a centerline connecting the ostium and the endpoint of a branch. These centerlines were overlapped on the coronary trunk, resulting in low efficiency.

2.3 Optimization and Application Strategy

Optimization Procedure. For optimizing **the tracker**, given a 3D CCTA, training samples were accumulated by experience replay. Then, the tracker calculated a maximum state-action value based on the current neural network Q among all twenty-six actions. When the agent reached the last point of the ground truth centerline G, current episode was terminated. Then, the tracker would be optimized in the next episode in a new CCTA. We defined the loss function of the tracker as

$$L_{track} = Q(I_t, a_t) - (r_t + \gamma Q'(I_{t+1}, \arg\max_a Q(I_{t+1}, a))) \tag{3}$$

where Q is the current network, Q' is the network from some previous iteration and γ is a reward decay coefficient. r_t is a scalar reward measuring the transition from patch I_t to I_{t+1} through action a_t.

For optimizing **the detector**, training samples were extracted along the ground truth centerline. The positive samples for the detector were 3D patches centered at branch points. And the negative samples were extracted at other location on the centerline. The radius label is the ground truth radius of the current location. During optimizing, we used the Adam optimizer to update the network parameters

$$L_{detect} = L_{branch} + \lambda L_{radius} \tag{4}$$

where L_{branch} is the binary cross-entropy loss for branch detection task, L_{radius} is the squared error loss for radius regression task. λ is a weight coefficient.

Application Procedure. Given a 3D CCTA, a start point p_0 was set at the coronary ostium. We fed a patch I_{p_0} centered at p_0 into our BACCE model. Firstly, the detector identified the radius of the coronary artery at p_0. Then, by the neural network Q, the tracker selected an action a_0 which moved the agent to p_1. With the movement of the agent, a new patch was updated as I_{p_1}. Then the detector identified the radius at p_1 and outputted a branch confidence that p_1 was a branch point. When the confidence was higher than a threshold K, we would use the ray-burst sampling at point p_1 to detect branches and the tracker would keep tracing detected branches. This process was repeated until the agent moved out of the CCTA or reached a point already visited before.

3   Experiments and Results

3.1   Network Architecture and Implementation

The experiments were conducted on CAT08 dataset [19]. The CAT08 dataset was resampled to voxel spacing $v = 0.5$. The patch shape was $w \times w \times w$ and w was 19. The number N of rays on the hemisphere was 100. To augment data, we used random rotation and translation. The experiments were implemented using Tensorflow with an NVIDIA RXT2080ti GPU.

For the tracker, Q and Q' had the same architecture, which was the same as that in [15]. As shown in Fig. 2, it consisted of two conv layers and three fully connected layers and state-action value outputs for twenty-six actions. The detailed architecture is introduced in [15]. The parameters of Q' were frozen and updated every 5,000 iterations. We used a set of 50,000 samples as experience replay. The batch size was 8 during optimizing. γ was 0.9. We trained the tracker up to 300 epochs and initial learning rate was 0.0005.

For the detector, the architecture was the same as that in [17]. As shown in Fig. 2, it consisted of four dilated conv layers and three conv layers and two outputs: radius and branch confidence. The detailed architecture is introduced in [17]. In the loss functions L_{detect}, λ was 15. We set the threshold $K = 0.5$ for the branch confidence. We trained the detector up to 200 epochs and initial learning rate was 0.001.

3.2   Experimental Results on CAT08 Dataset

For clinical applications, quantitative evaluation of experiment results is very important [20,21]. According to the Rotterdam Coronary Artery Algorithm Evaluation Framework [19], we evaluated our results on time-cost, overlap (OV), overlap until first error (OF), overlap of clinical relevant part (OT) and average inside accuracy (AI) metrics. In CAT08 dataset, there are 8 training CCTA scans, each of which contains four major vessels with point-wise centerlines and radii annotations. To evaluate our BACCE on the training CCTA scans, we adopted leave-one-out cross-validation. We compared our BACCE with several state-of-the-art methods as shown in Table 1.

Table 1. Performance comparison between the proposed method and state-of-the-art methods.

Method	Time (s) (per data)	OV (%)			OF (%)			OT (%)			AI (mm)		
		Min.	Max.	Avg.	Min.	Max.	Avg.	Min.	Max.	Avg	Min.	Max.	Avg.
HOT [10]	<30	66.6	100.0	97.3	10.0	100.0	85.0	67.2	100.0	97.7	0.22	0.51	0.34
MDC [8]	60	68.7	99.9	92.4	17.1	100.0	80.6	69.7	100.0	93.4	**0.12**	0.47	0.21
CNNTracker [16]	10	61.3	100.0	95.7	12.9	100.0	87.1	62.2	100.0	97.1	0.14	0.43	0.23
MHT [22]	360	**94.5**	100.0	99.3	23.9	100.0	94.6	**94.5**	100.0	**99.5**	0.16	0.37	0.24
DCAT [17]	10	83.1	100.0	**99.5**	**71.6**	100.0	**99.1**	83.1	100.0	**99.5**	0.15	**0.24**	**0.19**
BACCE (Ours)	**7**	67.2	**100.0**	96.2	13.5	**100.0**	88.3	67.7	**100.0**	96.5	0.17	0.41	0.21

As reported in Table 1, our BACCE had achieved state-of-the-art results in terms of OV, OF, OT and AI. In aspect of time cost, our method only took 7 s to extract the entire coronary tree, it was the fastest of all the public methods. Before this, the fastest method CNNTracker [16] needed 10 s to extract the entire coronary tree. MHT [22] needed 6 min, ModelDrivenCenterline [8] needed 1 min and HOTtractography [10] needed less than 30 s. Moreover, compared with other methods, our BACCE only needed one seed to extract the entire coronary tree, it was more flexible and can reduce human labor effectively. As a comparison, MHT [22] needed 2–5 points per vessel. CNNTracker [16] and DCAT [17] needed some seed points on the branches to extract the entire coronary tree.

Table 2. The performance of the proposed method with different ablation configurations. BACCE-6 denotes the proposed method with 6 adjacent voxels action space, BACCE-26 denotes the proposed method with 26 adjacent voxels action space.

Method	time (s) (per data)	OV (%)			OF (%)			OT (%)			AI (mm)		
		Min.	Max.	Avg.	Min.	Max.	Avg.	Min.	Max.	Avg.	Min.	Max.	Avg.
BACCE-6	12	66.3	100.0	94.5	13.1	100.0	87.3	66.1	100.0	94.3	0.19	0.43	0.23
BACCE-26	7	**67.2**	**100.0**	**96.2**	**13.5**	**100.0**	**88.3**	**67.7**	**100.0**	**96.5**	**0.17**	**0.41**	**0.21**

We also conducted experiments to show that the twenty-six adjacent voxels based movement strategy could speed up centerline tracing and get better extraction results. We evaluated the six-actions movement strategy and our movement strategy on CAT08 dataset, respectively. As illustrated in Table 2, BACCE-26 surpassed BACCE-6 in terms of all metrics. Moreover, we also demonstrated our results on qualitative evaluation. As shown in Fig. 4(a), our BACCE could extract the centerline accurately. Benefiting from learning based nature, our method could extract the centerline successfully and accurately, even in the presence of stenosis and plaque. Moreover, as shown in Fig. 4(b), the proposed BACCE also can achieve excellent 3D reconstruction of the coronary artery based on the accurate centerline extraction and radius estimation.

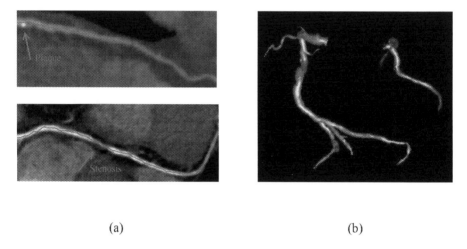

(a)   (b)

Fig. 4. The accurate coronary artery tracing and reconstruction a) The qualitative results on multiple 2D views. b) The qualitative results of coronary artery reconstruction on 3D space.

4   Conclusion

In this paper, we introduced a branch-aware coronary centerline extraction method (BACCE) via DDQN and 3D dilated CNN. The BACCE only needed one seed to extract the entire coronary tree. In addition, we proposed a twenty-six adjacent voxels based movement strategy and a dot product based reward calculation, which made BACCE be the fastest coronary tracing method of all the public methods. Moreover, benefiting from the learning-based nature, our BACCE can extract centerline successfully and accurately, even in the presence of stenosis and plaque. At last, the results of qualitative evaluation also illustrated the robustness and accuracy of our method.

Acknowledgements. This work was supported by China Postdoctoral Science Foundation under Grant 2020M670911.

References

1. Leipsic, J., et al.: SCCT guidelines for the interpretation and reporting of coronary CT angiography: a report of the society of cardiovascular computed tomography guidelines committee. J. Cardiovasc. Comput. Tomogr. **8**(5), 342–358 (2014)
2. Dong, S., et al.: Deep atlas network for efficient 3D left ventricle segmentation on echocardiography. Med. Image Anal. **61**, 101638 (2020)
3. Luo, G., Dong, S., Wang, K., Zuo, W., Cao, S., Zhang, H.: Multi-views fusion CNN for left ventricular volumes estimation on cardiac MR images. IEEE Trans. Biomed. Eng. **65**(9), 1924–1934 (2017)
4. Wink, O., Frangi, A.F., Verdonck, B., Viergever, M.A., Niessen, W.J.: 3D MRA coronary axis determination using a minimum cost path approach. Magn. Reson. Med. Official J. Int. Soc. Magn. Resonance in Medicine **47**(6), 1169–1175 (2002)
5. Krissian, K., Bogunovic, H., Pozo, J., Villa-Uriol, M., Frangi, A.: Minimally interactive knowledge-based coronary tracking in CTA using a minimal cost path. Insight J. **2018**, 1–8 (2008)
6. Tetteh, G., et al.: Deepvesselnet: Vessel segmentation, centerline prediction, and bifurcation detection in 3-d angiographic volumes. arXiv preprint arXiv:1803.09340 (2018)
7. Yang, G., et al.: Automatic centerline extraction of coronary arteries in coronary computed tomographic angiography. The Int. J. Cardiovasc. Imaging **28**(4), 921–933 (2012)
8. Zheng, Y., Tek, H., Funka-Lea, G.: Robust and accurate coronary artery centerline extraction in CTA by combining model-driven and data-driven approaches. In: Mori, K., Sakuma, I., Sato, Y., Barillot, C., Navab, N. (eds.) MICCAI 2013. LNCS, vol. 8151, pp. 74–81. Springer, Heidelberg (2013). https://doi.org/10.1007/978-3-642-40760-4_10
9. Zhou, C., et al.: Automated coronary artery tree extraction in coronary CT angiography using a multiscale enhancement and dynamic balloon tracking (MSCAR-DBT) method. Comput. Med. Imaging Graph. **36**(1), 1–10 (2012)
10. Cetin, S., Unal, G.: A higher-order tensor vessel tractography for segmentation of vascular structures. IEEE Trans. Med. Imaging **34**(10), 2172–2185 (2015)
11. Lesage, D., Angelini, E.D., Funka-Lea, G., Bloch, I.: Adaptive particle filtering for coronary artery segmentation from 3D CT angiograms. Comput. Vis. Image Underst. **151**, 29–46 (2016)
12. Yin, Y., Adel, M., Bourennane, S.: Retinal vessel segmentation using a probabilistic tracking method. Pattern Recogn. **45**(4), 1235–1244 (2012)
13. Xiao, R., Yang, J., Li, T., Liu, Y.: Ridge-based automatic vascular centerline tracking in x-ray angiographic images. In: Yang, J., Fang, F., Sun, C. (eds.) IScIDE 2012. LNCS, vol. 7751, pp. 793–800. Springer, Heidelberg (2013). https://doi.org/10.1007/978-3-642-36669-7_96
14. Friman, O., Hindennach, M., Kühnel, C., Peitgen, H.O.: Multiple hypothesis template tracking of small 3d vessel structures. Med. Image Anal. **14**(2), 160–171 (2010)
15. Zhang, P., Wang, F., Zheng, Y.: Deep reinforcement learning for vessel centerline tracing in multi-modality 3D volumes. In: Frangi, A.F., Schnabel, J.A., Davatzikos, C., Alberola-López, C., Fichtinger, G. (eds.) MICCAI 2018. LNCS, vol. 11073, pp. 755–763. Springer, Cham (2018). https://doi.org/10.1007/978-3-030-00937-3_86
16. Wolterink, J.M., van Hamersvelt, R.W., Viergever, M.A., Leiner, T., Išgum, I.: Coronary artery centerline extraction in cardiac CT angiography using a CNN-based orientation classifier. Med. Image Anal. **51**, 46–60 (2019)

17. Yang, H., Chen, J., Chi, Y., Xie, X., Hua, X.: Discriminative coronary artery tracking via 3D CNN in cardiac CT angiography. In: Shen, D., et al. (eds.) MICCAI 2019. LNCS, vol. 11765, pp. 468–476. Springer, Cham (2019). https://doi.org/10.1007/978-3-030-32245-8_52

18. Ming, X., et al.: Rapid reconstruction of 3D neuronal morphology from light microscopy images with augmented rayburst sampling. PloS One **8**(12), e84557 (2013)

19. Schaap, M., et al.: Standardized evaluation methodology and reference database for evaluating coronary artery centerline extraction algorithms. Med. Image Anal. **13**(5), 701–714 (2009)

20. Luo, G., Wang, W., Tam, C., Wang, K., Li, S.: Dynamically constructed network with error correction for accurate ventricle volume estimation. Med. Image Anal. **64**, 101723 (2020)

21. Luo, G., Dong, S., Wang, W., Wang, K., Li, S.: Commensal correlation network between segmentation and direct area estimation for bi-ventricle quantification. Med. Image Anal. **59**, 101591 (2019)

22. Friman, O., Kühnel, C., Peitgen, H.O.: Coronary centerline extraction using multiple hypothesis tracking and minimal paths. In: Proceedings of the MICCAI, vol. 42 (2008)

Automatic CAD-RADS Scoring Using Deep Learning

Felix Denzinger[1,2]([⊠]), Michael Wels[2], Katharina Breininger[1],
Mehmet A. Gülsün[2], Max Schöbinger[2], Florian André[3], Sebastian Buß[3],
Johannes Görich[3], Michael Sühling[2], and Andreas Maier[1]

[1] Pattern Recognition Lab, Universität Erlangen-Nürnberg, Erlangen, Germany
`felix.denzinger@fau.de`
[2] Computed Tomography, Siemens Healthcare GmbH, Forchheim, Germany
[3] Das Radiologische Zentrum - Radiology Center,
Sinsheim-Eberbach-Erbach-Walldorf-Heidelberg, Heidelberg, Germany

Abstract. Coronary CT angiography (CCTA) has established its role
as a non-invasive modality for the diagnosis of coronary artery disease
(CAD). The CAD-Reporting and Data System (CAD-RADS) has been
developed to standardize communication and aid in decision making
based on CCTA findings. The CAD-RADS score is determined by manual assessment of all coronary vessels and the grading of lesions within
the coronary artery tree.

We propose a bottom-up approach for fully-automated prediction of
this score using deep-learning operating on a segment-wise representation of the coronary arteries. The method relies solely on a prior fully-
automated centerline extraction and segment labeling and predicts the
segment-wise stenosis degree and the overall calcification grade as auxiliary tasks in a multi-task learning setup.

We evaluate our approach on a data collection consisting of 2,867
patients. On the task of identifying patients with a CAD-RADS score
indicating the need for further invasive investigation our approach
reaches an area under curve (AUC) of 0.923 and an AUC of 0.914 for
determining whether the patient suffers from CAD. This level of performance enables our approach to be used in a fully-automated screening
setup or to assist diagnostic CCTA reading, especially due to its neural
architecture design – which allows comprehensive predictions.

Keywords: Coronary Artery Disease · Coronary CT Angiography ·
Deep learning · Data representation · CAD-RADS

1 Introduction

Coronary Artery Disease (CAD), which may lead to major adverse events like
cardiac infarction or significantly decrease quality of life in the form of coronary ischemia, remains the most common cause of death [7]. Most kinds of CAD

© Springer Nature Switzerland AG 2020
A. L. Martel et al. (Eds.): MICCAI 2020, LNCS 12266, pp. 45–54, 2020.
https://doi.org/10.1007/978-3-030-59725-2_5

result from atherosclerotic plaque deposits aggregating in the vessel wall creating a stenosis, hence narrowing the vessel and obstructing the blood flow. The plaque lesions are categorized by the degree of stenosis into no (0%), minimal (1–24%), mild (25–49%), moderate (50–69%), severe stenosis (70–99%), and occluded vessel (100%) [2].

Coronary CT Angiography (CCTA) is a common non-invasive rule-out modality for CAD due to its high negative predictive value. In order to standardize communication and guide patient management, the CAD-RADS score based on above mentioned stenosis grades was introduced [2]. It ranges between 0 and 5 and is strongly influenced by the degree of the severest stenosis within a patient. Additionally, this score is influenced by the location of the lesion and includes qualitative assessments based on the experience of the physician, especially in edge-cases.

From a high-level perspective for the case of stable CAD, the resulting patient management decision can be divided into three options: the patient has no CAD and does not need any treatment in the direction of CAD (0), the patient has a non-obstructive CAD (1–2) without need for further investigation, or the patient has an obstructive CAD and should undergo a further functional investigation or direct intervention (3–5).

Therefore, at least these clinical questions need to be answered by an assisting image analysis tool: in the rule-out case, the CAD-RADS 0 score needs to be differentiated from 1–5, and in the hold-out case, the CAD-RADS scores 0–2 need to be differentiated from 3–5. However, prediction on an even finer scale is necessary when the exact required action needs to be identified.

In clinical practice, the assessment of the CAD-RADS score is cumbersome, since the whole coronary tree needs to be assessed and the severest lesion is graded manually based on experience and eyeballing, which is prone to error. Therefore, approaches to ease the workflow and help to detect and grade stenotic lesions have been developed in recent years. Previous approaches focus on detection and quantification of stenoses and are based on the segmentation of the entire coronary tree [5,10], which is time consuming and often needs manual correction [12].

Recently, deep-learning approaches [6] without the need for a prior segmentation were introduced [1,11,14]. These methods operate on multi-planar reformatted (MPR) image stacks which are extracted by interpolating orthogonal planes for each centerline point of the vessel. Approaches for this task include a recurrent convolutional neural network (RCNN) [14], a 2D texture-based multiview [11], and a 3D CNN approach [1]. A 2D CNN approach, which classifies the whole CCTA volume scaled down and placed in a 2D grid, is described in [8], but might have optimistic results since the training and test splits are described not to be patient-wise.

However, most of the above approaches have the disadvantage of determining the patient score based on single lesions, again introducing a large amount of potential error sources, with no global context incorporated into the decision.

To overcome these pitfalls, we propose a bottom-up approach to directly predict the patient-level CAD-RADS score using a deep-learning based approach that leverages a task-specific hierarchical data representation building up on the coronary tree segments as defined by the American Health Association (AHA) norm. By having the segment-wise stenosis degree as an additional output and by utilizing a global max pooling operation, which identifies the most relevant features across the whole coronary tree, the network is designed to be comprehensive. Additionally, since all steps in the workflow of our approach can be automated, it can be used for patient screening as well as a preprocessing utility to ease and speed up the clinical workflow.

2 Data

We train and evaluate our methods on a data set consisting of CCTA scans from 2,867 patients collected at a single site.

For each patient, labels regarding the stenosis degree were given on a segment-level as no-stenosis, minimal, mild, moderate, severe or occluded with frequencies of 3,625, 34,889, 4,565, 2,324, 722 and 70 and on patient-level with frequencies of 53, 940, 861, 611, 352 and 50. Furthermore, the CAD-RADS score was annotated on the patient-level with categories 0–5 with frequencies of 436, 584, 873, 568, 348 and 58 [2]. The difference between the patient-wise stenosis degree and the CAD-RADS score can be explained by edge-cases and is especially severe in the CAD-RADS 0 case, since lesions with very minor wall irregularities were classified as minimal according to literature [2]. Additionally, for a subset of 2,828 patients, the Agatston scores were annotated based on additional calcium scoring scans, which were utilized in a binned version according to Rumberger et al. [9] as no, minimal, mild, moderate and severe calcifications with frequencies of 911, 317, 649, 491 and 460.

The data collection did not include patients with stents or bypass grafts. It was split into two parts with two thirds (1,899) used for training and one third (968) used for testing.

3 Methods

Preprocessing. For each patient, centerlines are automatically extracted using the algorithm described by Zheng et al. [13] and assigned to the AHA segments [4]. The extracted AHA segment centerlines are used to create MPR image stacks, which are then resized to the mode segment length resulting in a subvolume of size $128 \times 32 \times 32$ for each segment according to Denzinger et al. [3]. Subsequently, the Hounsfield Unit (HU) value range is clipped between -324 and 1,176 HU and normalized to a value range of $[0, 1]$. In order to focus on the more important sections and prevent error propagation from mislabeled AHA-segments, we only select a subset of AHA-segments (RCA_p, RCA_m, RCA_d, LM, LAD_p, LAD_m, LAD_d, LAD_D1, CX_m, CX_d, RAMUS), which were more robustly labeled according to Gülsün et al. [4]. We confirmed to reach

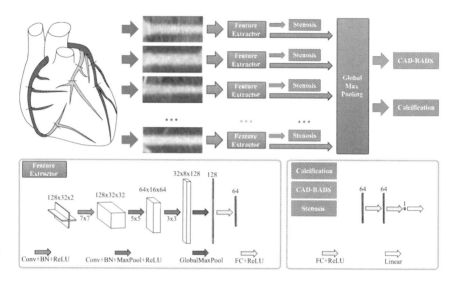

Fig. 1. Model overview (Conv = 2D Convolutional Layer; BN = Batch Normalization Layer; MaxPool = 2D MaxPooling Layer; ReLU = Rectified Linear Unit; FC = Fully-Connected Layer).

similar performance with this subset compared to utilizing all segments in preliminary experiments.

Neural Architecture Design. The general workflow of the proposed method is outlined in Fig. 1. Since our classes are ordered, we reformulate our classification task as a regression problem. This carries the benefit that misclassifications are penalized stronger depending on the class distance, which is convenient since misclassifications between neighboring classes are not as severe for our task. In order to reformat the whole coronary tree in a reasonable representation for neural network training, we divided the whole coronary tree into its sub-segments and extracted straightened MPR volumes. Since we assume all segments to be able to contribute equally, we utilize a feature extractor block with shared weights across all segments to extract spatial features. The feature extractor blocks work on a 2.5D representation utilizing a simple convolutional neural network (CNN) [3]. This architecture choice is motivated by the fact that we strived for simple building blocks to reduce the overall computational effort. Furthermore, we validated in prior experiments that adding additional views or having a feature extractor block similar to the method of [14] did not improve the performance. Since we do not want our model to depend on the location of the stenosis within one segment we choose to decouple the spatial features using a global max pooling operation. A fully-connected layer is used as the last layer of the feature extractor block in order to weight and combine the features such that our different targets can influence each other in the multi-task learning setup. The output of the feature extractor block is then either processed by a stenosis regression block with shared weights across segments to predict the stenosis degree of each segment or the maximum feature responses across all segments are extracted by a global max pooling layer. These global maximum feature responses are then fed into two further regression blocks for the CAD-RADS and calcification score prediction. This architecture choice is motivated by the definition of the CAD-RADS score as being heavily influenced by the severest lesion. Furthermore, the use of global max pooling allows the network to be more comprehensible since the regions with the highest activations as determined by the network can be displayed to the physician.

Evaluation. In order to evaluate the effectiveness of the use of multi-task learning, we evaluated our approach on three different configurations: directly regressing the CAD-RADS score (CAD-RADS), additionally regressing the segment-wise stenosis scores (CAD-RADS $+ \asymp$) and also regressing the calcification score (CAD-RADS $+ \asymp + $ Ca). Furthermore, to verify whether the global context introduced by our architecture improves the performance, we also evaluate the combination of the feature extractor block and the stenosis regression block with the severest prediction being propagated to the patient-level (Patient-level $\asymp$), which is as close as we can get to related work algorithms with our given labels. The training set is split into five folds of actual training and validation data (80%/20%). The model with the overall lowest loss on the validation set is used as a checkpoint for later evaluation. We choose the Adam optimizer with a learning rate of 0.0001, a batch size of 32 and mean squared error loss for all targets. Furthermore, we utilize data augmentation in the form of rotations around the centerline and minor shifts in x and y direction. In all experiments involving the segment-wise stenosis grade, the feature extractor block is pretrained on the stenosis grade on segment-level before getting integrated into the full model. This is done to condition the feature extractor block towards learning relevant features for the prediction of the stenosis degree. In order to convert our regressed predictions back into classes, we enforce the binned predictions to have the same class distribution as the ground truth labels. The thresholds used for this are calculated on the training set and propagated to the test set.

4   Results

As mentioned in Sect. 1, most reference approaches perform the classification of the severeness on a per-lesion-level with only Zreik et al. [14] performing an evaluation on the patient-level. However, the severest lesion per patient is not equivalent to the CAD-RADS score and differs especially often in the CAD-RADS 0 case (see Sect. 2), hence complicating a direct comparison.

CAD-RADS Performance. Before analyzing the clinical tasks at hand (rule-out/hold-out), we want to analyze the performance of our approach under different configurations for all six classes. Results for our baseline (severest lesion

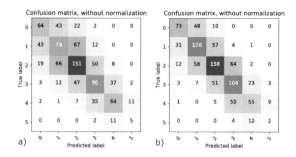

Fig. 2. Example confusion matrices of a single fold on the target of predicting the CAD-RADS using the maximum segment-wise prediction (a) and our proposed multi-task learning procedure (b).

score as patient score) approach and our full model are given in Fig. 2 and Table 1. By leveraging multi-task learning we are able to boost the performance of our approach incrementally (Table 1) from an accuracy of 0.810 to 0.840. While the baseline approach performs better compared to direct CAD-RADS scoring without auxiliary targets, we manage to outperform it in the multi-task setup. The biggest performance difference in comparison to the baseline are the lower CAD-RADS scores since in these cases overestimation of single-segment stenoses degrees are especially severe. As displayed in Fig. 2, the hardest class to identify was CAD-RADS 5. An explanation for this is the fact that the centerline extraction fails in the case of occluded vessels. Our method has a low specificity due to the high class imbalance for the single class metrics. Apart from this, most misclassifications are within one class distance, especially in our multi-task learning setup, which is a good feature with respect to the confidence in the network decision.

Table 1. Mean performance on the six class problem of the baseline approach and the three different multi-task learning network configurations ($\bowtie$ = segment-wise stenosis grade; Ca = patient-wise calcification grade; MCC = Matthews Correlation Coefficient).

Approach/Metric	Accuracy	Sensitivity	Specificity	MCC
Patient-level $\bowtie$	0.825	0.895	0.476	0.371
CAD-RADS	0.810	0.886	0.430	0.316
CAD-RADS + $\bowtie$	0.832	0.899	0.496	0.395
CAD-RADS + $\bowtie$ + Ca	**0.840**	**0.904**	**0.520**	**0.424**

Rule-Out. On the task of classifying whether a patient suffers from CAD, we see incremental improvements in the performance of our method with each auxiliary target from an AUC of 0.860 to 0.894 to 0.914 (Fig. 3a and Table 2). The

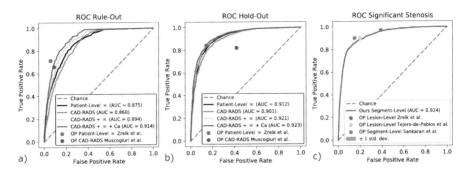

Fig. 3. Results: a) Mean receiver operating characteristic (ROC) curves for the rule-out case. The operating points (OP) of Zreik et al. [14] and Muscogiuri et al. [8] refer to metrics calculated on their data set with Zreik et al. operating on the related task of classifying the severest stenosis degree. b) Mean ROC curves for the hold-out case. c) Mean ROC curve for the classification of significant stenoses. Prediction in Zreik et al. [14] and Tejero-de-Pablos et al. [11] is performed on a per-lesion level and Sankaran et al. [10] utilize the vessel segmentation as additional preprocessing.

performance boost of utilizing the calcification grade can be explained by the fact that patients without CAD should not exhibit any calcifications in the coronary arteries. Also the baseline approach of propagating the severest segment-prediction to the patient-level only reaches an AUC of 0.875 compared to the 0.914 of our full model. Furthermore, there is a severe gap between sensitivity and specificity due to class imbalance. However, as the ROC curve (Fig. 3a) indicates an operating point with both sensitivity and specificity above 0.800 – which is often times required in a clinical setting – can be selected.

Table 2. Results for the rule-out case (predicting CAD-RADS 0 vs 1–5). Results of Zreik et al. [14] refer to the related but different task of predicting the severest stenosis degree on a different data set (abbreviations as in Table 1).

Approach/Metric	Patients	AUC	Accuracy	Sensitivity	Specificity	MCC
Patient-level ⋉	955	0.875	0.865	0.508	0.921	0.430
CAD-RADS	955	0.860	0.849	0.489	0.907	0.384
CAD-RADS + ⋉	955	0.894	0.875	0.510	0.933	0.456
CAD-RADS + ⋉ + Ca	955	**0.914**	**0.888**	**0.532**	**0.945**	**0.504**
Zreik et al. [14]	65	–	0.892	0.714	0.941	0.674
Muscogiuri et al. [8]	284	0.89	0.863	0.660	0.909	0.558

Hold-Out. In the hold-out case, the use of auxiliary tasks did not boost the performance as much as for the other targets (Fig. 3b and Table 3), with the biggest gain caused by adding the segment-wise stenosis degree. However, we outperform our baseline with an AUC, accuracy and MCC of 0.923, 0.860 and 0.692.

Table 3. Results for the hold-out case (predicting CAD-RADS 0–2 vs 3-5). (abbreviations as in Table 1)

Approach/Metric	Patients	AUC	Accuracy	Sensitivity	Specificity	MCC
Patient-level ⋈	955	0.912	0.850	0.885	0.781	0.666
CAD-RADS	955	0.901	0.838	0.879	0.759	0.640
CAD-RADS + ⋈	955	0.921	0.858	**0.895**	0.787	0.684
CAD-RADS + ⋈ + Ca	955	**0.923**	**0.860**	0.891	**0.802**	**0.692**
Zreik et al. [14]	65	–	0.846	0.841	0.857	0.671
Muscogiuri et al. [8]	284	0.78	0.711	0.822	0.583	0.420

Auxiliary Targets. For the target of predicting the stenosis degree on a segment-wise level, we reach results comparable to state-of-the-art methods when looking at the binary case of predicting significant stenosis ($>50\%$) (Fig. 3c). It should be noted that competing methods are evaluated on different data sets and use labels on lesion-level with defined start and end points, which require a remarkable amount of effort for annotation. Furthermore, our performance on this level enables that segments with the highest score are highlighted in order to aid physicians in their decision making process.

On the task of predicting our calcification grade (as defined in Sect. 1) we are able to reach a mean accuracy of 0.878.

5   Conclusion

In clinical practice, a standardized way to report CAD from CCTA scans is the CAD-RADS score. To the best of our knowledge – this work presents and evaluates the first approach to directly predict the six class CAD-RADS score using a deep-learning based algorithm. By leveraging two auxiliary tasks – the prediction of the segment-wise stenosis grade and a patient-wise calcification grade – we boosted the performance of our method. The method only relies on a prior centerline extraction and AHA segment label but not on the segmentation of the coronary tree, which is time-consuming to obtain and may need manual correction. Our approach is able to robustly identify patients suffering from CAD (AUC 0.914) or requiring further clinical investigation (AUC 0.923). Segments with severe lesions can be identified by our approach due to the neural architecture design and since we predict segment-wise stenosis with the same network. We validated our approach on a data set of 2,867 patients, a data set considerably larger compared to what has been reported in related work.

Still, the used 2.5D data representation of the single segments may omit some 3D information. We expect this to be successfully addressed by using test augmentation or utilizing additional views in future work. Within this study, it was not possible to apply algorithms defined in related work to our data set, since our stenosis degree labels were segment-wise and not on a per lesion-level. Still, with our experimental design we address this issue in order to allow for a fair comparison. Furthermore, the definition of CAD-RADS also includes report modifiers related to high-risk plaques, stents and bypass grafts which will be addressed in future work.

Disclaimer. The methods and information here are based on research and are not commercially available.

References

1. Candemir, S., et al.: Coronary Artery Classification and Weakly Supervised Abnormality Localization on Coronary CT Angiography with 3-Dimensional Convolutional Neural Networks. arXiv preprint arXiv:1911.13219 (2019)
2. Cury, R.C., et al.: Coronary artery disease-reporting and data system (CAD-RADS): an expert consensus document of SCCT, ACR and NASCI: endorsed by the ACC. JACC CI **9**(9), 1099–1113 (2016)
3. Denzinger, F., et al.: Deep learning algorithms for coronary artery plaque characterisation from CCTA scans. Bildverarbeitung für die Medizin 2020. I, pp. 193–198. Springer, Wiesbaden (2020). https://doi.org/10.1007/978-3-658-29267-6_42
4. Gülsün, M.A., Funka-Lea, G., Zheng, Y., Eckert, M.: CTA coronary labeling through efficient geodesics between trees using anatomy priors. In: Golland, P., Hata, N., Barillot, C., Hornegger, J., Howe, R. (eds.) MICCAI 2014. LNCS, vol. 8674, pp. 521–528. Springer, Cham (2014). https://doi.org/10.1007/978-3-319-10470-6_65
5. Kirişli, H., et al.: Standardized evaluation framework for evaluating coronary artery stenosis detection, stenosis quantification and lumen segmentation algorithms in computed tomography angiography. Med. Image Anal. **17**(8), 859–876 (2013)
6. Maier, A., Syben, C., Lasser, T., Riess, C.: A gentle introduction to deep learning in medical image processing. Zeitschrift für Medizinische Physik **29**(2), 86–101 (2019)
7. Mendis, S., Davis, S., Norrving, B.: Organizational update: the World Health Organization global status report on noncommunicable diseases 2014. Stroke **46**(5), e121–e122 (2015)
8. Muscogiuri, G., et al.: Performance of a deep learning algorithm for the evaluation of CAD-RADS classification with CCTA. Atherosclerosis **294**, 25–32 (2020)
9. Rumberger, J., Kaufman, L.: A rosetta stone for coronary calcium risk stratification: agatston, volume, and mass scores in 11,490 individuals. Am. J. Roentgenol. **181**(3), 743–748 (2003)
10. Sankaran, S., Schaap, M., Hunley, S.C., Min, J.K., Taylor, C.A., Grady, L.: HALE: healthy area of lumen estimation for vessel stenosis quantification. In: Ourselin, S., Joskowicz, L., Sabuncu, M.R., Unal, G., Wells, W. (eds.) MICCAI 2016. LNCS, vol. 9902, pp. 380–387. Springer, Cham (2016). https://doi.org/10.1007/978-3-319-46726-9_44

11. Tejero-de-Pablos, A., et al.: Texture-based classification of significant stenosis in CCTA multi-view images of coronary Arteries. In: Shen, D., et al. (eds.) MICCAI 2019. LNCS, vol. 11765, pp. 732–740. Springer, Cham (2019). https://doi.org/10.1007/978-3-030-32245-8_81
12. Wels, M., Lades, F., Hopfgartner, C., Schwemmer, C., Sühling, M.: Intuitive and accurate patient-specific coronary tree modeling from cardiac computed-tomography angiography. In: The 3rd Interactive MIC Workshop, pp. 86–93 (2016)
13. Zheng, Y., Tek, H., Funka-Lea, G.: Robust and accurate coronary artery centerline extraction in CTA by combining model-driven and data-driven approaches. In: Mori, K., Sakuma, I., Sato, Y., Barillot, C., Navab, N. (eds.) MICCAI 2013. LNCS, vol. 8151, pp. 74–81. Springer, Heidelberg (2013). https://doi.org/10.1007/978-3-642-40760-4_10
14. Zreik, M., et al.: A recurrent CNN for automatic detection and classification of coronary artery plaque and stenosis in coronary CT angiography. IEEE Trans. Med. Imaging **38**(7), 1588–1598 (2018)

Higher-Order Flux with Spherical Harmonics Transform for Vascular Analysis

Jierong Wang[✉] and Albert C. S. Chung

Lo Kwee-Seong Medical Image Analysis Laboratory,
Department of Computer Science and Engineering,
The Hong Kong University of Science and Technology, Kowloon, Hong Kong
{jwangdh,achung}@cse.ust.hk

Abstract. In this paper, we present a novel flux-based method to robustly identify the vasculature structure in the angiography, where the curvilinear geometry is delineated by the higher-order tensor computed in the spherical frequency domain. We first modify the vesselness measurement derived from the oriented flux and introduce an antisymmetry measurement to generate the curvilinear responses. We then extend the responses to the cylindrical model and fit them into spherical harmonics transform to perform high-order tensor analysis, in which fiber orientation distribution function is utilized. A graphical framework based on the random walker is applied for vascular segmentation. It is experimentally demonstrated that the proposed method can achieve accurate and stable segmentation performance with various noise levels, demonstrating the proposed method can deliver reliable curvilinear structure responses.

Keywords: Oriented flux · Spherical harmonics · Fiber ODF

1 Introduction

Curvilinear structure segmentation is of importance in computer-aided diagnosis, which is a fundamental pre-processing step and the foundation of a range of applications in medical image analysis. One major category of the vascular network extraction methods is designed based on the order-2 tensor, targeting designing filters with high responses at the curvilinear structures. The Frangi filter [6] is one of the well-known vascular filters exploiting the second-order statistics of the image intensity in a multi-scale scenario. The Hessian matrix is constructed by an isotropic Gaussian function, which will limit the detection performance due to bifurcations, intensity inhomogeneity and changing of curvature in the clinical practice. Recently, [13] proposed a steerable curvilinear Gaussian filterbank to compute vascular connectivity map based on the Hessian matrix and Laplacian. Considering second-order based methods are susceptive to intensity fluctuation and adjacent structures, a flux-based method, which exploits

© Springer Nature Switzerland AG 2020
A. L. Martel et al. (Eds.): MICCAI 2020, LNCS 12266, pp. 55–65, 2020.
https://doi.org/10.1007/978-3-030-59725-2_6

the first-order derivative, was presented [15]. Followed by the flux geometry, an oriented-flux descriptor was proposed [10], namely Optimal Oriented Flux (OOF), by projecting the gradient on a local sphere and computing the minimal inward flux. The order-2 tensor has limitations of orientation estimation and modeling abnormal cross-sections. Therefore, higher-order Cartesian tensor (HOT) has raised research interests in the community. For vascular analysis, [3] constructed HOTs by least-square estimation, while [14] generated the tensors directly by outer product. In the high angular resolution diffusion imaging (HARDI), HOT can be fit to HARDI data by spherical harmonics (SH). Comparing to HOT, the determination of the SH coefficients is more efficient, which is analogous to a Fourier decomposition restricted on a unit sphere. On the other hand, although HOT is hypersymmetry, the non-redundant components need to be decided manually, which is not convenient to extend the orders of the tensor.

In this paper, we first propose a gradient symmetry quantification based on a new vesselness measurement derived from OOF and an antisymmetry measurement based on the derivatives of the Gaussian function. Such quantification is then modified in a cylindrical scheme, where the cylindrical length is determined by the scale detected from the vesselness measurement and the size of the input volume. We then fit the curvilinear responses with spherical harmonics series using least-square estimation and estimate the fiber orientation distribution function as the final curvilinear descriptor of the proposed method. A random walker based graphical framework is applied to perform vascular segmentation.

2 Antisymmetry Optimal Oriented Flux

2.1 Background of Optimal Oriented Flux (OOF)

The optimal oriented flux (OOF) in [10] reflects the minimal amount of projected gradient going through a local sphere S_r in the 3D space. Given a gradient field $\boldsymbol{v}$ of an image I and a direction of interest $\hat{\rho}$, the projected flux is defined as,

$$f(\boldsymbol{x}; r, \hat{\rho}) = \frac{1}{4\pi r^2} \int_{\partial S_r} ((\boldsymbol{v}(\boldsymbol{x} + \boldsymbol{h}) \cdot \hat{\rho})\hat{\rho}) \cdot \hat{n} dA = \hat{\rho}^T \boldsymbol{Q}_{r,\boldsymbol{x}} \hat{\rho}, \qquad (1)$$

where $\hat{n}$ is the outward unit normal of the spherical surface ∂S_r, $\boldsymbol{h}$ and dA are the position vector and infinitesimal area on ∂S_r. $\boldsymbol{Q}_{r,\boldsymbol{x}}$ is a matrix with the entry $q_{r,\boldsymbol{x}}^{i,j} = \int_{\partial S_r} v_i(\boldsymbol{x} + \boldsymbol{h})n_j dA$ at ith row and jth column ($i, j \in \{1, 2, 3\}$). Finding the outward optima of Eq. 1 can be solved as a generalized eigenvalue problem of the order-2 tensor $\boldsymbol{Q}_{r,\boldsymbol{x}}$ s.t. $\|\hat{\rho}\| = 1$. Let $\lambda_i(\boldsymbol{x}; r)$ and $\omega_i(\boldsymbol{x}; r)$ be the eigenvalues and eigenvectors of $\boldsymbol{Q}_{r,\boldsymbol{x}}$ ($i \in \{1, 2, 3\}$). $\lambda_3(\cdot) \leq \lambda_2(\cdot) \ll \lambda_1(\cdot) \approx 0$ when $\boldsymbol{x}$ is on the centerline and the radius of S equals to the scale of the structure. $\omega_2(\cdot)$ and $\omega_3(\cdot)$ span the normal plane and $\omega_1(\cdot)$ represents the orientation of the structure at $\boldsymbol{x}$. The values of $q_{r,\boldsymbol{x}}^{i,j}$ can be easily acquired by just convolving the input image with a group of filters $\psi_{r,i,j}(\boldsymbol{x}) = (b_r * g_{\hat{a}_i, \hat{a}_j, \sigma})(\boldsymbol{x})$, where $b_r(\boldsymbol{x})$ is the spherical step function with values equal to $(4\pi r^2)^{-1}$ within the sphere $\|\boldsymbol{x}\| \leq r$

(0 otherwise) and $g_{\hat{a}_i, \hat{a}_j, \sigma}$ is the second-order derivative of Gaussian with the scale σ. The convolution can be calculated efficiently in the frequency domain, where multiplication will be used to replace convolution. Denote $\Psi_{r,i,j}(\boldsymbol{u})$ as the corresponding Fourier transform of $\psi_{r,i,j}$, where $\boldsymbol{u} = (u_1, u_2, u_3)^T$ is the position vector in the frequency domain. Then, in the frequency domain, we have,

$$\mathscr{F}\{q_{r,\boldsymbol{x}}^{i,j}\} = \mathscr{F}\{\psi_{r,i,j} * I\} = \Psi_{r,i,j}(\boldsymbol{u}) \cdot \mathscr{F}\{I\} = \mathscr{F}\{b_r\} \cdot \mathscr{F}\{g_{\hat{a}_i, \hat{a}_j, \sigma}\} \cdot \mathscr{F}\{I\}. \quad (2)$$

Different from the responses computed in the original paper, the vesselness measurement we use are defined by,

$$\Lambda_{23}(\boldsymbol{x}; r) = \pi(\lambda_2^2 + \lambda_3^2)/r. \quad (3)$$

2.2 Gradient Symmetry Quantification

Inspired by [1], we propose a gradient symmetry measurement based on the derivatives of Gaussian function and Hankel transform. Combing the Hessian and the directional derivatives along the gradient of the image, define the gradient antisymmetry measurement as below,

$$I_{ww}(\boldsymbol{x}; r) = \frac{1}{r(I_x^2 + I_y^2 + I_z^2)} \sum_{i \in \{x,y,z\}} (I_x I_{ix} + I_y I_{iy} + I_z I_{iz}) I_i. \quad (4)$$

The second-order derivative in Eq. 4 can be obtained directly in OOF,

$$\mathscr{F}\{I_{ij}\} = \mathscr{F}\{\psi_{r,i,j} * I(x)\} = \Psi_{r,i,j}(\boldsymbol{u}) \cdot \mathscr{F}\{I\} \quad (i, j \in \{x, y, z\}), \quad (5)$$

$$\Psi_{r,i,j}(\boldsymbol{u}) = 4\pi r u_i u_j e^{-2(\pi \|\boldsymbol{u}\| \sigma)^2} \frac{1}{\|\boldsymbol{u}\|} (\cos(2\pi \|\boldsymbol{u}\| r) - \frac{\sin(2\pi \|\boldsymbol{u}\| r)}{2\pi \|\boldsymbol{u}\| r}). \quad (6)$$

And the first-order statistic parts can be acquired in a similar manner,

$$\mathscr{F}\{I_i\} = \mathscr{F}\{\Gamma_{r,i} * I(x)\} = \mathscr{F}\{b_r\} \cdot \mathscr{F}\{g_{i,\sigma}\} \cdot \mathscr{F}\{I\} \quad (i \in \{x, y, z\}), \quad (7)$$

$$\mathscr{F}\{\Gamma_{r,i}(\boldsymbol{u})\} = (-\frac{r}{\pi \|\boldsymbol{u}\|} (\cos(2\pi \|\boldsymbol{u}\| r) - \frac{\sin(2\pi \|\boldsymbol{u}\| r)}{2\pi \|\boldsymbol{u}\| r}))(-\mathbf{j} 2\pi u_i e^{-2(\pi \|\boldsymbol{u}\| \sigma)^2}), \quad (8)$$

where $\mathbf{j}$ is the imaginary unit in Eq. 8. Finally, the gradient symmetry quantification for spherical harmonics analysis is defined as,

$$\mathcal{M}(\boldsymbol{x}) = \max(0, \max_{r \in R}(\Lambda_{23}(\boldsymbol{x}; r)) - \gamma \max_{r \in R}(I_{ww}(\boldsymbol{x}; r))), \quad \gamma \in [0, 1], \quad (9)$$

where γ is the metric parameter and it was set to 0.5 in our experiments. Figure 1 provides an illustration of Λ_{23}, I_{ww} and $\mathcal{M}(\boldsymbol{x})$. It is clear that $\Lambda_{23} \gg I_{ww}$ inside the curvilinear structure and $\Lambda_{23} \ll I_{ww}$ at the object boundary. The second row exhibits the response maps of the same image disturbed by a high-level random noise generated by the operator Λ_{23} and $\mathcal{M}(\boldsymbol{x})$ respectively. It is obvious that combining with the gradient antisymmetric measurement, the new curvilinear descriptor can delineate the vascular structures better under noisy environment.

3   Spherical Harmonics Transform

We propose to construct the higher-order tensor in the frequency domain by spherical harmonics transform. The designed symmetry quantification in Sect. 2 is employed to mimic the diffusion signal required in HARDI. The higher-order tensor construction in the spherical frequency domain will be remarkably more efficient than that in the spatial domain [3,16] since rank-1 decomposition is not required. We then focus on the spherical harmonics coefficients evaluated by solving a linear system for vessel segmentation. The proposed method is different from the Ring Pattern Detector in [12] where the high-order tensors are also constructed in the spatial domain using normalized gradient and the spherical harmonics coefficients are computed by the spherical harmonics defined with Cartesian coordinates. Our framework is actually an analog of fitting high-order tensors to HARDI data using spherical harmonics transform and focuses on calculating the spherical harmonics coefficients for vessel segmentation.

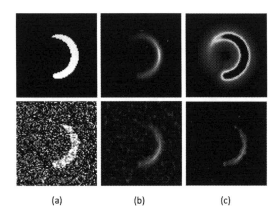

(a)   (b)   (c)

Fig. 1. An illustration for λ_{23}, I_{ww} and $\mathcal{M}(\boldsymbol{x})$. (a-c) 1st row: A synthetic tube, Λ_{23} and I_{ww}; 2nd row: The same tube disturbed by random noise, Λ_{23} and $\mathcal{M}(\boldsymbol{x})$.

3.1   Fitting Curvilinear Responses with Spherical Harmonics Series

Any real function parameterized by the spherical coordinates (θ, ϕ) can be written in a Spherical Harmonics (SH) series,

$$x(\theta, \phi) = \sum_{l=0}^{\infty} \sum_{m=-l}^{l} c_l^m Y_l^m(\theta, \phi), \tag{10}$$

where l is the order, $Y_l^m(\theta, \phi) = \sqrt{\frac{2l+1}{4\pi} \frac{(l-m)!}{(l+m)!}} P_l^m(\cos \theta) e^{jm\phi}$ are the SH series and $P_l^m(\cdot)$ is a Legendre polynomial. In HARDI, the continuous function $x(\cdot)$ is replaced by the discrete measured diffusion signal $X(\boldsymbol{g}) = X(\theta, \phi)$ with the number of SH: $R = \frac{(l+1)(l+2)}{2}$ for approximation. In the usual practice, a modified SH basis is employed to compute the coefficients c_l^m of the SH series [4], where $Y_j = \frac{\sqrt{2}}{2}((-1)^m Y_l^m + Y_L^{-m})$ if $-l \leq m < 0$; $Y_j = \frac{\sqrt{-2}}{2}((-1)^{m+1} Y_l^m + Y_L^{-m})$ if $0 < m \leq l$ and $Y_j = Y_l^0$ for $m = 0$, in which $j(l,m) = \frac{l^2+l+2}{2} + m$, l is even and $m = -l, ..., 0, ...l$. With the above modification, the SH transform can be represented with only one summation and finite terms,

$$X(\theta, \phi) = \sum_{j=1}^{R} c_j Y_j(\theta, \phi). \tag{11}$$

Define $X_p = (X_p(\theta_1, \phi_1), X_p(\theta_2, \phi_2), ..., X_p(\theta_M, \phi_M))^T$ as the measured diffusivities at the voxel p, where M is the number of sampled directions, $C_p = (c_{1_p}, c_{2_p}, ..., c_{R_p})$ as the coefficients of SH series at p and $Y_p = (\tilde{Y}_p(1), \cdots, \tilde{Y}_p(M))^T$ as the SH series matrix at p, where $\tilde{Y}_p(i) = (Y_{1_p}(\theta_i, \phi_i), \cdots, Y_{R_p}(\theta_i, \phi_i))$. The coefficients c_j can be obtained by least-square estimation: $C_p = (Y_p^T Y_p)^{-1} Y_p^T X_p$.

Consider the whole 3D volume with the size of $H \times W \times D$, the diffusivity matrix is denoted as X, which is a $M \times (HWD)$ matrix and the coefficient matrix of SH series is represented by C, which is a $R \times (HWD)$ matrix. Since the SH series are not related with the location p over the whole image, we have $Y = Y_p$. Thus, the coefficients of each voxel can be computed simultaneously,

$$C = (Y^T Y + \beta L)^{-1} Y^T X, \tag{12}$$

where $L = \text{diag}(l_1^2(l_1+1)^2, ..., l_R^2(l_R+1)^2)$ is the Laplace-Beltrami regularization term and l_j is the order corresponding to jth SH basis (e.g., $l_j = 0$ for $j = 1$).

To fit the curvilinear responses with SH series, we modify the gradient symmetry quantification in Sect. 2 with a cylindrical manner. Given a direction $g_i := (\theta_i, \phi_i)$, the diffusivity gradient symmetry quantification is defined as,

$$\mathcal{M}(x; \theta_i, \phi_i) = \mathcal{M}(x; g_i) = \sum_{l=0,...,h_x} \mathcal{M}(x_l), \quad x_l = x + gl, \tag{13}$$

where $h_x = f(\arg\max_{r \in R} \Lambda_{23}(x; r))$ and $f(r) = \frac{\log(1+r)\log(HWD/3)}{2}$. The first term $\log(1 + r)$ allows small curvilinear structures to have a larger h_x and the second term takes the size of the volume into consideration during traveling. Thus, the cylinder constructed for simulating the diffusion signal is adaptive.

3.2   Fiber Orientation Distribution Function

The vesselness response we propose is determined via the orientation distribution function (ODF) computed by the SH coefficients. In HARDI, diffusion ODF (dODF) is usually employed for fiber tracking, which is computed by multiplying the SH coefficients c_j with $2\pi P_{l_j}(0)$, where $P_{l_j}(\cdot)$ is a Legendre polynomial. Comparing to the dODF, fiber ODF (fODF) removes the smooth parts by spherical deconvolution transform and thus becomes sharper along the orientations of fibers [5]. Since the smooth parts in dODF could be the disturbance for detecting the vessels especially on the bifurcations, we utilize fODF in our work,

$$\Psi_{sharp}(\theta, \phi) = \sum_{j=1}^{R} 2\pi P_{l_j}(0)\frac{c_j}{f_j}Y_j(\theta, \phi), \text{ where } f_j = 2\pi \int_{-1}^{1} P_{l_j}(t)R(t)dt. \quad (14)$$

$R(\cdot)$ is the dODF kernel, which exhibits a prolate tensor profile, and l_j is the even order corresponding to jth SH basis. The term $\frac{1}{f_j}$ is used to transform the dODF into fODF. In [5], analytical expressions of $R(\cdot)$ and f_j are given by,

$$R(t) := \frac{1}{Z}((\frac{e_2}{e_1} - 1)t^2 + 1)^{\frac{1}{2}} \quad \text{and} \quad f_j = \frac{2\pi}{Z}\int_{-1}^{1} P_{l_j}(t)((\frac{e_2}{e_1} - 1)t^2 + 1)^{\frac{1}{2}}dt, \quad (15)$$

where $Z = \int_{-1}^{1}((\frac{e_2}{e_1}-1)t^2+1)^{\frac{1}{2}}dt$ is the normalization term and the eigenvalues of the tensor profile are assumed to be $\{e_1, e_2, e_2\}(e_1 \gg e_2)$. In our computation, we use the order-2 tensor computed from OOF with the corresponding top-k largest $\mathcal{M}(x)$ values and set $e_1 = \sum(|\lambda_2| + |\lambda_3|)/k$, $e_2 = \sum(|\lambda_1|)/k$, where $k = (H * W * D)^{\frac{1}{3}}$. Then, the final curvilinear responses, namely Spherical Harmonics flux, are obtained by averaging summation among M fiber orientation distribution functions,

$$\Psi_f = \frac{1}{M}\sum_{i}^{M}\Psi_{sharp}(\theta_i, \phi_i). \quad (16)$$

3.3   Spherical Harmonics Connectivity Enhanced Random Walks

Random walks [7] is a segmentation framework formulated on a weighted graph. Given an undirected graph $G_o = (V_o, E_o)$ with vertices $v \in V_o$ and edges $e \subseteq E_o = V_o \times V_o$, denote w_{ij} as the weight of e_{ij}, v_b as the background seeds and v_s as the foreground seeds. Based on the random walks framework, we present a vasculature enhanced graphical framework with the Spherical Harmonics flux,

$$x = \arg\min_{x} \sum_{v_i \in V_o \backslash \{v_s, v_b\}} (\Psi_f(x_i)(x_i - 1)^2 + (1 - \Psi_f(x_i))(x_i - 0)^2) +$$
$$\sum_{e_{ij} \in E_o} w_{ij}(x_i - x_j)^2, \quad \text{s.t.} \quad x(v_s) = 1, x(v_b) = 0, \quad (17)$$

where $x = (x_1, ..., x_n)^T$ is the probability for a walker starting from a vertex j that first arrives at a foreground seed and $w_{ij} = e^{-\beta(g_i - g_j)^2}$, g_i is the intensity value at voxel i and β is the only parameter of random walks, which equals to 100 in all experiments. Define $SHG = (V_{shg}, E_{shg})$ with $V_{shg} = V_o\backslash\{v_s, v_b\}$ and $E_{shg} = V_{shg} \times \{v_s, v_b\}$, Eq. 17 can be rewritten as,

$$x = \arg\min_{x} \sum_{e_{ij} \in E_o} w_{ij} \cdot (x_i - x_j)^2 + \sum_{e_{ij} \in E_{shg}} s_{ij} \cdot (x_i - x_j)^2,$$
$$\text{s.t.} \quad x(v_s) = 1, x(v_b) = 0, \quad (18)$$

where $s_{ij} = \Psi_f(x_i) \cdot x_j + (1 - \Psi_f(x_i)) \cdot (1 - x_j)$. With the construction of SHG, Eq. 18 can be solved efficiently and easily by the algorithm in [7].

4 Experiments

We have performed the validation on the synthetic images and clinical images. For the 1st group of synthetic experiments, the evaluation was performed on the synthetic helix-tube whose radii were from 1 to 4 voxels and the synthetic tree with radii from 2 to 4 voxels (Fig. 2). The intensities of the helix-tube varied from 0.5 on the top to 1 on the bottom. These two synthetic images were designed to simulate bifurcation challenges and shrinking problem for real vessel images. A subset of Vascusynth [8,9] was used to perform the 2nd group of experiments, in which there were 12 synthetic data with distinct bifurcation numbers. Prior to performing segmentation, different levels of random noise were added to these synthetic images. The clinical experiments were performed on a set of MRA images from TubeTK [2].

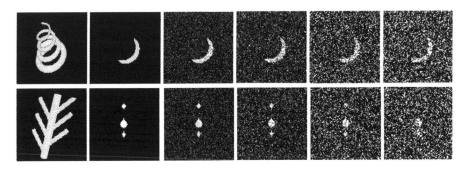

Fig. 2. First row: helix-tubes. Second row: synthetic trees. $2^{nd}-6^{th}$ columns: 2D view with noise levels $\mathcal{N} \in \{0, 0.04, 0.1, 0.3, 0.5\}$ respectively.

The proposed method (Ψ_f) was compared with classical random walks (RW), OOF and the Hessian [6] with direction coherence enforcement framework in [17] (OOF_{Coh}, HES_{Coh}) and the ranking orientation responses of path operators ($RORPO$) [11]. We also compared the gradient symmetry quantification ($\mathcal{M}(\boldsymbol{x})$) to see the effects of SH transform. For the proposed method, we set the order of the SH series to 4 and the number of sampled directions was equal to 64. The radii for Ψ_f, OOF_{Coh} and HES_{Coh} were in the range of 1 to 10 with steps equal to 0.5 in the synthetic experiments, while there were in the range of 1 to 20 with steps equal to 0.5 in the clinical tests. For $RORPO$, the length of the minimum path, the factor and the number of scales were set to 2, 1.5 and 10, respectively for synthetic data and for the clinical parts, they were set to 2, 1.5 and 20, respectively. We optimized the thresholding values in the scope of 0.02 to 0.8 with steps equal to 0.02 for binarization. Dice score was employed for statistic comparison, which is given by $\frac{2TP}{2TP+FP+FN}$.

Table 1. Dice on Helix-tubes and synthetic trees

Synthetic data	Methods	Noise level								
		0	0.02	0.04	0.06	0.1	0.2	0.3	0.4	0.5
Helix-tubes	$\mathcal{M}(x)$	99.79%	97.15%	**96.54%**	95.71%	94.74%	90.93%	86.52%	82.30%	77.09%
	Ψ_f	99.20%	96.82%	96.33%	**95.88%**	**94.79%**	**92.75%**	**89.81%**	**86.79%**	**83.84%**
	$RORPO$	**100.00%**	**97.80%**	88.17%	76.75%	57.44%	0.33%	0.00%	0.00%	0.00%
	OOF_{Coh}	98.78%	92.23%	90.08%	90.50%	91.47%	89.68%	86.09%	80.79%	75.10%
	HES_{Coh}	98.84%	95.94%	95.26%	94.35%	92.13%	82.92%	73.49%	62.79%	53.29%
	RW	**100.00%**	51.70%	51.10%	36.72%	48.26%	49.07%	30.23%	53.92%	36.44%
Synthetic trees	$\mathcal{M}(x)$	99.70%	93.33%	92.66%	91.10%	90.90%	89.60%	87.08%	84.23%	80.69%
	Ψ_f	99.62%	92.27%	91.55%	91.41%	**90.99%**	**89.77%**	**88.17%**	**86.48%**	**84.79%**
	$RORPO$	99.97%	**99.83%**	**95.80%**	86.18%	67.16%	0.74%	0.00%	0.00%	0.00%
	OOF_{Coh}	99.77%	94.79%	93.74%	**92.76%**	90.42%	89.55%	86.54%	85.18%	81.27%
	HES_{Coh}	98.73%	94.51%	93.39%	92.22%	90.45%	85.14%	75.09%	59.89%	47.01%
	RW	**100.00%**	77.80%	46.93%	56.01%	26.12%	58.11%	41.76%	40.00%	54.50%

The numerical results are listed in Table 1 and Table 2. It is clear that both $\mathcal{M}(x)$ and Ψ_f exhibit an accurate and stable performance on both types of synthetic data with different noise levels. When the noise level is low, $\mathcal{M}(x)$ performs slightly better than Ψ_f due to the antipodally-symmetry enforced by the even order tensor. However, when the noise level is high, the difference between $\mathcal{M}(x)$ and Ψ_f becomes distinct and Ψ_f presents a better ability on curvilinear structure analysis thanks to the SH deconvolution in the computation of fODF. Despite the high performance when noise level is 0, RW suffers from the inhomogeneous intensity caused by noise. Similar to RW, $RORPO$ achieves high scores with no noise and low noise levels but it fails in the cases of high noise levels. In the clinical experiments, we will see that noise of the clinical data will still be retained by the feature map of $RORPO$.

Table 2. Dice on Group 1 of Vascusynth

Noise level	0			0.04			0.1		
Bif num.	16	36	56	16	36	56	16	36	56
Ψ_{fODF}	93.37%	95.02%	92.73%	83.82%	**88.42%**	84.09%	**79.27%**	**85.58%**	**80.64%**
$RORPO$	**98.70%**	98.19%	95.04%	73.92%	71.45%	65.14%	45.40%	40.12%	36.29%
OOF_{Coh}	97.84%	**98.22%**	**97.61%**	77.92%	81.67%	78.37%	75.01%	78.93%	76.05%
HES_{Coh}	94.13%	91.45%	94.30%	**84.17%**	84.28%	**84.73%**	77.12%	80.48%	78.85%

For the clinical experiments, Ψ_f was compared with OOF_{Coh} and $RORPO$ qualitatively. Segmentation results of Ψ_f are shown in Fig. 3 with different thresholding values (0.04 and 0.14). The comparison of vascular maps generated from the above-mentioned methods is presented in Fig. 4. It is noted that promising segmentation can be achieved by Ψ_f despite high noise levels and intensity inhomogeneity, while vessels of small scales can be obtained under low thresholding value. Although both Ψ_f and OOF_{Coh} use the similar graphical framework, OOF_{Coh} fails outside the region of the Circle of Willis. $RORPO$ does generate relative good vascular responses but as shown in the zoom region of the green rectangle, the noise of background and some unwanted segments are retained.

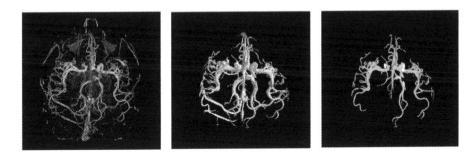

Fig. 3. From left to right: original image and two segmentation results with distinct thresholding values (0.04 and 0.14).

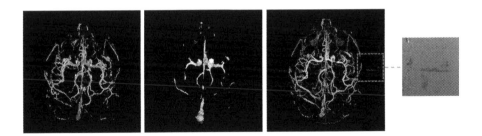

Fig. 4. From left to right: the vascular maps of Ψ_f, OOF_{Coh} and $RORPO$.

5 Conclusion

We have introduced a spherical harmonics framework for curvilinear structure analysis, which can be regarded as the higher-order tensor analysis in the frequency domain. Comparing to the higher-order tensor, the spherical harmonics, combining with the orientation distribution function, is more efficient and convenient to use. We have also presented a novel gradient symmetry quantification and fit it into the spherical harmonics framework. Three groups of experiments have been carried out and the results demonstrate that our method is robust to noise.

References

1. Zheng, Y., Tek, H., Funka-Lea, G.: Robust and accurate coronary artery centerline extraction in CTA by combining model-driven and data-driven approaches. In: Mori, K., Sakuma, I., Sato, Y., Barillot, C., Navab, N. (eds.) MICCAI 2013. LNCS, vol. 8151, pp. 74–81. Springer, Heidelberg (2013). https://doi.org/10.1007/978-3-642-40760-4_10

2. Aylward, S.R., Bullitt, E.: Initialization, noise, singularities, and scale in height ridge traversal for tubular object centerline extraction. IEEE Trans. Med. Imaging **21**(2), 61–75 (2002)

3. Cetin, S., Unal, G.: A higher-order tensor vessel tractography for segmentation of vascular structures. IEEE Trans. Med. Imaging **34**(10), 2172–2185 (2015)

4. Descoteaux, M., Angelino, E., Fitzgibbons, S., Deriche, R.: Apparent diffusion coefficients from high angular resolution diffusion imaging: estimation and applications. Magn. Reson. Med. **56**(2), 395–410 (2006)

5. Descoteaux, M., Deriche, R., Knosche, T.R., Anwander, A.: Deterministic and probabilistic tractography based on complex fibre orientation distributions. IEEE Trans. Med. Imaging **28**(2), 269–286 (2009)

6. Frangi, A.F., Niessen, W.J., Vincken, K.L., Viergever, M.A.: Multiscale vessel enhancement filtering. In: Wells, W.M., Colchester, A., Delp, S. (eds.) MICCAI 1998. LNCS, vol. 1496, pp. 130–137. Springer, Heidelberg (1998). https://doi.org/10.1007/BFb0056195

7. Grady, L.: Random walks for image segmentation. IEEE Transaction on Pattern Analysis and Machine Intelligence **28**(11), 1768–1783 (2006)

8. Hamarneh, G., Jassi, P.: Vascusynth: simulating vascular trees for generating volumetric image data with ground-truth segmentation and tree analysis. Compute. Med. Imaging Graph. **34**(8), 605–616 (2010)

9. Jassi, P., Hamarneh, G.: Vascusynth: vascular tree synthesis software. Insight J. (2011)

10. Law, M.W.K., Chung, A.C.S.: Three dimensional curvilinear structure detection using optimally oriented flux. In: Forsyth, D., Torr, P., Zisserman, A. (eds.) ECCV 2008. LNCS, vol. 5305, pp. 368–382. Springer, Heidelberg (2008). https://doi.org/10.1007/978-3-540-88693-8_27

11. Merveille, O., Talbot, H., Najman, L., Passat, N.: Curvilinear structure analysis by ranking the orientation responses of path operators. IEEE Trans. Pattern Anal. Mach. Intell. **40**(2), 304–317 (2018)

12. Moreno, R., Smedby, Ö.: Vesselness estimation through higher-order orientation tensors. In: 2016 IEEE 13th International Symposium on Biomedical Imaging (ISBI), pp. 1139–1142. IEEE (2016)

13. Moriconi, S., Zuluaga, M.A., Jäger, H.R., Nachev, P., Ourselin, S., Cardoso, M.J.: VTrails: inferring vessels with geodesic connectivity trees. In: Niethammer, M., Styner, M., Aylward, S., Zhu, H., Oguz, I., Yap, P.-T., Shen, D. (eds.) IPMI 2017. LNCS, vol. 10265, pp. 672–684. Springer, Cham (2017). https://doi.org/10.1007/978-3-319-59050-9_53

14. Schultz, T.: Towards resolving fiber crossings with higher order tensor inpainting. In: Laidlaw, D., Vilanova, A. (eds.) New Developments in the Visualization and Processing of Tensor Fields. MATHVISUAL. Springer, Heidelberg (2012)

15. Vasilevskiy, A., Siddiqi, K.: Flux maximizing geometric flows. IEEE Trans. Pattern Anal. Mach. Intelligence **24**(12), 1565–1578 (2002)

16. Wang, J., Chung, A.C.S.: High-order oriented cylindrical flux for curvilinear structure detection and vessel segmentation. In: Chung, A.C.S., Gee, J.C., Yushkevich, P.A., Bao, S. (eds.) IPMI 2019. LNCS, vol. 11492, pp. 479–491. Springer, Cham (2019). https://doi.org/10.1007/978-3-030-20351-1_37

17. Zhang, Q., Chung, A.C.S.: 3D vessel segmentation using random walker with oriented flux analysis and direction coherence. In: Zheng, G., Liao, H., Jannin, P., Cattin, P., Lee, S.-L. (eds.) MIAR 2016. LNCS, vol. 9805, pp. 281–291. Springer, Cham (2016). https://doi.org/10.1007/978-3-319-43775-0_25

Cerebrovascular Segmentation in MRA via Reverse Edge Attention Network

Hao Zhang[1,2,5,6,7], Likun Xia[1,5,6,7]($\boxtimes$), Ran Song[3], Jianlong Yang[2],
Huaying Hao[2], Jiang Liu[4], and Yitian Zhao[2]($\boxtimes$)

[1] College of Information Engineering, Capital Normal University, Beijing, China
xlk@cnu.edu.cn
[2] Cixi Institute of Biomedical Engineering, Ningbo Institute of Industrial
Technology, Chinese Academy of Sciences, Ningbo, China
yitian.zhao@nimte.ac.cn
[3] School of Control Science and Engineering, Shandong University, Jinan, China
[4] Department of Computer Science and Engineering, Southern University of Science
and Technology, Shenzhen, China
[5] International Science and Technology Cooperation Base of Electronic System
Reliability and Mathematical Interdisciplinary, Capital Normal University,
Beijing, China
[6] Laboratory of Neural Computing and Intelligent Perception,
Capital Normal University, Beijing, China
[7] Beijing Advanced Innovation Center for Imaging Theory and Technology,
Capital Normal University, Beijing, China

Abstract. Automated extraction of cerebrovascular is of great importance in understanding the mechanism, diagnosis, and treatment of many cerebrovascular pathologies. However, segmentation of cerebrovascular networks from magnetic resonance angiography (MRA) imagery continues to be challenging because of relatively poor contrast and inhomogeneous backgrounds, and the anatomical variations, complex geometry and topology of the networks themselves. In this paper, we present a novel cerebrovascular segmentation framework that consists of image enhancement and segmentation phases. We aim to remove redundant features, while retaining edge information in shallow features when combining these with deep features. We first employ a Retinex model, which is able to model noise explicitly to aid removal of imaging noise, as well as reducing redundancy within an image and emphasizing the vessel regions, thereby simplifying the subsequent segmentation problem. Subsequently, a reverse edge attention module is employed to discover edge information by paying particular attention to the regions that are not salient in high-level semantic features. The experimental results show that the proposed framework enables the reverse edge attention network to deliver a reliable cerebrovascular segmentation.

Keywords: Cerebrovascular · 3D segmentation · Attention · Learning

© Springer Nature Switzerland AG 2020
A. L. Martel et al. (Eds.): MICCAI 2020, LNCS 12266, pp. 66–75, 2020.
https://doi.org/10.1007/978-3-030-59725-2_7

1   Introduction

Cerebrovascular diseases, such as stroke, aneurysm, and arteriovenous malformation, are some of the most common fatal diseases threatening human health worldwide [1]. MRA is a common imaging technique for observing the cerebrovascular system, and an accurate detection of the cerebrovascular from MRA imagery is essential for many clinical applications to support early diagnosis, optimal treatment, and neurosurgery planning for vascular-related diseases. However, manual annotation of cerebrovascular networks is an exhausting task even for experts, and existing computer-aided systems cannot reliably extract and segment these networks, due to the high degree of anatomical variation.

In the last two decades, we have witnessed the rapid development of vessel segmentation methods for different medical imaging modalities, as evidenced by extensive reviews [2–5]. However, the extraction of cerebrovascular networks is a less explored topic. Most techniques tend to over-segment or mis-segment, primarily due to the complex geometry involved, as well as problems posed by varying scales of noise, or imbalanced illumination within an image, as well as low image contrast and spatial resolution. As a result, it is desirable to design a fully automated cerebrovascular segmentation method.

Conventional cerebrovascular segmentation approaches have been based on handcrafted features, including vessel intensity distribution, gradient features, morphological features and many others [6]. For example, the cerebrovascular were segmented by means of active contour models [7] and geometric models [8]. In addition, a variety of filtering methods have been proposed, including Hessian matrix-based filters [9], a symmetry filter [4], and a tensor-based filter [10]. These approaches aim to remove undesired intensity variations in the image, and suppress background structures and image noise. However, these methods require elaborate design, which depends heavily on the user's domain knowledge and expertise, and errors can be propagated and accumulated, especially in the case of small vessels, due to poor imaging quality.

In recent years, several deep learning-based methods have been proposed for cerebrovascular segmentation. Phellan et al. [11] explored a relatively shallow Convolutional Neural Network (CNN) on MRA scans to segment the vessels at 2D slice level. Livne et al. [12] utilized the U-Net deep learning framework [13] to segment the vessel region on each slice of an MRA volume. Nevertheless, 2D CNN in taken from an MRA slice necessarily discards valuable three-dimensional(3D) context information that is crucial for tracking curvilinear structures. Sanches et al. [14] presented a Uception model for segmentation of an arterial cerebrovascular network. Zhang et al. [15] presented an efficacious framework that applies deep 3D CNN to automatic cerebrovascular segmentation with sparsely labeled data: however, they neglected to prioritize the extraction of blood vessel edges.

The aforementioned methods have not yet completely addressed the issues posed by the high degree of anatomical variation across the population, and poor contrast and varying scales of noise within an image. In this paper, we propose a novel cerebrovascular segmentation framework that hybridizes image enhancement and segmentation steps. First, we employ a Retinex model to improve

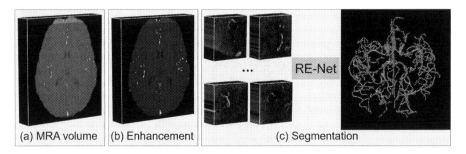

Fig. 1. Pipeline of the proposed framework. A given MRA volume (a) is firstly enhanced by a Retinex model (b), and a RE-Net is then introduced to segment the cerebrovascular (c) from the enhanced MRA volume.

image contrast and also model noise explicitly to aid the removal of imaging noise. Subsequently, we make use of a reverse attention mechanism, and introduce a Reverse Edge attention Network (RE-Net) capable of discovering the missing edge features and residual details effectively. This leads in turn to a significant improvement in the cerebrovascular segmentation.

2   Methodology

In this section, we detail the proposed cerebrovascular segmentation method. Figure 1 illustrates the pipeline of the proposed framework.

2.1   Image Enhancement via Retinex Model

In general, it is observed that laminar flow within cerebrovascular vessels causes velocity variations in blood flow, which leads to highly varying contrast distribution in MRA imagery. In addition, degrading noise is usually inherited from the image acquisition process. To this end, the enhancement of MRA images is essential so as to obtain a more precise segmentation result. In this section, we introduce a novel noise-suppression Retinex model to enhance MRA images.

The classic Retinex model [16] assumes that an image $\mathbf{S}$ can be decomposed into two components, the reflectance $\mathbf{R}$ and the illumination $\mathbf{L}$, with the range $(0, \infty)$: $\mathbf{S} = \mathbf{L} \cdot \mathbf{R}$, and it follows that $\mathbf{S} \leq \mathbf{L}$. By removing the influence of illumination $\mathbf{L}$, the resulting $\mathbf{R}$ is able to reveal the reflectance of the object of interest more objectively, and it can thus be regarded as the enhanced image. In this work, we utilized the Retinex model proposed by Elad et al. [17] for MRA volume enhancement.

2.2   Cerebrovascular Segmentation via RE-Net

In this section, we introduce the proposed Reverse Edge Attention network for cerebrovascular segmentation. The proposed RE-Net consists of three phases: the

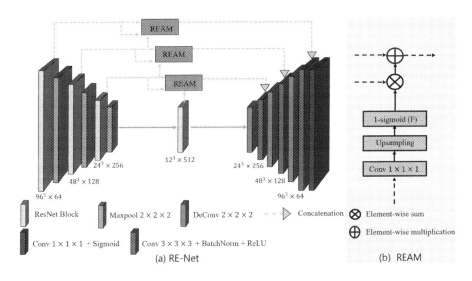

Fig. 2. Proposed RE-Net architecture and REAM module.

encoder module, the Reverse Edge Attention Module (REAM), and the decoder module, as shown in Fig. 2.

The encoder module contains four encoder stages based on the ResNet Block [18]. For each stage, the inputs are first fed into a stack of $3 \times 3 \times 3 - 3 \times 3 \times 3$ convolutional layers, and are then summed with the shortcut of inputs to generate the final outputs, followed by a max-pooling of $2 \times 2 \times 2$ to increase the receptive field for better extraction of global features. With the residual connection, the model can avoid the gradient vanishing and accelerate the network convergence. In the decoder module, each of the three stages consists of a deconvolution of $2 \times 2 \times 2$ with a stride of 2, followed by two $3 \times 3 \times 3$ convolutions activated with ReLU. To assist the decoding process, skip connections copy feature maps generated by the encoder module to the decoder module. Afterwards, a REAM is embedded in a skip connection to extract edge information from encoder layers.

Reverse Edge Attention Module: Consecutive pooling and striding convolutional operations may enlarge the receptive field, and obtain more global information. However, they lead to the loss of edge information. To maintain such information, the features generated by the deep layers in the decoder module are usually concatenated with features in the encoder module. Unfortunately, these features encoded by the shallow layers not only contain edges, but also keep textural features, which might then constitute interference factors compromising the robustness of high-level features [19].

Inspired by the reverse attention model in [20], we propose to make use of a reverse edge attention module, as shown in Fig. 2 (b), to extract the edge information from the feature maps generated by the encoding layers. This edge information is then fused with the features maps generated by the decoder for the corresponding positions, so that the vessel edge features will be enhanced and the original image edge details will be more accurately restored.

$X_i \in R^{h \times w \times d \times c_x}$, $2 \leq i \leq 4$ denotes the features generated in the ith encoder stage. A $1 \times 1 \times 1$ convolution is first used to fuse the features into a single channel. The features are then upsampled to the same resolution as the outputs of the $(i-1)$th encoder stage, thus generating $F \in R^{h \times w \times d \times c}$. The corresponding weight A_{i-1} in the $(i-1)$th encoder stage is simply generated by subtracting the upsampled prediction of the ith stage from 1, as below:

$$A_{i-1} = 1 - \mathrm{sigmoid}(F) = 1 - \frac{1}{1 + e^{-F}}. \tag{4}$$

In REAM, the edge information was discovered by paying attention to the regions that are not salient in high-level semantic features. Let $X_{i-1} \in R^{h \times w \times d \times c_x}$ denote features generated in the $(i-1)$th encoder stage. Then edge feature E_{i-1} can be captured by element-wise multiplication, expressed as follows:

$$E_{i-1} = X_{i-1} * A_{i-1}. \tag{5}$$

The edge feature is fused into the features of the decoder at the corresponding position by skip connection after summing with X_{i-1}. Comparably, the feature map after concatenation at the same position has more edge information with the aid of REAM, thus improving the segmentation accuracy.

3   Experimental Results

Our RE-Net was implemented on a PyTorch framework with a single GPU (TITAN RTX). Adaptive moment estimation (Adam) was employed for network optimization. The initial learning rate was set to 0.0001, with a weight decay of 0.0005. A poly learning rate policy [21] with power 0.9 was used, and the maximum epoch was 4000. Dice loss was adopted as the loss function. The batch size was set to 4 during training, because of the limitations of GPU memory.

Data and metrics: a publicly available dataset[1] with 42 time-of-flight MRA volumes was used in our work. These images were captured by a 3 T unit under standardized protocols, with a voxel size of $0.5 \times 0.5 \times 0.8$ mm^3, and reconstructed with a $448 \times 448 \times 128$ matrix. Manual annotations were available online, and the cerebrovascular labels were obtained via an open source toolkit - TubeTK [22]. In the experiment, patches of size $96 \times 96 \times 96$ were randomly cropped from the MRA volume, followed by a 90-° rotation for data augmentation. To facilitate better observation and objective evaluation of the cerebrovascular segmentation method, the following metrics were calculated: sensitivity (Sen) = TP/ (TP + FN); specificity (Spe) = TN/ (TN + FP); Precision (Pre) = TP/(TP + FP); and Dice similarity coefficient (DSC) = 2*TP/ (FP + FN + 2*TP). In addition, the Average Hausdorffs Distance (AHD) was adopted, because it is sensitive to the edge of segmentation results.

---

[1] https://public.kitware.com/Wiki/TubeTK/Data.

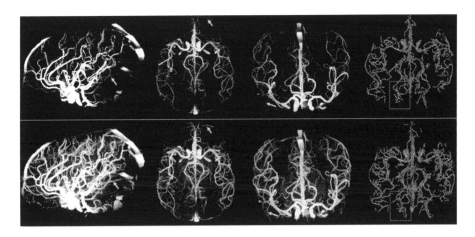

Fig. 3. MIP views and segmentation results of raw (top) and enhanced data by our NSR method (bottom). From left to right are the sagittal, axial, coronal views of the volume, and segmentation results produced by using 3D U-Net.

Table 1. Segmentation results obtained using 3D U-Net with different enhancement methods.

Enhancement Methods	Sen	Spe	Pre	DSC	AHD[mm]
Raw data	63.26%	99.85%	61.34%	62.29%	1.2466
Gamma Correction	63.75%	99.85%	60.89%	62.29%	1.2470
Guided Filter [23]	64.71%	99.87%	64.43%	64.53%	1.1216
NSR	**66.06%**	**99.87%**	**65.05%**	**65.53%**	**1.1188**

3.1 Evaluation of Image Enhancement Performance

Firstly, we analyze the effect of the Retinex (NSR) pre-processing step in the proposed method.

Figure 3 illustrates the sagittal, axial, and coronal views of maximum intensity projection (MIP) of a sample MRA, before and after the NSR method was applied. Overall, our method provides similar performance with raw data on the large vessels. However, when scrutinizing the smaller vessels, we can see that the proposed method provides relatively stronger responses; i.e., it has successfully enhanced some small vessels, as indicated by the green arrows. This is because our NSR enhancement is able to reduce noise, and normalize the entire background to a similar level, so as to increase the contrast between the vessel regions and their background. In addition to visual inspection of the enhancement results, objective evaluation was also undertaken by comparing the segmentation results with a state-of-the-art 3D segmentation model, 3D U-Net [24]. Table 1 shows the segmentation results in comparison with two classic image enhancement methods, i.e. gamma correction, and guided filter [23]. The proposed method improves the segmentation of the original images (i.e. raw data)

Table 2. Segmentation results obtained using different methods on NSR enhanced data.

Segmentation Methods	Sen	Spe	Pre	DSC	AHD[mm]
3D U-Net [24]	66.06%	99.87%	65.05%	65.53%	1.1188
V-Net [25]	64.84%	99.84%	60.70%	62.67%	1.2173
Uception [14]	61.12%	99.87%	64.40%	62.68%	1.3776
Backbone	67.30%	99.88%	67.07%	67.18%	1.1101
RE-Net	**69.57%**	**99.88%**	**68.59%**	**69.05%**	**1.0449**

by 2.80% and 3.24% in terms of *Sen* and *DSC*. By contrast, relatively more significant margins of segmentation results have been shown, when compared with two alternative enhancement methods, which indicates that our enhancement method demonstrates larger improvement than its competitors. It can be seen that our NSR method achieves the best performance in terms of all metrics.

3.2   Evaluation of Cerebrovascular Segmentation Performance

In this section, we compare performance in segmenting cerebrovascular networks using different 3D segmentation models. Three state-of-the-art 3D segmentation networks - including 3D U-Net [24], V-Net [25] and Uception [14] - as well as the proposed RE-Net, and RE-Net without REAM (backbone) were used for comparison. Figure 4 illustrates the segmentation results on two MRA volumes. It should be noted that, for purposes of fair comparison, 3D U-Net, V-Net, and Uception were applied to the same enhanced data. Overall, all methods demonstrate a similar performance on large vessels. However, the proposed RE-Net produces better segmentation results than other models in preserving small vessels, and maintaining vascular continuity, as highlighted by the green arrows with close-ups shown in the green rectangles in Fig. 4. RE-Net is capable of discovering lost edge features and residual details effectively, which leads to a significant improvement on detecting small vessels.

The above findings are reinforced by the evaluation scores in Table 2. The RE-Net outperforms 3D U-Net, V-Net, and Uception by 3.51%, 4.73%, and 8.45% in terms of *Sen* respectively. In addition, the *DSC* score of the Backbone method (RE-Net without REAM) is approximately 1.87% lower than that of RE-Net, which constitutes additional proof that the reverse attention mechanism benefits the performance of cerebrovascular segmentation. Overall, the proposed RE-Net yields the best segmentation results in terms of all metrics.

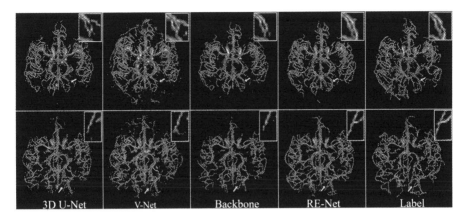

Fig. 4. Illustrative of cerebrovascular segmentation results of different methods on two sample MRA volumes.

4   Conclusion

Cerebrovascular segmentation is an important step for the diagnosis of many cerebrovascular diseases. In this paper, we have proposed a novel cerebrovascular segmentation framework, which consists of Retinex-based method for MRA image enhancement and a newly designed 3D segmentation network. The former takes into account a noise term in conventional Retinex model, so as to suppress noise in MRA imagery and further ease the segmentation task. The RE-Net implements a self-attention mechanism, and pays more attention to regions that are not salient in high-level global features. The experimental results based on a publicly available dataset demonstrate that the proposed method significantly improves cerebrovascular segmentation.

Acknowledgment. This work was supported by Beijing Natural Science Foundation (4202011), National Natural Science Foundation of China (61572076), Key Research Grant of Academy for Multidisciplinary Studies of CNU (JCKXYJY2019018), Zhejiang Provincial Natural Science Foundation of China (LZ19F010001), and Ningbo "2025 S&T Megaprojects" (2019B10033, 2019B10061).

References

1. Yasugi, M., Hossain, B., Nii, M., Kobashi, S.: Relationship between cerebral aneurysm development and cerebral artery shape. J. Adv. Comput. Intell. Intell. Inform. **22**(2), 249–255 (2018)
2. Fraz, M., et al.: Blood vessel segmentation methodologies in retinal images - a survey. Comput. Methods Programs Biomed. **108**(1), 407–433 (2012)
3. Zhao, Y., Rada, L., Chen, K., Harding, S.P., Zheng, Y.: Automated vessel segmentation using infinite perimeter active contour model with hybrid region information with application to retinal images. IEEE Trans. Med. Imaging **34**(9), 1797–1807 (2015)

4. Zhao, Y., et al.: Automatic 2-D/3-D vessel enhancement in multiple modality images using a weighted symmetry filter. IEEE Trans. Med. Imaging **37**(2), 438–450 (2017)
5. Zhao, Y., et al.: Retinal artery and vein classification via dominant sets clustering-based vascular topology estimation. In: Frangi, A.F., Schnabel, J.A., Davatzikos, C., Alberola-López, C., Fichtinger, G. (eds.) MICCAI 2018. LNCS, vol. 11071, pp. 56–64. Springer, Cham (2018). https://doi.org/10.1007/978-3-030-00934-2_7
6. Zhao, Y., et al.: Automated tortuosity analysis of nerve fibers in corneal confocalmicroscopy. In: IEEE Transactions on Medical Imaging (2020)
7. Yang, X., Cheng, K.T., Chien, A.: Geodesic active contours with adaptive configuration for cerebral vessel and aneurysm segmentation. In: 2014 22nd International Conference on Pattern Recognition, pp. 3209–3214. IEEE (2014)
8. Forkert, N.D., et al.: 3D cerebrovascular segmentation combining fuzzy vessel enhancement and level-sets with anisotropic energy weights. Magn. Reson. Imaging **31**(2), 262–271 (2013)
9. Frangi, A.F., Niessen, W.J., Vincken, K.L., Viergever, M.A.: Multiscale vessel enhancement filtering. In: Wells, W.M., Colchester, A., Delp, S. (eds.) MICCAI 1998. LNCS, vol. 1496, pp. 130–137. Springer, Heidelberg (1998). https://doi.org/10.1007/BFb0056195
10. Cetin, S., Unal, G.: A higher-order tensor vessel tractography for segmentation of vascular structures. IEEE Trans. Med. Imaging **34**(10), 2172–2185 (2015)
11. Phellan, R., Peixinho, A., Falcão, A., Forkert, N.D.: Vascular segmentation in TOF MRA images of the brain using a deep convolutional neural network. In: Cardoso, M.J., et al. (eds.) LABELS/CVII/STENT -2017. LNCS, vol. 10552, pp. 39–46. Springer, Cham (2017). https://doi.org/10.1007/978-3-319-67534-3_5
12. Livne, M., et al.: A u-net deep learning framework for high performance vesselsegmentation in patients with cerebrovascular disease. Frontiers Neurosci. **13**, 97 (2019)
13. Ronneberger, O., Fischer, P., Brox, T.: U-Net: convolutional networks for biomedical image segmentation. In: Navab, N., Hornegger, J., Wells, W.M., Frangi, A.F. (eds.) MICCAI 2015. LNCS, vol. 9351, pp. 234–241. Springer, Cham (2015). https://doi.org/10.1007/978-3-319-24574-4_28
14. Sanchesa, P., Meyer, C., Vigon, V., Naegel, B.: Cerebrovascular network segmentation of mra images with deep learning. In: IEEE 16th International Symposium on Biomedical Imaging (ISBI 2019), pp. 768–771. IEEE (2019)
15. Zhang, B., et al.: Cerebrovascular segmentation from TOF-MRA using model-and data-driven method via sparse labels. Neurocomputing, (2019)
16. Land, E.H., McCann, J.J.: Lightness and retinex theory. J. Opt. Soc. Am. **61**(1), 1–11 (1971)
17. Elad, M.: Retinex by two bilateral filters. In: Kimmel, R., Sochen, N.A., Weickert, J. (eds.) Scale-Space 2005. LNCS, vol. 3459, pp. 217–229. Springer, Heidelberg (2005). https://doi.org/10.1007/11408031_19
18. He, K., Zhang, X., Ren, S., Sun, J.: Deep residual learning for image recognition. In: Proceedings of the IEEE conference on computer vision and pattern recognition, pp. 770–778. IEEE (2016)
19. Pang, Y., Li, Y., Shen, J., Shao, L.: Towards bridging semantic gap to improve semantic segmentation. In: Proceedings of the IEEE International Conference on Computer Vision, pp. 4230–4239. IEEE (2019)
20. Chen, S., Tan, X., Wang, B., Hu, X.: Reverse attention for salient object detection. In: Proceedings of the European Conference on Computer Vision (ECCV), pp. 234–250 (2018)

21. Zhao, H., Shi, J., Qi, X., Wang, X., Jia, J.: Pyramid scene parsing network. In: Proceedings of the IEEE conference on computer vision and pattern recognition, pp. 2881–2890. IEEE (2017)
22. Aylward, S.R., Bullitt, E.: Initialization, noise, singularities, and scale in height ridge traversal for tubular object centerline extraction. IEEE Trans. Med. Imaging **21**(2), 61–75 (2002)
23. He, K., Sun, J., Tang, X.: Guided image filtering. IEEE Trans. Pattern Anal. Mach. Intell. **35**, 1397–1409 (2013)
24. Çiçek, Ö., Abdulkadir, A., Lienkamp, S.S., Brox, T., Ronneberger, O.: 3D U-Net: learning dense volumetric segmentation from sparse annotation. In: Ourselin, S., Joskowicz, L., Sabuncu, M.R., Unal, G., Wells, W. (eds.) MICCAI 2016. LNCS, vol. 9901, pp. 424–432. Springer, Cham (2016). https://doi.org/10.1007/978-3-319-46723-8_49
25. Milletari, F., Navab, N., Ahmadi, S.A.: V-net: fully convolutional neural networks for volumetric medical image segmentation. In: 2016 Fourth International Conference on 3D Vision (3DV), pp. 565–571. IEEE (2016)

Automated Intracranial Artery Labeling Using a Graph Neural Network and Hierarchical Refinement

Li Chen[1] ⓘ, Thomas Hatsukami[2], Jenq-Neng Hwang[1], and Chun Yuan[3](✉)

[1] Department of Electrical and Computer Engineering, University of Washington,
Seattle, WA, USA
{cluw,hwang}@uw.edu
[2] Department of Surgery, University of Washington, Seattle, WA, USA
tomhat@uw.edu
[3] Department of Radiology, University of Washington, Seattle, WA, USA
cyuan@uw.edu

Abstract. Automatically labeling intracranial arteries (ICA) with their anatomical names is beneficial for feature extraction and detailed analysis of intracranial vascular structures. There are significant variations in the ICA due to natural and pathological causes, making it challenging for automated labeling. However, the existing public dataset for evaluation of anatomical labeling is limited. We construct a comprehensive dataset with 729 Magnetic Resonance Angiography scans and propose a Graph Neural Network (GNN) method to label arteries by classifying types of nodes and edges in an attributed relational graph. In addition, a hierarchical refinement framework is developed for further improving the GNN outputs to incorporate structural and relational knowledge about the ICA. Our method achieved a node labeling accuracy of 97.5%, and 63.8% of scans were correctly labeled for all Circle of Willis nodes, on a testing set of 105 scans with both healthy and diseased subjects. This is a significant improvement over available state-of-the-art methods. Automatic artery labeling is promising to minimize manual effort in characterizing the complicated ICA networks and provides valuable information for the identification of geometric risk factors of vascular disease. Our code and dataset are available at https://github.com/clatfd/GNN-ART-LABEL.

Keywords: Artery labeling · Graph neural network · Hierarchical refinement · Intracranial artery

1 Introduction

Intracranial arteries (ICA) have complex structures and are critical for maintaining adequate blood supply to the brain. There are substantial variations in these arteries

---

Electronic supplementary material The online version of this chapter (https://doi.org/10.1007/978-3-030-59725-2_8) contains supplementary material, which is available to authorized users.

A. L. Martel et al. (Eds.): MICCAI 2020, LNCS 12266, pp. 76–85, 2020.
https://doi.org/10.1007/978-3-030-59725-2_8

among individuals that are associated with vascular disease and cognitive functions [1–3]. Comprehensive characterization of ICA including labeling each artery segment with its anatomical name (Fig. 1 (b)) is desirable for both clinical evaluation and research. The center of the ICA is the Circle of Willis (CoW, normally incorporating nine artery segments forming a ring shape), which connects the left and right hemispheres, as well as anterior and posterior circulations. It has been reported that only 52% of the population has a complete CoW [4]. Many natural variations of CoW exist, including those missing one or multiple arterial segments [5]. In addition, disease related changes within the complex network of ICA are also challenging for automated labeling. For example, stenosis may cause decreased cerebral flow, reflected as reduced blood signal in arterial images; collateral flow forms near the severe stenosis, leading to abnormal structures in the ICA. These situations make automated artery labeling challenging. A simplified graph illustration of ICA is shown in Fig. 1 (c).

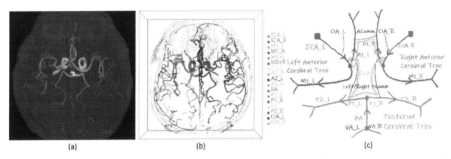

Fig. 1. (a) Time of flight (ToF) Magnetic Resonance Angiography (MRA) of cerebral arteries. (b) ICA labeled in different colors. (c) Illustration of CoW (yellow), left (blue) and right (red) anterior circulation, posterior circulation (red) and optional artery branches (black) with their anatomical names. When there are ICA variations, not only the optional artery branches, but also A1, M1, P1 segments may be missing. See supplementary material for abbreviations. (Colour figure online)

There have been continuous efforts in automating ICA labeling, using either private datasets with a limited number of scans or the publicly available UNC dataset with 50 cerebral Magnetic Resonance Angiography (MRA) images [6]. Takemura et al. [7] built a template of the CoW on five subjects, then arteries were labeled by template alignment and matching on fifteen scans. A more complete artery atlas was built from a population-based cohort of 167 subjects by Dunås et al. [8, 9] using a similar matching approach, and arteries were labeled in 10 clinical cases. Bilgel et al. [10] considered connection probability within the cerebral network using belief propagation for labeling 30 subjects but the method was limited to anterior circulations. Using the UNC dataset, in the serial work from Bogunović et al. [11–13], eight typical ICA graph templates were used to represent ICA with variations, and bifurcations of interest (BoI) were defined and classified so that vessels were labeled indirectly. However, more variations exist beyond the eight typical types. Using the same dataset, by combining artery segmentation along with the labeling, Robben et al. [14] simultaneously optimized the artery centerlines and their labels from an over complete graph. However, their computation involved thousands of variables and constraints, and takes as long as 510 s per case. In summary, while previous works have shown success in labeling relatively small datasets with limited variations in mostly healthy populations, prior knowledge about the global artery structures and relations has not been fully explored. Furthermore, labeling efficiency has not been considered for a large number of scans, which will be needed for clinical applications.

The Graph neural network (GNN) is an emerging network structure recently attracting significant interest [15, 16], including applications on vasculature [17, 18]. By passing information between nodes and edges within the graph, useful properties for the graph can be predicted. Considering the graph topology in anatomical structures of ICA, in this work, we propose a GNN model with hierarchical refinement (HR), aiming to overcome the challenges in arterial labeling by training with large and diversely labeled datasets (more than 500 scans in the training set from multiple sources) and applying refinements after network predictions to combine prior knowledge on ICA. In addition to its superior performance compared with methods described in the literature, our work shows robustness and generalizability on various challenging anatomical variations.

2   Methods

2.1   Intracranial Artery Labeling

The definition of arteries and their abbreviations follows [19] (also in supplementary material). All visible ICA in MRA are traced and labeled using a validated semi-automated artery analysis tool [19, 20] by experienced reviewers with the same labeling criteria, then examined by a peer-reviewer to control quality. Arteries not connected in the main artery tree are excluded from labeling, such as the arteries outside the skull.

2.2   Graph Neural Network (GNN) for Node and Edge Probabilities

The ICA network is represented as the centerlines of arteries, each with consecutively connected 3D points with radius. Centerlines in one MRA scan are constructed as an attributed relational graph $G = (V, E)$. $V = \{v_i\}$ represents all unique points in the centerlines with node features of v_i, and $E = \{e_k, r_k, s_k\}$ represents all point connections where edge k connects between the node index r_k, s_k with edge features of e_k. $e(i)_{1,...,D(i)}$ are all the edges connected with node i ($r_k = i$ or $s_k = i$). $D(i)$ is the degree (number of neighbor nodes) of node i.

Features for node v_i include p_i for x, y, z coordinates, r_i for radius and b_i for the directional embedding of the node. Due to the uncertain number of edges connected to the node, direction features cannot be directly used as an input in GNN. Here we use the multi-label binary encoding to represent direction features. First, 26 major directions in the 3D space are defined as $n_u = (x_u, y_u, z_u)_{u=1,...,26}$, with 45 degrees apart in each axis, excluding duplicates.

$$\begin{cases} x = \sin(45° * a) * \cos(45° * b) \\ y = \cos(45° * a) * \cos(45° * b), a \in \{0, \ldots, 7\}, b \in \{-2, .., 2\} \\ \quad z = \sin(45° * b) \end{cases} \quad (1)$$

Then each edge direction (x_v, y_v, z_v) originating from the node is matched with the major directions with $dir_v = argmax_u(x_u x_v + y_u y_v + z_u z_v)$. $\boldsymbol{b_i}$ is the 26-dimensional feature with encoded direction for all dir_v. (an example in supplementary figure).

Features for edges $\boldsymbol{e_k}$ include edge direction $\boldsymbol{n_k} = \left(\boldsymbol{p_{s_k}} - \boldsymbol{p_{r_k}}\right)$, which is then normalized (and inverted) so that $\|\boldsymbol{n_k}\| = 1$ and $z_k > 0$; distance between nodes at two ends $d_k = \left\|\boldsymbol{p_{s_k}} - \boldsymbol{p_{r_k}}\right\|$; and mean radius at two nodes $\bar{r}_k = \left(r_{s_k} + r_{r_k}\right)/2$.

With similar purpose of BoI [11–13], we remove all nodes with a degree of 2 to reduce the graph size, as nodes requiring labeling are usually at bifurcations or ending points. If the remaining nodes are correctly predicted as one of the 21 possible bifurcation/ending types, then the ICA (edges) can be labeled based on their connections.

We implemented the GNN based on the message passing GNN framework proposed in [16, 21] to predict the types for each node and edge. The GNN takes a graph with node and edge features as input and returns a graph as output with additional features for node and edge types. The input features of edges and nodes in the graph are encoded to an embedding in the encoder layer. Then the core layer passes messages for 10 rounds by concatenating the encoder's output with the previous output of the core layer. The embedding is restored to edge and node features in the decoder layer with additional label features. Computation in each graph block is shown in the Eq. 2. The edge attributes are updated through the per-edge "update" function $\emptyset^e$, and features for edges connected to the same node are "aggregated" through function $\rho^{e \rightarrow v}$ to update node features through the per-node "update" function $\emptyset^v$. The network structure is shown in Fig. 2.

$$
\begin{cases}
updated\,edge\,attributes\ \boldsymbol{e}'_k = \emptyset^e\left(\boldsymbol{e_k}, \boldsymbol{v_{r_k}}, \boldsymbol{v_{s_k}}\right) \\
updated\,edge\,attributes\,per\,node\ \bar{\boldsymbol{e}}'_i = \rho^{e \rightarrow v}\left(\left\{\left(\boldsymbol{e}'_k, r_k, s_k\right)\right\}_{r_k=i}\right) \\
updated\,node\,attributes\ \boldsymbol{v}'_i = \emptyset^v\left(\bar{\boldsymbol{e}}'_i, \boldsymbol{v_i}\right)
\end{cases} \quad (2)
$$

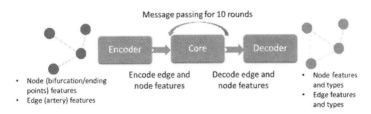

Fig. 2. GNN structure used in this study.

Probability $P_{nt}(i)$ for node i being bifurcation/ending type $nt \in \{0 : Non\_Type, 1 : ICA\_Root\_L, \ldots, 20 : ICA\_PComm\}$ is calculated using a softmax function of GNN output $O_{nt}(i)$. The predicted node type $T_n(i)$ is then identified by selecting the node type with the maximum probability.

$$
P_{nt}(i) = \frac{e^{O_{nt}(i)}}{\sum_{nti=0}^{20} e^{O_{nti}(i)}} \quad (3)
$$

$$
T_n(i) = argmax_{nt}(P_{nt}(i)) \quad (4)
$$

Similar for edges, $et \in \{0 : Non\_Type, 1 : ICA\_L, \ldots, 22 : OA\_R\}$, the edge probability and predicted edge type are

$$P_{et}(k) = \frac{e^{O_{et}(k)}}{\sum_{eti=0}^{22} e^{O_{eti}(k)}} \qquad (5)$$

$$T_e(k) = argmax_{et}(P_{et}(k)) \qquad (6)$$

Ground truth types for nodes and edges are $G(i)$, $G(k)$.

The GNN was trained using combined weighted cross entropy losses in both nodes and edges, with weights inverse proportional to frequencies of the node and edge types. Batch size of 32 graphs was used in training the GNN. Adam optimizer [22] was used for controlling the learning rate. Positions of nodes from different datasets were normalized based on the imaging resolution, and a random translation of positions (within 10%) was used as the data augmentation method.

2.3 Hierarchical Refinement (HR)

Predictions from the GNN might not be perfect, as end-to-end training cannot easily learn global ICA structures and relations. Human reviewers are likely to subdivide ICA into three sub-trees (i.e., left/right anterior, posterior cerebral trees), find key nodes (such as the bifurcation for ICA/MCA/ACA) in sub-trees, then add additional branches which are less important and more prone to variations (such as PComm, AComm). Enlighted by the sequential behavior during manual labeling, a hierarchical refinement (HR) framework based on GNN outputs is proposed to further improve the labeling. Starting from the most confident nodes, the three-level refinement is shown in Fig. 3.

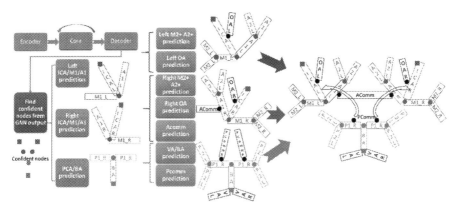

Fig. 3. Workflow of HR framework. In the first level (blue box), confident nodes (circle and square dots) are identified from the GNN outputs. In the second level (orange boxes), confident nodes as well as their inter-connected edges in the left (blue lines)/right (red lines) anterior, posterior (green) sub-trees are identified. In the third level (grey boxes), optional nodes and edges (black lines) are added to each of the three sub-trees to form a complete artery tree. (Colour figure online)

Level One Labeling. We consider nodes as confident if the predicted node type fits the predicted edge types in edges they are connected with.

$$F\left(\{T_e\left(e(i)_{1,...,D(i)}\right)\}\right) = T_n(i) \tag{7}$$

F is a lookup table for all valid pairs of edge types and node types. For example, $F(P1\_L, P1\_R, BA) = PCA/BA$, $F(ICA\_L) = ICA\_ROOT\_L$

Level Two Labeling. From confident nodes, three sub-trees are built, and major-branch nodes are predicted in each sub-tree individually. Major node i is defined as ICA/MCA/ACA (for anterior trees) and PCA/BA (for posterior trees), and branch nodes j are defined as ICA_Root, M1/2, A1/2 (for anterior trees) and BA/VA, P1/2 (for posterior trees). If major nodes are not confident nodes in each sub-tree, they are predicted with type $argmax_i(P_{nt_i}(i)|D(i) \neq 1)$ with additional constraints if branch nodes j are confident ($i \notin j$ and i must be in the path between any pair of j). Then from the major node, all unconfident branch nodes are predicted using the target function of

$$\begin{cases} argmax_j\left[P_{nt_j}(j) + P_{et_{e(i)}}\left(e(i)_{1,...,D(i)}\right)_{r_{e(i)}=j,s_{e(i)}=i}\right], if\ P_{nt_i}(i) > Thres \\ argmax_j\left(P_{nt_j}(j)\right), if\ P_{nt_i}(i) < Thres \end{cases} \tag{8}$$

On rare occasions, when the major nodes have a probability lower than a certain threshold *Thres* (when there are anatomical variations where major nodes do not exist), branch nodes are predicted without edge probability.

If certain distance between the optimal i and j is beyond the mean plus 1.5 standard deviation of $G(k)|r_k = j, s_k = i$ from the training set, labeling on j will be skipped and a node with a degree of 2 will be labeled so that its distance to node i is closest to the mean distance of $G(k)$. This happens when there are missing Acomm or Pcomm.

Level Three Labeling. Optional branches are added to three sub-trees. M2+, A2+, and P2+ edges are assigned for all distal neighbors of M1/2, A1/2, P1/2 nodes. Based on node probabilities, OAs are identified on the path between ICA_Root and ICA/MCA/ACA nodes, Acomm is assigned if there is a connection between A1/2_L and A1/2_R, Pcomm is assigned if there is connection between P1/2 and ICA/MCA/ACA, VA_Root is predicted from neighbors of BA/VA.

2.4 Experiments

Datasets. Five datasets from our previous research [23] were used to train and evaluate our method, then the generalizability was assessed on the public UNC dataset with/without further training. Details for the datasets are in supplementary material.

Our five datasets were collected with different resolutions from different scanner manufacturers. Subjects enrolled in the datasets include both healthy (no recent or chronic vascular disease) and with various vascular related diseases, such as recent stroke events and hypertension. All the datasets were randomly divided into a training set (508 scans), a validation set (116 scans) and a testing set (105 scans). If the subject had multiple scans, these scans were divided into the same set. All scans from the UNC dataset (https://public.kitware.com/Wiki/TubeTK/Data, healthy volunteers) with publicly available artery traces (N = 41) were used. Generally, our dataset has more ICA variations and more challenging anatomies than the UNC dataset.

Evaluation Metrics. As our purpose is to label the ICA, the accuracy of predicted node labels is the primary metric for evaluation (Node_Acc). In addition, we also used number of wrongly predicted nodes per scan (Node_Wrong), edge accuracy (Edge_Acc) and the percent of scans with CoW nodes (ICA/MCA/ACA, PCA/BA, A1/2, P1/2/PComm, PComm/ICA), all nodes and all edges correctly predicted (CoW_Node_Solve, Node_Solve, Edge_Solve). For detailed analysis of detection performance on each bifurcation type, the detection accuracy, precision and recall for 7 major bifurcation types (ICA-OA, ICA-M1, ICA-PComm, ACA1-AComA, M1-M2, VBA-PCA1, PCA1-PComA) were calculated. The processing time was also recorded. Due to the lack of criteria for labeling nodes with degrees of 2, nodes such as A1/2 without AComm were excluded from the evaluation.

Comparison Methods. With the same artery traces of our dataset, three artery labeling methods [7, 9, 19] were used to compare the performance. Due to the unavailability of two methods using the UNC dataset, we only cite evaluation results from their publications. Direction features and HR were sequentially added to our baseline model to evaluate the contribution of different features and the effectiveness of the HR framework.

As the ablation study, GNN without HR predicts node and edge types directly from the GNN outputs $T_n(i)$ and $T_e(k)$. We further tested the removal of direction features.

3   Results

In the testing set of our dataset, 1035 confident nodes (9.86/scan) were identified, and 5 of them (0.5%, none are major or branch nodes) were predicted wrongly, showing labeling of confident nodes is reliable, so that labeling in the following up levels in HR was meaningful. *Thres* was chosen as 1e-10 from the validation set. Examples of correctly labeling challenging cases are shown in Fig. 4. Our method was robust, even with artificial noise branches added in the M1 branch shown in Fig. 4 (d).

The comparison with other artery labeling algorithms and the ablation study is shown in Table 1. Our method demonstrates a better node accuracy of 97.5% with 3.0 wrong nodes/scan. Our method is the only one with cases where all nodes and edges were predicted correctly with the minimum processing time (less than 0.1 s). With direction features and the HR added to the baseline model, the performance is further improved. ICA-OA is the most accurately detected bifurcation type with detection accuracy of 96.2% while the challenging M1/2 has an accuracy of 68.1%. Mean detection accuracy is 83.1%, precision is 91.3%, recall is 83.8%.

Our method showed good generalizability on the UNC dataset (Table 2). Even without additional training on the UNC dataset, the node accuracy was 99.03% (2.0 wrong nodes/scan) with 56% of cases solved. Mean detection accuracy for all node types was 92%. As a reference with methods using leave-one-out cross validation trained and evaluated using the same dataset, 58% of cases were solved [13]. The mean detection accuracies were 94% and 95% in [13, 14], respectively. If trained in combination with the UNC dataset using three fold cross validation, our method outperforms [13, 14].

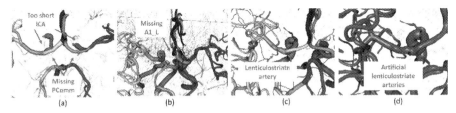

Fig. 4. Examples of challenging anatomical variations where our method predicted all arteries correctly. (a) A subject with Parkinson's disease. Occlusions cause both right and left internal carotid arteries to be partially invisible. In addition, Pcomms are missing. (b) A hypertensive subject with rare A1_L artery missing, which is not among the 8 anatomical types and thus not solvable in [13]. (c) Some lenticulostriate arteries are visible in our dataset with higher resolution, an additional challenge for labeling, our method predicted it correctly as a non-type. (d) With more artificial lenticulostriate arteries added in the M1_L segment, our method is still robust to these additional noise branches.

Table 1. Comparison with existing methods and the ablation study on our testing set (N = 105).

Method	Node_Acc↑	Node_Wrong↓	Node_Solve↑	CoW_Node_Solve↑	Edge_Acc↑	Edge_Solve↑	Process time (s) ↓
MAP [19]	0.9153	10.0	0	0.0476	0.3304	0	1.075
Template [7]	0.7316	31.6	0	0.0476	0.7934	0	5.057
Atlas [9]	0.8856	13.5	0	0.0095	0.7010	0	9.253
GNN(Pos)	0.9553	5.3	0.0286	0.3524	0.9099	0	**0.017**
GNN(Pos + Dir)	0.9637	4.3	0.0381	0.4286	0.9223	0	0.020
GNN(Pos + Dir) + HR	**0.9746**	**3.0**	**0.3238**	**0.6381**	**0.9246**	**0.3238**	0.092

Table 2. Performance of detection accuracy (A), precision (P) and recall (R) for each bifurcation type, compared with previous methods using the UNC dataset. Note that our method is trained with our dataset with (GNN + HR + UNC) and without (GNN + HR) further training in the UNC dataset, but other methods were trained and evaluated by leave-one-out cross validation on the UNC dataset alone (more likely to overfit on the UNC dataset).

Method	GNN + HR (Ours)			GNN + HR + UNC (Ours)			Robben [14]			Bogunović [13]		
	A	P	R	A	P	R	P	R	R	A	P	R
ICA-OA	97	100	97	100	100	100	99	99	100	99	100	99
ICA-M1	90	96	93	95	97	97	100	100	100	99	99	100
ICA-PComm	88	97	89	95	97	97	98	100	98	93	94	96
ACA1-AComA	95	100	95	96	100	96	96	100	95	92	93	97
M1-M2	90	95	95	95	97	97	78	78	100	89	89	100
VBA-PCA1	94	97	97	90	94	94	90	98	91	94	100	93
PCA1-PComm	90	97	92	92	98	94	95	100	92	96	100	94
Mean	92	**97**	94	**95**	**97**	**97**	94	96	**97**	**95**	96	**97**

4   Discussion and Conclusion

We have developed a GNN approach to label ICA with HR on our comprehensive ICA dataset (729 scans). Four contributions and novelties in our work are worth highlighting. 1) The dataset includes more diverse and challenging ICA variations compared with the existing UNC dataset, which is better suited to evaluate labeling performance. 2) The GNN and HR framework is an ideal method to learn from the graph representation of ICA and incorporate prior knowledge about ICA structure. 3) With accurate predictions of 20 node and 22 edge types covering all major artery branches visible in MRA, this method can automatically provide comprehensive features for detailed analysis of cerebral flow and structures in less than 0.1 s. 4) It should also be noted that our GNN and HR framework is not only applicable to ICA, but also to any graph structures where sequential labeling helps. For example, major lower extremity arteries and branches can be labeled ahead of labeling the collateral arteries.

Accurate ICA labeling using our method relies on reliable artery tracing, which is one of our limitations, although this is a lesser concern compared with non-graph based methods, where some artery tracing mistakes (such as centerlines off-center or zigzags in the path) can be avoided through a simplified representation of the ICA through graph constructions.

Acknowledgements. This work was supported by National Institute of Health under grant R01-NS092207. We are grateful for the collaborators who provided the datasets for this study, including the CROP and BRAVE investigators, and researchers from the University of Arizona, USA, Beijing Anzhen hospital, China, and Tsinghua University, China and the public data from The University of North Carolina at Chapel Hill (distributed by the MIDAS Data Server at Kitware Inc.). We acknowledge NIVIDIA for providing the GPU used for training the neural network model.

Our code and dataset are available at https://github.com/clatfd/GNN-ART-LABEL.

References

1. Kayembe, K.N., Sasahara, M., Hazama, F.: Cerebral aneurysms and variations in the circle of Willis. Stroke **15**, 846–850 (1984)
2. Alpers, B.J., Berry, R.G., Paddison, R.M.: Anatomical studies of the circle of willis in normal brain. Arch. Neurol. Psychiatry. **81**, 409–418 (1959)
3. Chen, L., et al.: Quantitative intracranial vasculature assessment to detect dementia using the intra-Cranial Artery Feature Extraction (iCafe) technique. In: Proc. Annu. Meet. Int. Soc. Magn. Reson. Med. Palais des congrès Montréal, Montréal, QC, Canada May, pp. 11–16 (2019)
4. Alpers, B.J., Berry, R.G.: Circle of willis in cerebral vascular disorders. Anat. Struct. Arch. Neurol. **8**, 398–402 (1963)
5. Ustabaşıoğlu, F.E.: Magnetic resonance angiographic evaluation of anatomic variations of the circle of willis. Med. J. Haydarpaşa Numune Training Res. Hosp. **59**, 291–295 (2018)
6. Bullitt, E., et al.: Vessel tortuosity and brain tumor malignancy: a blinded study. Acad. Radiol. **12**, 1232–1240 (2005)
7. Takemura, A., Suzuki, M., Harauchi, H., Okumura, Y.: Automatic anatomical labeling method of cerebral arteries in MR-angiography data set. Japanese J. Med. Phys. **26**, 187–198 (2006)

8. Dunås, T., Wåhlin, A., Ambarki, K., Zarrinkoob, L., Malm, J., Eklund, A.: A stereotactic probabilistic atlas for the major cerebral arteries. Neuroinformatics **15**(1), 101–110 (2016). https://doi.org/10.1007/s12021-016-9320-y

9. Dunås, T., et al.: Automatic labeling of cerebral arteries in magnetic resonance angiography. Magn. Reson. Mater. Phys., Biol. Med. **29**(1), 39–47 (2015). https://doi.org/10.1007/s10334-015-0512-5

10. Bilgel, M., Roy, S., Carass, A., Nyquist, P.A., Prince, J.L.: Automated anatomical labeling of the cerebral arteries using belief propagation. Med. Imaging 2013 Image Process. **8669**, 866918 (2013)

11. Bogunović, H., Pozo, J.M., Cárdenes, R., Frangi, A.F.: Automatic identification of internal carotid artery from 3DRA images. In: 2010 Annual International Conference of the IEEE Engineering in Medicine and Biology, pp. 5343–5346. IEEE (2010)

12. Bogunović, H., Pozo, J.M., Cárdenes, R., Frangi, A.F.: Anatomical labeling of the anterior circulation of the circle of willis using maximum a posteriori classification. In: Fichtinger, G., Martel, A., Peters, T. (eds.) MICCAI 2011. LNCS, vol. 6893, pp. 330–337. Springer, Heidelberg (2011). https://doi.org/10.1007/978-3-642-23626-6_41

13. Bogunović, H., Pozo, J.M., Cardenes, R., Roman, L.S., Frangi, A.F.: Anatomical labeling of the circle of willis using maximum a posteriori probability estimation. IEEE Trans. Med. Imaging. **32**, 1587–1599 (2013)

14. Robben, D., et al.: Simultaneous segmentation and anatomical labeling of the cerebral vasculature. Med. Image Anal. **32**, 201–215 (2016)

15. Zhou, J., et al.: Graph neural networks: a review of methods and applications, pp. 1–22 (2018)

16. Battaglia, P.W., et al.: Relational inductive biases, deep learning, and graph networks. arXiv: 1806.01261, pp. 1–40 (2018)

17. Zhai, Z., et al.: Linking convolutional neural networks with graph convolutional networks: application in pulmonary artery-vein separation. In: Zhang, D., Zhou, L., Jie, B., Liu, M. (eds.) GLMI 2019. LNCS, vol. 11849, pp. 36–43. Springer, Cham (2019). https://doi.org/10.1007/978-3-030-35817-4_5

18. Wolterink, J.M., Leiner, T., Išgum, I.: Graph convolutional networks for coronary artery segmentation in cardiac CT angiography. In: Zhang, D., Zhou, L., Jie, B., Liu, M. (eds.) GLMI 2019. LNCS, vol. 11849, pp. 62–69. Springer, Cham (2019). https://doi.org/10.1007/978-3-030-35817-4_8

19. Chen, L., et al.: Development of a quantitative intracranial vascular features extraction tool on 3D MRA using semiautomated open-curve active contour vessel tracing. Magn. Reson. Med. **79**, 3229–3238 (2018)

20. Chen, L., et al.: Quantification of morphometry and intensity features of intracranial arteries from 3D TOF MRA using the intracranial artery feature extraction (iCafe): a reproducibility study. Magn. Reson. Imaging **57**, 293–302 (2018)

21. Gilmer, J., Schoenholz, S.S., Riley, P.F., Vinyals, O., Dahl, G.E.: Neural message passing for quantum chemistry. arXiv:1704.01212v2, (2017)

22. Pinto, A., Alves, V., Silva, C.A.: Brain tumor segmentation using convolutional neural networks in MRI images. IEEE Trans. Med. Imaging **35**, 1240–1251 (2016)

23. Chen, L., et al.: Quantitative assessment of the intracranial vasculature in an older adult population using iCafe (intraCranial Artery Feature Extraction). Neurobiol. Aging **79**, 59–65 (2019)

Time Matters: Handling Spatio-Temporal Perfusion Information for Automated TICI Scoring

Maximilian Nielsen[1,2], Moritz Waldmann[3], Thilo Sentker[1,2], Andreas Frölich[3], Jens Fiehler[3], and René Werner[1,2(✉)]

[1] Department of Computational Neuroscience, University Medical Center Hamburg-Eppendorf, Martinistr. 52, 20246 Hamburg, Germany
r.werner@uke.de
[2] Center for Biomedical Artificial Intelligence (bAIome), University Medical Center Hamburg-Eppendorf, Martinistr. 52, 20246 Hamburg, Germany
[3] Department of Diagnostic and Interventional Neuroradiology, University Medical Center Hamburg-Eppendorf, Martinistr. 52, 20246 Hamburg, Germany

Abstract. X-ray digital subtraction angiography (DSA) imaging is the backbone of diagnosis and therapy response assessment in cerebral ischemic stroke. To evaluate and document the success of endovascular interventions, the spatio-temporal DSA image information and perfusion dynamics are visually assessed by a clinical expert and reperfusion rated using the so-called TICI (treatment in cerebral ischemia) score. Although clinical standard, it is well known that TICI scoring is time-consuming, observer-dependent and not practicable especially in larger clinical studies. Automated TICI scoring has, however, been considered beyond the scope of machine learning capabilities, due to the complexity of the classification task (eg. heterogeneity of clinical DSA data and a complex dependence between TICI score and perfusion dynamics). The present work describes the first study that tackles automated TICI scoring using deep spatio-temporal learning. It thereby defines the first corresponding benchmark. Methodically, we build on gated recurrent unit networks (GRUs) and integrate knowledge about the perfusion and TICI scoring process into loss functions and network training to increase prediction robustness. Differences between GRU-predicted mTICI scores and routine mTICI scores are in the order of literature-reported interrater variability of human expert-based TICI scoring.

Keywords: Spatio-temporal imaging · Digital subtraction angiography (DSA) · Gated recurrent unit (GRU) networks · Ischemic stroke · TICI

Electronic supplementary material The online version of this chapter (https://doi.org/10.1007/978-3-030-59725-2_9) contains supplementary material, which is available to authorized users.

A. L. Martel et al. (Eds.): MICCAI 2020, LNCS 12266, pp. 86–96, 2020.
https://doi.org/10.1007/978-3-030-59725-2_9

1   Introduction

Ischemic stroke is one of the leading causes of neurological mortality and morbidity [1,2], causing considerable burden for patients and relatives as well as significant costs in the health care system [3]. Clinical gold standard for acute ischemic stroke diagnosis and treatment response evaluation is X-ray digital subtraction angiography (DSA) imaging, which provides 2D time series data of high spatio-temporal resolution at low risk and costs. DSA-based diagnosis and treatment response assessment is performed using the so-called TICI (treatment in cerebral ischemia) grading system [4,5], which is based on the extent of tissue reperfusion after intervention. The common mTICI maps the spatio-temporal image information onto the 5-level score detailed in Table 1. Corresponding DSA image examples are shown in Fig. 1.

Table 1. Definition of classes of the mTICI grading system according to [5] and number of DSA data sets available for the individual classes in the present study.

Grade	Definition	#Data
TICI 0	No perfusion	167
TICI 1	Antegrade reperfusion past initial occlusion, but limited distal branch filling with little/slow reperfusion	25
TICI 2a	Antegrade reperfusion of *less* than half of the previously occluded target artery ischemic territory	71
TICI 2b	Antegrade reperfusion of *more* than half of the previously occluded target artery ischemic territory	98
TICI 3	Complete antegrade reperfusion of the previously occluded target artery ischemic territory, with absence of visualized occlusion in all distal branches	99

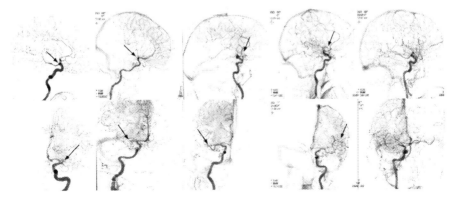

Fig. 1. Temporal minimum intensity projections of DSA time series data. Top: sagittal views. Bottom: coronal views. Left to right: mTICI scores 0, 1, 2a, 2b, and 3. Arrows indicate location of occlusion. By definition, mTICI 3 contains no occlusion.

The TICI score correlates to favorable clinical response [6] and therefore represents an early indicator of intervention success and an appropriate end point of clinical studies for, eg., comparison of different treatment modalities and endovascular devices. If the score is objectively assessed, it could also be applied as quality assurance tool for endovascular interventions.

At the moment, TICI grading is done by visual review of the DSA image time series by a neuroradiologist – which is time consuming, error prone and subject to interrater variability [7,8]. Objective TICI score assessment therefore requires automatizing the grading process. However, in 2016, Scalzo and Liebeskind stated that "automatic TICI [...] scores are still beyond the capabilities of current methods" [8]. In line with this statement, we are not aware of any approaches to automated TICI scoring, except for [9]. In [9], standard image classification CNN architectures were applied; the resulting classification accuracy was relatively poor (eg. 44% accuracy for mTICI 3).

Our hypothesis was that this was in parts due to the authors' methodical approach, which does not fit the specific problem and data: for standard CNNs, temporal aspects like slow or retrograde perfusion are hard to learn. The aim of our study was therefore

[**Goal 1**] to develop appropriate deep spatio-temporal learning approaches, and

[**Goal 2**] to compare the classification performance to reported TICI grading interrater variability – with the ultimate goal to obtain a classification accuracy that is, in terms of interrater-variability, comparable to human experts.

1.1   Interrater Variability of TICI Grading: Literature Overview

Exact values for interrater variability of TICI grading differ, but they are mainly of the same order of magnitude. Suh *et al.* reported an intra-class coefficient (ICC) of 0.67 for mTICI scoring [10]. Mair *et al.* evaluated the interrater variability by means of Krippendorff's Alpha (K-Alpha) for three groups: multicenter angiography panel experts (K-Alpha of 0.60), neuroradiology trainees (0.63) and non-experts (0.39) [11]. They also reported expert intrarater variability, which was in the same order than the interrater variability (0.60). Furthermore, Tung *et al.* evaluated interrater variability for two radiologists and obtained an ICC of 0.77 and an overall Cohen's Kappa of 0.58 for oTICI (class border between 2a and 2b at 67%) and 0.62 for mTICI [12]. The use of the different variability measures exacerbates direct comparison across the studies, but Kappa, ICC and K-Alpha reside on the same scale and can be assumed to be broadly similar [12].

1.2   Novelty and Contribution of the Study

The present study is the first study that tackles automated TICI scoring by deep spatio-temporal learning. As methodical contributions,

[**Contribution 1**] we propose using gated recurrent unit networks (GRUs) to address the complexity and nature of the classification task, and

[**Contribution 2**] we suggest problem-tailored loss function modifications and a model-based adaptation of the GRU learning process.

Our study therefore represents the first benchmark for automated mTICI scoring using deep spatio-temporal learning. From an application perspective,

[**Contribution 3**] we illustrate our knowledge-driven approaches to improve classification accuracy and

[**Contribution 4**] demonstrate that we indeed end up with residual differences between automated mTICI and routine mTICI scoring that are in the order of literature-reported interrater variability values.

2   Methods and Material

DSA data are simultaneously acquired as sagittal and coronal views (Fig. 1). Subsequently, DSA temporal image series of a patient are denoted as $I_{cor}, I_{sag} \in$ Img, $I_{cor}, I_{sag} : \Omega \times \mathcal{T} \subset \mathbb{Z}^2 \times \mathbb{Z} \to \mathbb{Z}$ with $(x, t) \in \Omega \times \mathcal{T}$ representing pixel index x in time frame t of the image series. Img $\equiv$ Img $(\Omega \times \mathbb{Z})$ denotes the space of DSA image series. For each pair (I_{cor}, I_{sag}), a corresponding mTICI label $y \in \mathcal{Y} = \{`0`, `1`, `2a`, `2b`, `3`\}$ exists. The goal of automated TICI scoring is to learn a mapping $f_{TICI} :$ Img $\times$ Img $\to \mathcal{Y}$, i.e. a classification that assigns a correct mTICI label $\hat{y} = f_{TICI} (I_{cor}, I_{sag})$ to an DSA image series pair (I_{cor}, I_{sag}).

2.1   Network Architectures and Loss Functions

Standard CNN-Based TICI Learning. To define a baseline, standard CNN-based image classification was implemented, similar to [9] building on a MinIP approach: The DSA time series I_{cor}, I_{sag} were condensed to temporal minimum intensity projections $I_{(\cdot)}^{MinIP} (x) := \min \{I_{(\cdot)} (x, t) \mid t \in \mathcal{T}\}$, $(\cdot) \in \{cor, sag\}$, to highlight maximum contrast agent concentration (associated with low intensity values) per pixel over time. The idea is that pixel-by-pixel representation of the maximum contrast agent concentration allows assessing the entire perfused area in a static image that can be efficiently processed; this, however, neglects by design aspects like slow or retrograde perfusion (Sect. 1).

I_{cor}^{MinIP} and I_{sag}^{MinIP} were processed by a two-path EfficientNet-B0 network, pretrained on ImageNet [13]. We used and compared two different loss functions:

– **Standard CE loss**: The network was first trained using the standard categorical cross entropy (CE) loss

$$\mathcal{L}_{CE} (\mathbf{y}, \hat{\mathbf{y}}) = -\langle \mathbf{y}, \log \hat{\mathbf{y}} \rangle \tag{1}$$

with $\mathbf{y} \in \{0,1\}^K$ as one-hot-encoded ground truth (GT) label, $\hat{\mathbf{y}} \in \mathbb{R}^K$ as network (softmax) estimate and K as number of classes (here: $K = 5$).

– **Uncertainty-driven TICI loss**: Standard CE loss and an one-hot-encoded GT label neglect the inherent uncertainty of manual TICI scoring. Therefore, similar to [14], we propose label smoothing to encourage the network to be less confident. We do, however, not consider modeling label uncertainty by an additive uniform distribution to be appropriate. Due to the roughly ordinal character of the TICI score, we assumed that labeling uncertainties lead to TICI labels that, if wrong, are close to the true score. Thus, let y_l denote the component of $\mathbf{y}$ that corresponds to the GT label; we then replaced $\mathbf{y}$ by a smoothed version $\tilde{\mathbf{y}}$ with non-zero elements $\tilde{y}_l = 1 - \epsilon_L$ and $\tilde{y}_{l\pm1} = \epsilon_L/2$. $\epsilon_L \in [0, 0.5]$ is referred to as labeling uncertainty factor.

Yet, label smoothing is not directly compatible with standard CE. To nevertheless benefit from the advantageous CE convergence behavior, we introduced a loss with gradient properties similar to CE. In detail, we transformed the network softmax output $\hat{\mathbf{y}}$ into an auxiliary vector $\hat{\mathbf{y}}^*$ by

$$\hat{\mathbf{y}}^* = \tilde{\mathbf{y}} - \mathrm{ReLU}\left(\tilde{\mathbf{y}} - \hat{\mathbf{y}}\right). \tag{2}$$

Based on $\hat{\mathbf{y}}^*$, the loss function was defined as

$$\mathcal{L}_{\tilde{\mathrm{CE}}}\left(\tilde{\mathbf{y}}, \hat{\mathbf{y}}\right) = \sum_i \mathcal{L}_{\tilde{\mathrm{CE}},}\left(\hat{y}_i^*\right), \quad \mathcal{L}_{\tilde{\mathrm{CE}}}\left(\hat{y}_i^*\right) \propto -\log\left(\hat{y}_i^* + \sum_{k \neq i} \hat{y}_k^*\right) \tag{3}$$

with coefficients of proportionality $\left(\mathrm{ReLU}\left(\tilde{y}_i - \hat{y}_i^*\right)\right) / \left(\sum_k \mathrm{ReLU}\left(\tilde{y}_k - \hat{y}_k^*\right)\right)$. To further penalize large TICI differences between GT and prediction, we also introduced a mean squared error (MSE) loss variant

$$\mathcal{L}_{\tilde{\mathrm{MSE}}}\left(\tilde{\mathbf{y}}, \hat{\mathbf{y}}\right) = d_l\left(\tilde{\mathbf{y}}, \hat{\mathbf{y}}\right)\left\langle\left(\tilde{\mathbf{y}} - \hat{\mathbf{y}}\right), \left(\tilde{\mathbf{y}} - \hat{\mathbf{y}}\right)\right\rangle \tag{4}$$

with d_l as an index distance: if $\tilde{y}_l$ is the largest element of $\tilde{\mathbf{y}}$ and $\hat{y}_{l'}$ is the largest element of $\hat{\mathbf{y}}$, $d_l\left(\tilde{\mathbf{y}}, \hat{\mathbf{y}}\right)$ is given by $|l - l'|$. The applied TICI loss function was eventually defined as

$$\mathcal{L}_{\mathrm{TICI}}\left(\tilde{\mathbf{y}}, \hat{\mathbf{y}}\right) = \frac{\mathcal{L}_{\tilde{\mathrm{CE}}}\left(\tilde{\mathbf{y}}, \hat{\mathbf{y}}\right)}{\alpha_{\tilde{\mathrm{CE}}}} + \frac{\mathcal{L}_{\tilde{\mathrm{MSE}}}\left(\tilde{\mathbf{y}}, \hat{\mathbf{y}}\right)}{\alpha_{\tilde{\mathrm{MSE}}}} \tag{5}$$

with the normalization factors being the loss values for $\tilde{y}_i = 1/K$ for all i.

GRU-Based TICI Learning. To evaluate if explicit handling of the DSA temporal information helps improving automated TICI scoring, we further implemented TICI learning by recurrent neural networks (RNNs). RNNs allow operating on frame-level information and information integration over time. We focussed on gated recurrent unit networks (GRUs), which have been shown to be more resource efficient at similar accuracy levels than long short-term memory (LSTM) networks in, eg., sequence modeling and video representation learning [15,16]. Like [17], we also worked on TICI learning by convolutional GRUs; the performance was, however, not superior compared to standard GRUs.

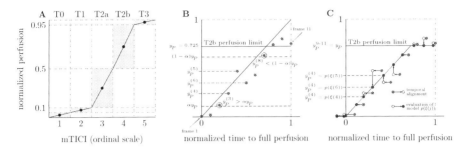

Fig. 2. A – Assumed piecewise linear relation between mTICI and normalized perfusion. The dots denote the mTICI interval centers and corresponding y_P values. Back-transforming the derived targets $y_P^{(j)}$ to $y_T^{(j)}$ is also based on the shown relation (e.g., a $y_P^{(j)} = 0.4$ corresponds to $y_T^{(j)} = 3.25$). **B** – First step of the temporal alignment of the GRU-predicted frame-specific perfusion data $\hat{y}_P^{(j)}$, $j \in \{1, \dots, n\}$, with the linear model p. $\hat{y}_P^{(1)}$ is aligned with $t = 0$ and $\hat{y}_P^{(n)}$ with $t = 1$. The first frames k, m with $\hat{y}_P^{(k)} \geq \alpha y_P$ and $\hat{y}_P^{(m)} \geq (1 - \alpha) y_P$, respectively, are sought (here: $y_P = 0.725$ for GT label TICI 2b, $k = 3$, $m = 8$). **C** – Final alignment $\xi(j)$ and computation of frame-specific targets $p(\xi(j))$. The alignment is based on afore-mentioned $\hat{y}_P^{(k)}$ and $\hat{y}_P^{(m)}$, which are shifted onto the perfusion model line $p(t) = t$. Thereon based, temporal shifts of the other data points are linearly interpolated and the sought targets $y_P^{(j)} = p(\xi(j))$ derived.

The original DSA image size allowed, however, no direct image-based GRU training (at least for standard GPUs). Here, the MinIP-pretrained EfficientNet-B0 was used as frame-level feature extractor and the encoder bottleneck information as GRU input. The entire system (encoder and GRU) was trained end-to-end. Let $n = |\mathcal{T}|$ denote the number of a training DSA image series and $\hat{\mathbf{y}}^{(n)}$ the GRU softmax output for the final DSA frames after forward pass of the entire DSA series; temporal backpropagation was then performed based on $\mathcal{L}_{\text{TICI}}(\hat{\mathbf{y}}^{(n)}, \tilde{\mathbf{y}})$.

GRU-Based TICI Learning Using Frame-Specific Targets. While GRU training via minimization of $\mathcal{L}_{\text{TICI}}(\hat{\mathbf{y}}^{(n)}, \tilde{\mathbf{y}})$ already allowed incorporation of DSA time information, we hypothesized that integration of prior knowledge about the contrast medium (CM) inflow and perfusion process into the GRU training could further improve TICI learning. In detail, we addressed three aspects:

(A) TICI scoring is directly linked to CM inflow, which is visually perceived as perfusion. Thus, if evaluated based on only the first DSA frame(s), the predicted score has to be close to TICI 0.

(B) The final TICI score should be determined latest by making use of the last DSA frame; however, strong variations and a jump of the score during observation of the final DSA frames are most probably physiologically related to retrograde perfusion.

(C) Due to its link to CM inflow, the temporal development of predicted TICI scores over time is assumed to be a monotonically increasing function.

To enforce the network to adhere to (A)–(C), we integrated frame-specific targets into the GRU learning process. Let $\hat{\mathbf{y}}^{(j)}, j \in \{1, \ldots, n\}$ be the GRU softmax output after having processed the information of the frames $I_{\text{cor}}(\cdot, j'), I_{\text{sag}}(\cdot, j'), j' \in \{1, \ldots, j\}$. The goal was now to define corresponding $\tilde{\mathbf{y}}^{(j)}$ for all j that fulfill (A)–(C) and that could be used to evaluate $\mathcal{L}_{\text{TICI}}(\hat{\mathbf{y}}^{(j)}, \tilde{\mathbf{y}}^{(j)})$ for additional guidance during temporal backpropagation after a completed forward pass of $I_{\text{cor}}, I_{\text{sag}}$.

To define the targets, we built on the (simplistic) assumption that CM-based perception of the perfusion process between first CM inflow and full patient-specific (re-)perfusion behind the occlusion site can be modeled as a linear function. Therefore, each $\hat{\mathbf{y}}^{(j)}$ was first transformed into a continuous version of the TICI score by $\hat{y}_T^{(j)} = \sum_i \hat{y}_i^{(j)} i$ with the sum covering the index of the largest vector element and its neighboring elements. $\hat{y}_T^{(j)}$ was afterwards transformed into a relative pseudo-perfusion value $\hat{y}_P^{(j)}$ under the assumption of a piecewise linear relation between $\hat{y}_T$ and $\hat{y}_P$. Similarly, the GT TICI was converted into a perfusion value $y_P \in [0, 1]$; see Fig. 2A. We then introduced a perfusion model $p : \mathbb{R} \to [0, 1]$ with $p(t \leq 0) = 0$, $p(0 < t \leq y_P) = t$, $p(t > y_P) = y_P$. Based on the first frame k with $\hat{y}_P^{(k)} \geq \alpha y_P$ and the first frame m with $\hat{y}_P^{(m)} \geq (1 - \alpha) y_P$, the frame-specific GRU predictions $\hat{y}_P^{(i)}$ were aligned with p via a linear map ξ of the frame number to the temporal domain of p and using $k \mapsto \hat{y}_P^{(k)}$ and $m \mapsto \hat{y}_P^{(m)}$ as anchor points; see Fig. 2B–C. After establishing ξ, frame-specific perfusion targets were computed by $y_P^{(j)} = p(\xi(j))$ and back-transformed into continuous TICI targets $y_T^{(j)}$, again exploiting the assumed piecewise linear relation. To convert $y_T^{(j)}$ into the desired vector $\tilde{\mathbf{y}}^{(j)}$, label smoothing was conducted as described for the labeling uncertainty factor ϵ_L. However, as $y_T^{(j)}$ does not necessarily correspond to an integer value, the part $\epsilon_T = y_T^{(j)} - \text{INT}[y_T^{(j)}]$, with $\text{INT}[\cdot]$ mapping a real number to the nearest integer, had also to be accounted for. The corresponding procedure is detailed in the supplementary materials.

2.2 Data Set and Experiments

Data Set. The study built on 460 routine DSA image series pairs $(I_{\text{cor}}, I_{\text{sag}})$ of size 1024×1024 pixel (resolution $0.19 \times 0.19\,\text{mm}^2$) and varying length (5–30 frames; time resolution $0.33\,\text{s}$) and corresponding mTICI scores (see Table 1 for class frequencies). Infarct area was the M1 segment of the middle cerebral artery.

Data Preprocessing and Parameter Settings. Due to GPU memory constraints, the original image series were down-sampled to 224×224 pixel. The intensity range of the data was rescaled to the interval $[-1, 1]$ (note that a DSA intensity distribution has a U-shape and standard z-transformation is not appropriate). Class imbalance was countered by adjusting the loss values inversely proportional to the class frequency. The labeling uncertainty ϵ_L was 0.4 and α 0.15. Data augmentation was based on random rotation, translation and scaling of the input images, contrast variation and frame dropout (ie. single frame removed from a DSA series). Optimization was conducted using the ADAM optimizer (learning rate 0.001, weight decay 0.1, default β_1 and β_2; GRU: dropout factor 0.5, EfficientNet: default parameters; training for 40 epochs, batch size of 2 for GRU and 24 for MinIP-based classification) and the implementation based on PyTorch. Training and testing was performed using a standard PC (Intel Xeon CPU E5-2620 v4, 64 GB RAM, 2×GeForce GTX 1080 Ti).

Evaluation. Evaluation was performed by 5-fold cross validation (CV), with the folds kept consistent for the different models. Classification performance was assessed by exact accuracy and a sliding window accuracy (deviation of ± 1 scores from the GT mTICI value also considered correct) [18]. The latter was performed to account for mTICI labeling uncertainty (Sect. 1.1). To be comparable to literature values on TICI scoring interrater variability, we also computed Cohen's Kappa between the GT mTICI labels and the network-predicted scores. We evaluated the overall Kappa values as well as Kappa values for joint consideration of mTICI 2b and 3 values and Kappa for mTICI 3 cases only. The latter was clinically motivated: At the moment, either mTICI 2b and 3 or only mTICI 3 are considered indicating success of endovascular interventions [19,20].

Table 2. Summary of automated TICI scoring evaluation (sl-win = sliding window; T2b∪T3 = combined consideration of mTICI 2b and mTICI 3 cases; T3 = mTICI 3). For each model, the first row represents the median values of the five folds; the second row denotes mean ± standard deviation values. [†]Cohen's Kappa values for interrater variability assessment were taken from [12].

| Model | Accuracy | | Cohen's Kappa | | |
	Exact	sl-win	Overall	T2b∪T3	T3
CNN, $\mathcal{L}_{CE}$	0.51	0.82	0.40	0.73	0.38
	0.52 ± 0.03	0.82 ± 0.02	0.41 ± 0.04	0.71 ± 0.07	0.34 ± 0.09
CNN, $\mathcal{L}_{TICI}$	0.50	0.89	0.45	0.77	0.27
	0.49 ± 0.02	0.90 ± 0.02	0.46 ± 0.04	0.76 ± 0.04	0.29 ± 0.09
GRU, $\mathcal{L}_{TICI}$	0.63	0.95	0.56	0.79	0.42
	0.61 ± 0.04	0.95 ± 0.03	0.55 ± 0.04	0.79 ± 0.04	0.46 ± 0.11
GRU, per-mod	0.63	0.95	0.59	0.80	0.58
	0.62 ± 0.03	0.95 ± 0.03	0.58 ± 0.02	0.82 ± 0.08	0.54 ± 0.06
Interrater var.[†]	–	–	0.58/0.62	n/a	0.47

3   Results

Accuracy and Cohen's Kappa evaluation results are summarized in Table 2. The GRU models outperformed the CNN-based models for all evaluation criteria.

Using the CNN models as baseline models, the difference between using $\mathcal{L}_{CE}$ and $\mathcal{L}_{TICI}$ was, eg., reflected by an increased sliding window accuracy for $\mathcal{L}_{TICI}$. This means that a reduced number of test cases was assigned to a completely wrong TICI score (ie. not to the neighboring score of the GT TICI label) when using $\mathcal{L}_{TICI}$. Exact accuracy values of 0.51 and 0.50 (although being in the order of the results in [9]) and, e.g., Cohen's Kappa values for TICI 3 cases appeared nevertheless rather low for the CNN-based classification systems.

Training the RNN systems (here: GRU variants) resulted in a jump in accuracy. Still, the observed exact accuracy of 0.63 appears not high – at least at first glance. However, given the TICI scoring-inherent uncertainty (and, thus, the uncertainty of our GT mTICI labels), we consider the exact accuracy to underestimate the network performance. Indeed, the high sliding window accuracy values directly translated into Cohen's Kappa values that are in the same order of and higher than reported literature values for comparison of human expert-based TICI grading (note that corresponding numbers for non-experts are much smaller). Integration of the perfusion model into GRU training further payed off, especially with regard to Cohen's Kappa for mTICI 3 scoring.

4   Conclusions

The present study tackled automated DSA time series classification by means of the mTICI scoring system that is routinely used for therapy response assessment in cerebral ischemic stroke. The study demonstrates for the fist time that automated TICI scoring with an accuracy and residual misclassification frequency (with respect to the available ground truth labels) on par with literature-reported expert-based interrater variability is feasible. This was achieved by means of deep spatio-temporal learning, problem-tailored loss function definition and integration of task-specific knowledge into the network training process.

Adherence to MICCAI Commitment to Reproducible Research. Trained models and example data are provided at github.com/autoTICI.

Acknowledgements. This work was supported by the European Fund for Regional Development (ERDF) and the Free and Hanseatic City of Hamburg.

References

1. Go, A.S., Mozaffarian, D., Roger, V.L., Benjamin, E.J., Berry, J.D., Blaha, M.J., et al.: Heart disease and stroke statistics-2014 update: a report from the American Heart Association. Circulation **129**, e28–e292 (2014)
2. Organisation, European Stroke, (ESO) Executive Committee; ESO Writing Committee.: Guidelines for management of ischaemic stroke and transient ischaemic attack. Cerebrovasc. Dis. **25**(2008), 457–507 (2008)
3. Taylor, T.N., Davis, P.H., Torner, J.C., Holmes, J., Meyer, J.W., Jacobsen, M.F.: Lifetime cost of stroke in the United States. Stroke **27**, 1459–1466 (1996)

4. Higashida, R.T., Furlan, A.J., Roberts, H., Tomsick, T., Connors, B., Barr, J., et al.: Trial design and reporting standards for intra-arterial cerebral thrombolysis for acute ischemic stroke. Stroke **34**, e109–e137 (2003)

5. Zaidat, O.O., Yoo, A.J., Khatri, P., Tomsick, T.A., von Kummer, R., Saver, J.L., et al.: Recommendations on angiographic revascularization grading standards for acute ischemic stroke: a consensus statement. Stroke **44**, 2650–2663 (2013)

6. Marks, P.M., Lansberg, M.G., Mlynash, M., Kemp, S., McTaggart, R.A., Zaharchuk, G., et al.: Angiographic outcome of endovascular stroke therapy correlated with MR findings, infarct growth, and clinical outcome in the DEFUSE 2 trial. Int. J. Stroke **9**, 860–865 (2014)

7. Drewer-Gutland, F., et al.: CTP-based tissue outcome: promising tool to prove the beneficial effect of mechanical recanalization in acute ischemic stroke. In: RoFo, vol. 187, pp. 459–466 (2015)

8. Scalzo, F., Liebeskind, D.S.: Perfusion angiography in acute ischemic stroke. Comput. Math. Methods. Med. 2478324 (2016)

9. Nielsen, M., Waldmann, M., Frölich, A., Fiehler, J., Werner, R.: Machbarkeitsstudie zur CNN-basierten Identifikation und TICI-Klassifizierung zerebraler ischämischer Infarkte in DSA-Daten. In: Bildverarbeitung für die Medizin 2019, pp. 200–205. Springer, Wiesbaden (2019). https://doi.org/10.1007/978-3-658-25326-4_45

10. Suh, S.H., Cloft, H.J., Fugate, J.E., Rabinstein, A.A., Liebeskind, D.S., Kallmes, D.F.: Clarifying differences among thrombolysis in cerebral infarction scale variants. Stroke **44**, 1166–1168 (2013)

11. Mair, G., et al.: Observer reliability of CT angiography in the assessment of acute ischaemic stroke: data from the Third International Stroke Trial. Neuroradiology **57**(1), 1–9 (2014). https://doi.org/10.1007/s00234-014-1441-0

12. Tung, E.L., McTaggart, R.A., Baird, G.L., Yaghi, S., Hemendinger, M., Dibiasio, E.L., et al.: Rethinking thrombolysis in cerebral infarction 2b: which thrombolysis in cerebral infarction scales best define near complete recanalization in the modern thrombectomy era? Stroke **48**, 2488–2493 (2017)

13. Tan, M., Le, Q.: EfficientNet: rethinking model scaling for convolutional neural networks. In: Chaudhuri, K., Salakhutdinov, R. (eds) 36th International Conference on Machine Learning, Proceeding of Machine Learning Research (PMLR) 97, pp. 105–6114 (2019)

14. Szegedy, C., Vanhoucke, V., Ioffe, S., Shlens, J., Wojna, Z.: Rethinking the inception architecture for computer vision. In: IEEE Conference on Computer Vision and Pattern Recognition (CVPR), pp. 2818–2826. IEEE (2016)

15. Chung, J., Gulcehre, C., Cho, K., Bengio, Y.: Empirical evaluation of gated recurrent neural networks on sequence modeling. In: NIPS 2014 Workshop on Deep Learning (2014)

16. Ballas, N., Yao, L., Pal, C., Courville, A.C.: Delving deeper into convolutional networks for learning video representations. In: Bengio, Y., LeCun, Y. (eds) 4th International Conference on Learning Representations, (ICLR) (2016). http://arxiv.org/abs/1511.06432

17. Shi, X., Chen, Z., Wang, H., Yeung, D.-Y., Wong, W.-K., Woo, W.-C.: Convolutional LSTM network: a machine learning approach for precipitation nowcasting. In: NIPS'15: Proceeding of 28th International Conference on Neural Information Processing Systems, pp. 802–810 (2015)

18. Forkert, N., Verleger, T., Cheng, B., Thomalla, G., Hilgetag, C.C., Fiehler, J.: Multiclass support vector machine-based lesion mapping predicts functional outcome in ischemic stroke patients. PLoS ONE **10**, e0129569 (2015)

19. Fugate, J.E., Klunder, A.M., Kallmes, D.F.: What is meant by "TICI"? AJNR Am. J. Neuroradiol. **34**, 1792–1797 (2013)
20. Dargazanli, C., Consoli, A., Barral, M., Labreuche, J., Redjem, H., Ciccio, G., et al.: Impact of modified TICI 3 versus modified TICI 2b reperfusion score to predict food outcome following endovascular therapy. AJNR Am. J. Neuroradiol. **38**, 90–96 (2017)

ID-Fit: Intra-Saccular Device Adjustment for Personalized Cerebral Aneurysm Treatment

Romina Muñoz[1(✉)], Ana Paula Narata[2(✉)], and Ignacio Larrabide[1(✉)]

[1] Instituto PLADEMA - CONICET, Universidad Nacional del Centro de la Provincia de Buenos Aires, Tandil, Argentina
rominalucianam@gmail.com, larrabide@exa.unicen.edu.ar
[2] Neuroradiology Department, University Hospital of Southampton, Southampton, UK
apnarata@gmail.com

Abstract. Intrasaccular devices, like Woven EndoBridge (WEB), are novel braided devices employed for the treatment of aneurysms with a complex shape and location, mostly terminal aneurysms. Such aneurysms are often challenging or impossible to treat with other endovascular techniques such as coils, stents, flow diverter stents. The selection of an appropriate endosaccular device size is crucial for a successful treatment and strongly depends of the final configuration that the device adopts when it adapts to the aneurysm sac morphology. This is frequently a problem during the intervention, leading to replacement of the device, reopening of the aneurysm or a need for re-treatment. A technique that allows predicting the released WEB configuration before intervention will provide a powerful computational tool to aid the interventionist during device selection. We propose a technique based on device design and aneurysm morphology that, by virtually deploying a WEB, will enable the assessment of different device sizes before the device implantation. This technique was tested on 6 MCA aneurysm cases and the simulation results were compared to the size of the deployed device on the patient, using post-treatment images.

Keywords: Intracranial aneurysm · Endovascular treatment · Woven endobridge

1 Introduction

An intracranial aneurysm is an abnormal dilation in the wall of a cerebral artery. The aneurysm wall weakens by adverse physiological and hemodynamic factors and the risk of rupture increases, which would provoke hemorrhagic ischemia, leading to severe consequences. Given this risk, the physician evaluates which intervention is most appropriate: surgical or endovascular treatment. Endovascular treatment with braided devices has been growing in last years due its

© Springer Nature Switzerland AG 2020
A. L. Martel et al. (Eds.): MICCAI 2020, LNCS 12266, pp. 97–105, 2020.
https://doi.org/10.1007/978-3-030-59725-2_10

efficacy and versatility. Further, it is a low risk non-invasive technique compared to surgical treatment [1].

One of the most recent devices in the market is the Woven EndoBridge (WEB, Microvention, Aliso Viejo, CA), an intrasacular device designed to treat intracranial aneurysms with unfavorable anatomy (wide-neck, large or giant by their size, bifurcation or terminal by their location) [6]. The WEB is an intrasaccular flow disrupter that is inserted inside the aneurysm, unlike stents and flow diverters, which are the current alternatives and are inserted in the parent vessel. The WEB, will disrupt and redirect the flow from the aneurysm neck to the parent vessel [4]. When released into the aneurysm sac, the WEB suffers a deformation while adapting to the aneurysm morphology. The final device configuration strongly depends on the local aneurysm morphology and, therefore, the interventionist is unable to accurately assess a priori how one specific device size will fit a particular aneurysm.

At the moment, two orthogonal X-Ray angiography projections are used in the size selection of the WEB [11]. Width, height and neck of the aneurysm are measured on those projections to choose the device size. WEB diameter is oversized by 1 mm with respect to the average aneurysm diameter measured on the projections, and its height is undersized 1 mm to the average height. Such heuristics proved to fail, leading to an unsuitable device size selection most of the times. Choosing the wrong device can lead to a bad positioning or migration of the device after its deployment [3]. When the wrong device size is detected during the procedure, it must be removed and replaced by one that appropriately fits the aneurysm. These devices are expensive and, therefore, when one or more devices are discarded the costs increase dramatically. In the works of Mut et al. and Cebral et al., a tool to manually place a WEB device inside a vascular model was used [2,5]. However this technique does not consider the specific design or behavior of each device according to their size and is based on deformation through an expansion spring model. Having a tool that accurately predicts the final configuration of specific device size and shape inside patients' own anatomy, thus allowing to assess the fit of different alternatives, will considerably reduce such risks.

In this work we present ID-Fit, a computational method to model and select the size of a WEB-like device. The method presented is based on the analysis of the aneurysm local morphology. Specifically, it analyzes how the device will adapt to the local morphology of the aneurysm. The morphological description of the aneurysm and the descriptive specifications of the device design (height, diameter, number of wire turns, etc.) are taken into account during the deployment process. This tool allows virtually deploying and assessing different device sizes on the 3D model of the aneurysm being treated, to obtain an accurate representation of the final disposition of the selected device within of the aneurysm and observe if the chosen size is the optimum. With ID-Fit, different WEB sizes can be quickly and safely tested on a 3D model of the patient's anatomy and achieve a suitable device selection for the case to be treated.

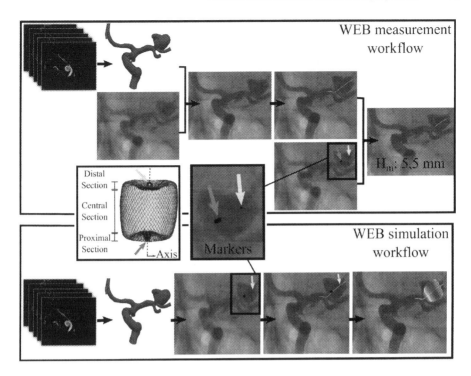

Fig. 1. Upper: Processing steps followed for measuring the height of the WEB after being implanted, using pre-treatment and post-treatment 3D angiographies. Distal marker (green arrow) and proximal marker (red arrow) were identified and projected on the centerline to obtain the WEB actual height. Middle: WEB device picture indicating longitudinal sections (distal, central and proximal). Lower: WEB simulation workflow. Distal marker (green arrow) was projected on the centerline and used as reference to start the deployment. (Color figure online)

2   Methods

The WEB is a braided wire intravascular device with a "barrel" shape. The wires that conform the device are joined at both ends, at a point called hub on the axis of the device. The device is tubular in shape with its ends closed. We propose a methodology to virtually deploy a WEB device inside the anatomy of an intracranial aneurysm patient considering how it will deform and adapt to the local morphology of the treated vessel. To this end, the device was divided in three sections, as indicated in Fig. 1 (Middle).

The WEB device is inserted into an aneurysm radially compressed inside a micro-catheter and progressively released, allowing it to expand. The distal section is the first portion released from the catheter when the interventionist starts pushing the device inside the aneurysm sac. The central section is then released out of the catheter and finally the proximal section, fully liberating the device inside the aneurysm.

Cerebral aneurysms are irregular in shape, forcing the device to deform and adapt to it. The aneurysm morphology is, therefore, an input of the method. Aneurysm and vessel morphology is obtained from their centerline, which is derived from a 3D mesh of the vessel lumen. To properly represent the behaviour of each particular device, its design specifications are needed. These parameters are number of turns of the threads, nominal dimensions (diameter and heigth) and constrained dimensions at two different configurations (constrained diameter and resulting height). This information is used in the representation of the central section of the device.

2.1   Model Generation

Three-dimensional rotational angiography (3DRA) images before treatment of 6 patients (5 women and 1 man), aged 33–66 years (mean 55.17 ± 12.42) with unruptured aneurysms in the middle cerebral artery (MCA) were used to obtain the 3D anatomical model necessary to validate the proposed method. A 3D model of the lumen is obtained by the segmentation of the three-dimensional angiography of the treated vessel using AngioLab software [10]. The centerline of the aneurysm and parent vessel is obtained and their morphological descriptors are computed. Four anatomical landmarks are manually identified to build the centerline: the aneurysm dome, the aneurysm center, the aneurysm neck and the parent vessel center. A spline that interpolates between these four points is obtained, identified as the centerline and divided in small segments of equal length. The device will be deployed following the aneurysm centerline along the proximal direction. Using the centerline, the 3D model was divided in cross sections perpendicular to it, thus obtaining morphological descriptors of the aneurysm and parent vessel at each point.

2.2   WEB Device Deployment

We recall the vessel centerline as $\mathbf{C}$, and for each point $\mathbf{p}^i \in \mathbf{C}$ we compute the perimeter of the cross section obtained by cutting the model with a plane passing by that point. The equivalent radius r^i to the cross section perimeter $p(s^i)$ is computed as $r^i = p(s^i)/2\pi$.

The device deployment is done in 3 stages. At the first stage, a distal point $\mathbf{p}^d$ on the centerline is selected for starting the deployment of the distal section of the device. This point is typically inside the aneurysm. The method then runs $\forall \mathbf{p}^i \in \mathbf{C}$ towards the proximal direction, calculating the distance h^i as follows:

$$h^i = \frac{r^i}{\sin(\arctan(\frac{r^i}{d_{di}}))} \tag{1}$$

being d_{di} the Euclidean distance between $\mathbf{p}^d$ and $\mathbf{p}^i$. This first part of the deployment ends at the centerline point $\mathbf{p}^m$ where $h^m \geq r_N$.

In second stage, the point $\mathbf{p}^m$ is used to start the release of the central section, which is similar in shape to a braided flow diverter. To this end, the deployment of the central section is done following the algorithm described in [7]. A function of the height with respect to the perimeter of each cross section perpendicular to the centerline of the aneurysm was obtained in a similar way as in previous studies [7,9]. The section height for equivalent radius r^i at each point is calculated and subtracted to the nominal height h_n of the device central section at each step until $h_n \approx 0$. At that point $\mathbf{p}^c$, the central section release ends.

In the final stage, the proximal section is released searching for the point $\mathbf{p}^p$, the proximal end of the device along the centerline. The deployment begins at the point $\mathbf{p}^c$ of the centerline where the central section deployment ended. Similarly as in the first stage, for each point along the centerline, the distance h^i is calculated as:

$$h^i = \frac{r^c}{\sin(\arctan(\frac{r^c}{d_{ci}}))}. \tag{2}$$

being d_{ci} the distance between $\mathbf{p}^c$ and $\mathbf{p}^i$. When $h^p \geq r_N$ the algorithm ends and the point $\mathbf{p}^p \in \mathbf{C}$ is returned as the position of the proximal end of the device.

When the three sections are deployed, the distance between $\mathbf{p}^d$ and $\mathbf{p}^p$ along the centerline is computed, obtaining the simulated height of the virtually deployed WEB device. Algorithm 1 succinctly describes the stages of the device deployment.

2.3   WEB Representation

A visual representation of the simulated WEB device was created using VTK and VMTK tools [8]. Between $\mathbf{p}_c$ and $\mathbf{p}_a$ a tube was created that follows the centerline. This tube has as diameter that of the virtually deployed device for the central section. Distal and proximal sections have a sinusoidal-like shape. The function

$$\frac{\sin(\frac{\pi k}{2} - 2\pi)}{\frac{\pi k}{2} - 2\pi}$$

$k = 1, 2, ..., 6$ was used to obtain six points in the plane generated by vectors $\overrightarrow{t}$ and $\overrightarrow{\mathbf{p}_d\mathbf{p}_i}$, where $\overrightarrow{t}$ is the centerline tangent vector at the distal point $\mathbf{p}_d$ and $\overrightarrow{\mathbf{p}_d\mathbf{p}_i}$ is the vector originated in $\mathbf{p}_d$ and ended at the point $\mathbf{p}_i$, $i = 1, 2, ..., N$ (N total number of wires) belonging to the distal extreme of the tubular geometry representative of the central section. A spline that passes by the 6 points is built and N splines with sinusoidal shape are obtained to represent the distal section. In a similar way, the sinusoidal geometry representative of the proximal section is built.

Distal, central and proximal geometries are simultaneously displayed allowing to visualize a representation of the final disposition of the selected WEB device inside the aneurysm, observing its expansion, the occlusion of the neck and the position of the device ends (hubs).

Algorithm 1. Virtual deployment.

    Input   : Aneurysm geometry, aneurysm centerline, WEB geometrical
                description.
    Output: Centerline point $\mathbf{p}_p$, Simulated height.

1 Select distal point, $\mathbf{p}_d$;   // Distal section release

2 Calculate h_e at $\mathbf{p}_d$

3 **while** $h_e/r_n \leq 1$ **do**

4   **for** $\mathbf{p}_w \in centerline$ **do**

5 | Calculate h_e at $\mathbf{p}_w$

6   **end for**

7 **end while**

8 **return** $\mathbf{p}_w$

9 Calculate h_r at $\mathbf{p}_w$;   // Central section release

10 **while** $h_n > 0$ **do**

11   **for** $\mathbf{p}_w \in centerline$ **do**

12 | Calculate h_r at $\mathbf{p}_w$

13   **end for**

14 **end while**

15 **return** $\mathbf{p}_w$

16 Calculate h_e at $\mathbf{p}_w$;   // Proximal section release

17 **while** $h_e/r_n \leq 1$ **do**

18   **for** $\mathbf{p}_w \in centerline$ **do**

19 | Calculate h_e at $\mathbf{p}_w$

20   **end for**

21 **end while**

22 **return** $\mathbf{p}_p$

2.4   Validation

Two-dimensional post-treatment images were used to identify distal and proximal ends of the WEB device after implantation in real patients. Device height was measured on 2D projections from angiographic images. Post-treatment 2D images with visible proximal and distal hubs, were used. Table 1 shows the WEB size device used in each patient.

Table 1. WEB size used for treated 6 MCA aneurysms, measured height (H_m), simulated height (H_s) in each case and difference between them.

Patient	WEB size	Measured height (mm)	Simulated height (mm)	$H_s - H_m$ (mm)
1	5×3	4.04	3.7	-0.34
2	5×3	3.1	3.6	0.5
3	8×6	10.8	11.7	0.96
4	7×6	5.5	6.3	0.8
5	5×3	3.8	4.3	0.5
6	7×3	5.3	5.6	0.3

Deployed Device Height Measurement. 3D models were manually oriented in the same position as the 2D post-treatment images and mutually aligned. The WEB distal and proximal hubs were projected on the centerline, and the distance between them was measured along the centerline (Fig. 1, upper). The results of these measures are presented in Table 1 (Measured height (mm)).

Simulated Height. The proposed method was used to simulate the virtual deployment of the WEB device on the 3D anatomical models of the 6 studied cases. The distal position for the WEB simulation was obtained from the post-treatment images, identifying the distal hub and projecting it on the centerline (Fig. 1, lower). The proposed algorithm returns the proximal hub position and the simulated height. This height was compared with the measurement obtained from the post-treatment images to assess the accuracy of the method. Table 1 shows the simulated heights.

3   Results

A total of 6 intracranial aneurysm cases located at the middle cerebral artery and treated with WEB devices were used for the validation. Table 1 shows the WEB size used on each patient. The mean time required to process and obtain the model is 13 min and the computational time required to simulate the deployment of each device size is around 2.5 s.

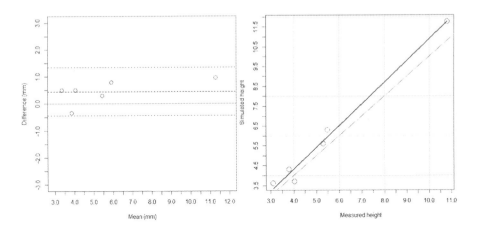

Fig. 2. Left: Bland-Altman plot showing differences between simulated height and measured height. The mean difference is 0.45 mm. Right: Scatterplot showing positive correlation between measures ($\rho = 0.94$).

After processing the complete data-set for geometry reconstruction and simulation of the WEB device, the error between the simulated height and the measured height was assessed calculating the difference between them ($H_s - H_m$).

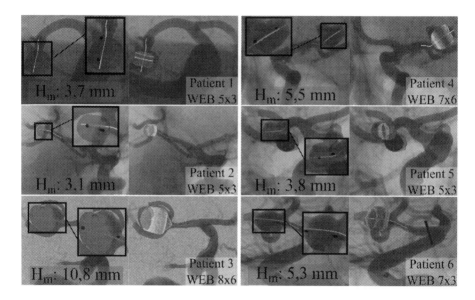

Fig. 3. Height measurement of the implanted device in each patient and the virtual resulting geometry of the same device.

Table 1 shows the measured height, simulated height and the difference between them for each case.

The Wilcoxon signed-rank paired test showed that there are no statistically significant differences between simulated height and measured height ($p > 0.05$). The concordance between measures was analyzed with Bland-Altman method. Figure 2 (left) shows the differences between measures with a mean of 0.45 mm and a standard deviation of 0.89 mm. Measured height and simulated height were plotted in a scatterplot (Fig. 2, right), which shows a positive correlation ($\rho = 0.94$, $p < 0.05$).

Simulated height obtained from the virtual deployment and measured height for each patient are showed in Fig. 3. The final disposition of the WEB geometries resulting of the simulations are also shown. Color scale indicates the expansion of the device released into the aneurysm, red for maximum expansion, corresponding to the nominal size, and blue for minimum expansion, corresponding to sections diameters smaller than the nominal diameter of the device.

4   Conclusion

A fast method to simulate intra-saccular devices inside the patient anatomy is proposed which allows to know, in a quick and precise way, how a specific device size adapts to the aneurysm morphology. The method presents small errors in relation to the device size and in less than 15 min (considering anatomy processing and simulation time) it will allow assessing a complete data base of device sizes to identify the one that best fits the patient anatomy.

References

1. Pierot, L., Wakhloo, A.K.: Endovascular treatment of intracranial aneurysms - current status. Stroke **44**, 2046–2054 (2013)
2. Cebral, J.R., et al.: Analysis of flow dynamics and outcomes of cerebral aneurysms treated with intrasaccular flow-diverting devices. Am. J. Neuroradiol. **40**(9), 1511–1516 (2019)
3. König, I., Weber, A., Weber, W., Fischer, S.: Dislocation of a WEB device into the middle cerebral artery - removal with the alligator retrieval device. Clin Neuroradiol. **29**, 361–364 (2019)
4. Pierot, L., et al.: Intrasaccular flow-disruption treatment of intracranial aneurysms: preliminary results of a multicenter clinical study. Am. J. Neuroradiol. **33**(7), 1232–1238 (2012)
5. Mut, F., et al.: Image-based modeling of blood flow in cerebral aneurysms treated with intrasaccular flow diverting devices. Numer. Methods Biomed. Eng. **35**(6), e3202 (2019)
6. Asnafi, S., Rouchaud, A., Pierot, L., Brinjikji, W., Murad, M.H., Kallmes, D.F.: Efficacy and safety of the Woven Endobridge (WEB) device for the treatment of inracranial aneurysms: a systematic review and meta-analysis. Am J Neuroradiol. **37**, 2287–2292 (2016)
7. Larrabide, I.: Method for determining the final length of stents before the positioning thereof. US Patent App. 14/911,938, August 11 2016
8. Piccinelli, M., Veneziani, A., Steinman, D.A., Remuzzi, A., Antiga, L.: A framework for geometric analysis of vascular structures: application to cerebral aneurysms. IEEE Trans. Med. Imaging **28**(8), 1141–1155 (2009)
9. Fernandez, H., et al.: Computation of the change in length of a braided device when deployed in realistic vessel models. Int. J. Comput. Assist. Radiol. Surg. **10**(10), 1659–1665 (2015). https://doi.org/10.1007/s11548-015-1230-1
10. Larrabide, I., et al.: Angiolab-a software tool for morphological analysis and endovascular treatment planning of intracranial aneurysms. Comput. Methods Programs Biomed. **108**(2), 806–819 (2012)
11. Popielski, J., Berlis, A., Weber, W., Fisher, S.: Two-Center experience in the endovascular treatment of ruptured and unruptured intracranial aneurysms using the WEB device: a retrospective analysis. Am. J. Neuroradiol. **39**(1), 111–117 (2018)

JointVesselNet: Joint Volume-Projection Convolutional Embedding Networks for 3D Cerebrovascular Segmentation

Yifan Wang[1], Guoli Yan[1], Haikuan Zhu[1], Sagar Buch[2], Ying Wang[2],

Ewart Mark Haacke[2], Jing Hua[1], and Zichun Zhong[1(✉)]

[1] Department of Computer Science, Wayne State University, Detroit, MI, USA
`zichunzhong@wayne.edu`
[2] Department of Radiology, Wayne State University, Detroit, MI, USA

Abstract. In this paper, we present an end-to-end deep learning method, *JointVesselNet*, for robust extraction of 3D sparse vascular structure through embedding the image composition, generated by maximum intensity projection (MIP), into the 3D magnetic resonance angiography (MRA) volumetric image learning process to enhance the overall performance. The MIP embedding features can strengthen the local vessel signal and adapt to the geometric variability and scalability of vessels. Therefore, the proposed framework can better capture the small vessels and improve the vessel connectivity. To our knowledge, this is the first time that a deep learning framework is proposed to construct a joint convolutional embedding space, where the computed joint vessel probabilities from 2D projection and 3D volume can be integrated synergistically. Experimental results are evaluated and compared with the traditional 3D vessel segmentation methods and the state-of-the-art in deep learning, by using both public and real patient cerebrovascular image datasets.

Keywords: Deep neural network · 3D cerebrovascular segmentation · Maximum intensity projection (MIP) · Joint embedding

1   Introduction

A growing body of evidence in animal and human studies shows that micro cerebrovascular abnormalities are the source of many vascular diseases and neurologic disorders, e.g., hypertension, arteriosclerosis, cerebral amyloid angiopathy, diabetes [1–3]. There is a pressing need for better extracting and understanding of vascular abnormalities in vivo at the micro-level [4]. However, compared with traditional organ segmentations, segmenting the cerebrovascular structure from magnetic resonance angiography (MRA) is very challenging due to difficulties, such as complex geometry and topology variations as well as data sparseness (artifacts, noises, low signal-to-noise ratio).

In recent decades, there have been many automatic model-driven vessel segmentation approaches proposed, such as multiscale filtering [5], region growing techniques [6], active contours [7], statistical and shape models [8,9], particle filtering [10],

© Springer Nature Switzerland AG 2020
A. L. Martel et al. (Eds.): MICCAI 2020, LNCS 12266, pp. 106–116, 2020.
https://doi.org/10.1007/978-3-030-59725-2_11

geometric flow [11], path tracing [12], level-set approach [13], etc. However, these approaches are easily overwhelmed by tons of low-level handcrafted features and complicated manual parameter adjustments to overcome aforementioned difficulties and subject variations.

Recently, data-driven approaches have been proposed to robustly investigate the correlations between different objects/instances without relying on hard-coded metrics. In medical image processing, several deep learning-based methods have been proposed to extract vessels from 2D retinal images, such as DeepVessel [14], cross-modality learning approach [15], multi-level deep supervised networks [16], DNN-based method [17], unified convolutional neural network (CNN) and graph neural network (GNN) [18], etc. These methods can well perform 2D vessel segmentation tasks, but are far from success in 3D vessel scenarios. There are still very few dedicated deep learning architectures for 3D vessel segmentation. For instance, Uception [19] presents a network inspired by 3D U-Net [20] and Inception modules [21]. DeepVesselNet [22] and VesselNet [23] propose 2D orthogonal cross-hair filters in three planes on each voxel to obtain the 3D contextual information at a reduced computational burden and less memory. The major challenges of 3D cerebrovascular segmentation are complicated vessel geometry and topology variations, high sparseness of vessel data in a large-sized 3D volume, and the limited resource of 3D vasculature datasets. Existing methods are not specifically designed for solving these challenges.

To fill the gap in the high-fidelity 3D cerebrovascular segmentation, we present an end-to-end deep learning method, *JointVesselNet*. A multi-stream CNN framework is adopted to effectively learn the 3D volume and multislice composited 2D maximum intensity projection (MIP) [24] feature vectors respectively. It then explores the inter-dependencies between 3D and 2D embedded feature vectors in a joint volume-projection embedding space by backprojecting the 2D feature vectors, learned from MIP, into the 3D volume embedding space. MIP is a widely-used scientific method for visualizing and analyzing 3D vasculature structure in MRA diagnosis by domain scientists. It can enhance the local vessel signal by canceling out the random noises and adapt to geometric variability and scalability. Through the novel integrative learning of both 3D volume and 2D MIPs, the proposed framework can better capture the small vessels and improve the subtle vessel connectivity. To our knowledge, this is the first time that a deep learning framework is proposed to construct a joint convolutional embedding space, where the computed joint vessel probabilities from 2D projection and 3D volume can be integrated synergistically. The key motivation of the proposed network is to integrate the trustworthy auxiliary from learned 2D MIP features into the 3D volume segmentation network, instead of using more complicated networks empirically. Experimental results are evaluated and compared with the traditional 3D vessel segmentation methods and the state-of-the-art in deep learning by using both public and real patient cerebrovascular image datasets. The application of this accurate segmentation of sparse and complicated 3D vascular structure facilitated by our method demonstrates the potential in improving MRA diagnosis of vascular diseases.

2   Method

In this section, we introduce the components of the JointVesselNet model: dataset preparation and generation, network architecture, and loss function.

2.1   Dataset Preparation and Generation

In this work, we use two different real patient datasets to evaluate our proposed JointVesselNet method. The first dataset is the public TubeTK Toolkit MRA dataset from University of North Carolina at Chapel Hill[1], acquired by a Siemens Allegra head-only 3T MR system. There are 42 patient cases in the whole dataset, which have the manual-labeled vessel segmentation masks. The voxel spacing of the MRA images is $0.5 \times 0.5 \times 0.8$ mm^3 with a volume size of $448 \times 448 \times 128$ voxels.

The second dataset is provided by domain experts, who are neurologists and radiologists under our collaboration. 11 healthy volunteers were scanned in midbrain regions with a dual echo susceptibility weighted imaging (SWI) sequence at four time points: the first was acquired pre-contrast and the remaining three were acquired post-contrast during a gradual increase in dose delivered over the time frame of 20 mins (final concentration = 4 mg/kg); with the imaging parameters: echo time (TE)1/echo time (TE)2/repetition time (TR) = 7.5/22.5/27 ms, bandwidth = 180 Hz/pxl, flip angle = 15° (pre-contrast and final post-contrast data) and 20° (first and second post-contrast data). The voxel spacing is $0.22 \times 0.22 \times 1$ mm^3 with a volume size of $1024 \times 832 \times 96$ voxels. This protocol enables short and long TE magnitude data to produce MRA. Then, the MRA is calculated using a non-linear subtraction (MRAnls) method [25], which is employed for selective MRA enhancement utilizing the flow rephrased and dephased images. Finally, the ground truth vessel labels are obtained by integrating an enhanced angiography map from the computed MRAnls to set a more reasonable threshold for the initial mask, followed by domain experts' post-manual labeling refinement using a developed cerebrovascular labeling and visualization tool.

2.2   Network Architecture

The proposed JointVesselNet mainly consists of a dual-stream component (i.e., a 3D volume segmentation stream and a 2D composited MIP segmentation stream) and the bi-directional operations between these two streams (i.e., 3D-to-2D projection and 2D-to-3D backprojection). The overall architecture is demonstrated in Fig. 1.

In this work, the two-stream segmentation component learns vessel feature vectors in 3D volume and corresponding multiple 2D MIPs (enhanced and dense depiction of 3D relationships via a 3D-to-2D projection computation) contexts, respectively. We use a 3D U-Net [20] as the 3D volume segmentation branch and a half 2D U-Net [26] (in terms of feature channel numbers) as the 2D composited MIP segmentation branch, respectively. Since U-Net-like networks are the most commonly-used and robust medical imaging segmentation neural networks across different data modalities for varying

---
[1] https://public.kitware.com/Wiki/TubeTK/Data.

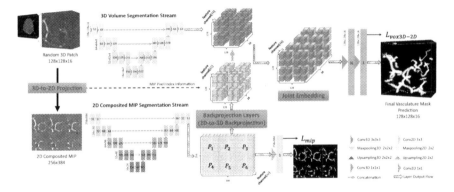

Fig. 1. The architecture of JointVesselNet. The major procedure includes compositing MIPs via 3D-to-2D projection, dual-stream segmentation learning for 3D volume and 2D composited MIP feature vectors, mapping 2D composited MIP feature vectors back to 3D volume feature space via 2D-to-3D backprojection, building a joint embedding for learning the final vasculature mask.

organ/tissue geometries, it is suitable for our work to justify the benefits from the 2D-to-3D backprojection and joint embedding of 3D volume and 2D composited MIP. A U-Net-like network is essentially a convolutional encoder-decoder network. In Fig. 1, the layer output feature channel numbers are denoted in the corresponding blocks and layer input spatial dimensions are shown in the horizontal levels of every block.

Due to the limited data availability and the large volume size in cerebrovascular image datasets, we choose to train the network patch-wisely. Specifically, from the observation that most brain MRAs have much higher resolutions in axial plane than other planes, we adaptively train our network using none-cubic patches, which have larger dimension sizes across axial plane, instead of resizing the data into the uniform voxel spacing through an interpolation before the network training, to avoid potential segmentation inaccuracy. As shown in Fig. 1, the key step in our network is the effective integration of the features from two different streams/domains. Accordingly, two main challenges need to be overcome in this work. The first one is the effective format of the corresponding 2D composited MIPs from a randomly-extracted 3D volume patch that is suitable for simultaneous dual-stream learning design. The second one is the effective approach for backprojecting the feature vectors extracted from the composited MIP image plane pixels (in a dimension-reduced 2D space) back to the corresponding 3D volume spacing voxels. More details are introduced in the following.

3D-to-2D Projection in Dual-Stream Design. The major motivation for projecting the 3D volume space into the 2D MIP space is to enhance the local vessel probability (sparseness) as well as the signal-to-noise ratio. Given a randomly-extracted 3D volume patch V of the size $K_1 \times K_2 \times K_3$ (e.g., we use $128 \times 128 \times 16$ in our experiments) and K_3 along vertical axis, we compute s-sliced (e.g., $s = 5$ in our experiments as suggested by domain experts) MIPs of V along vertical axis with overlapping coverage every t slice interval. Consequently we can get a set of m consecutive MIPs, i.e., $\mathbf{P} = \{P_1, P_2, \ldots, P_k, \ldots, P_{m-1}, P_m\}$, in which P_k is the MIP across the

$[(k-1)t+1]^{th}$ slice to the $[(k-1)t+s]^{th}$ slice in V. It is noted that in a 2D MIP, only one voxel with the maximum intensity among the s voxels along the vertical axis in V will be recorded, which is prone to an information loss, considering the segmentation task needing information of every voxel. Consequently, we set $t = 2$ as a trade-off between computation cost and information completeness/denseness. We can get m MIPs of size $K_1 \times K_2$ for V, where the MIP number m is computed as:

$$m = \left\lfloor \frac{1}{t}(K_3 - s) \right\rfloor + 1. \tag{1}$$

A MIP conveys denser vessel information and is naturally suitable for 2D convolution. However, we now have m different MIPs and need to feed them to our network in the MIP stream, in company with the 3D volume stream V as an input pair to our entire network. The information from the m MIPs is equally important. In order to avoid stacking them to a $K_1 \times K_2 \times m$ volume such that the 2D CNN would essentially treat it as a 2D input of a spatial dimension $K_1 \times K_2$ with m different properties (feature channels), we convert the m MIPs to a tiled MIP with a larger 2D spatial size, such as $0.5\,mK_1 \times 2K_2$. In this case, the 2D convolution operates equally across the 2D composited MIP plane domain. The slice indices from where the MIP pixels are selected in the original V are also recorded so as to effectively restore the pixel-wise information extracted from MIP to the 3D volume space, which will be used in the 2D-to-3D back-projection transformation in the following process. The format of the 2D composited MIP (e.g., six consecutive MIPs) computed from a 3D patch is shown in Fig. 2(a).

2D-to-3D Backprojection for Joint Embedding. Once the 3D volume and 2D MIP streams learn their segmentation features respectively, we intend to integrate them in a unified joint hidden feature embedding space to yield the final 3D segmentation prediction. In order to achieve this, we conduct several operations within our network to backproject the pixel features extracted from the composited MIP back to their corresponding 3D voxel feature space.

The final-stage hidden feature from 2D composited MIP segmentation branch has the size $0.5\,m\,K_1 \times 2K_2$ with C_1 channels ($C_1 = 32$ as shown in Fig. 1), which is the input of the backprojection layers. We first disassemble it to restore m C_1-channel features for the corresponding MIPs (e.g., $P_1, P_2, \ldots, P_{m-1}, P_m$, where $m = 6$ as illustrated in Fig. 2 b). Then we use the recorded index information to map the MIP pixel features back to where they are selected from V during the 2D composited MIP generation. Figure 2 (b) shows how the feature vectors of two consecutive MIPs (P_1 and P_2) are disassembled from the composited MIP. They project their pixel feature space (P_{m-j}, i.e., the j-th slice among 5-sliced MIP P_m, $1 \leq j \leq 5$) back to the voxel feature space (S_n, i.e., the n-th slice in the input 3D patch, $1 \leq n \leq 16$). It is noted that the feature dimension is reduced from 3D to 2D for a convenient illustration in Fig. 2 (b) (i.e., without considering the 32 feature channels).

For the features of overlapping slices (from the consecutive MIPs) derived from the enhanced vessel probability/features, which are covered by multiple MIPs, we take the element-wise maximum value across the overlapping restoration through the feature

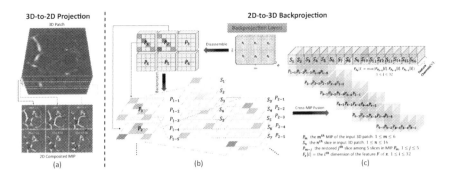

Fig. 2. (a) Illustration of the 3D-to-2D projection in the spatial domain for computing a 2D composited MIP from a 3D volume patch. (b) and (c) Illustration of the detailed 2D-to-3D restorations within backprojection layers in the embedded feature domain. Backprojected pixel-voxel feature pair examples can be traced via the pairing colors (e.g., green, orange, purple, and blue) in (b). (Color figure online)

channels, i.e., $F_{S_n}[i] = \max(F_{P_{1-1}}[i], \dots, F_{P_{6-5}}[i]), 1 \leq i \leq 32$, where $F_{S_n}[i]$ represents the i-th dimensional feature (channel) F at the n-th slice in the 3D patch. For example, the feature F_{S_9} is computed across the overlapping slices of $P_{3-5}, P_{4-3}, P_{5-1}$ as highlighted in the pink color in Fig. 2 (c). The whole process of the cross-MIP fusion in the feature channels of the 3D volume feature space is shown in Fig. 2 (c) in detail. Now, in this 3D volume feature space, the backprojected 2D MIP features and 3D volume features from two streams are integrated together through concatenation, constructing a unified high-dimensional joint convolutional embedding for predicting the final vessel segmentation.

2.3 Loss Function

The major learning objective of our JointVesselNet network is to extract the sparse 3D vasculature structure from the 3D MRA volume image using a 3D segmentation network supplemented by information from the denser and more connected multiple 2D MIPs. Consequently the network loss function consists of two terms:

$$L = L_{vox_{3D-2D}} + \lambda L_{mip}, \qquad (2)$$

where $L_{vox_{3D-2D}}$ is the joint 3D-2D segmentation Dice loss defined as:

$$L_{vox_{3D-2D}} = -\frac{2\Sigma_{x \in V} p(x)g(x) + \delta}{\Sigma_{x \in V} p(x) + \Sigma_{x \in V} g(x) + \delta}, \qquad (3)$$

where $p(x)$ and $g(x)$ are the predicted voxel-wise vessel probability maps and ground truth binary labels within the query volume patch V, respectively. δ is a smooth constant. L_{mip} acts as a regularization term during training, which is also a Dice loss function defined (similarly to $L_{vox_{3D-2D}}$) within the 2D composited MIP and supervised by the ground truth MIP vessel binary labels. λ is the constant coefficient of L_{mip}, which is set to be 0.2 for our best experiment performance.

3 Experiments and Results

For both datasets, we first apply the MR-based skull-stripping method [27] to extract the pure brain from each MRA image. 3D training patches with the imbalanced dimensions are randomly-extracted with overlapping focusing on the brain area in a whole 3D MRA, e.g., 80 patches for each TubeTK case and 440 patches for each collaborative clinical case. The random training/validation/testing case split is 33/3/6 and 6/2/3 for TubeTK dataset and the clinical dataset, respectively. The testing accuracy is computed in the full brain patched with no overlap.

Our JointVesselNet network adopts the Adam optimizer with an initial learning rate of 0.0001 with 0.5 as the learning decay factor and 10 epochs as the learning patience. The network is implemented with the TensorFlow framework and the total training time is about 10 h on two NVIDIA GeForce GTX 1080 GPUs with 8 GB GDDR5X memory. The source code of our method and datasets will be made available later.

We first compare our JointVesselNet performance on TubeTK dataset with four state-of-the-art deep learning based methods (i.e., 3D U-Net [20], 2D U-Net [26], DeepVesselNet [22], and Uception [19]) and one classical parametric intensity-based method (i.e., vesselness algorithm [5,11]) in 3D vessel segmentation. All deep learning methods in comparison are trained until convergence by using the same dataset split or using the results reported from their original publication (such as Uception). For 2D U-Net, we train it with 128×128 2D patches, whose amount is over 10 times of the amount of 3D patches extracted for the 3D CNN based methods with on-the-fly data augmentation for a fair data acquisition. For DeepVesselNet, we have tried different combinations of their data pre-processing processes and chosen the image intensity clipping for obtaining an optimal performance on TubeTK dataset. For the evaluation on the clinical dataset from our medical collaboration, we use the state-of-the-art model-driven MRAnls method [25] mentioned in Sect. 2.1 to extract the vessels for comparison.

Four quantitative metrics, i.e., Dice Similarity (Dice), Sensitivity, Precision, and False Positive Rate (FPR), are used for numerical evaluation. The performance comparison of different methods on TubeTK dataset is shown in Table 1. '$-$' means 'not applicable' due to the lack of their implementations or results. The best results in the table are shown in bold font. From Table 1, we can see that our JointVesselNet has the best overall performance among all the methods on TubeTK dataset. With the 2D composited MIP feature integration, our network performs better than a pure 3D U-Net [20] over all four different metrics. The qualitative comparison of MIP-wise (e.g., 5-sliced) segmentation results and 3D global vessel segmentation results between our JointVesselNet and 3D U-Net (one of the most robust state-of-the-art deep learning based methods for biomedical image segmentation) is shown in Fig. 3. With the 2D composited MIP complementary information, the final vessel segmentation shows better connectivity and better small vessel capturing compared to 3D U-Net, as marked in red circles (3D global vessel segmentation visualization) and green circles (2D MIP vessel segmentation visualization). Another observation is that 3D U-Net greatly outperforms 2D U-Net [26] since the former method is able to capture the cross-slice continuity. That is why 3D CNN should be involved in sparse 3D object segmentation with complex topology. DeepVesselNet [22] fails to yield a good performance as reported in their dataset, which could result from the lack of the pre-training procedure and the instability of their loss function, as well as over-simplified network (e.g., five convolutional layers). Our method also performs better than the best Uception result in [19] on TubeTK dataset with even less data pre-processing procedure. We also employ the vesselness algorithm [5,11], a widely-used approach to segment cylindrical vessel structures in the medical field, as a traditional benchmark method for comparison. Last but not least, all deep learning methods greatly outperform vesselness method on TubeTK dataset.

Table 1. Quantitative performance evaluation of different methods.

Metrics/Methods	Ours	3D U-Net	Uception	DeepVesselNet	2D U-Net	Vesselness
Dice (%) ↑	**71.81**	71.01	67.01	64.12	65.10	37.71
Sensitivity (%) ↑	**79.33**	75.99	66.02	63.20	73.93	65.34
Precision (%) ↑	**76.66**	74.00	–	63.75	70.05	47.69
FPR (%) ↓	**0.0821**	0.0958	–	0.1465	0.1041	0.1393

Note: Accuracy metric is not included here, since it is always very high (e.g., ≥99%) in highly sparse vessel data.

As mentioned in Sect. 2.1, our collaborative medical domain experts collect clinical MRA datasets and use the state-of-the-art non-linear subtraction (MRAnls) method [25] to extract the clean midbrain vessels, which tend to be a good physical indicator of several cerebral diseases. MRAnls requires different data modalities and tedious manual parameter-tuning; however, as shown in Fig. 3, its segmentation result still fails to be free from location-dependent interference, such as superior sagittal sinus (red dotted circles in 3D visualization and green dotted circles in MIP visualization) and some random noises (red solid circles). In order to improve the segmentation performance with limited data inventory (11 cases in total), we fine-tune our network pretrained on TubeTK dataset with six cases in the clinical dataset. Our network only needs TE1 pre-contrast SWI (a single-modal MRA) data as input. The quantitative comparison results between ours and MRAnls method are **82.98%**/80.60%, **83.77%**/82.26%, **83.69%**/82.07%, **0.0337%**/0.0354% for Dice, Sensitivity, Precision, and FPR, respectively. Our numeric evaluation outperforms MRAnls method on all metrics. Figure 3 further shows that our vessel segmentation results are more continuous and cleaner than MRAnls results.

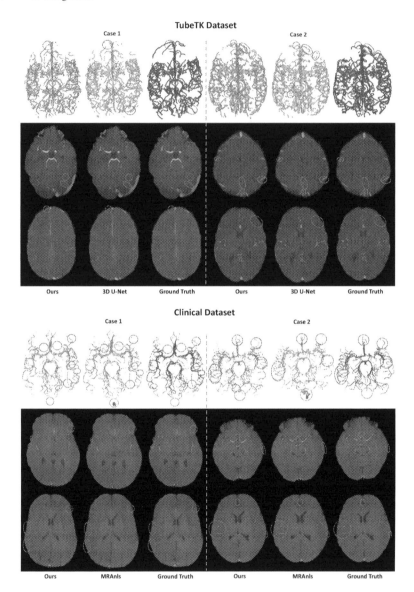

Fig. 3. Some qualitative comparison results from two datasets. 3D global vessel segmentations are shown from superior direction. Two MIP segmentations are visualized by 5-sliced MRA images.

4   Conclusion

In this work, we have proposed a deep neural network method, JointVesselNet, to segment high-fidelity 3D cerebrovascular structure from MRA images. By backprojecting the learned multislice composited 2D MIP feature vectors into the 3D volume embedding space, the proposed framework can strengthen the sparse 3D vascular representation by better capturing the small vessels as well as improving the vessel connectivity, which outperforms the state-of-the-art classical and deep learning based methods. In medical practice, this work can be used as the key functions for real-time segmentation and visualization of sparse and complicated 3D vasculature to improve MRA diagnosis.

Acknowledgements. We would like to thank the reviewers for their valuable comments. We are grateful to Yongsheng Chen from Neurology for the early discussion of this work, Pavan K. Jella from Radiology for preparing and collecting the clinical datasets, and Michelle Hua from Cranbrook Schools for pre-processing the datasets and proofreading. This work was partially supported by the NSF under Grant Numbers IIS-1816511, CNS-1647200, OAC-1657364, OAC-1845962, OAC-1910469, the Wayne State University Subaward 4207299A of CNS-1821962, NIH 1R56AG060822-01A1, NIH 1R44HL145826-01A1, ZJNSF LZ16F020002, and NSFC 61972353.

References

1. Brown, W., Thore, C.: Cerebral microvascular pathology in ageing and neurodegeneration. Neuropathol. Appl. Neurobiol. **37**(1), 56–74 (2011)
2. Dorr, A., Sahota, B., et al.: Amyloid-β-dependent compromise of microvascular structure and function in a model of Alzheimer's disease. Brain **135**(10), 3039–3050 (2012)
3. Gouw, A., Seewann, A., Van Der Flier, W., Barkhof, F., et al.: Heterogeneity of small vessel disease: a systematic review of MRI and histopathology correlations. J. Neurol. Neurosurg. Psychiatry **82**(2), 126–135 (2011)
4. Mott, M., Pahigiannis, K., Koroshetz, W.: Small blood vessels: big health problems: national institute of neurological disorders and stroke update. Stroke **45**(12), e257–e258 (2014)
5. Frangi, A., Niessen, W., et al.: Multiscale vessel enhancement filtering. In: International Conference on Medical Image Computing and Computer-Assisted Intervention, pp. 130–137 (1998)
6. Martínez-Pérez, M., et al.: Retinal blood vessel segmentation by means of scale-space analysis and region growing. In: International Conference on Medical Image Computing and Computer-Assisted Intervention, pp. 90–97 (1999)
7. Nain, D., Yezzi, A., Turk, G.: Vessel segmentation using a shape driven flow. In: International Conference on Medical Image Computing and Computer-Assisted Intervention, pp. 51–59 (2004)
8. Chung, A., et al.: Statistical 3D vessel segmentation using a Rician distribution. In: International Conference on Medical Image Computing and Computer-Assisted Intervention, pp. 82–89 (1999)
9. Liao, W., Rohr, K., Wörz, S.: Globally optimal curvature-regularized fast marching for vessel segmentation. In: International Conference on Medical Image Computing and Computer-Assisted Intervention, pp. 550–557 (2013)
10. Florin, C., Paragios, N., Williams, J.: Globally optimal active contours, sequential Monte Carlo and on-line learning for vessel segmentation. In: European Conference on Computer Vision, pp. 476–489 (2006)
11. Descoteaux, M., Collins, D., Siddiqi, K.: A geometric flow for segmenting vasculature in proton-density weighted MRI. Med. Image Anal. **12**(4), 497–513 (2008)
12. Wang, S., et al.: Sequential Monte Carlo tracking for marginal artery segmentation on CT angiography by multiple cue fusion. In: International Conference on Medical Image Computing and Computer-Assisted Intervention, pp. 518–525 (2013)

13. Forkert, N., et al.: 3D cerebrovascular segmentation combining fuzzy vessel enhancement and level-sets with anisotropic energy weights. Magn. Reson. Imaging **31**(2), 262–271 (2013)

14. Fu, H., Xu, Y., Lin, S., Wong, D., Liu, J.: DeepVessel: retinal vessel segmentation via deep learning and conditional random field. In: International Conference on Medical Image Computing and Computer-Assisted Intervention, pp. 132–139 (2016)

15. Li, Q., Feng, B., Xie, L., et al.: A cross-modality learning approach for vessel segmentation in retinal images. IEEE Trans. Med. Imaging **35**(1), 109–118 (2015)

16. Mo, J., Zhang, L.: Multi-level deep supervised networks for retinal vessel segmentation. Int. J. Comput. Assist. Radiol. Surg. **12**(12), 2181–2193 (2017)

17. Liskowski, P., Krawiec, K.: Segmenting retinal blood vessels with deep neural networks. IEEE Trans. Med. Imaging **35**(11), 2369–2380 (2016)

18. Shin, S., Lee, S., Yun, I., Lee, K.: Deep vessel segmentation by learning graphical connectivity. Med. Image Anal. **58**, 101556 (2019)

19. Sanchesa, P., et al.: Cerebrovascular network segmentation of MRA images with deep learning. In: IEEE International Symposium on Biomedical Imaging, pp. 768–771 (2019)

20. Çiçek, Ö., et al.: 3D U-Net: learning dense volumetric segmentation from sparse annotation. In: International Conference on Medical Image Computing and Computer-Assisted Intervention, pp. 424–432 (2016)

21. Szegedy, C., et al.: Inception-v4, inception-resnet and the impact of residual connections on learning. In: The AAAI Conference on Artificial Intelligence (2017)

22. Tetteh, G., et al.: DeepVesselNet: vessel segmentation, centerline prediction, and bifurcation detection in 3D angiographic volumes. arXiv preprint arXiv:1803.09340 (2018)

23. Kitrungrotsakul, T., et al.: VesselNet: a deep convolutional neural network with multi pathways for robust hepatic vessel segmentation. Comput. Med. Imaging Graph. **75**, 74–83 (2019)

24. Napel, S., et al.: CT angiography with spiral CT and maximum intensity projection. Radiology **185**(2), 607–610 (1992)

25. Ye, Y., Hu, J., Wu, D., Haacke, E.: Noncontrast-enhanced magnetic resonance angiography and venography imaging with enhanced angiography. J. Magn. Reson. Imaging **38**(6), 1539–1548 (2013)

26. Ronneberger, O., Fischer, P., Brox, T.: U-Net: convolutional networks for biomedical image segmentation. In: International Conference on Medical Image Computing and Computer-Assisted Intervention, pp. 234–241 (2015)

27. Jenkinson, M., et al.: BET2: MR-based estimation of brain, skull and scalp surfaces. In: Eleventh Annual Meeting of the Organization for Human Brain Mapping (2005)

Classification of Retinal Vessels into Artery-Vein in OCT Angiography Guided by Fundus Images

Jianyang Xie[1], Yonghuai Liu[2], Yalin Zheng[3], Pan Su[1], Yan Hu[4],
Jianlong Yang[1], Jiang Liu[4(✉)], and Yitian Zhao[1(✉)]

[1] Cixi Institute of Biomedical Engineering,
Ningbo Institute of Materials Technology and Engineering,
Chinese Academy of Sciences, Ningbo, China
yitian.zhao@nimte.ac.cn

[2] Department of Computer Science, Edge Hill University, Ormskirk, UK

[3] Department of Eye and Vision Science, University of Liverpool, Liverpool, UK

[4] Department of Computer Science and Engineering,
Southern University of Science and Technology, Shenzhen, China
liuj@sustech.edu.cn

Abstract. Automated classification of retinal artery (A) and vein (V) is of great importance for the management of eye diseases and systemic diseases. Traditional colour fundus images usually provide a large field of view of the retina in color, but often fail to capture the finer vessels and capillaries. In contrast, the new Optical Coherence Tomography Angiography (OCT-A) images can provide clear view of the retinal microvascular structure in gray scale down to capillary levels but cannot provide A/V information alone. For the first time, this study presents a new approach for the classification of A/V in OCT-A images, guided by the corresponding fundus images, so that the strengths of both modalities can be integrated together. To this end, we first estimate the vascular topologies of paired color fundus and OCT-A images respectively, then we propose a topological message passing algorithm to register the OCT-A onto color fundus images, and finally the integrated vascular topology map is categorized into arteries and veins by a clustering approach. The proposed method has been applied to a local dataset contains both fundus image and OCT-A, and it reliably identified individual arteries and veins in OCT-A. The experimental results show that despite lack of color and intensity information, it produces promising results. In addition, we will release our database to the public.

Keywords: OCT-A · Message passing · Artery/vein classification

1 Introduction

Vascular morphological changes in the retina are frequently associated with a variety of diseases such as diabetic retinopathy (DR), age-related macular

© Springer Nature Switzerland AG 2020
A. L. Martel et al. (Eds.): MICCAI 2020, LNCS 12266, pp. 117–127, 2020.
https://doi.org/10.1007/978-3-030-59725-2_12

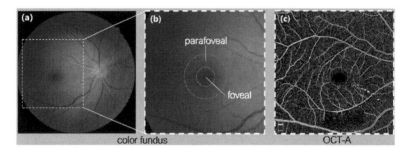

Fig. 1. Example of a retinal color fundus image and its corresponding macula-centred $6 \times 6\,\mathrm{mm}^2$ OCT-A image of the same eye. (Color figure online)

degeneration (AMD), pathological myopia and other systemic diseases such as cardiovascular diseases and hypertension [1,2]. A number of studies [3,4] suggested that different systematic diseases and their severity could affect arteries and veins differently. For example, a low artery-vein ratio (AVR) is a direct characteristic sign of DR [5] whilst a high AVR has been linked to high level of cholesterol [6,7]. However, the calculation of AVR requires the detection and classification of retinal vessels into arteries (A) and veins (V). Manual annotation of A/V is time consuming and prone to human errors [8]. There is increasing demand on automated methods for the A/V detection and classification.

In the past decade, extensive effort has been made to automate the process of A/V detection and classification in color fundus images. Vazquez et al. [9] utilized a tracking strategy based on the minimal path approach to support the A/V classification. Dashtbozorg et al. [7] proposed an A/V classification method based on graph nodes and intensity features. Estrada et al. [1] employed a graph method to estimate the vessel topology with domain-specific knowledge to classify the A/V. Huang et al. [10] introduced an A/V classification framework by using a linear discriminate analysis classifier. Zhao et al. [11] formalized the A/V classification as a pairwise clustering problem. Ma et al. [12] proposed a multi-task neural networks for retinal vessel segmentation and A/V classification.

However, all the aforementioned methods are only applicable to color fundus images. Color fundus imagery are not able to capture fine vessels as well as capillaries, which is prominent in the center of the retina, also known as fovea and parafovea, see Fig. 1(a–b). Fluorescein angiography (FA) can resolve the whole retinal vasculature including capillaries, but it is invasive and also has side effects [13,14]. In contrast, optical coherence tomography angiography (OCT-A) is a new emerging non-invasive imaging modality that enables observation of microvascular details up to capillary level, as shown in Fig. 1(c). It thus opens up a new avenue to study the relation between retinal vessels and various vessel-related diseases. In particular, the microvasculature distributed within the parafovea is of great interest as the abnormality there often indicates different diseases such as the early stage of DR and hypertension [4,15,16].

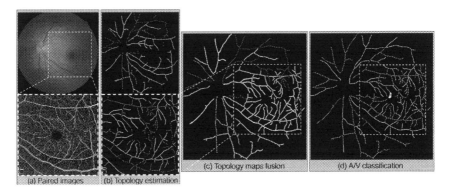

Fig. 2. Overview of the proposed method. (a) A pair of color fundus and OCT-A images from the same eye. (b) Estimated vascular topology maps. (c) Fused topology map. (d) A/V classification results in the fused image. (Color figure online)

OCT-A images provide a huge potential for the quantification of fine vessels and capillaries for improved diagnosis and monitoring of disease. However, existing methods cannot discriminate arteries from veins in OCT-A for two reasons. First, most of the vessels have relatively small calibres which is a challenging task for current methods to accurately segment them. Second, A/V classification is impossible due to the lack of color and intensity features, on which most of A/V classification algorithms are based [17,18].

To overcome the above issues, we propose a novel A/V classification method for OCT-A images guided by the color information in fundus images. Our method consists of three phases: vessel topology estimation, registration of vessel structures extracted from OCT-A and color fundus images, and A/V classification on the registered vasculature. This paper makes three main contributions. **1)** This is the first attempt to use color fundus images to guide the A/V classification in OCT-A, and thus exploit their complementary information for the task; **2)** We use the topological message passing approach to achieve multi-modal image registration, so as to obtain a comprehensive vascular topology map; **3)** We made our dataset publicly available, which contains paired fundus and OCT-A images, manual annotations of vessel topology and A/V classification.

2 Method

The proposed new framework comprises three steps: (1) Estimation of vascular topology in both colour fundus and corresponding OCT-A images, (2) Topology maps fusion using message passing; and (3) Classification of AV in OCT-A images. Figure 2 provides an overview of the proposed method.

2.1 Vascular Topology Estimation

Figure 3 illustrates the pipeline of the topology estimation in an OCT-A image. We first segment the blood vessels in both fundus and OCT-A images, respec-

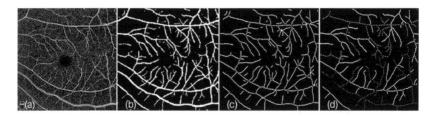

Fig. 3. Pipeline of the topology estimation. (a) Original image. (b) Vessel segmentation result. (c) Significant points detection. (d) Topology estimation result.

tively by using a pre-trained CS-Net [19], which has been validated for the segmentation of vessels on both fundus and OCT-A images. An undirected graph G with weighted edges is represented as $G = (V, E, \omega)$, where V is a set of nodes, edge set $E \subseteq V \times V$ indicates all the connections of the relevant nodes, and ω represents the similarity among the nodes in V. A $|V| \times |V|$ symmetric matrix $A = \{a_{ij}\}$ is used to represent the weighted graph G, which is named an adjacency matrix. The value of a_{ij} is derived by a similarity measure defined in the feature space of the nodes [20]. Here, we define $a_{ij} = 0$ for $i = j$, which indicates that the generated graph G does not include self-loop. The Dominant Sets are then employed to detect which nodes (and the segments they are representing) belong to the same vessel branch. A dominant set can be determined by solving a standard quadratic program below:

$$\max f(\mathbf{x}) = \mathbf{x}^\top A \mathbf{x}$$
$$\text{s.t. } \mathbf{x} \in \mathbb{R}^{|V|}, \sum_{i=1}^{|V|} x_i = 1, \text{ and } x_i \geq 0 \text{ for all } i = 1, \cdots, |V|, \tag{1}$$

where A is the adjacency matrix of G. $\mathbf{x}^*$ is a strict local solution of Eq. (1). If the i-th element of the $\mathbf{x}^*$ is larger than zero, it means that the i-th node of G is in the dominant set S identified by $\mathbf{x}^*$. Effective optimization approaches for solving Eq. (1) can be found in [21,22]. The nodes in the dominant set S of G (and the segments represented by them) are identified as they are in the same vessel branch. The identification of other vessel branches in G is carried out by iteratively solving the dominant set problem on the updated graph G defined as $G = G \setminus S$. Figure 3(d) shows the vascular network with topological information.

2.2   Topology Maps Fusion via Message Passing

The proposed classification of A/V on OCT-A requires the guidance of topology information of the fundus image. As such we map the vascular topology of OCT-A to that from fundus image so as to form a comprehensive topology map. This involves registration and topological consistency enforcement processes.

We first use a rigid registration algorithm [23] for mapping the significant points between OCT-A and fundus images. We consider two significant points from different image modalities within a six-pixel tolerance as a matched pair. However, this coarse registration approach does not guarantee spatial consistency, and may lead to a large portion of potential mismatches because of the

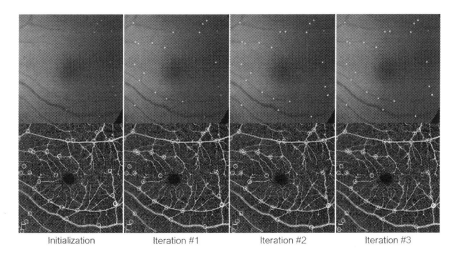

| Initialization | Iteration #1 | Iteration #2 | Iteration #3 |

Fig. 4. Significant points matching at different iterations. Highly similar corresponding points are selected to instantiate a GPR that will map the remaining red points in fundus (top) to the locations (red) in the OCT-A (bottom) image. The yellow circles denote the covariances associated to these locations. Those points in cyan on the OCT-A that are close enough to the red locations and correlate well with the original red points in fundus are taken to form new corresponding set for a new iteration. (Color figure online)

non-linear deformation between OCT-A and fundus. To this end, we propose a novel method to remove correspondences inconsistent with a locally smooth deformation model, inspired by Gaussian Process regression (GPR) [24].

In practice, to find a geometrically consistent set of correspondences S_n between fundus image I^F and OCT-A I^O, we first define a set of correspondences $S_n^0 = \{\mathbf{p}_l^F \leftrightarrow \mathbf{p}_l^O\}_{1 \leq l \leq L}$ of the L overlapped significant points. In the example of Fig. 4 (Iteration #1), the selected $\mathbf{p}_l^F$ points are shown in green. We treat S_F^0 as being a reliable set and use the GPR to estimate the mean $m_{S_n^0}(\mathbf{p}^F)$ and covariance $\sigma_{S_n^0}^2(\mathbf{p}^F)$ of the location of a point $\mathbf{p}^F$ in I_O, and it is computed by

$$m_{S_n^0}(\mathbf{p}^F) = \mathbf{k}' \Gamma_{S_n^0}^{-1} \mathbf{P}_{S_n^0}^O,$$

$$\sigma_{S_n^0}^2(\mathbf{p}^F) = \mathbf{k}(\mathbf{p}^F, \mathbf{p}^F) + \beta^{-1} - \mathbf{k}' \Gamma_{S_n^0}^{-1} \mathbf{k}, \tag{2}$$

where $\mathbf{k}$ is the vector $[k(\mathbf{p}_1^F, \mathbf{p}^F), ..., k(\mathbf{p}_L^F, \mathbf{p}^F)]^T$, and k is a kernel function which defines a mapping composed of an affine and a non-linear transformation as in [24,25], β^{-1} denotes a measurement noise variance, $\Gamma_{S_n^0}$ is the $L \times L$ symmetric matrix with elements $\Gamma_{i,j} = k(\mathbf{p}_i^F, \mathbf{p}_j^F) + \beta^{-1}\delta_{i,j}$, and $\mathbf{P}_{S_n^0}^O$ is the $L \times D$ matrix $[\mathbf{p}_1^O, ..., \mathbf{p}_L^O]^T$, where D is the dimension of the image.

Afterwards, all correspondences that are consistent with this GPR are added to S_n^0. A correspondence is treated as valid if the Mahalanobis distance between the corresponding points between $\mathbf{p}^O$ and $m_{s_n^0}(\mathbf{p}^F)$ is small enough. This gives us an augmented correspondence set S_n^1, such as the one depicted by Fig. 4 (Iteration #2). The process using S_n^1 to compute the regression of Eq. 2 is then repeated until the set stabilizes, typically after 3 to 4 iterations, as shown in Fig. 4 (Iteration #3). Finally, it yields two sets of image points $(\mathbf{p}_i^F, \mathbf{p}_i^O)$ and of geometrically consistent correspondences S_n. Typically, all the significant points $\mathbf{p}^F$ in the fundus image would be contained in the stabilized set, so i is usually equal to the number of the significant points in fundus image I_F.

We now turn to enforcing topological consistency constraint by message passing on the result introduced in Sect. 2.1. Topological consistency means that if one subtree of fundus vessel network contains a set of significant points $\mathbf{p}_m^F$, and there is a edge set $E^F = \{E_{ij}^F, i, j \in m\}$, then we can find the corresponding point set $\mathbf{p}_m^O$ and edge set $E^O = \{E_{ij}^O, i, j \in m\}$ in OCT-A. For each subtree in fundus, the corresponding subtree can be found by searching for those containing all the edges in E^O in OCT-A. The topological message passing is achieved by linking corresponding subtree pair, as show in Fig. 2(c).

2.3 A/V Classification in OCT-A

A topology map is obtained after mapping the topology results from an OCT-A to its corresponding fundus - the complete vessel network is divided into several subgraphs each with a label. The final goal is to differentiate the labels into artery and vein categories.

Again, the dominant sets-based classification method (DOS) introduced in Sect. 2.1 is employed to classify these individual branches into two clusters, A and B. Note, the features suggested in [7] are utilized to compute the weights ω. For each subgraph i, the probability of its being cluster A: P_A^i, is computed by the number of vessel pixels classified by DOS as A: $P_A^i = n_A(i)/(n_A^i + n_B^i)$, where n_A^i is the number of pixels classified as A, and n_B^i is the number of pixels classified as B. In this work, the subgraphs are determined as 'artery' if the average intensity value in the fundus image is larger than 0.48, otherwise, the subgraphs are assigned as 'vein'.

3 Experiments

Datasets: The proposed A/V classification method was evaluated on 22 pairs of OCT-A and fundus images from 22 eyes with pathological myopia. OCT-A images were acquired using an Angiovue Spectral-domain-OCT angiography system (Optovue Fremont, USA), with resolutions of 400×400 pixels. All analyzed OCT-A images were $6 \times 6\,\text{mm}^2$ scans. Color fundus images were captured using Canon CR-2 (Cannon, Japan) with a $45°$ field-of-view and a resolution of 1623×1642 pixels. One senior image expert was invited to manually label the vascular topology (each single tree is marked with a distinct color), arteries and veins on both image types. We made our dataset publicly available [https://imed.nimte.ac.cn/ACRO.html].

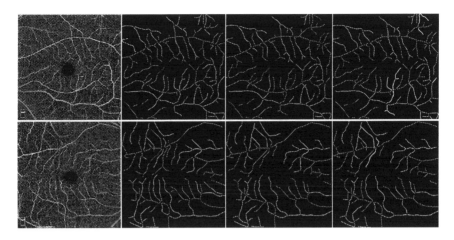

Fig. 5. A/V classification performances in OCT-A. From left to right: two example OCT-A images, A/V manual annotations, A/V classification results by our method with incorrectly identified vessels highlighted (**yellow** indicate the incorrectly classified veins, **green** indicate the incorrectly identified arteries). (Color figure online)

Table 1. Performances of different A/V classification methods.

	OCT-A		INSPIRE		DRIVE		VICAVR		WIDE	
	Se	Acc	Se	Acc	Se	Acc	Se	Acc	Se	Acc
Dashbozorg [7]	–	–	0.910	0.885	0.900	0.870	–	0.898	–	–
Estrada [28]	–	–	0.915	0.909	0.930	0.941	–	–	0.910	0.910
Huang [10]	–	–	–	0.851	–	–	–	0.906	–	–
Zhao [11]	–	–	0.918	0.910	0.919	–	–	0.910	–	–
Ma [12]	–	–	0.924	0.916	0.934	**0.945**	–	–	–	–
Proposed	**0.906**	**0.890**	**0.968**	**0.964**	**0.942**	0.935	**0.954**	**0.946**	**0.962**	**0.952**

3.1 Evaluation of A/V Classification

Figure 5 demonstrates the results of A/V classification on two example OCT-A images by using the proposed method. Overall, our method correctly distinguishes most of the arteries (red) and veins (blue), when compared with the corresponding manual annotations. In order to demonstrate the effectiveness of the proposed A/V classification more objectively, Table 1 reports the sensitivity (Se), and balanced accuracy (Acc) scores in terms of centerline pixel-level. Se illustrates the ability of a given method to detect arteries, and Acc indicates the overall classification performance, and thus reflects the trade-off between sensitivity and specificity. Our method obtains promising performance of A/V classification on OCT-A, with Se = 0.906 and Acc = 0.890, respectively.

We have also evaluated our method against the state-of-the-art A/V classification methods over four public retinal vessel datasets: INSPIRE [26], VICAVR [9], DRIVE [27], and WIDE [1], to validate the superiority our method

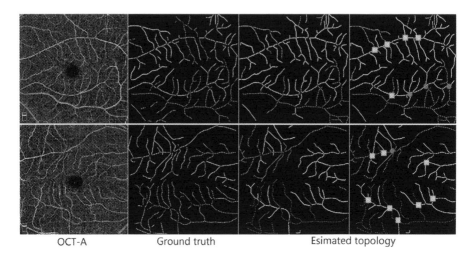

OCT-A   Ground truth   Esimated topology

Fig. 6. Example topology estimation results of the proposed method. From left to right: OCT-A image, manual annotations, estimated topology where green cubes and red circles indicate correctly and incorrectly traced points respectively. (Color figure online)

in A/V classification. It is clearly shown that the proposed method obtains competing results in terms of Se and Acc. The results of the existing methods were quoted from the relevant papers.

3.2 Evaluation of Topology Estimation

Our A/V classification relies on the prior results of topology reconstruction, and false topology estimation may cause incorrect A/V classification. Therefore, it is important to establish the performance on topology estimation. The last two columns of Fig. 6 illustrates the results of vascular topology estimation on OCT-A. It can be seen that our method is able to trace most vascular structures correctly: only a few significant points were incorrectly traced in red circles, as shown in the last column of Fig. 6. These errors were located at crossovers since it may suffer from failures at the tiny vessel segmentation and skeletonization stages, leading to misrepresentation of the topological structures. The percentage of the correctly identified (True Positive, TP) relevant significant points, i.e., bifurcation (BIF) and crossovers (CRO), is presented in Table 2. Compare to color fundus, although there is a lack of color and intensity information in OCT-A, our results showed that our method still achieves promising results in OCT-A. Th accuracy of significant point identification is 87.1% in OCT-A images.

Table 2. Performance of the proposed method on topology reconstruction at significant points over different datasets.

	OCT-A	INSPIRE	VICAVR	DRIVE	WIDE
# BIF	1549	1998	2478	4955	3678
# TP	1360	1945	2288	4799	3551
# CRO	260	778	832	1421	1230
# TP	216	697	728	1182	1107
Accuracy	87.1%	95.1%	91.1%	93.8%	94.9%

4   Conclusion

An automated method for classification of vessels as arteries or veins in OCT-A is indispensable, to understand disease progression and facilitate for the management of many diseases. In this paper, we have demonstrated a novel A/V classification method in OCT-A, guided by the color and intensity information from the corresponding fundus images. We utilized the observation that paired fundus and OCT-A images share partial vascular topological networks. After estimating the vascular topology of both color fundus and OCT-A images, Gaussian process regression was applied to register and translate A/V classification results from color fundus image to OCT-A images. The proposed algorithm accurately estimated topology and classified vessel types in OCT-A. The significance of our method is that it is the first attempt to classify A/Vs in OCT-A. Future work will focus on testing the proposed technique for the diagnosis of eye-related disease in clinical settings.

Acknowledgment. This work was supported by Zhejiang Provincial Natural Science Foundation of China (LQ20F030002, LZ19F010001), Ningbo "2025 S&T Megaprojects" (2019B10033, 2019B10061), National Natural Science Foundation of China (61906181).

References

1. Estrada, R., Tomasi, C., Schmidler, S., Farsiu, S.: Tree topology estimation. IEEE Trans. Pattern Anal. Mach. Intell. **37**(8), 1688–1701 (2015)
2. Zheng, Y., et al.: Automatic 2-D/3-D vessel enhancement in multiple modality images using a weighted symmetry filter. IEEE Trans. Med. Imaging **37**(2), 438–450 (2017)
3. Vázquez, S., et al.: Improving retinal artery and vein classification by means of a minimal path approach. Mach. Vis. Appl. **24**(5), 919–930 (2013)
4. Alam, M., Toslak, D., Lim, J.I., Yao, X.: Color fundus image guided artery-vein differentiation in optical coherence tomography angiography. Invest. Ophthalmol. Vis. Sci. **59**(12), 4953–4962 (2018)
5. Estrada, R., Tomasi, C., Schmidler, S.C., Farsiu, S.: Tree topology estimation. IEEE Trans. Pattern Anal. Mach. Intell. **37**(8), 1688–1701 (2014)

6. Zhao, Y., Rada, L., Chen, K., Harding, S.P., Zheng, Y.: Automated vessel segmentation using infinite perimeter active contour model with hybrid region information with application to retinal images. IEEE Trans. Med. Imaging **34**(9), 1797–1807 (2015)

7. Dashtbozorg, B., Mendonça, A.M., Campilho, A.: An automatic graph-based approach for artery/vein classification in retinal images. IEEE Trans. Image Process. **23**(3), 1073–1083 (2013)

8. Zhao, Y., et al.: Retinal vascular network topology reconstruction and artery/vein classification via dominant set clustering. IEEE Trans. Med. Imaging **39**(2), 341–356 (2020)

9. Vázquez, S., Cancela, B., Barreira, N., Saez, M.: Improving retinal artery and vein classification by means of a minimal path approach. Mach. Vis. Appl. **24**(5), 919–930 (2013)

10. Huang, F., Dashtbozorg, B., Romeny, B.M.H.: Artery/vein classification using reflection features in retina fundus images. Mach. Vis. Appl. **29**(1), 23–34 (2017). https://doi.org/10.1007/s00138-017-0867-x

11. Zhao, Y., et al.: Retinal artery and vein classification via dominant sets clustering-based vascular topology estimation. In: Frangi, A.F., Schnabel, J.A., Davatzikos, C., Alberola-López, C., Fichtinger, G. (eds.) MICCAI 2018. LNCS, vol. 11071, pp. 56–64. Springer, Cham (2018). https://doi.org/10.1007/978-3-030-00934-2_7

12. Ma, W., Yu, S., Ma, K., Wang, J., Ding, X., Zheng, Y.: Multi-task neural networks with spatial activation for retinal vessel segmentation and artery/vein classification. MICCAI 2019. LNCS, vol. 11764, pp. 769–778. Springer, Cham (2019). https://doi.org/10.1007/978-3-030-32239-7_85

13. Talu, S., Calugaru, D.M., Lupascu, C.A.: Characterisation of human non-proliferative diabetic retinopathy using the fractal analysis. Int. J. Ophthalmol. **8**(4), 770 (2015)

14. Zhao, Y., et al.: Intensity and compactness enabled saliency estimation for leakage detection in diabetic and malarial retinopathy. IEEE Trans. Med. Imaging **36**(1), 51–63 (2017)

15. Zahid, S., et al.: Fractal dimensional analysis of optical coherence tomography angiography in eyes with diabetic retinopathy. Invest. Ophthal. Vis. Sci. **57**(11), 4940–4947 (2016)

16. Zhao, Y., et al.: Automated tortuosity analysis of nerve fibers in corneal confocal microscopy. IEEE Trans. Med. Imaging **39**, 2725–2737 (2020)

17. Niemeijer, M., et al.: Automated measurement of the arteriolar-to-venular width ratio in digital color fundus photographs. IEEE Trans. Med. Imaging **30**(11), 1941–1950 (2011)

18. Xie, J., Zhao, Y., Zheng, Y., Su, P., Liu, J., Wang, Y.: Retinal vascular topology estimation via dominant sets clustering. In: International Symposium on Biomedical Imaging, pp. 1458–1462. IEEE (2018)

19. Mou, L., et al.: CS-Net: channel and spatial attention network for curvilinear structure segmentation. In: Shen, D., et al. (eds.) MICCAI 2019. LNCS, vol. 11764, pp. 721–730. Springer, Cham (2019). https://doi.org/10.1007/978-3-030-32239-7_80

20. Xie, J., et al.: Topology reconstruction of tree-like structure in images via structural similarity measure and dominant set clustering. In: Conference on Computer Vision and Pattern Recognition, vol. 10, pp. 8505–8513 (2019)

21. Pavan, M., Pelillo, M.: Dominant sets and pairwise clustering. IEEE Trans. Pattern Anal. Mach. Intell. **29**(1), 167–172 (2006)

22. Zemene, E., Pelillo, M.: Interactive image segmentation using constrained dominant sets. In: Leibe, B., Matas, J., Sebe, N., Welling, M. (eds.) ECCV 2016. LNCS, vol. 9912, pp. 278–294. Springer, Cham (2016). https://doi.org/10.1007/978-3-319-46484-8_17

23. Chen, J., Tian, J., Lee, N., Zheng, J., Smith, R.T., Laine, A.F.: A partial intensity invariant feature descriptor for multimodal retinal image registration. IEEE Trans. Biomed. Eng. **57**(7), 1707–1718 (2010)

24. Quiñonero-Candela, J., Rasmussen, C.E.: A unifying view of sparse approximate gaussian process regression. J. Mach. Learn. Res. **6**(Dec), 1939–1959 (2005)

25. Serradell, E., Glowacki, P., Kybic, J., Moreno-Noguer, F., Fua, P.: Robust non-rigid registration of 2D and 3D graphs. In: Conference on Computer Vision and Pattern Recognition, pp. 996–1003. IEEE (2012)

26. http://webeye.ophth.uiowa.edu/component/k2/item/270

27. Qureshi, T., Habib, M., Hunter, A., Al-Diri, B.: A manually-labeled, artery/vein classified benchmark for the drive dataset. In: 2013 IEEE 26th International Symposium on Computer-Based Medical Systems, pp. 485–488 (2013)

28. Estrada, R., Allingham, M.J., Mettu, P.S., Cousins, S.W., Tomasi, C., Farsiu, S.: Retinal artery-vein classification via topology estimation. IEEE Trans. Med. Imaging **34**(12), 2518–2534 (2015)

Vascular Surface Segmentation for Intracranial Aneurysm Isolation and Quantification

Žiga Bizjak[✉], Boštjan Likar, Franjo Pernuš, and Žiga Špiclin

Faculty of Electrical Engineering, University of Ljubljana,
Tržaška cesta 25, 1000 Ljubljana, Slovenia
ziga.bizjak@fe.uni-lj.si

Abstract. Predicting rupture risk and deciding on optimal treatment plan for intracranial aneurysms (IAs) is possible by quantification of their size and shape. For this purpose the IA has to be isolated from 3D angiogram. State-of-the-art methods perform IA isolation by encoding neurosurgeon's intuition about former non-dilated vessel anatomy through principled approaches like fitting a cutting plane to vasculature surface, using Gaussian curvature and vessel centerline distance constraints, by deformable contours or graph cuts guided by the curvature or restricted by Voronoi surface decomposition and similar. However, the large variability of IAs and their parent vasculature configurations often leads to failure or non-intuitive isolation. Manual corrections are thus required, but suffer from poor reproducibility. In this paper, we aim to increase the accuracy, robustness and reproducibility of IA isolation through two stage deep learning based segmentation of vascular surface. The surface was represented by local patches in form of point clouds, which were fed into first stage multilayer neural network (MNN) to obtain descriptors invariant to point ordering, rotation and scale. Binary classifier as second stage MNN was used to isolate surface belonging to the IA. Method validation was based on 57 3D-DSA, 28 CTA and 5 MRA images, where cross-modality-validation showed high segmentation sensitivity of 0.985, a substantial improvement over 0.830 obtained for the state-of-the-art method on the same datasets. Visual analysis of IA isolation and its high accuracy and reliability consistent across CTA, MRA and 3D-DSA scans confirmed the clinical applicability of proposed method.

Keywords: Intracranial aneurysm · Angiographic images · 3D-DSA · CTA · MRA · Point cloud · Multi-stage deep learning · Cross-modality-validation

1   Introduction

Intracranial aneurysms (IAs) are abnormal bulges mainly located at vessel bifurcations and arising from weakened vessel wall. Despite being highly prevalent

© Springer Nature Switzerland AG 2020
A. L. Martel et al. (Eds.): MICCAI 2020, LNCS 12266, pp. 128–137, 2020.
https://doi.org/10.1007/978-3-030-59725-2_13

(~3.2% of population), most IAs do not rupture throughout patient's lifetime and are not associated to any symptoms. In case of rupture, however, the subsequent subarachnoid hemorrhage (SAH) has 50% fatality rate. Imaging studies have shown that aneurysm size and shape are one of the key factors for prediction rupture risk and deciding on optimal treatment plan for unruptured IAs [10]. For instance, in ELAPSS score [2] the IA size can amount to 55% of the final score, whereas ELAPSS also takes into consideration earlier SAH, location of aneurysm, age, population and shape of aneurysm.

Clinical treatment of small and medium IAs is yet not well defined. Note that size groups of IAs are small (0–4.9 mm in diameter), medium (5 to 9.9 mm), large (10 to 25 mm) and giant (>25 mm) [10]. Studies have shown that risk of complications during procedure is larger than risk of spontaneous rupture [3, 4, 8, 10, 23]. There are many cases, where risk of spontaneous rupture is less likely than the risk due to treatment and where follow-up imaging is considered to be the best course of treatment. Follow-up imaging may involve several modalities like 3D-DSA, CTA and MRA. Therefore, an automatic and reproducible modality-independent process of aneurysm size and shape measurement is essential to provide best achievable treatment for patient and to correctly measure change of aneurysm morphology between consecutive imaging sessions.

Aneurysm size and shape quantification is possible through its segmentation and isolation from parent vessels. Most neurosurgeons still use manual methods to isolate IAs [5]. Due to high intra- and inter-rater variability of manual aneurysm segmentation, and the time-consuming and cumbersome 3D image manipulation, there is a need for fast, accurate and consistent computer-assisted segmentation methods.

In this paper we propose surface point classification (SPC) method to segment and isolate the IA. Once the point cloud is extracted from 3D image, for instance by simple interactive thresholding, the procedure to isolate IAs remains the same for all modalities. Our segmentation is based on the point cloud data and is thus modality-independent. This was successfully verified on a total of 100 IA cases, extracted from 57 3D-DSA, 28 CTAs and 5 MRA scans. The proposed method achieved consistent performance across different image modalities and also improved segmentation over the state-of-the-art.

2   Related Work

An approach mimicking manual IA isolation is by positioning a cutting plane. In past clinical practice the cutting plane was chosen manually with the help of the 3D visualization software [17]. Jerman et al. [11] developed method that automatically positions the cutting plane (ACP). This method works well on most of the aneurysms and is considered as current state-of-the-art. However, if the aneurysm is small or if the aneurysm is blended with surrounding vessels the ACP may fail. Furthermore, if the shape of IA's neck does not lie in a plane, the method may again fail to isolate aneurysm correctly as shown in Fig. 1.

Non-planar curve for separation of aneurysm from the surrounding vessels seem a better option. Ruben et al. [6] used automatic approach based on the

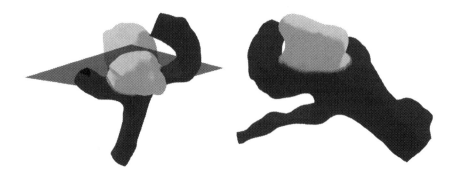

Fig. 1. Two examples of aneurysm segmentation. Current state-of-the-art-method (ACP) fails at isolate aneurysm with plane (*grey surface on the left*), while our SPC method (*yellow-colored surface*) is very close to reference manual segmentation (*red curve*). (Color figure online)

computation of a minimal path along a scalar field obtained on the vessel surface. In order to assure correct topology of the aneurysms bulb, the neck computation was constrained with a region defined by surface Voronoi diagram. The method was evaluated on 26 real cases against manual aneurysm isolation using cutting plane. This method works well for standard cases, but can fail if the configuration of the arteries is complex. Marina et al. [18] proposed a method also based on Voronoi diagram with the combination of so-called tube function. The method was evaluated on 30 cases. In 20 cases, the manual and computed separation curves were considered as being very similar and both acceptable. In five cases manual curve was better then the computed, but still acceptable. In five cases the method did not produce similar curves compared to manual ones. Hence, robustness remains an issue with non-planar separation curves.

Sylivia et al. [20] proposed a semi-automatic separation curve reconstruction algorithm for automatic extraction of morphological parameters. The first drawback of this approach is that it requires a pronounced aneurysm neck to work properly. The second drawback is that the user has to click on the aneurysm to initialize the method's parameters. Lawonn et al. [16] proposed a geometric optimization approach for the detection and segmentation of multiple aneurysms in two stages. First, a set of aneurysm candidate regions was identified by segmenting regions of the vessels. Second stage was a classifier, that identified the actual aneurysms among the candidates. While the method was capable of detecting aneurysm location automatically, the user still had to use the brush tool to get correct aneurysm neck curve. To address this problem they proposed a smoothing algorithm based on ostium curve extraction to improve upon previous results. While smoothing algorithm worked fine, additional manual user input was still necessary for some aneurysm cases.

All of the aforementioned methods mostly provide an acceptable result for large saccular IAs that have a well defined aneurysm neck (the cross section of the inlet is significantly smaller than cross-section of the aneurysm). For other aneurysm size and shapes most of these methods fails to deliver correct result.

To our knowledge, there is still no fully automatic method that would work well for IAs of various sizes and shapes, regardless of the configuration of surrounding vessels.

3   Materials and Methods

Current clinical detection and measurement of IAs is performed on angiographic images like 3D-DSA, CTA and MRA (Sect. 3.1). Procedures used to extract 3D surface mesh from angiographic images are explained in Sect. 3.2, data augmentation and training process in Sect. 3.3 and the inference on clinical images in Sect. 3.4.

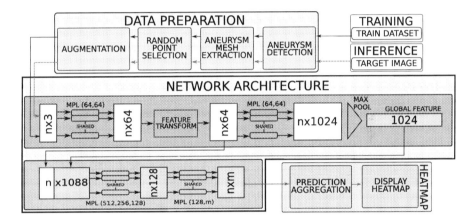

Fig. 2. Flowchart of the proposed aneurysm segmentation method.

3.1   Case Information

This study was performed with the approval of the institutional review board. A total of 90 patient angiographic 3D scans, including 57 3D-DSA images containing 70 aneurysm, 28 CTA images containing 25 aneruysms and 5 MRA images containing 5 aneurysms, of the cerebrovascular region were obtained for the purpose of this study. All images were acquired at University Medical Centre Ljubljana using standard imaging protocols used in clinical routine. Per patient incidence of IAs in our dataset was from zero to three. We found a total of 100 IAs, from which 27 are considered small ($<5\,$mm), 49 medium sized ($5 < $ size $ < 10\,$mm) and 24 are large ($>25\,$mm). The median diameter of the aneurysms was 6.99 mm.

3.2   Mesh Extraction from 3D Angiographic Image

First step in extracting the IA mesh from 3D angiographic image is IA detection. The search for IA is mainly still performed by skilled surgeon without any special computer-assisted tools. Many algorithms for automatic aneurysm detection were previously proposed [9,12,13,24], but all have the difficulty of presenting with to many false positives or are not able to detect small aneurysms. For our purposes detection was carried out manually using visual image inspection.

Following detection, a region of interest (ROI) containing the aneurysm and surrounding vessels was determined. Segmentation of vascular structures was performed by using interactive thresholding, followed by the application of marching cubes and smooth non-shrinking algorithms [7,15]. In this way, the corresponding 3D surface mesh of aneurysm and surrounding vessels was reconstructed. The following procedures ware performed using image analysis software (RealGUIDE Software version 5.0; 3DIEMME Srl). Example mesh containing an aneurysm and surrounding vessels is shown in Fig. 1.

3.3   Training

On every input mesh the aneurysm was manually isolated from surrounding vessels by skilled neurosurgeon by drawing a closed surface curve (see Fig. 1 for examples of manual reference curve for IA isolation). Aneurysm points were labeled according to the neurosurgeon's segmentation. Surgeon was not restricted to the specific shape of aneurysm neck. The obtained manual IA segmentation isolation were used for training and validation of multi-layer neural network (MNN) model [19].

To train MNN model, with network architecture as presented on Fig. 2, that is robust to point ordering, rotations and scaling of input data we replicated each mesh n times ($n =$ number of replications). Rotations were randomly applied with angles from 0 to $360°$ in all three axes. Scaling factor was randomly set between values 0.5 and 1. From each augmented mesh we randomly picked m points ($m =$ number of points) and saved them as point cloud. Augmented point clouds of one aneurysm appeared only in train, test or validation set.

3.4   Inference - Using Trained Model for Isolation

To further suppress MNN model's sensitivity to rotation, we created a so-called multiple partial voting algorithm. Similar to the augmentation process before training, we used the same augmentation procedures for the target mesh until each point of the target mesh was used at least 20 times, each time with different rotation. For every augmented instance we used m random points from target mesh. Each augmented point cloud contributed one vote to segmentation for each point. The sum of all votes was divided by number of votes for each point. The resulting heatmaps represented the output segmentation, which was compared to the other methods as shown on Fig. 4.

To determine which points on heatmap form the aneurysm and which the surrounding vessels we used simple thresholding at 0.5. We also verified our decision on such threshold by plotting the ROC (receiver operating characteristic) curve.

4   Experiment

The training of MNN segmentation model was executed only once. As train dataset we used 70% of all available data, i.e. 70 aneurysm meshes. Each mesh was reproduced with different rotation about 100 times ($n = 100$) and 3000 points ($m = 3000$) were randomly chosen after each rotation. For test dataset we used 25% of available data, i.e. 25 aneurysm meshes, which were also rotated n times and m points were chosen. We used 5% of data or 5 aneurysm meshes for validation purposes. Those aneurysm meshes were used to create a valid heatmap. The number of epochs was 100 and the learning rate was 0.001 for fist 20 epochs, 0.0005 for epochs between 21–40, 0.00025 for epochs between 40–60, 0.000125 for epochs between 61–80 and 0.000063 for the rest of the epochs. Batch size was set to 16.

To fairly compare our results with the state-of-the-art method proposed by Jerman et al. [11] we also calculated SPC with plane (SPCP) as follows: using least-square fit a cutting plane was positioned onto border points between aneurysm and vessel.

The model was trained on the Linux based computer with 32 GB RAM, Intel Core 8700 K 6 Core 4,7 GHz processor and 11 GB NVIDIA Graphic Card.

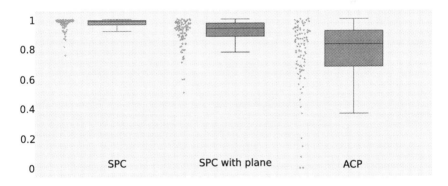

Fig. 3. Point-jitter and box plots of sensitivity values for all meshes for three tested methods.

5   Results

After training our model on 70 meshes, we tested the performance on all datasets. Point-wise predictions were rounded to the nearest integer (0 or 1) for all n rotation of each mesh, the aggregated heatmap values of each target mesh were normalized to interval $[0, 1]$. For each mesh we computed true positive rate (TPR; sensitivity). Table 1 shows sensitivity evaluation of all data subsets. Median sensitivity on learning dataset was 0.985, 0.983 on test dataset and 0.994 on validation dataset. If we fit plane to the SPC model result we achieve median

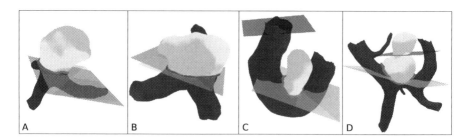

Fig. 4. Images **A–D** depict four examples of our isolation compared to gold standard and previous work. Red line represents gold standard aneurysm neck, grey plane represents the ouput of ACP algorithm, orange plane was fitted from boarder points of SPC. Yellow surface is our result of aneurysm isolation from vessels. Dark purple surface is predicted vessel surface. (Color figure online)

Table 1. Aneurysm segmentation evaluation across all 100 datasets and across the training, testing and validation cases.

Method	TRAIN			TEST			VAL		
	SPC	SPCP	ACP	SCP	SPCP	ACP	SCP	SPCP	ACP
min TPR	**0.759**	0.507	0	**0.918**	0.730	0.074	**0.982**	0.892	0.709
max TPR	1	1	1	1	1	0.985	**0.999**	0.972	0.972
median TPR	**0.985**	0.938	0.830	**0.983**	0.924	0.775	**0.994**	0.947	0.846

sensitivity 0.938 on training dataset, 0.924 on test dataset and 0.947 on validation dataset. Current state-of-the-art method achieved sensitivity of 0.830 on training, 0.775 on test and 0.846 on validation dataset, with median sensitivity of 0.830. State-of-the-art method failed to isolate aneurysm from the vessels in 8% of all cases, while our proposed method succeeded in accurately isolating aneurysm from the vessel across all cases. The distribution of sensitivity for all three tested methods is presented in box plots on Fig. 3.

Average time to segment and isolate one aneurysm from the vessels with the current state-of-the-art method was approximately 11 min. On the other hand, the isolation of using novel method executed in under 23 s per input image.

6   Discussion and Conclusion

We successfully developed and validated a novel MNN based method for aneurysm segmentation and isolation and compared its performances to current state-of-the-art ACP method [11]. To our knowledge our method is the first that works on all aneurysm cases regardless of their own shape or shape of the surrounding vessels. At the same time, the ACP method fails to provide correct results in approximately 10% of the cases and provide partially satisfactory segmentations on 30% of our cases. The computational times, compared to approximately 11 min for the ACP, were less then 23 s for the proposed method.

The proposed method achieved 15% better median sensitivity than the ACP. For comparison we use the least-square minimization to fit a plane to the original output of our method and still achieved 10% better median sensitivity than ACP. We succeeded to isolate aneurysm from vessel for all cases with the sensitivity of at least 0.759 (median sensitivity for train dataset was 0.985, while for test dataset the sensitivity was 0.983). The 5 validation cases were not used in training and were extracted only from CTA images. All aneurysms in test dataset were extracted from MRA images, while aneurysms extracted from 3D-DSA images were used for training. Results showed that our model is not sensitive to the imaging modality used for extracting aneurysm meshes.

Figure 4 shows 2 cases where all methods work and 2 cases where current state-of-the-art ACP method fails. Red curve is manual reference neck curve determined by skilled neurosurgeon, yellow/purple areas on meshes represents our isolation of aneurysm from vessel, black surface is a manually positioned cutting plane, while the orange surface represents the plane fit on border points between aneurysm and vessel of our method's output. In sections **A** and **B** of Fig. 4 all approaches worked well, while in sections **C** the ACP method failed, while the plane acquired from output of our method still provided good results, in section **D** only our method provided meaningful results.

We have successfully surpassed the sensitivity of MNN to rotation and scaling of input data with random rotations and scaling of training data. The additional rotation of target mesh contributed to better and more robust prediction of our MNN model. The output in the form of heatmap allows us to see the probability of each point being part of aneurysm or vessel. Probability map also emphasizes the area between aneurysm or vessel (the values smaller than one, but bigger than zero).

We have also shown that point cloud based approach is not sensitive to different input image modalities and artifacts associated to them. To create our data we used all common angiography modalities, namely the CTA, 3D-DSA and MRA.

Accurate and fast prediction of aneurysm neck is essential for computing aneurysm features that are based on neck curve determination, such as volume, surface area, sphericity index, surface are to volume ratio, etc. Those measurements are useful in a variety of clinical and research applications related to intracranial aneurysms. An important treatment procedure called coiling involves coil embolization for treating unruptured aneurysms. The quantity of inserted coil depends on the measured aneurysm volume. Reliable volume calculations can prevent complications during and after the procedure, for instance, caused by inaccurate amount of inserted coil that can cause aneurysm lumen reopening and eventually result in rupture [14,21].

Surface area and volume have both been demonstrated to be important indicators of aneurysm rupture risk [1,22]. As shown on Fig. 4 the segmentation and isolation with cutting plane does not seem robust enough to isolate aneurysm from the vessels regardless of the position and the shape of aneurysm. Though prior work on isolation with non-planar separation curves showed some promising results [16,20], none of them achieved clinically acceptable results and robust enough performance. Small aneurysms with curved veins proved to be extremely challenging. Our method showed high accuracy and sensitivity, even with a relatively small dataset of 100 aneurysms. The method was able to predict, which points belong to the aneurysm automatically with sensitivity of 0.985. We feel that the robustness of our model will be reconfirmed even on larger volumes of training and test data.

The proposed deep learning approach to aneurysm isolation can identify and label unordered aneurysm points in 3D vascular mesh automatically and with great sensitivity. This approach enables robust, repeatable and fast isolation of aneurysms from the surrounding vessels, thus the method seems readily usable in research and clinical applications.

References

1. Austin, G.M., Austin, G.M., Schievink, W., Williams, R.: Controlled pressure-volume factors in the enlargement of intracranial aneurysms. Neurosurgery **24**(5), 722–730 (1989)
2. Backes, D., et al.: ELAPSS score for prediction of risk of growth of unruptured intracranial aneurysms **88**(17), 1600–1606 (2017). https://doi.org/10.1212/WNL. 0000000000003865, http://europepmc.org/abstract/med/28363976
3. Brinjikji, W., et al.: Risk factors for growth of intracranial aneurysms: a systematic review and meta-analysis **37**(4), 615–620 (2016). https://doi.org/10.3174/ajnr. A4575, http://www.ajnr.org/content/37/4/615
4. Brown, R.D., Broderick, J.P.: Unruptured intracranial aneurysms: epidemiology, natural history, management options, and familial screening **13**(4), 393–404 (2014). https://doi.org/10.1016/S1474-4422(14)70015-8, http://www. sciencedirect.com/science/article/pii/S1474442214700158
5. Cárdenes, R., Larrabide, I., San Román, L., Frangi, A.F.: Performance assessment of isolation methods for geometrical cerebral aneurysm analysis. Med. Biol. Eng. Comput. **51**(3), 343–352 (2013)
6. Cárdenes, R., Pozo, J.M., Bogunovic, H., Larrabide, I., Frangi, A.F.: Automatic aneurysm neck detection using surface voronoi diagrams. IEEE Trans. Med. Imaging **30**(10), 1863–1876 (2011)
7. Cebral, J.R., Löhner, R.: From medical images to anatomically accurate finite element grids. Int. J. Numeric. Methods Eng. **51**(8), 985–1008 (2001)
8. Chien, A., et al.: Unruptured intracranial aneurysm growth trajectory: occurrence and rate of enlargement in 520 longitudinally followed cases **1**, 1–11 (2019). https://doi.org/10.3171/2018.11.JNS181814, https://thejns.org/view/ journals/j-neurosurg/aop/article-10.3171-2018.11.JNS181814.xml
9. Duan, H., Huang, Y., Liu, L., Dai, H., Chen, L., Zhou, L.: Automatic detection on intracranial aneurysm from digital subtraction angiography with cascade convolutional neural networks. Biomed. Eng. Online **18**(1), 110 (2019)
10. Ishibashi, T., et al.: Unruptured intracranial aneurysms **40**(1), 313–316 (2009). https://doi.org/10.1161/STROKEAHA.108.521674, https://doi.org/10. 1161/STROKEAHA.108.521674
11. Jerman, T., Chien, A., Pernus, F., Likar, B., Spiclin, Z.: Automated cutting plane positioning for intracranial aneurysm quantification. IEEE Trans. Biomed. Eng. **67**(2), 577–587 (2019)

12. Jerman, T., Pernus, F., Likar, B., Špiclin, Ž.: Aneurysm detection in 3d cerebral angiograms based on intra-vascular distance mapping and convolutional neural networks. In: 2017 IEEE 14th International Symposium on Biomedical Imaging (ISBI 2017), pp. 612–615. IEEE (2017)

13. Jin, H., Yin, Y., Hu, M., Yang, G., Qin, L.: Fully automated unruptured intracranial aneurysm detection and segmentation from digital subtraction angiography series using an end-to-end spatiotemporal deep neural network. In: Medical Imaging 2019: Image Processing, vol. 10949, p. 109491I. International Society for Optics and Photonics (2019)

14. Kai, Y., Hamada, J.I., Morioka, M., Yano, S., Kuratsu, J.I.: Evaluation of the stability of small ruptured aneurysms with a small neck after embolization with guglielmi detachable coils: correlation between coil packing ratio and coil compaction. Neurosurgery **56**(4), 785–792 (2005)

15. Larrabide, I., et al.: Three-dimensional morphological analysis of intracranial aneurysms: a fully automated method for aneurysm sac isolation and quantification. Med. Phys. **38**(5), 2439–2449 (2011)

16. Lawonn, K., Meuschke, M., Wickenhöfer, R., Preim, B., Hildebrandt, K.: A geometric optimization approach for the detection and segmentation of multiple aneurysms. In: Computer Graphics Forum, vol. 38, pp. 413–425. Wiley Online Library (2019)

17. Ma, B., Harbaugh, R.E., Raghavan, M.L.: Three-dimensional geometrical characterization of cerebral aneurysms. Ann. Biomed. Eng. **32**(2), 264–273 (2004)

18. Piccinelli, M., Steinman, D.A., Hoi, Y., Tong, F., Veneziani, A., Antiga, L.: Automatic neck plane detection and 3D geometric characterization of aneurysmal sacs. Ann. Biomed. Eng. **40**(10), 2188–2211 (2012)

19. Qi, C.R., Su, H., Mo, K., Guibas, L.J.: Pointnet: deep learning on point sets for 3D classification and segmentation. In: Proceedings of the IEEE Conference on Computer Vision and Pattern Recognition, pp. 652–660 (2017)

20. Saalfeld, S., Berg, P., Niemann, A., Luz, M., Preim, B., Beuing, O.: Semiautomatic neck curve reconstruction for intracranial aneurysm rupture risk assessment based on morphological parameters. Int. J. Comput. Assist. Radiol. Surg. **13**(11), 1781–1793 (2018). https://doi.org/10.1007/s11548-018-1848-x

21. Slob, M.J., Sluzewski, M., van Rooij, W.J.: The relation between packing and reopening in coiled intracranial aneurysms: a prospective study. Neuroradiology **47**(12), 942–945 (2005)

22. Valencia, A., Morales, H., Rivera, R., Bravo, E., Galvez, M.: Blood flow dynamics in patient-specific cerebral aneurysm models: the relationship between wall shear stress and aneurysm area index. Med. Eng. Phys. **30**(3), 329–340 (2008)

23. Wiebers, D.O.: Unruptured intracranial aneurysms: natural history, clinical outcome, and risks of surgical and endovascular treatment **362**(9378), 103–110 (2003). https://doi.org/10.1016/S0140-6736(03)13860-3, http://www.sciencedirect.com/science/article/pii/S0140673603138603

24. Zhou, M., Wang, X., Wu, Z., Pozo, J.M., Frangi, A.F.: Intracranial aneurysm detection from 3D vascular mesh models with ensemble deep learning. In: Shen, D., et al. (eds.) MICCAI 2019. LNCS, vol. 11767, pp. 243–252. Springer, Cham (2019). https://doi.org/10.1007/978-3-030-32251-9_27

Breast Imaging

Deep Doubly Supervised Transfer Network for Diagnosis of Breast Cancer with Imbalanced Ultrasound Imaging Modalities

Xiangmin Han[1], Jun Wang[1], Weijun Zhou[2], Cai Chang[3], Shihui Ying[4], and Jun Shi[1($\boxtimes$)]

[1] Shanghai Institute for Advanced Communication and Data Science, School of Communication and Information Engineering, Shanghai University, Shanghai, China
junshi@shu.edu.cn
[2] Department of Ultrasound, The First Affiliated Hospital of Anhui Medical University, Hefei, China
[3] Department of Ultrasound, Department of Oncology, Shanghai Medical College, Fudan University Shanghai Cancer Center, Fudan University, Shanghai, China
[4] Department of Mathematics, School of Science, Shanghai University, Shanghai, China

Abstract. Elastography ultrasound (EUS) provides additional bio-mechanical information about lesion for B-mode ultrasound (BUS) in the diagnosis of breast cancers. However, joint utilization of both BUS and EUS is not popular due to the lack of EUS devices in rural hospitals, which arouses a novel modality imbalance problem in computer-aided diagnosis (CAD) for breast cancers. Current transfer learning (TL) pay little attention to this special issue of clinical modality imbalance, that is, the source domain (EUS modality) has fewer labeled samples than those in the target domain (BUS modality). Moreover, these TL methods cannot fully use the label information to explore the intrinsic relation between two modalities and then guide the promoted knowledge transfer. To this end, we propose a novel doubly supervised TL network (DDSTN) that integrates the Learning Using Privileged Information (LUPI) paradigm and the Maximum Mean Discrepancy (MMD) criterion into a unified deep TL framework. The proposed algorithm can not only make full use of the shared labels to effectively guide knowledge transfer by LUPI paradigm, but also perform additional supervised transfer between unpaired data. We further introduce the MMD criterion to enhance the knowledge transfer. The experimental results on the breast ultrasound dataset indicate that the proposed DDSTN outperforms all the compared state-of-the-art algorithms for the BUS-based CAD.

Keywords: Ultrasound imaging · Breast cancer · Deep doubly supervised transfer learning · Support vector machine plus · Maximum mean discrepancy

1 Introduction

B-mode ultrasound (BUS) is a clinical routine imaging tool to diagnose breast cancers. With the fast development of artificial intelligence technology, the BUS-based computer-aided diagnosis (CAD) has attracted considerable attention in recent years [1]. However,

© Springer Nature Switzerland AG 2020
A. L. Martel et al. (Eds.): MICCAI 2020, LNCS 12266, pp. 141–149, 2020.
https://doi.org/10.1007/978-3-030-59725-2_14

142 X. Han et al.

BUS only provides diagnostic information related to the lesion structure and internal echogenicity, which limits the performance of CAD to a certain extent.

Elastography ultrasound (EUS) imaging has emerged as an effective imaging technology for the diagnosis of breast cancers, which shows information pertaining to the biomechanical and functional properties of a lesion [2]. Joint utilization of both BUS and EUS provides complementary information for breast cancers to promote diagnostic accuracy [3]. However, the EUS devices are generally scarce in rural hospitals, which makes EUS not popular in diagnosing breast cancers in clinical practice.

Transfer learning (TL) aims to improve a learning model in the target domain by transferring knowledge from the related source domains [4, 5]. TL has achieved great success in various classification tasks, including CAD [6, 7]. Therefore, the performance of a single-modal imaging-based CAD model can be effectively promoted by transferring knowledge from other related imaging modalities or diseases [6].

It is worth noting that modality imbalance is a common phenomenon in clinical practice. That is, there are not only some paired BUS and EUS images with shared labels but also additional single-modal labeled BUS images in this work. Therefore, the source domain (EUS modality) has fewer samples than those in the target domain (BUS modality) in our work, which is contrary to the conventional TL applications. The inadequate data in the source domain also increase the difficulty for TL since it cannot provide enough supervision for TL. The conventional TL methods can handle this transfer task by performing the feature- or classifier-level transfer [4, 6, 8, 9]. However, these TL methods have no constraints on the labels of both the source and target domains, and therefore cannot fully use the label information to explore the intrinsic relation between two modalities and then guide the promoted knowledge transfer.

Learning using privileged information (LUPI) is a newly proposed TL paradigm developed on the paired data in source and target domains with shared labels [10]. Support vector machine plus (SVM+) is a typical classifier under the LUPI paradigm, which generally outperforms the conventional TL classifiers due to the supervision of the shared labels [10]. However, SVM+ cannot conduct TL for unpaired or imbalanced data also due to the limitation of the LUPI paradigm.

On the other hand, convolutional neural network (CNN) based TL methods generally achieve superior performance to the conventional TL approaches in many classification tasks [5]. Although the source domain generally includes a large number of labeled data while the target domain only has a few labeled data, most of these works focus on the knowledge transfer between unpaired data in an unsupervised way[11].

Therefore, it is necessary to develop a new TL paradigm that can effectively address the issue of TL for imbalanced medical modalities in a supervised way. To this end, we propose a novel deep doubly supervised transfer network (DDSTN) for the BUS-based CAD of breast cancers. As shown in Fig. 1, this new TL paradigm doubly transfers knowledge between both the paired and unpaired data between the source and target domains in a unified framework. Specifically, the SVM+ classifier performs the transfer for the paired ultrasound data with shared labels, while the two-channel CNNs conduct another supervised transfer for the unpaired labeled data by MMD criterion. The double transfer mechanism can effectively adopt both shared and unshared labels to mine the intrinsic transferred information, and then guide the knowledge transfer from the limited samples in the source domain.

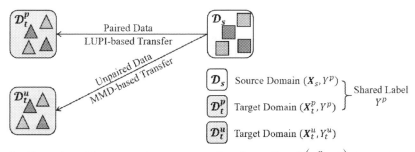

Fig. 1. Illustration of the proposed DDSTN paradigm. (X_s, Y^p) and $\left(X_t^p, Y^p\right)$ denote the paired data with shared labels Y^p in the source domain $\mathcal{D}_s$ and target domain $\mathcal{D}_t^p$, respectively. (X_t^u, Y_t^u) is the additional single-modal data in the target domain $\mathcal{D}_t^u$.

The main contributions are twofold as follows:

1) We propose a new doubly supervised TL paradigm to address the issue of TL for imbalanced modalities with labeled data, which can not only make full use of the shared labels to effectively guide knowledge transfer, but also perform additional information transfer between unpaired data. Therefore, more transferred knowledge promotes the classification performance.

2) We develop a novel DDSTN algorithm to perform the doubly supervised TL from fewer EUS samples in the source domain to the BUS-based CAD for breast cancers. Specifically, DDSTN integrates the SVM+ paradigm for the TL of paired data and the deep TL network for transfer between the unpaired data into a unified framework. The experimental results show its effectiveness on the BUS-based CAD for breast cancers.

2 Method

2.1 Network Architecture of DDSTN

Figure 2 shows the flowchart of our proposed DDSTN, which consists of two components, namely, LUPI-based supervised TL module for paired data and MMD-based supervised TL module for unpaired data. There are two independent CNNs for the source and target domains, respectively, which mainly learn feature representation, and also perform knowledge transfer for both paired and unpaired data.

In this work, BUS and EUS imaging work as target domain and source domain, respectively. We define (X_s, Y^p) and $\left(X_t^p, Y^p\right)$ to be the paired data with shared labels Y^p in the source domain $\mathcal{D}_s$ and target domain $\mathcal{D}_t^p$, respectively, and (X_t^u, Y_t^u) is the additional single-modal data in the target domain $\mathcal{D}_t^u$. The superscript p and u denote paired and unpaired, the subscript s and t mean the source and target domain.

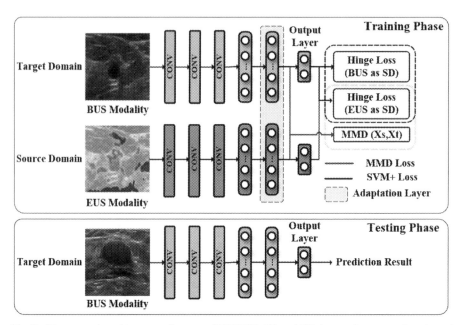

Fig. 2. The network architecture of proposed DDSTN. SD and TD denote the source domain and the target domain, respectively. The black dotted box represents the loss of LUPI, and the light green dotted box represents the supervised MMD criterion. (Color figure online)

The LUPI-based supervised TL module performs knowledge transfer under the guidance of shared labels to promote the classifier in the target domain. As shown in Fig. 2, the loss function of LUPI contains coupled SVM+ loss. We optimize both the two-channel networks simultaneously with this coupled loss.

The MMD-based supervised TL module shares the same network with the LUPI-based supervised TL module. We integrate the MMD learning criterion and hinge loss into a uniform supervised architecture. MMD is used to minimize the distribution imparity between two domains, and the hinge loss in SVM can help to learn a strong classifier. Since we introduce the label information, the unpaired data can be trained in a supervised way. Moreover, it is worth noting that the hinge loss is just the same as the LUPI-based supervised TL module.

In the training phase, the source and target networks are optimized under an overall objective function, while in the testing phase, only the learned target networks (BUS modality network) is used to predict the results.

2.2 Doubly Supervised Transfer Learning

We propose a doubly supervised transfer strategy to perform knowledge transfer across the imbalance modality. The overall object function incorporates two loss parts for transferring the paired and unpaired data, respectively, into the following formula:

$$\mathcal{L} = \mathcal{L}_{paired} + \mathcal{L}_{unpaired} \tag{1}$$

where $\mathcal{L}_{paired}$ is the LUPI paradigm for the TL of paired data and $\mathcal{L}_{unpaired}$ is the MMD learning criterion for the TL of unpaired data.

LUPI Paradigm for TL of Paired Data. The LUPI paradigm is adopted to perform transfer for the paired data with shared labels [10]. Here, the typical SVM+ classifier is used with the objective function as following:

$$\mathcal{L}_{paired} = \min \frac{1}{2}\left(\|W_t\|^2 + \lambda_1 \|W_s\|^2\right) + C_1 \sum_{i=1}^{n^p} \left[\langle W_s, x_{s_i}^p \rangle + b_s\right] \qquad (2)$$

$$\text{s.t. } y_i\left[\langle W_t, x_{t_i}^p \rangle + b_t\right] \geq 1 - \left[\langle W_s, x_{s_i}^p \rangle + b_s\right], \quad i = 1, \ldots, n^p$$
$$\text{and } \langle W_s, x_{s_i}^p \rangle + b_s \geq 0, \quad i = 1, \ldots, n^p$$

where $y_i \in Y^p$, $\{W_t, W_s\}$ and $\{b_t, b_s\}$ denote the weight matrices and bias vectors of the last layer in both target and source domains, respectively. $\lambda_1 > 0$ is a hyperparameter that restricts the correcting capacity, $C_1 > 0$ is a coefficient that balances the hinge loss term and the regularization term,n^p is the number of paired data, and $\|\cdot\|$ denotes the L2-norm of the weight matrix.

As shown in Fig. 2, the LUPI paradigm for the paired data has a coupled loss for two domains. Thus, EUS and BUS modalities are alternately taken as the source domain data to perform TL to improve the network of the target domain.

MMD Criterion for TL of Unpaired Data. To conduct knowledge transfer between unpaired data, MMD is introduced to minimize the distribution imparity between two domains. By considering labels as the supervision to further improve the learning performance of the classifier, we design a new loss function for unpaired data:

$$\mathcal{L}_{unpaired} = \min(\frac{1}{2}\|W\|_t^2 + C_2 \sum_{j=1}^{n_t^u} \max\left(0, 1 - y_{t_j}^u\left[\langle W_t, x_{t_j}^u \rangle + b_t\right]\right)$$
$$+ \lambda_2 \left\| \frac{1}{n_s^u} \sum_{k=1}^{n_s^u} \phi(x_{s_k}) - \frac{1}{n_t^u} \sum_{j=1}^{n_t^u} \phi(x_{t_j}^u) \right\|_{\mathcal{H}}) \qquad (3)$$

where $x_{s_k} \in X_s$, $x_{t_j}^u \in X_t^u$, $y_{t_j}^u \in Y_t^u$, λ_2 is non-negative hyperparameter of MMD. $\phi(\cdot)$ is a feature mapping function, we aim to find an optimal $\phi(\cdot)$ that can train a robust classifier, n_t^u, n_s^u are the number of BUS imaging and EUS imaging, respectively.

In order to minimize Eq. (3), we perform the domain adaption on the penultimate layer to transfer the knowledge from the source domain to the target domain for the unpaired features [14]. The supervised domain fusion loss makes the domains indistinguishable in the process of representation learning.

Doubly Supervised TL Strategy. The final objective function for doubly supervised TL is formulated by combining the $\mathcal{L}_{paired}$ and $\mathcal{L}_{unpaired}$ as following:

$$\mathcal{L} = \min \frac{1}{2}\left(\|W_t\|^2 + \lambda_1 \|W_s\|^2\right)$$
$$+ C_1 \sum_{i=1}^{n^p} \max\left(0, 1 - y_i\left[\langle W_s, x_{s_i}^p \rangle + b_t\right]\right) + C_2 \sum_{j=1}^{n_t^u} \max\left(0, 1 - y_{t_j}^u\left[\langle W_t, x_{t_j}^u \rangle + b_t\right]\right)$$
$$+ \lambda_2 \left\| \frac{1}{n_s^u} \sum_{k=1}^{n_s^u} \phi(x_{s_k}) - \frac{1}{n_t^u} \sum_{j=1}^{n_t^u} \phi(x_{t_j}^u) \right\|_{\mathcal{H}}$$
$$(4)$$

where C_1 and C_2 are non-negative constants of LUPI paradigm and distance metric loss, respectively, λ_1 restrict the correcting capacity of the classifier, and λ_2 is non-negative hyperparameter of MMD.

The overall objective function is optimized by stochastic gradient descent [12]. As shown in Fig. 2, only the learned target domain network is used to predict the results. The objective function is given by:

$$\hat{Y} = WX_t + b \tag{5}$$

where $X_t \subset \{X_t^p, X_t^u\}$, W and b are the learned parameters in the training stage.

3 Experiments

3.1 Data Processing

We evaluated the proposed DDSTN algorithm on a bimodal breast ultrasound dataset sampled by one of the authors, in which 106 patients (54 benign tumors patients and 51 malignant cancer patients) have both BUS and EUS modalities, while the other 159 patients (81 benign tumors patients and 78 malignant cancer patients) only have BUS data. The approval from the ethics committee of the hospital was obtained, and all patients had signed informed consent.

The bimodal ultrasound images were acquired by the Mindray Resona7 ultrasound scanner with the L11-3 probe by an experienced sonologist. All the malignant cancers have been proved by the pathological diagnosis. A region of interest (ROI) including the lesion region was selected by an experienced sinologist from each ultrasound image. Noting that for the paired BUS and EUS images, only the ROI in BUS image was manually selected, and the same location of ROI was then automatically mapped to EUS imaging to obtain the ROI.

3.2 Experimental Setup

The proposed DDSTN was compared with the following related or state-of-the-art TL algorithms.

1) CNN-SVM: CNN-SVM is a single-channel CNN which is compared as a baseline, we selected ResNet18 as the classification network for single-modality BUS and replace the softmax classifier with SVM.
2) CNN-SVM+ [13]: CNN-SVM+ is another baseline which consists of two-channel CNNs and an SVM+ classifier. BUS is considered as the diagnostic modality, while EUS is the source domain.
3) DDC [14]: DDC is a typical deep TL algorithm which uses the MMD criterion as the distribution distance metric.

4) DAN [15]: Deep adaptation networks (DAN) is an improved DDC algorithm that replaces the MMD with multi-kernel MMD and then calculates the multiple layer losses.

5) Deep CORAL [16]: Deep correlation alignment (Deep CORAL) is a deep TL algorithm based on correlation alignment, which learns a second-order feature transformation to minimize the feature distance between the source and the target domain.

The 3-fold cross-validation was adopted to evaluate all the algorithms. Specifically, the 106 paired data were always fixed as training data for the LUPI-based TL module, and the 159 additional BUS data were divided into three groups. We selected two of three groups of additional BUS data and all the EUS images from the 106 paired data to form another training set for the MMD-based TL module, while the remaining one BUS group was set as testing data. The experiment repeated three times. The final results were presented with the format of the mean ± SD (standard deviation).

The commonly used classification accuracy (ACC), sensitivity (SEN), specificity (SPE) and Youden index (YI) were selected evaluation indices. Moreover, the receiver operating characteristic (ROC) curve and the area under ROC curve (AUC) were also adopted for evaluation.

3.3 Experimental Results

Table 1 shows the classification results of different algorithms. It can be found that the proposed DDSTN outperforms all the compared algorithms with the best accuracy of 86.79 ± 1.54%, sensitivity of 86.45 ± 1.44%, specificity of 87.31 ± 4.37%, and YI of 73.77 ± 3.17%. DDSTN improves at least 1.92%, 2.04%, 0.4%, and 3.85% on accuracy, sensitivity, specificity and YI, respectively compared with other algorithms (Table 1).

Table 1. Classification results of different algorithms

	ACC (%)	SEN (%)	SPE (%)	YI (%)
CNN-SVM	82.34 ± 5.67	81.56 ± 4.22	84.56 ± 1.22	66.12 ± 3.51
CNN-SVM+	84.87 ± 2.85	84.41 ± 4.45	85.28 ± 1.41	69.69 ± 5.63
DDC	83.33 ± 1.44	81.53 ± 3.02	85.40 ± 4.68	66.93 ± 2.97
DAN	84.85 ± 1.11	83.01 ± 2.61	86.91 ± 3.58	69.92 ± 2.27
Deep CORAL	84.47 ± 2.31	84.07 ± 3.12	85.66 ± 2.14	69.73 ± 2.25
DDSTN (proposed)	**86.79 ± 1.54**	**86.45 ± 1.44**	**87.31 ± 4.37**	**73.77 ± 3.17**

The experiments show that CNN-SVM+ achieves superior performance to CNN-SVM, which indicates the effectiveness of transferring information from EUS for the BUS-based CAD by LUPI paradigm. It also can be found that DDSTN improves at least 1.94% on accuracy, 2.38% on sensitivity, 0.40% on specificity and 3.85% on YI compared with DDC, DAN and Deep CORAL, which indicates the effectiveness of our doubly supervised TL paradigm. Moreover, DDSTN improves 1.92%, 2.04%, 2.03%, and 4.08% on accuracy, sensitivity, specificity and YI, respectively, over CNN-SVM+, suggesting the positive effect of TL between unpaired data for learning an effective classifier.

Figure 3 shows the ROC curves and the corresponding AUC values of different algorithms. DDSTN again achieves the best AUC value of 0.871, which improves at least 0.028 over all the other algorithms.

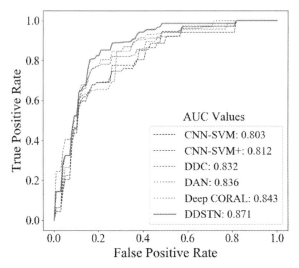

Fig. 3. ROC curves of different algorithms with the corresponding AUC values.

4   Conclusion

In summary, we propose a novel doubly supervised TL paradigm to address the issue of TL between imbalanced modalities with labeled data. The proposed DDSTN algorithm effectively performs the double supervised transfer between both the paired data with shared labels by the SVM+ paradigm and the unpaired data with different labels by the MMD criterion in a unified framework. The experimental results indicate that DDSTN outperforms all the compared algorithms on the BUS-based CAD for breast cancers.

In current work, we adopt MMD as the distribution distance metric for TL, and therefore we select DDC, DAN and Deep CORAL for comparison, since all these algorithms are developed based on MMD or MMD related criterion. In our future work, we will further improve the doubly supervised transfer network by studying other TL methods instead of MMD. Moreover, we will try to integrate the advantages of adversarial domain adaption networks in this new doubly supervised TL paradigm.

Acknowledgements. This work is supported by the National Natural Science Foundation of China (81830058, 81627804), the Shanghai Science and Technology Foundation (17411953400,

18010500600), the 111 Project (D20031) and the Nanjing Science and Technology Commission (201803027).

References

1. Cheng, H.D., Shan, J., Ju, W., Guo, Y., et al.: Automated breast cancer detection and classification using ultrasound images: a survey. Pattern Recogn. **43**(1), 299–317 (2010)
2. Sigrist, R.M., et al.: Ultrasound elastography: review of techniques and clinical applications. Theranostics **7**(5), 1303–1329 (2017)
3. Ara, S.R., et al.: Bimodal multiparameter-based approach for benign–malignant classification of breast tumors. Ultrasound Med. Biol. **41**(7), 2022–2038 (2015)
4. Pan, S.J., Yang, Q.: A survey on transfer learning. IEEE Trans. Knowl. Data Eng. **22**(10), 1345–1359 (2009)
5. Tan, C., et al.: A survey on deep transfer learning. In: ICANN, pp. 270–279 (2018)
6. Cheplygina, V., de Bruijne, M., Pluim, J.P.: Not-so-supervised: a survey of semi-supervised, multi-instance, and transfer learning in medical image analysis. Med. Image Anal. **54**, 280–296 (2019)
7. Lu, S., Lu, Z., Zhang, Y.D.: Pathological brain detection based on AlexNet and transfer learning. J. Comput. Sci. **30**, 41–47 (2019)
8. Zheng, X., Shi, J., Ying, S., Zhang, Q., Li, Y.: Improving single-modal neuroimaging based diagnosis of brain disorders via boosted privileged information learning framework. In: Wang, L., Adeli, E., Wang, Q., Shi, Y., Suk, H.-I. (eds.) MLMI 2016. LNCS, vol. 10019, pp. 95–103. Springer, Cham (2016). https://doi.org/10.1007/978-3-319-47157-0_12
9. Zheng, X., et al.: Improving MRI-based diagnosis of Alzheimer's disease via an ensemble privileged information learning algorithm. In: ISBI, pp. 456–459 (2017)
10. Vapnik, V., Vashist, A.: A new learning paradigm: learning using privileged information. Neural Netw. **22**(5–6), 544–557 (2009)
11. Pan, S.J., Tsang, I.W., Kwok, J.T.: Domain adaptation via transfer component analysis. IEEE TNN **22**(2), 199–210 (2011)
12. Kingma, D.P., Ba, J.: Adam: a method for stochastic optimization. In: ICLR (2015)
13. Li, W., Dai, D., Tan, M., et al.: Fast algorithms for linear and kernel SVM+ . In: CVPR, pp. 2258–2266 (2016)
14. Tzeng, E., et al.: Deep domain confusion: maximizing for domain invariance. arXiv preprint arXiv:1412.3474 (2014)
15. Long, M., et al.: Learning transferable features with deep adaptation networks. In: ICML, pp. 97–105 (2015)
16. Sun, B., Saenko, K.: Deep coral: correlation alignment for deep domain adaptation. In: ECCV, pp. 443–450 (2016)

2D X-Ray Mammogram and 3D Breast MRI Registration

Hossein Soleimani$^{(\boxtimes)}$ and Oleg V. Michailovich

Department of Electrical and Computer Engineering,
University of Waterloo, Waterloo, Canada
{h3soleim,olegm}@uwaterloo.ca

Abstract. X-ray mammography and breast Magnetic Resonance Imaging (MRI) are two principal imaging modalities which are currently used for detection and diagnosis of breast disease in women. Since these imaging modalities exploit different contrast mechanisms, establishing spatial correspondence between mammograms and volumetric breast MRI scans is expected to aid the assessment and quantification of different type of breast malignancies. Finding such correspondence is, unfortunately, far from being a trivial problem – not only that the images have different contrasts and dimensionality, they are also acquired under vastly different physical conditions. As opposed to many complex standard methods relying on patient-specific bio-mechanical modelling, we developed a new simple approach to find the correspondences. This paper introduces a two-stage computational scheme which estimates the global (compression dependent) part of the spatial transformation first, followed by estimating the residual (tissue dependent) part of the transformation of much smaller magnitude. Experimental results on a clinical data-set, containing 10 subjects, validated the efficiency of the proposed approach. The average Target Registration Error (TRE) on the data-set is 5.44 mm with a standard deviation of 3.61 mm.

Keywords: Breast cancer · Registration · Mammogram · Breast MRI

1 Introduction

Breast cancer is the most common malignancy diagnosed in women worldwide, with about 1.1 million cases of breast cancer diagnosed each year and the annual fatality costing 400,000 lives worldwide [1]. Since its establishment as a method of population-wise screening, X-ray mammography has helped to slash the mortality rates to a remarkable extent. The relatively low sensitivity of mammography limits its efficacy in patients with relatively dense composition of breast tissue. For this reason, in the cases of newly diagnosed breast cancer as well as in patients considered to be at an elevated risk of developing breast disease, it is nowadays a standard practice to warrant MRI examination [2]. In such cases, to improve the specificity of MRI findings as well as to facilitate biopsy,

© Springer Nature Switzerland AG 2020
A. L. Martel et al. (Eds.): MICCAI 2020, LNCS 12266, pp. 150–159, 2020.
https://doi.org/10.1007/978-3-030-59725-2_15

it is often necessary to locate the *same* lesion in the MRI and mammography scans concurrently. Establishing such correspondence requires one to find a spatial transformation that relates the coordinates of breast tissue in its pendulous and compressed states based on the imaging data alone. This problem can be conveniently formulated as problems of *image registration*, which, in the case at hand, can be further characterized as being both *cross-modal* and *cross dimensional* (CMCD). Moreover, the expected ill-posedness of CMCD formulation is further exacerbated by the effect of mechanical compression of the breast during mammography examination along with the fact that, as opposed to MRI scans, mammographic images are, in fact, projective. Hence, it hardly comes as a surprise that the range of approaches to the problem of 3D breast MRI to 2D mammography (MRI/MMG) registration remains comparatively limited, while the drawbacks of existing solutions hamper their widespread adoption into clinical practice.

The problem of MRI/MMG registration have been addressed in several studies using a range of different approaches. At a conceptual level, the solutions proposed hitherto differ in how they: a) deal with the change in the dimensionality of imaging data, b) model and estimate the geometric transformation, and c) assess image similarity. Thus, for example, to address the cross-dimensional aspects of the registration problem at hand, [3] relied on landmark-based registration of mammographic scans with 2D MRI images derived from their associated 3-D MRI volumes via radiographic projection. It was proposed in [4] to restrict all admissible transformations to a low-dimensional space of para-metric models, affine transformation, which is commonly used to describe image deformation due to shifts, rotations, scaling and shearing of spatial coordinates. Moreover, assuming the breast volume to be invariant under the change of coordinates allowed the authors to reduce the number of unknown transformation parameters (from 12 to 11), while improving the stability of overall estimation to a substantial degree. However, by its very nature, affine transformation is incapable of describing curvilinear displacements of matter, which are likely to take place in the breast under deformation.

The most promising results have been thus far obtained with the help of biomechanical Finite Element Models (FEMs), which can be used to predict the deformation of breast tissue due to compression [6,7]. While different in their finer details, all FEM-based methods share a common algorithmic structure consisting of four principal stages, *viz.*: MRI image segmentation, material modelling, computation of the displacement, and registration [5]. Note that, in this case, image segmentation is a key step required to discriminate between different types of breast tissue (e.g., adipose and fibro-glandular tissue, skin, etc.), which is critical for accuracte material modelling. Following the "compression stage", the pre-warped MRI volumes are reduced to their 2D projections by means of ray-tracing [4], followed by estimating a transformation between the latter and their associated mammograms. Thus, for example, [8] relied on a rigid-body transformation model, with Normalized Cross Correlation (NCC) used a similarity measure between the "simulated" and real mammograms.

A fully automated method has been proposed in [9] which performs a complete registration of MRI volumes and X-ray images in both directions, i.e. from MRI to mammogram and from mammogram to MRI. In [10], in contrary to other FEM-approaches, it was proposed to perform registration on the density maps extracted from both MRI and mammography scans by means of Vollpara software suit [11]. In [12], the same group of authors proposed to define the similarity measure using intensity gradients, which was shown to be much less sensitive to the difference in imaging contrasts between MRI and mammography.

One of the main disadvantages of using FEM-based methods are due to their computational complexity, high sensitivity to the results of image segmentation as well as their dependency on third-party numerical solvers. Moreover, the biomechanical models used by FEM are build individually for each subject, thus ignoring the common characteristics of the compression-related displacement of breast tissue which are likely to be shared between different cases. To overcome some of these drawbacks, this paper introduces a new approach to the problem of MRI/MMG registration that does not require the use of FEM-based modelling to account for the large-amplitude component of breast deformation. To this end, given a mammography scan, the breast boundary is used to build a 3D *reference* surface that predicts the shape of compressed breast. Subsequently, the MRI volume is registered to the reference shape (thus accounting for the major portion of breast motion), followed by estimating the residual deformation in a non-rigid intensity-based registration setting. The proposed approach has been observed to be both numerically straightforward and accurate, as supported by a series of our experiments conducted on clinical datasets.

2   Method

Let f and g denote a 3D breast MRI volume and a 2D mammography scan defined over a rectangular domain Ω, respectively. Also, let $\mathcal{P}$ denote a projection operator such that $\mathcal{P}(f)$ is defined over Ω as well. Then, given an appropriate *distance* $d(\cdot, \cdot)$ between two (planar) images, the problem of MRI/MMG registration can be formulated as one of finding the optimal spatial transformation $\phi^* \in \Phi$ which solves

$$\phi^* = \arg\min_{\phi \in \Phi} d(\mathcal{P}(f \circ \phi), g), \tag{1}$$

with $(f \circ \phi)(\mathbf{r}) = f(\phi(\mathbf{r}))$, where $\mathbf{r} \in \mathbb{R}^3$. Note that, in the above formulation, all admissible transformations are limited to the set Φ which could be, e.g., the set of topology preserving homeomorphisms.

Let ϕ be an admissible deformation which brings $\mathcal{P}(f \circ \phi)$ and g into a close correspondence w.r.t. the chosen metric d. In this work, the entire deformation ϕ is decomposed into two constituents, namely $\phi = \phi_{\text{glb}} \circ \phi_{\text{res}}$, with ϕ_{glb} and ϕ_{res} being the global and residual components of ϕ. Each of the two components is then estimated separately according to the algorithmic steps depicted in flowchart shown in Fig. 1. In the preprocessing step, the breast geometry is extracted from MRI volume, and then MRI voxels are segmented to either to

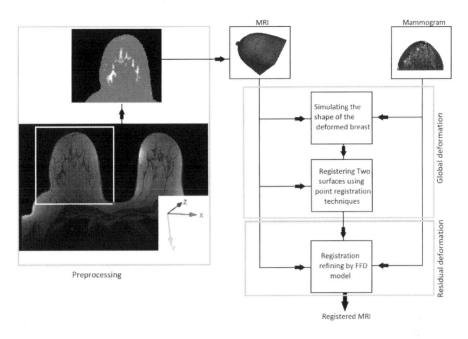

Fig. 1. Scheme of the proposed method.

adipose (fat) or fibroglandular [13]. The particular methods for estimation of ϕ_{glb} and ϕ_{res} are described next.

2.1 Estimation of Global Deformation

The global deformation of the breast during mammographic compression has many properties and characteristics which appear to be common to subjects within different breast geometry and composition. Thus, in particular, the boundary of a compressed breast can be closely approximated by a super-quadratic [18] of the form

$$\left|\frac{x}{a}\right|^r + \left|\frac{y}{b}\right|^s + \left|\frac{z}{c}\right|^t = 1, \quad y \geq 0, \tag{2}$$

where the x, y and z coordinates are aligned with the left-right, posterior-anterior and inferior-superior directions, respectively. Note that the condition $y \geq 0$ is added to keep the anterior part of the surface only.

The parameter $\theta = \{a, b, c, r, s, t\}$ control the shape of the super-quadratic and, hence, they need to be properly defined. To this end, we first notice that the projection of the super-quadratic onto the (x, y) plane is described by a simplified equation of the form $|x/a|^r + |y/b|^s = 1$. This shape can be reasonably expected to be aligned with the mammographic boundary of the breast. Consequently, the parameters $\{a, b, c, r\}$ can be estimated from mammographic data using, e.g., the heuristic optimization algorithm of [14].

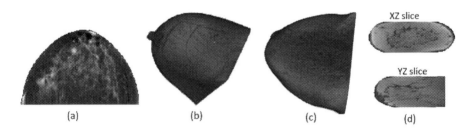

Fig. 2. Deformation of breast MRI under ϕ_{glb}. (a) Digital mammogram, (b) Original breast MRI boundary, (c) Deformed breast MRI boundary, (d) Examples of coronal (XZ) and sagittal (YZ) slices of the deformed MRI.

To estimate the remaining parameters of the super-quadratic (i.e., c and t), we take advantage of the fact that information about the distance between compression paddles is always indicated in the header of mammography (DICOM) files. Consequently, denoting this distance by h, it is straightforward that $c = h/2$. Finally, to estimate t, the compression is assumed to be volume preserving. Thus, given an estimate of the breast volume derived from 3D MRI, one can simply find a value of t yielding a super-quadratic (2) of the equal volume.

Given the original boundary of the breast (as observed in MRI scans) and its referenced "compressed" boundary (as represented by the fitted super-quadratic), the final step in estimation of ϕ_{glb} consists in finding a spatial transformation that aligns these surfaces. Note that, in practical computations, the surfaces are represented by sets of discrete point coordinates. Consequently, in this work, we took advantage of the Coherent Point Drift (CPD) point registration algorithm [15], which is a set-point registration technique allowing one to determine a spatial transformation that brings the two sets of discrete (surface) points into close correspondence with each other. This method was chosen because of the simplicity of its algorithmic structure that requires neither preprocessing nor special initialization.

It should be noted that the above method of surface registration can only be used to predict the motion of the boundary points, while remaining oblivious to what happens inside the breast mass. To overcome this problem, we extrapolate the boundary motion *inside* the breast volume by means of Thin Plate Spline (TPS) interpolation [19]. Note that this type of interpolation is guaranteed to find a spatial transformation of minimum possible bending energy, which agrees well with the general tendency of soft biological tissue to deform in the most "ergonomic" way. Figure 2 provides an illustration of the above-described process.

It should be noted that the proposed method for estimating ϕ_{glb} does not take into consideration the actual composition of breast tissue, effectively assuming it to be homogeneous. As a result, it would be unreasonable to expect ϕ_{glb} thus obtained to be sufficient to explain the real displacement. This brings us to the problem of estimation of the residual transformation ϕ_{res}, which is detailed next.

2.2   Estimation of Residual Deformation

Let the estimated global transformation be denoted by ϕ^*_{glb}. Then, the residual deformation is estimated by registering $\mathcal{P}(f \circ \phi^*_{\text{glb}})$ to g, with a proper choice of the projection operator $\mathcal{P}$. It goes without saying, such a registration task is far from being trivial on the account of vast differences in the contrast mechanisms of the images being registered. One way to overcome this problem is to subject the images to an *intensity transformation* which would make them appear as if they have been unimodal. In particular, in this work, we compare two types of such transformations which are described below.

The first type of intensity transformation was based on the method proposed in [12], which can be used to transform the intensities of $f \circ \phi^*_{\text{glb}}$ to emulate the contrast of X-ray images (before applying the projection transform $\mathcal{P}$). The second approach, on the other hand, used the thickness of dense (fibroglandular) tissue as a "contrast independent" measure of image content. In particular, the thickness measurements are straightforward to extract from MRI volumes based on the results of image segmentation, while in the case of mammograms, these measurements are straightforward to compute using the method in [16]. Note that, in this work, $\mathcal{P}$ was assumed to be a parallel-ray radiological projection, as suggested in [17]. In both cases, we applied a free-form deformation model to describe ϕ_{res}. In particular, the latter has been modelled as a linear combination of separable cubic b-splines, with their knots distributed uniformly across Ω.

Mutual Information (MI) [20,21] and Cross Correlation (CC) [22] are two common metrics being commonly used in the literature. However, since the intensities of MRI and mammogram have been normalized to be comparable, we used the Sum of Square Distance (SSD) criterion as a similarity measure in order to quantitatively asses the alignment of the images under registration. Formally, the problem of finding an optimal ϕ_{res} can be formulated as given by

$$\phi^*_{\text{res}} = \arg\min_{\phi_{\text{res}}} \frac{1}{2} \iint_{\Omega} \left| \int_z \left(f(\mathbf{r} + \phi_{\text{res}}(\mathbf{r}\,|\,\mu)) \, dz \right) - g(x,y) \right|^2 dx\,dy, \qquad (3)$$

where $\mathbf{r} = (x,y,z)$ and $\phi(\mathbf{r}|\mu) = \sum_{i=1}^M \mu_i \beta^3(\mathbf{r} - \mathbf{r}_i)$. Note that the projection operator is incorporated in (3) explicitly, i.e. as an integration over z. Also note that the separability of b-splines implies that $\beta^3(\mathbf{r}) = \beta^3(x)\beta^3(y)\beta^3(z)$.

The image deformation is defined on a sparse, regular grid of control points $(\mathbf{r}_i)$ placed over the Ω. Each control point $\mathbf{r}_i$ has an associated three-element deformation coefficient, describing the x-, y-, and z-components of the deformation. The aim of the registration is to find optimal $\mu = \{\mu_i\}_{i=1}^M$ which minimizes the SSD criterion. Differentiating the latter w.r.t. the spline parameters produces the gradient vector $\nabla S(\mu) = [\frac{\partial S(\mu)}{\partial \mu_1}, \frac{\partial S(\mu)}{\partial \mu_2}, \ldots, \frac{\partial S(\mu)}{\partial \mu_M}]^T$, whose components are given by

$$\frac{\partial S(\mu)}{\partial \mu_i} = \iint_{\Omega} \left\{ \left[\int_z f(\mathbf{r} + \phi(\mathbf{r})\,dz - g(x,y) \right] \cdot \int_z \nabla f(\mathbf{r} + \phi(\mathbf{r}\,|\,\mu))^T \frac{\partial \phi(\mathbf{r})}{\partial \mu_i} dz \right\} dx\,dy.$$
$$(4)$$

Subsequently, the minimization of SSD is carried out by means of the Gradient Descent algorithm, the iteration of which are given by

$$\mu^{(t+1)} = \mu^{(t)} - \tau \nabla S(\mu^{(t)}), \qquad (5)$$

where $\tau > 0$ is a predefined step-size. In order to avoid local optima and decrease computation time, we used a multi-grid optimization scheme, where the registration is initiated at a coarse resolution level, followed by its gradual refinement at finer resolutions. By increasing the number of control points in multi-grid scheme, the deformation may not remain locally smooth (i.e. the deformation is not feasible). To further stabilize the numerical behaviour of image registration, the SSD cost function has been augmented by an additional (regularization) term given by

$$R(\mu) = \int_{\mathbf{r} \in \Omega} \left(\|\nabla \phi_x(\mathbf{r}\,|\,\mu)\|_2^2 + \|\nabla \phi_y(\mathbf{r}\,|\,\mu)\|_2^2 + \|\nabla \phi_z(\mathbf{r}\,|\,\mu)\|_2^2 \right) d\mathbf{r}, \qquad (6)$$

where $\phi_x(\mathbf{r}|\mu)$, $\phi_y(\mathbf{r}|\mu)$ and $\phi_z(\mathbf{r}|\mu)$ indicate deformation in x, y and z directions, respectively. Again as before, one needs to take gradient of $R(\mu)$ respect to μ_i to compute the gradient of R and, eventually, the gradient of the augmented cost.

In Eq. (4) we need to compute the gradient of $f(\mathbf{x})$ at any $\mathbf{x} \in \Omega$. However, values of f are only available at a relatively small number of grid points. Therefore, an interpolation procedure is required in order to compute f at $\mathbf{x}$. To this end, we again used cubic B-splines to define f *continuously* over Ω as $f(\mathbf{x}) = \sum_j \alpha_j \beta^{(3)}(\mathbf{x} - \mathbf{x}_j)$, where α_j are the spline coefficients. To compute the gradient of f w.r.t. the latter, one needs to take the derivatives of f respect to x, y and z, i.e. $\nabla f = [\frac{\partial f}{\partial x}, \frac{\partial f}{\partial y}, \frac{\partial f}{\partial z}]^T$. Thus, for instance, the first-order derivative of f w.r.t. x is computed as

$$\frac{\partial f(\mathbf{r})}{\partial x} = \sum_j \alpha_j \left(\left. \frac{d\beta^{(3)}(u)}{du} \right|_{u=x-x_j} \beta^{(3)}(y - y_j) \beta^{(3)}(z - z_j) \right), \qquad (7)$$

where $\frac{d\beta^{(3)}(u)}{du} = \beta^{(2)}\left(u + \frac{1}{2}\right) - \beta^{(2)}\left(u - \frac{1}{2}\right)$.

3   Experimental Results

To validate our proposed framework, we used a clinical dataset containing 10 clinical cases from 10 different subjects. Each case consisted of one MRI volume and two mammographic images (i.e., two projections in the cranio-caudal (CC) and medio-lateral oblique directions). All the subjects in our database had a unilateral breast lesion. The images were acquired approximately at the same time point to avoid significant change of tissues inside the breast. Breast MRI scans were acquired at the Princess Margaret Cancer Center (Toronto, Canada) with a 3T $Signa^{TM}$ Premier MRI scanner (GE Healthcare, Inc.). MRI volumes had a size of $[448 \times 448 \times 210]$ voxels and $[0.76 \times 0.76 \times 1.2]\,\text{mm}^3$ per voxel. Mammograms were, acquired at the same center, composed of $[2294 \times 1914]$

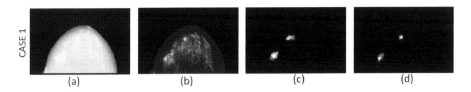

CASE 1

(a)   (b)   (c)   (d)

Fig. 3. Registration results for case 1. (a) is the mammogram, (b) is the thickness image computed from mammogram, (c) shows aligned tumors using pseudo-CT image and (d) shows aligned tumors using fibroglandular thickness images (Color figure online)

pixels, with $[0.094 \times 0.094]\,\mathrm{mm}^2$ per pixel. Both MRI and mammograms were re-sampled to 1 mm resolution. Furthermore, histogram equalization was applied to increase the contrast of the glandular tissue. In the experiments reported in this paper, we focused only on registering breast MRI to CC view.

The TRE criterion has been used to quantitatively assess the accuracy of the proposed framework. To compute TRE, it is necessary to have reference points in both images, which usually landmarks are used in the literature. We used the lesion centres as landmarks to compute registration error. All the lesions have been delineated by a trained radiologist, with the resulting segmentations used as a reference for error computation. The TRE figures have been computed as the Euclidean distance between the centroid of the 2D lesion in the mammogram and the centroid of the same lesion in the projected MRI.

As it was mentioned before, we applied two approaches to transform the intensity of breast MRI images to be comparable with mammograms: a) building "emulated" X-ray images (pseudo-CT) from MRI volumes and projecting then onto the x-y plane, and b) computing the fibroglandular thickness from both MRI images and mammograms. Figure 3 shows the registration results for case 1. Lesions are approximately shown by a red circle in thickness image. deformed and projected lesions are shown with green colour, while purple colour shows the lesion mask annotated from the mammogram. The overlap between the two lesions is white. As it can be seen, in both cases, the projected lesions from MRI appear to be very close to the lesions visible in the mammograms.

Table 1. Target registration error, in millimeters, for the 10 CC-mammograms using pseudo-CT and thickness image

	C 1	C 2	C 3	C 4	C 5	C 6	C 7	C 8	C 9	C 10	Mean	Std
pseudo-CT	11.63	8.15	13.4	5.4	2.45	6.24	6.74	12.01	5.33	3.58	7.49	3.72
Thickness	0.54	8.27	11.21	10.66	3.29	2.44	6.91	3.13	3.78	4.25	5.44	3.61

Table 1 summarizes the registration results on all 10 cases in dataset. The obtained TRE for registration by thickness images is 5.44 ± 3.61 mm and it is 7.49 ± 3.72 mm for registration using pseudo-CT images. These values are comparable with TRE obtained from FEM-based method (shown in Table 2) in the

Table 2. TRE of recent FEM-based methods in the literature

Method	Garcia [12]	Garcia [10]	Mertz [8]	Sloves [9]	Mertz [17]
TRE	$9.02 \pm 4.28\,\text{mm}$	$5.65 \pm 2.78\,\text{mm}$	$11.6 \pm 3.8\,\text{mm}$	$4.2 \pm 1.9\,\text{mm}$	$12.7\,\text{mm}$

literature. All of these methods use image intensity to optimize the objective function over registration parameters. Note that there is no standard database and one should be cautious comparing the results provided by previous works. However, comparing the TRE figures obtained by our approach as well as by the FEM-based methods, we can see that results are of the same order of accuracy. On the other hand, the computational time of the proposed registration (including segmentation, surface fitting and registration refinement) using MATLAB on CPU (Intel(R) CORE (TM) -i7 6500U) is about 40 min on average. At the same time, FEM-based methods can complete a single registration in an hour or few hours, even using numerical accelerations by means of GPUs. Finally, the proposed approach is fully automatic and can be executed on a standard PC at a reasonable time.

4   Conclusion

In this paper, we introduced a new registration framework to align breast MRIs and X-ray mammograms. We used surface registration and the FDD model to estimate the breast deformation in mammography. The proposed solution is simpler than FEM-based methods and requires less computational resources. The average target registration error of the presented registration approach was less than 6 mm which is an assumable error in the clinical practice with the aim of localizing susceptible areas within the MRI or the mammogram. Our approach is automatic and can run in an acceptable time with regular CPUs in hospitals.

References

1. Breast Cancer homepage. https://www.breastcancer.org/symptoms/understand-bc/statistics. Last Accessed 13 February 2019
2. Monticciolo, D.L., Newell, M.S., Moy, L., Niell, B., Monsees, B., Sickles, E.A.: Breast cancer screening in women at higher-than-average risk: recommendations from the ACR. J. Am. Coll. Radiol. **15**(3), 408–414 (2018)
3. Behrenbruch, C.P., et al.: Fusion of contrast-enhanced breast MR and mammographic imaging data. Med. Image Anal. **7**(3), 311–340 (2003)
4. Mertzanidou, T., et al.: MRI to X-ray mammography registration using a volume-preserving affine transformation. Med. Image Anal. **16**(5), 966–975 (2012)
5. García, E., et al.: A step-by-step review on patient-specific biomechanical finite element models for breast MRI to x-ray mammography registration. Med. Phys. **45**(1), e6–e31 (2018)
6. Ruiter, N.V., Stotzka, R., Muller, T.O., Gemmeke, H., Reichenbach, J.R., Kaiser, W.A.: Model-based registration of X-ray mammograms and MR images of the female breast. IEEE Trans. Nuclear Sci. **53**(1), 204–211 (2006)

7. Lee, A.W., Rajagopal, V., Gamage, T.P.B., Doyle, A.J., Nielsen, P.M., Nash, M.P.: Breast lesion co-localisation between X-ray and MR images using finite element modelling. Med. Image Anal. **17**(8), 1256–1264 (2013)

8. Mertzanidou, T.: MRI to X-ray mammography intensity-based registration with simultaneous optimisation of pose and biomechanical transformation parameters. Med. Image Anal. **18**(4), 674–683 (2014)

9. Solves-Llorens, J. A., Rupérez, M. J., Monserrat, C., Feliu, E., García, M., Lloret, M.: A complete software application for automatic registration of x-ray mammography and magnetic resonance images. Med. Phys. **41**(8Part1), 081903 (2014)

10. Garcia, E., et al.: Multimodal breast parenchymal patterns correlation using a patient-specific biomechanical model. IEEE Trans. Med. Imaging **37**(3), 712–723 (2017)

11. Vollpara package. https://volparasolutions.com/science-hub/breast-density/measuring-breast-density

12. García, E., et al.: Breast MRI and X-ray mammography registration using gradient values. Med. Image Anal. **54**, 76–87 (2019)

13. Soleimani, H., Rincon, J., Michailovich, O.V.: Segmentation of breast MRI scans in the presence of bias fields. In: Karray, F., Campilho, A., Yu, A. (eds.) ICIAR 2019. LNCS, vol. 11662, pp. 376–387. Springer, Cham (2019). https://doi.org/10.1007/978-3-030-27202-9_34

14. Rao, R.V., Savsani, V.J., Vakharia, D.P.: Teaching-learning-based optimization: a novel method for constrained mechanical design optimization problems. Comput. -Aided Des. **43**(3), 303–315 (2011)

15. Myronenko, A., Song, X.: Point set registration: coherent point drift. IEEE Trans. Pattern Anal. Mach. Intell. **32**(12), 2262–2275 (2010)

16. Highnam, R., Brady, S.M., Yaffe, M.J., Karssemeijer, N., Harvey, J.: Robust breast composition measurement - Volpara^{TM}. In: Martí, J., Oliver, A., Freixenet, J., Martí, R. (eds.) IWDM 2010. LNCS, vol. 6136, pp. 342–349. Springer, Heidelberg (2010). https://doi.org/10.1007/978-3-642-13666-5_46

17. Mertzanidou, T., Hipwell, J., Han, L., Huisman, H., Karssemeijer, N., Hawkes, D.: MRI to X-ray mammography registration using an ellipsoidal breast model and biomechanically simulated compressions. In MICCAI Workshop on Breast Image Analysis, pp. 161–168 (2011)

18. Barr, A.H.: Superquadrics and angle-preserving transformations. IEEE Comput. Graph. Appl. **1**(1), 11–23 (1981)

19. Bookstein, F.L.: Principal warps: thin-plate splines and the decomposition of deformations. IEEE Trans. Pattern Anal. Mach. Intell. **11**(6), 567–585 (1989)

20. Soleimani, H., Khosravifard, M.A.: Reducing interpolation artifacts for mutual information based image registration. J. Med. Sig. Sensors **1**(3), 177–183 (2011)

21. Dowson, N., Kadir, T., Bowden, R.: Estimating the joint statistics of images using nonparametric windows with application to registration using mutual information. IEEE Trans. Pattern Anal. Mach. Intell. **30**(10), 1841–1857 (2008)

22. Luo, J., Konofagou, E.E.: A fast normalized cross-correlation calculation method for motion estimation. IEEE Trans. Ultrason. Ferroelectr. Freq. Control **57**(6), 1347–1357 (2010)

A Second-Order Subregion Pooling Network for Breast Lesion Segmentation in Ultrasound

Lei Zhu[1,5], Rongzhen Chen[2], Huazhu Fu[3], Cong Xie[2], Liansheng Wang[2(✉)], Liang Wan[1,4], and Pheng-Ann Heng[5]

[1] College of Intelligence and Computing, Tianjin University, Tianjin, China
[2] Department of Computer Science at School of Informatics, Xiamen University, Xiamen, China
lswang@xmu.edu.cn
[3] Inception Institute of Artificial Intelligence (IIAI), Abu Dhabi, UAE
[4] Medical College of Tianjin University, Tianjin , China
[5] Department of Computer Science and Engineering, The Chinese University of Hong Kong, Shatin, Hong Kong

Abstract. Breast lesion segmentation in ultrasound images is a fundamental task for clinical diagnosis of the disease. Unfortunately, existing methods mainly rely on the entire image to learn the global context information, which neglects the spatial relation and results in ambiguity in the segmentation results. In this paper, we propose a novel second-order subregion pooling network (S^2P-Net) for boosting the breast lesion segmentation in ultrasound images. In our S^2P-Net, an attention-weighted subregion pooling (ASP) module is introduced in each encoder block of segmentation network to refine features by aggregating global features from the whole image and local information of subregions. Moreover, in each subregion, a guided multi-dimension second-order pooling (GMP) block is designed to leverage additional guidance information and multiple feature dimensions to learn powerful second-order covariance representations. Experimental results on two datasets demonstrate that our proposed S^2P-Net outperforms state-of-the-art methods.

Keywords: Ultrasound image · Breast lesion segmentation · Second-order subregion pooling

1 Introduction

Breast cancer is a leading cause of women death in the world [17]. Ultrasound imaging is an useful tool for breast cancer detection in clinical due to its versatility, safety and high sensitivity [18]. Segmenting breast lesions from ultrasound imaging is an important step of computer-aided diagnosis systems, which

---

L. Zhu and R. Chen—Joint first authors of this work.

© Springer Nature Switzerland AG 2020
A. L. Martel et al. (Eds.): MICCAI 2020, LNCS 12266, pp. 160–170, 2020.
https://doi.org/10.1007/978-3-030-59725-2_16

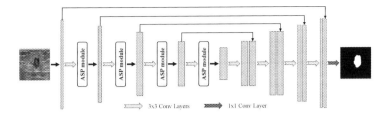

Fig. 1. Schematic illustration of the proposed breast ultrasound lesion segmentation network. Please see Fig. 2 for the ASP module. Best viewed in color. (Color figure online)

assists radiologists in the ultrasound-based breast cancer diagnosis [22]. However, accurate segmentation in ultrasound images is challenging due to the missing/ambiguous boundary, the inhomogeneous intensity distribution of breast ultrasound image, the similar visual appearance between lesions and non-lesion backgrounds, as well as the irregular shapes and complex variants of breast lesions [10,22].

Early methods [2,16,20] mainly examine hand-crafted features for inferring breast lesion boundaries in ultrasound images. Madabhushi [14] incorporated empirical domain knowledge from radiologists and low-level image features (e.g., texture, intensity and directional gradient) into a deformable shape-based model for the segmentation. Gómez-Flores et al. [6] segmented breast lesions by analyzing textures of ultrasound images. Later, several methods based on convolution neural networks (CNNs) [10,21] have achieved superior breast lesion performance than early methods by learning discriminative features from annotated images. Yap et al. [22] investigated the patch-based LeNet, U-Net, and transfer learning with a pre-trained FCN-AlecNet for segmenting breast ultrasound lesions. Lei et al. [10] designed a boundary regularized encoder-decoder network for predicting segmentation maps. Unfortunately, due to diverse ambiguous boundaries and complex shape variances, existing CNN-based methods mainly rely on the entire ultrasound image to learn the global context information, which neglects the spatial relation and results in ambiguity in the segmentation results.

In this work, we develop a second-order subregion pooling network (S^2P-Net) for boosting breast lesion segmentation performance by aggregating the multi-context information from both the whole image and multiple subregions. The contributions of this work could be summarized as: **1)** A new second-order subregion pooling network is proposed for boosting breast lesion segmentation in ultrasound images. **2)** The ASP module is utilized to attentively aggregate global and multiple subregion representations for inferring breast lesion regions. **3)** The GMP block is designed to leverage additional guidance information and all three feature dimensions for modeling higher-order statistics for more discriminative image representations. **4)** Moreover, experimental results on two datasets

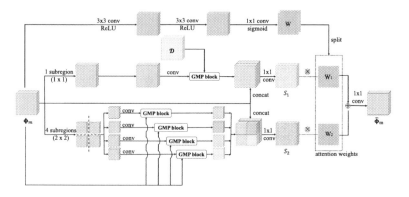

Fig. 2. Schematic illustration of the ASP module at the m-th CNN layer. Φ_m is the feature map at the m-th CNN layer. Please see Fig. 3 for the details of the GMP block. Note that W has 2 channels. The first channel is W_1 while the second channel is W_2. $\mathcal{D}$ is an unit matrix (all elements are 1). Best viewed in color. (Color figure online)

demonstrate that our network sets a new state-of-the-art performance on breast lesion segmentation in ultrasound images.

2   Proposed Method

Figure 1 illustrates the architecture of the proposed breast lesion segmentation network, which takes a breast lesion ultrasound image as the input and produces a segmentation resultant image of breast lesions. In our method, the U-Net architecture [15] is utilized as the backbone, which consists of the encoder and decoder paths. Each CNN block of the encoder contains two 3×3 convolutional layers and a max-pooling layer (stride $= 2$), and decoder blocks have two 3×3 convolutional layers and a upsampling operation. After each CNN block of the encoder, we introduce an attention-weighted subregion pooling (ASP) module (see Fig. 2) to refine the CNN features by learning second-order statistics from multiple subregions. In each subregion, a guided multi-dimension second-order pooling (GMP) block (see Fig. 3) is designed to leverage guidance information for learning second-order covariance matrices. Then, we iteratively merge two adjacent layers by first upsampling the low-resolution feature map and then applying two 3×3 convolutional layers. Finally, the feature map with the largest spatial resolution is used to predict the final segmentation result.

2.1   Attention-Weighted Subregion Pooling (ASP) Module

Existing methods mainly rely on the entire ultrasound image to learn a global information for inferring breast ultrasound lesions, which loses the spatial relations and tends to contain non-lesion regions or loss parts of breast lesions in the segmentation result. We develop an ASP module (see Fig. 2) to fuse the

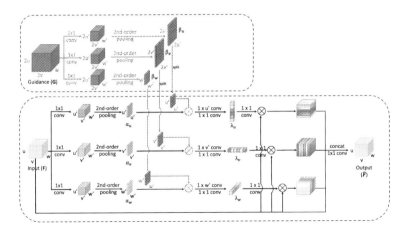

Fig. 3. Schematic illustration of the GMP block. Best viewed in color. (Color figure online)

global second-order features from the whole image and the local second-order features from multiple subregions together. Intuitively, image subregions have less non-breast-lesion details than the whole input image, and thus reduce the interference from non-lesion regions, resulting in a superior segmentation performance via our ASP modules.

Figure 2 shows the detailed architecture of the ASP module at m-th CNN layer. It refines the feature map (denoted as $\mathbf{\Phi}_m$) at m-th CNN layer by aggregating features learned from different subregion separation branches. The first branch of the ASP module uses a 1×1 separation on $\mathbf{\Phi}_m$. It passes the whole $\mathbf{\Phi}_m$ to a GMP block to learn a second-order feature map, concatenates the resultant features with $\mathbf{\Phi}_m$, and uses a 1×1 convolutional layer to produce a new feature map (denoted as S_1). The second branch separates the input $\mathbf{\Phi}_m$ into 4 (2×2) subregions and then extracts second-order features from each subregion by feeding its feature map into a GMP block. The resultant second-order features from the four subregions are then combined together to form a feature map, which is merged with $\mathbf{\Phi}_m$ by using a concatenation and a 1×1 convolution to produce a feature map (denoted as S_2). To fuse the two feature maps, we generate an attention map (denoted as W; 2 channels) by employing two successive convolutional layers (with 3×3 kernels) followed by a ReLU non-linear operation [9,26,27], and the third layer (with 1×1 kernels and a sigmoid activation layer). Finally, we split W into two maps (W_1 and W_2), multiply them with S_1 and S_2, add two multiplication results, and apply a 1×1 convolutional layer on the addition result to produce the output feature map (denoted as $\hat{\mathbf{\Phi}}_m$) of the ASP module. Mathematically, we can compute $\hat{\mathbf{\Phi}}_m$ as:

$$\hat{\mathbf{\Phi}}_m = f_{conv}(W_1 \times S_1 + W_2 \times S_2) , \tag{1}$$

where f_{conv} is the 1×1 convolutional parameters.

2.2   Guided Multi-dimension Second-Order Pooling (GMP) Block

Breast lesion segmentation in ultrasound images is challenging due to the ambiguous boundary, the inhomogeneous intensity distribution, and the complex variants of breast lesions. CNN features in existing methods almost are first-order, and tend to produce unsatisfactory results of the challenging breast lesion segmentation in ultrasound. To enhance the segmentation accuracy, our work introduces the second-order information into an end-to-end breast lesion segmentation network. In this regard, we devise a novel guided multi-dimension second-order pooling (GMP) block to leverage an entire feature map as a guidance to provide more spatial information from other subregions for helping to learn second-order covariances in each subregion of the entire features. Moreover, instead of considering only channel dimension, GMP further augments the resulting second-order covariances by exploring element statistical dependencies from all three feature dimensions. Hence, GMP enables our method to better identify breast lesions than original second-order pooling [5]; see the ablation study results in Table 3.

Given a 3D CNN feature map $\mathbf{F}$ (size: $u \times v \times w$), original second-order pooling block [5] learns a $w \times w$ region covariance matrix from $\mathbf{F}$ by reformulating $\mathbf{F}$ as a set of points $\{P_{i,j}, 1 \leq i \leq u; 1 \leq j \leq v\}$, and each point $P_{i,j}$ is a vector with w elements. The resulting covariance matrix represents the pair-wise channel correlations among all w elements, and has a clear physical meaning, i.e., its k-row ($\{1 \leq k \leq w\}$) indicates the statistical dependencies of the k-th element (channel) with all w elements (channels). We argue that only exploring element dependencies along the channel dimension suffers from a limited capability to capture the second-order statistics, due to ignoring other two feature dimensions. On the other hand, as presented in classical guided filters [8,23], incorporating additional information from a guidance image generates a better filtering behavior.

In this regard, our GMP block leverages all three dimensions to learn the second-order covariance matrix by considering additional information from another feature map (we call it "guidance feature map"). Before going into details of GMP block, we first present which information is employed as the guidance feature map. As shown in Fig. 2, our ASP module uses two branches to divide the input features ($\mathbf{\Phi}_m$). In the 1×1 subregion branch, we set the guidance feature map (denoted as $\mathcal{D}$ in Fig. 2) as an all-1 3D feature map (all elements are 1) to bypass the filtering, while the guidance feature map for all the four subregions in the 2×2 subregion branch is set as $\mathbf{\Phi}_m$. The reason behind is that $\mathbf{\Phi}_m$ has information of all the four subregions, and thus can provide additional spatial relations in other three subregions for each GMP block to better learn second-order statistics.

Figure 3 shows the architecture of our GMP block, which takes the feature map $\mathbf{F}$ (size: $u \times v \times w$) and guidance feature map $\mathbf{G}$ (size: $2u \times 2v \times w$) as two inputs and produces a refined feature map $\hat{\mathbf{F}}$. Specifically, we first apply a 1×1 convolution layer on $\mathbf{F}$ to reduce its feature dimension size into $u^{'} \times v^{'} \times w^{'}$ ($u^{'} < u; v^{'} < v; w^{'} < w$) for saving the computational cost, and take three copies of the resultant features for learning three region covariance representations along u, v and w dimensions: a $u' \times u'$ covariance matrix (α_u), a $v' \times v'$ covariance matrix (α_v), and a $w' \times w'$ covariance matrix (α_w). To do so, apart from considering the w dimension in original second-order pooling [5], we reformulate $\mathbf{F}$ into $v' \times w'$ points $\{Q_{i,j}, \text{ where } 1 \leq i \leq v'; 1 \leq j \leq w'\}$, and the vector $Q_{i,j}$ has u' elements. After that, we learn a $u' \times u'$ covariance matrix to capture the statistical dependencies among all u' elements in the vector $Q_{i,j}$. Moreover, we decompose $\mathbf{F}$ into $u' \times w'$ vectors $\{R_{i,j}, \text{ where } 1 \leq i \leq u'; 1 \leq j \leq w'\}$, and each $R_{i,j}$ has v' elements. Then, we learn a $v' \times v'$ covariance matrix to compute the second-order statistics of all v' elements of $R_{i,j}$.

Similarly, we can learn three region covariance matrices from $\mathbf{G}$ along its three dimensions. To this end, we first reduce the dimension of $\mathbf{G}$ to $2u' \times 2v' \times w'$ by applying a 1×1 convolutional layer on $\mathbf{G}$ and produce three covariance matrices (i.e., a $2u' \times 2u'$ covariance matrix (β_u), a $2v' \times 2v'$ covariance matrix (β_v), and a $w' \times w'$ covariance matrix (β_w)) from three copies of the resized feature map. Once obtaining three covariance matrices from $\mathbf{F}$ and $\mathbf{G}$ respectively, we multiply them (e.g., $\alpha_w \times \beta_w$) together to integrate the guidance features for learning guided second-order covariance matrices. Note that β_u $(2u' \times 2u')$ and β_v $(2v' \times 2v')$ have not the same size of α_u $(u' \times u')$ and α_v $(v' \times v')$. Hence, according to the 2×2 subregion partition manner in Fig. 2, we similarly split β_u into four $u' \times u'$ subregion matrices, select the corresponding one (subregion) with the size of $u' \times u'$, and multiply it with α_u. The same operations are applied to β_v and α_v. After that, following [5], we use a row-wise convolution, a 1×1 convolution, and a sigmoid activation function on each resultant covariance matrix to produce three statistic weight vector $(\lambda_u, \lambda_v, \text{ and } \lambda_w)$, which are then multiplied with the input $\mathbf{F}$ for scaling different channels, in order to emphasize useful channel and suppress bad channel information for detecting breast lesion boundaries. Finally, we concatenate the scaled feature maps from three dimensions and the input $\mathbf{F}$, followed by a 1×1 convolution, to generate the output feature map $(\hat{\mathbf{F}})$ of the developed GMP block.

Implementation Details. We train our network from scratch and all the network parameters are initialized by a normal distribution. All the training images are randomly rotated, cropped, and horizontally flipped for data augmentation. The focal loss [12] is employed to compute the total loss of our network, and we utilize the Adam optimizer to minimize our total loss for training the whole framework with $15,000$ iterations. The learning rate is initialized as 0.0001 and then reduced to 0.00001 at $3,000$ iterations. We implement our network on Keras and train it on two GPUs with a mini-batch size of 4. Our method In our experiments, we empirically set $u'=24$, $v'=24$, and $w'=128$,

3 Experiments

Datasets. We used two breast ultrasound image datasets to evaluate the effectiveness of the proposed network. The first one is a public dataset, BUSI [1],

Table 1. The results (mean ± variance) of different methods on our dataset.

Method	#paras (M)	Dice ↑	ADB ↓	Jaccard ↑	Precision ↑	Recall ↑
FCN [13]	513	0.8289±0.0007	7.1340±1.4860	0.7461±0.0008	0.8600±0.0005	0.8465±0.0002
U-Net [15]	30	0.7907±0.0003	14.428± 3.2128	0.7020±0.0004	0.8212± 0.0004	0.8168±0.0003
U-Net++ [25]	35	0.7933±0.0003	13.9142±3.0879	0.7036±0.0005	0.8300±0.0004	0.8101±0.0002
FPN [11]	52	0.8336±0.0007	7.6704±4.8599	0.7548±0.0009	0.8742±0.0009	0.8438±0.0001
DeeplabV3+ [4]	159	0.8180±0.0007	8.5677±3.4262	0.7318±0.0009	0.8533±0.0008	0.8271±0.0002
GSoP [5]	48	0.8361±0.0005	8.4699±5.6814	0.7540±0.0006	0.8640±0.0012	0.8473±0.0004
ConvEDNet [10]	310	0.8428±0.0003	6.2834±1.3013	0.7652±0.0004	0.8847±0.0005	0.8485±0.0003
Our method	96	**0.8967±0.0001**	**5.5967±0.7930**	**0.8311±0.0001**	**0.9072±0.0003**	**0.9047±0.0002**

Table 2. The results (mean ± variance) of different methods on the BUSI dataset.

Method	#paras (M)	Dice ↑	ADB ↓	Jaccard ↑	Precision ↑	Recall ↑
FCN [13]	513	0.8259±0.0120	16.4322±0.7647	0.7412±0.0089	0.8551± 0.0093	0.8435±0.0173
U-Net [15]	30	0.7660±0.0041	28.8543±4.0009	0.6710±0.0058	0.8326±0.0148	0.7799±0.0060
U-Net++ [25]	35	0.7621±0.0053	33.0030±2.1888	0.6664±0.0042	0.8267±0.0060	0.7802±0.0059
FPN [11]	52	0.7993±0.0033	22.5294±1.4860	0.7012±0.0083	0.8352±0.0171	0.8254±0.0154
DeeplabV3+ [4]	159	0.8210±0.0056	14.6643±1.3872	0.7321±0.0024	0.8648±0.0113	0.8266±0.0100
GSoP [5]	48	0.8309±0.0062	13.0799±1.0415	0.7455±0.0050	0.8745±0.0109	0.8311±0.0044
ConvEDNet [10]	310	0.8254±0.0015	14.6643±1.8361	0.7386±0.0056	0.8408±0.0136	0.8516±0.0109
Our method	96	**0.8470±0.0094**	**11.1760±0.9436**	**0.7639±0.0107**	**0.8762±0.0081**	**0.8551±0.0100**

from the Baheya Hospital for Early Detection & Treatment of Womens' Cancer (Cairo, Egypt). It has a total of 780 tumor images from 600 female patients (25–75 years old). Second, we collected 632 breast ultrasound images from Shenzhen People's Hospital to build the second dataset for evaluation, and informed consent forms were obtained from all patients. We invited experienced clinicians to manually annotate the breast lesion regions of each image. Moreover, we further adopt the five-folder cross-validation to statistically test different segmentation methods on the two datasets.

Evaluation Metrics. We employ widely-used segmentation metrics for quantitatively comparing different methods. They are Dice Similarity Coefficient (Dice), Average Distance of Boundaries (ADB, in pixel), Jaccard, Precision, and Recall; see [3,7,19,24] for details of these five metrics. A better segmentation result shall have smaller ADB and larger values for all other four metrics.

3.1   Segmentation Performance

We validate our segmentation network by comparing with seven state-of-the-art methods, including the fully convolutional network (FCN) [13], U-Net [15], U-Net++ [25], feature pyramid network (FPN) [11], DeeplabV3+ [4], a very recent second-order method (i.e., GSoP [5]), and a recent ultrasound breast lesion segmentation method (i.e., ConvEDNet [10]). For a fair comparison, we obtain the segmentation results of all the competitors by exploiting its public implementations or implementing them by ourselves, and fine-tuning the network training parameters for best segmentation results.

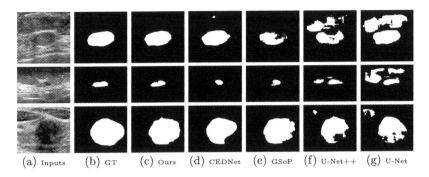

(a) Inputs   (b) GT   (c) Ours   (d) CEDNet   (e) GSoP   (f) U-Net++   (g) U-Net

Fig. 4. Comparing segmentation maps produced by different methods. (a) Breast ultrasound lesion images. (b) Ground truths (denoted as GT). (c)–(g): Segmentation results produced by our method, ConvEDNet [10] (denoted as CEDNet), GSoP [5], U-Net++ [25], and U-Net [15], respectively.

Table 3. Metric results of different components on the BUSI dataset.

Method	Dice ↑	ADB ↓	Jaccard ↑	Precision ↑	Recall ↑
w/o-GMP	0.801 + 0.0024	21.1114 + 3.5251	0.7113 + 0.0034	0.8436 + 0.0143	0.8199 + 0.0116
w/o-subregions	0.8315 + 0.0023	13.7236 + 1.8757	0.7463 + 0.0036	0.8635 + 0.0055	0.8397 + 0.0080
Ours-channelRC	0.8341 + 0.0151	12.9171 + 2.4227	0.7488 + 0.0170	0.8672 + 0.0145	0.8422 + 0.0171
w/o-guidance	0.8410 + 0.0098	14.2395 + 3.4063	0.7547 + 0.0148	0.8709 + 0.0193	0.8485 + 0.0186
Our method	**0.8470 ± 0.0094**	**11.1760 ± 0.9436**	**0.7639 ± 0.0107**	**0.8762 ± 0.0081**	**0.8551±0.0100**

Quantitative Comparisons. Tables 1 and 2 report the metric results of different segmentation methods on our collected dataset and BUSI dataset, respectively. Apparently, our method consistently and stably has the superior performances of the mean and variance values of all the five metrics over all the competitors. It indicates that our method has more accurate segmentation results than all the competitors.

Visual Comparisons. Figure 4 visualizes the segmentation results produced by different methods. Apparently, the compared methods tend to neglect some details of the breast lesion regions or include other non-lesion regions into their predicted segmentation results, while our method can more accurately detect the blurry breast lesion boundaries for the input images (1st row), and can better detect the whole breast lesion regions for the images (last two rows) with multiple intensity distributions. The superior segmentation results of our method show that our subregion based second-order features have more discriminative capabilities in inferring breast lesion regions from ultrasound images. Apparently, our network is more complex than standard UNet, but the GMP blocks first downsample the resolutions of the input feature map and the guidance features a lot for computing second-order statistics. Thus our network does not increase the inference time too much. As shown in Table 1, our method only has about a half model size of the Deeplabv3+, but achieves a better breast lesion segmentation performance.

3.2   Ablation Study

We conduct the ablation study experiments to verify the major components in our network design. Here, we consider four baseline networks, and report their quantitative results on the BUSI dataset in Table 3. (i) The first baseline (denoted as "w/o-GMP") removes all the GMP modules (see Fig. 3) from our network, while (ii) the second baseline (denoted as 'w/o-subregions') removes the 2×2 subregions (the second branch of Fig. 2) from our network, meaning that we do not model the second-order representation from any subregion. (iii) The third baseline (denoted as "Ours-1channelRC") is to replace our GMP block (see Fig. 3) with the original second-order block [5], which uses only the feature channel dimension. (iv) The last baseline (denoted as "w/o-guidance") is to remove the guidance features from our GMP block (see Fig. 3).

Table 3 shows the comparison results. Apparently, our method has superior metric results than "w/o-GMP" and "w/o-subregions", which demonstrates that both ASP module and GMP block have the contributions to the superior segmentation results of our method. Our method can also more accurately segment breast lesions than "Ours-channelRC" and "w/o-guidance", showing that both guidance feature and the multiple feature dimension help to learn more powerful second-order features.

4   Conclusion

This paper presents a second-order subregion network for the breast lesion segmentation from an ultrasound image by harnessing second-order statistics from multiple feature subregions. Our key idea is to develop an ASP module at each CNN layer to aggregate global features from the whole image and local high-order features from multiple subregions, and a GMP block in each subregion to leverage additional guidance information and all the three feature dimensions for learning powerful second-order covariance features. Experiments on two datasets demonstrate that our method clearly outperforms state-of-the-art methods. In addition, the proposed segmentation network has the potential for other similar medical image segmentation tasks, e.g., the prostate segmentation. Although our method obtained best performance in two datasets (i.e., the Dice score is 0.847 for BUSI, and 0.8967 for our dataset), there is a large room to further improve the segmentation accuracy.

Acknowledgements. This work was supported by National Natural Science Foundation of China (Project No. 61902275, 61671399), the Fundamental Research Funds for the Central Universities (Grant No. 20720190012), and Hong Kong Innovation and Technology Fund (GHP/002/13SZ and GHP/003/11SZ). We thank Yunzhu Wu for her efforts of data collection and annotations.

References

1. Al-Dhabyani, W., Gomaa, M., Khaled, H., Fahmy, A.: Dataset of breast ultrasound images. Data Brief **28**, 104863 (2020)

2. Boukerroui, D., Basset, O., Guerin, N., Baskurt, A.: Multiresolution texture based adaptive clustering algorithm for breast lesion segmentation. Eur. J. Ultrasound **8**(2), 135–144 (1998)

3. Chang, H.H., Zhuang, A.H., Valentino, D.J., Chu, W.C.: Performance measure characterization for evaluating neuroimage segmentation algorithms. Neuroimage **47**(1), 122–135 (2009)

4. Chen, L.C., Zhu, Y., Papandreou, G., Schroff, F., Adam, H.: Encoder-decoder with atrous separable convolution for semantic image segmentation. In: ECCV, pp. 801–818 (2018)

5. Gao, Z., Xie, J., Wang, Q., Li, P.: Global second-order pooling convolutional networks. In: CVPR, pp. 3024–3033 (2019)

6. Gómez-Flores, W., Ruiz-Ortega, B.A.: New fully automated method for segmentation of breast lesions on ultrasound based on texture analysis. Ultrasound Med. Biol. **42**(7), 1637–1650 (2016)

7. Gu, Z., et al.: CE-Net: context encoder network for 2D medical image segmentation. IEEE Trans. Med. Imaging **38**(10), 2281–2292 (2019)

8. He, K., Sun, J., Tang, X.: Guided image filtering. IEEE Trans. Pattern Anal. Mach. Intell. **35**(6), 1397–1409 (2012)

9. Krizhevsky, A., Sutskever, I., Hinton, G.E.: ImageNet classification with deep convolutional neural networks. In: Advances in Neural Information Processing Systems (NIPS), pp. 1097–1105 (2012)

10. Lei, B., et al.: Segmentation of breast anatomy for automated whole breast ultrasound images with boundary regularized convolutional encoder-decoder network. Neurocomputing **321**, 178–186 (2018)

11. Lin, T.Y., Dollár, P., Girshick, R., He, K., Hariharan, B., Belongie, S.: Feature pyramid networks for object detection. In: CVPR, pp. 2117–2125 (2017)

12. Lin, T.Y., Goyal, P., Girshick, R., He, K., Dollár, P.: Focal loss for dense object detection. In: ICCV, pp. 2980–2988 (2017)

13. Long, J., Shelhamer, E., Darrell, T.: Fully convolutional networks for semantic segmentation. In: CVPR, pp. 3431–3440 (2015)

14. Madabhushi, A., Metaxas, D.N.: Combining low-, high-level and empirical domain knowledge for automated segmentation of ultrasonic breast lesions. IEEE Trans. Med. Imaging **22**(2), 155–169 (2003)

15. Ronneberger, O., Fischer, P., Brox, T.: U-Net: convolutional networks for biomedical image segmentation. In: Navab, N., Hornegger, J., Wells, W.M., Frangi, A.F. (eds.) MICCAI 2015. LNCS, vol. 9351, pp. 234–241. Springer, Cham (2015). https://doi.org/10.1007/978-3-319-24574-4_28

16. Shan, J., Cheng, H., Wang, Y.: Completely automated segmentation approach for breast ultrasound images using multiple-domain features. Ultrasound Med. Biol. **38**(2), 262–275 (2012)

17. Siegel, R.L., Miller, K.D., Jemal, A.: Cancer statistics. CA: Cancer J. Clin. **67**(1), 7–30 (2017)

18. Stavros, A.T., Thickman, D., Rapp, C.L., Dennis, M.A., Parker, S.H., Sisney, G.A.: Solid breast nodules: use of sonography to distinguish between benign and malignant lesions. Radiology **196**(1), 123–134 (1995)

19. Wang, Y., et al.: Deep attentional features for prostate segmentation in ultrasound. In: Frangi, A.F., Schnabel, J.A., Davatzikos, C., Alberola-López, C., Fichtinger, G. (eds.) MICCAI 2018. LNCS, vol. 11073, pp. 523–530. Springer, Cham (2018). https://doi.org/10.1007/978-3-030-00937-3_60

20. Xian, M., Zhang, Y., Cheng, H.D.: Fully automatic segmentation of breast ultrasound images based on breast characteristics in space and frequency domains. Pattern Recogn. **48**(2), 485–497 (2015)

21. Xu, Y., Wang, Y., Yuan, J., Cheng, Q., Wang, X., Carson, P.L.: Medical breast ultrasound image segmentation by machine learning. Ultrasonics **91**, 1–9 (2019)

22. Yap, M.H., et al.: Automated breast ultrasound lesions detection using convolutional neural networks. IEEE J. Biomed. Health Inform. **22**(4), 1218–1226 (2018)

23. Zhang, Q., Shen, X., Xu, L., Jia, J.: Rolling guidance filter. In: Fleet, D., Pajdla, T., Schiele, B., Tuytelaars, T. (eds.) ECCV 2014. LNCS, vol. 8691, pp. 815–830. Springer, Cham (2014). https://doi.org/10.1007/978-3-319-10578-9_53

24. Zhang, Z., Fu, H., Dai, H., Shen, J., Pang, Y., Shao, L.: ET-Net: A generic edge-aTtention guidance network for medical image segmentation. In: Shen, D., et al. (eds.) MICCAI 2019. LNCS, vol. 11764, pp. 442–450. Springer, Cham (2019). https://doi.org/10.1007/978-3-030-32239-7_49

25. Zhou, Z., Rahman Siddiquee, M.M., Tajbakhsh, N., Liang, J.: UNet++: a nested U-Net architecture for medical image segmentation. In: Stoyanov, D., et al. (eds.) DLMIA/ML-CDS -2018. LNCS, vol. 11045, pp. 3–11. Springer, Cham (2018). https://doi.org/10.1007/978-3-030-00889-5_1

26. Zhu, L., et al.: Aggregating attentional dilated features for salient object detection. IEEE Trans. Circ. Syst. Video Technol. **PP**(99), 1 (2019)

27. Zhu, L., et al.: Bidirectional feature pyramid network with recurrent attention residual modules for shadow detection. In: Proceedings of the European Conference on Computer Vision (ECCV), pp. 121–136 (2018)

Multi-scale Gradational-Order Fusion Framework for Breast Lesions Classification Using Ultrasound Images

Zhenyuan Ning[1,2], Chao Tu[1,2], Qing Xiao[1,2], Jiaxiu Luo[1,2], and Yu Zhang[1,2(✉)]

[1] School of Biomedical Engineering, Southern Medical University,
Guangzhou 510515, China
yuzhang@smu.edu.cn
[2] Guangdong Provincial Key Laboratory of Medical Image Processing,
Southern Medical University, Guangzhou 510515, China

Abstract. Predicting malignant potential of breast lesions based on breast ultrasound (BUS) images is crucial for computer-aided diagnosis (CAD) system for breast cancer. However, since breast lesions in BUS images have various shapes with relatively low contrast and the textures of breast lesions are often complex, it still remains challenging to predict the malignant potential of breast lesions. In this paper, a novel multi-scale gradational-order fusion (MsGoF) framework is proposed to make full advantages of features from different scale images for predicting malignant potential of breast lesions. Specifically, the multi-scale patches are first extracted from the annotated lesions in BUS images as the multi-channel inputs. Multi-scale features are then automatically learned and fused in several fusion blocks that armed with different fusion strategies to comprehensively capture morphological characteristics of breast lesions. To better characterize complex textures and enhance non-linear modeling capability, we further propose isotropous gradational-order feature module in each block to learn and combine different-order features. Finally, these multi-scale gradational-order features are utilized to perform prediction for malignant potential of breast lesions. The major advantage of our framework is embedding the gradational-order feature module into a fusion block, which is used to deeply integrate multi-scale features. The proposed model was evaluated on an open dataset by using 5-fold cross-validation. The experimental results demonstrate that the proposed MsGoF framework obtains the promising performance when compared with other deep learning-based methods.

Keywords: Multi-scale · Gradational-order fusion · Breast lesions classification

1 Introduction

Breast cancer is the most frequently diagnosed cancer with high incidence and mortality among women worldwide, and early detection and diagnosis are crucial

---

Z. Ning and C. Tu—Equally contribute to this paper.

© Springer Nature Switzerland AG 2020
A. L. Martel et al. (Eds.): MICCAI 2020, LNCS 12266, pp. 171–180, 2020.
https://doi.org/10.1007/978-3-030-59725-2_17

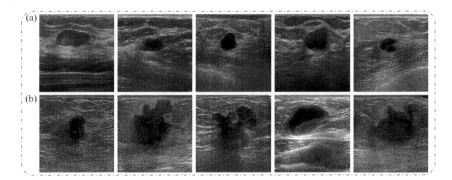

Fig. 1. Some samples of BUS images are presented: (a) benign lesions. (b) malignant lesions. They have the characteristics of various low-contrast shapes and complex textures.

for successful treatment of breast cancer [1,2]. Among many imaging technologies, breast ultrasound (BUS) has been widely used to detect and diagnose breast lesions due to its high-efficiency, non-invasiveness, and the capability to improve the sensitivity of screening [3]. Reviewing hundreds of images produced by BUS requires a large amount of time, even for experienced physicians [4]. To reduce the reviewing time, many learning-based computer-aided diagnosis (CAD) approaches have thus been proposed to support physicians' decision-making [5,6]. However, since breast lesions in BUS images have the characteristics of various low-contrast shapes and complex textures (as shown in Fig. 1), it still remains challenging to predict malignant potential of breast lesions.

Recently, many deep learning-based methods, especially convolutional neural networks (CNNs), have been proposed to learn task-oriented features from BUS images for breast classification-related task.

For example, Cheng et al. [7] explored stacked denoising autoencoders to build the deep-learning-based computer-aided diagnosis (CADx) for the differential diagnosis of benign and malignant nodules in BUS images, and got the state-of-the-art performance when compared with two traditional machine learning-based methods. Yap et al. [8] investigated and compared three different deep learning-based models (including Patch-based LeNet, U-Net, and FCN-AlexNet) for breast ultrasound lesions detection. In [9], Byra et al. introduced a matching layer to convert the grayscale BUS images to red, green, blue (RGB) to more efficiently utilize the discriminative power of the CNN for improving lesions classification of BUS images. However, most of these existing methods typically utilized singe-scale BUS images for prediction and thus neglected that multi-scale images can complementally provide more morphological information than singe-scale images [10]. Besides, studies have indicated that higher-order feature representations can better characterize complex textures and enhance non-linear modeling capability [11,12]. Therefore, it is intuitively desirable to explore a special strategy to dig and integrate different-order features from multi-scale images for improving the performance of breast lesions classification.

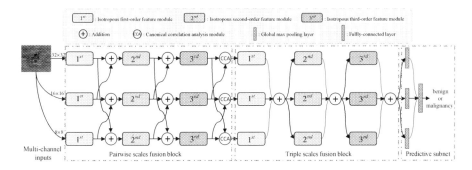

Fig. 2. The overview of MsGoF framework for predicting malignant potential of breast lesions, including multi-channel inputs (i.e., three scales), several fusion blocks (i.e., pairwise scales fusion block and triple scales fusion block) and a predictive subnet.

To this end, a novel multi-scale gradational-order fusion (MsGoF) framework is proposed to make full advantages of features from different scale images for predicting malignant potential of breast lesions. Specifically, the multi-scale patches are first extracted from the annotated lesions in BUS images as the multi-channel inputs. The strategy of using images patches rather than the whole BUS images as the inputs can partly alleviate the problem of limited data. Multi-scale features are then automatically learned and fused in several fusion blocks that armed with different fusion strategies to comprehensively capture morphological characteristics of breast lesions. For each fusion block, we further propose gradational-order feature module to characterize and combine different-order features, which can better characterize complex textures and effectively enhance non-linear modeling capability of the framework. Finally, these multi-scale gradational-order features are utilized to perform prediction for malignant potential of breast lesions. The experimental results on an open dataset demonstrate that the proposed MsGoF framework obtains the promising performance when compared with other deep learning-based methods.

2 Method

2.1 Overview of MsGoF Framework

The architecture of the proposed MsGoF framework is shown in Fig. 2. The framework basically consists of multi-channel inputs, several multi-scale fusion blocks and a predictive subnet. In particular, each fusion block arms with special fusion strategy and gradational-order feature module to deeply integrate multi-scale features.

2.2 Multi-channel Inputs Using Multi-scale Patches

Direct feeding the whole BUS images into a framework may introduce some bias, as the whole BUS images commonly include some irrelevant information (such as

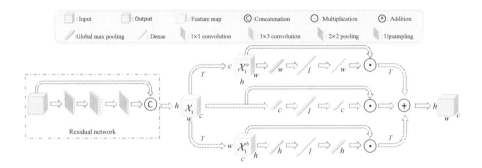

Fig. 3. The architecture of the first-order feature module. T:transpose operation.

noise and artifact). In addition, for small dataset, using the whole BUS images as input would bring difficulties in network training due to limited number of training samples. Motivated by [13], we adopt patch-based strategy to implement data augmentation. Specifically, we first extract patches with the size of 32×32 from region of interest (ROI) annotated by radiologists. These patches are then resized into several scales, and used as the multi-channel inputs of the framework. To decrease computational burden, we just generate three scales, i.e., 32×32, 16×16, and 8×8 in this work (as shown in Fig. 2). It's worthy mentioning that the framework can be extended to more scales when computational resource is sufficient and the patch size of original scale is large.

2.3   Multi-scale Fusion Block

The fusion block is designed according to the number of scales. To match with the three scales, two fusion blocks are developed to deeply integrate multi-scale features, including pairwise scales fusion (PsF) block and triple scales fusion (TsF) block, as illustrated in Fig. 2. Both of PsF block and TsF block arm with the same gradational-order feature module and their respective fusion strategies.

Pairwise Scales Fusion (PsF) Block. The PsF block, mainly consisting of isotropous first-order, second-order, and third-order feature modules, aims to learn and integrate low- and high-order features extracted from any two scales (as shown in Fig. 2). Let $i = \{1, 2, 3\}$ denote the channels of 32×32, 16×16, and 8×8 image patches, respectively. For the i-th channel, the output of each feature module keeps the same size with the input. For the 1-st channel, the output of each-order module is averaged with the upsampled version of output generated by the corresponding module of the 2-nd channel. Similar operation is used for other channels. Notably, the downsampling layer takes the place of the upsampling layer while the 3-rd channel is combined with the 1-st channel. In addition, at the end of PsF block, we further use canonical correlation analysis (CCA) [14] operator to obtain the high-correlated information for any two scales.

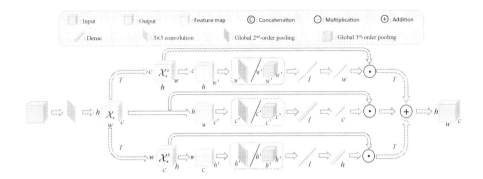

Fig. 4. The architecture of the higher-order feature module, including second-order and triple-order feature modules. T:transpose operation.

1) The First-Order Feature Module. For each channel, the isotropous first-order feature module is first developed to extract first-order feature representations (as the Fig. 3 shown). Specifically, the input is fed into a residual network (RN) which contains a 2×2 pooling layer, three convolutional layers with the kernel size of 1×1, 3×3, and 1×1, a 2×2 upsampling layer and skip-connection. Let $\mathcal{X}_1$ denote the feature maps generated by the RN, with the size of $h \times w \times c$. To extract isotropic character, we first transpose $\mathcal{X}_1$ to get the feature maps $\mathcal{X}_1^a$ and $\mathcal{X}_1^b$ with the shape of $c \times h \times w$ and $w \times c \times h$, respectively. Subsequently, a global max pooling layer and two dense layers are used to generate $1 \times 1 \times c$, $1 \times 1 \times w$, and $1 \times 1 \times h$ weighted vectors which are scaling (multiplication) of the input one along the channel dimension for $\mathcal{X}_1$, $\mathcal{X}_1^a$ and $\mathcal{X}_1^b$, respectively. Finally, the weighted feature maps $\mathcal{X}_1^a$ and $\mathcal{X}_1^b$ are transposed back to the original direction, and integrated into the weighted $\mathcal{X}_1$ to obtain the first-order feature representations.

2) The second-Order Feature Module. In statistics, traditional pooling operator (such as max or mean pooling) represents first order statistical moment of feature map, which can not capture the correlation among features [12]. It have been proven that correlation can better characterize complex patterns of subjects and enhance non-linear modeling capability [11]. Therefore, we plug a second-order feature module into the block to further explore the high-order information exited in features. Specifically, as illustrated in Fig. 4, the first-order feature representations from different scales are first fed into the 3×3 convolutional layer with 8 kernels. Let $\mathcal{X}_2$ denote the feature maps with the shape of $h \times w \times c$, which is generated by the convolutional layer. Similar to the first-order feature module, we get the feature maps $\mathcal{X}_2^a$ and $\mathcal{X}_2^b$ with shape of $c \times h \times w$ and $w \times c \times h$, respectively, by transposing $\mathcal{X}_2$. Then, a 1×1 convolutional layer is utilized to reduce the number of channels from c to c' for $\mathcal{X}_2$ (from w to w' for $\mathcal{X}_2^a$; from h to h' for $\mathcal{X}_2^b$), which decreases the computational cost of the following operations. Subsequently, a global second-order pooling layer is used to compute pairwise-channel correlations as the second-order statistics, obtaining a $c' \times c'$

covariance matrix for $\mathcal{X}_2$ ($w' \times w'$ for $\mathcal{X}_2^a$ and $h' \times h'$ for $\mathcal{X}_2^b$). In particular, the coordinate position of (i, j) in the covariance matrix indicates statistical dependency between the i-th channel and j-th channel of feature maps. To maintain the structural information, a $l \times l$ convolutional layer with l kernels is placed to get a $1 \times 1 \times l$ feature vector, followed by a dense layer to generate $1 \times 1 \times c$, $1 \times 1 \times w$, and $1 \times 1 \times h$ weighted vectors which are performed dot product with the input one along the channel dimension for $\mathcal{X}_2$, $\mathcal{X}_2^a$, and $\mathcal{X}_2^b$, respectively. At the end of the second-order feature module, the second-order feature representations are obtained by transposing the weighted feature maps $\mathcal{X}_2^a$ and $\mathcal{X}_2^b$ back to the original direction and integrating them into the weighted $\mathcal{X}_2$.

3) The Third-Order Feature Module. To learn more higher-order feature representations, we further develop an isotropous third-order feature module, as shown in Fig. 4. This module shares similar structure with the second-order feature module. The difference between them is that a global third-order pooling layer takes place of the global second-order pooling layer. Specifically, we compute correlations of three channels as the third-order statistics in the global third-order pooling layer, obtaining a relevance tensor, whose coordinate position of (i, j, k) presents statistical dependency among three feature maps along the channel dimension.

The Triple Scales Fusion (TsF) Block. The structure of TsF block is similar to that of PsF block, consisting of isotropous first-order, second-order, and third-order feature modules. The difference between them is aggregative way for different scales. Different from the PsF, we straightly integrate three scales by downsampling (upsampling) the feature maps at the 1-st (3-rd) channel into feature maps with the shape of 16×16 and averaging with the feature maps of 2-nd channel, and then resize back to three scales again, as shown in Fig. 2.

2.4   The Predictive Subnet

The multi-scale gradational-order features learned by the PsF and TsF blocks only focus on local patches, which would ignore global information. Inspired by [13], we perform a global max pooling along patches to integrate local features into global features. Finally, by leveraging these features, a fully-connected layer is used to perform malignant potential prediction for each patient.

3   Experiment and Result Analysis

Data and Experimental Setup. An open BUS images dataset was used in this study, which was collected from the UDIAT Diagnostic Centre of the Parc Taulí Corporation, Sabadell (Spain) with a Siemens ACUSON Sequoia C512 system 17L5 HD linear array transducer (8.5 MHz) [8]. The dataset consists of a total of 163 breast BUS images (53 malignant and 110 benign lesions) with a mean image size of 760×570 pixels and the range from 307×233 to 791×641.

Table 1. The ablation experiments evaluating the effectiveness of multi-scale inputs (mean ± standard deviation).

Method	Accuracy	Sensitivity	Specificity	AUC
SsGoF	0.877 ± 0.104	0.871 ± 0.072	0.880 ± 0.140	0.880 ± 0.111
PsGoF	0.897 ± 0.135	0.909 ± 0.115	0.891 ± 0.145	0.905 ± 0.131
Our method	**0.909 ± 0.032**	**0.927 ± 0.106**	**0.900 ± 0.044**	**0.939 ± 0.031**

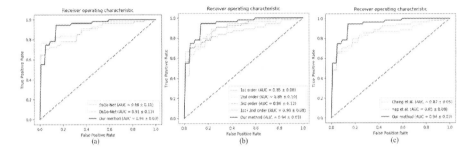

Fig. 5. The receiver operating characteristic curves for the experiments: (a) The ablation experiments evaluating the effectiveness of multi-scale inputs. (b) The ablation experiments evaluating the effectiveness of gradational-order feature module. (c) The contrast experiments with other existing state-of-the-art methods.

All lesion regions in the BUS images were delineated by experienced radiologists. To evaluate the proposed model, we performed a 5-fold cross-validation strategy, where 80% images were used for training, and the remaining 20% for testing. For multi-scale patch extraction, we totally extracted 47055 patches (the ratio of benign and malignant patches is about 1.9 : 1) for training and testing, and each patch was ensured to include substantial tumor tissue (>300 tumor voxels). We compared the predictive performance in terms of accuracy (ACC), sensitivity (SEN), and specificity (SPE). Considering imbalance and distribution variation between classes, we also calculated the area under the receiver operating characteristics curve (AUC). The Adam solver was used as the optimizer with an initial learning rate of 0.001, a L_2 weight decay of 1×10^{-4}. The batch-size was set to 128. We used Python3.7 to implement all the experiments and evaluations and Tensorflow to build the deep learning network architecture. All computationally intensive calculations were offloaded to a 12 GB NVIDIA Pascal Titan X GPU.

Effectiveness of Multi-scale Inputs. We perform ablation experiments on the number of input scale. To be specific, total two variants of MsGoF framework are built, including 1) single-scale gradational-order fusion (SsGoF): by removing all multi-scale fusion strategies and remaining the first channel of MsGoF, 2) pairwise-scale gradational-order fusion (PsGoF): by removing triple scales fusion stream and the third channel of MsGoF. The results are listed in Table 1 and the ROC curves are shown in Fig. 5(a). From the Table 1 and Fig. 5(a), we can

Table 2. The ablation experiments evaluating the effectiveness of gradational-order feature module (mean ± standard deviation).

Method	Accuracy	Sensitivity	Specificity	AUC
First-order	0.842 ± 0.081	0.800 ± 0.145	0.863 ± 0.090	0.849 ± 0.084
Second-order	0.877 ± 0.107	0.887 ± 0.109	0.871 ± 0.133	0.887 ± 0.101
Third-order	0.895 ± 0.114	0.907 ± 0.058	0.889 ± 0.146	0.902 ± 0.117
First-order + second-order	0.883 ± 0.102	0.889 ± 0.089	0.880 ± 0.116	0.901 ± 0.092
Our method	**0.909 ± 0.032**	**0.927 ± 0.106**	**0.900 ± 0.044**	**0.939 ± 0.031**

Table 3. The performance of the proposed breast lesions classification method comparing with other existing state-of-the-art methods on the same dataset (mean ± standard deviation). * indicates that there is a significant difference between this method and our method by Delong test. [a] indicates directly comparing the result of [9] based on the same dataset.

Method	Accuracy	Sensitivity	Specificity	AUC
*Cheng et al. [7]	0.853 ± 0.021	0.795 ± 0.137	0.881 ± 0.073	0.869 ± 0.052
*Yap et al. [8]	0.846 ± 0.087	0.813 ± 0.105	0.862 ± 0.104	0.847 ± 0.086
[a]Byra et al. [9]	0.840 ± 0.024	0.851 ± 0.042	0.834 ± 0.030	0.893 ± 0.030
Our method	**0.909 ± 0.032**	**0.927 ± 0.106**	**0.900 ± 0.044**	**0.939 ± 0.031**

observe that our proposed MsGoF, with triple scales as input, achieves better performance on all metrics when compared with other models.

Effectiveness of Isotropous Gradational-Order Feature Module. To validate the effectiveness of isotropous gradational-order feature module, we conduct a set of variants with different combinations of feature modules. From Table 2 and Fig. 5(b), we can observe that:1) for these models with single-order feature module, higher-order feature module obtains better performance than lower-order ones; 2) the third-order feature module achieves comparable results with the combination of first-order and second-order feature modules; 3)utilizing features learned from all three orders is beneficial to further enhancing the classification performance.

Comparison with Other Methods. In this section, we compare the predictive performance of our proposed method with other existing state-of-the-art methods [7–9] on the same dataset. Specifically, we directly use the results (no ROC curve given) from Byra et al. [9] based on the same dataset. And we reimplement the methods from Cheng et al. [7] and Yap et al. [8], and maximize performance by fine tuning. In [8], we modify the Patch-based LeNet detection algorithm to address breast ultrasound lesion classification by averaging outputs of the same sample patches as the new output. Table 3 presents the mean results of 5-fold cross-validation for all methods. As shown in Table 3 and Fig. 5(c), our method obtains the best performance among all compared methods.

3.1 Conclusion

In this paper, we introduce a novel deep learning-based framework, i.e., multi-scale gradational-order fusion (MsGoF) framework, for predicting malignant potential of breast lesions using ultrasound images. Multi-scale features are then automatically learned and fused in several fusion blocks that armed with different fusion strategies to comprehensively capture morphological characteristics of breast lesions. Besides, we further propose isotropous gradational-order feature module in each block to learn and combine different-order features for better characterizing complex textures and enhancing non-linear modeling capability. Finally, by utilizing these multi-scale gradational-order features, a predictive sub-net is built to perform prediction for malignant potential of breast lesions. The experimental results demonstrate that the proposed MsGoF framework obtains the promising performance when compared with other deep learning-based methods.

Acknowledgements. This work was supported in part by the National Natural Science Foundation of China [61971213, 61671230], and in part by the Basic and Applied Basic Research Foundation of Guangdong Province [2019A1515010417].

References

1. DeSantis, C., Ma, J., Bryan, L., Jemal, A.: Breast cancer statistics 2013. CA Cancer J. Clin. **64**(1), 52–62 (2014)
2. Siegel, R.L., Miller, K.D., Jemal, A.: Cancer statistics 2019. CA Cancer J. Clin. **69**(1), 7–34 (2019)
3. Wang, Y., et al.: Deeply-supervised networks with threshold loss for cancer detection in automated breast ultrasound. IEEE Trans. Med. Imaging **39**(4), 833–879 (2019)
4. Qi, X., et al.: Automated diagnosis of breast ultrasonography images using deep neural networks. Med. Image Anal. **52**, 185–198 (2019)
5. Huang, Q., Zhang, F., Li, X.: Machine learning in ultrasound computer-aided diagnostic systems: a survey. BioMed Res. Int. **2018**(7), 1–10 (2018)
6. Litjens, G., et al.: A survey on deep learning in medical image analysis. Med. Image Anal. **42**, 60–88 (2017)
7. Cheng, J.Z., et al.: Computer-aided diagnosis with deep learning architecture: applications to breast lesions in us images and pulmonary nodules in CT scans. Sci. Rep. **6**(1), 1–13 (2016)
8. Yap, M.H., et al.: Automated breast ultrasound lesions detection using convolutional neural networks. IEEE J. Biomed. Health Inform. **22**(4), 1218–1226 (2017)
9. Byra, M., et al.: Breast mass classification in sonography with transfer learning using a deep convolutional neural network and color conversion. Med. Phys. **46**(2), 746–755 (2019)

10. Lee, H., Park, J., Hwang, J.Y.: Channel attention module with multi-scale grid average pooling for breast cancer segmentation in an ultrasound image. IEEE Trans. Ultrason. Ferroelectr. Freq. Control **67**(7), 1344–1353 (2020)
11. Zoumpourlis, G., Doumanoglou, A., Vretos, N., Daras, P.: Non-linear convolution filters for CNN-based learning. In: Proceedings of the IEEE International Conference on Computer Vision, pp. 4761–4769 (2017)
12. Gao, Z., Xie, J., Wang, Q., Li, P.: Global second-order pooling convolutional networks. In: Proceedings of the IEEE Conference on Computer Vision and Pattern Recognition, pp. 3024–3033 (2019)
13. Ning, Z., et al.: Pattern classification for gastrointestinal stromal tumors by integration of radiomics and deep convolutional features. IEEE J. Biomed. Health Inform. **23**(3), 1181–1191 (2018)
14. Härdle, W.K., Simar, L.: Canonical correlation analysis. In: Applied Multivariate Statistical Analysis, pp. 443–454. Springer, Heidelberg (2015). https://doi.org/10.1007/978-3-662-45171-7_16

Computer-Aided Tumor Diagnosis in Automated Breast Ultrasound Using 3D Detection Network

Junxiong Yu[1,2], Chaoyu Chen[1,2], Xin Yang[1,2], Yi Wang[1,2], Dan Yan[3], Jianxing Zhang[3], and Dong Ni[1,2(✉)]

[1] National-Regional Key Technology Engineering Laboratory for Medical Ultrasound, School of Biomedical Engineering, Health Science Center, Shenzhen University, Shenzhen, China
nidong@szu.edu.cn
[2] Medical UltraSound Image Computing (MUSIC) Lab, Shenzhen University, Shenzhen, China
[3] Guangdong Province Traditional Chinese Medical Hospital, GuangZhou, China

Abstract. Automated breast ultrasound (ABUS) is a new and promising imaging modality for breast cancer detection and diagnosis, which could provide intuitive 3D information and coronal plane information with great diagnostic value. However, manually screening and diagnosing tumors from ABUS images is very time-consuming and overlooks of abnormalities may happen. In this study, we propose a novel two-stage 3D detection network for locating suspected lesion areas and further classifying lesions as benign or malignant tumors. Specifically, we propose a 3D detection network rather than frequently-used segmentation network to locate lesions in ABUS images, thus our network can make full use of the spatial context information in ABUS images. A novel similarity loss is designed to effectively distinguish lesions from background. Then a classification network is employed to identify the located lesions as benign or malignant. An IoU-balanced classification loss is adopted to improve the correlation between classification and localization task. The efficacy of our network is verified from a collected dataset of 418 patients with 145 benign tumors and 273 malignant tumors. Experiments show our network attains a sensitivity of 97.66% with 1.23 false positives (FPs), and has an area under the curve(AUC) value of 0.8720.

Keywords: Automated Breast Ultrasound (ABUS) · 3D detection network · Similarity loss

1 Introduction

For women all around the world, breast cancer is the most commonly diagnosed type of cancer. Early detection through screening and advances in treatment have been shown significantly reduced the mortality rates.

J. Yu and C. Chen—Contribute equally to this work.

© Springer Nature Switzerland AG 2020
A. L. Martel et al. (Eds.): MICCAI 2020, LNCS 12266, pp. 181–189, 2020.
https://doi.org/10.1007/978-3-030-59725-2_18

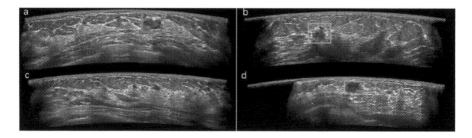

Fig. 1. Example ABUS images, the area in the box is the lesion marked by the doctor, where a, c are benign lesions, and b, d are malignant tumors.

Due to the advantages of non-invasive and convenient, ultrasound has become the most commonly used screening tool in the diagnosis of breast cancer, among which hand-held ultrasound (HHUS) is the most widely used. However, HHUS has a high dependence on the diagnosticians and a low repeatability. In contrast, automated breast ultrasound (ABUS) can make up for these shortcomings by providing more intuitive three-dimensional information and coronal plane information with great diagnostic value.

Although ABUS images have many advantages, they also inevitably increase the workload of doctors. Generally a typical ABUS exam has at least three volumes to complete coverage of the entire unilateral breast. Even for senior doctors, it is very time-consuming to manually screen tumors and overlook of abnormalities may happen. Therefore, the development of efficient and accurate computer-aided diagnosis is of great significance for reducing the workload of doctors, improving the tumor screening rate of ABUS images and promoting the early diagnosis of breast cancer.

Nevertheless, developing computer-aided diagnosis (CAD) schemes for ABUS images remains challenging. As shown in Fig. 1, 1) compared with other imaging modality, ultrasound imaging quality is relatively poor, thus making the boundary labeling difficult; 2) in most cases the proportion of lesion areas is less than 1%, at the same time, the high similarity of benign and malignant lesions makes the classification task difficult; 3) the reconstructed ABUS images have approximately 800 frames, which requires huge computing resources.

In order to improve the efficiency of reviewing ABUS images, researchers have been developed many CAD systems. Tan *et al.* [10] proposed an ensemble of neural network classifiers which obtains sensitivity of 64% at 1 false positives (FPs) per image. Lo *et al.* [4] proposed a CAD system based on watershed transform, achieving sensitivity of 100%, 90% and 80% with FPs of 9.44, 5.42, and 3.33, respectively. Moon *et al.* [6] proposed a CAD system based on quantitative tissue clustering algorithm to identify tumors, achieving sensitivity of 89.19% with 2.0 FPs per volume. Wang *et al.* [12] employed convolutional neural networks (CNNs) with threshold loss for cancer detection, obtaining a sensitivity of 95.12% with 0.84 FPs per volume. Chiang *et al.* [2] applied 3D CNN and prioritized candidate aggregation, achieving sensitivities of 95%, 90%, 85% and 80%

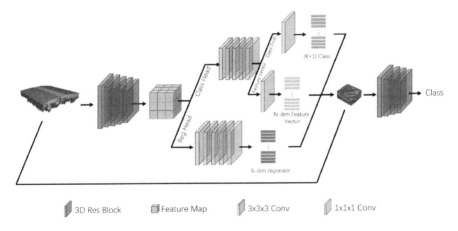

Fig. 2. An overview of our proposed two-stage 3D detection network. A 3D region proposal network (RPN) is employed to locate suspected lesions, and then a classification network is used to identify benign or malignant tumors.

with 14.03, 6.92, 4.91, and 3.62 FPs per volume, respectively. Moon *et al.* [5] proposed a 3D CNN with focal loss and ensemble learning, obtaining a sensitivity of 95.3% with 6.0 FPs. Wang *et al.* [11] proposed a CNN model which employs a multi-view strategy to classify breast lesions, obtaining AUC value of 0.95 with the sensitivity of 86.6% and specificity of 87.6%.

It can be found that most of the deep neural networks are based on U-Net [9] architecture. However, U-Net architecture consumes a lot of computing resource in the decode stage, which means only the small patch can be input into the network. Therefore we propose a 3D detection network to make full use of the spatial context information in ABUS images. In most traditional detection networks, regression and classification are two branches in parallel. Many paper [7,13] have proved that enhancing the relationship between regression and classification will improve the performance of the network. We propose to use IoU-balanced classification loss to make those anchors with high scores and good regression contributing more to the network. To better distinguish the lesions from the background areas, we employ the similarity loss to increase intra-category correlation and inter-category discrimination. After locating the lesions, we use a classification network to predict the class of lesions.

2   Method

In this section, we present the proposed two-stage 3D detection network. Figure 2 provides an overview of our method, which leverages the backbone to extract the feature maps from the input ABUS images. Then the feature maps are input into a 3D region proposal network (RPN) [8] to locate suspected lesions, finally a classification network is employed to predict the class of these suspected areas.

2.1   3D RPN

To extract representative features, we use a 3D CNN consisting of 5 Res-blocks as the backbone. The last feature map is input into a 3D RPN to locate suspected lesion areas. After analyzing the size of lesions, we use 5 basic sizes (i.e., 8, 16, 28, 40 and 55) to generate 125 different anchors at each feature cell. 3D RPN is comprised of classification and regression branches, both of which contain four $3 \times 3 \times 3$ convolution layers. At the head of regression branch, a $1 \times 1 \times 1$ convolution layer is employed to predict a set of six real-valued numbers representing bounding-box positions of the classes. We use a $1 \times 1 \times 1$ convolution layer to predict the probability of (K+1) classes of each anchor, meanwhile using a $1 \times 1 \times 1$ convolution layer to encode the corresponding area of each anchor to generate an n-dim (i.e., 32, 64, 128) feature vector.

2.2   IoU-Balanced Classification Loss

In ABUS images, the lesion area often only accounts for a small proportion of the entire image, and is very similar with the background area. Therefore, strengthening the classification weight of those anchors that regress well also helps to improve the performance of the network. Thus, we propose to use IoU-balanced loss as the classification loss:

$$L_{cls} = \sum_{i \in Pos}^{N} \omega_i(iou_i) * CE(p_i, \widehat{p}_i) + \sum_{i \in Neg}^{M} CE(p_i, \widehat{p}_i), \tag{1}$$

$$\omega_i(iou_i) = iou_i^{\eta} * \frac{\sum_i^n CE(p_i, \widehat{p}_i)}{\sum_i^n iou_i^{\eta} * CE(p_i, \widehat{p}_i)}, \tag{2}$$

In Eq. 1, iou is the Intersection-over-Union between positive proposal and its corresponding ground truth, CE means cross entropy loss where p_i is the predicted probability vector and $\widehat{p}_i$ is the real distribution, $\omega_i(iou_i)$ is the IoU weight from Eq. (2). The parameter η can regulate IoU-balanced classification loss to focus on samples with high IoU and suppress the ones with low IoU. When η is assigned to 0, the IoU-balanced classification loss is equivalent to cross entropy classification loss.

We calculate the IoU between the positive regression bounding boxes and their corresponding ground truth boxes as the weight coefficient, which acts on the classification loss. On the one hand, the relationship between the regression branch and classification branch is strengthened; on the other hand, compared with the traditional cross-entropy loss for positive samples, IoU-balanced classification loss will get a higher weight coefficient for those positive samples which get higher IoU, thus when the network is updated, those positive samples with good regressions are more inclined to obtain higher classification scores. At the same time, those samples with poor regressions would get smaller weight coefficients, which suppress the impact of those samples with high classifications scores but poor regressions on the network.

2.3   Similarity Loss

Because of the characteristics of wide intra-class differences and small inter-class differences in ABUS images, in order to better locate and classify the lesions, an essential goal of our model is to learn the common characteristics of the same category as much as possible and expand the differences between different categories.

Inspired by [1], we propose a similarity loss. After encoding the area corresponding to the anchor into N-dimensional feature vectors, we select several feature vectors that match predetermined conditions. The specific condition is that the IoU between each anchor and its corresponding ground truth needs to be greater than a certain threshold (i.e., 0.3), and the IoU between anchors also need to be greater than a certain threshold (i.e., 0.2). After selecting these feature vectors that meet above conditions, we use Eq. (3) to calculate the cosine similarity between Z_i and Z_j. Specifically, we calculate the cosine similarity between these selected vectors as $Sim_{pos,pos}$. Then we randomly select the same number of negative sample feature vectors and calculate the similarity between the negative sample feature vectors and the positive vectors as $Sim_{pos,neg}$. Our goal is to maximize the similarity between positive samples and reduce the similarity between negative samples, thus our loss function is as follows:

$$Sim_{i,j} = \frac{Z_i^T Z_j}{\|Z_i\| * \|Z_j\|}, \tag{3}$$

$$L_{sim} = \frac{2 - \log\left(e^{Sim_{pos,pos}}/e^{Sim_{pos,neg}}\right)}{4}. \tag{4}$$

The total 3D RPN loss is then summarized as

$$L_{rpn} = L_{reg} + L_{cls} + \lambda * L_{sim}, \tag{5}$$

where the regression loss is smooth L1 loss, λ ($= 0.7$ in our implementation) balances the importance between L_{cls} and L_{sim}.

2.4   Lesion Classification

We observe that the classification of the lesion is still insufficient if only using the output of the classification branch of the 3D RPN network. Therefore, in order to predict the lesion category more accurately, we input the predicted candidates into a trained classification network similar to backbone, to predict its possible category. We then weigh the score of the detection network and the classification network to ascertain the final category.

3   Experimental Results

3.1   Materials and Implementation Details

Our experimental data were acquired from Sun Yat-Sen University Cancer Center. Our institutional review board approved the consent process. There are

Table 1. Quantitative evaluation of our proposed framework.

Method	mIoU (%)	FPs	Sensitivity (%)
2D-U-Net	**44.34**	3.18	85.59
3D-U-Net	41.77	2.73	87.04
RPN	36.49	1.22	94.55
RPN-IoU	37.25	1.37	96.11
RPN-IoU-Sim	41.47	**1.24**	**97.66**

totally 145 benign patients and 273 malignant patients involved in this study. Each patient was scanned about 6 to 12 volumes with the voxel resolution of 0.511 mm, 0.082 mm, 0.200 mm in the transverse, sagittal and coronal direction, respectively. We randomly divided the dataset: 250 patients in the training set, 84 patients in the validation set, and 84 patients in the test set. In test set, 84 patients have 251 volumes with 257 lesions (144 maligant and 113 benign). All lesions were manually annotated by an experienced clinician. Since ABUS data itself has a large scale (i.e., $800 \times 200 \times 800$) thus is limited by the size of a single GPU memory, we down-sampled the raw volume to $\frac{1}{8}$ of its original size (i.e., $400 \times 100 \times 400$).

During the training phase, we firstly randomly cropped a volume of $400 \times 98 \times 360$ around the lesion and then randomly cropped a volume of $320 \times 96 \times 320$ into the network. Such operation can ensure that the input image maintains a high resolution, and can contain as many lesion areas as possible. We specified an anchor as positive if it had the highest IoU with the ground truth or its IoU with ground truth was above 0.2. An anchor was considered as negative if its IoU with every ground truth was less than 0.1. Other anchors would be ignored in this study.

During testing phase, we got 4 patches of size $320 \times 96 \times 320$ from ABUS volume through regular crop. For each patch, we only output the three boxes with the highest scores after non maximum suppression (NMS), and then the relative coordinates of the output box were converted into absolute coordinate. The oversized or undersized prediction bounding-box were removed through post-processing. The final prediction was retained after the NMS operation, and the maximum IoU is calculated for the reserved bounding-box.

The evaluation metrics consist of mIoU (the mean IoU across all categories), FPs (the number of false positives in a single data which the IoU between the ground truth is 0), and sensitivity.

3.2 Performance Evaluation

Detection Results. Table 1 shows quantitative comparison between the proposed framework (RPN-IoU-Sim, "IoU" denotes IoU-balanced classification loss; "Sim" denotes similarity loss) and other methods. Compared with the traditional segmentation algorithm 2D U-net and 3D U-net [3], the proposed class-specific

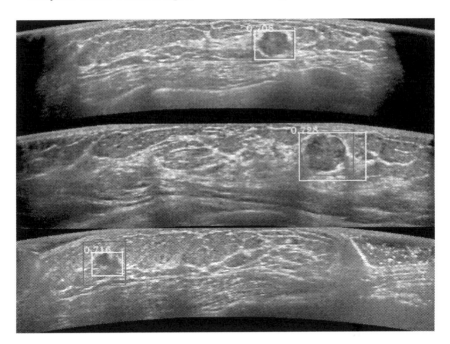

Fig. 3. Example of lesion detection results. The areas in green boxes are the labels by doctors. The yellow boxes are model predicted with prediction probability values. (Color figure online)

Table 2. Quantitative evaluation of our classification network.

Method	Accuracy	Sensitivity	Specificity	AUC
RPN	0.7341	0.8241	0.6667	0.8154
RPN-IoU	0.8016	0.9259	0.7083	0.8628
RPN-IoU-Sim	0.8016	0.9537	0.6875	0.8720

RPN method has achieved better results, with a hit rate of 94.55% and mIoU of 36.49%. While enhancing the relationship between classification and regression, RPN-IoU improves the performance of the RPN network, with a sensitivity of 96.55% and mIoU of 37.25%. After using IoU-balanced classification loss, our detection performance has obviously improved in sensitivity and mIoU. Finally, by combining IoU-balanced classification loss and similarity loss, our RPN-IoU-Sim network achieves a sensitivity of 97.66% and mIoU of 41.47%. Experimental results show that the proposed 3D detection scheme can achieve superior performance when using both IoU-balanced classification loss and similarity loss.

Figure 3 shows three lesion detection results. Figure 4 shows the sensitivity of our network to different sizes of lesions. For lesions smaller than $2\,cm^3$, our network achieved a sensitivity above 95%; and when the lesion sizes was larger than $4\,cm^3$, the sensitivity is 100%

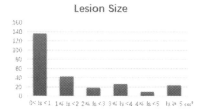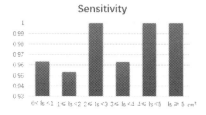

Fig. 4. Left: the lesion size distribution of all lesions. Right: the detection sensitivities of different lesions.

Classification Results. Table 2 shows the quantitative comparison of classification results. The RPN-IoU-Sim network outperformed the basic 3D RPN with respect to all evaluation metrics.

4   Conclusion

In this paper, we propose a 3D detection network for locating suspected lesions and classifying lesions as benign or malignant. In the proposed network, we use a 3D detection network rather than frequently-used segmentation network to locate lesions in ABUS images. By handling larger input patch, our network can make full use of the spatial context information in ABUS images. Furthermore, IoU-balanced classification loss is employed to improve the sensitivity greatly by leveraging the correlation between classification and localization tasks. Meanwhile, similarity loss is designed to effectively distinguish lesions from background. Experimental results show our network obtains a sensitivity of 97.66% with 1.23 FPs per ABUS volume and with an AUC value of 0.8720.

Acknowledgements. This work was supported in part by the National Key R&D Program of China (No. 2019YFC0118300), in part by the National Natural Science Foundation of China under Grant 61701312, in part by the Guangdong Basic and Applied Basic Research Foundation (2019A1515010847), in part by the Medical Science and Technology Foundation of Guangdong Province (B2019046), in part by the Natural Science Foundation of SZU (No. 860-000002110129), and in part by the Shenzhen Peacock Plan (KQTD2016053112051497).

References

1. Chen, T., Kornblith, S., Norouzi, M., Hinton, G.: A simple framework for contrastive learning of visual representations. arXiv preprint arXiv:2002.05709 (2020)
2. Chiang, T.C., Huang, Y.S., Chen, R.T., Huang, C.S., Chang, R.F.: Tumor detection in automated breast ultrasound using 3-D CNN and prioritized candidate aggregation. IEEE Trans. Med. Imaging **38**(1), 240–249 (2018)
3. Çiçek, Ö., Abdulkadir, A., Lienkamp, S.S., Brox, T., Ronneberger, O.: 3D U-Net: learning dense volumetric segmentation from sparse annotation. In: Ourselin, S., Joskowicz, L., Sabuncu, M.R., Unal, G., Wells, W. (eds.) MICCAI 2016. LNCS, vol. 9901, pp. 424–432. Springer, Cham (2016). https://doi.org/10.1007/978-3-319-46723-8_49

4. Lo, C.M., et al.: Multi-dimensional tumor detection in automated whole breast ultrasound using topographic watershed. IEEE Trans. Med. Imaging **33**(7), 1503–1511 (2014)

5. Moon, W.K., et al.: Computer-aided tumor detection in automated breast ultrasound using a 3-D convolutional neural network. Comput. Methods Programs Biomed. **190**, 105360 (2020)

6. Moon, W.K., et al.: Tumor detection in automated breast ultrasound images using quantitative tissue clustering. Med. Phys. **41**(4), 042901 (2014)

7. Pang, J., Chen, K., Shi, J., Feng, H., Ouyang, W., Lin, D.: Libra R-CNN: towards balanced learning for object detection. In: Proceedings of the IEEE Conference on Computer Vision and Pattern Recognition, pp. 821–830 (2019)

8. Ren, S., He, K., Girshick, R., Sun, J.: Faster R-CNN: towards real-time object detection with region proposal networks. In: Advances in Neural Information Processing Systems, pp. 91–99 (2015)

9. Ronneberger, O., Fischer, P., Brox, T.: U-Net: convolutional networks for biomedical image segmentation. In: Navab, N., Hornegger, J., Wells, W.M., Frangi, A.F. (eds.) MICCAI 2015. LNCS, vol. 9351, pp. 234–241. Springer, Cham (2015). https://doi.org/10.1007/978-3-319-24574-4_28

10. Tan, T., Platel, B., Mus, R., Tabar, L., Mann, R.M., Karssemeijer, N.: Computer-aided detection of cancer in automated 3-D breast ultrasound. IEEE Trans. Med. Imaging **32**(9), 1698–1706 (2013)

11. Wang, Y., Choi, E.J., Choi, Y., Zhang, H., Jin, G.Y., Ko, S.B.: Breast cancer classification in automated breast ultrasound using multiview convolutional neural network with transfer learning. Ultrasound Med. Biol. (2020)

12. Wang, Y., et al.: Deeply-supervised networks with threshold loss for cancer detection in automated breast ultrasound. IEEE Trans. Med. Imaging **39**(4), 866–876 (2019)

13. Wu, S., Li, X.: Iou-balanced loss functions for single-stage object detection. arXiv preprint arXiv:1908.05641 (2019)

Auto-weighting for Breast Cancer Classification in Multimodal Ultrasound

Jian Wang[1], Juzheng Miao[2], Xin Yang[1], Rui Li[1], Guangquan Zhou[2],
Yuhao Huang[1], Zehui Lin[1], Wufeng Xue[1], Xiaohong Jia[3],
Jianqiao Zhou[3], Ruobing Huang[1(✉)], and Dong Ni[1]

[1] Medical UltraSound Image Computing (MUSIC) Lab,
School of Biomedical Engineering, Shenzhen University, Shenzhen, China
ruobing.huang@szu.edu.cn
[2] School of Biological Science and Medical Engineering, Southeast University,
Nanjing, China
[3] Department of Ultrasound Medicine, Ruijin Hospital, School of Medicine,
Shanghai Jiaotong University, Shanghai, China

Abstract. Breast cancer is the most common invasive cancer in women. Besides the primary B-mode ultrasound screening, sonographers have explored the inclusion of Doppler, strain and shear-wave elasticity imaging to advance the diagnosis. However, recognizing useful patterns in all types of images and weighing up the significance of each modality can elude less-experienced clinicians. In this paper, we explore, for the first time, an automatic way to combine the four types of ultrasonography to discriminate between benign and malignant breast nodules. A novel multimodal network is proposed, along with promising learnability and simplicity to improve classification accuracy. The key is using a weight-sharing strategy to encourage interactions between modalities and adopting an additional cross-modalities objective to integrate global information. In contrast to hardcoding the weights of each modality in the model, we embed it in a Reinforcement Learning framework to learn this weighting in an end-to-end manner. Thus the model is trained to seek the optimal multimodal combination without handcrafted heuristics. The proposed framework is evaluated on a dataset contains 1616 sets of multimodal images. Results showed that the model scored a high classification accuracy of 95.4%, which indicates the efficiency of the proposed method.

Keywords: Ultrasound · Breast cancer · Multi-modality

1 Introduction

Breast cancer is one of the leading causes of cancer death in women [2]. Early and accurate diagnosis is crucial for better prognosis and the improvement of survival

---

J. Wang and J. Miao—Contribute equally to this work.

© Springer Nature Switzerland AG 2020
A. L. Martel et al. (Eds.): MICCAI 2020, LNCS 12266, pp. 190–199, 2020.
https://doi.org/10.1007/978-3-030-59725-2_19

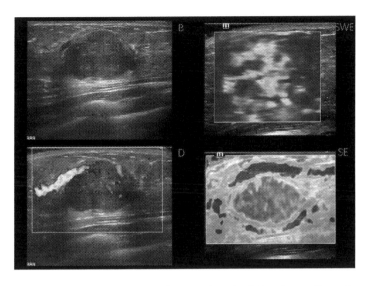

Fig. 1. Multimodal ultrasound images (top to down). *Left column*: B-mode and color doppler images. *Right column*: Shear wave and strain elastography images.

rate [1]. Mammography is the most common tool for breast cancer screening, while it could be less sensitive for patients with dense breasts [10], which might cause unnecessary biopsies or missed diagnosis [6]. Being radiation-free, non-invasive, and relatively cheap, Ultrasound (US) has shown great potential in breast cancer diagnosis. To assist the most widely adopted B-mode ultrasound, researchers have adopted Color Doppler, Share Wave Elastography (SWE) and Strain Elastography (SE) images to improve the diagnosis accuracy (see Fig. 1), as they provide valuable supplementary information such as vascularity and tissue stiffness. However, accurate diagnosis using rich multimodal images is extremely challenging and highly depends on the operators' experience.

Computer-Aided Diagnosis (CAD) methods using single-modality breast images have been developed. In [8], a multi-task learning (MTL) framework was proposed to classify breast B-mode images and predict tumor malignancy. Zhang et al. [15] classified the SWE images into benign or malignant using a deep belief network. Recently, CAD methods of multimodal breast images have been studied. Sultan et al. [11] employed traditional regression methods to differentiate breast lesion in the manually cropped B-mode and Doppler images. Zhang et al. [14] further proposed a two-stage method to segment tumors from B-mode and SWE images and classify them into benign and malignant by a deep polynomial network. The above CAD methods only consider two modalities and can be promoted greatly by making use of state-of-the-art multimodal deep learning techniques. Good multimodal representations are crucial for the performance of machine learning models. They are categorized into joint and coordinated representations in [3]. While the former project different modalities into the same representation space, the latter enforces the representations of different modalities to be more complementary. Ge et al. [4] explored three representations based on a MTL framework, namely sole-net, share-net, and triple-net, for skin cancer diagnosis. The sole-net and share-net are joint and coordinated representations, respectively and the triple-net achieved best diagnosis accuracy because it made use of the advantages of both representations. Despite showing promising results, existing methods usually treat different modalities equally, neglecting the fact that some of them should contribute more to the final decision-making. How to assign proper weight to each of the modalities remains an open question.

In this paper, we propose a novel multi-modality classification framework, termed as Auto-weighting Multimodal (AWMM), to address the challenging problem in breast cancer diagnosis. Our contribution is threefold: 1) jointly employ all four types of ultrasonography, namely B-mode, Doppler, SWE and SE for the first time, which are used clinically for breast nodule identification. 2) we cast the task as a multi-task learning problem and adopt a weight-sharing mechanism to avoid overfitting with a mixed loss to integrate extracted representations; 3) to automatically obtain the optimal weighting between modalities, we embed the model into a Reinforcement Learning (RL) framework to learn the weighting in an end-to-end fashion. Experiment results showed that the trained model is able to process rich multimodal information and identify the breast nodules accurately.

2   Methodology

In order to better mimic experts behaviour in diagnosing, we explore the advantages of linking information from four types of ultrasound imaging modalities through a multi-branch neural network model. The bespoke model is equipped with a weight-sharing mechanism and mixed fusion losses. The model is trained under a RL framework to achieve automatic branch (modality) weighting. As shown in Fig. 2, the proposed model replicates ResNet-18 base model four times to construct a weight sharing MTL model to extract multimodal representations while suppress overfitting. Besides independent loss for each stream, an additional fusion loss is added to encourage competition. As opposed to manually select the weight of each branch, the model leverages an RL based framework to learn the optimal weighting between different modalities. This design can exempt the hand-crafted heuristics to weight different task as common multimodal approaches. More importantly, it allows progressively interaction across modalities that contain different information and should contribute differently to the final prediction. Next, we explain the details of the framework.

2.1   Learning Multimodal Representation

Representing data in a format that a computational model can work with has always been a challenge in machine learning. It is especially true for multimodal models, which need to process and relate information from heterogeneous

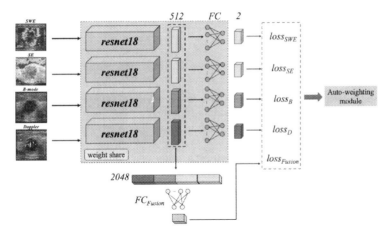

Fig. 2. Schematic of the proposed multimodal learning model. It consists of four streams, each corresponds to one investigated modality (B-mode, Doppler, SE, SWE). A fusion loss (yellow cube) that merges all branches and other four losses are combined together. (Color figure online)

sources. Inspired by Share-Net [4], the proposed model builds on four identical networks with shared-weights, each of which is able to make its own prediction. We adopt the Resnet-18 as the backbone that has been widely utilized in image recognition [5]. We argue that this architecture is less prone to overfitting since the network is forced to extract features that are functional across different modalities while reducing the number of learnable parameters. On the other hand, each stream has its own objective which enable more flexibility in learning and might contribute to accuracy improvement. At the output level, a fully connected layer (FC_{AWMM}) with a size of 2048×2 is added to project multimodal features into the shared space (see Fig. 2). Then, this joint representation is used to compute a fusion loss as an additional objective. It poses extra regularization that can help generalization (similar finding reported in [12]). Therefore, the proposed multimodal learning framework is further promoted at the output level, where both complementary and joint information can be learned using the combination of these loss functions. The model parameters are optimized by applying a stochastic gradient descent on the global loss, defined as: $\sigma_i^{t,j}$

$$Loss_G = \sum \alpha_{mod} Loss_{mod}, \tag{1}$$

where $mod \in SWE, SE, B, D, Fusion$. $Loss_{Fusion}$ is the loss computed from the joint representation, and $Loss_{mod}, mod \in SWE, SE, B, D$ is the loss computed only using the mod modality. The weights, α_{mod}, are the coefficients to setup the trade-off between each modality and the cross-modality learning, which satisfy $\sum \alpha_{mod} = 1$. Our AWMM model allows the back-propagation of both modality-specific and universal signals for parameter update, thus helps to coordinate the learning process and further suppress the overfitting.

2.2 Auto-weighting Modality Using Reinforcement Learning Framework

Similar to the behavior of clinical experts in making a diagnosis, an ideal model should make predictions using all the available information but puts emphasis on one or two of the imaging modalities. This could be incorporated into a multi-task model as weights associated with different tasks. It is impractical to define these weights manually, as even the most experienced experts have difficulty deciding if B-mode mode is 1.2 or 0.8 times more important than the elastic images. Inspired by the work [7], we enable the model to learn the optimal set of weighting using an RL framework in an end-to-end manner, avoiding the introduction of subjective bias.

In the classic setting of RL, an agent in its current state S, interacts with the environments E by making successive actions $\alpha \in A$ that maximize the expectation of reward. In this study, the environments E is the ultrasound images of different modalities. The state S is the set of weights, each of its elements corresponds to the weight of each task. These weights, represented by $\{\alpha_{mod}\}$, can be calculated by applying softmax to five real-valued agent output $\{\beta_{mod}\}$: $\alpha_{mod} = \frac{exp(\beta_{mod})}{\sum exp(\beta_{mod})}$. As the agent interacts with E to maximize the reward, the system can continuously adjust the weighting between different tasks. The action space $\{\Delta\beta_{mod}\}$ is defined as $[-0.2, 0, +0.2]$. Each valid action gets a scalar reward R calculated based on the classification accuracy, which indicates whether the agent is moving towards the preferred target.

The learning process is a standard bi-level optimization problem, in which we aim to maximize the reward w.r.t. the controller parameters $\{\theta\}$, as well as minimizing the loss of the network w.r.t. model parameters $\{\omega\}$. $\{\theta\}$ and $\{\omega\}$ are updated alternately. As illustrated in Fig. 3, at the inner level, we sample K times from the parameter space to generate K different sets of task weights $\{\alpha_{mod}^{t,1:k}\}$ for K identical networks (e.g. the whole model shown in Fig. 2). These networks are then trained independently using the training set for one epoch to update their network parameters ω^k, respectively. In the outer level of the optimization, the model that has the maximum validation accuracy is selected to update $\{\theta_i\}$ for each controller. Each controller corresponds to one modality and is independent to each other. They are updated iteratively using the REINFORCE rule [13]. In specific, the ith controller is updated following:

$$
\begin{aligned}
\Theta_i^{t+t'} &= \Theta_i^t + \eta \frac{1}{K} \sum_{j=1}^{K} R_j \cdot \nabla_\theta log\left(g\left(\alpha^{t,j}\right)\right) \\
&= \Theta_i^t + \eta \frac{1}{K} \sum_{j=1}^{K} R_j \cdot \nabla_\theta \sum log\left(p\left(\alpha_{mod}^{t,j}\right)\right),
\end{aligned}
\tag{2}
$$

where η is the learning rate, and $g\left(\alpha^{t,j}\right) = \prod p\left(\alpha_{mod}^{t,j}\right)$ is the joint probability distribution of four single-modality losses and the fusion loss. The reward signal, R_j, is calculated based on the validation accuracy of the corresponding jth network. After training the over-parameterized network, an optimal set of the parameters of the multimodal CNN model and a desirable weights between tasks are simultaneously obtained.

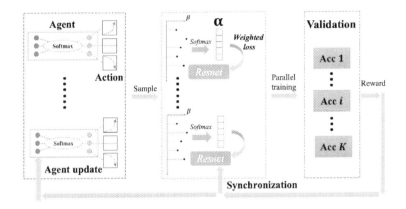

Fig. 3. Learning process of the proposed framework.

3   Materials and Experiments

We validate our solution on a dataset of 1616 sets of multimodal ultrasound images of breast nodules. Approved by the local Institutional Review Board, the images were acquired from 835 different patients. Each set has a total of four images (each of B-mode, Color Doppler, SWE, and SE modality) collected from the same patient. We randomly split the dataset into 60% training, 20% validation, and 20% test at the patient level. All images are resized to 544×320. Biopsies were carried out for each subject to classified whether the nodule is benign or malignant, and the result is used as the ground truth in this experiment.

To validate the effectiveness of the proposed model, comparison experiments were carried out using single-modality models and various popular multimodal approaches, including Share-Net [4], Centralnet [12], and Voting Schemes [9]. We termed our Manual-weighting MultiModal framework with or without sharing network parameters as MWMM-SN, MWMM respectively and its Auto-weighting version as AWMM-SN. All experiments are implemented in PyTorch with a GeForce RTX 2080Ti GPU.

Different augmentation strategies were applied, including scaling, rotation (up to 30°), flipping, and mixup. Weights pre-trained from ImageNet were used for initializing. We use the Adam optimizer with a learning rate of $1e^{-4}$. For the MWMM framework, the weights of each task's loss are equal ($\alpha_{mod} = 0.2$). Meanwhile, we set the sample number $K = 10$ and all controllers value β_{mod} to 1 for employing the RL based weight tuning framework. The controller parameters Θ is trained using Adam optimizer with learning rate $1e^{-3}$.

Table 1. Performance evaluation of single-modality and multi-modality methods. Sensitivity (SEN), specificity (SPE), overall accuracies (ACC), precision with the cut-off value of 0.5 (PRE), and F-score metric (F1-score) are used.

	Methods	ACC(%)	SEN(%)	SPE(%)	PRE(%)	F1-score(%)
Single-modality	SWE	91.46	91.03	91.86	91.03	91.03
	SE	89.02	88.46	89.53	88.46	88.46
	B-mode	86.28	83.97	88.37	86.75	85.34
	Doppler	86.59	82.05	90.70	88.89	85.33
Multi-modality	Voting	91.16	91.03	91.28	90.45	90.74
	Centralnet	92.99	91.67	94.19	93.46	92.56
	Sharenet	93.29	96.79	90.12	89.88	93.21
	MWMM	92.99	92.95	93.02	92.36	92.65
	MWMM-SN	93.60	92.95	94.19	93.55	93.25
	AWMM-SN	95.43	96.15	94.77	94.34	95.24

4   Results and Discussion

We first compare the performance of single modality models, each of which only had access to one of the four investigated modalities. As shown in Table 1, the SWE model achieved the best performance in Accuracy (ACC) = 91.46%, Sensitivity (SEN) = 91.03%, Specificity (SPE) = 91.86%, Precision (PRE) = 91.03%, and F1-score = 91.03%. It indicated that the stiffness of nodule tissue might be a strong indicator of its malignancy. On the contrary, the vascularity of nodule seemed to be less important, as the single modality model using only Doppler images achieved relatively lower accuracy (row 4 in Table 1). This is in accord with our clinical collaborators' finding in the real world situation, as a malignant breast nodule can have either rich or scarce blood flow.

Compared with single modality models, multi-modalities models showed better performance overall, whichever combination strategies they used. This proves that there is indeed some complementary information hidden in different types of ultrasonography, and their combination could lead to a more accurate diagnosis. Furthermore, it can be observed that both the proposed models and the Share-Net outperformed the CentralNet and the Voting Schemes approaches. This might result from the weight sharing strategy which forces the model to extract more universal features and also helps to reduce the overfitting by introducing fewer parameters. The last two rows in Table 1 showed that the necessity of the RL-based auto-weighting process as AWMM-SN (the proposed method) outperformed the MWMM-SN method, which weights each modality equally. In specific, the automatically learned weights for each branch are: SWE: 0.15, SW:0.19, B-mode: 0.13, Doppler: 0.19, and Fusion: 0.34. This further validates the importance of incorporating all the modalities in decision-making and verifies our model design. Another interesting observation is that the SWE did not gain the highest weight as expected. We conjecture that there might exist certain competition between SWE and SW modalities that lead to this phenomenon. Note that our model does not require the presence of all four modalities to predict and still works well when the modalities are incomplete. The proposed model achieved Acc of 91.2% (SWE), 86.0% (SE), 84.8% (B-mode), and 86.6% (Doppler) when only one modality is available during the test. The key challenge here is to extract and combine multi-modal information efficiently.

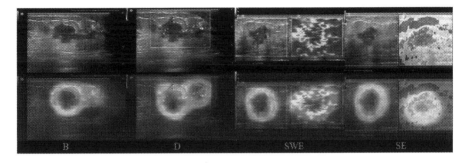

Fig. 4. Grad-CAM Visualization of AWMM-SN for one sample. Note that the network focused on the nodule region despite no location information was given in training.

Ablation Study. As an ablation study, we tested whether removing the weight sharing scheme would effect the model performance. As shown in Table 1, the MWMM model achieved better classification accuracy than that of the single-modality methods, but were inferior to the proposed method. This result indicates that the multimodal information is beneficial to accuracy improvement, while joint representation plays a vital role in the multimodal representations.

We further employed Gradient-weighted Class Activation Mapping (Grad-CAM) to visualize the classification decisions of AWMM-SN model (Fig. 4). It demonstrated that our model focused on the nodule regions during classification despite no location information was given for any of the modes. Note that no preprocessing was done to align the images, while the model automatically focused on the corresponding regions. It should be noted that clinical experts usually rely on the shape prior in the B-mode image to guide the diagnosis of SWE or SE. We, therefore, keep the B-mode region in the SWE or SE modality.

5   Conclusions

In this paper, we proposed a novel multimodal framework that is able to process four types of ultrasound modalities for automatic classification of breast nodules. The model design encourages interactions between different modalities while allowing each contributes to the final prediction. It is trained under a Reinforcement Learning framework to automatically find the optimal weighting across modalities to increase accuracy. As the design of this framework is general, it could be applied to other multimodal applications in the future.

Acknowledgement. This work was supported by NationalKey R&D Program of China (No. 2019YFC 0118300); Shenzhen Peacock Plan (No. KQTD2016053112051497, KQJSCX2018 0328095606003); Medical Scientific Research Foundation of Guangdong Province, China (No. B2018031); National Natural Science Foundation of China (Project No. NSFC61771130).

References

1. Ciatto, S., Cecchini, S., Lossa, A., Grazzini, G.: Category and operable breast cancer prognosis. Tumori J. **75**(1), 18–22 (1989)
2. World Health Organization (WHO). breast cancer. www.who.int/cancer/prevention/diagnosis-screening/breastcancer/. Accessed 2019
3. Baltrušaitis, T., Ahuja, C., Morency, L.P.: Multimodal machine learning: a survey and taxonomy. IEEE Trans. Pattern Anal. Mach. Intell. **41**(2), 423–443 (2018)
4. Ge, Z., Demyanov, S., Chakravorty, R., Bowling, A., Garnavi, R.: Skin disease recognition using deep saliency features and multimodal learning of dermoscopy and clinical images. In: Descoteaux, M., Maier-Hein, L., Franz, A., Jannin, P., Collins, D.L., Duchesne, S. (eds.) MICCAI 2017. LNCS, vol. 10435, pp. 250–258. Springer, Cham (2017). https://doi.org/10.1007/978-3-319-66179-7_29
5. He, K., Zhang, X., Ren, S., Sun, J.: Deep residual learning for image recognition. In: Proceedings of the IEEE Conference on Computer Vision and Pattern Recognition, pp. 770–778 (2016)
6. Jesneck, J.L., Lo, J.Y., Baker, J.A.: Breast mass lesions: computer-aided diagnosis models with mammographic and sonographic descriptors. Radiology **244**(2), 390–398 (2007)
7. Li, C., et al.: AM-LFS: AutoML for loss function search. In: Proceedings of the IEEE International Conference on Computer Vision, pp. 8410–8419 (2019)
8. Liu, J., et al.: Integrate domain knowledge in training CNN for ultrasonography breast cancer diagnosis. In: Frangi, A.F., Schnabel, J.A., Davatzikos, C., Alberola-López, C., Fichtinger, G. (eds.) MICCAI 2018. LNCS, vol. 11071, pp. 868–875. Springer, Cham (2018). https://doi.org/10.1007/978-3-030-00934-2_96
9. Morvant, E., Habrard, A., Ayache, S.: Majority vote of diverse classifiers for late fusion. In: Fränti, P., Brown, G., Loog, M., Escolano, F., Pelillo, M. (eds.) S+SSPR 2014. LNCS, vol. 8621, pp. 153–162. Springer, Heidelberg (2014). https://doi.org/10.1007/978-3-662-44415-3_16
10. Murtaza, G., et al.: Deep learning-based breast cancer classification through medical imaging modalities: state of the art and research challenges. Artif. Intell. Rev. **53**, 1–66 (2019)
11. Sultan, L.R., Cary, T.W., Sehgal, C.M.: Machine learning to improve breast cancer diagnosis by multimodal ultrasound. In: 2018 IEEE International Ultrasonics Symposium (IUS), pp. 1–4. IEEE (2018)
12. Vielzeuf, V., Lechervy, A., Pateux, S., Jurie, F.: CentralNet: a multilayer approach for multimodal fusion. In: Leal-Taixé, L., Roth, S. (eds.) ECCV 2018. LNCS, vol. 11134, pp. 575–589. Springer, Cham (2019). https://doi.org/10.1007/978-3-030-11024-6_44
13. Williams, R.J.: Simple statistical gradient-following algorithms for connectionist reinforcement learning. Mach. Learn. **8**(3–4), 229–256 (1992). https://doi.org/10.1007/BF00992696

14. Zhang, Q., Song, S., Xiao, Y., Chen, S., Shi, J., Zheng, H.: Dual-mode artificially-intelligent diagnosis of breast tumours in shear-wave elastography and b-mode ultrasound using deep polynomial networks. Med. Eng. Phys. **64**, 1–6 (2019)
15. Zhang, Q., et al.: Deep learning based classification of breast tumors with shear-wave elastography. Ultrasonics **72**, 150–157 (2016)

MommiNet: Mammographic Multi-view Mass Identification Networks

Zhicheng Yang[1], Zhenjie Cao[1], Yanbo Zhang[1], Mei Han[1], Jing Xiao[2],
Lingyun Huang[2], Shibin Wu[2], Jie Ma[3], and Peng Chang[1(✉)]

[1] Ping An Tech, US Research Lab, Palo Alto, USA
pengchang@gmail.com
[2] Ping An Technology, Shenzhen, China
[3] Shenzhen People's Hospital, Shenzhen, China

Abstract. Most Deep Neural Networks (DNNs) based approaches for mammogram analysis are based on single view. Some recent DNN-based multi-view approaches can perform either bilateral or ipsilateral analysis, while in practice, radiologists use both to achieve the best clinical outcome. In this paper, we present the first DNN-based tri-view mass identification approach (MommiNet), which can simultaneously perform end-to-end bilateral and ipsilateral analysis of mammogram images, and in turn can fully emulate the radiologists' reading practice. Novel network architectures are proposed to learn the symmetry and geometry constraints, to fully aggregate the information from all views. Extensive experiments have been conducted on the public DDSM dataset and our in-house dataset, and state-of-the-art (SOTA) results have been obtained in terms of mammogram mass detection accuracy.

Keywords: Mammogram · Deep learning · Multi-view · Mass

1 Introduction

Mammography is widely used as a cost-effective early detection method for breast cancer. A computer-aided diagnosis (CAD) system has the promise to detect the abnormal regions on digitized mammogram images. Significant progress has recently been made in the performance of CAD systems, especially with the advance of DNN-based methods. Nonetheless, mammographic abnormality detection remains challenging, largely due to the high accuracy requirement set by the clinical practice. In this paper, we focus on mass detection in mammograms.

A standard mammography screening procedure acquires two low-dose X-ray projection views for each breast, a craniocaudal (CC) view and a mediolateral

---

Z. Yang and Z. Cao—Equal contribution.

Electronic supplementary material The online version of this chapter (https://doi.org/10.1007/978-3-030-59725-2_20) contains supplementary material, which is available to authorized users.

A. L. Martel et al. (Eds.): MICCAI 2020, LNCS 12266, pp. 200–210, 2020.
https://doi.org/10.1007/978-3-030-59725-2_20

oblique (MLO) view. Radiologists routinely use all views in breast cancer diagnosis. The ipsilateral analysis refers to the diagnosis based on the CC and MLO views of the same breast, while the bilateral analysis combines the findings from the same views of the two breasts. For example, the radiologists may cross-check the lesion locations through the ipsilateral analysis, and use the symmetry information from the bilateral analysis to improve the decision accuracy.

Many of the previous work on mammographic lesion detection focus on one view, therefore unable to capture the rich information from the multiple view analysis. Recently several DNN-based dual-view approaches have been proposed, performing either ipsilateral or bilateral analysis. In this paper, we present MommiNet (MammOgraphic Multi-view Mass Identification NETworks [maa·mee·net]), the first DNN-based architecture to perform tri-view based mass detection. Our main contributions include: 1) the first tri-view DNN architecture to perform joint end-to-end ipsilateral and bilateral analysis, to fully aggregate information from all views; 2) a novel relation network designed in tandem with a DNN-based nipple detector to incorporate geometry constraint across views; and 3) SOTA Free-Response Operating Characteristic (FROC) performance has been achieved on both DDSM and in-house datasets.

2   Related Work

The importance of information fusion from multi-view mammograms has been recognized previously. A detailed review of multi-view information fusion is given in [11], covering mostly traditional methods, which rely on handcrafted features and fusion rules. In this paper, we focus on DNN-based approaches due to its wide success in medical image processing and computer vision in general.

Deep learning has been applied to mammographic mass detection, and most of the work focus on single view based approaches [1,3,4,14,15,29,30]. Recently, multi-view based approaches have gained popularity. In [5,21,22], different DNN-based approaches have been presented for ipsilateral analysis of multi-view mammograms. However, most of these approaches do not model the geometry relation across views explicitly. In [18], a cross-view relation network is added to the Siamese Networks for mass detection. However, this approach uses the same geometric features and embedding for the relation network as in [9], which was designed for single view object detection. In [7,17], DNN-based approaches are presented for bilateral analysis. Other related work includes an RNN-based multi-view approach for mass classification [25], and DNN-based multi-view approaches for breast cancer screening [19,28].

In our approach, a Faster-RCNN Network with Siamese input module and a DeepLab Network with Siamese input module are working in parallel to perform the ipsilateral and bilateral analysis simultaneously. Unlike [18], our relation network is explicitly designed to encode the mass to nipple distance for the ipsilateral analysis, in tandem with a DNN-based nipple detector. The mass to nipple distance has been used before [23], and our approach is the first to explicitly embed this prior knowledge into a DNN architecture.

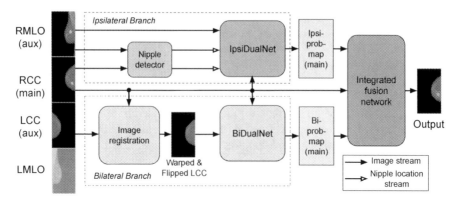

Fig. 1. Framework of MommiNet.

3   Materials and Methods

3.1   Datasets

Public Dataset. We leverage the widely used Digital Database for Screening Mammography (DDSM) [13] as our public dataset. DDSM has 2,620 patient cases, each of which has four views of mammograms (two views of each breast: left CC & MLO, right CC & MLO). Excluding some defective/corrupted cases, 2,578 cases (10,312 images in total) are used in this work. All cases are randomly divided into the training, validation, and test sets by approximately 8:1:1, resulting in 8,256, 1,020 and 1,036 images in the respective sets.

In-House Dataset. We are collaborating with a hospital to validate our proposed methods. Our in-house mammogram dataset has 2,749 cases, consisting of normal, cancer, and benign cases, which are close to the practical distribution. Lesion regions are first annotated by two radiologists and then reviewed by a senior radiologist. Same as the division strategy on the public DDSM dataset, all cases are randomly split by 8:1:1 and the training, validation and test sets have 8,988, 1,120, and 1,120 images, respectively. Note that some cases have mammograms taken at multiple dates.

3.2   System Overview

Figure 1 outlines the framework of MommiNet. At first one view is selected as the main image and its corresponding ipsilateral view and bilateral view are selected as the auxiliary views, together as input into MommiNet. As in Fig. 1, the main image ("RCC (main)") and the corresponding ipsilateral view ("RMLO (aux)") are input together into the ipsilateral branch. In parallel, the main image and the bilateral view ("LCC (aux)") are input into the bilateral branch. Each branch generates a probability map of the main image ("Ipsi-prob map" and "Bi-prob map") and the probability maps are then input into the integrated fusion network

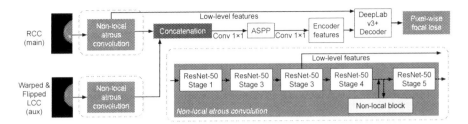

Fig. 2. Architecture of BiDualNet. *ASPP*: Atrous Spatial Pyramid Pooling. The blue dashed subfigure enlarges the "Non-local atrous convolution", where the only difference is that the main stream outputs low-level features at Stage 3. (Color figure online)

along with the main image, to generate the final output. Inside the ipsilateral branch, a DNN-based nipple detector is added to extract the nipple locations on both views ("RCC" and "RMLO") before input into *IpsiDualNet* along with the two views for ipsilateral analysis. Inside the bilateral branch, the bilateral view ("LCC") is first registered before input along with the main image into *BiDualNet*. This combined ipsilateral and bilateral analysis can be applied to any given view as the main image. In practice, we apply it to all views available to generate mass detection on each view.

3.3   Image Pre-processing

Image Registration. To facilitate the DNN-based learning of the symmetry constraint from the bilateral images, we register the input pair of the same view images (e.g. two CC view images or two MLO view images). The auxiliary image is horizontally flipped, and then warped toward the main image according to the breast contours. In particular, nipple locations are used to roughly align the two MLO images before warping. A warped CC view example is shown in Fig. 1.

Nipple Detection. Nipple locations are required in image registration and IpsiDualNet. A Faster-RCNN based keypoint detection framework [8] is used to identify the nipple locations with satisfactory accuracy. For example, there is only one incorrect nipple prediction in our in-house dataset (total 11,228 images).

3.4   BiDualNet

Most women have roughly symmetric breasts in terms of density and texture [6]. This property is well leveraged by radiologists to identify the abnormalities in mammograms. Hinging on a bilateral dual-view, radiologists are able to locate a mass based on its distinct morphologic appearance and relative position compared to its corresponding area in the other lateral image.

To incorporate this diagnostic prior information and facilitate the learning of the symmetry constraint, we develop *BiDualNet* (*Bi*lateral *Dual*-view *Net*work) as illustrated in Fig. 2. It is derived from a DeepLab v3+ structure, enhanced

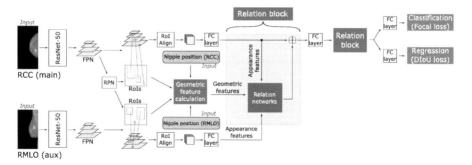

Fig. 3. Architecture of IpsiDualNet.

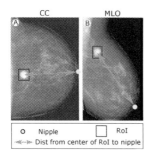

Fig. 4. Similarity of RoI-to-nipple distances

Table 1. Impact of different geometric features on prediction performance on our in-house dataset.

Geometric features	R@0.5	R@1.0	R@2.0
Shape and location of RoI (i.e., Eq. (2) in [18])	0.86	0.90	0.93
Dummy nipple point (Central point of every image)	0.80	0.85	0.89
RoI-to-nipple distance (Ours, in IpsiDualNet)	**0.88**	**0.92**	**0.96**

with a Siamese input module [21], a non-local (NL) module [26] and the pixel-wise focal loss (PWFL) function [16]. The Siamese input module consists of two NL atrous convolution modules. They share the same weights, extracting features from the bilateral images in the same manner. The auxiliary feature map is then assumed as a reference and concatenated with the main feature map, and in turn the feature difference at the same location can highlight the abnormality. After a 1×1 convolution and the ASPP module, deeplab v3+ decoder finally generates a segmentation result for the main image.

3.5   IpsiDualNet

Ipsilateral images provide information of the same breast from two different views. Hence, a mass in the ipsilateral images usually presents similar brightness, shapes, sizes and distances to the nipple. This prior knowledge is essential to help radiologists make a decision. To incorporate this prior diagnostic knowledge, we design *IpsiDualNet* (*Ipsi*lateral *Dual*-view *Net*work) as illustrated in Fig. 3. It is built on the Faster-RCNN detection architecture, by adding the Siamese input

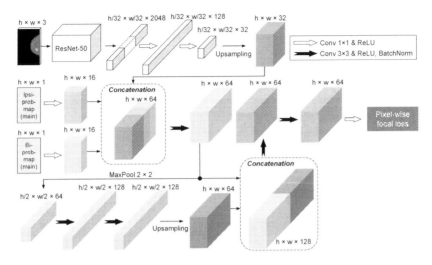

Fig. 5. Integrated fusion network

module, Feature Pyramid Networks (FPN) module [8], and the designed relation block. The Siamese input module with FPN enables the two input branches to share the same weights and extract the features from the two ipsilateral views in the same way. In turn the appearance similarity and the geometry constraint between RoIs from the two branches are computed using the proposed relation block (described in the next paragraph). Finally, the mass regions in the main image are detected and converted into a probability map. In addition, focal loss (FL) [16] and Distance-IoU loss (DIoU) [31] are used to improve the performance of *IpsiDualNet*, and training with negative samples (normal cases) is enabled.

Relation networks [9] model the attention-based relationships [24] between two RoIs in single image based on the similarity of their appearance and geometric features, leading to an improvement of detection accuracy. Inspired by [9], we propose a new relation block that emphasizes the appearance and geometric similarities of a lesion RoI in two ipsilateral images. Equation (1) defines the appearance similarity weight ω_A^{ij} between the ith RoI $\mathbf{f}_m^i$ in main image and the jth RoI $\mathbf{f}_a^j$ in auxiliary image. Regarding the geometric similarity, Eq. (2) considers RoIs' geometric factors $\mathbf{g}_t^k = \{d_t^k, w_t^k, h_t^k\}$, including the RoI-to-nipple distance, RoI width and height, where the subscript "t" indicates "m" or "a" (the main or auxiliary image). Other variables in Eqs. (1) and (2) have the same meaning as described in [9]. The lesion-to-nipple distance is routinely leveraged by radiologists for estimating a lesion, since this distance is approximately the same in both CC and MLO views [10,27]. Figure 4 shows an example of similarity of RoI-to-nipple distances.

$$\omega_A^{ij} = \frac{dot(W_a\mathbf{f}_a^j, W_m\mathbf{f}_m^i)}{\sqrt{D}}, \qquad i \in \{1,2,...,I\}, j \in \{1,2,...,J\}, \qquad (1)$$

Table 2. Ablation study on ipsilateral and bilateral branches on DDSM.

Networks	Ipsilateral (Recall@FPI)			Bilateral (Recall@FPI)		
	R@0.5	R@1.0	R@2.0	R@0.5	R@1.0	R@2.0
BiDualNet	0.67	0.81	**0.89**	**0.78**	**0.84**	**0.89**
IpsiDualNet w/o Relation Blocks	0.68	0.77	0.84	0.73	0.79	0.84
IpsiDualNet	**0.76**	**0.83**	0.88	0.65	0.75	0.82

Table 3. Performance comparison of various methods on the DDSM dataset. *CVR-RCNN*: Cross-View Relation Region-based Convolutional Neural Network; *CBN*: Contrasted Bilateral Network

View	Method	DDSM	Recall@FPI		
		(train/val/test)	R@0.5	R@1.0	R@2.0
Single	Campanini et al. [2]	1400/_/512	∼0.54	∼0.74	∼0.86
	Nazaré Silva et al. [20]	349/150/100	n/a	∼0.8033	n/a
	Faster-RCNN [17]	80%/10%/10%	0.6610	0.7246	0.7839
	Mask-RCNN [17]		0.6441	0.7458	0.8178
	DeepLab+NL+PWFL	8256/1020/1036	0.68	0.78	0.83
	Faster-RCNN+FPN+FL+DIoU		0.74	0.82	0.88
Dual	CVR-RCNN [18]	410/_/102	n/a	n/a	∼0.88
	CBN [17]	80%/10%/10%	0.6907	0.7881	0.8559
	BiDualNet	8256/1020/1036	0.78	0.84	0.89
	IpsiDualNet		0.76	0.83	0.88
Tri	MommiNet	8256/1020/1036	**0.80**	**0.85**	**0.89**

$$\mathcal{E}(\mathbf{g}_m^i, \mathbf{g}_a^j) = \mathcal{E}([\log(\frac{d_m^i}{d_a^j}), \log(\frac{w_m^i}{w_a^j}), \log(\frac{h_m^i}{h_a^j})]^T), \tag{2}$$

3.6　Integrated Fusion Network

To integrate the outputs of both ipsilateral and bilateral learning, we design the novel integrated fusion network that accepts three inputs: the main image and the two probability maps from the IpsiDualNet and BiDualNet (shown in Fig. 1). These two probability maps are attentions of the comprehensive information from both bilateral and ipsilateral images. This strategy is also applied for mammogram cancer screening as in [28]. Figure 5 describes the architecture of integrated fusion network. Note that the ResNet-50 backbone comes from IpsiD-ualNet and is frozen during the training process. The final prediction result is generated by this integrated fusion network.

4 Experimental Results

In this section, we perform extensive experiments on the DDSM and our in-house datasets to validate our proposed methods. The model on each dataset is independently trained based on the pre-trained ImageNet model [12]. FROC is selected as our evaluation metric to compare with previous work. A mass is assumed as successfully identified if the Intersection-over-Union (IoU) of a prediction output and the ground truth is greater than 0.2.

4.1 Ablation Study on Ipsilateral and Bilateral Learning

We train and test BiDualNet, IpsiDualNet and its degraded version "IpsiDual-Net w/o Relation Blocks" (the two feature streams from the main and auxiliary images are directly concatenated after the RoI align stage) on both ipsilateral and bilateral images. Table 2 shows their results on DDSM. It can be observed that BiDualNet always achieves the highest recall scores on bilateral images, and IpsiDualNet has generally better performance on ipsilateral images. It is also clear that IpsiDualNet outperforms IpsiDualNet w/o Relation Blocks on ipsilateral images which suggests that the designed relation module remarkably enhances IpsiDualNet. Thus, BiDualNet and IpsiDualNet are respectively applied to bilateral and ipsilateral analysis in MommiNet. We believe segmentation network maintains spatial information therefore more appropriate for symmetry learning, while detection network is more suitable for RoI-based relation learning.

Table 4. Performance comparison of various methods on the in-house dataset.

View type	Method	R@0.5	R@1.0	R@2.0
Single-view	DeepLab+NL+PWFL	0.81	0.84	0.90
	Faster-RCNN+FPN+FL+DIoU	0.82	0.89	0.91
Dual-view	BiDualNet	0.87	0.93	0.95
	IpsiDualNet	0.88	0.92	0.96
Tri-view	MommiNet	**0.90**	**0.94**	**0.96**

We also study the impact of different geometric features on IpsiDualNet, including the feature in [18], the dummy nipple, and our RoI-to-nipple-distance based feature as in Table 1. It demonstrates that the RoI-to-nipple-distance based geometric features generate the best performance of IpsiDualNet.

4.2 Results on the DDSM and In-House Datasets

Table 3 compares the performance of various single-view, dual-view and tri-view methods on DDSM. The approaches in references [2,17,18,20], reported evaluation on DDSM with normal patients data using the FROC metric, are selected as competing methods. Among the single-view methods, our Faster-RCNN with the modules of FPN, FL, and DIoU loss is superior to others. Compared to other recent dual-view methods, our BiDualNet and IpsiDualNet achieve better recall results at the respective FPI. Finally, our proposed tri-view MommiNet surpasses all the methods above. To the best of our knowledge, MommiNet achieves the highest recall score on DDSM among all published works.

Various methods are also tested on the in-house dataset as shown in Table 4. Our designed two dual-view networks are constantly better than single-view methods, and the tri-view MommiNet again achieves the best recall result at all FPIs. Furthermore, due to the DR images' better quality, the proposed method achieves remarkably higher recall on the in-house dataset than DDSM. It is worth noting that our system is capable of accepting mammograms with incomplete views (please refer to Table A3 in Supplementary Material for our additional evaluation results in this situation).

5   Conclusions

In this paper, we proposed a novel multi-view DNN architecture MommiNet to perform joint ipsilateral and bilateral analysis on mammograms for high precision mass detection. By carefully designing the DNN architecture, MommiNet can effectively learn the geometry constraint and symmetry constraint from the ipsilateral and bilateral analysis respectively. Its efficacy can be further verified by our extensive experiment results and the SOTA FROC performance achieved on both the DDSM dataset and our in-house dataset. Our future work includes multimodal learning to further improve mass detection accuracy.

References

1. Agarwal, R., Diaz, O., Lladó, X., Yap, M.H., Martí, R.: Automatic mass detection in mammograms using deep convolutional neural networks. J. Med. Imaging **6**(3), 1–9 (2019)
2. Campanini, R., et al.: A novel featureless approach to mass detection in digital mammograms based on support vector machines. Phys. Med. Biol. **49**(6), 961 (2004)
3. Cao, Z., et al.: Deep learning based mass detection in mammograms. In: GlobalSIP, pp. 1–5 (2019)
4. Cao, Z., et al.: DeepLIMa: deep learning based lesion identification in mammograms. In: Proceedings of the IEEE International Conference on Computer Vision Workshops (2019)
5. Carneiro, G., Nascimento, J., Bradley, A.P.: Automated analysis of unregistered multi-view mammograms with deep learning. IEEE Trans. Med. Imaging **36**(11), 2355–2365 (2017)
6. Cunningham, D.: The Ups and Downs of Breasts, Physicians & Midwives (2013). https://physiciansandmidwives.com/2013/12/11/ups-and-downs-of-breasts/

7. Diniz, J.O.B., Diniz, P.H.B., Valente, T.L.A., Silva, A.C., de Paiva, A.C., Gattass, M.: Detection of mass regions in mammograms by bilateral analysis adapted to breast density using similarity indexes and convolutional neural networks. Comput. Methods Programs Biomed. **156**, 191–207 (2018)

8. Facebook: Fast, modular reference implementation of Instance Segmentation and Object Detection algorithms in PyTorch (2019). https://github.com/facebookresearch/maskrcnn-benchmark

9. Hu, H., Gu, J., Zhang, Z., Dai, J., Wei, Y.: Relation networks for object detection. In: 2018 IEEE/CVF Conference on Computer Vision and Pattern Recognition, pp. 3588–3597 (2017)

10. Ikeda, D., Miyake, K.K.: Breast Imaging: The Requisites E-Book. Elsevier Health Sciences (2016)

11. Jouirou, A., Baâzaoui, A., Barhoumi, W.: Multi-view information fusion in mammograms: a comprehensive overview. Inf. Fusion **52**, 308–321 (2019)

12. Krizhevsky, A., Sutskever, I., Hinton, G.E.: ImageNet classification with deep convolutional neural networks. In: Advances in Neural Information Processing Systems, pp. 1097–1105 (2012)

13. Lee, R.S., Gimenez, F., Hoogi, A., Miyake, K.K., Gorovoy, M., Rubin, D.L.: A curated mammography data set for use in computer-aided detection and diagnosis research. Sci. Data **4**, 170177 (2017)

14. Li, H., Chen, D., Nailon, W.H., Davies, M.E., Laurenson, D.: A deep dual-path network for improved mammogram image processing. In: ICASSP 2019–2019 IEEE International Conference on Acoustics, Speech and Signal Processing (ICASSP), pp. 1224–1228. IEEE (2019)

15. Li, Y., Chen, H., Zhang, L., Cheng, L.: Mammographic mass detection based on convolution neural network. In: 2018 24th International Conference on Pattern Recognition (ICPR), pp. 3850–3855. IEEE (2018)

16. Lin, T.Y., Goyal, P., Girshick, R., He, K., Dollár, P.: Focal loss for dense object detection. In: Proceedings of the IEEE International Conference on Computer Vision, pp. 2980–2988 (2017)

17. Liu, Y., et al.: From unilateral to bilateral learning: detecting mammogram masses with contrasted bilateral network. In: Shen, D., et al. (eds.) MICCAI 2019. LNCS, vol. 11769, pp. 477–485. Springer, Cham (2019). https://doi.org/10.1007/978-3-030-32226-7_53

18. Ma, J., et al.: Cross-view relation networks for mammogram mass detection. arXiv abs/1907.00528 (2019)

19. McKinney, S., et al.: International evaluation of an AI system for breast cancer screening. Nature **577**, 89–94 (2020)

20. de Nazaré Silva, J., de Carvalho Filho, A.O., Silva, A.C., De Paiva, A.C., Gattass, M.: Automatic detection of masses in mammograms using quality threshold clustering, correlogram function, and SVM. J. Digit. Imaging **28**(3), 323–337 (2015)

21. Perek, S., Hazan, A., Barkan, E., Akselrod-Ballin, A.: Siamese network for dual-view mammography mass matching. In: Stoyanov, D., et al. (eds.) RAMBO/BIA/TIA -2018. LNCS, vol. 11040, pp. 55–63. Springer, Cham (2018). https://doi.org/10.1007/978-3-030-00946-5_6

22. Ren, Y., et al.: Multiview mammographic mass detection based on a single shot detection system. In: Medical Imaging 2019: Computer-Aided Diagnosis, vol. 10950, p. 109500E. International Society for Optics and Photonics (2019)

23. Sahiner, B., et al.: Joint two-view information for computerized detection of microcalcifications on mammograms. Med. Phys. **33**(7Part1), 2574–2585 (2006)

24. Vaswani, A., et al.: Attention is all you need. In: Advances in Neural Information Processing Systems, pp. 5998–6008 (2017)
25. Wang, H., et al.: Breast mass classification via deeply integrating the contextual information from multi-view data. Pattern Recogn. **80**, 42–52 (2018)
26. Wang, X., Girshick, R., Gupta, A., He, K.: Non-local neural networks. In: Proceedings of the IEEE Conference on Computer Vision and Pattern Recognition, pp. 7794–7803 (2018)
27. Wei, J., et al.: Computer-aided detection of breast masses on mammograms: dual system approach with two-view analysis. Med. Phys. **36**(10), 4451–4460 (2009)
28. Wu, N., et al.: Deep neural networks improve radiologists' performance in breast cancer screening. IEEE Trans. Med. Imaging **39**(4), 1184–1194 (2019)
29. Xi, P., Shu, C., Goubran, R.: Abnormality detection in mammography using deep convolutional neural networks. In: 2018 IEEE International Symposium on Medical Measurements and Applications (MeMeA), pp. 1–6. IEEE (2018)
30. Zhang, F., et al.: Cascaded generative and discriminative learning for microcalcification detection in breast mammograms. In: Proceedings of the IEEE Conference on Computer Vision and Pattern Recognition, pp. 12578–12586 (2019)
31. Zheng, Z., Wang, P., Liu, W., Li, J., Ye, R., Ren, D.: Distance-IOU loss: faster and better learning for bounding box regression. arXiv preprint arXiv:1911.08287 (2019)

Multi-site Evaluation of a Study-Level Classifier for Mammography Using Deep Learning

Dustin Sargent[1(✉)], Sun Young Park[1], Amod Jog[2], Aly Mohamed[3], and David Richmond[3]

[1] IBM Watson Health, San Diego, CA 92121, USA
dsargent@us.ibm.com
[2] IBM Watson Health, San Jose, CA 95120, USA
[3] IBM Watson Health, Cambridge, MA 02142, USA

Abstract. We present a computer-aided diagnosis algorithm for mammography trained and validated on studies acquired from six clinical sites. We hold out the full dataset from a seventh hospital for testing to assess the algorithm's ability to generalize to new sites. Our classifiers are convolutional neural networks that take multiple input images from a mammography study and produce classifications for the study. The studies are globally labeled as normal, biopsy benign, high risk or biopsy malignant. We report on experimental results from several network variants, including study-level and breast-level models, single- and multiple-output models, and a novel model architecture that incorporates prior studies. Each model variation includes an image-level classifier that is pre-trained with per-image labels and is used as a feature extractor in our study-level models. Our best study-level model achieves 0.85 area under the ROC curve for normal vs malignant classification on the held-out test site. In comparison with other recent work, we achieve a similar level of classification sensitivity and specificity on a dataset with greater site and vendor variation. Additionally, our test performance is demonstrated on a held-out site to more accurately assess how the model would perform when deployed in the field.

Keywords: Convolutional neural networks · Computer-aided diagnosis · Breast cancer

1 Introduction

Breast cancer is the leading cause of new cancer cases among women in the US according to the American Cancer Society's Cancer Facts and Figures for 2019 [1]. Despite this, it is only the third leading cause of cancer-related deaths among women, which suggests that outcomes for breast cancer patients are significantly improved by early detection and treatment. Mammography is the primary and most effective [2,3] screening method for breast cancer, but the high

A. L. Martel et al. (Eds.): MICCAI 2020, LNCS 12266, pp. 211–219, 2020.
https://doi.org/10.1007/978-3-030-59725-2_21

occurrence rate of breast cancer creates a large demand and workload that is difficult for mammographers to keep up with. Additionally, diagnosis by a consensus assessment between two radiologists is very effective in improving screening and reducing unnecessary patient recalls. This is standard practice in England [4] and improves patient outcomes, but it also increases the number of exams that must be read.

Fortunately, advances in deep learning for medical imaging have shown that artificial intelligence is a promising candidate to provide additional mammography readers to alleviate this burden. Several recent publications have showcased applications of deep learning models to mammography image classification that achieve accuracy comparable to radiologists. Some examples include image-level classification of digital mammograms explored in [5,6], breast density classification in [7], and local methods that incorporate annotations of masses and calcifications such as [8]. These and most of the other existing work operate at the image or patch level, which does not take advantage of looking at the entire study to perform classification.

Recent papers have started to explore the challenging task of study-level assessment; for example, in [9], the authors present a study-level classifier that produces a benign vs malignant classification for each breast. Their method also takes advantage of local annotations to include heatmaps as input to the classifier, achieving classification accuracy similar to the study annotators. Their results were demonstrated on a very large dataset with a limited amount of site variation. A recent paper from Google [10] demonstrated superior diagnostic accuracy over six readers on a small dataset. Finally, Lunit [11] also presented a deep learning classifier for mammography with diagnostic accuracy rivaling that of experienced radiologists on a large dataset taken from a limited geographical area. These and other contributions have demonstrated the potential of deep learning to read a subset of cases, reducing the workload of radiologists, or to serve as second readers, flagging potential missed cancers.

Table 1. Number of studies per clinical site.

Site	A	B	C	D	E	F	G (test)	Total
Studies	10,623	2,048	10,503	5,089	18,940	2,248	1,068	50,519

However, as proven in [12], deep learning models trained on mammography datasets from a limited number of clinical sites often fail to generalize to unseen data. Even if the size of the dataset is large, mammography images acquired from a small number of clinical sites tend to feature similar patient demographics and lack variety in device vendors and settings. The datasets used in previous studies have had this characteristic to some degree, which creates some concern as to the generalization capabilities of their models. We address this issue by training deep learning classifiers on a large mammography dataset taken from a greater variety of clinical sites than previous work in this area. Our dataset includes studies collected from multiple sites across the United States and in the UK. In particular, we ensure that our classifiers can generalize to unseen data by holding out an entire site's data for testing our models. We evaluate several model architectures and demonstrate that our best model does not suffer a significant decrease in performance when applied to the unseen site data.

Our main contributions are: (1) a mammography classifier that looks at the whole study rather than images or patches, (2) an analysis of multiple model architectures to determine which generalizes best in a practical scenario, (3) a best-performing model that outperforms the recent work on our test set, and (4) a demonstration that the model performs well on the screening assessment task, and could therefore serve as a second opinion for breast cancer screening.

2   Methods

2.1   Data

Our algorithms are trained on a dataset of mammography studies taken from seven clinical sites in the US and UK. Each study contains at least one each of the four standard mammography views (left and right craniocaudal (CC) and mediolateral oblique (MLO)). As many cases also include non-standard diagnostic views, we allow CC-like (XCCM, XCCL, etc.) and MLO-like (ML, LM, etc.) views to substitute for their respective standard views throughout our work in order to process all images in a study. Each image is labeled globally as either normal, benign, or malignant. These labels correspond to cases that were determined to be normal during screening, cases that have at least one biopsy-confirmed benign finding (and no malignant findings), and cases that have at least one biopsy-confirmed malignant finding (along with possible benign findings). A large number of the studies in our dataset al.so have at least one associated prior study. Table 1 reports the number of available studies per clinical site.

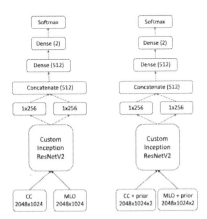

Fig. 1. Side-based model architectures: (left) single-channel model, and (right) two-channel model with corresponding prior images as the second input channel.

2.2   Model Architecture

Each of our models contain a single pre-trained image-level classifier that takes a single view and produces a (normal, benign) versus (malignant) classification for that view. This image-level classifier consists of a single-channel Inception-ResNetV2 [13] model cut off at an intermediate layer and followed by two fully-connected layers terminating in a two-class output layer, with shared weights for all input images. This allows us to avoid the GPU memory issues that arise from handling high resolution mammography images, without sacrificing accuracy compared to the full InceptionResNetV2 model. Our baseline model, included for comparison, classifies a left and right CC- and MLO-like view from a study at the image level, averages the CC and MLO results from each side, and returns the maximum of the two sides. The image-level modules in all of our study-level models were initialized with the pre-trained weights from this baseline model, which increased accuracy and sped up training convergence.

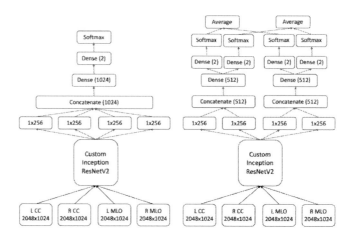

Fig. 2. Study-based model architectures: (left) one output for the whole study, and (right) one output per side.

We evaluate four variations of study-level mammography classifiers: a side-based model, a study-based model that produces a single label for the whole study, a study-based model that produces a label for each side, and a side-based model that takes as input corresponding images from the current and first prior exams stacked as two-channel images. The architectures of these models are shown in Fig. 1 and 2. Each model features a single instance of the image-level classifier that processes all of the input images with shared weights. For the side-based models, a study-level label is produced by taking the maximum result between the two sides. We use the maximum finding as ground truth for each study; that is, if any image in the study contains a malignant finding, the entire study is labeled as malignant; otherwise, the study is labeled as (normal, benign).

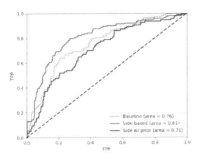 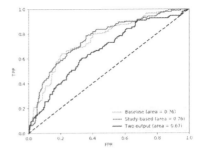

Side-based models vs baseline   Study-based models vs baseline

Fig. 3. Test ROC curves for different models on the held-out site. (a) Comparison of side-based models: ROC-AUC of 0.81 (side-based), 0.71 (side-based w/prior). (b) Comparison of study-based models: ROC-AUC of 0.76 (study-based), and 0.67 (two output). Baseline ROC-AUC is 0.76 ((a) and (b)).

2.3   Training

We filter our dataset by removing images containing implants, spot compression images, and magnified or partial views. We then center the tissue region in the image, and resize images to 2048 rows and 1024 columns. As our dataset has many more normal cases than cases with malignant biopsies, we use augmentation during training and ensure that each batch has a balance of positive and negative cases. We apply random rotation, translation, zoom, brightness shift and horizontal flipping to each image. Finally, the 16-bit input images are rescaled to the range $(-1, 1)$, and we train with a batch size of 4. Our batch generator is implemented such that each negative case is sampled once per epoch, which means that each positive case is sampled several times per epoch with different augmentation. Our models are implemented in Keras [16] with Tensorflow [17] as the backend, and we use the Adam optimizer with an initial learning rate of 0.0001. The best version of each model is selected by choosing the epoch with the best validation ROC-AUC.

3   Results

We created several training/test/validation splits in which one of our seven clinical sites was held out as the test set, and the remaining six sites were split up with 80% for training and 20% for validation. The data from one particular site was held out for testing by quality assurance, so we feel that the results from that site are the most representative of the model's generalization capability. This results section mainly reports on performance from that site. Performance was also verified on the other sites, and followed the same trends reported here.

We compared the accuracy of our four model variations to the baseline on the most difficult held-out test site (Fig. 3). All test ROC-AUC scores were obtained from the same study-level test sets, combining the left and right breast output by

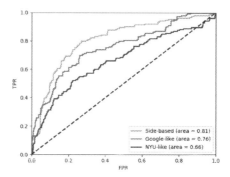

Fig. 4. Comparison of our best model vs previously published models. ROC-AUC on held-out site: 0.81 (ours), 0.76 (Google-like), and 0.66 (NYU-like).

max operations when necessary. As the figure shows, the side-based model was the only model to significantly outperform the baseline. This result repeated itself over multiple tests with different random splits and held-out sites. We also applied our best-performing model to a pipeline including inference-time augmentation and filtering out very high density, implant and low quality images. This produced a 4% improvement in the test results, bringing the ROC-AUC score up from 0.81 to 0.85.

The test results summarized in Table 2 show that the more complex models often achieved better validation performance than the side-based model, but performed worse in testing. This suggests that the negative effect of overfitting caused by greater model complexity was stronger than the benefit gained by providing more input data to the model. This may also be related to the distribution and labeling of our data, which we discuss in Sect. 4. The side-based model with priors achieved 0.93 validation but only 0.71 test ROC-AUC, displaying the same trend as the complex models.

Table 2. Comparison of different models on validation and held-out test set.

	Validation ROC-AUC	Test ROC-AUC
Proposed architectures		
Baseline (image-based)	–	0.76
Side-based	0.85	**0.81**
Study-based	0.77	0.76
Study-based (two-output)	**0.92**	0.67
Published architectures		
Google-like	0.88	0.76
NYU-like	0.70	0.66

We also compared our best-performing model against our implementations of recent models reported in [9] and [10], which we refer to as the 'NYU-like' and 'Google-like' models. For a fair comparison, we replaced the ResNet18 and ResNet50 [14,15] in those models with our pre-trained InceptionResNetV2, and modified the input and output layers accordingly. Our NYU-like model is identical to their view-wise model other than these changes. Due to GPU memory constraints, the post-image CNN component in our Google-like model is not as deep as in [10]; we maintain the same architecture but reduce the number of bottleneck blocks.

Figure 4 compares these results to our best model, which outperforms both on this test set. The Google-like model achieves a result similar to our study-based model, and the NYU-like model is similar to our two output model. The overall trend is the same as above, in which simple models outperform complex models the test set. We repeated these tests on three other test sites, obtaining test AUC values of (0.85, 0.89, 0.83) for our model, (0.81, 0.84, 0.79) for the Google-like model and (0.73, 0.78, 0.72) for the NYU-like model.

Like our models, the Google-like model achieved a high validation score that dropped significantly on the test set, as shown in Table 2. The Google-like model used a single image-level model to process all four images, outperforming the NYU-like model that used a separate image-level model for the CC and MLO images. The inclusion of separate image-level components for CC and MLO or for left and right consistently reduced performance on our dataset no matter which variant of the NYU-like model we tested. This also held when we tested our study-level models with two image-level components.

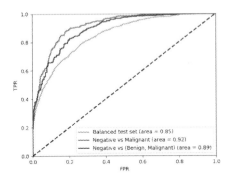

Fig. 5. Performance comparison on different tasks. ROC-AUC of our best model: 0.85 on balanced test set, 0.92 on negative vs malignant, and 0.89 on negative vs (benign, malignant).

Finally, we evaluated our best-performing side-based model on an additional training/validation/test data split, and three different classification tasks. We generated a balanced dataset in which the training, validation, and test splits all contain a balanced mix of data from our seven clinical sites, rather than the test set being formed exclusively from one site. This test set provides performance estimates that are more comparable to other work versus the lower values obtained from our most difficult single-site test set. We re-trained our best model on this dataset for the same (negative, benign) vs malignant classification task as above. Next, we found that benign biopsy cases in our data were the most difficult to classify correctly. Therefore, we re-trained our best model on a dataset in which only the negative and malignant biopsy cases were included (negative vs malignant). We also re-trained our best model for the negative vs (benign, malignant) classification task. That is, any cases that were sent for biopsy were labeled as positive cases. In other words, this classification task mimics the screening diagnosis rather than the biopsy results. The results of these additional comparisons are shown in Fig. 5. These results show that our best model compares favorably with the recent work on our test data, with test ROC-AUC rising from 0.81 to 0.85 when evaluated on a balanced test set. The high performance on the negative versus benign task confirms our observation that benign cases are the hardest to classify. This is also shown by the model's strong performance on the screening diagnosis task (0.89 test ROC-AUC), in which benign biopsy cases are added to the positive class. This result indicates that our model is a suitable candidate to serve as a second reader in mammography reading to increase sensitivity and reduce false positives.

4   Conclusion

We have presented study of multiple deep learning models that perform breast cancer classification at the study level, rather than the image or patch level as in much of the existing work. Our best model outperforms recently-published methods on test datasets from clinical sites that are totally withheld from training and validation. This test scenario demonstrates that our algorithm is capable of generalizing to new data when deployed in a realistic clinical setting. We have also demonstrated that the same algorithm can be applied to negative versus high-risk screening and other classification tasks with high accuracy. This demonstrates the potential of our classifier to serve as a second reader, to help boost the sensitivity and specificity of mammography reading and alleviate burdens for radiologists.

One interesting conclusion of our work was that a side-based model using the inter-side maximum as the final classification outperformed models that looked at all four images together. We believe that this was because our data is annotated on a per-side basis, and our data does not have many cases with asymmetrical findings that would benefit from simultaneous assessment. The prior model may also have been affected by the smaller training set available due to being limited to patients with priors. For our future work, we are in the process of gathering additional site data and annotations for new cases. We intend to use this data to improve our results and further investigate the discrepancy in test accuracy between side-, study- and prior-based models found in this work.

References

1. American Cancer Society: Cancer facts & figures 2019. American Cancer Society, Atlanta (2019)
2. Duffy, S.W., et al.: The impact of organized mammography service screening on breast carcinoma mortality in seven Swedish country. Cancer **95**(3), 458–469 (2002)
3. Kopans, D.B.: Beyond randomized controlled trials: organized mammographic screening substantially reduces breast carcinoma mortality. Cancer **94**(2) (2002)
4. NHS Breast Screening Programme: Clinical guidance for breast cancer screening assessment. Public Health England, London (2016)
5. Geras, K.J., et al.: High-resolution breast cancer screening with multi-view deep convolutional neural networks. arXiv:1703.07047 (2017)
6. Lotter, W., Sorensen, G., Cox, D.: A multi-scale CNN and curriculum learning strategy for mammogram classification. In: Cardoso, M.J., et al. (eds.) DLMIA/ML-CDS -2017. LNCS, vol. 10553, pp. 169–177. Springer, Cham (2017). https://doi.org/10.1007/978-3-319-67558-9_20
7. Wu, N., et al.: Breast density classification with deep convolutional neural networks. In: ICASSP (2018)
8. Ribli, D., Horvath, A., Unger, Z., Pollner, P., Csabai, I.: Detecting and classifying lesions in mammograms with deep learning. Sci. Rep. **8**(1) (2018)
9. Wu, N., et al.: Deep neural networks improve radiologists' performance in breast cancer screening. IEEE Trans. Med. Imaging **1**(1), 99 (2019)
10. McKinney, S.M., et al.: International evaluation of an AI system for breast cancer screening. Nature **577**(7788), 89–94 (2020)
11. Kim, E.K., et al.: Applying data-driven imaging biomarker in mammography for breast cancer screening: preliminary study. Sci. Rep. (2018)
12. Wang, X., Liang, G., Zhang, Y., Blanton, H., Bessinger, Z., Jacobs, N.: Inconsistent performance of deep learning models on mammogram classification. J. Am. Coll. Radiol. **17**(6), 796–803 (2020)
13. Szegedy, C., Ioffe, S., Vanhoucke, V., Alemi, A.: Inception-v4, Inception-ResNet and the impact of residual connection on learning. arXiv:1602.07261 (2016)
14. He, K., Zhang, X., Ren, S., Sun, J.: Deep residual learning for image recognition. arXiv:1512.03385 (2015)
15. He, K., Zhang, X., Ren, S., Sun, J.: Identity mappings in deep residual networks. arXiv:1603.05027 (2016)
16. Tensorflow. www.tensorflow.org. Accessed 15 Mar 2020
17. Keras. keras.io. Accessed 15 Mar 2020

The Case of Missed Cancers: Applying AI as a Radiologist's Safety Net

Michal Chorev[1]([⊠]) [iD], Yoel Shoshan[1], Ayelet Akselrod-Ballin[2], Adam Spiro[1], Shaked Naor[1], Alon Hazan[1], Vesna Barros[1], Iuliana Weinstein[3], Esma Herzel[4], Varda Shalev[4], Michal Guindy[3], and Michal Rosen-Zvi[1]

[1] Department of Healthcare Informatics, IBM Research, IBM R&D Labs, University of Haifa Campus, Mount Carmel, 3498825 Haifa, Israel
{michalc,yoels,rosen}@il.ibm.com
[2] Zebra Medical Vision, Shefayim, Israel
[3] Department of Imaging, Assuta Medical Centers, Tel Aviv, Israel
michalgu@assuta.co.il
[4] MaccabiTech, MKM, Maccabi Healthcare Services, Tel Aviv, Israel

Abstract. We investigate the potential contribution of an AI system as a safety net application for radiologists in breast cancer screening. As a safety net, the AI alerts on cases suspected to be malignant which the radiologist did not recommend for a recall. We analyzed held-out data of 2,638 exams enriched with 90 missed cancers. In screening mammography settings, we show that a system alerting on 11 out of every 1,000 cases, could detect up to 10.7% of the radiologists' missed cancers. Thus, significantly increasing radiologist's sensitivity to 80.3%, while only slightly decreasing their specificity to 95.3%. Importantly, the safety net demonstrated a significant contribution to their performance even when radiologists utilized both mammography and ultrasound images. In those settings, it would have alerted 8.5 times per 1,000 cases, and detected 11.7% of the radiologists' missed cancers. In an analysis of the missed cancers by an expert, we found that most of the cancers detected by the AI were visible post-hoc. Finally, we performed a reader study with five radiologists over 120 exams, 10 of which were originally missed cancers. The AI safety net was able to assist 3 out of the 5 radiologists in detecting missed cancers without raising any false alerts.

Keywords: Computer-aided diagnosis · Deep learning · Breast imaging

1 Introduction

1.1 Radiologists Performance in Screening Digital Mammography

Breast cancer (BC) is the most commonly diagnosed cancer among women worldwide, and the second leading cause of cancer-related deaths. As treatment options improve,

---

The original version of this chapter was revised: the list of coauthors was corrected. The correction to this chapter is available at https://doi.org/10.1007/978-3-030-59725-2_79

A. L. Martel et al. (Eds.): MICCAI 2020, LNCS 12266, pp. 220–229, 2020.
https://doi.org/10.1007/978-3-030-59725-2_22

early detection may have a larger impact on morbidity and mortality. Presently, digital mammography (DM) is the most common method of screening being used globally. Women undergo a DM exam every 1–3 years depending on their familial history and national policy. These exams are then interpreted by radiologists based on the Breast Imaging Report and Data System (BIRADS). According to the breast cancer surveillance consortium (BCSC) benchmark for radiologists in DM screening [1], the average radiologist's sensitivity and specificity are 87% and 89%, respectively. While 97.1% of radiologists are within the acceptable range of sensitivity $\geq75\%$, only 63.0% met the acceptable range for specificity of 88%–95%. Indeed, analyzing mammograms is a challenging task. Previous works have shown the agreement between radiologists to be slight to moderate at best [2–4]. A second reading of mammograms by an additional radiologist has been proven to increase sensitivity and specificity [5, 6]. However, lack of trained radiologists, budget, and time limitations often make it inapplicable to the standard screening procedure [7]. AI systems may help close the gap of readily available second readers, but their real-world efficacy is still a matter of debate [8–10].

1.2 AI Performance in Screening Digital Mammography

Since June 2019, six different papers [11–16] reported results of AI models trained on retrospective screening mammography data for the detection of BC within 12 months. Most studies were based on large-scale data (ranging from an order of 15K to 150K women) and reported impressive results, often at the range of the radiologists' performance or even surpassing it. The radiologists' performance on the held-out data in the above papers (when reported) is presented in Fig. 1 by the full circles. The reported numbers vary between sensitivity of 77% (readers in Italy and the Netherlands) to 98.5% (UK consensus reader) and specificity between 67% (Italian readers) and 97.4% (Israeli readers), which is consistent with the performance of radiologists derived by BCSC [1]. The performance of AI models is illustrated by empty circles.

The ability to achieve radiologists' level in specificity and sensitivity indicates that the technology could have assisted radiologists. Unfortunately, it is almost impossible to compare the AI models. Not only that each paper is leveraging different datasets from different geographies (USA, South Korea, Israel, and multiple countries in western Europe), but also different evaluation methods and performance measures were used for reporting the results. Most importantly, some key elements that could demonstrate the usefulness of these models as part of a screening routine were often missing from the report. Namely, their expected performance in real-life settings, including what composition of cases was considered real-life settings, and the existence of false negative (FN) cases by radiologists (definition, count, and performance). Moreover, two of the six papers conducted reader studies [13, 14] that only shed light on part of the picture. Kim et al. only tested the different readers on cancers that were detectable by the original radiologist by DM or ultrasound (US). McKinney et al. did not explicitly report the number of missed cancer cases in their reader study.

In this work, we focused on a safety net application, where the aim of the technology was to alert on FN cases within 12 months from the index exam, while maintaining a low number of false alerts. Reducing AI's false alarms is key, as computer-aided diagnosis systems have been shown in the past to generate a large number of false positive findings, slowing the radiologist's work without contributing to their performance [8]. False positive recalls may induce extra costs, unnecessary anxiety and additional procedures for a healthy population [17, 18]. For that purpose, our system worked in a high specificity operation point, in which it was expected to identify normal cases with high confidence. However, high specificity operation point is not necessarily optimal for the detection of all cancer cases, let alone those that were missed by the first reader.

Here, we analyzed the overall contribution of the AI system in a safety net application to the radiologist's performance; first, on a held-out set based on the original radiologists' performance, and second, in a multi-reader study to examine its contribution to individual readers. Both the held-out data and the reader study were enriched with radiologists' FN cases.

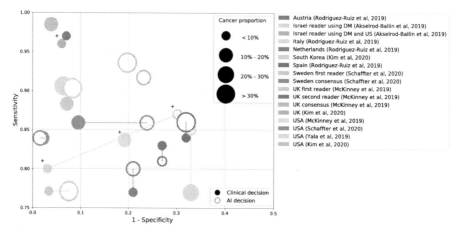

Fig. 1. Performance of radiologists and AI systems within a 12-month follow-up period as reported in recent publications. Straight lines correspond to studies who reported performance of both AI and radiologists on the same cohort. $^{+}$Indicates adjusted results for screening scenario. Results with specificity below 50% and sensitivity below 75% are not shown.

2   Methods

This work was approved by the research ethics review board of Assuta Medical Centers (AMC), who also waived the need to obtain a written informed consent. Data were collected, managed and anonymized by Maccabi Health Services (MHS).

2.1   AI as a Safety Net Application

The AI system being used in this work was trained on the index exam DM images and detailed clinical history of the women. Its architecture consisted of an ensemble of three deep learning (DL) models on the image level, these models were used to produce a malignancy score and as feature extractors. Another machine learning model utilized the DL models' output and incorporated the clinical information (including imaging history, gynecological and familial history, medications, diagnoses, and lab results) into a final malignancy score on the breast-level. For a study-level decision, the maximum between the prediction for each side was taken. For more details see [11].

The model's output is a continuous score in the range between 0 and 1. For the system to provide a final binary decision, thresholds were set at specific operation points using a validation set from a previous study [11] without overlap with this work's datasets. One operation point was 87% sensitivity, consistent with the average reader sensitivity according to BCSC benchmark [1]. Here, we focused mostly on a second operation point of 99% specificity in a safety net application. In such an application, the AI system analyzes the cases independently of the radiologist and is activated after the radiologist's BI-RADS score assignment. When the AI system deems a case malignant, only then it checks its original BI-RADS category. If the case was assigned BI-RADS 1–2, it raises an alarm for a recall. A 99% specificity operation point was selected to reduce the number of false alarms.

2.2   Retrospective Held-Out Set

The dataset was collected from five AMC imaging facilities and from MHS database. The cohort was composed of women who underwent at least one DM examination between 2013 and 2017 consecutively and had at least one year of clinical history. Most women in AMC undergo a screening mammography biannually, with the exception of women with familial history who are offered an annual exam. Mammograms are typically read by a single fellowship-trained reader and interpreted using the BI-RADS scale. We excluded exams of women with a history of breast cancer; exams post breast operations (e.g., lumpectomy or mammoplasty); and exams of a single breast. Exams were considered positive if there was an indication of a biopsy positive for cancer or they appeared in the cancer registry within 12 months. Exams were considered negative if there was an indication of a negative biopsy within 12 months or they had a BI-RADS 1–3 index exam and a completely clean follow-up for at least two years (i.e. all follow-up exams in that period were BI-RADS 1–2, without biopsy recommendations or procedures). For each woman, the first exam meeting the inclusion/exclusion criterion was selected as the index examination. All FN in the retrospective test set were reviewed post-hoc by a breast radiology specialist with more than 20 years of experience. The retrospective test set analyzed in this work was never used to train or tune the AI system.

2.3   Reader Study

In the reader study, five AMC board-certified radiologists with breast mammography fellowship have interpreted 120 exams. Two of the readers had 20 years of experience, two had 10 years of experience, and one had more than a year of experience reading breast mammograms. Another >20 years experienced reader conducted a post-hoc review of the FN cases in the study. Readers used their regular system and screens. Readers and AI model were exposed to the index exam images and the entire set of clinical data. They had no access to previous exams or to other modalities taken in the index examination date. However, we excluded cases with high BI-RADS due only to high US BI-RADS (i.e., the retrospective DM required a recall). Additionally, when an exam had a negative biopsy indicating a healthy tissue, we made sure that there isn't a follow-up positive biopsy exam or a record in the cancer registry. Here the definition of FN was loosened in comparison to the retrospective test set, to introduce cancers that were diagnosed within a two years window. We made no distinction whether a finding was visible in DM, US, or neither. Cases were assigned in a random order to each reader, and each has covered the entire set of cases.

2.4   Bootstrapping and Statistical Analysis

Ideally, AI system's performance and contribution to a human reader should be assessed in conditions as close to real-world prevalence as possible. However, in most studies, this is not the case. Roughly 98% of mammograms in a screening population are normal. The manner in which AI models are trained often results in datasets enriched in abnormal cases. Here too, the retrospective test set is not reflective of real-world prevalence of breast cancer as reported by AMC or the BCSC benchmark [1]. The data is enriched with biopsy cases (880/2,638, 33%) and especially FN (90/2,638, 3%).

For this purpose, we utilized the entire set of clinical data in MHS (69,149 cases) to estimate real-world prevalence in the population. Performance of the original radiologist was estimated once according to DM alone (TP: 1,444/69,149, 2.08%; FN: 405/69,149, 0.59%; TN: 64,832/69,149, 93.76%; FP: 2,468/69,149, 3.57%), and once according to DM and US (TP: 1,692/69,149, 2.45%; FN: 157/69,149, 0.23%; TN: 61,667/69,149, 89.17%; FP: 5,633/69,149, 8.15%). Using these proportions, we bootstrapped with replacements a sample set of 1,000 cases in 1,000 iterations and calculated the reader's and AI's performance on the retrospective dataset. We report mean and 95% confidence interval (CI) for each measure.

Fisher exact test and Wilcoxon signed-rank test were used to evaluate significant differences in performance. For multiple hypotheses we used Benjamini-Hochberg adjustment. Inter-reader agreement was estimated using Cohen's Kappa statistic. P-values less than 0.05 were considered statistically significant.

3   Results and Discussion

3.1   Safety Net Application on Held-Out Retrospective Data

A held-out set of 2,638 individual exams was collected (age 55 [47–63], BMI 26 [23–30], median and interquartile range). Each exam in the dataset included the four standard mammography images as well as detailed clinical history of the women. The dataset consisted of: 1,688/2,638 (64%) BI-RADS 1–2 cases with a clean follow-up, 70/2,638 (3%) BI-RADS 3 cases with a clean follow-up, 501/2,638 (19%) negative-biopsy cases, and 379/2,638 (14%) positive-biopsy cases. The dataset was intentionally enriched in FN cases, with 24% (90/379) of cancer cases originally missed by the radiologist. A screening US exam was performed in 75% (1,967/2,638) of the cases, and BI-RADS were reported separately for DM and US. Performance of the original radiologist was estimated twice; based on DM alone and based on both modalities, when US was available.

The AI system's performance was estimated at an operation point of 99% specificity as a safety net application (see Sect. 2.1). We used a bootstrapping analysis, mean and CI, to estimate performance on real-word prevalence (see Sect. 2.4).

Based on DM alone, the radiologists obtained a sensitivity of 77.9% [60.9%–92.6%] (or 20.8 cancers detected per 1,000 cases) and specificity of 96.4% [95.1%–97.5%]. With the addition of US, their sensitivity increased to 91.3% [79.2%–100.0%] (or 24.3 cancers detected per 1,000 cases) and their specificity decreased to 91.7% [89.8%–93.3%]. The AI system obtained a sensitivity of 47.0% [27.3%–66.7%] at a specificity of 98.7% [98.0%–99.4%]. We then simulated a safety net application of the AI system. In this scenario, the AI system analyzed the cases independently after the radiologist work is done. For cases it deemed likely to be missed cancer, it raised an alert (see Sect. 2.1).

For radiologists reading only DM, the AI system would have raised 10.68 [4.00–20.00] alerts for every 1,000 cases. From the alerts, 0.63 would be valid. In other words, for an average of 5.87 [1.00–13.00] radiologists' FN per 1,000 cases, the AI could have identified 10.73% [0.00%–23.07%]. The remaining 10.05/1000 of the alerts would be false alarms. Hence, the safety net application is able to increase the reader's sensitivity significantly (80.3% vs. 77.9%, p-value $= 1.0 \times 10^{-78}$), with a slight but significant decrease in specificity (95.3% vs. 96.4%, p-value $= 3.3 \times 10^{-165}$). When the reader's interpretation was based on both DM and US, the AI's contribution was significant still. It would have raised an alarm on 8.51 [3.00–16.00] out of every 1,000 cases, of which 0.27 [0.00–2.00] would have been valid. For an average of 2.30 [0.00–7.00] radiologists' FN, the AI could have identified 11.73% [0.00%–28.57%] for every 1,000 cases. The remaining 8.24 alarms would have been false. The AI safety net was still able to increase sensitivity (92.3% vs. 91.3% p-value $= 5.6 \times 10^{-39}$) with a slight decrease in specificity (90.8% vs 91.7%, p-value $= 3.3 \times 10^{-165}$).

From the 90 cases that were missed by the radiologist, the AI safety net was able to identify 11. We asked an expert breast radiologist to review all FN cases and determine whether the malignant lesion was visible in the index exam's DM or not. The expert reader analyzed first the index exam, and only then utilized any other US or follow-up examinations images as well as pathology reports to localize the malignant finding. The AI was able to identify a larger proportion of visible cancers than not (Fisher exact test, p-value $= 1.76 \times 10^{-2}$; see Table 1).

Table 1. False negative cases' post-hoc visibility vs. identification by AI safety net

	Visible	Not visible	Total
AI identified	8	3	11
AI missed	27	52	79
Total	35	55	90

3.2 Reader Study

We then continued to examine the AI safety net application in a separate reader study with five certified breast radiologists from AMC. The study consisted of 120 cases (age 53 [47–65], BMI 26 [23–29], median and interquartile range), including 36/120 (30%) normal cases (original BI-RADS 1–2 with a clean two years follow-up), 13/120 (11%) original BI-RADS 3 cases without biopsy and a clean two years follow-up, 35/120 (29%) with a negative biopsy with one year, and 36/120 (30%) with a positive biopsy. The cancer cases further included 10/36 (28%) FN. The original BI-RADS assigned to the FN was either 1 or 2.

Readers were asked to assign each case a BI-RADS 1–5 score. Their answers were compared to the ground truth and individual sensitivity/specificity measures were calculated. We compared the readers performance to the AI system in two operation points: 1) 87% sensitivity, and 2) 99% specificity, for the safety net application (Fig. 2a, see Sect. 2.3).

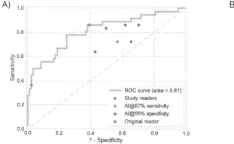
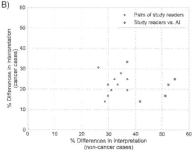

Fig. 2. Performance of the readers and AI system in the reader study. A) ROC curve of the AI system (AUC = 0.81). Sensitivity/specificity of each study reader (in blue), the original reader (purple) and the AI at 87% sensitivity and 99% specificity (green and red, respectively). B) Differences in interpretation (recall/no-recall) of cancer and non-cancer cases between pairs of readers (blue dots) and readers and AI (green squares). (Color figure online)

At an operation point of 87% sensitivity, the AI system has exceeded the average radiologist performance (sensitivity 86.1% vs. 78.3%; specificity 60.7% vs. 41.9%). Interestingly, the readers' average sensitivity matched the retrospective average sensitivity estimated for AMC radiologists on DM (see Sect. 2.4). However, the retrospective specificity was much higher than the one obtained in the reader study (96% vs. 42%). Indeed, in regular settings, the radiologist could compare the index exam to previous exams of the same woman or to the US, if either existed. Moreover, the reader study is enriched with negative-biopsy cases, which are more contestable. The average level of agreement between radiologists based on BI-RADS score was only fair (Cohen's Kappa of 0.34 [0.28–0.42]), with a slight increase when it was based on recall/no-recall bins instead (0.37 [0.30–0.47]). In general, the radiologists tended to agree more between themselves on cancer cases than on non-cancer cases (Fig. 2b). The average agreement with the AI system was even lower (0.19 [0.02–0.35], Table 2). Similarly, most of the disagreement between readers and the AI system was rooted in the non-cancer cases (which the AI more often classified correctly) rather than the cancer-cases (Fig. 2b).

At an operation point of 99% specificity (sensitivity of 36.1% at specificity of 97.6%), the AI system was still able to detect four missed cancers (two cases for reader #3, one case for reader #2, and one case for reader #5). Importantly, the safety net application in the reader study did not add additional false alarms to any of the readers (Table 2).

In a post-hoc visibility analysis of then 10 FN cases (see Sect. 2.3), an independent expert has determined that 6 out of 10 FN were not visible at the index exam. Moreover, there was no association between the FN cases each reader has suspected to be malignant and their post-hoc visibility (p-value > 0.05, Fisher exact test). As such, the suspected lesions identified by the readers in those cases were most likely benign, and if biopsied, would have returned negative.

Table 2. Individual readers performance in the reader study.

	Reader 1	Reader 2	Reader 3	Reader 4	Reader 5
Specificity	50.0%	38.1%	57.1%	29.8%	34.5%
Sensitivity	83.3%	86.1%	63.9%	86.1%	72.2%
TP (with AI)	30 (30)	31 (32)	23 (25)	31 (31)	26 (27)
FP (with AI)	42 (42)	52 (52)	36 (36)	59 (59)	55 (55)
Mean κ with other readers	0.41 [0.25–0.57]	0.39 [0.22–0.56]	0.39 [0.24–0.54]	0.34 [0.17–0.50]	0.35 [0.18–0.52]
κ with AI	0.32 [0.16–0.49]	0.16 [−0.01–0.33]	0.28 [0.11–0.46]	0.10 [−0.06–0.26]	0.06 [−0.11–0.23]

Note—TP = true positives out of the 36 cancer cases. In parentheses – TP identified with AI safety net. Similarly, FP = false positives, in parentheses with AI safety net. κ refers to Cohen's Kappa agreement statistic, first with other readers (mean and CI), and then with the AI at sensitivity 87% (κ and CI).

4   Conclusions

In this work, we evaluated an AI system as a radiologist's safety net application. The safety net has contributed significantly to reader's sensitivity, especially when the analysis was based on DM alone, but also when combined with US. In a reader study, we demonstrated that even in a challenging dataset enriched with biopsy and FN cases, a safety net application could have benefited the readers. When the AI operated at the average sensitivity level of radiologists according to the BCSC, it had a low agreement with the readers, and as such, was in a better position to give useful insights, especially when there were no prior images or US available, such as in the case of women undergoing their first exam.

This analysis was not without limitations. Data originated from five different facilities of a single provider in one country, using a single mammography vendor (Hologic). In some cases, US was performed prior to the mammogram, and the radiologist may have been aware of the US report before analyzing the DM. Even so, according to the data, the existence of an US did not guarantee that the DM's BI-RADS was equal to or higher than the US's. In the reader study, the readers have operated in their regular environment, but did not have access to prior images or US, both essential tools in their daily work known to have impact on their performance. The AI system did not use those either. To account for the lack of US, cases with original high BI-RADS due only to US were excluded from the study. Readers and AI had access to the same clinical data.

The AI safety net application was designed to interfere as little as possible with the radiologist routine; analyzing cases independently and raising a minimal amount of alarms as a second reader. Even under those restrictions, it demonstrated useful abilities. This is only one possible application of AI systems, but one we believe to be practical for immediate use.

Funding. This project has received funding from the European Union's Horizon 2020 research and innovation programme under the Marie Sklodowska-Curie grant agreement No. 813533.

References

1. Lehman, C.D., et al.: National performance benchmarks for modern screening digital mammography: update from the Breast Cancer Surveillance Consortium. Radiology **283**, 49–58 (2016)
2. Antonio, A.L.M., Crespi, C.M.: Predictors of interobserver agreement in breast imaging using the Breast Imaging Reporting and Data System. Breast Cancer Res. Treat. **120**, 539–546 (2010). https://doi.org/10.1007/s10549-010-0770-x
3. Nishikawa, R.M., Comstock, C.E., Linver, M.N., Newstead, G.M., Sandhir, V., Schmidt, R.A.: Agreement between radiologists' interpretations of screening mammograms. In: Tingberg, A., Lång, K., Timberg, Pontus (eds.) IWDM 2016. LNCS, vol. 9699, pp. 3–10. Springer, Cham (2016). https://doi.org/10.1007/978-3-319-41546-8_1
4. Katalinic, A., Bartel, C., Raspe, H., Schreer, I.: Beyond mammography screening: quality assurance in breast cancer diagnosis (The QuaMaDi Project). Br. J. Cancer **96**, 157 (2007)
5. Karssemeijer, N.: Effect of independent double and multiple reading of screening mammograms by breast density. 1136 words (2014). https://doi.org/10.1594/ecr2014/c-0358
6. Taylor-Phillips, S., Jenkinson, D., Stinton, C., Wallis, M.G., Dunn, J., Clarke, A.: Double reading in breast cancer screening: cohort evaluation in the CO-OPS trial. Radiology **287**, 749–757 (2018). https://doi.org/10.1148/radiol.2018171010
7. Leivo, T., et al.: Incremental cost-effectiveness of double-reading mammograms. Breast Cancer Res. Treat. **54**, 261–267 (1999). https://doi.org/10.1023/A:1006136107092
8. Lehman, C.D., Wellman, R.D., Buist, D.S.M., Kerlikowske, K., Tosteson, A.N.A., Miglioretti, D.L.: Breast cancer surveillance consortium: diagnostic accuracy of digital screening mammography with and without computer-aided detection. JAMA Intern. Med. **175**, 1828–1837 (2015). https://doi.org/10.1001/jamainternmed.2015.5231
9. Cole, E.B., Zhang, Z., Marques, H.S., Edward Hendrick, R., Yaffe, M.J., Pisano, E.D.: Impact of computer-aided detection systems on radiologist accuracy with digital mammography. Am. J. Roentgenol. **203**, 909–916 (2014). https://doi.org/10.2214/AJR.12.10187
10. Gao, Y., Geras, K.J., Lewin, A.A., Moy, L.: New frontiers: an update on computer-aided diagnosis for breast imaging in the age of artificial intelligence. AJR Am. J. Roentgenol. **212**, 300–307 (2019). https://doi.org/10.2214/AJR.18.20392

11. Akselrod-Ballin, A., et al.: Predicting breast cancer by applying deep learning to linked health records and mammograms. Radiology **292**, 331–342 (2019). https://doi.org/10.1148/radiol. 2019182622
12. Yala, A., Lehman, C., Schuster, T., Portnoi, T., Barzilay, R.: A deep learning mammography-based model for improved breast cancer risk prediction. Radiology **292**, 60–66 (2019). https:// doi.org/10.1148/radiol.2019182716
13. McKinney, S.M., et al.: International evaluation of an AI system for breast cancer screening. Nature **577**, 89–94 (2020). https://doi.org/10.1038/s41586-019-1799-6
14. Kim, H.-E., et al.: Changes in cancer detection and false-positive recall in mammography using artificial intelligence: a retrospective, multireader study. Lancet Digit. Health **2**, e138–e148 (2020). https://doi.org/10.1016/S2589-7500(20)30003-0
15. Schaffter, T., et al.: Evaluation of combined artificial intelligence and radiologist assessment to interpret screening mammograms. JAMA Netw. Open **3**, e200265 (2020). https://doi.org/ 10.1001/jamanetworkopen.2020.0265
16. Rodriguez-Ruiz, A., et al.: Stand-alone artificial intelligence for breast cancer detection in mammography: comparison with 101 radiologists. J. Natl Cancer Inst. **111**, 916–922 (2019). https://doi.org/10.1093/jnci/djy222
17. Siu, A.L.: Screening for breast cancer: U.S. preventive services task force recommendation statement. Ann. Intern. Med. **164**, 279 (2016). https://doi.org/10.7326/M15-2886. On behalf of the U.S. Preventive Services Task Force
18. Alcusky, M., Philpotts, L., Bonafede, M., Clarke, J., Skoufalos, A.: The patient burden of screening mammography recall. J. Womens Health **23**, S-11 (2014). https://doi.org/10.1089/ jwh.2014.1511

Decoupling Inherent Risk and Early Cancer Signs in Image-Based Breast Cancer Risk Models

Yue Liu[1,2]([✉]), Hossein Azizpour[1], Fredrik Strand[3,4], and Kevin Smith[1,2]

[1] KTH Royal Institute of Technology, Stockholm, Sweden
yue3@kth.se
[2] Science for Life Laboratory, Solna, Sweden
[3] Karolinska Institutet, Stockholm, Sweden
[4] Karolinska University Hospital, Stockholm, Sweden

Abstract. The ability to accurately estimate risk of developing breast cancer would be invaluable for clinical decision-making. One promising new approach is to integrate image-based risk models based on deep neural networks. However, one must take care when using such models, as selection of training data influences the patterns the network will learn to identify. With this in mind, we trained networks using three different criteria to select the positive training data (*i.e.* images from patients that will develop cancer): an *inherent risk* model trained on images with no visible signs of cancer, a *cancer signs* model trained on images containing cancer or early signs of cancer, and a *conflated* model trained on all images from patients with a cancer diagnosis. We find that these three models learn distinctive features that focus on different patterns, which translates to contrasts in performance. Short-term risk is best estimated by the cancer signs model, whilst long-term risk is best estimated by the inherent risk model. Carelessly training with all images conflates inherent risk with early cancer signs, and yields sub-optimal estimates in both regimes. As a consequence, conflated models may lead physicians to recommend preventative action when early cancer signs are already visible.

Keywords: Mammography · Risk prediction · Deep learning

1 Introduction

Breast cancer is the most commonly occurring type of cancer worldwide for women [1]. An effective method to reduce breast cancer mortality is to detect it early while it is still curable. Population-wide mammographic screening is

Electronic supplementary material The online version of this chapter (https://doi.org/10.1007/978-3-030-59725-2_23) contains supplementary material, which is available to authorized users.

A. L. Martel et al. (Eds.): MICCAI 2020, LNCS 12266, pp. 230–240, 2020.
https://doi.org/10.1007/978-3-030-59725-2_23

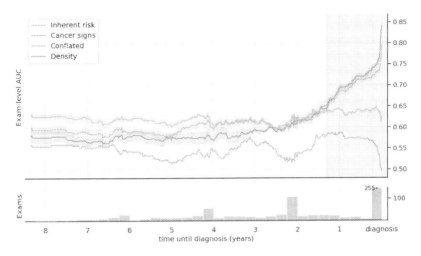

Fig. 1. If trained carelessly, *conflated* (blue) neural networks confound long-term *inherent risk* (orange) and early *cancer signs* (red). Top: Test AUC of the models (based on 20% of the positive samples and all negative samples), computed every 7 days over an 8-year period using a sliding window of varying width (shaded areas show 95% confidence interval). Bottom: Number of positive exams in the test set (women who will develop cancer) vs. time until diagnosis. Spikes correspond to scheduled screenings. The right-most bin contains 255 exams. The gray region shows when the sliding window contains samples ≤30 days to diagnosis, which likely corresponds to screen-detected cancers. The *conflated* model, trained on all images, is decoupled into *inherent risk* and *cancer signs*. All three models outperform the mammographic density baseline. Short-term risk is best estimated by the cancer signs model, which is unsurprising as it was trained like a cancer detector. Long-term risk is best estimated by the inherent risk model, whose AUC remains constant even near diagnosis. Although the conflated model was trained with more data, it is sub-optimal in both the short- and long-term. (Color figure online)

proven to have a positive effect in this regard, and has been implemented across many developed countries [2]. However, studies have shown that mammographic screening has limited sensitivity for some women [3]. Cancers that could potentially be found with more sensitive screening methods are routinely missed. For example, adding MRI or ultrasound screening would improve early detection, but are too costly to offer to the whole population. A reliable method to estimate breast cancer risk would allow hospitals to offer more personalized care to high-risk women, including enhanced screening and other preventive measures.

Breast cancer risk prediction approaches include questionnaire-based models such as Gail and Tyrer-Cuzick models [4,5] and breast density models. A new state-of-the-art in breast cancer risk estimation was recently established using deep neural networks trained on mammograms [6,7]. These risk models represent a paradigm shift towards learned features, and have been shown to substantially outperform prior models. Based on these successes, we anticipate that risk assessment research will shift towards deep learning approaches.

The key message of this work is a warning that, if care is not taken when selecting the training data and designing the training procedure, *neural networks trained to estimate breast cancer risk may conflate actual risk prediction and cancer detection*. Conflated models purport to perform long-term risk prediction, but in reality are highly sensitive to cancer signs. This yields sub-optimal long-term risk estimation, and could cause cancers to go undetected if physicians believe women have high long-term risk when in fact they exhibit cancer signs.

Through a series of experiments, we illustrate the phenomenon of risk conflation both qualitatively and quantitatively, and measure how it impacts the performance of risk prediction over time. Code to reproduce our work is available at https://github.com/yueliukth/decoupling_breast_cancer_risk.

2 Related Works

Breast cancer prevention demands accurate and individualized risk assessment for decision-making. Over the last decades, many models for estimating individual breast cancer risk have been developed. The Gail model [4] is a questionnaire-based method for estimating 5-year and lifetime risk of developing invasive breast cancer. It considers risk factors such as a woman's age and family history. Tyrer–Cuzick [5], another commonly-used risk model, incorporates more detailed family history. Glynn *et al.* recently compared questionnaire-based models and found that their practical usefulness is limited by performance [8].

Breast density, aside from age, is one of the strongest risk factors for breast cancer [9]. Density measures if a breast is more fatty or contains more fibroglandular tissue, can be obtained from mammographic screens, and has been shown to improve questionnaire-based models [10]. Density is often defined by a few statistics obtained either through ad-hoc [11] or learning-based approaches [12]. In general, methods for quantifying density lack consistency [13] and tend to over-simplify image data, limiting their general application.

In the era of deep learning, most research in mammography has focused on computer-aided diagnosis (CAD) [14–16]. A handful of studies have addressed risk prediction, though most have been restricted to small datasets and short-term prediction. Two such studies [17,18] considered a few hundred negative screening samples, and predicted which would be positive at the next screening. He *et al.* used a multi-modal approach to combine mammographic screenings, ultrasound images, patient demographics, and language from clinical reports to predict if a patient with an abnormal mammogram should be sent for biopsy [19].

Two recent breakthrough studies showed substantial improvements in long-term risk prediction using neural networks on large population-level cohorts. Yala *et al.* showed mammogram-based deep learning models outperform the Tyrer–Cuzick model for five-year risk prediction [6]. Dembrower *et al.* similarly showed that five-year risk predictions from a neural network surpass density-based predictions [7]. In this study, we consider the same cohort as Dembrower *et al.*, but our focus is not to push performance, rather to raise awareness of the dangers of conflating long-term risk and cancer signs in risk models.

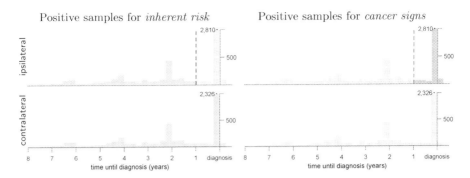

Fig. 2. We decouple *inherent risk* (orange) and *cancer signs* (red) from the *conflated* model by splitting the positive training data in two separate parts. Top: Histogram of positive ipsilateral images over the study period (ipsilateral is the breast that develops cancer). A cutoff of 1 year from diagnosis (dashed line) separates images with no visible cancer signs (orange) from those with possible cancer signs (red). Bottom: Positive contralateral images, from the other breast that is usually confirmed cancer-free. As it has been exposed to the same environmental and genetic risk factors as the ipsilateral, it is included in the inherent risk model. The conflated risk model is trained with all images. The cancer signs model is trained with red-marked positive examples, while the inherent risk model is trained with orange-marked positive examples. (Color figure online)

3   Decoupling Breast Cancer Risk

A straightforward approach to train a network to predict breast cancer risk from mammograms is to provide all images from cancer patients as positive examples. Several prior works have trained models in this manner. The problem with this approach is that the images recorded near the date of diagnosis are included in the positive set, and are likely to include signs of actual cancer. We can imagine separating the positive training images with no visible cancer signs from those containing cancer signs by drawing an arbitrary cutoff within one screening interval, *e.g.* at one year from diagnosis (Fig. 2). From this perspective, the data contains two different classification problems: *inherent risk vs. healthy* and *cancer signs vs. healthy*. When we train using all the data, we conflate them into a single binary classification task, *at-risk vs. healthy*.

This is problematic for the learning process, since recognizing long-term risk is more difficult than detecting cancer signs. Networks are known to converge faster with easier examples [20,21], and if it focuses too strongly on increasing confidence of the easy samples [22] learning on the harder long-term risk problem may be crippled.

Consequently, we hypothesize that the conflated model will perform worse at long-term risk prediction than a model trained exclusively with images acquired before onset of early cancer signs. This effect will be more acute when a substantial portion of the positive data contains cancer signs, which is typical for population datasets (for CSAW [23], up to 31% of the positive samples may contain cancer signs). In order to test this hypothesis, we decomposed the conflated

Table 1. Model performance of risk prediction on test set

	AUC (95% CI)			
	31d–1 year	>1 year	>2 years	>5 years
Inherent risk	0.62 (0.62, 0.63)	**0.62** (0.61, 0.62)	**0.62** (0.61, 0.62)	**0.61** (0.60, 0.62)
Cancer signs	**0.71** (0.68, 0.73)	0.59 (0.58, 0.60)	0.59 (0.58, 0.59)	0.56 (0.55, 0.57)
Conflated	**0.72** (0.69, 0.75)	**0.61** (0.60, 0.62)	0.60 (0.59, 0.61)	0.58 (0.56, 0.59)
Density	0.61 (N/A)	0.54 (N/A)	0.54 (N/A)	0.55 (N/A)

model by dissecting the data and training models on those splits. We trained an *inherent risk* model using data with no visible cancer signs, and a *cancer signs* model using data that contains a substantial number of cancer signs. The *conflated* model was provided with all available data.

Details of the data selection strategy are provided in Fig. 2. For ipsilateral – breasts that will develop cancer – we selected a cutoff of one year prior to diagnosis to separate inherent risk and cancer signs (dashed line). The contralateral breast is usually confirmed cancer-free in patients with breast cancer. It reflects actual risk without revealing any cancer cues, as it has been exposed to the same environmental and genetic risk factors. Therefore, we included the contralateral breast in the inherent risk model but not the cancer signs model.

Using these models, we conducted a series of experiments to understand the phenomenon of risk conflation. We address the following questions:

1. *How does the conflated model compare to the decoupled models over time?*
2. *Does the conflated model identify the same at-risk women as the inherent risk/cancer signs models?*
3. *Do the inherent risk/cancer signs models recognize the same patterns?*

4   Experimental Setup

Dataset. The dataset used in our study is extracted from CSAW, a population-based screening cohort containing millions of mammographic images [23]. Mammograms of multiple views were collected every 18 to 24 months from women aged 40 to 74. Outcome and date of diagnosis was determined through the Regional Cancer Center Registry. The data was curated by excluding images from patients with implants, biopsy images, or other issues such as aborted exposure. We randomly assigned the participants to the training, validation and test set. Negative exams were randomly sampled among women with at least two years' cancer-free follow-up. A flowchart describing the data curation is given in Supplementary Fig. 1. The resulting training set contains 138,032 mammograms from 15,558 women, the validation set contains 3,008 mammograms from 332 women, and the test set contains 6,436 mammograms from 731 women.

(a) >2 years   (b) 31 days to 2 years   (c) <31 days

Fig. 3. Venn diagrams showing *true positive rates* of various models, given their top-5% predictions. The *inherent risk* (orange) model consistently identifies decidedly different sets of at-risk women, compared to the *cancer signs* (red) model. The *conflated* (blue) model identifies nearly the exact same positive images as the cancer signs model near the date-of-diagnosis. But farther from diagnosis it overlaps both decoupled models. (Color figure online)

Preprocessing. The source images are in standard DICOM format. Using DICOM metadata, we flip images horizontally to make all breasts left-posed. We rescale the intensity to the range defined in the acquisition metadata [24], and we detect and correct inverted contrast images using the photometric attribute. We perform a rough alignment of each image using a distance transform to locate the center of mass. Zero-padding is applied to ensure all images have uniform size, then images are resized to 632×512. This ensures each image retains relative scale and aspect ratio. Finally, the images are converted to 16-bit PNG format.

Implementation Details. We use the same architecture and training setup for all models. In particular, we use ResNet50 [25] with group normalization [26], and replace standard ReLU activation with Leaky-ReLU [27]. We use binary cross-entropy loss and batch size of 32 with a stochastic gradient descent with momentum (SGDM) optimizer. All models were initialized with ImageNet pre-trained weights [28]. We employ standard data augmentation including random rotation, crops, brightness and contrast. Hyperparameters detailed below were selected using grid search. The initial learning rate for the cancer signs and conflated models is 0.0001, and 0.001 for the inherent risk model. The inherent risk and conflated models were run for 50 epochs, and the learning rate was lowered by a factor of 10 at epoch 20. The cancer signs model was run for 100 epochs, with a similar learning rate drop at epoch 50. Dropout [29] with a rate of 0.5 was applied after the last fully connected layer in the inherent risk model.

We repeated each experiment five times and report the mean, unless otherwise specified. As a baseline, we provide risk estimation results using mammographic density (breast dense area) from publicly available software, LIBRA [30].

5   Results and Discussion

Through a series of experiments based on the setup described above, we address the questions raised in Sect. 3.

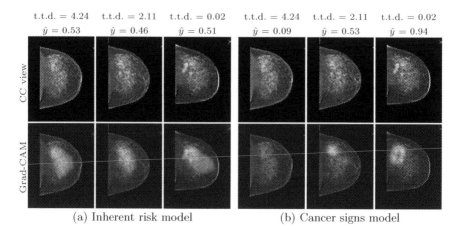

t.t.d. = 4.24   t.t.d. = 2.11   t.t.d. = 0.02   t.t.d. = 4.24   t.t.d. = 2.11   t.t.d. = 0.02
$\hat{y} = 0.53$ $\hat{y} = 0.46$ $\hat{y} = 0.51$ $\hat{y} = 0.09$ $\hat{y} = 0.53$ $\hat{y} = 0.94$

(a) Inherent risk model   (b) Cancer signs model

Fig. 4. Grad-CAM visualizations suggest that the inherent risk model and cancer signs model base their decisions on different image cues. Top: CC views of a breast that develops cancer over 4 years. An expert cancer annotation (red region) appears in the most recent image, and the cancer-developing location is identified in the prior image (red dot). Time-to-diagnosis (t.t.d., in years) and risk prediction $\hat{y}$ are provided for each image. Bottom: Grad-CAM visualizations. The activation maps are weighted by the prediction score. The inherent risk model appears to rely on a broad range of image cues, while the cancer signs model concentrates activations on tumor-like patterns. (Color figure online)

Conflated Risk Model vs. Decoupled Models. We find that the conflated model is a weakened hybrid of the inherent risk and cancer signs models. It underperforms the decoupled models in both short- and long-term risk prediction. In Fig. 1 we plot the exam-level AUC for our three models along with the density baseline. The x-axis shows how performance varies with time-to-diagnosis using a sliding window. Exam-level predictions are the maximum breast risk score; breast scores are the average score of both views. Near diagnosis, the cancer signs model is the best risk estimator. This is unsurprising because it was trained like a tumor detector, and many of the positive mammograms within the first year, especially within the first 30 days, are screen-detected cancers with visible tumors. Long-term risk is best estimated by the inherent risk model, whose AUC remains constant, even in the first year. This suggests that the inherent risk model has the desirable property of *ignoring early cancer signs*[1] *and focusing on cues correlated with long-term risk, which do not change near time-of-diagnosis.*

Similar conclusions can be drawn from Table 1, where we break down risk prediction by short-term and long-term outlooks. Inherent risk performs best at long-term risk prediction, while the cancer signs/conflated models show similar performance in the short-term (bold values indicate significant improvements; statistical tests can be found in Supplementary Table 1).

---

[1] Cancer detection is the purview of established screening routines or CAD systems.

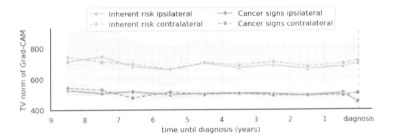

Fig. 5. Localization of gradient-weighted class activation maps (Grad-CAM) from Fig. 4, computed over all positive test images (x-axis indicates time until diagnosis). Lines are the mean computed from five models, shaded areas indicate the 95% CI. The cancer signs (red) model exhibits more localized activations, measured by total variation (TV) norm of Grad-CAMs. The inherent model (orange) consistently covers a larger area, indicating that the two models concentrate on different patterns. A sharp dip near time-of-diagnosis in the ipsilateral cancer signs model suggests that it concentrates on tumor-like patterns, as these images are likely to contain visible cancer. (Color figure online)

Identifying At-Risk Women. An important clinical question is: do these models identify the same at-risk women? To test this, we consider images identified by the top-5% predictions of each model – a number chosen to reflect the capacity of a healthcare system for additional screening. In Fig. 3 we compare positive-identified images from all three models. The inherent risk model consistently identifies different images than the cancer signs model, supporting our hypothesis that it focuses on different cues. Near the date-of-diagnosis, the conflated model highly overlaps with the cancer signs model, but farther from diagnosis it overlaps both decoupled models. Its proportion of novel at-risk findings is consistently low, suggesting it could be replaced by the decoupled models.

Image Cues that Indicate Risk. The final question we address is: do the decoupled models recognize different patterns? This is a difficult question to answer conclusively, but we can gain some insight by understanding and quantifying where the network pays attention.

In Fig. 4 we visualize how gradient-weighted class activation maps (Grad-CAM) [31] of the inherent risk and cancer signs models evolve over time. Qualitatively, we can see that the cancer signs model exhibits sharp activations localized to the tumor, whereas the inherent risk model has broad activations in the center of the breast. We empirically confirm this trend over the entire positive test set by computing the total variation of the Grad-CAM heatmaps in Fig. 5, and using multi-scale blob detection [32] in Supplementary Fig. 2. Based on these results, we surmise that the inherent risk model relies on a broader range of image cues than the cancer signs model, which appears to concentrate activations near tumor-like patterns.

6   Conclusions

Our key finding is that risk estimation models conflate inherent risk and cancer signs if care is not taken during training. We demonstrate that conflated models can be decoupled by selecting appropriate training data, and that the decoupled models consistently outperform the conflated model, even though it is trained with more data. In particular, short-term risk (≤ 1 year in our study) should rely on cancer sign models. Long-term risk models should be trained exclusively on images with no visible cancer signs, or use other strategies to mitigate model conflation. When models are put to clinical use, it is important to state which type of model is used, or to somehow assist in the interpretation of conflated models – otherwise physicians may believe that a woman has high long-term risk when, in fact, her images already exhibit cancer signs. Our hope is that this work will provide valuable insights for the development and clinical translation of deep neural networks for cancer risk estimation.

Acknowledgements. This work was partially supported by Region Stockholm HMT 20170802, the Swedish Innovation Agency (Vinnova) 2017-01382, the Wallenberg Autonomous Systems Program (WASP), and the Swedish Research Council (VR) 2017-04609.

References

1. Bray, F., Ferlay, J., Soerjomataram, I., et al.: GLOBOCAN estimates of incidence and mortality worldwide for 36 cancers in 185 countries. CA Cancer J. Clin. **68**(6), 394–424 (2018)
2. Duffy, S.W., Tabár, L., Chen, H.H., et al.: The impact of organized mammography service screening on breast carcinoma mortality in seven Swedish counties: a collaborative evaluation. Cancer Interdisc. Int. J. Am. Cancer Soc. **95**(3), 458–469 (2002)
3. Kolb, T.M., Lichy, J., Newhouse, J.H.: Comparison of the performance of screening mammography, physical examination, and breast US and evaluation of factors that influence them: an analysis of 27,825 patient evaluations. Radiology **225**(1), 165–175 (2002)
4. Gail, M.H.: Personalized estimates of breast cancer risk in clinical practice and public health. Stat. Med. **30**(10), 1090–1104 (2011)
5. Tyrer, J., Duffy, S.W., Cuzick, J.: A breast cancer prediction model incorporating familial and personal risk factors. Stat. Med. **23**(7), 1111–1130 (2004)
6. Yala, A., Lehman, C., Schuster, T., et al.: A deep learning mammography-based model for improved breast cancer risk prediction. Radiology **292**(1), 60–66 (2019)
7. Dembrower, K., Liu, Y., Azizpour, H., et al.: Comparison of a deep learning risk score and standard mammographic density score for breast cancer risk prediction. Radiology **294**(2), 265–272 (2019). https://doi.org/10.1148/radiol.2019190872
8. Glynn, R.J., Colditz, G.A., Tamimi, R.M., et al.: Comparison of questionnaire-based breast cancer prediction models in the nurses' health study. Cancer Epidemiol. Prev. Biomark. **28**(7), 1187–1194 (2019)
9. Boyd, N.F., Guo, H., Martin, L.J., et al.: Mammographic density and the risk and detection of breast cancer. New Engl. J. Med. **356**(3), 227–236 (2007)

10. Brentnall, A.R., Harkness, E.F., Astley, S.M., et al.: Mammographic density adds accuracy to both the Tyrer-Cuzick and Gail breast cancer risk models in a prospective UK screening cohort. Breast Cancer Res. **17**(1) (2015). Article number: 147. https://doi.org/10.1186/s13058-015-0653-5

11. Rauh, C., Hack, C., Häberle, L., et al.: Percent mammographic density and dense area as risk factors for breast cancer. Geburtshilfe Frauenheilkd. **72**(08), 727–733 (2012)

12. Keller, B.M., Nathan, D.L., Wang, Y., et al.: Estimation of breast percent density in raw and processed full field digital mammography images via adaptive fuzzy c-means clustering and support vector machine segmentation. Med. Phys. **39**(8), 4903–4917 (2012)

13. Amir, E., Freedman, O.C., Seruga, B., et al.: Assessing women at high risk of breast cancer: a review of risk assessment models. JNCI: J. Natl. Cancer Inst. **102**(10), 680–691 (2010)

14. Geras, K.J., Wolfson, S., Shen, Y., et al.: High-resolution breast cancer screening with multi-view deep convolutional neural networks. arXiv preprint arXiv:1703.07047 (2017)

15. Shen, L., Margolies, L.R., Rothstein, J.H., et al.: Deep learning to improve breast cancer detection on screening mammography. Sci. Rep. **9**(1), 1–12 (2019)

16. McKinney, S.M., Sieniek, M., Godbole, V., et al.: International evaluation of an AI system for breast cancer screening. Nature **577**(7788), 89–94 (2020)

17. Sun, W., Tseng, T.-L.B., Zheng, B., Qian, W.: A preliminary study on breast cancer risk analysis using deep neural network. In: Tingberg, A., Lång, K., Timberg, P. (eds.) IWDM 2016. LNCS, vol. 9699, pp. 385–391. Springer, Cham (2016). https://doi.org/10.1007/978-3-319-41546-8_48

18. Qiu, Y., Wang, Y., Yan, S., et al.: An initial investigation on developing a new method to predict short-term breast cancer risk based on deep learning technology. In: Medical Imaging 2016: Computer-Aided Diagnosis, vol. 9785, p. 978521. International Society for Optics and Photonics (2016)

19. He, T., Puppala, M., Ezeana, C.F., et al.: A deep learning-based decision support tool for precision risk assessment of breast cancer. JCO Clin. Cancer Inform. **3**, 1–12 (2019)

20. Bengio, Y., Louradour, J., Collobert, R., et al.: Curriculum learning. In: Proceedings of the 26th Annual International Conference on Machine Learning (2009)

21. Weinshall, D., Cohen, G., Amir, D.: Curriculum learning by transfer learning: theory and experiments with deep networks. In: International Conference on Machine Learning (2018)

22. Guo, C., Pleiss, G., Sun, Y., Weinberger, K.Q.: On calibration of modern neural networks. In: Proceedings of the 34th International Conference on Machine Learning, vol. 70, pp. 1321–1330. JMLR.org (2017)

23. Dembrower, K., Lindholm, P., Strand, F.: A multi-million mammography image dataset and population-based screening cohort for the training and evaluation of deep neural networks-the cohort of screen-aged women (CSAW). J. Digit. Imaging **33**, 408–413 (2020). https://doi.org/10.1007/s10278-019-00278-0

24. Clunie, D.A.: DICOM implementations for digital radiography. RSNA **2003**, 163–172 (2003)

25. He, K., Zhang, X., Ren, S., et al.: Deep residual learning for image recognition. In: Proceedings of the IEEE Conference on Computer Vision and Pattern Recognition (2016)

26. Wu, Y., He, K.: Group normalization. In: Proceedings of the European Conference on Computer Vision (ECCV), pp. 3–19 (2018)

27. Maas, A.L., Hannun, A.Y., Ng, A.Y.: Rectifier nonlinearities improve neural network acoustic models. In: Proceedings of ICML, vol. 30, p. 3 (2013)
28. Deng, J., Dong, W., Socher, R., et al.: ImageNet: a large-scale hierarchical image database. In: 2009 IEEE Conference on Computer Vision and Pattern Recognition. IEEE (2009)
29. Hinton, G.E., Srivastava, N., Krizhevsky, A., et al.: Improving neural networks by preventing co-adaptation of feature detectors. arXiv preprint arXiv:1207.0580 (2012)
30. Keller, B.M., Chen, J., Daye, D., et al.: Preliminary evaluation of the publicly available laboratory for breast radiodensity assessment (LIBRA) software tool: comparison of fully automated area and volumetric density measures in a case-control study with digital mammography. Breast Cancer Res. 17(1), 117 (2015)
31. Selvaraju, R.R., Cogswell, M., Das, A., et al.: Grad-CAM: visual explanations from deep networks via gradient-based localization. In: Proceedings of the IEEE International Conference on Computer Vision (2017)
32. Lindeberg, T.: Feature detection with automatic scale selection. Int. J. Comput. Vis. 30(2), 79–116 (1998). https://doi.org/10.1023/A:1008045108935

Multi-task Learning for Detection and Classification of Cancer in Screening Mammography

Maria V. Sainz de Cea[1], Karl Diedrich[1], Ran Bakalo[2], Lior Ness[2], and David Richmond[1(✉)]

[1] IBM Watson Health, Cambridge, MA 02142, USA
david.richmond@ibm.com
[2] IBM Research, Haifa University Campus, Mount Carmel, 3498825 Haifa, Israel

Abstract. Breast screening is an effective method to identify breast cancer in asymptomatic women; however, not all exams are read by radiologists specialized in breast imaging, and missed cancers are a reality. Deep learning provides a valuable tool to support this critical decision point. Algorithmically, accurate assessment of breast mammography requires both detection of abnormal findings (object detection) and a correct decision whether to recall a patient for additional imaging (image classification). In this paper, we present a multi-task learning approach, that we argue is ideally suited to this problem. We train a network for both object detection and image classification, based on state-of-the-art models, and demonstrate significant improvement in the recall vs no recall decision on a multi-site, multi-vendor data set, measured by concordance with biopsy proven malignancy. We also observe improved detection of microcalcifications, and detection of cancer cases that were missed by radiologists, demonstrating that this approach could provide meaningful support for radiologists in breast screening (especially non-specialists). Moreover, we argue that this multi-task framework is broadly applicable to a wide range of medical imaging problems that require a patient-level recommendation, based on specific imaging findings.

Keywords: Decision support · Deep learning · RetinaNet · ResNet

1 Introduction

Breast cancer is the leading cause of cancer death in women world wide [19]. Screening aims to increase early detection by identifying suspicious findings in asymptomatic women, and has been shown to reduce the risk of dying from breast cancer [17]. However, radiologists struggle to keep up with the volume of breast screening, increasing the risk of burn-out and missed cancer [13]. Thus, computer algorithms that can assist radiologists in reading mammography exams have the potential for a significant impact on women's health.

© Springer Nature Switzerland AG 2020
A. L. Martel et al. (Eds.): MICCAI 2020, LNCS 12266, pp. 241–250, 2020.
https://doi.org/10.1007/978-3-030-59725-2_24

The task of screening mammography is to decide whether or not to recall a patient for additional work-up. This clinical decision is made on the basis of abnormal findings within the breast, but may also be influenced by the patient's risk profile, which can be inferred from the mammogram's overall appearance [21]. Despite the dual nature of this problem, requiring both object detection and image classification, and the vast literature on breast screening decision support, we are aware of only one publication utilizing multi-task learning (MTL) [9].

Previous methods based on image classification have produced competitive results [22], somewhat surprisingly, given the fact that suspicious findings may occupy only 1% of a mammogram. Detection-based methods, on the other hand, have the clear advantage that they are trained on highly discriminative findings, and have demonstrated good results on detection of masses and calcifications [1,2,7]. However, local annotations are inherently subjective, and are typically not as consistent as outcome-based ground truth used by classification. Furthermore, we observe that detection-only approaches potentially overlook ancillary findings, which can be unlabeled. There are many ancillary findings that have significant implication for determining malignancy, such as skin thickening, nipple retraction, neovascularity, adenopathy and multifocal disease.

A few notable publications have combined image classification and detection. In [8,12], the authors initialize a classification model from a patch-based model pretrained on locally annotated data. However, weight sharing in this approach is sequential, offering much less flexibility than MTL. In addition, similar to classification methods mentioned above, these models can only provide indirect evidence for their decision via saliency maps. In [4], the authors address detection, segmentation and classification; however, similar to Mask R-CNN [5] and RetinaMask [3], their method applies only to detection ROIs, and does not address classification of the full image, a central focus of our paper. In [20], the authors input heatmaps from a sliding-window classifier as an additional channel to a study-level classification model, whereas in [18], the authors combine a sliding-window classifier and image classifier using a Random Forest, and in [14], the authors employ ensemble averaging over multiple classification and detection models; however, none of these approaches benefit from weight sharing across tasks. The paper that is methodologically most similar to ours is [9]. This work applies MTL of segmentation and image classification; however, they achieve a smaller performance improvement, and only evaluate their model on DDSM.

In this paper, we combine both image classification and object detection in a single multi-class MTL framework to derive the benefits of strong outcome-based ground truth information, global image features, and highly discriminative, albeit possibly noisy, local annotations. Using this flexible approach, we observe a significant performance boost in classification of malignant vs non-malignant images, and an improvement in detection of malignant calcifications.

2   Methods

2.1   Baseline RetinaNet Detector

We use the state-of-the-art RetinaNet model [11] as our baseline detection algorithm. RetinaNet is a single-shot object detector that utilizes a novel focal loss to counteract background-foreground imbalance, and has been used for object detection in several fields, including medical [7,23]. The overall architecture is composed of a ResNet34-Feature Pyramid Network (FPN) backbone, and two sub-networks performing (i) classification, and (ii) coordinate regression, for each of the candidate detections.

2.2   Proposed Multi-task Algorithm

Building from the baseline RetinaNet model, we added an image classification subnet to ResNet34, to perform full image classification. In this way, the ResNet34 weights were shared between the image classification and detection tasks (Fig. 1). The image classification subnet matches the architecture of the published ResNet34 classification model [6], consisting of global pooling followed by a fully-connected layer. We experimented with additional Conv blocks followed by multiple fully-connected layers; however, it did not further improve performance. The final loss is a combination of categorical cross-entropy for image classification, and the focal loss and regression loss for object detection. We used a relative weighting factor, λ, to balance the two tasks. Best results were obtained for $\lambda = 0.2$, and $\gamma = 2$ and $\alpha = 0.5$ in the focal loss [11].

$$L_{\text{multi-task}} = L_{\text{focal}} + L_{\text{regression}} + \lambda L_{\text{cross entropy}}$$

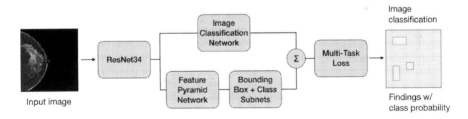

Fig. 1. Architecture of our proposed algorithm. Starting from the state-of-the-art RetinaNet algorithm, we added a classification task with shared weights (ResNet34) and multi-task loss. The trained algorithm outputs both detected findings, each with their own classification (e.g., soft-tissue lesion vs calcification) and probability of malignancy, as well as the probability of malignancy for the full image.

We also made multiple changes to the standard RetinaNet model underlying our algorithm. All changes were applied consistently, for fair comparison of the baseline and proposed algorithms.

First, we addressed the issue that the RetinaNet model is initialized from pre-trained ImageNet weights containing 3 color channels. The typical approach in medical imaging is to replicate the grayscale image 3-fold to match the expected input of ImageNet models. We observe that there is a much simpler and more efficient solution. We take the ImageNet model, and sum each of the first layer kernels over its channel dimension, reducing it to a single channel input. It is trivial to show that this is mathematically equivalent to replicating the input image, due to the linearity of the convolution operation.

Additional changes were made as follows: (i) To address the fact that findings such as microcalcification clusters may have irregular shapes, yielding a low IoU with the generated anchor boxes, we use the best matching policy of [3]. (ii) We also use a wider range of aspect ratios {1:3, 1:2, 1:1, 2:1, 3:1} for the anchor boxes to account for elongated findings. (iii) We modified the FPN architecture by removing the highest resolution level, because the majority of anchors from the highest level were "easy" background anchors that made negligible contribution to the focal loss. In practice, we found that small findings could be adequately explained with anchors from the second highest level, leading to a more efficient network, without loss in performance. (iv) In the focal loss, we normalized positive and negative anchors separately, whereas the original implementation normalized both positive and negative anchors by the number of positive anchors. By normalizing positive and negative anchors separately, we gave a greater relative weight to positive anchors in the loss computation, which we found to improve performance.

2.3   Model Training

Training experiments were conducted on hardware with 80 Intel(R) CPUs and 8 T V100-SXM2 GPUs with 32 GB memory per GPU. The training architecture was developed in Python 3.7.6 using packages: Pytorch 1.3.0, Torchvision 0.4.2, and Apex 0.1 automatic mixed precision (AMP) from Nvidia. Using AMP enabled training with a batch size of 4 full-sized images per GPU. The training environment was deployed in Docker 18.06.2-ce containers based on Ubuntu Linux 18.04.2 LTS. Training time ranged from 3–6 days for up to 60 epochs. At inference time, the model can be run with single-image batches, and fits on a consumer-grade GPU card with 12 GB memory.

2.4   Dataset and Performance Evaluation

Model training and evaluation was done on a multi-site, multi-vendor in-house research data set containing 8613 images (5825 negative, 2788 positive) from 2699 patients (1351 negative, 1348 positive) at 4 geographically distinct sites within the USA. Images were acquired on Hologic (70%), Siemens (25%), and GE (3%) machines (2% Fuji or undefined). Data was split 80/10/10 at the patient level for training, validation and test, respectively, ensuring that images from the same patient weren't included in multiple splits. Images were scaled to a pixel size of $100 \times 100\,\mu$m, cropped to eliminate background, padded to match the largest image in the batch (approx. 2400×1200 pixels), and then normalized to $[-1, 1]$. Training time augmentation was used, including zoom.

Patient-level ground truth was defined as follows. Positive cases were screening exams with biopsy proven malignancy within 12 months. Negative exams were either screening negative (BI-RADS 1 or 2 with 24–48 months of normal follow up), diagnostic negative (recalled with negative diagnostic exam), or biopsy negative (screening exams with biopsy proven benign findings within 12 months). For positive cases, only the biopsied breast was used in training and test, the contralateral breast was discarded due to lack of follow up.

Finding-level ground truth was generated by expert annotation of all biopsy positive findings. No additional findings were annotated. During training and test, annotated findings were assigned to two classes: (i) soft-tissue lesions (masses and asymmetries; 1655 in train, 165 in test), and (ii) calcifications (980 in train, 102 in test). Architectural distortions were not included in this study.

To generate test results that are representative of performance on a screening population [10], inverse weighting [16] was applied to simulate the following prevalence: screening negative (88.4%), diagnostic negative (9.9%), biopsy negative (1.2%) and biopsy positive (0.5%).

We also evaluated our algorithm on a separate, small data set containing 24 interval cancer cases, where the screening exam was assessed as negative by a radiologist, and the patient returned with cancer within 12 months. These cases were selected from a larger pool of interval cancers, on the basis that in each case, the malignant finding is visible upon retrospective investigation.

3 Results

Below we demonstrate that our method improves both image classification performance (Sect. 3.1), and detection performance (Sect. 3.2) over strong baseline algorithms. All results are summarized in Table 1.

3.1 Image Classification Analysis

We compared five methods of image classification: (i) ResNet34, a popular classification architecture and the backbone of RetinaNet, (ii) RetinaNet, using the max detection score as the image classification score, as in [15] (iii) an ensemble of ResNet34 and the max detection score from RetinaNet, (iv) our proposed method, using the max detection score from the detection head as the image classification score, and (v) our proposed method, using the classification head.

Using the classification head from our proposed method increases the AUC for image classification by 0.058 (p-value $< 10^{-4}$), from 0.851 (95% CI (0.824–0.876)) for the second-best performing algorithm (ensemble of ResNet34 and RetinaNet) to 0.909 (0.890–0.927). Compared to the naive method of using the maximum detection score from RetinaNet [15], we see an even greater improvement of 0.195 (p-value $< 10^{-4}$) from an AUC of 0.714 (0.679–0.750).

We show all ROC curves with corresponding AUCs in Fig. 2. Our method yields better performance for a continuum of operating points. For example, at an operating point with sensitivity of 0.80, our method increases the specificity from 0.471 using the maximum finding output in RetinaNet, 0.495 using the maximum finding output with our proposed architecture, 0.712 using ResNet34, and 0.723 using an ensemble of ResNet and RetinaNet, to 0.876 with the classification head of our method (p-value $< 10^{-4}$ in all cases).

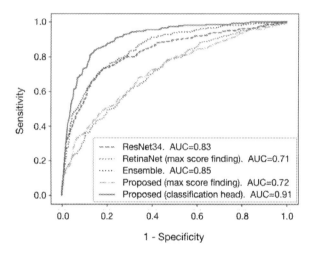

Fig. 2. ROC analysis. ROC curves obtained for (i) ResNet34, (ii) baseline RetinaNet using the max-detection score, (iii) ensemble of ResNet34 and the max-detection score from baseline RetinaNet, (iv) our proposed method using the max-detection score, and (v) our proposed method using the classification head.

3.2   Finding Detection Analysis

As reported in Sect. 2.4, we train the detection algorithm with two different kinds of annotated findings: soft tissue lesions (masses and asymmetries) and calcifications. We report detection performance for each finding type individually, using Free Response ROC Curve (FROC) analysis (Fig. 3).

In Table 1, we show the trade-off between sensitivity and the number of FPPIs at different operating points for soft tissue lesions, calcifications, and both finding types combined (average). For example, at a sensitivity of 0.7, the number of FPPIs is reduced from 9.04 to 2.09 (77% reduction; $p < 10^{-4}$) for calcifications, from 1.38 to 1.39 (0.72% increase; not significant) for masses, and from 3.08 to 1.65 (46% reduction; $p < 10^{-4}$) for both finding types combined. Similarly, at a fixed number of 2 FPPIs, the sensitivity increases from 0.58 to 0.7 for calcifications (21% increase; p 0.04), from 0.74 to 0.77 for masses (4%

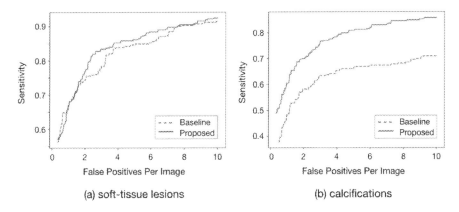

(a) soft-tissue lesions   (b) calcifications

Fig. 3. FROC analysis. FROC curves show sensitivity against the average number of false positives per image (FPPI) for detection of (a) soft-tissue lesions, and (b) calcifications. Baseline: RetinaNet; proposed: detection head of our multi-task network.

Table 1. Detection and classification results for baseline and proposed algorithms. p-values are reported as follows: *($p < 0.05$); **($p < 0.01$); ***($p < 10^{-3}$)

	Resnet34	RetinaNet	Ensemble	Proposed
Classification				
$AUC_{classification}$	0.83	–	0.85	**0.91*****
$AUC_{max\ detection}$	–	0.71	–	0.72
Detection average soft-tissue lesions and calcifications				
Se @ FPPI = 2.0	–	0.67	–	**0.76*****
Se @ FPPI = 3.0	–	0.74	–	**0.80****
FPPI @ Se = 0.6	–	1.51	–	**0.89*****
FPPI @ Se = 0.7	–	3.08	–	**1.65*****
Detection calcifications				
Se @ FPPI = 2.0	–	0.58	–	**0.70***
Se @ FPPI = 3.0	–	0.63	–	**0.77****
FPPI @ Se = 0.6	–	2.66	–	**1.12*****
FPPI @ Se = 0.7	–	9.04	–	**2.09*****
Detection soft-tissue lesions				
Se @ FPPI = 2.0	–	0.74	–	**0.77**
Se @ FPPI = 3.0	–	0.77	–	**0.83***
FPPI @ Se = 0.6	–	**0.54**	-	0.69
FPPI @ Se = 0.7	–	**1.38**	-	1.39

increase; not significant), and from 0.67 to 0.76 for both findings combined (13% increase; $p < 10^{-3}$).

We also evaluated our algorithm on a separate set of 24 interval cancer cases, where the screening exam was assessed as negative by a radiologist, and the patient returned with cancer within 12 months. At a threshold of 0.85, our algorithm achieved a sensitivity of 0.67 with 1.4 FPPI for soft tissue lesions, and a sensitivity of 0.5 with 3 FPPI for calcifications on the interval cancer cases. In Fig. 4, we show example detections from our algorithm. Figure 4(a) is a case of malignancy that was detected during screening, and Fig. 4(b) is a case of an interval cancer.

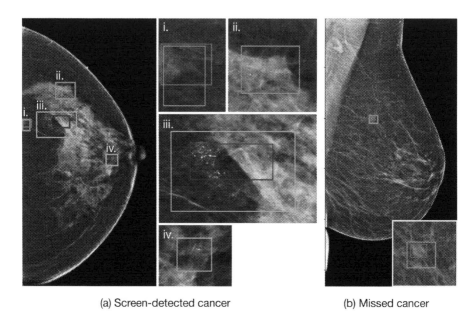

(a) Screen-detected cancer   (b) Missed cancer

Fig. 4. Example detections from our proposed method. Predicted boxes obtained with a detection threshold of 0.85 are shown for soft-tissue lesions (prediction: red; ground truth: green) and calcifications (prediction: yellow; ground truth: blue). (a) A mammogram containing lesions detected during breast screening, with algorithm output overlaid. Enlarged candidate detections show: (i) a true positive detection of a malignant mass, (ii) a false positive detection of a malignant mass, (iii) a true positive detection of malignant calcifications, and (iv) a false positive detection of malignant calcifications (the calcifications are benign). (b) Detection, by our algorithm, of a malignant mass that was missed by a radiologist. (Color figure online)

4   Conclusion

We have presented a flexible and efficient single-shot, multi-class MTL algorithm that takes as input a screening mammogram and returns a probability of malignancy for the image, as well as detected findings. This method leverages both subjective expert annotations and non-subjective outcome-based ground truth.

We tested our method on a simulated screening population and achieved an AUC of 91% for image classification, an improvement of 6% ($p < 10^{-4}$) over the second-best performing method. We also observed an improvement in the detection of malignant microcalcifications, but not for soft-tissue lesions. This may be due to the fact that the baseline performance was much higher for soft-tissue lesions, leaving less room for improvement. Compared to other published results on public datasets [1,2,7], we achieved a lower detection performance; however, this may be due to the challenging multi-site, multi-vendor nature of our data.

In future, we plan to apply methods such as RetinaMask [3], to address the issue of low IOU between findings and RetinaNet's bounding box representation. We anticipate this will further improve the detection performance. We will also focus on detecting biopsy negative findings, which currently have a 10x higher false positive rate compared to screening negative cases when evaluated at an operating point with sensitivity of 80%.

We demonstrated the potential impact of our algorithm by using it to successfully detect cancer that was missed by radiologists during screening but visible retrospectively. Moreover, we feel that this approach will be useful for a wide range of medical imaging problems, where a clinical decision is made at a patient or organ level, but finding-level information confers significant advantage, both during training, and as a form of direct explanatory output at run time.

References

1. Akselrod-Ballin, A., Karlinsky, L., Alpert, S., Hasoul, S., Ben-Ari, R., Barkan, E.: A region based convolutional network for tumor detection and classification in breast mammography. In: Carneiro, G., et al. (eds.) LABELS/DLMIA -2016. LNCS, vol. 10008, pp. 197–205. Springer, Cham (2016). https://doi.org/10.1007/978-3-319-46976-8_21

2. Akselrod-Ballin, A., et al.: Deep learning for automatic detection of abnormal findings in breast mammography. In: Cardoso, M.J., et al. (eds.) DLMIA/ML-CDS -2017. LNCS, vol. 10553, pp. 321–329. Springer, Cham (2017). https://doi.org/10.1007/978-3-319-67558-9_37

3. Fu, C.Y., Shvets, M., Berg, A.C.: RetinaMask: learning to predict masks improves state-of-the-art single-shot detection for free. arXiv preprint arXiv:1901.03353 (2019)

4. Gao, F., Yoon, H., Wu, T., Chu, X.: A feature transfer enabled multi-task deep learning model on medical imaging. Expert Syst. Appl. **143**, 112957 (2020)

5. He, K., Gkioxari, G., Dollár, P., Girshick, R.: Mask R-CNN. In: International Conference on Computer Vision (2017)

6. He, K., Zhang, X., Ren, S., Sun, J.: Deep residual learning for image recognition. In: CVPR, pp. 770–778 (2016)

7. Jung, H., et al.: Detection of masses in mammograms using a one-stage object detector based on a deep convolutional neural network. PLoS ONE **13**(9), e0203355 (2018)

8. Kim, H.E., et al.: Changes in cancer detection and false-positive recall in mammography using artificial intelligence: a retrospective, multireader study. The Lancet Digit. Health **2**(3), e138–e148 (2020)
9. Le, T.L.T., Thome, N., Bernard, S., Bismuth, V., Patoureaux, F.: Multitask classification and segmentation for cancer diagnosis in mammography. In: International Conference on Medical Imaging with Deep Learning - Extended Abstract Track, London, UK (2019)
10. Lehman, C.D., et al.: National performance benchmarks for modern screening digital mammography: update from the breast cancer surveillance consortium. Radiology **283**(1), 49–58 (2017)
11. Lin, T.Y., Goyal, P., Girshick, R., He, K., Dollár, P.: Focal loss for dense object detection. In: Proceedings of the IEEE International Conference on Computer Vision, pp. 2980–2988 (2017)
12. Lotter, W., Sorensen, G., Cox, D.: A multi-scale CNN and curriculum learning strategy for mammogram classification. In: Cardoso, M.J., et al. (eds.) DLMIA/ML-CDS -2017. LNCS, vol. 10553, pp. 169–177. Springer, Cham (2017). https://doi.org/10.1007/978-3-319-67558-9_20
13. Mainiero, M.B., Parikh, J.R.: Recognizing and overcoming burnout in breast imaging. J. Breast Imaging **1**(1), 60–63 (2019)
14. McKinney, S.M., et al.: International evaluation of an AI system for breast cancer screening. Nature **577**(7788), 89–94 (2020)
15. Ribli, D., Horváth, A., Unger, Z., Pollner, P., Csabai, I.: Detecting and classifying lesions in mammograms with deep learning. Sci. Rep. **8**(1), 1–7 (2018)
16. Seaman, S.R., White, I.R.: Review of inverse probability weighting for dealing with missing data. Stat. Methods Med. Res. **22**(3), 278–295 (2013)
17. Tabár, L., et al.: The incidence of fatal breast cancer measures the increased effectiveness of therapy in women participating in mammography screening. Cancer **125**(4), 515–523 (2019)
18. Teare, P., Fishman, M., Benzaquen, O., Toledano, E., Elnekave, E.: Malignancy detection on mammography using dual deep convolutional neural networks and genetically discovered false color input enhancement. J. Digit. Imaging **30**(4), 499–505 (2017). https://doi.org/10.1007/s10278-017-9993-2
19. International Agency for Research on Cancer: World Health Organization: Global cancer observatory database (2018)
20. Wu, N., et al.: Deep neural networks improve radiologists' performance in breast cancer screening. IEEE Trans. Med. Imaging **39**(4), 1184–1194 (2020)
21. Yala, A., Lehman, C., Schuster, T., Portnoi, T., Barzilay, R.: A deep learning mammography-based model for improved breast cancer risk prediction. Radiology **292**(1), 60–66 (2019)
22. Yala, A., Schuster, T., Miles, R., Barzilay, R., Lehman, C.: A deep learning model to triage screening mammograms: a simulation study. Radiology **293**(1), 38–46 (2019)
23. Zlocha, M., Dou, Q., Glocker, B.: Improving RetinaNet for CT lesion detection with dense masks from weak RECIST labels. In: Shen, D., et al. (eds.) MICCAI 2019. LNCS, vol. 11769, pp. 402–410. Springer, Cham (2019). https://doi.org/10.1007/978-3-030-32226-7_45

Colonoscopy

Adaptive Context Selection for Polyp Segmentation

Ruifei Zhang[1], Guanbin Li[1(✉)], Zhen Li[2], Shuguang Cui[2], Dahong Qian[3], and Yizhou Yu[4]

[1] School of Data and Computer Science, Sun Yat-sen University, Guangzhou, China
liguanbin@mail.sysu.edu.cn
[2] Shenzhen Research Institute of Big Data, The Chinese University of Hong Kong, Shenzhen, Guangdong, China
[3] Institute of Medical Robotics, Shanghai Jiao Tong University, Shanghai, China
[4] Deepwise AI Lab, Beijing, China

Abstract. Accurate polyp segmentation is of great significance for the diagnosis and treatment of colorectal cancer. However, it has always been very challenging due to the diverse shape and size of polyp. In recent years, state-of-the-art methods have achieved significant breakthroughs in this task with the help of deep convolutional neural networks. However, few algorithms explicitly consider the impact of the size and shape of the polyp and the complex spatial context on the segmentation performance, which results in the algorithms still being powerless for complex samples. In fact, segmentation of polyps of different sizes relies on different local and global contextual information for regional contrast reasoning. To tackle these issues, we propose an adaptive context selection based encoder-decoder framework which is composed of Local Context Attention (LCA) module, Global Context Module (GCM) and Adaptive Selection Module (ASM). Specifically, LCA modules deliver local context features from encoder layers to decoder layers, enhancing the attention to the hard region which is determined by the prediction map of previous layer. GCM aims to further explore the global context features and send to the decoder layers. ASM is used for adaptive selection and aggregation of context features through channel-wise attention. Our proposed approach is evaluated on the EndoScene and Kvasir-SEG Datasets, and shows outstanding performance compared with other state-of-the-art methods. The code is available at https://github.com/ReaFly/ACSNet.

1 Introduction

Colorectal cancer is a serious threat to human health, with the third highest morbidity and mortality among all cancers [15]. As one of the most critical precursors of this disease, polyp localization and segmentation play a key role in the early diagnosis and treatment of colorectal cancer. At present, colonoscopy is the most commonly used means of examination, but this process involves manual and thus expensive labor, not to mention its higher misdiagnosis rate [16].

© Springer Nature Switzerland AG 2020
A. L. Martel et al. (Eds.): MICCAI 2020, LNCS 12266, pp. 253–262, 2020.
https://doi.org/10.1007/978-3-030-59725-2_25

Therefore, automatic and accurate polyp segmentation is of great practical significance. However, polyp segmentation has always been a challenging task due to the diversity of polyp in shape and size. Some examples of polyp segmentation are displayed in Fig. 1.

In recent years, with the prevalence of deep learning technology, a series of convolutional neural network variants have been applied to polyp segmentation and have made breakthrough progress. Early fully convolutional neural networks [1,2,9,12] replaced the fully connected layers of the neural network with convolutional ones. In order to enlarge the receptive field of the neurons, the neural network gradually reduces the scale of the feature map and finally generates the prediction with very low resolution, resulting in a rough segmentation result and prone to inaccurate boundaries. Later, UNet [14] based structure was proposed, which adopts a stepwise upsample learning to restore the feature map resolution while maintaining the relatively large receptive field of the neurons. At the same time, the skip connection is used to enhance the fusion of shallow and deep features to improve the original FCN, greatly improving the segmentation performance and boundary localization of the specific organs or diseased regions. SegNet [19] is similar to UNet, but utilizes the max pooling indices to achieve up-sample operation in the decoder branch. SFANet [3] incorporates a sharing encoder branch and two decoder branches to detect polyp regions and boundaries respectively, and includes a new boundary-sensitive loss to mutually improve both polyp region segmentation and boundary detection. In addition, by adopting the upward concatenation to fuse multi-level features and embedding the selective kernel module to learn multi-scale features, the model is further enhanced and achieves competitive results. However, most of the methods have not taken proper measures to deal with the shape and size variance of polyps regions.

In this paper, we propose the Adaptive Context Selection Network (ACSNet). Inspired by [4], we believe that the global context features are helpful for the segmentation of large polyps, while the local context information is crucial for the identification of small ones. Therefore, the intent of our designed network is to adaptively select context information as contrast learning and feature enhancement based on the size of the polyp region to be segmented. Specifically, our ACSNet is based on the encoder-decoder framework, with Local Context Attention (LCA) module, Global Context Module (GCM), and Adaptive Selection Module (ASM). LCAs and GCM are responsible for mining local and global context features and sending them to the ASM modules in each decoder layer. Through channel-wise attention, ASM well achieves adaptive feature fusion and selection. In summary, the contributions of this paper mainly include: (1) Our designed ACSNet can adaptively attend to different context information to better cope with the impact of the diversity of polyp size and shape on segmentation. (2) Our tailored LCA and GCM modules can achieve more consistent and accurate polyp segmentation through complementary selection of local features and cross-layer enhancement of global context. (3) ACSNet achieves new state-of-the-art results on two widely used public benchmark datasets.

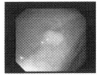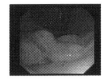

Fig. 1. Two examples of polyp segmentation

2 Method

The architecture of our ACSNet is shown in Fig. 2, which can be regarded as an enhanced UNet [14] or Feature Pyramid Network (FPN) [10]. We utilize ResNet34 [5] as our encoder, which contains five blocks in total. Accordingly, the decoder branch also has five blocks. Each decoder block is composed of two Conv-BN-ReLU combinations, and generates one prediction map with different resolution, which is supervised by the down-sampled ground truth respectively.

The GCM is placed on top of the encoder branch, which captures the global context information and densely concatenates to the ASM of each layer in the decoder path. At the same time, each skip-connection between the encoder and decoder paths of UNet [14] is replaced by the LCA module, which gives each positional feature column of every decoding layer a local context enhancement of different receptive field and at the same time delicately leverages the prediction confidence of the previous layer as a guidance to force the current layer to focus on harder regions. Finally, we utilize the ASM modules to integrate the features output from each previous decoder block, the LCA module and the GCM, based on a channel-wise attention scheme for context selection.

2.1 Local Context Attention Module (LCA)

LCA is designed as a kind of spatial attention scheme, which aims to incorporate hard sample mining when merging shallow features and pay more attention to the uncertain and more complex area to achieve layer-wise feature complementation and prediction refinement. As shown in Fig. 3, the attention map of each LCA module is determined by the prediction map generated from the upper layer of the decoder stream. Specifically, the attention map of the ith LCA module is denoted as $Att_i \in \mathbb{R}^{1 \times H_i \times W_i}$, in which H_i, W_i are the height and width of the attention map respectively. The value of position $j \in [1, 2, \cdots, H_i \times W_i]$, denoted as Att_i^j can be calculated as follows:

$$Att_i^j = 1 - \frac{\left| p_{i+1}^j - T \right|}{\max \left(T, 1 - T \right)}, \tag{1}$$

where $P_{i+1}^j \in (0, 1)$ is the jth location value of the prediction map $P_{i+1} \in \mathbb{R}^{1 \times H_i \times W_i}$ which is generated by the $(i+1)^{th}$ decoder block. T is the threshold to determine whether the specific position belongs to foreground or background.

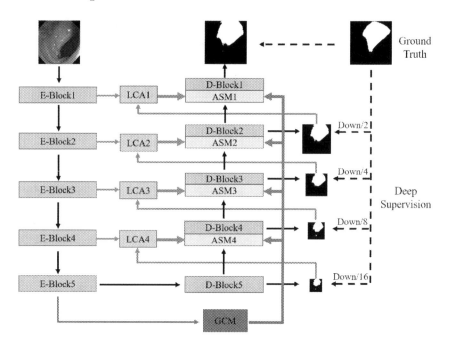

Fig. 2. Overview of our proposed ACSNet

We calculate the absolute difference between the prediction value and threshold T, and limit it to the range of 0 to 1 by dividing the maximum difference. We believe that the closer the predicted value is to the threshold T, the more uncertain the prediction of the corresponding position is, so it should be given a larger attention weight in the forwarding layer, and vice versa. Finally, we multiply the features by the attention values, and then sum with the original features to get the output of this module. For simplicity, T is set to 0.5 in our experiments.

2.2 Global Context Module (GCM)

We borrow the idea from pyramid pooling [6,11,20] to design our GCM and also put it as an independent module for global context inferring on top of the encoder branch. Meanwhile, GCM forwards the output to each ASM module to compensate the global context which is gradually diluted during layer-wise refinement.

As shown in Fig. 4, GCM contains four branches to extract context features at different scales. Specifically, this module is composed of a global average pooling branch, two adaptive local average pooling branches, and outputs three feature maps of spatial size 1×1, 3×3, 5×5, respectively. It also contains an identity mapping branch with non local operation [18] to capture the long range dependency while maintaining the original resolution. We introduce a non-

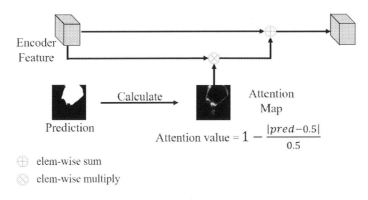

Fig. 3. Local Context Attention Module (LCA)

local operation based feature representation here to finely capture the global dependency of each positional feature to enhance the output of the encoder. In the end, we up-sample the above four feature maps and concatenate them to obtain the resulted global context feature of this module, which will be densely fed to each designed ASM module in the decoder stream.

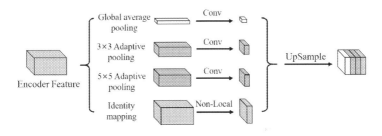

Fig. 4. Global Context Module (GCM)

2.3 Adaptive Selection Module (ASM)

We believe that local context and global context have different reference values for the segmentation of polyp regions with different appearances, sizes, and feature contrasts. Therefore, we attach an adaptive context selection module (ASM) to each block in the decoder stream. Based on the local context features generated by the LCA, the global context features from the GCM, and the output features of previous decoder block as inputs, it learns to adaptively select context feature for aggregation in each block.

As shown in Fig. 5, we incorporate a "Squeeze-and-Excitation" block [7] to adaptively recalibrate channel-wise feature responses for feature selection. Specifically, ASM takes the concatenated feature as input, and employs global average pooling to squeeze the feature map to a single vector which is further fed to a fully connected layer to learn the weight of each channel. After sigmoid operation, the attention weight is limited to the range of 0 to 1. Through multiplying the original feature maps with the attention values, some informative context features can be picked out while those not conducive to improving discrimination will be suppressed. Noted that we also apply non local operation [18] to the features output from previous decoder block before concatenation to enhance the decoder features with long range dependency.

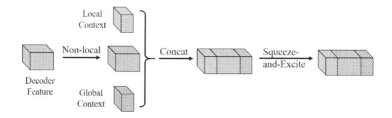

Fig. 5. Adaptive Selection Module (ASM)

3   Experiments

3.1   Datasets

We evaluate our proposed method on two benchmark colonoscopy image datasets, collected from the examination of colorectal cancer. The first is the EndoScene Dataset [17], which contains 912 images and each of which has at least one polyp region. It is divided into the training set, validation set and test set, with 547, 183, and 182 images respectively. For simplicity, we resize the images to 384×288 uniformly in our experiments. The second is Kvasir-SEG Dataset [8] containing 1000 images with polyp regions. We randomly use 60% of the dataset as training set, 20% as validation set, and the remaining 20% as test set. Since the image resolution of this dataset varies greatly, we refer to the setting of [8] and set all images to a fixed size of 320×320.

3.2   Implementation Details and Evaluation Metrics

In the training stage, we use data augmentation to enlarge the training set, including random horizontal and vertical flips, rotation, zoom and shift. All the images are randomly cropped to 224×224 as input. We set batch size to 4, and use SGD optimizer with a momentum of 0.9 and a weight decay of 0.0005 to optimize the model. A poly learning rate police is adopted to adjust the initial learning rate, which is $lr = init\_lr \times (1 - \frac{epoch}{nEpoch})^{power}$, where $init\_lr = 0.001$, $power = 0.9$, $nEpoch = 150$. We utilize the combination of a binary cross

Images   FCN8s   UNet   Unet++   SegNet   SFANet   Ours   GT   Ours-HRM

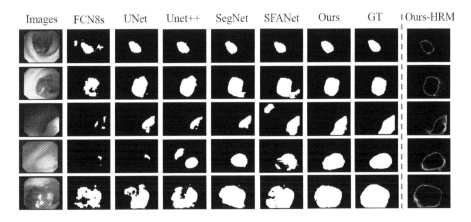

Fig. 6. Visual comparison of polyp region segmentation from state-of-the-art methods. The ground truth (GT) is shown in the penultimate column. Our proposed method consistently produces segmentation results closest to the ground truth. The hard region mining result is shown in the rightmost column.

entropy loss and a dice loss as the loss function. Our model is implemented using PyTorch [13] framework.

As in [3], we use eight metrics to evaluate the segmentation performance, including "Recall", "Specificity", "Precision", "Dice Score", "Intersection-over-Union for Polyp (IoUp)", "IoU for Background (IoUb)", "Mean IoU (mIoU)" and "Accuracy".

Table 1. Comparison with other state-of-the-art methods on the EndoScene dataset

Methods	Rec	$Spec$	$Prec$	$Dice$	$IoUp$	$IoUb$	$mIoU$	Acc
FCN8s [1]	60.21	98.60	79.59	61.23	48.38	93.45	70.92	93.77
UNet [14]	85.54	98.75	83.56	80.31	70.68	95.90	83.29	96.25
UNet++ [21]	78.90	99.15	86.17	77.38	68.00	95.48	81.74	95.78
SegNet [19]	86.48	99.04	86.54	82.67	74.41	96.33	85.37	96.62
SFANet [3]	85.51	98.94	86.81	82.93	75.00	96.33	85.66	96.61
Ours	**87.96**	**99.16**	**90.99**	**86.59**	**79.73**	**96.86**	**88.29**	**97.11**

3.3   Results on the EndoScene Dataset

We compare our ACSNet with FCN8s [1], UNet [14], UNet++ [21], SegNet [19] and SFANet [3] on the test set. As shown in Table 1, our method achieves the best performance over all metrics, with $Dice$ of 86.59%, a 3.66% improvement over the second best algorithm. Some visualization results are shown in Fig. 6 (Col.1-8), as can be seen that our algorithm is very robust to some complex situations such as polyp region sizes and image brightness changes. At the same time, due to the introduction of the effective context selection module and especially the hard region mining (abbr.HRM) mechanism, the algorithm is significantly more accurate for polyp boundary positioning. In the rightmost column of Fig. 6, it can be observed that the hard regions mined by our method are usually located in the border area of polyps, which is worthy of attention during prediction refinement.

3.4 Results on the Kvasir-SEG Dataset

On this dataset, we compare our ACSNet with UNet [14], UNet++ [21], Seg-Net [19], ResUNet [8] and SFANet [3]. The results are listed in Table 2. Similarly, our method achieves the best performance and outperforms others by large margins, further demonstrating the robustness and effectiveness of our method.

Table 2. Comparison with other state-of-the-art methods and Ablation study on the Kvasir-SEG dataset

Methods	Rec	Spec	Prec	Dice	IoUp	IoUb	mIoU	Acc
UNet [14]	87.89	97.69	83.89	82.85	73.95	94.73	84.34	95.65
UNet++ [21]	88.67	97.49	83.17	82.80	73.74	94.49	84.11	95.42
ResUNet [8]	81.25	98.31	87.88	81.14	72.23	94.00	83.11	94.90
SFANet [3]	91.99	97.05	82.95	84.68	77.06	94.83	85.94	95.71
SegNet [19]	90.03	98.13	87.51	86.43	79.11	95.90	87.51	96.68
Ours	**93.14**	98.55	**91.59**	**91.30**	**85.80**	**97.00**	**91.40**	**97.64**
Baseline	89.53	98.63	90.32	88.21	81.59	96.27	88.93	96.99
Baseline+LCAs	91.79	98.39	89.15	89.00	82.47	96.41	89.44	97.15
Baseline+LCAs+GCM	92.18	**98.72**	90.90	90.28	84.35	96.88	90.62	97.52

3.5 Ablation Study

To validate the effectiveness and necessity of each of the three modules in our proposed method, we compare ACSNet with its three variants in Table 2. Specifically, the baseline model refers to the original U-shape encoder-decoder framework with skip-connections, and we gradually add LCAs, GCM, and ASMs to it, denoted as Baseline+LCAs, Baseline+LCAs+GCM and Ours, respectively. As shown in the table, with the progressive introduction of LCAs, GCM, and ASMs, our algorithm has witnessed a certain degree of performance improvement, boosting Dice by 0.79%, 1.28%, 1.02% respectively.

4 Conclusion

In this paper, we believe that an efficient perception of local and global context is essential to improve the performance of polyps region localization and segmentation. Based on this, we propose an adaptive context selection based encoder-decoder framework which contains the LCA module for hard region mining based local context extraction, the GCM module for global feature representation and enhancement in each decoder block, and the ASM component for contextual information aggregation and selection. Extensive experimental results and ablation studies have demonstrated the effectiveness and superiority of the proposed method.

Acknowledgement. This work is supported in part by the Guangdong Basic and Applied Basic Research Foundation (No. 2020B1515020048), in part by the National Natural Science Foundation of China (No. 61976250 and No. 61702565), in part by the ZheJiang Province Key Research & Development Program (No. 2020C03073) and in part by the Key Area R&D Program of Guangdong Province (No. 2018B030338001).

References

1. Akbari, M., et al.: Polyp segmentation in colonoscopy images using fully convolutional network. In: 2018 40th Annual International Conference of the IEEE Engineering in Medicine and Biology Society, pp. 69–72 (2018)
2. Brandao, P., et al.: Fully convolutional neural networks for polyp segmentation in colonoscopy. In: Medical Imaging 2017: Computer-Aided Diagnosis, vol. 10134, p. 101340F. International Society for Optics and Photonics (2017)
3. Fang, Y., Chen, C., Yuan, Y., Tong, K.: Selective feature aggregation network with area-boundary constraints for polyp segmentation. In: Shen, D., et al. (eds.) MICCAI 2019. LNCS, vol. 11764, pp. 302–310. Springer, Cham (2019). https://doi.org/10.1007/978-3-030-32239-7_34
4. Fu, J., et al.: Adaptive context network for scene parsing. In: Proceedings of the IEEE International Conference on Computer Vision, pp. 6748–6757 (2019)
5. He, K., Zhang, X., Ren, S., Sun, J.: Deep residual learning for image recognition. In: Proceedings of the IEEE Conference on Computer Vision and Pattern Recognition, pp. 770–778 (2016)
6. He, X., Yang, S., Li, G., Li, H., Chang, H., Yu, Y.: Non-local context encoder: robust biomedical image segmentation against adversarial attacks. In: Proceedings of the AAAI Conference on Artificial Intelligence, vol. 33, pp. 8417–8424 (2019)
7. Hu, J., Shen, L., Sun, G.: Squeeze-and-excitation networks. In: Proceedings of the IEEE Conference on Computer Vision and Pattern Recognition, pp. 7132–7141 (2018)
8. Jha, D., et al.: Kvasir-SEG: a segmented polyp dataset. In: Ro, Y.M., et al. (eds.) MMM 2020. LNCS, vol. 11962, pp. 451–462. Springer, Cham (2020). https://doi.org/10.1007/978-3-030-37734-2_37
9. Li, G., Yu, Y.: Contrast-oriented deep neural networks for salient object detection. IEEE Trans. Neural Netw. Learn. Syst. **29**(12), 6038–6051 (2018)
10. Lin, T.Y., Dollár, P., Girshick, R., He, K., Hariharan, B., Belongie, S.: Feature pyramid networks for object detection. In: Proceedings of the IEEE Conference on Computer Vision and Pattern Recognition, pp. 2117–2125 (2017)

11. Liu, J.J., Hou, Q., Cheng, M.M., Feng, J., Jiang, J.: A simple pooling-based design for real-time salient object detection. In: Proceedings of the IEEE Conference on Computer Vision and Pattern Recognition, pp. 3917–3926 (2019)
12. Long, J., Shelhamer, E., Darrell, T.: Fully convolutional networks for semantic segmentation. In: Proceedings of the IEEE Conference on Computer Vision and Pattern Recognition, pp. 3431–3440 (2015)
13. Paszke, A., et al.: PyTorch: an imperative style, high-performance deep learning library. In: Advances in Neural Information Processing Systems, pp. 8026–8037 (2019)
14. Ronneberger, O., Fischer, P., Brox, T.: U-Net: convolutional networks for biomedical image segmentation. In: Navab, N., Hornegger, J., Wells, W.M., Frangi, A.F. (eds.) MICCAI 2015. LNCS, vol. 9351, pp. 234–241. Springer, Cham (2015). https://doi.org/10.1007/978-3-319-24574-4_28
15. Siegel, R.L., Miller, K.D., Jemal, A.: Cancer statistics, 2020. CA: A Cancer J. Clin. **70**(1), 7–30 (2020)
16. Van Rijn, J.C., et al.: Polyp miss rate determined by tandem colonoscopy: a systematic review. Am. J. Gastroenterol. **101**(2), 343–350 (2006)
17. Vázquez, D., et al.: A benchmark for endoluminal scene segmentation of colonoscopy images. J. Healthc. Eng. **2017** (2017)
18. Wang, X., Girshick, R., Gupta, A., He, K.: Non-local neural networks. In: Proceedings of the IEEE Conference on Computer Vision and Pattern Recognition, pp. 7794–7803 (2018)
19. Wickstrøm, K., Kampffmeyer, M., Jenssen, R.: Uncertainty and interpretability in convolutional neural networks for semantic segmentation of colorectal polyps. arXiv preprint arXiv:1807.10584 (2018)
20. Zhao, H., Shi, J., Qi, X., Wang, X., Jia, J.: Pyramid scene parsing network. In: Proceedings of the IEEE Conference on Computer Vision and Pattern Recognition, pp. 2881–2890 (2017)
21. Zhou, Z., Rahman Siddiquee, M.M., Tajbakhsh, N., Liang, J.: UNet++: a nested U-Net architecture for medical image segmentation. In: Stoyanov, D., et al. (eds.) DLMIA/ML-CDS -2018. LNCS, vol. 11045, pp. 3–11. Springer, Cham (2018). https://doi.org/10.1007/978-3-030-00889-5_1

PraNet: Parallel Reverse Attention Network for Polyp Segmentation

Deng-Ping Fan[1], Ge-Peng Ji[2], Tao Zhou[1], Geng Chen[1], Huazhu Fu[1(✉)], Jianbing Shen[1(✉)], and Ling Shao[1,3]

[1] Inception Institute of Artificial Intelligence, Abu Dhabi, UAE
{huazhu.fu,jianbing.shen}@inceptioniai.org
[2] School of Computer Science, Wuhan University, Wuhan, Hubei, China
[3] Mohamed bin Zayed University of Artificial Intelligence, Abu Dhabi, UAE
https://github.com/DengPingFan/PraNet

Abstract. Colonoscopy is an effective technique for detecting colorectal polyps, which are highly related to colorectal cancer. In clinical practice, segmenting polyps from colonoscopy images is of great importance since it provides valuable information for diagnosis and surgery. However, accurate polyp segmentation is a challenging task, for two major reasons: (i) the same type of polyps has a diversity of size, color and texture; and (ii) the boundary between a polyp and its surrounding mucosa is not sharp. To address these challenges, we propose a parallel reverse attention network (*PraNet*) for accurate polyp segmentation in colonoscopy images. Specifically, we first aggregate the features in high-level layers using a parallel partial decoder (PPD). Based on the combined feature, we then generate a global map as the initial **guidance area** for the following components. In addition, we mine the **boundary cues** using the reverse attention (RA) module, which is able to establish the relationship between areas and boundary cues. Thanks to the recurrent cooperation mechanism between areas and boundaries, our *PraNet* is capable of calibrating some misaligned predictions, improving the segmentation accuracy. Quantitative and qualitative evaluations on five challenging datasets across six metrics show that our *PraNet* improves the segmentation accuracy significantly, and presents a number of advantages in terms of generalizability, and real-time segmentation efficiency (~**50 fps**).

Keywords: Colonoscopy · Polyp segmentation · Colorectal cancer

1 Introduction

Colorectal cancer (CRC) is the third most common type of cancer around the world [23]. Therefore, preventing CRC by screening tests and removal of preneoplastic lesions (colorectal adenomas) is very critical and has become a worldwide

Electronic supplementary material The online version of this chapter (https://doi.org/10.1007/978-3-030-59725-2_26) contains supplementary material, which is available to authorized users.

public health priority. Colonoscopy is an effective technique for CRC screening and prevention since it can provide the location and appearance information of colorectal polyps, enabling doctors to remove these before they develop into CRC. A number of studies have shown that early colonoscopy has contributed to a 30% decline in the incidence of CRC [14]. Thus, in a clinical setting, accurate polyp segmentation is of great importance. It is a challenging task, however, due to two major reasons. First, the polyps often vary in appearance, *e.g.*, size, color and texture, even if they are of the same type. Second, in colonoscopy images, the boundary between a polyp and its surrounding mucosa is usually blurred and lacks the intense contrast required for segmentation approaches. These issues result in the inaccurate segmentation of polyps, and sometimes even cause the missing detection of polyps. Therefore, an automatic and accurate polyp segmentation approach capable of detecting all possible polyps at an early stage is of great significance in the prevention of CRC [17].

Among the various polyp segmentation methods, the early learning-based methods rely on extracted hand-crafted features [18,24], such as color, texture, shape, appearance, or a combination of these features. These methods are usually trained a classifier to distinguish a polyp from its surroundings. However, these models often suffer from a high miss-detection rate. The main reason is that the representation capability of hand-crafted features is quite limited when it comes to dealing with the high intra-class variations of polyps and low inter-class variations between polyps and hard mimics [30]. Recently, numerous deep learning based methods have been developed for polyp segmentation [30,32]. Although progress has been made by these methods, they only detect polyps using a bounding boxes, thus failing to locate accurate boundaries of polyps. To address this issue, Brandao *et al.* [3] employed an FCN with a pre-trained model to identify and segment polyps. Akbari *et al.* [1] utilized a modified version of FCN to improve the accuracy of polyp segmentation. Inspired by the success of the U-Net [22] applied in biomedical image segmentation, U-Net++ [39] and ResUNet++ [16] were employed for polyp segmentation and obtained promising performance. These methods focus on segmenting the whole area of the polyp, but they ignore the area-boundary constraint, which is very critical for enhancing the segmentation performance. To this end, Psi-Net [19] utilized area and boundary information simultaneously in polyp segmentation, but the relationship between the area and boundary was not fully captured. Besides, Fang *et al.* [10] proposed a three-step selective feature aggregation network with area and boundary constraints for polyp segmentation. This method *explicitly* considers the dependency between areas and boundaries and obtains good results with additional edge supervision; however, it is time-consuming (>20 h) and easily corrupted with over-fitting.

In this paper, we propose a novel deep neural network, called **P**arallel **R**everse **A**ttention **Net**work (*PraNet*), for the polyp segmentation task. Our motivation stems from the fact that, during polyp annotation, clinicians first roughly locate a polyp and then accurately extract its silhouette mask according to the local features. We therefore argue that the area and boundary are two key characteristics

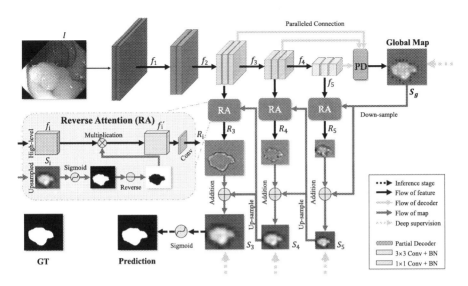

Fig. 1. Overview of the proposed *PraNet*, which consists of three reverse attention modules with a parallel partial decoder connection. See Sect. 2 for details.

that distinguish normal tissues and polyps. Different from [10], we first predict coarse areas and then *implicitly* model the boundaries by means of reverse attention. There are three advantages to this strategy, including better learning ability, improved generalization capability, and higher training efficiency. Please refer to our experiments (Sect. 3) for more details. In a nutshell, our contributions are threefold. (1) We present a novel deep neural network for real-time and accurate polyp segmentation. By aggregating features in high-level layers using a parallel partial decoder (PPD), the combined feature takes contextual information and generates a global map as the initial *guidance area* for the subsequent steps. To further mine the *boundary cues*, we leverage a set of recurrent reverse attention (RA) modules to establish the relationship between areas and boundary cues. Due to this recurrent cooperation mechanism between areas and boundaries, our model is capable of calibrating some misaligned predictions. (2) We introduce several novel evaluation metrics for polyp segmentation and present a comprehensive benchmark for existing SOTA models that are publicly available. (3) Extensive experiments demonstrate that the proposed *PraNet* outperforms most cutting-edge models and advances the SOTAs by a large margin, on five challenging datasets, with real-time inference and shorter training time.

2   Method

Figure 1 shows our *PraNet*, which utilizes a parallel partial decoder to generate the high-level semantic global map and a set of reverse attention modules for accurate polyp segmentation from the colonoscopy images. Each component will be elaborated as follows.

2.1 Feature Aggregating via Parallel Partial Decoder

Current popular medical image segmentation networks usually rely on a U-Net [22] or a U-Net like network (*e.g.*, U-Net++ [39], ResUNet [35], *etc*). These models are essentially encoder-decoder frameworks, which typically aggregate *all* multi-level features extracted from CNNs. As demonstrated by Wu *et al.* [29], compared with high-level features, low-level features demand more computational resources due to their larger spatial resolutions, but contribute less to performance. Motivated by this observation, we propose to aggregate high-level features with a ***parallel partial decoder*** component. More specifically, for an input polyp image I with size $h \times w$, five levels of features $\{\mathbf{f}_i, i = 1, ..., 5\}$ with resolution $[h/2^{k-1}, w/2^{k-1}]$ can be extracted from Res2Net-based [12] backbone network. Then, we divide $\mathbf{f}_i$ features into low-level features $\{\mathbf{f}_i, i = 1, 2\}$ and high-level features $\{\mathbf{f}_i, i = 3, 4, 5\}$. We introduce the partial decoder $p_d(\cdot)$ [29], a new SOTA decoder component, to aggregate the high-level features with a paralleled connection. The partial decoder feature is computed by $\mathbf{PD} = p_d(f_3, f_4, f_5)$, and we can obtain a global map $\mathbf{S}_g$.

2.2 Reverse Attention Module

In a clinical setting, doctors first roughly locate the polyp region, and then carefully inspect local tissues to accurately label the polyp. As discussed in Sect. 2.1, our global map $\mathbf{S}_g$ is derived from the deepest CNN layer, which can only capture a relatively rough location of the polyp tissues, without structural details (see Fig. 1). To address this issue, we propose a principle strategy to progressively mine discriminative polyp regions through an erasing foreground object manner [4,27]. Instead of aggregating features from all levels like in [4,13,33,36], we propose to adaptively learn the ***reverse attention*** in three parallel high-level features. In other words, our architecture can sequentially mine complementary regions and details by erasing the existing estimated polyp regions from high-level side-output features, where the existing estimation is up-sampled from the deeper layer.

Specifically, we obtain the output reverse attention features R_i by multiplying (element-wise $\odot$) the high-level side-output feature $\{f_i, i = 3, 4, 5\}$ by a reverse attention weight A_i, as below:

$$R_i = f_i \odot A_i. \tag{1}$$

The reverse attention weight A_i is de-facto for salient object detection in the computer vision community [4,34], and can be formulated as:

$$A_i = \ominus(\sigma(\mathcal{P}(S_{i+1}))), \tag{2}$$

where $\mathcal{P}(\cdot)$ denotes an up-sampling operation, $\sigma(\cdot)$ is the Sigmoid function, and $\ominus(\cdot)$ is a reverse operation subtracting the input from matrix $\mathbf{E}$, in which all the elements are 1. Figure 1 (RA) shows the details of this process. It is worth noting that the erasing strategy driven by reverse attention can eventually refine the imprecise and coarse estimation into an accurate and complete prediction map.

2.3   Learning Process and Implementation Details

Loss Function. Our loss function is defined as $\mathcal{L} = \mathcal{L}_{IoU}^w + \mathcal{L}_{BCE}^w$, where $\mathcal{L}_{IoU}^w$ and $\mathcal{L}_{BCE}^w$ represent the weighted IoU loss and binary cross entropy (BCE) loss for the global restriction and local (pixel-level) restriction. Different from the standard IoU loss, which has been widely adopted in segmentation tasks, the weighted IoU loss increases the weights of hard pixels to highlight their importance. In addition, compared with the standard BCE loss, $\mathcal{L}_{BCE}^w$ pays more attention to hard pixels rather than assigning all pixels equal weights. The definitions of these losses are the same as in [21,26] and their effectiveness has been validated in the field of salient object detection. Here, we adopt deep supervision for the three side-outputs (*i.e.*, S_3, S_4, and S_4) and the global map S_g. Each map is up-sampled (*e.g.*, S_3^{up}) to the same size as the ground-truth map G. Thus the total loss for the proposed *PraNet* can be formulated as: $\mathcal{L}_{total} = \mathcal{L}(G, S_g^{up}) + \sum_{i=3}^{i=5} \mathcal{L}(G, S_i^{up})$.

Implementation Details. We implement our model in PyTorch, which is accelerated by an NVIDIA TITAN RTX GPU. All the inputs are uniformly resized to 352×352 and employ a multi-scale training strategy $\{0.75, 1, 1.25\}$ rather than data augmentation. We employ the Adam optimization algorithm to optimize the overall parameters with a learning rate of $1e - 4$. The whole network is trained in an end-to-end manner, which takes 32 min to converge over 20 epochs with a batch size of 16. Our final prediction map S_p is generated by S_3 after a sigmoid operation.

3   Experiments

3.1   Experiments on Polyp Segmentation

In this section, we compare our *PraNet* with existing methods in terms of learning ability, generalization capability, complexity, and qualitative results.

Datasets and Baselines. Experiments are conducted on five polyp segmentation datasets: ETIS [23], CVC-ClinicDB/CVC-612 [2], CVC-ColonDB [24], EndoScene [25], and Kvasir [15]. The first four are standard benchmarks, and the last one is the largest-scale challenging dataset, recently released. We compare our *PraNet* with four SOTA medical image segmentation methods: U-Net [22], U-Net++ [39], ResUNet-mod [35], and ResUNet++ [16]. We also report the cutting edge polyp segmentation model, *i.e.*, SFA [10]. The segment results of SFA are generated by the released code with default settings.

Training Settings and Metrics. Unless otherwise noted, we follow the same training settings as in [16], *i.e.*, the images from Kvasir, and CVC-ClinicDB are randomly split into 80% for training, 10% for validation, and 10% for testing. We employ two metrics (*i.e.*, mean Dice and mean IoU) for quantitative evaluation, similar to [15,16]. To provide deeper insight into the model performance, we further introduce four other metrics which are widely used in the field of object detection [7,8,11,31,37,38]. The weighted Dice metric

Table 1. Quantitative results on Kvasir [15] and CVC-612 [2] datasets. 'n/a' denotes that the results are not available. '†' represents evaluation scores from [16].

	Methods	mean Dice	mean IoU	F_β^w	S_α	E_ϕ^{max}	MAE
Kvasir	U-Net (MICCAI'15) [22]	0.818	0.746	0.794	0.858	0.893	0.055
	U-Net++ (TMI'19) [39]	0.821	0.743	0.808	0.862	0.910	0.048
	ResUNet-mod† [35]	0.791	n/a	n/a	n/a	n/a	n/a
	ResUNet++† [16]	0.813	0.793	n/a	n/a	n/a	n/a
	SFA (MICCAI'19) [10]	0.723	0.611	0.670	0.782	0.849	0.075
	PraNet (**Ours**)	**0.898**	**0.840**	**0.885**	**0.915**	**0.948**	**0.030**
CVC-612	U-Net (MICCAI'15) [22]	0.823	0.755	0.811	0.889	0.954	0.019
	U-Net++ (TMI'19) [39]	0.794	0.729	0.785	0.873	0.931	0.022
	ResUNet-mod† [35]	0.779	n/a	n/a	n/a	n/a	n/a
	ResUNet++† [16]	0.796	0.796	n/a	n/a	n/a	n/a
	SFA (MICCAI'19) [10]	0.700	0.607	0.647	0.793	0.885	0.042
	PraNet (**Ours**)	**0.899**	**0.849**	**0.896**	**0.936**	**0.979**	**0.009**

F_β^w is used to amend the "Equal-importance flaw" in Dice. The MAE metric is utilized to evaluate the pixel-level accuracy. To evaluate pixel-level and global-level similarity, we adopt the recently released enhanced-alignment metric E_ϕ^{max} [6]. Since F_β^w and MAE are based on a pixel-wise evaluation system and ignore structural similarities, S_α [5] is adopted to assess the similarity between predictions and ground-truths. The evaluation toolbox is available at https://github.com/DengPingFan/PraNet.

Learning Ability. In this section, we conduct two experiments to validate our model's learning ability on two *seen* datasets, *i.e.*, Kvasir and CVC-612. *Kvasir* is a recently released challenging dataset that contains 1,000 images selected from a sub-class (polyp class) of the Kvasir dataset [20]. *CVC-ClinicDB*, also called *CVC-612*, includes 612 open-access images from 31 colonoscopy clips. As shown in Table 1, our *PraNet* outperforms all SOTAs by a large margin (mean Dice: about > 7%), across both datasets, in all metrics. This suggests that our model has a strong learning ability to effectively segment polyps.

Generalization Capability. We conduct three experiments to test the model's generalizability. The three *unseen* datasets have their own challenging situations and properties. *CVC-ColonDB* is a small-scale database which contains 380 images from 15 short colonoscopy sequences. All images are used as our testing set. *ETIS* is an early established dataset which has 196 polyp images for early diagnosis of colorectal cancer. *EndoScene* is a combination of CVC-612 and CVC300. We follow Fang *et al.* [10] and split it into training, validation, and testing subsets. We only use the testing set of EndoScene-CVC300 in this experiment, since part of CVC-612 may be seen in the training stage. *PraNet* again outperforms existing classical medical segmentation baselines (*i.e.*, U-Net, U-Net++), as well as SFA, with significant improvements (see Table 2) on all three unseen datasets. One notable finding is that SFA drops dramatically on these unseen datasets, partially demonstrating that the model generalizability is poor.

Table 2. Quantitative results on CVC-ColonDB [24], ETIS [23], and test set (CVC-T) of EndoScene [25] datasets. SFA [10] results are generated using the released code.

	Methods	mean Dice	mean IoU	F_β^w	S_α	E_ϕ^{max}	MAE
ColonDB	U-Net(MICCAI'15) [22]	0.512	0.444	0.498	0.712	0.776	0.061
	U-Net++(TMI'19) [39]	0.483	0.410	0.467	0.691	0.760	0.064
	SFA (MICCAI'19) [10]	0.469	0.347	0.379	0.634	0.765	0.094
	PraNet (Ours)	**0.709**	**0.640**	**0.696**	**0.819**	**0.869**	**0.045**
ETIS	U-Net (MICCAI'15) [22]	0.398	0.335	0.366	0.684	0.740	0.036
	U-Net++ (TMI'19) [39]	0.401	0.344	0.390	0.683	0.776	0.035
	SFA (MICCAI'19) [10]	0.297	0.217	0.231	0.557	0.633	0.109
	PraNet (Ours)	**0.628**	**0.567**	**0.600**	**0.794**	**0.841**	**0.031**
CVC-T	U-Net (MICCAI'15) [22]	0.710	0.627	0.684	0.843	0.876	0.022
	U-Net++ (TMI'19) [39]	0.707	0.624	0.687	0.839	0.898	0.018
	SFA (MICCAI'19) [10]	0.467	0.329	0.341	0.640	0.817	0.065
	PraNet (Ours)	**0.871**	**0.797**	**0.843**	**0.925**	**0.972**	**0.010**

Table 3. Training and inference analysis (same platform) on CVC-ClinicDB [2] dataset. We record the #epochs when the model converges. Lr = learning rate.

	Methods	Epoch	Lr	Training	Inference	mean Dice
CVC-612	U-Net (MICCAI'15) [22]	30	3e−4	~40 min	~8 fps	0.823
	U-Net++ (TMI'19) [39]	30	3e−4	~45 min	~7 fps	0.794
	SFA (MICCAI'19) [10]	500	1e−2	>20 h	~40 fps	0.700
	PraNet (Ours)	**20**	1e−4	**~30 min**	**~50 fps**	**0.899**

Qualitative Results. In Fig. 2, we provide the polyp segmentation results of our *PraNet* on the Kvasir test set. Our model can precisely locate and segment the polyp tissues in many challenging cases, such as varied size, homogeneous regions, different kinds of texture, *etc.*

Training and Inference Analysis. In Table 3, we present the training time, and inference time of *PraNet* and current SOTA approaches. The running times of all compared models are tested on an Intel i9-9820X CPU and a TITAN RTX GPU with 24 GB memory. As shown, our model achieves convergence with only 20 epochs (~0.5 h) of training. One reason is that the parallel structure of our *PraNet* provides a short connection way to back-propagate the loss to the early layer in the decoder path (red flow of map in Fig. 1). Moreover, the side-outputs also relieve the vanishing gradient problem and guide the early layer training. Note that our *PraNet* runs at a real-time speed of ~50 fps for a 352 × 352 input, which guarantees our method can be implemented in colonoscopy video.

3.2 Ablation Study

In this section, we test each component of our *PraNet* on the *seen* and *unseen* datasets to provide deeper insight into our model.

Table 4. Ablation study for *PraNet* on the CVC-612 and CVC300 datasets.

Settings	CVC-612 (*seen*)			CVC300 (*unseen*)		
	mean Dice	mean IoU	S_α	mean Dice	mean IoU	S_α
Backbone (No. 1)	0.747	0.668	0.735	0.726	0.631	0.670
PPD + Backbone (No. 2)	0.865	0.798	0.902	0.824	0.734	0.893
RA + Backbone (No. 3)	0.888	0.845	0.912	0.871	**0.800**	0.888
PPD + RA + Backbone (No. 4)	**0.899**	**0.849**	**0.936**	**0.871**	0.797	**0.925**

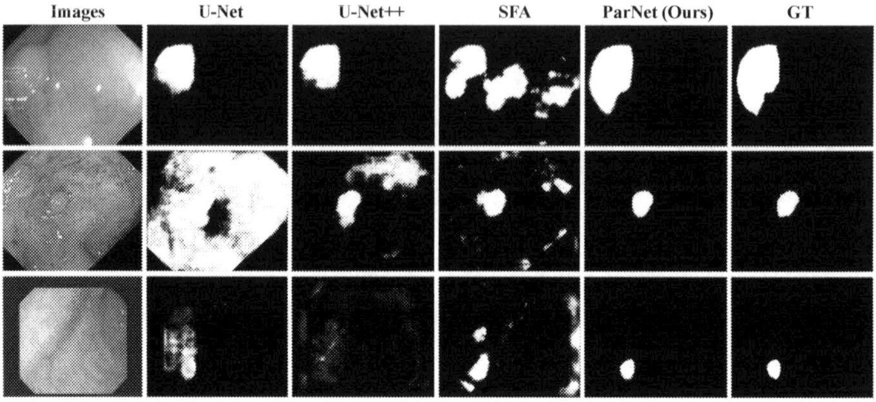

Fig. 2. Qualitative results of different methods.

Effectiveness of PPD. We investigate the importance of the cascaded mechanism (parallel partial decoder, PPD). From Table 4, we observe that No. 2 (backbone + PPD) outperforms No. 1 (backbone), clearly showing that the cascaded mechanism is necessary for increasing performance. Note that our PPD is only deployed on the high-level features, which greatly reduces the training time (See Table 3, *Inference* = ∼50 fps) of the model.

Effectiveness of RA. We further investigate the contribution of the reverse attention. The results are listed in the first and third column of Table 4. We observe that No. 3 improves the backbone (No. 1) performance on the CVC-612, increasing the mean Dice from 0.747 to 0.888 and the structure measure S_α from 0.735 to 0.912. These improvements suggest that introducing reverse attention component can enable our model to accurately distinguish true polyp tissues.

Effectiveness of PPD and RA. To assess the combination of the PPD and RA modules, we test the performance of No. 4 (PPD + RA + Backbone). As shown in Table 4, our *PraNet* (No. 4) is generally better than other settings (No. 1–No. 3). In addition, *PraNet* outperforms four SOTA models on all datasets tested, with significant improvements (>5%), making it a robust, unified architecture that can help promote future research in polyp segmentation.

4 Conclusion

We have presented a novel architecture, *PraNet*, for automatically segmenting polyps from colonoscopy images. Extensive experiments demonstrated that *PraNet* consistently outperforms all state-of-the-art approaches by a large margin (>5%) across five challenging datasets. Furthermore, *PraNet* achieves a very high accuracy (mean Dice = 0.898 on Kvasir dataset) without any pre-/post-processing. Another advantage is that *PraNet* is universal and flexible, meaning that more effective modules can be added to further improve the accuracy. Compared with current top-ranked SFA models, *PraNet* can achieve strong learning, generalization ability, and real-time segmentation efficiency. We hope this study will offer the community an opportunity to explore more powerful models on the related topics such as lung infection segmentation [9]/classification [28], or even on the upstream task, *etc.*

References

1. Akbari, M., et al.: Polyp segmentation in colonoscopy images using fully convolutional network. In: IEEE EMBC, pp. 69–72 (2018)
2. Bernal, J., Sánchez, F.J., Fernández-Esparrach, G., Gil, D., Rodríguez, C., Vilariño, F.: WM-DOVA maps for accurate polyp highlighting in colonoscopy: validation vs. saliency maps from physicians. CMIG **43**, 99–111 (2015)
3. Brandao, P., et al.: Fully convolutional neural networks for polyp segmentation in colonoscopy. In: Medical Imaging 2017: Computer-Aided Diagnosis, vol. 10134, p. 101340F (2017)
4. Chen, S., Tan, X., Wang, B., Hu, X.: Reverse attention for salient object detection. In: Ferrari, V., Hebert, M., Sminchisescu, C., Weiss, Y. (eds.) ECCV 2018. LNCS, vol. 11213, pp. 236–252. Springer, Cham (2018). https://doi.org/10.1007/978-3-030-01240-3_15
5. Fan, D.P., Cheng, M.M., Liu, Y., Li, T., Borji, A.: Structure-measure: a new way to evaluate foreground maps. In: IEEE ICCV, pp. 4548–4557 (2017)
6. Fan, D.P., Gong, C., Cao, Y., Ren, B., Cheng, M.M., Borji, A.: Enhanced-alignment measure for binary foreground map evaluation. In: IJCAI (2018)
7. Fan, D.P., Ji, G.P., Sun, G., Cheng, M.M., Shen, J., Shao, L.: Camouflaged object detection. In: IEEE CVPR (2020)
8. Fan, D.-P., Cheng, M.-M., Liu, J.-J., Gao, S.-H., Hou, Q., Borji, A.: Salient objects in clutter: bringing salient object detection to the foreground. In: Ferrari, V., Hebert, M., Sminchisescu, C., Weiss, Y. (eds.) ECCV 2018. LNCS, vol. 11219, pp. 196–212. Springer, Cham (2018). https://doi.org/10.1007/978-3-030-01267-0_12
9. Fan, D.P., et al.: Inf-Net: automatic COVID-19 lung infection segmentation from CT images. IEEE TMI (2020)
10. Fang, Y., Chen, C., Yuan, Y., Tong, K.: Selective feature aggregation network with area-boundary constraints for polyp segmentation. In: Shen, D., et al. (eds.) MICCAI 2019. LNCS, vol. 11764, pp. 302–310. Springer, Cham (2019). https://doi.org/10.1007/978-3-030-32239-7_34
11. Fu, K., Fan, D.P., Ji, G.P., Zhao, Q.: JL-DCF: Joint learning and densely-cooperative fusion framework for RGB-D salient object detection. In: IEEE CVPR, pp. 3052–3062 (2020)

12. Gao, S.H., Cheng, M.M., Zhao, K., Zhang, X.Y., Yang, M.H., Torr, P.: Res2net: a new multi-scale backbone architecture. IEEE TPAMI 1 (2020)
13. Gu, Z., et al.: CE-Net: context encoder network for 2D medical image segmentation. IEEE TMI **38**(10), 2281–2292 (2019)
14. Haggar, F.A., Boushey, R.P.: Colorectal cancer epidemiology: incidence, mortality, survival, and risk factors. Clin. Colon Rectal Surg. **22**(04), 191–197 (2009)
15. Jha, D., et al.: Kvasir-SEG: a segmented polyp dataset. In: Ro, Y.M., et al. (eds.) MMM 2020. LNCS, vol. 11962, pp. 451–462. Springer, Cham (2020). https://doi.org/10.1007/978-3-030-37734-2_37
16. Jha, D., et al.: Resunet++: an advanced architecture for medical image segmentation. In: IEEE ISM, pp. 225–2255 (2019)
17. Jia, X., Xing, X., Yuan, Y., Xing, L., Meng, M.Q.H.: Wireless capsule endoscopy: a new tool for cancer screening in the colon with deep-learning-based polyp recognition. Proc. IEEE **108**(1), 178–197 (2019)
18. Mamonov, A.V., Figueiredo, I.N., Figueiredo, P.N., Tsai, Y.H.R.: Automated polyp detection in colon capsule endoscopy. IEEE TMI **33**(7), 1488–1502 (2014)
19. Murugesan, B., Sarveswaran, K., Shankaranarayana, S.M., Ram, K., Joseph, J., Sivaprakasam, M.: Psi-Net: shape and boundary aware joint multi-task deep network for medical image segmentation. In: IEEE EMBC, pp. 7223–7226 (2019)
20. Pogorelov, K., et al.: Kvasir: a multi-class image dataset for computer aided gastrointestinal disease detection. In: ACM MSC, pp. 164–169 (2017)
21. Qin, X., Zhang, Z., Huang, C., Gao, C., Dehghan, M., Jagersand, M.: BASNet: boundary-aware salient object detection. In: IEEE CVPR, pp. 7479–7489 (2019)
22. Ronneberger, O., Fischer, P., Brox, T.: U-Net: convolutional networks for biomedical image segmentation. In: Navab, N., Hornegger, J., Wells, W.M., Frangi, A.F. (eds.) MICCAI 2015. LNCS, vol. 9351, pp. 234–241. Springer, Cham (2015). https://doi.org/10.1007/978-3-319-24574-4_28
23. Silva, J., Histace, A., Romain, O., Dray, X., Granado, B.: Toward embedded detection of polyps in WCE images for early diagnosis of colorectal cancer. Int. J. Comput. Assist. Radiol. Surg. **9**(2), 283–293 (2014)
24. Tajbakhsh, N., Gurudu, S.R., Liang, J.: Automated polyp detection in colonoscopy videos using shape and context information. IEEE TMI **35**(2), 630–644 (2015)
25. Vázquez, D., et al.: A benchmark for endoluminal scene segmentation of colonoscopy images. J. Healthcare Eng. **2017** (2017)
26. Wei, J., Wang, S., Huang, Q.: F3Net: fusion, feedback and focus for salient object detection. In: AAAI (2020)
27. Wei, Y., Feng, J., Liang, X., Cheng, M.M., Zhao, Y., Yan, S.: Object region mining with adversarial erasing: a simple classification to semantic segmentation approach. In: IEEE CVPR, pp. 1568–1576 (2017)
28. Wu, Y.H., et al.: JCS: An Explainable COVID-19 Diagnosis System by Joint Classification and Segmentation. arXiv preprint arXiv:2004.07054 (2020)
29. Wu, Z., Su, L., Huang, Q.: Cascaded partial decoder for fast and accurate salient object detection. In: IEEE CVPR, pp. 3907–3916 (2019)
30. Yu, L., Chen, H., Dou, Q., Qin, J., Heng, P.A.: Integrating online and offline three-dimensional deep learning for automated polyp detection in colonoscopy videos. IEEE JBHI **21**(1), 65–75 (2016)
31. Zhang, J., et al.: UC-Net: uncertainty inspired RGB-D saliency detection via conditional variational autoencoders. In: IEEE CVPR (2020)
32. Zhang, R., Zheng, Y., Poon, C.C., Shen, D., Lau, J.Y.: Polyp detection during colonoscopy using a regression-based convolutional neural network with a tracker. Pattern Recogn. **83**, 209–219 (2018)

33. Zhang, S., et al.: Attention guided network for retinal image segmentation. In: Shen, D., et al. (eds.) MICCAI 2019. LNCS, vol. 11764, pp. 797–805. Springer, Cham (2019). https://doi.org/10.1007/978-3-030-32239-7_88

34. Zhang, Z., Lin, Z., Xu, J., Jin, W., Lu, S.P., Fan, D.P.: Bilateral attention network for RGB-D salient object detection. arXiv preprint arXiv:2004.14582 (2020)

35. Zhang, Z., Liu, Q., Wang, Y.: Road extraction by deep residual U-net. IEEE Geosci. Remote Sens. Lett. **15**(5), 749–753 (2018)

36. Zhang, Z., Fu, H., Dai, H., Shen, J., Pang, Y., Shao, L.: ET-Net: a generic edge-aTtention guidance network for medical image segmentation. In: Shen, D., et al. (eds.) MICCAI 2019. LNCS, vol. 11764, pp. 442–450. Springer, Cham (2019). https://doi.org/10.1007/978-3-030-32239-7_49

37. Zhao, J.X., Cao, Y., Fan, D.P., Cheng, M.M., Li, X.Y., Zhang, L.: Contrast prior and fluid pyramid integration for RGBD salient object detection. In: IEEE CVPR, pp. 3927–3936 (2019)

38. Zhao, J.X., Liu, J.J., Fan, D.P., Cao, Y., Yang, J., Cheng, M.M.: EGNet: edge guidance network for salient object detection. In: IEEE ICCV, pp. 8779–8788 (2019)

39. Zhou, Z., Siddiquee, M.M.R., Tajbakhsh, N., Liang, J.: UNet++: a nested u-net architecture for medical image segmentation. In: IEEE TMI, pp. 3–11 (2019)

Few-Shot Anomaly Detection for Polyp Frames from Colonoscopy

Yu Tian[1,3(✉)], Gabriel Maicas[1], Leonardo Zorron Cheng Tao Pu[2,4], Rajvinder Singh[2], Johan W. Verjans[1,2,3], and Gustavo Carneiro[1]

[1] Australian Institute for Machine Learning, The University of Adelaide, Adelaide, Australia
yu.tian01@adelaide.edu.au
[2] Faculty of Health and Medical Sciences, The University of Adelaide, Adelaide, Australia
[3] South Australian Health and Medical Research Institute, Adelaide, Australia
[4] Department of Gastroenterology and Hepatology, Nagoya University, Nagoya, Japan

Abstract. Anomaly detection methods generally target the learning of a normal image distribution (i.e., inliers showing healthy cases) and during testing, samples relatively far from the learned distribution are classified as anomalies (i.e., outliers showing disease cases). These approaches tend to be sensitive to outliers that lie relatively close to inliers (e.g., a colonoscopy image with a small polyp). In this paper, we address the inappropriate sensitivity to outliers by also learning from inliers. We propose a new few-shot anomaly detection method based on an encoder trained to maximise the mutual information between feature embeddings and normal images, followed by a few-shot score inference network, trained with a large set of inliers and a substantially smaller set of outliers. We evaluate our proposed method on the clinical problem of detecting frames containing polyps from colonoscopy video sequences, where the training set has 13350 normal images (i.e., without polyps) and less than 100 abnormal images (i.e., with polyps). The results of our proposed model on this data set reveal a state-of-the-art detection result, while the performance based on different number of anomaly samples is relatively stable after approximately 40 abnormal training images. Code is available at https://github.com/tianyu0207/FSAD-Net.

Keywords: Machine learning · Anomaly detection · Few-shot learning · Weakly-supervised learning · Polyp detection · Colonoscopy

This work was partially supported by Australian Research Council grant DP180103232.

Electronic supplementary material The online version of this chapter (https://doi.org/10.1007/978-3-030-59725-2_27) contains supplementary material, which is available to authorized users.

A. L. Martel et al. (Eds.): MICCAI 2020, LNCS 12266, pp. 274–284, 2020.
https://doi.org/10.1007/978-3-030-59725-2_27

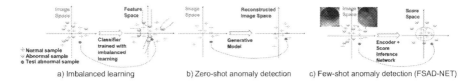

a) Imbalanced learning   b) Zero-shot anomaly detection   c) Few-shot anomaly detection (FSAD-NET)

Fig. 1. Depiction of the three different approaches to handle few-shot and zero-shot anomaly detection. Our proposed FSAD-NET demonstrate better deviations between normal and abnormal samples

1   Introduction

Classification of rare events is a common problem in medical image analysis [12], e.g., disease detection in medical screening tests such as colonoscopy. In this scenario, normal images generally come from healthy patients, while abnormal images are from unhealthy ones, where the proportion of normal images in the training set tends to be substantially larger than the abnormal ones. One possible way to address such problems is through the design of training methods that can deal with imbalanced learning problems [10,11] (Fig. 1-(a)). Even though they are often effective, these approaches still need a fairly high number of abnormal training images. Alternatively, zero-shot anomaly detection methods [4,14,15,31] tackle this problem using a training set containing only normal images to train a conditional generative model that can reconstruct normal images, and anomalies are detected based on the reconstruction errors of testing images (Fig. 1-(b)). Unfortunately, in practice these methods can misclassify outliers that lie relatively close to inliers (e.g., when cancer tissue occupies a small area of the image). Therefore, we propose a middle ground between these two approaches to address the issues of requiring a relatively large annotated data set and misclassifying challenging outliers.

In this paper, we propose a few-shot anomaly detection method network (FSAD-NET) that is trained with a highly imbalanced training set, containing a large number of normal images (more than $10,000$) and few abnormal images (less than 100) – Fig. 1-(c). The method first learns a feature encoder that is trained with normal images to maximise the mutual information (MI) between the training images and feature embeddings [7]. Next, we train a score inference network (SIN) [19] that pulls the feature embeddings of normal images close together toward a particular region of the feature space and pushes the embeddings of abnormal images away from that region of normal features.

In practice, FSAD-NET needs significantly less abnormal training images than typical imbalanced learning problems [10,11]. Moreover, given that we access a few abnormal training images, FSAD-NET has the potential to be more effective at correctly classifying challenging outliers compared to typical zero-shot anomaly detection methods [4,14,15,31]. To the best of our knowledge, our method is the first medical image analysis work to explore few-shot anomaly detection with a feature encoder that maximises MI between training images and embeddings, and explicitly optimises anomaly scores. We evaluate

FSAD-NET on the detection of colonoscopy video frames that contain polyps with a training set of more than 10000 normal images (without polyps) and less than 100 abnormal images. Results show that our FSAD-NET is more accurate than previous zero-shot anomaly detection approaches, which allows us to conclude that incorporating few abnormal cases into the training process improves the performance of anomaly detection methods. Our approach also shows better accuracy than imbalanced learning methods, suggesting that FSAD-NET is more effective at dealing with very small training sets of abnormal images.

2   Related Work

Colorectal cancer is considered to be one of the most harmful cancers [23,29]. One effective method for screening patients for colorectal cancer is colonoscopy, where the goal is to detect polyps that are malignant or pre-malignant using a camera that is inserted into the bowel. Accurate early detection of polyps may improve the 5-year survival rate to over 90% [27]. Unfortunately, the accuracy and speed of manual polyp detection can be affected by human factors, such as fatigue and expertise [22,23,30]. Therefore, automated polyp detection systems could help doctors improve polyp detection accuracy during a colonoscopy [23,29]. Traditional systems to detect polyps are based on a supervised two-class classifier [9,29] trained with large training sets of images without polyps (i.e. normal) and images containing polyps (i.e. abnormal). Annotation of such training sets is unfortunately difficult because the vast majority of colonoscopy video frames contain normal images, making the manual search for images that contain polyps challenging. Imbalanced learning solutions can therefore be used in this context [10,11], but its extension to polyp detection may not be effective without a relatively large number of abnormal images in the training set. Because of this limitation, zero-shot anomaly detection methods have been studied [13,14,16,19,21,25,26], where the idea is to learn a distribution of normal images in a particular feature space, to subsequently test samples that do not fit well in this distribution and are then classified as an outlier that may contain a polyp.

Zero-shot anomaly detection methods assume that the conditional generative model [4,14,15,21,25,26,31]) can only reconstruct normal data. Hence, when presented with an abnormal test image, the model produces a large reconstruction error. However, using an image reconstruction error for training is an indirect optimisation of the anomaly score, which can lead to a sub-optimal training process. For example, an abnormal image with a small polyp may have a low reconstruction error because the small area affected by the polyp and can be wrongly classified as normal. We advocate that the performance of zero-shot anomaly detection methods can improve with the use of a small set of abnormal training images (less than 100). Such imbalance learning problem has been tackled by few-shot classification approaches before. However, our problem has a different setup compared to problems handled by traditional few-shot learning methods that generally have many few-shot balanced multi-class problems for training [3,17,28], while ours has only one few-shot highly imbalanced binary problem for training. Hence, we can only compare our method with baseline approaches that handle imbalance learning [11,24]. For instance, Ren et al. [24] propose a learning algorithm for highly imbalanced learning problems that weights training samples using a balanced validation set – the need for this validation set is a disadvantage of this approach. The focal loss approach [11] is effective at handling imbalanced learning, but it may still need a large number of samples from both classes.

Few-shot anomaly detection has been shown in a non-medical image analysis context with the method SIN [18] that is designed to directly optimise an anomaly score for normal and abnormal images. The main challenge to train SIN lies in the high dimensionality of the images [18]. Therefore, one way to alleviate this challenge is to introduce a dimensionality reduction before training SIN. Recently, deep infomax (DIM) [7] has been shown to be an effective dimensionality reduction approach. In our paper, we propose a method that uses DIM to learn a low-dimensionality feature embedding that is then used by SIN to classify anomalies.

3 Data Set and Method

3.1 Data Set

The data set is obtained from 18 colonoscopy videos from 15 patients. Video frames containing blurred visual information are removed using the variance of Laplacian method [6]. We then sub-sample consecutive frames by taking one of every five frames because the correlation between them makes training ineffective. We also remove frames containing feces and water to reduce the need for a very large normal training set (we plan to handle such distractors in future work). This data set is defined by $\mathcal{D} = \{(\mathbf{x}, d, y)_i\}_{i=1}^{|\mathcal{D}|}$, where $\mathbf{x} : \Omega \to \mathbb{R}^3$ denotes a colonoscopy frame (Ω represents the frame lattice), $d \in \mathbb{N}$ represents patient identification[1], $y \in \mathcal{Y} = \{Normal, Abnormal\}$ denotes the normal (without polyp) and abnormal (with polyp) classes. The distribution of this data set is as follows: 1) Training set: a set of 13250 normal images (without polyps), denoted by $\mathcal{D}_N \subset \mathcal{D}$, and a set containing between 10 and 80 abnormal images, denoted by $\mathcal{D}_A \subset \mathcal{D}$; 2) Validation set: 100 normal images and 100 abnormal images for model selection; and 3) Testing set: 967 images, with 217 (25% of the set) abnormal images and 700 (75% of the set) normal images. The patients in the testing set do not appear in the training/validation sets and vice versa. This abnormality proportion (on the testing set) is commonly defined in other anomaly detection literature [21,26]. These frames were obtained with the Olympus ®190 dual focus endoscope.

---

[1] Note that the data set has been de-identified, so d is useful only for splitting $\mathcal{D}$ into training, testing and validation sets in a patient-wise manner.

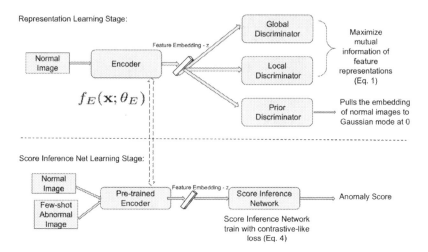

Fig. 2. The first stage of FSAD-NET training consists of modelling the encoder by maximising the MI between normal training images and embeddings in a global and local manner and by minimising the divergence of embeddings and a prior distribution [7]. The embeddings produced by the encoder are then used to train the SIN using a contrastive-like loss [19].

3.2   Method

The training process of our proposed FSAD-NET (Fig. 3) is divided into two stages: 1) pre-training of a feature encoder $\mathbf{z} = f_E(\mathbf{x}; \theta_E)$ (θ_E is the encoder parameter and $\mathbf{z} \in \mathbb{R}^Z$) to learn an image embedding that maximises the mutual information (MI) between normal images $\mathbf{x} \in \mathcal{D}_N$ and their embeddings $\mathbf{z}$ [7]; and 2) training of the SIN $f_S(f_E(\mathbf{x}; \theta_E); \theta_S)$ [19], parameterised by θ_S, with a contrastive-like loss that uses $\mathcal{D}_N$ and $\mathcal{D}_A$ to achieve the goal $f_S(f_E(\mathbf{x} \in \mathcal{D}_A; \theta_E); \theta_S) > f_S(f_E(\mathbf{x} \in \mathcal{D}_N; \theta_E); \theta_S)$.

More specifically, the training of the encoder to maximise the MI between the normal samples $\mathbf{x} \in \mathcal{D}_N$ and their feature embeddings $\mathbf{z} = f_E(\mathbf{x} \in \mathcal{D}_N; \theta_E)$ [7] is achieved with

$$\theta_E^*, \theta_G^*, \theta_L^* = \arg \max_{\theta_E, \theta_G, \theta_L} \left(\alpha \hat{I}_{\theta_G}(\mathbf{x}; f_E(\mathbf{x}; \theta_E)) + \frac{\beta}{|\mathcal{M}|} \sum_{\omega \in \mathcal{M}} \hat{I}_{\theta_L}(\mathbf{x}(\omega); f_E(\mathbf{x}(\omega); \theta_E)) \right)$$
$$+ \gamma \arg \min_{\theta_E} \arg \max_{\phi} \hat{D}_\phi(\mathbb{V} || \mathbb{U}_{\mathbb{P}, \theta_E}) \tag{1}$$

where α, β, γ are the model hyperparameters, the functions $\hat{I}_G(.)$ and $\hat{I}_L(.)$ denote an MI lower bound based on the Donsker-Varadhan representation of the Kullback-Leibler (KL)-divergence [7], defined by

$$\hat{I}_{\theta_G}(\mathbf{x}; f_E(\mathbf{x}; \theta_E)) = \mathbb{E}_{\mathbb{J}}[f_G(\mathbf{x}, f_E(\mathbf{x}; \theta_E); \theta_G)] - \log \mathbb{E}_{\mathbb{M}}[e^{f_G(\mathbf{x}, f_E(\mathbf{x}; \theta_E); \theta_G)}], \tag{2}$$

with $\mathbb{J}$ denoting the joint distribution between images $\mathbf{x}$ and their respective embeddings $\mathbf{z} = f_E(\mathbf{x}; \theta_E)$, $\mathbb{M}$ representing the product of the marginals of the images and embeddings, and $f_G(\mathbf{x}, f_E(\mathbf{x}; \theta_E); \theta_G)$ being a discriminator parameterised by θ_G. Also in (1), the function $\hat{I}_{\theta_L}(\mathbf{x}(i); f_E(\mathbf{x}(i); \theta_E))$, defined similarly as (2) for the discriminator $f_L(\mathbf{x}(\omega), f_E(\mathbf{x}(\omega); \theta_E); \theta_L)$, is the local MI between image regions $\mathbf{x}(\omega)$ ($\omega \in \mathcal{M} \subset \Omega$) and respective local embeddings $f_E(\mathbf{x}(\omega), \theta_E)$. Moreover in (1),

$$\arg\min_{\theta_E} \arg\max_{\phi} \hat{D}_\phi(\mathbb{V} || \mathbb{U}_{\mathbb{P}, \theta_E}) = \mathbb{E}_{\mathbb{V}}[\log d(\mathbf{z}; \phi)] + \mathbb{E}_{\mathbb{P}}[\log(1 - d(f_E(\mathbf{x}; \theta_E)); \phi))],$$
(3)

with $\mathbb{V}$ denoting a prior distribution for the embeddings $\mathbf{z}$ ($\mathbb{V}$ is assumed to be a normal distribution $\mathcal{N}(.; \mu_{\mathbb{V}}, \Sigma_{\mathbb{V}})$, with mean $\mu_{\mathbb{V}}$ and covariance $\Sigma_{\mathbb{V}}$), $\mathbb{P}$ the distribution of the embeddings $\mathbf{z} = f_E(\mathbf{x} \in \mathcal{N}_N; \theta_E)$, and $d(.; \phi)$ is a discriminator modelled with adversarial training to estimate the likelihood that the input is sampled from $\mathbb{V}$ or $\mathbb{P}$. This objective function pulls the feature embeddings of the normal images toward $\mathcal{N}(.; \mu_{\mathbb{V}}, \Sigma_{\mathbb{V}})$.

The next step of the learning process consists of computing the embeddings of normal and abnormal images with $\mathbf{z} = f_E(\mathbf{x} \in \mathcal{D}_A \bigcup \mathcal{D}_N; \theta_E^*)$ to train $f_S(\mathbf{z}; \theta_S)$ using a contrastive-like loss to directly optimise the anomaly score [19]. More specifically, the constrastive loss for each training sample is defined as:

$$\ell_S = \mathbb{I}(y \text{ is } Normal)|s(f_S(\mathbf{z}; \theta_S))| + \mathbb{I}(y \text{ is } Abnormal) \max(0, a - s(f_S(\mathbf{z}; \theta_S))),$$
(4)

where $\mathbb{I}(.)$ is an indicator function that is equal to one when the condition in the parameter is true, and zero otherwise, $s(x) = \frac{x - \mu_S}{\sigma_S}$ with $\mu_S = 0$ and $\sigma_S = 1$ representing the mean and standard deviation of the prior distribution for the anomaly scores for normal images, and a is the minimum margin between μ_S and the anomaly scores of abnormal images [19]. The loss in (4) pulls the scores from normal images to μ_S and pushes the scores of abnormal images away from μ_S with a margin of at least a.

During inference, we take a test image $\mathbf{x}$, compute the feature embedding with $f_E(\mathbf{x}; \theta_E)$ and then compute the score with $s = f_S(\mathbf{z}; \theta_S)$ – the score result s is then compared to a threshold τ to determine if the test image is normal or abnormal. We considered the score s as the estimation of the notion of closeness which is related to the likelihood that the embedding of a colonoscopy image is classified as belonging to the set of normal images.

4 Experiment

4.1 Experimental Setup

The original colonoscopy images are resized from initial resolution $1072 \times 1072 \times 3$ to $64 \times 64 \times 3$ to reduce the training and inference computational costs. We found that $64 \times 64 \times 3$ is the minimum size that we can use without a negative impact on AUC. We note the polyps are still visible at such resolution, as shown in Fig. 4. The model selection (to select optimiser, learning rate and model structure) is done using the validation set mentioned in Sect. 3.1. We use Adam [1] optimiser during training with a learning rate of 0.0001 for the encoder and SIN learning.

Table 1. Comparison between our proposed FSAD-NET and other state of the art zero-shot and few-shot anomaly detection methods.

	Methods	AUC
Zero-Shot	DAE [16]	0.6384
	VAE [2]	0.6528
	OC-GAN [21]	0.6137
	f-AnoGAN(ziz) [26]	0.6629
	f-AnoGAN(izi) [26]	0.6831
	f-AnoGAN(izif) [26]	0.6997
	ADGAN [14]	0.7391
Few-Shot	Densenet121 [8] (40 abnormal samples)	0.8231
	cross-entropy (30 abnormal samples)	0.6826
	cross-entropy (40 abnormal samples)	0.7115
	Focal loss (30 abnormal samples)	0.7038
	Focal loss (40 abnormal samples)	0.7235
	without RL (40 abnormal samples)	0.6011
	Learning to Reweight [24] (40 abnormal samples)	0.7862
	AE network (30 abnormal samples)	0.819
	AE network (40 abnormal samples)	0.835
	FSAD-NET (30 abnormal samples)	0.855
	FSAD-NET (40 abnormal samples)	**0.9033**

We adopt batch normalisation for both stages. We make sure our method uses a similar backbone architecture as other competing approaches in Table 1. In particular, the encoder $f_E(.;\theta_E)$ uses four convolution layers (with 64, 128, 256, 512 filters of size 4×4). The global discriminator $f_G(.;\theta_G)$ has three convolutional layers (with 128, 64, 32 filters of size 3×3). The local discriminator $f_G(.;\theta_G)$ has three convolutional layers (with 192, 512, 512 filters of size 1×1). The prior discriminator $d(.;\phi)$ has three linear layers with 1000, 200, 1 nodes per layer). We also use the validation set to estimate $a = 6$ in (4). In (1), we follow the DIM paper for setting the hyper-parameters as follows [7]: $\alpha = 0.5$, $\gamma = 1$, $\beta = 0.1$. For the prior distribution for the embeddings in (3), we set $\mu_{\mathbb{V}} = \mathbf{0}$ (i.e., a Z-dimensional vector of zeros), and $\Sigma_{\mathbb{V}}$ is a $Z \times Z$ identity matrix. To train the model, we first train the encoder, local, global and prior discriminator (representation learning stage) for 6000 epochs with a mini-batch of 64 samples. We then train SIN for 1000 epochs, with a batch size of 64, while fixing the parameters of encoder, local, global and prior discriminator. We implement our method using Pytorch [20].

The detection results are measured with the area under the receiver operating characteristic curve (AUC) on the test set [21,26], computed by varying the inference threshold τ for the score result s.

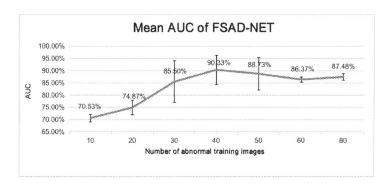

Fig. 3. AUC mean and standard deviation of FSAD-NET computed over different number of abnormal training images.

4.2 Anomaly Detection Results

The test set AUC results shown in Table 1 are divided into zero-shot and few-shot. The zero-shot rows show results obtained from the following zero-shot anomaly detection methods[2]: ADGAN [14], OCGAN [21], f-anogan and its variants [26] that involve image-to-image mean square error (MSE) loss (izi), Z-to-Z MSE loss (ziz) and its hybrid version (izif). Our FSAD-NET model outperforms all zero-shot learning methods by a large margin, showing the importance of using a few abnormal samples for training. For the few-shot results, we consider the cases where we have 30 and 40 abnormal training images, and we test several variants of the FSAD-NET. We use between 30 and 40 abnormal training images because that is the range, where we observe that our FSAD-NET produces stable AUC results. As a baseline approach, we train Densenet121 [8] using high levels of data augmentation to deal with the training imbalance issue. However, our FSAD-Net outperforms Densenet121 by a large margin. The variants of FSAD-NET are designed to test the importance of each stage of our method. The methods labelled as 'Cross entropy' and 'Focal loss' replace the contrastive loss in (4) by the cross entropy loss (commonly used in classification problems) [5] and the focal loss (robust to imbalanced learning problems) [11], respectively. FSAD-NET shows substantially better results, indicating the importance of using a more appropriate loss function for few-shot anomaly detection. To show the importance of representation learning (RL) in FSAD-Net, we tested FSAD-Net without it, which shows much lower AUC results than competing approaches. Also, we compared our method with a few-shot learning baseline [24], which proposes a learning algorithm for highly imbalanced learning problems. When used to train FSAD-Net, it achieved 78.62% of mean AUC when training with 40 abnormal training samples. Hence our model shows more accurate results than that approach. Furthermore, we test the importance of DIM to train the encoder in (1) by replacing it by the deep auto-encoder [16] (labelled as AE network) –

---

[2] Codes were downloaded from the authors' Github pages and tuned for our problem.

results show that FSAD-NET is more accurate, indicating the effectiveness of using MI and prior distribution for learning the feature embeddings in (1).

We further investigate the performance of our proposed FSAD-NET as a function of the number of abnormal training images that can vary from 10 to 80. For each number of abnormal training images, we train our model three times, using different training sets each time, and we compute the mean and standard deviation of the AUC results. The result of this experiment in Fig. 3 shows that: 1) the performance stabilises between 85%–90% when feeding the model 30 or more abnormal training images; and 2) our method is robust to extremely small training sets of abnormal images. We show a few true positive, true negative, false positive and false negative results produce by FSAD-NET in Fig. 4.

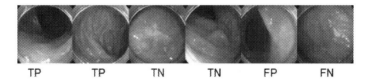

Fig. 4. True positive (TP), true negative (TN), false positive (FP) and false negative(FN) results produce by FSAD-NET (Negative = frame with polyp).

5   Conclusion

We propose the first few-shot anomaly detection framework, named as FSAD-NET, for medical image analysis applications. FSAD-NET consists of an encoder trained to maximise the mutual information between normal images and respective embeddings and a score inference network that classifies between normal and abnormal colonoscopy frames. Results show that our method achieves state-of-the-art anomaly detection performance on our colonoscopy data set, compared to previous zero-shot anomaly detection methods and imbalanced learning methods. In the future, we expect to extend our approach to polyp localisation and to work with colonoscopy frames containing distractors, like feces and water.

References

1. Kingma, D.P., Ba, J.: Adam: a method for stochastic optimization. ICLR (2015)
2. Doersch, C.: Tutorial on variational autoencoders. arXiv preprint arXiv:1606.05908 (2016)
3. Finn, C., Abbeel, P., Levine, S.: Model-agnostic meta-learning for fast adaptation of deep networks. arXiv preprint arXiv:1703.03400 (2017)
4. Gong, D., et al.: Memorizing normality to detect anomaly: memory-augmented deep autoencoder for unsupervised anomaly detection. In: Proceedings of the IEEE International Conference on Computer Vision, pp. 1705–1714 (2019)

5. Goodfellow, I., Bengio, Y., Courville, A.: Deep Learning. MIT Press, Cambridge (2016)
6. He, X., Cai, D., Niyogi, P.: Laplacian score for feature selection. In: Advances in Neural Information Processing Systems, pp. 507–514 (2006)
7. Hjelm, R.D., et al.: Learning deep representations by mutual information estimation and maximization. arXiv preprint arXiv:1808.06670 (2018)
8. Huang, G., Liu, Z., Van Der Maaten, L., Weinberger, K.Q.: Densely connected convolutional networks. In: Proceedings of the IEEE Conference on Computer Vision and Pattern Recognition (2017)
9. Korbar, B., et al.: Deep learning for classification of colorectal polyps on whole-slide images. J. Pathol. Inform. **8** (2017)
10. Li, Z., Kamnitsas, K., Glocker, B.: Overfitting of neural nets under class imbalance: analysis and improvements for segmentation. In: Shen, D. et al. (eds.) MICCAI 2019, pp. 402–410. Springer, Heidelberg (2019). https://doi.org/10.1007/978-3-030-32248-9_45
11. Lin, T.Y., Goyal, P., Girshick, R., He, K., Dollár, P.: Focal loss for dense object detection. IEEE Trans. Pattern Anal. Mach. Intell. (2018)
12. Litjens, G., et al.: A survey on deep learning in medical image analysis. Med. Image Anal. **42**, 60–88 (2017)
13. Liu, W., Luo, W., Lian, D., Gao, S.: Future frame prediction for anomaly detection-a new baseline. In: Proceedings of the IEEE Conference on Computer Vision and Pattern Recognition, pp. 6536–6545 (2018)
14. Liu, Y., et al.: Photoshopping colonoscopy video frames. In: 2020 IEEE 17th International Symposium on Biomedical Imaging (ISBI), pp. 1–5 (2020)
15. Makhzani, A., Shlens, J., Jaitly, N., Goodfellow, I., Frey, B.: Adversarial autoencoders. arXiv preprint arXiv:1511.05644 (2015)
16. Masci, J., Meier, U., Cireşan, D., Schmidhuber, J.: Stacked convolutional autoencoders for hierarchical feature extraction. In: Honkela, T., Duch, W., Girolami, M., Kaski, S. (eds.) ICANN 2011. LNCS, vol. 6791, pp. 52–59. Springer, Heidelberg (2011). https://doi.org/10.1007/978-3-642-21735-7_7
17. Nichol, A., Achiam, J., Schulman, J.: On first-order meta-learning algorithms. arXiv preprint arXiv:1803.02999 (2018)
18. Pang, G., Cao, L., Chen, L., Liu, H.: Learning representations of ultrahigh-dimensional data for random distance-based outlier detection. In: Proceedings of the 24th ACM SIGKDD International Conference on Knowledge Discovery & Data Mining, pp. 2041–2050 (2018)
19. Pang, G., Shen, C., van den Hengel, A.: Deep anomaly detection with deviation networks. In: Proceedings of the 25th ACM SIGKDD International Conference on Knowledge Discovery & Data Mining, pp. 353–362 (2019)
20. Paszke, A., et al.: Automatic differentiation in pytorch (2017)
21. Perera, P., Nallapati, R., Xiang, B.: OCGAN: One-class novelty detection using GANs with constrained latent representations. In: Proceedings of the IEEE Conference on Computer Vision and Pattern Recognition, pp. 2898–2906 (2019)
22. Pu, L., et al.: Prospective study assessing a comprehensive computer-aided diagnosis for characterization of colorectal lesions: results from different centers and imaging technologies. J. Gastroenterol. Hepatol. **34**, 25–26 (2019)
23. Pu, L.Z.C.T., et al.: Computer-aided diagnosis for characterisation of colorectal lesions: a comprehensive software including serrated lesions. Gastrointestinal Endosc. (2020)
24. Ren, M., Zeng, W., Yang, B., Urtasun, R.: Learning to reweight examples for robust deep learning. arXiv preprint arXiv:1803.09050 (2018)

25. Schlegl, T., Seeböck, P., Waldstein, S.M., Schmidt-Erfurth, U., Langs, G.: Unsupervised anomaly detection with generative adversarial networks to guide marker discovery. In: Niethammer, M., Styner, M., Aylward, S., Zhu, H., Oguz, I., Yap, P.-T., Shen, D. (eds.) IPMI 2017. LNCS, vol. 10265, pp. 146–157. Springer, Cham (2017). https://doi.org/10.1007/978-3-319-59050-9_12

26. Schlegl, T., Seeböck, P., Waldstein, S.M., Langs, G., Schmidt-Erfurth, U.: f-AnoGAN: fast unsupervised anomaly detection with generative adversarial networks. Med. image anal. **54**, 30–44 (2019)

27. Siegel, R., DeSantis, C., Jemal, A.: Colorectal cancer statistics, 2014. CA: Cancer J. Clin. **64**(2), 104–117 (2014)

28. Sung, F., Yang, Y., Zhang, L., Xiang, T., Torr, P.H., Hospedales, T.M.: Learning to compare: relation network for few-shot learning. In: Proceedings of the IEEE Conference on Computer Vision and Pattern Recognition, pp. 1199–1208 (2018)

29. Tian, Y., Pu, L.Z., Singh, R., Burt, A.D., Carneiro, G.: One-stage five-class polyp detection and classification. In: 2019 IEEE 16th International Symposium on Biomedical Imaging (ISBI 2019), pp. 70–73. IEEE (2019)

30. Van Rijn, J.C., Reitsma, J.B., Stoker, J., Bossuyt, P.M., Van Deventer, S.J., Dekker, E.: Polyp miss rate determined by tandem colonoscopy: a systematic review. Am. J. Gastroenterol. **101**(2), 343–350 (2006)

31. Zong, B., et al.: Deep autoencoding Gaussian mixture model for unsupervised anomaly detection. In: International Conference on Learning Representations (2018)

PolypSeg: An Efficient Context-Aware Network for Polyp Segmentation from Colonoscopy Videos

Jiafu Zhong[1], Wei Wang[1], Huisi Wu[1(✉)], Zhenkun Wen[1], and Jing Qin[2]

[1] College of Computer Science and Software Engineering, Shenzhen University, Shenzhen, China
hswu@szu.edu.cn
[2] Centre for Smart Health, School of Nursing, The Hong Kong Polytechnic University, Hung Hom, Hong Kong

Abstract. Polyp segmentation from colonoscopy videos is of great importance for improving the quantitative analysis of colon cancer. However, it remains a challenging task due to (1) the large size and shape variation of polyps, (2) the low contrast between polyps and background, and (3) the inherent real-time requirement of this application, where the segmentation results should be immediately presented to the doctors during the colonoscopy procedures for their prompt decision and action. It is difficult to develop a model with powerful representation capability, yielding satisfactory segmentation results in a real-time manner. We propose a novel and efficient context-aware network, named *PolypSeg*, in order to comprehensively address these challenges. The proposed *PolypSeg* consists of two key components: adaptive scale context module (ASCM) and semantic global context module (SGCM). The ASCM aggregates the multi-scale context information and takes advantage of an improved attention mechanism to make the network focus on the target regions and hence improve the feature representation. The SGCM enriches the semantic information and excludes the background noise in the low-level features, which enhances the feature fusion between high-level and low-level features. In addition, we introduce the deep separable convolution into our *PolypSeg* to replace the traditional convolution operations in order to reduce parameters and computational costs to make the *PolypSeg* run in a real-time manner. We conducted extensive experiments on a famous public available dataset for polyp segmentation task. Experimental results demonstrate that the proposed *PolypSeg* achieves much better segmentation results than state-of-the-art methods with a much faster speed.

Keywords: Polyp segmentation · Context-aware · Deep learning · Colo-noscopy video analysis

© Springer Nature Switzerland AG 2020
A. L. Martel et al. (Eds.): MICCAI 2020, LNCS 12266, pp. 285–294, 2020.
https://doi.org/10.1007/978-3-030-59725-2_28

1   Introduction

Colorectal cancer (CRC) is the second-highest cause of cancer-related deaths in the United States. According to the American Cancer Society's estimation, there are about 145600 new cases of colon cancer and over 50000 deaths in 2019 [15]. Fortunately, colon cancer can be cured if it could be detected at early stages [9]. It is reported that the advanced stages of colon cancer have poor five-year survival rate of 10% while the survival rate is 90% in its early stages. To the end, early diagnosis and intervention play a crucial role in colon cancer prevention and treatment.

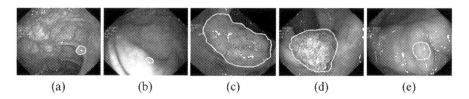

| (a) | (b) | (c) | (d) | (e) |

Fig. 1. The challenges of automatic polyp segmentation from colonoscopy videos: (a)–(b) small polyps, (c)–(d) large polyps, and (e) low contrast between polyp and the background.

Colorectal polyps are masses that bulge on the surface of the colorectum; it is the precursor to colon cancer. Early diagnosis and intervention of colon cancer are achievable if colorectal polyps can be identified and removed. Currently, colonoscopy is the primary method for colon cancer screening and prevention, in which a tiny camera is navigated into the colon in order to locate and remove polyps. However, according to the study reported in [10], during the colonoscopy procedures, one out of every four polyps was missed. In addition, it is also a difficult task for physicians to identify the polyps' boundaries from colonoscopy videos to precisely remove them, particularly in some challenging cases. In this regard, a real-time and effective automatic approach to segmenting polyps from colonoscopy videos is highly demanded in clinical practice.

However, automatic polyp segmentation is still a challenging task because (1) colorectal polyps vary significantly in size and shape (see Fig. 1 (a)-(d)), (2) it is difficult to distinguish them from the background due to the low intensity contrast (see Fig. 1 (e)), and (3) in order to fulfill the requirements of clinical practice, the algorithm should run in a real-time manner so that it can immediately alarm the doctors to take necessary action during the colonoscopy procedures. A lot of effort has been dedicated to overcoming these challenges. Early studies attempted to employ fuzzy clustering and template matching methods to segment the polyp from colonoscopy videos [6] [17]. However, the accuracy of these methods are far from sufficient for clinical practice. Recently, with the rise of deep learning, convolutional neural networks (CNNs) have shown powerful capability in image segmentation tasks. Among the deep models for segmentation, the fully convolutional network (FCN) [13] is a milestone and has been widely used in various segmentation applications. Based on the FCN, some models have been developed for polyp segmentation [1,2,11,16]. Compare with traditional methods using hand-crafted features, these deep learning models are able to extract more representative features and hence achieve much better results. However, existing models are still incapable of meeting above-mentioned challenges. First, these methods cannot adapt well to the large variation of polyp size. While Sun et al. [16] propose to widen the receptive field by introducing dilated convolution [18], it is difficult to find suitable receptive fields to capture the appropriate context. ResUNet++ [8] attempts to address this problem by introducing atrous spatial pyramid pooling (ASPP) [4] to capture multi-scale context information, but it is still insufficient to deal with polyps with different sizes, especially small polyps. Second, most of the existing methods ignore the low contrast between polyps and the background, and hence bring a lot of noise to the feature maps when performing feature fusion, which limit the segmentation accuracy. Third, it is still difficult for these models to achieve real-time performance to fulfill the clinical requirement.

In this paper, we propose a novel and efficient context-aware network for segmenting polyps from colonoscopy videos in a real-time manner; we call it *PolypSeg*. The proposed network has two key components. First, we design an adaptive scale context module (ASCM) to tackle the size variations among polyps. It aggregates multi-scale context information and takes the advantage of an attention mechanism [7] to make our model focus on the target regions and hence improve the feature representation capability. Second, we propose a semantic global context module (SGCM) to enhance the fusion between high-level and low-level features, and hence enrich the semantic information in the features as well as exclude the background noise in low-level features. SGCM is helpful to deal with the low contrast between polyps and the background and further improve the segmentation accuracy. Furthermore, in order to reduce parameters and calculations to make our model be capable of running in a real-time manner, we replaced the general convolution in the decoder with recently proposed deep separable convolution [5]. We conducted extensive experiments on a famous colonoscopy videos dataset to evaluate the effectiveness of the proposed *PolypSeg*. Experimental results demonstrate that the proposed method outperforms the state-of-the-art methods by a large margin with a much faster speed.

2   Method

Figure 2 illustrates the pipeline of the proposed network *PolypSeg*, which employs the U-Net as the backbone network. In the encoder, we extract high-level features by using multiple convolutions and downsampling operations, then harness the proposed ASCM to extract multi-scale context features and leverage an attention mechanism to make our model focus on the target regions. In the decoder, we employ the SGCM to enhance the feature fusion between high-level and low-level features and alleviate the effect of background noise.

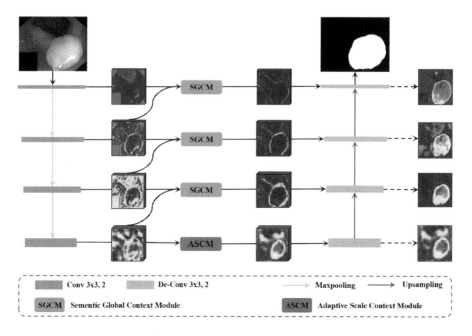

Fig. 2. An overview of the proposed *PolypSeg*.

2.1   Adaptive Scale Context Module

In order to tackle the large size variation of polyps, we introduce ASCM to (1) extract multi-scale context information and (2) drive the network to focus on regions of interest especially small polyps. The mechanism of ASCM is shown in Fig. 3. Given an input feature $\mathbf{X} \in \mathbb{R}^{C \times H \times W}$, we first feed it into three dilation convolutions with different dilation rates (1, 3 and 5) to generate three new feature maps $\mathbf{X_1}$, $\mathbf{X_2}$ and $\mathbf{X_3}$, respectively, where $\{\mathbf{X_1}, \mathbf{X_2}, \mathbf{X_3}\} \in \mathbb{R}^{C/3 \times H \times W}$, then we concatenate these three new feature maps together to generate a new feature map $\mathbf{X}' \in \mathbb{R}^{C \times H \times W}$, which hence contains multi-scale context information. Next, we feed the feature $\mathbf{X}'$ into a regular convolution block to generate feature map $\mathbf{X}'' \in \mathbb{R}^{C \times H \times W}$, aiming at further fusing the extracted multi-scale context information via the convolution operation.

After that, a channel attention mechanism is utilized to make our network focus on target regions. Specifically, We first mapping feature $\mathbf{X}''$ to feature $\mathbf{U} = [\mathbf{u_1}, \mathbf{u_2}, ..., \mathbf{u_c}] \in \mathbb{R}^{C \times 1 \times 1}$ by using global average pooling (GAP) [12]. The c-th element of U is c^{th} channel's descriptor of the feature $\mathbf{X}''$; it is calculated by:

$$\mathbf{u_c} = \frac{1}{H \times W} \sum_{i=1}^{H} \sum_{j=1}^{W} x_c(i, j) \tag{1}$$

where $x_c(i, j)$ represents the pixel value at position (i, j) in the cth channel of the feature $\mathbf{X}''$. To make use of information aggregated in the GAP operation, we

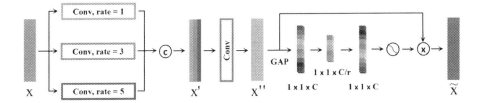

Fig. 3. An overview of the proposed ASCM.

feed the c elements of the feature $\mathbf{U}$ into two fully-connected (FC) layers with ReLU function, and then employ a sigmoid activation to generate weights $\mathbf{T}$ for each feature channels. This process can be described as follows:

$$\mathbf{T} = \sigma\left(\mathbf{W}_2\delta\left(\mathbf{W}_1\mathbf{u} + b_1\right) + b_2\right) \tag{2}$$

where $\mathbf{W}_1 \in \mathbb{R}^{C\times C/r}$, $\mathbf{W}_2 \in \mathbb{R}^{C/r\times C}$, b_1 and b_2 refers to the bias, σ refers to the ReLU function and δ refers to the Sigmoid function. We set the r to 16 to tradeoff the model complexity and accuracy. After two FC layers and Sigmoid function, the feature $\mathbf{U}$ was mapped to feature maps $\mathbf{T} \in \mathbb{R}^{C\times 1\times 1}$. In $\mathbf{T}$, channel features with target region will be assigned large weight. Finally, we get the $\tilde{\mathbf{X}} \in \mathbb{R}^{C\times H\times W}$ by multiplying the feature $\mathbf{X}''$ and the feature $\mathbf{T}$. To this end, compared with feature $\mathbf{X}''$, context features in the target region will be strengthened, and context features with background will be weakened.

2.2 Semantic Global Context Module

The extracted features from ASCM has rich semantic information, but lack of details on the boundaries of polyps. In this regard, we fuse the high-level and low-level features to restore detailed information in the decoder. Low-level features are rich in details, but it lack of semantic information, and contain too much background noise because the low contrast between the polyps and the background. We thus propose SGCM to enrich semantic information and exclude the background noise from low-level features. First, we embed the semantic information from the high-level features into the low-level features, and this process can be described as follows:

$$I^F\left(X_l\right) = I^L\left(X_l\right) \otimes U_{psample}\left(I^H\left(X_{l+1}\right)\right) \tag{3}$$

where I^L, I^H denote low-level features and high-level features, respectively. The $\otimes$ refers to the element-wise multiplication and l refers to the number of feature levels.

Second, we attempt to exclude background noise by introducing global context information for each pixel. To model the global contexts for I^F, we first apply a global attention pooling introduced by GC [3] to generate global attention map Z, and this process can be described as follows:

$$\mathbf{Z} = \sum_{j=1}^{N_p} \frac{e^{\mathbf{W}_k I^F \left(x_l^j\right)}}{\sum_{m=1}^{N_p} e^{\mathbf{W}_k I^F \left(x_l^m\right)}} \tag{4}$$

where $\mathbf{W}_k$ denotes the linear transformation matrices, $N_p = H \cdot W$ is the number of positions in feature map I^F and x_l^i denote the pixel at the ith position in feature map X_l.

Then we apply a bottleneck transform $\varphi(\cdot) = \mathbf{W}_{v2} ReLU \left(BN \left(\mathbf{W}_{v1}(\cdot)\right)\right)$ for global attention map $\mathbf{Z}$ to capture channel-wise dependencies, aiming to model the global contexts $O = \varphi(Z)$. Finally, we aggregate global contexts to each pixel, and the SGCM can be defined as follows:

$$I^O \left(x_l^i\right) = I^F \left(x_l^i\right) \oplus O \tag{5}$$

where $\oplus$ is broadcast element-wise addition, and $I^O \left(x_l\right)$ is the output of SGCM which excludes background noise with only a slight increase in computation cost.

3   Experiments and Results

3.1   Data Description

We conduct extensive experiments on a famous dataset Kvasir-SEG[1] to validate the performance of the proposed network. The Kvasir-SEG dataset contains $1,000$ RGB polyp images, which are all obtained from real colonoscopy videos. The resolution of each image is different and most of them are about 500 x 500 pixels.

3.2   Implementation Details

We train and test the model on a single NVIDIA RTX 2080Ti. The 1,000 images are randomly divided into 600 training images, 200 validating images, and 200 for the test. All the images are resized to a size 384×384 and we set batchsize to 8. Here, we use Adam with an initial learning rate of 0.001 as our optimization algorithm, and adopted the weight decay strategy when the loss on the validation set has not dropped by 5 epochs. Note that we did not use any data augmentation and image post-processing operations. Each experiment was trained from scratch for 80 epochs.

3.3   Results

We first compared our method with several state-of-the-art methods, including Akbari [1], U-Net [14], Sun [16] and ResUNet++ [8]. We employ Dice coefficient

---

[1] https://datasets.simula.no/kvasir-seg/.

Table 1. Quantitative comparison of segmentation results on Kvasir-SEG.

Method	AUC(%)	IoU(%)	Dice(%)	Params(M)	FLOPs(G)	FPS
U-Net [14]	91.48	62.72	73.41	**1.95**	**7.18**	**34**
Akbari et al. [1]	92.18	63.62	73.89	5.01	15.67	17
Sun et al. [16]	92.27	63.82	73.84	4.89	12.16	19
ResUNet++ [8]	92.56	64.87	74.18	5.65	25.27	12
Ours	**95.67**	**67.55**	**77.02**	3.61	8.55	30

(Dice), Intersection over Union (IoU) and Area Under ROC (AUC) as our evaluation metrics. In addition, we also figure out the parameters and GFLOPs for each method on a GPU card commonly equipped in a notebook, i.e., NVIDIA GTX 965M, to evaluate their efficiency. Table 1 shows the experimental results. From the results, it is observed that our PolypSeg outperforms all the other methods in these metrics. More importantly, it yielded the results with a much faster speed than other methods, achieving 30 frames per second on the commonly-used GPU card, which is sufficient to be applied to colonoscopy procedures for real-time segmentation of polyps. Particularly, compared with U-Net [14], PolypSeg has improved the three metrics from 91.48%, 62.72%, 73.41% to 95.67%, 67.55% and 77.02%, respectively, demonstrating that the proposed ASCM and SGCM is capable of greatly improving the segmentation performance. In addition, compared with other start-of-the-art methods, our PolypSeg achieved better results with much less parameters and computational resources.

Table 2. Ablation study for polyp segmentation.

Method	AUC(%)	IoU(%)	Dice(%)
Baseline	91.28	60.85	72.61
Baseline+ASCM	94.21	66.99	76.73
Baseline+SGCM	93.17	61.96	73.22
PolypSeg	**95.67**	**67.55**	**77.02**

Figure 4 visualizes six typical examples of segmentation results. It is observed that our results are closest to the ground truth. Compared with our competitors, the proposed method is capable of better dealing with many challenging cases of small polys (Fig. 4 the first and second rows), large polyps (Fig. 4 the third and fourth rows), large polys with low contrast to background (Fig. 4 fifth rows), and multiple small polyps (Fig. 4 sixth row). These examples intuitively demonstrate the proposed method's capability in taking full advantage of context information to overcome the barriers of existing polyp segmentation approaches and hence greatly improve the performance.

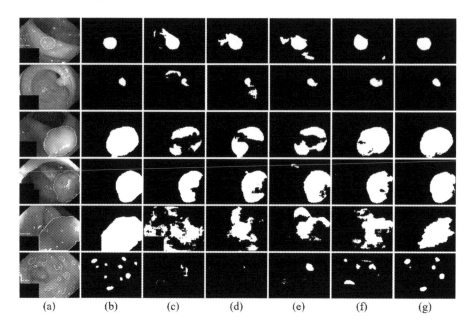

Fig. 4. Typical Examples of the segmentation results, from left to right: (a) Input images, (b) Ground truth, (c) U-Net [14], (d) Akbari et al. [1], (e) Sun et al. [16], (f) ResUNet++ [8], (g) our PolypSeg.

We further conduct ablation studies to demonstrate the effectiveness of the two key components in the proposed network: ASCM and SGCM. Also, we use U-Net [14] as our Baseline and the experimental results are shown in Table 2. It is observed in Table 2 that the Baseline network reached 91.28%, 60.85% and 72.61% in terms of AUC, IoU and Dice, respectively. When we add ASCM to the encoder, it reached 94.21%, 66.99%, 76.73% (AUC/IoU/Dice), outperforming the Baseline network by a large margin. When we add SGCM on Baseline, the Auc, Iou and Dice reach 93.17%, 61.96% and 73.22% respectively. When we employ both ASCM and SGCM to Baseline together, the Auc, Iou and Dice were further improved, achieving 95.67%, 67.55%, and 77.02%, which outperforms Baseline by 4.39%, 6.7%, and 4.41%, respectively. In addition, the two modules bring only small extra parameters and calculation costs to the Baseline network. These results demonstrate that the proposed ASCM and SGCM can boost the segmentation performance while well managing the extra costs in order to achieve real-time performance.

4   Conclusions

We present a novel and efficient context-aware network, namely *PolypSeg*, for accurate and real-time segmentation of polyps from colonoscopy videos. Extensive experiments demonstrate the proposed *PolypSeg* is capable of yielding much better segmentation results than existing methods with a much faster speed. Further investigations include validating it on more datasets and integrating it in clinical colonoscopy procedures.

Acknowledgement. This work was supported in part by grants from the National Natural Science Foundation of China (No. 61973221), the Natural Science Foundation of Guangdong Province, China (Nos. 2018A030313381 and 2019A1515011165), the Major Project or Key Lab of Shenzhen Research Foundation, China (Nos. JCYJ2016060 8173051207, ZDSYS201707311550233, KJYY201807031540021294 and JSGG201 805081520220065), the COVID-19 Prevention Project of Guangdong Province, China (No. 2020KZDZX1174), the Major Project of the New Generation of Artificial Intelligence (No. 2018AAA0102900) and the Hong Kong Research Grants Council (Project No. PolyU 152035/17E and 15205919).

References

1. Akbari, M., et al.: Polyp segmentation in colonoscopy images using fully convolutional network. In: 2018 40th Annual International Conference of the IEEE Engineering in Medicine and Biology Society (EMBC), pp. 69–72. IEEE (2018)
2. Brandao, P., et al.: Fully convolutional neural networks for polyp segmentation in colonoscopy. In: Medical Imaging 2017: Computer-Aided Diagnosis, vol. 10134, p. 101340F. International Society for Optics and Photonics (2017)
3. Cao, Y., Xu, J., Lin, S., Wei, F., Hu, H.: GCNet: non-local networks meet squeeze-excitation networks and beyond. In: Proceedings of the IEEE International Conference on Computer Vision Workshops (2019)
4. Chen, L.C., Papandreou, G., Kokkinos, I., Murphy, K., Yuille, A.L.: Deeplab: semantic image segmentation with deep convolutional nets, atrous convolution, and fully connected CRFs. IEEE Trans. Pattern Anal. Mach. Intell. **40**(4), 834–848 (2017)
5. Chollet, F.: Xception: Deep learning with depthwise separable convolutions. In: Proceedings of the IEEE Conference on Computer Vision and Pattern Recognition, pp. 1251–1258 (2017)
6. Gross, S., et al.: Polyp segmentation in nbi colonoscopy. In: Meinzer, H.P., Deserno, T.M., Handels, H., Tolxdorff, T. (eds.) Bildverarbeitung für die Medizin 2009, pp. 252–256. Springer, Heidelberg (2009). https://doi.org/10.1007/978-3-540-93860-6_51
7. Hu, J., Shen, L., Sun, G.: Squeeze-and-excitation networks. In: Proceedings of the IEEE Conference on Computer Vision and Pattern Recognition, pp. 7132–7141 (2018)
8. Jha, D., et al.: Resunet++: an advanced architecture for medical image segmentation. In: 2019 IEEE International Symposium on Multimedia (ISM), pp. 225–2255. IEEE (2019)
9. Kolligs, F.T.: Diagnostics and epidemiology of colorectal cancer. Visceral Med. **32**(3), 158–164 (2016)
10. Leufkens, A., Van Oijen, M., Vleggaar, F., Siersema, P.: Factors influencing the miss rate of polyps in a back-to-back colonoscopy study. Endoscopy **44**(05), 470–475 (2012)
11. Li, Q., et al.: Colorectal polyp segmentation using a fully convolutional neural network. In: 2017 10th International Congress on Image and Signal Processing, BioMedical Engineering and Informatics (CISP-BMEI), pp. 1–5. IEEE (2017)

12. Lin, M., Chen, Q., Yan, S.: Network in network. arXiv preprint arXiv:1312.4400 (2013)
13. Long, J., Shelhamer, E., Darrell, T.: Fully convolutional networks for semantic segmentation. In: Proceedings of the IEEE Conference on Computer Vision and Pattern Recognition, pp. 3431–3440 (2015)
14. Ronneberger, O., Fischer, P., Brox, T.: U-Net: convolutional networks for biomedical image segmentation. In: Navab, N., Hornegger, J., Wells, W.M., Frangi, A.F. (eds.) MICCAI 2015. LNCS, vol. 9351, pp. 234–241. Springer, Cham (2015). https://doi.org/10.1007/978-3-319-24574-4_28
15. Siegel, R.L., et al.: Colorectal cancer incidence patterns in the United States, 1974–2013. JNCI: J. Natl. Cancer Inst. **109**(8) (2017)
16. Sun, X., Zhang, P., Wang, D., Cao, Y., Liu, B.: Colorectal polyp segmentation by u-net with dilation convolution. arXiv preprint arXiv:1912.11947 (2019)
17. Yao, J., Miller, M., Franaszek, M., Summers, R.M.: Colonic polyp segmentation in CT colonography-based on fuzzy clustering and deformable models. IEEE Trans. Med. Imaging **23**(11), 1344–1352 (2004)
18. Yu, F., Koltun, V.: Multi-scale context aggregation by dilated convolutions. arXiv preprint arXiv:1511.07122 (2015)

Endoscopic Polyp Segmentation Using a Hybrid 2D/3D CNN

Juana González-Bueno Puyal[1,2(✉)], Kanwal K. Bhatia[2], Patrick Brandao[1,2], Omer F. Ahmad[1], Daniel Toth[2], Rawen Kader[1], Laurence Lovat[1], Peter Mountney[2], and Danail Stoyanov[1]

[1] Wellcome/EPSRC Centre for Interventional and Surgical Sciences (WEISS), University College London, London, UK
j.puyal@ucl.ac.uk
[2] Odin Vision, London, UK

Abstract. Colonoscopy is the gold standard for early diagnosis and pre-emptive treatment of colorectal cancer by detecting and removing colonic polyps. Deep learning approaches to polyp detection have shown potential for enhancing polyp detection rates. However, the majority of these systems are developed and evaluated on static images from colonoscopies, whilst applied treatment is performed on a real-time video feed. Non-curated video data includes a high proportion of low-quality frames in comparison to selected images but also embeds temporal information that can be used for more stable predictions. To exploit this, a hybrid 2D/3D convolutional neural network architecture is presented. The network is used to improve polyp detection by encompassing spatial and temporal correlation of the predictions while preserving real-time detections. Extensive experiments show that the hybrid method outperforms a 2D baseline. The proposed architecture is validated on videos from 46 patients. The results show that real-world clinical implementations of automated polyp detection can benefit from the hybrid algorithm.

Keywords: Colonoscopy · Polyp detection · Computer aided diagnosis

1 Introduction

Colorectal cancer (CRC) is the third most common cancer worldwide accounting for 10% of all forms of cancer [6] but early diagnosis and treatment can significantly improve the associated prognosis. Colonoscopy is the gold standard colon screening procedure for early detection, during which the bowel is visually inspected for polyps using an endoscope [17]. Unfortunately colonoscopy is highly operator dependent, with high reported polyp miss rates and associated interval cancers [20].

Electronic supplementary material The online version of this chapter (https://doi.org/10.1007/978-3-030-59725-2_29) contains supplementary material, which is available to authorized users.

Computer-aided polyp detection (CAD) systems aiming to assist endoscopists with automatic polyp identification from video have been researched for several decades but significant clinical progress has been reported only in recent years [8,21,22]. In particular, approaches based on Convolutional Neural Networks (CNNs) [4,5,19,22] have reported robust and promising results [1] and multiple clinical studies and randomised control trials have begun to evaluate CAD technology as well as to consider ethical and regulatory aspects [2,18]. One of the main challenges when developing such detection models is the limited availability of labelled data because full length colonoscopic videos are not usually recorded clinically, whereas still frames are stored in clinical reports enabling still image databases [3,12]. While most current CAD systems have been trained and evaluated on still images they are used in endoscopy units where real time videos are used to detect polyps. It is therefore necessary to demonstrate sound performance on videos and address model behavior stability in practical conditions with poor visibility and variability in polyp appearance that might lead to a lack of temporal coherence in consecutive frames yielding short, false predictions [3].

Yet temporal information in endoscopic video can be exploited for more temporally correlated predictions by extracting temporal representations. Recurrent neural networks (RNN), such as long short-term memory (LSTM), 3D CNNs, or two-stream models have demonstrated good results for temporal recognition tasks [7]. In endoscopic CAD, dense 3D networks have been explored, such as C3D to classify endoscopic frames containing polyps [10,11,14] and also a 3D Fully Convolutional Network for polyp segmentation [23]. A major challenge remains the problem of training a 3D CNN with a limited number of videos and various strategies have been proposed to overcome this limitation. For example including a module following a CNN's prediction to increase temporal coherence on consecutive frames or a hand-tuned false positive reduction stage appended to the polyp detection model [15]. Tracking algorithms can be combined with detection CNNs to temporally refine results but the re-initialisation of the tracker can be problematic [25]. Recently an approach fusing two CNN streams, one receiving the input frame, and the other one optical flow information, was reported but can suffer from errors in the optical flow estimation [24].

In this paper, we propose a novel hybrid 2D/3D architecture for polyp detection and segmentation in colonoscopic videos. The proposed architecture intrinsically learns spatio-temporal representations from videos. This increases the network's ability to generalize to challenging clinical endoscopic situations with lower quality data or temporally inconsistent data. A 2D neural network is used to extract spatial features and leverages large training databases through transfer learning. The 3D network component ensures temporal consistency in an efficient architecture designed for real time performance. The hybrid method has been quantitative and qualitatively evaluated and bench-marked against a 2D segmentation network. The results show an increase in performance with higher sensitivity, higher specificity, better spatial segmentation and more stable temporal segmentation.

2   Methods

A two-step temporal segmentation algorithm was developed (see Fig. 1). The proposed architecture was capable of learning a spatial representation of polyps through the 2D stage, allowing to apply transfer learning from larger 2D datasets. A 3D segmentation stage followed in order to generate temporally coherent polyp segmentation masks.

2.1   Hybrid Architecture

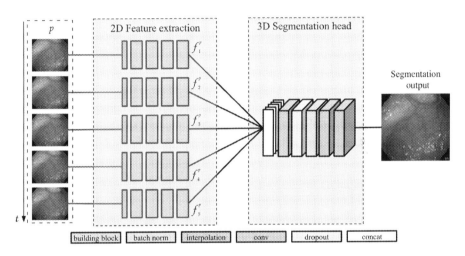

Fig. 1. Architecture of the proposed hybrid segmentation network. A polyp pseudo-batch as input and the corresponding output segmentation are presented.

The first part of the hybrid model corresponds to the 2D feature extraction. A Resnet-101 architecture was used as a backbone [9], which included a convolutional layer, followed by four sets of building blocks containing 3, 4, 23 and 3 residual blocks, sequentially. The last fully-connected layer was removed, the output then consisting of a set of 2048 feature maps per image.

The 3D segmentation stage is composed of two 3D convolutional layers, dropout, batch normalisation and an interpolation layer for upsampling (see Fig. 1). This structure is an inflated version of the segmentation head from a Fully Convolutional Network (FCN) [13]. The first convolutional layer reduces the number of features by four, applying $[d \times 3 \times 3]$ convolutions, where d is the depth, with a padding and a stride of $[1 \times 1 \times 1]$. The second convolutional layer uses the same stride, no padding, and a kernel of $[d - 2 \times 3 \times 3]$, outputting one channel per class. Note that, inherently, the temporal depth of the output maps is reduced by four filters in the segmentation head.

At train time, an input batch of N images was composed by P pseudo-batches of images. The different pseudo-batches were randomly sampled, but each contained d consecutive video frames, so that $N = d \times P$. The N images in the batch were passed through the backbone, extracting a set of spatial features f_i^p per image i. A set of features for an image had a shape of $[2048 \times w \times h]$, where w and h are the width and height of features, 32 times smaller than the original input images. The first layer of the segmentation head, namely the concatenation step, stacked the features f_i^p for the images in the same pseudo-batch $p \in [1, ..., P]$ into a batch of 3D features. Consequently, the input to the first convolutional layer had a shape of $[P \times 2048 \times d \times w \times h]$. After the last layer in the segmentation head, the network generated a probability map m_p per pseudo-batch, trained to predict the polyp in the image in the middle of the pseudo-batch.

During training, an additional auxiliary segmentation head receives features from the third backbone building block. These feature maps have undergone fewer pooling steps, and therefore are twice the size of the main backbone's feature maps. Thus, the corresponding segmentation output before the interpolation layer is double the resolution of the main segmentation output. A combination $\mathcal{L}_{aux} + \mathcal{L}_{main}$ of the losses computed from the two segmentation heads is used as the final loss $\mathcal{L}$, allowing to refine the spatial precision of the output.

At inference time, real-time performance can be achieved by running each video frame individually. The features f_i^p generated by the 2D backbone for each image can be computed once and be used by the 3D segmentation head during the following $d - 1$ frames, saving time and computational resources.

2.2   Training Strategy

A temporal window of $d = 5$ consecutive frames was selected to optimize the balance between new temporal information without sacrificing detection speed. Additionally, $d = 5$ intrinsically yields a single segmentation output from the model as explained in the previous section.

Random sampling of 5000 pseudo-batches was performed on each epoch to minimise overfitting, re-sampling at every new epoch. Data augmentation was applied in such a way as to guarantee identical augmentations within pseudo-batches. The augmentation operations consisted of random affine transformations (rotation, translation and scale) and random colour transformations (brightness, contrast and saturation). Finally, the images were preprocessed by cropping out the video borders, followed by resizing the images to 448 by 448 pixels, and an intensity normalization step.

All available positive images were used during training, and a data mining strategy was adopted for selecting the most beneficial negative images to the training set. An initial model was trained uniquely with positive samples and was used for inference on the available set of negative images (from training procedures), which was shuffled randomly. Images yielding false positives were selected until reaching 15% of the new training set.

Cross-entropy loss was used for the experiments and Adagrad for optimisation with a batch size N of 45, and pseudo-batch size P of 9. Two output classes were defined: polyp and no-polyp presence. The epoch with the highest pixel accuracy in the validation set was selected for testing. All models were trained with Pytorch on an NVIDIA Tesla V100 DGXS 32 GB GPU. The hybrid network was able to predict at 19 frames per second, ensuring real-time predictions.

3 Experimental Results

3.1 Datasets

The data was divided into two separate datasets: *Dataset V* composed of consecutive video frames and, *Dataset S* composed of static images.

Dataset V. A series of 95 videos, from 95 patients, was collected in University College London Hospital with an Olympus EVIS LUCERA endoscope under ethics REC reference 18/EE/0148. A total of 234 histologically confirmed polyps were extracted into single-polyp video sequences. The frames in these sequences were annotated by expert colonoscopists by drawing bounding boxes around each polyp. Only polyp, white light frames were included. The 25 full-length negative videos were added to the testing set, whereas the 70 procedures containing polyps were randomly split into training, validation and testing sets. The data was split on a per-procedure basis, where each procedure corresponded to a different patient, ensuring there was no patient data overlap between the sets of data. 51,426 frames from 173 polyps within 45 procedures were used for training and 2,152 frames from 8 polyps within 4 procedures were used for validation. 20,943 frames from 21 procedures and 53 polyps were used for testing, as well as 542,583 non-polyp frames from the 25 negative procedures.

Dataset S. Static polyp images were gathered from two sources: the publicly available Kvasir dataset [12] composed of 1,000 polyp images and corresponding masks, and a dataset containing 833 polyp images collected from reports from University College London Hospital under ethics REC reference 18/EE/0148 and annotated by expert colonoscopists by drawing bounding boxes around polyps. This set of 1,833 white light, polyp images was solely used for training purposes.

3.2 Comparison and Evaluation Metrics

In order to assess the temporal benefits of the model, its comparable 2D network, an FCN with a Resnet101 backbone, was implemented [13]. Whereas the backbone used was identical to the one in the hybrid model, the segmentation head was a deflated version of the hybrid one. In this case, 3D convolutional, batch normalisation and pooling layers were replaced by their 2D corresponding versions, maintaining all parameters. During training, an auxiliary segmentation head was used in the same manner as for the hybrid network. The training strategy and parameters for the baseline model were kept identical to the hybrid model, when possible, to ensure comparison fairness.

Object-wise metrics were used for evaluation, namely sensitivity, precision, and F1-score on videos with polyps, and specificity on non-polyp videos. Further implementation details are available in [3]. Polyp objects were denoted by a rectangle enclosing pixels classified as polyp at a threshold of 0.5. This allowed comparison with the ground truth annotations of rectangular bounding boxes. Dice score was reported on true positive frames to assess the quality of the overlap. Per-polyp sensitivity was also reported, considering a true positive when at least one frame was correctly detected for each polyp.

In order to determine the consistency of the predictions over consecutive frames, temporal coherence (TC) was computed as defined in [3]. Additionally, auto-correlation of masks was measured to assess both temporal and spatial correlation between two consecutive mask predictions. The auto-correlation for a given pixel position over a sequence of masks is defined as:

$$ r = \frac{\sum_{i=1}^{N-k}(Y_i - \bar{Y})(Y_{i+1} - \bar{Y})}{\sum_{i=1}^{N}(Y_i - \bar{Y})^2} \tag{1} $$

where Y_i is the value of a pixel in a certain position, and $\bar{Y}$ is the average of the pixel values in that position over the entire sequence. After obtaining a 2D vector with auto-correlation values per sequence, the average over the x and y axis was computed. The absolute difference with respect to the ground truth auto-correlation was computed, and mean and standard deviation were reported.

3.3   Results and Analysis

To establish a baseline, an FCN was trained initialising the backbone weights from ImageNet - referred to as *FCN (ImageNet)*. The training set consisted of images from the *Dataset V* training set and the full *Dataset S*. Furthermore, 10,000 negative images from the training procedures from *Dataset V* were added to the training set using the strategy previously described, by means of an FCN model formerly trained on positive images exclusively. Correspondingly, a hybrid model was trained on the training set from *Dataset V*, and 10,000 negative images, following the negative mining strategy. The hybrid network was trained using the weights from *FCN (ImageNet)* to initialise and freeze the backbone, only training the segmentation head (*Hybrid (FCN)*). Having a common backbone, it was then possible to evaluate the effect of the 3D segmentation head. Table 1 depicts the associated results when tested on the *Dataset V* testing set, where it can be observed that the incorporation of the 3D component caused a general increase in performance, particularly in terms of temporal consistency shown by the high temporal coherence and low distance between the predictions and ground truth auto-correlations.

The proposed model was also trained initialising the backbone from ImageNet weights and training the full network. This experiment will be referred to as *Hybrid (ImageNet)*. This showed the highest F1-score in Table 1, demonstrating that it is possible to train the hybrid architecture with limited amounts of data. As it can be observed, incorporating temporal components led to a

Table 1. Quantitative evaluation of baseline and proposed methods on Video *Dataset* V (*pp* and *pf* denote per-polyp and per-frame sensitivity, respectively)

Method	Sens (pp)(%)	Sens (pf)(%)	Spec (%)	Prec (%)	F1 (%)	Dice (%)	Δ A-corr (%)	TC (%)
FCN (ImageNet)	100.00	83.56	83.04	88.11	85.78	69.68	20.50 ± 16.16	79.55
Hybrid (FCN)	100.00	85.66	83.60	93.27	89.30	**74.08**	**11.92 ± 11.97**	84.24
Hybrid (ImageNet)	100.00	**86.14**	**85.32**	**93.45**	**89.65**	73.48	12.24 ± 11.74	**84.64**

reduction of false positives and negatives in both hybrid models. Figure 2 shows the per frame predictions on one of the non-polyp full colonoscopic withdrawals from the testing set, where it can be seen that the number of false positives is reduced throughout the procedure with the hybrid network compared to the FCN. Although the mapping to a 3D model of the colon is not fully realistic, it gives an indication of the importance of reducing the false positives. In addition to an improvement in sensitivity, the dice score increased considerably on detected polyps with the hybrid model, showing that the quality of the segmentation masks benefited from the temporal component. This can be supported by the increase in auto-correlation, which indicates that consecutive predicted masks presented a higher similarity. Finally, the temporal coherence benefited from the 3D component, suggesting a decrease of short false positive predictions.

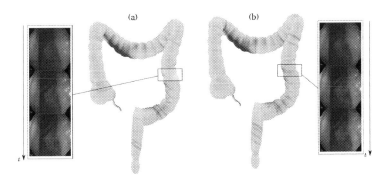

Fig. 2. Prediction timelines for a non-polyp procedure mapped onto a colon model for (a) Hybrid (ImageNet) and (b) FCN (ImageNet), where orange stripes denote false positives. Network outputs are shown as an overlay on a video section. (Color figure online)

Enhanced temporal correlation can also be observed on Fig. 3. The dice score over a sequence of polyp frames is more stable for the hybrid model, corroborating the increased dice score and auto-correlation. The segmentation examples in Fig. 3(bottom) show that both models generated similar outputs on good quality images. However, on blurry frames the FCN yielded false negatives while the Hybrid model successfully used information from surrounding frames.

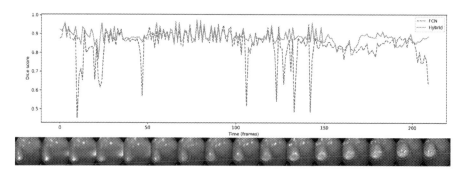

Fig. 3. Results on a polyp sequence showing (top) the dice overlap with the ground truth and (bottom) segmentation outputs for the FCN (blue) and Hybrid (green) (Color figure online)

Because the negative images used to train the models reported in Table 1 were not identical due to the negative mining strategy, a data benchmark was performed to show the closest comparison possible between models. Table 2 shows the performance of the models when trained uniquely on positive data. A 2D FCN was trained on *Dataset V* and *Dataset S*, initialising its backbone from ImageNet weights. A hybrid model was trained by initialising the backbone from this *FCN (ImageNet)*, and training exclusively the segmentation head. A hybrid model was also trained from ImageNet, without freezing any layers. It is important to note that these networks were trained exclusively on positive samples and false positives are to be expected. When compared to the FCN, which was also trained on *Dataset S*, the results presented in Table 2 show that the Hybrid model initialised from ImageNet yields poorer results. This can indicate that the model might be overfitting when training only on positive images, a common problem on 3D architectures. However, when initialised from an FCN trained on polyps, the proposed model improved the performance considerably in all aspects. Particularly, a 20.19% rise in specificity was achieved. This shows that the proposed architecture allows to pre-train on still images while benefiting from the temporal stability provided by the 3D segmentation head.

Table 2. Detailed evaluation of network performance when negative data is not included in the training set. Results are reported on the test set from *Dataset V* (*pp* and *pf* denote per-polyp and per-frame sensitivity, respectively).

Method	Sens (*pp*) (%)	Sens (*pf*) (%)	Spec (%)	Prec (%)	F1 (%)	Dice (%)	Δ A-corr (%)	TC (%)
FCN (ImageNet)	100.00	87.77	54.02	87.81	87.80	72.22	17.12 ± 15.18	84.58
Hybrid (FCN)	100.00	**88.88**	**74.18**	**92.73**	**90.76**	**75.06**	**9.77 ± 9.77**	**87.78**
Hybrid (ImageNet)	100.00	85.79	44.54	87.45	86.61	68.27	10.89 ± 11.25	84.42

4   Discussion and Conclusion

A novel hybrid 2D/3D segmentation CNN architecture for polyp detection in colonoscopic videos was developed. As a result of its 2D feature extraction, the hybrid network encompasses the benefits from a 2D architecture, namely spatial representation learning and the potential opportunity to apply transfer learning from a curated dataset of still images. This is particularly beneficial for clinical applications, where large video datasets are challenging to collect and are currently not widely available and hence still image data may be needed to provide strong and diverse representation of the spatial domain. In the proposed method, the 3D segmentation seamlessly incorporates temporal correlation in the results encapsulating learning of spatio-temporal information on smaller video datasets. The overall hybrid architecture is validated on videos from 46 patients, showing an increase in performance, along with higher quality segmentation potential. Future work includes optimization of the length of temporal information and potential handling of video aberration artefacts, as well as expanding testing to larger video datasets. Incorporation of depth and colon mapping information [16] can be incorporated to support full examination polyps location and identification.

Acknowledgments. The work was supported by the Wellcome/EPSRC Centre for Interventional and Surgical Sciences (WEISS) [203145Z/16/Z]; Engineering and Physical Sciences Research Council (EPSRC) [EP/P027938/1, EP/R004080/1, EP/P012841/1]; The Royal Academy of Engineering [CiET1819\2\36]; European Union's Horizon 2020 research and innovation programme under grant agreement No. 863146; Work carried out under a programme of and funded by the European Space Agency, the view expressed herein can in no way be taken to reflect the official opinion of the European Space Agency.

References

1. Ahmad, O.F., et al.: Artificial intelligence and computer-aided diagnosis in colonoscopy: current evidence and future directions. Lancet Gastroenterol. Hepatol. **4**(1), 71–80 (2019)

2. Ahmad, O.F., Stoyanov, D., Lovat, L.B.: Barriers and pitfalls for artificial intelligence in gastroenterology: ethical and regulatory issues. Tech. Gastrointestinal Endosc. 150636 (2019)

3. Bernal, J., et al.: Comparative validation of polyp detection methods in video colonoscopy: results from the MICCAI 2015 endoscopic vision challenge. IEEE Trans. Med. Imaging **36**(6), 1231–1249 (2017)

4. Brandao, P., et al.: Fully convolutional neural networks for polyp segmentation in colonoscopy. In: Medical Imaging 2017: Computer-Aided Diagnosis, vol. 10134, p. 101340F. International Society for Optics and Photonics (2017)

5. Brandao, P., et al.: Towards a computed-aided diagnosis system in colonoscopy: automatic polyp segmentation using convolution neural networks. J. Med. Robot. Res. **3**(02), 1840002 (2018)

6. Bray, F., Ferlay, J., Soerjomataram, I., Siegel, R.L., Torre, L.A., Jemal, A.: Global cancer statistics 2018: globocan estimates of incidence and mortality worldwide for 36 cancers in 185 countries. CA: Cancer J. Clin. **68**(6), 394–424 (2018)

7. Carreira, J., Zisserman, A.: Quo vadis, action recognition? A new model and the kinetics dataset. In: Proceedings of the IEEE Conference on Computer Vision and Pattern Recognition, pp. 6299–6308 (2017)

8. Hassan, C., et al.: New artificial intelligence system: first validation study versus experienced endoscopists for colorectal polyp detection. Gut, pp. gutjnl-2019 (2019)

9. He, K., Zhang, X., Ren, S., Sun, J.: Deep residual learning for image recognition. In: Proceedings of the IEEE Conference on Computer Vision and Pattern Recognition, pp. 770–778 (2016)

10. Itoh, H., et al.: Stable polyp-scene classification via subsampling and residual learning from an imbalanced large dataset. Healthcare Technol. Lett. **6**(6), 237–242 (2019)

11. Itoh, H., et al.: Towards automated colonoscopy diagnosis: binary polyp size estimation via unsupervised depth learning. In: Frangi, A.F., Schnabel, J.A., Davatzikos, C., Alberola-López, C., Fichtinger, G. (eds.) MICCAI 2018. LNCS, vol. 11071, pp. 611–619. Springer, Cham (2018). https://doi.org/10.1007/978-3-030-00934-2_68

12. Jha, D., et al.: Kvasir-SEG: a segmented polyp dataset. In: Ro, Y.M., et al. (eds.) MMM 2020. LNCS, vol. 11962, pp. 451–462. Springer, Cham (2020). https://doi.org/10.1007/978-3-030-37734-2_37. https://datasets.simula.no/kvasir-seg/

13. Long, J., Shelhamer, E., Darrell, T.: Fully convolutional networks for semantic segmentation. In: Proceedings of the IEEE Conference on Computer Vision and Pattern Recognition, pp. 3431–3440 (2015)

14. Misawa, M., et al.: Artificial intelligence-assisted polyp detection for colonoscopy: initial experience. Gastroenterology **154**(8), 2027–2029 (2018)

15. Qadir, H.A., Balasingham, I., Solhusvik, J., Bergsland, J., Aabakken, L., Shin, Y.: Improving automatic polyp detection using CNN by exploiting temporal dependency in colonoscopy video. IEEE J. Biomed. Health Inform. (2019)

16. Rau, A., et al.: Implicit domain adaptation with conditional generative adversarial networks for depth prediction in endoscopy. Int. J. Comput. Assist. Radiol. Surg. **14**(7), 1167–1176 (2019). https://doi.org/10.1007/s11548-019-01962-w

17. Rex, D.K., Johnson, D.A., Anderson, J.C., Schoenfeld, P.S., Burke, C.A., Inadomi, J.M.: American college of gastroenterology guidelines for colorectal cancer screening 2008. Am. J. Gastroenterol. **104**(3), 739–750 (2009)

18. Su, J.R., et al.: Impact of a real-time automatic quality control system on colorectal polyp and adenoma detection: a prospective randomized controlled study (with videos). Gastrointestinal Endosc. **91**(2), 415–424 (2020)

19. Tajbakhsh, N., Gurudu, S.R., Liang, J.: Automated polyp detection in colonoscopy videos using shape and context information. IEEE Trans. Med. Imaging **35**(2), 630–644 (2015)

20. Van Rijn, J.C., Reitsma, J.B., Stoker, J., Bossuyt, P.M., Van Deventer, S.J., Dekker, E.: Polyp miss rate determined by tandem colonoscopy: a systematic review. Am. J. Gastroenterol. **101**(2), 343–350 (2006)

21. Wang, P., et al.: Mo1712 automatic polyp detection during colonoscopy increases adenoma detection: an interim analysis of a prospective randomized control study. Gastrointestinal Endosc. **87**(6), AB490–AB491 (2018)

22. Wang, P., et al.: Development and validation of a deep-learning algorithm for the detection of polyps during colonoscopy. Nat. Biomed. Eng. **2**(10), 741–748 (2018)

23. Yu, L., Chen, H., Dou, Q., Qin, J., Heng, P.A.: Integrating online and offline three-dimensional deep learning for automated polyp detection in colonoscopy videos. IEEE J. Biomed. Health Inform. **21**(1), 65–75 (2016)
24. Zhang, P., Sun, X., Wang, D., Wang, X., Cao, Y., Liu, B.: An efficient spatial-temporal polyp detection framework for colonoscopy video. In: 2019 IEEE 31st International Conference on Tools with Artificial Intelligence (ICTAI), pp. 1252–1259. IEEE (2019)
25. Zhang, R., Zheng, Y., Poon, C.C., Shen, D., Lau, J.Y.: Polyp detection during colonoscopy using a regression-based convolutional neural network with a tracker. Pattern Recogn. **83**, 209–219 (2018)

Dermatology

A Distance-Based Loss for Smooth and Continuous Skin Layer Segmentation in Optoacoustic Images

Stefan Gerl[1], Johannes C. Paetzold[2,4(✉)], Hailong He[1,3,4], Ivan Ezhov[2,4], Suprosanna Shit[2,4], Florian Kofler[2,4], Amirhossein Bayat[2,4], Giles Tetteh[2,4], Vasilis Ntziachristos[1,3], and Bjoern Menze[2,4]

[1] Department of Electrical and Computer Engineering, Technische Universität München, Munich, Germany
[2] Department of Computer Science, Technische Universität München, Munich, Germany
johannes.paetzold@tum.de
[3] Institute of Biological and Medical Imaging (IBMI), Helmholtz Zentrum München, Neuherberg, Germany
[4] TranslaTUM Center for Translational Cancer Research, Munich, Germany

Abstract. Raster-scan optoacoustic mesoscopy (RSOM) is a powerful, non-invasive optical imaging technique for functional, anatomical, and molecular skin and tissue analysis. However, both the manual and the automated analysis of such images are challenging, because the RSOM images have very low contrast, poor signal to noise ratio, and systematic overlaps between the absorption spectra of melanin and hemoglobin. Nonetheless, the segmentation of the epidermis layer is a crucial step for many downstream medical and diagnostic tasks, such as vessel segmentation or monitoring of cancer progression. We propose a novel, shape-specific loss function that overcomes discontinuous segmentations and achieves smooth segmentation surfaces while preserving the same volumetric Dice and IoU. Further, we validate our epidermis segmentation through the sensitivity of vessel segmentation. We found a 20% improvement in Dice for vessel segmentation tasks when the epidermis mask is provided as additional information to the vessel segmentation network.

1 Introduction

Skin imaging plays an important role in dermatology; in both fundamental research and treatment of diverse diseases [10,20,31]. Optoacoustic (photoacoustic) mesoscopy offers unique opportunities in optical imaging, by bridging the

---

S. Gerl, J. C. Paetzold and H. He—Equal contribution.

Electronic supplementary material The online version of this chapter (https://doi.org/10.1007/978-3-030-59725-2_30) contains supplementary material, which is available to authorized users.

© Springer Nature Switzerland AG 2020
A. L. Martel et al. (Eds.): MICCAI 2020, LNCS 12266, pp. 309–319, 2020.
https://doi.org/10.1007/978-3-030-59725-2_30

gap between microscopic and macroscopic description of tissue and by enabling high-resolution visualizations which are deeper than optical microscopy [4,21]. Raster scan optoacoustic mesoscopy (RSOM) is a novel technique for noninvasive, high-resolution, and three-dimensional imaging of skin features based on optical absorption contrast [1,17]. Several studies using RSOM have recently demonstrated high resolution skin imaging by revealing different skin layers and the structure of the microvasculature [1,2]. RSOM imaging has been used for in-depth visual examination of psoriasis and analysis of vascularization of superficial tumors [1,18]. A critical first step for quantitative analysis of clinical RSOM images is to segment skin layers and vasculature in a rapid, reliable, and automated manner.

Previously, skin layers in RSOM images have been manually segmented by visual inspection of vasculature morphology; or automatically, based on signal intensity levels exploiting dynamic programming [15] and random forest [13]. Such procedures are slow, inaccurate, and unsuitable for processing larger numbers of patients, especially for making clinical decisions during the patient's visit. Manual segmentation is also subjective and hence compromises the reproducibility and robustness of RSOM skin image analysis. In addition to the rich, three-dimensional vascular information, RSOM images can be employed to compute biomarkers such as the total blood volume, vessel density, and complexity. These help to assess disease progression and identify skin inflammation. In current practice, the segmentation of RSOM images is thresholding-based and thus very sensitive to signal to noise ratio (SNR) variations. Therefore, there is a need

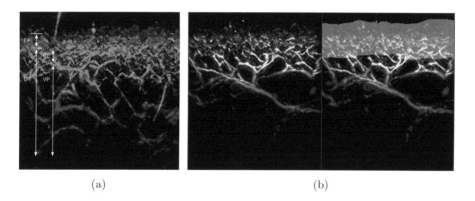

(a)   (b)

Fig. 1. *Visual problem definition:* (a) shows a maximum intensity projection (MIP) in y direction of a volumetric RSOM image of human skin. The anatomical structure is described by the white arrows: epidermis (EP) and dermis (DR); the dermis itself consists of the capillary loop layer (CL) on top of the vascular plexus (VP). Here, extracting an exact boundary is very difficult. (b) Our contribution: considering the RSOM image on the left of (b) we automated the epidermis segmentation (semitransparent white overlay on the right), which we use as a mask for the vascular segmentation of the vascular plexus (VP). The smoothness of the layer segmentation is crucial to input meaningful and reproducible images into the vessel segmentation.

to develop a reliable, automatic skin layer and vessel segmentation approach based on neural networks for rapid quantitative analysis of RSOM images.

Problem Definition: This paper sets out to develop a custom loss function for structured and smooth epidermis segmentation in RSOM images, which can be used in any segmentation network. This is difficult for the RSOM modality because melanin and hemoglobin overlap in their acoustic response. I.e. it is hard to distinguish between the melanin layer from the epidermis and the capillary loops from the dermis layer, because the absorption spectrum of melanin and hemoglobin is very close at the used laser wavelength (532 nm) of RSOM imaging [16]. First, segmentation of the epidermis is necessary to compute the average thickness of the epidermis layer, which is an important biomarker. Second, a vessel segmentation is less affected by the melanin signal, if it is smoothly masked out, see Fig. 1. Critically, the response of melanin is distributed irregularly and nonlinear, which increases the difficulty of epidermis segmentation. The smoothness is the key aspect where traditional segmentation networks trained on, e.g., BCE or soft-dice loss fail, because their segmentations lead to discontinuous surfaces, which miss parts of the epidermis, see Fig. 2. These "gaps" inevitably lead to false vessel segmentations because the melanin and hemoglobin signal cannot be distinguished, see Fig. 1 and Supplementary Fig. 1.

Methodologically, we overcome this by developing a custom loss function; previous works demonstrated how custom loss functions can be superior for difficult medical imaging tasks [3,8,14,26]. Regarding smoothness, previous approaches used post-processing steps to achieve smooth surfaces, e.g., filters [12]. More complex neural network approaches used topological concepts as priors for histology gland segmentation [5]. For general smooth shape segmentation, other approaches [9,22] successfully combined multiple fully convolutional networks, which incorporated arbitrary shape priors into the loss function of an additional network. Another successful approach used graph cuts [27]. Patino et al. implemented superpixel merging [20] and Li et al. graph theory to achieve smooth surfaces [11].

Our Contributions: At the core of our contribution is a new method to achieve an anatomically consistent and smooth epidermis layer segmentation in RSOM images. First, we introduce a custom loss term, which enforces smooth surfaces through a distance-based smoothness penalty. Next, we show that a combination of binary cross entropy loss and the custom smoothness loss optimizes epidermis segmentation. Conclusively, the resulting loss allows us to learn from very few examples, but well defined prior knowledge with very high accuracy, leading to the first automated RSOM epidermis segmentation algorithm, which preserves smooth layer structures. We validate the epidermis layer segmentation by evaluating the performance of RSOM vessel segmentation - a downstream image processing task - with the proposed segmentation algorithm and its alternatives.

2 Methodology

2.1 Loss Function

Our total loss function $\mathcal{L}_j : \mathbb{R}_{\geq 0}^{X \times Z} \to \mathbb{R}_{\geq 0}$ for a sample j consists of a per-pixel cross-entropy part $\mathcal{H}_j : \mathbb{R}_{\geq 0}^{X \times Z} \to \mathbb{R}_{\geq 0}^{X \times Z}$ and a smoothness penalty $\mathcal{S}_j : \mathbb{R}_{\geq 0}^{X \times Z} \to \mathbb{R}_{\geq 0}$, which is weighted by a constant parameter $s = \text{const}$. The total loss function is given in Eq. 1. The width and height of one 2D slice are X and Z (see also Sect. 3), consequently the summation term denotes the spatial average of $\mathcal{H}_j$.

$$\mathcal{L}_j = \frac{1}{XZ} \sum_{x=1}^{X} \sum_{z=1}^{Z} \mathcal{H}_j(x, z) + s \cdot \mathcal{S}_j \tag{1}$$

$\mathcal{H}_j$ is a per-pixel standard binary cross entropy. To incorporate prior knowledge about the shape of the epidermis, the smoothness cost function $\mathcal{S}_j$ is defined in Eq. 3. This concept is motivated by the clinical imaging setup, where the epidermis layer is always approximately parallel to the $x - y$ plane. The scenario of arbitrary orientations, non-parallel to some coordinate plane would complicate implementation, but is implausible, as the RSOM scan is acquired directly and directional on the skin surface. Firstly, we split the probability map $P_j \in \mathbb{R}_{\geq 0}^{X \times Z}$, in Z row vectors:

$$P_j = \left[\underline{p}_{1,j}^{\mathsf{T}}, \ \underline{p}_{2,j}^{\mathsf{T}}, \ \cdots, \ \underline{p}_{Z,j}^{\mathsf{T}} \right]. \tag{2}$$

Secondly, we perform a 1D convolution or correlation operation (denoted by $*$) of a vector $\underline{p}_{z,j}$ with kernel K, defined in Eq. 4. Note that this is a discrete convolution and K is a discrete kernel, and its weights are chosen to obey $\sum_{-\infty}^{\infty} K = 1$. Furthermore, convolving with K does not change the size of $p_{z,j}$. In Eq. 3, $\oslash$ is the Hadamard division [6], where $\mathbf{1}_X \in \mathbb{R}^X$ is a vector of ones. $|\cdot|$ denotes the element-wise absolute value.

$$\mathcal{S}_j = \sum_x \sum_z \left| \left(\left(\left(\underline{p}_{z,j} * K \right) + \mathbf{1}_X \right) \oslash \left(\underline{p}_{z,j} + \mathbf{1}_X \right) \right) - \mathbf{1}_X \right| \tag{3}$$

$$K(x) = \begin{cases} \frac{1}{5} & |x| \leq 2 \\ 0 & \text{else} \end{cases} \tag{4}$$

In the case of an equal prediction probability in x direction, $p_{z,j} = c_z \cdot \mathbf{1}_{1,X}$, with $c_z \in \mathbb{R}_{\geq 0, \leq 1}, c_z = \text{const}$ for all z. Consequently, $\mathcal{S}_j = 0$, which results in no smoothness penalty.

However, in the common case of an unequal prediction probability in x direction, it follows that $\mathcal{S}_j > 0$; and $\mathcal{S}_j$ contributes to the total loss function, i.e., penalizing a non-smooth layer in x direction. Note that $\mathcal{S}_j$ is differentiable with respect to the model weights, which is a necessary condition for any loss function. Due to incorporating the smoothness penalty, the model is directly taught to learn smooth representations, rather than requiring a manual post-processing step. Note that the computation of $\mathcal{S}_j$ is very inexpensive, as it requires only one 1D convolution, additions, and one division.

In practice, a perfectly segmented healthy epidermis is not rectilinear across a whole RSOM image, as the thickness of the skin layers deviates in spatially coarse patterns. This means that the thickness differences are smooth and coarse but not abrupt, see Fig. 3. Thus, $\mathcal{S}_j > 0$, while at the same time $\sum_x \sum_z \mathcal{H}_j(x, z) = 0$, resulting in an overall nonzero loss. Therefore, the scaling factor s in Eq. 1 must be tuned accordingly.

2.2 Network Architectures

We use two very general segmentation architectures to show that the novel loss function is agnostic to the network architecture. First, a U-Net [23] with dropout in all up-convolution blocks except the first one. Second, a fully convolutional network (FCN) with 7 layers depth and no dropout [28]. We train using the described 5 fold cross validation for all loss functions depicted in Table 1. All networks are implemented in Pytorch using the Adam optimizer.

3 Experiments and Discussion

Since our objective is to achieve a smooth epidermis and dermis surface segmentation, while maintaining the accuracy of traditional overlap and volumetric scores (Dice, Precision, IoU), we compare the segmentation from a pure BCE loss function to our combined loss function for a starkly varying smoothness loss term, weighted by s. We validate our epidermis segmentation by an additional sensitivity experiment. We run a standard CNN vessel segmentation on the masked image volume and show that the new and smooth epidermis segmentation is beneficial for vessel segmentation, thereby also for even further "downstream" tasks in clinical practice.

Dataset: The given RSOM dataset consists of two volumetric data channels with a size of $333 \times 171 \times 500$ pixels ($X \times Y \times Z$). Step sizes are $\Delta x = \Delta y = 12\,\mu m$ and $\Delta z = 3\,\mu m$, resulting in image volumes of $2 \times 4 \times 1\,mm^3$, where part of the data can represent voxels outside the skin. For the layer segmentation, data is processed in $x - z$ slices of 333×500 pixels. We split our dataset consisting of 31 3D volumes according to these in 25 volumes for training and validation (5-fold cross validation) and 6 volumes for testing. Next, we split all 3D volumes along the $x - z$ slices and shuffle the train and validation set across the 25 volumes. Thereby, we have a training set of 3420 2D images (20×171), a validation set of 855 2D images (5×171), and a completely unseen test set of 1026 2D images (6×171). The GT of the epidermis was labeled using the approach in [1,2] by experts familiar with RSOM images. In ambiguous situations, labels were discussed to reach consensus decisions.

Assessment of the Segmentation Smoothness: Dice score and IoU do not reveal detailed morphological information about the segmentation result. In order to quantify the epidermis and dermis surface smoothness, we calculate the arithmetic mean deviation in 1D, which is a common measure in material

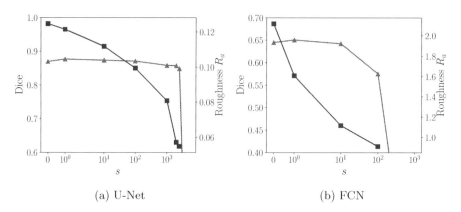

(a) U-Net   (b) FCN

Fig. 2. The resulting Dice (red) and Roughness R_a (blue) values for our Epidermis segmentation plotted against the scaling factor s between BCE and smoothness loss in Eq. (3). To calculate the roughness value, all 2D slices of the test set are accumulated; R_a is calculated by averaging over all x, for all slices for both, the epidermis and the dermis surfaces. For both the U-Net and the FCN, increasing the smoothness loss substantially improves the surface roughness, while maintaining a robust Dice score, for a wide range of s (1–2000). Please note that $s = 0$ represents a pure BCE loss function. We consider this a strong property of our loss term, as its robustness across log-scales is evident. (Color figure online)

science to assess the quality of a surface [7,30]. Where, $\tilde{\mu}_z(x)$, is the local mean at x over a moving window of 5 pixels.

$$R_a = \frac{1}{X} \sum_{x=1}^{X} |z(x) - \tilde{\mu}_z(x)| \qquad (5)$$

The local mean respects the coarse structure of the skin layers reducing its contribution to the roughness R_a to a minimum, all while the fine structure (high-frequency deviations) is reported. As an additional measure of the roughness, we use the angular distribution of surface normals, see Fig. 4.

Epidermis Segmentation: We train the U-Net and FCN architectures incorporating loss functions with differing smoothness terms. Inclusion of our novel loss term in any proportion improves the smoothness of the layer segmentation, as measured by R_a, independent of the network architecture, see Table 1. The same trend is visible in the distribution of surface normals, see Fig. 4. Dice scores are insensitive to the magnitude of the smoothness term, defined by s, until certain tipping points, where the networks fail to converge, see Fig. 2. Two-sided, paired Wilcoxon signed rank tests, comparing Dice scores for different smoothness factors to the pure BCE loss, support this observation by revealing no significant difference in underlying distributions for both U-Nets and FCNs with p-values >0.05 across the board (p-values > 0.4 for all U-Net models and > 0.06 for all FCN models). On the other hand, p-values for R_a show that our

Table 1. *Evaluation of the epidermis segmentation for the U-Net and FCN architecture for a varying s.* Overlap based scores, Dice, IoU, Precision, and Recall do not substantially differ for both U-Net and FCN when increasing the smoothness loss term s. In contrast, the surface roughness continuously improves with increasing s. Our U-Net outperforms the FCN in regards to overlap based scores and roughness as it is a substantially more complex model.

Network	Loss	Smoothness factor s	Dice	Precision	Recall	IoU	Roughn. R_a
U-Net	BCE	0	0.87 ± 0.09	0.87 ± 0.10	0.89 ± 0.13	0.78 ± 0.11	0.125
U-Net	BCE+S	1	0.88 ± 0.10	0.88 ± 0.08	0.90 ± 0.14	0.79 ± 0.12	0.122
U-Net	BCE+S	10	0.87 ± 0.10	0.87 ± 0.10	0.89 ± 0.13	0.79 ± 0.12	0.112
U-Net	BCE+S	100	0.87 ± 0.10	0.86 ± 0.10	0.90 ± 0.13	0.78 ± 0.12	0.099
U-Net	BCE+S	1000	0.86 ± 0.09	0.83 ± 0.11	0.90 ± 0.12	0.76 ± 0.11	0.081
U-Net	BCE+S	2000	0.86 ± 0.09	0.80 ± 0.09	0.94 ± 0.11	0.76 ± 0.11	0.057
U-Net	BCE+S	2500	0.85 ± 0.11	0.79 ± 0.11	0.94 ± 0.12	0.75 ± 0.13	0.055
7FCN	BCE	0	0.64 ± 0.23	0.85 ± 0.12	0.62 ± 0.32	0.52 ± 0.24	2.127
7FCN	BCE+S	1	0.65 ± 0.23	0.85 ± 0.12	0.63 ± 0.32	0.52 ± 0.23	1.607
7FCN	BCE+S	10	0.64 ± 0.24	0.85 ± 0.12	0.62 ± 0.32	0.52 ± 0.24	1.116
7FCN	BCE+S	100	0.58 ± 0.27	0.83 ± 0.14	0.55 ± 0.35	0.45 ± 0.26	0.909

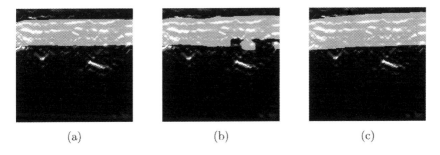

(a)   (b)   (c)

Fig. 3. *Magnified slice of an RSOM skin scan.* The epidermis layer is marked in white. (a) ground truth annotation (label), (b) segmentation result from the U-Net with BCE loss, (c) segmentation result from the U-Net with BCE and Smoothness Loss ($s = 2000$). Note that despite the highly unevenly distributed melanin response, the smoothness loss based prediction segments the epidermis layer superior with a very smooth surface, which is similar to the label.

models have significantly different R_a distributions across test samples (p-values < 0.05 for all FCN and all U-Net where s > 1). Overall, we achieve the best performance of around 87 % Dice and 0.06 R_a using the U-Net architecture. Across the samples (for the U-Net), the scores of the five-fold cross validation resulted in an agreement of the Dice scores of 0.985 ± 0.00068. From the very low standard deviation, we conclude that the statistical divergence between the training and validation set is very low. Visual inspection reveals that using our smooth loss indeed yields smooth and continuous epidermis and dermis surfaces, see Fig. 1 (b), Fig. 3 and Supplementary Fig. 3. The combined loss is robust for a

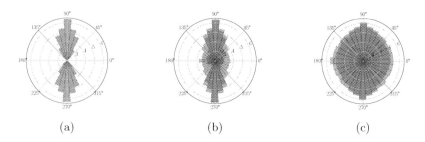

(a)   (b)   (c)

Fig. 4. *Histogram of the orientation of surface normals on the segmentation map* (logarithmic scale). (a) Ground truth. BCE loss (red), BCE and smoothness loss (green) for (b) U-Net ($s = 2000$), and (c) FCN ($s = 100$). The amount of wrongly orientated surface normals is reduced by one order of magnitude for the U-Net and less pronounced for the FCN, too. The anatomically desired smooth surface is better segmented using our smoothness loss term. For details on the calculation of the surface normals, please see the supplementary material. (Color figure online)

largely varying s. For the U-Net, the loss was stable for s ranging from 1 to 2800 and for the FCN for an s ranging from 1 to 1000. To be clear, our smoothness term is not a standalone loss, but works very well in combination with BCE; increasing the factor s too much leads to instabilities during training, see Fig. 2.

Table 2. *Vessel segmentation sensitivity experiment:* Here we report the performance of a standard vessel segmentation using DeepVesselNet [28]; without using our proposed epidermis layer segmentation and using our method. Numbers in bold indicate superior performance.

Configuration	Dice	Precision	Recall	IoU
No Mask	0.619 ± 0.187	0.673 ± 0.303	0.698 ± 0.160	0.474 ± 0.196
Our Method	$\mathbf{0.810 \pm 0.095}$	$\mathbf{0.883 \pm 0.100}$	$\mathbf{0.760 \pm 0.125}$	$\mathbf{0.690 \pm 0.117}$

Vessel Segmentation: We validate our epidermis segmentation via a sensitivity experiment of vessel segmentation, where we use the epidermis segmentation as a mask. Vessel segmentation in optoacoustic skin scans is of great clinical interest in order to characterize the vasculature of healthy human skin and in order to diagnose several disease cases, where vasculature and capillaries are altered or damaged; e.g., for the diagnosis of long term effects of diabetes on the patients' body. An established method [19,28,29] for vessel segmentation is used to verify the validity of the epidermis segmentation. Synthetically generated arterial trees [24,25] serve as training and validation data, see Supplementary Fig. 2. Testing is done on 32 annotated 3D RSOM volumes of size $166 \times 85 \times 250$. Epidermis segmentation increased the Dice similarity by more than 20% from 0.619 ± 0.187 to 0.810 ± 0.095, yielding high confidence of the validity and necessity of our epidermis segmentation approach. The complete results for vessel segmentation are given in Table 2.

4   Conclusion

In this paper, we introduced a novel, shape-specific loss function, for RSOM image skin layer segmentation. Our loss overcomes discontinuous segmentations and achieves smooth segmentation surfaces, while preserving the same volumetric segmentation performance, e.g. Dice. This is important because only meaningful and reproducible segmentation can be used for downstream tasks in medical practice, e.g. vessel segmentation for diagnostic purposes. We validate our epidermis segmentation through a sensitivity experiment, where we use our epidermis segmentation as a mask for vessel segmentation and improve their performance by more than 20% Dice.

References

1. Aguirre, J., et al.: Precision assessment of label-free psoriasis biomarkers with ultra-broadband optoacoustic mesoscopy. Nat. Biomed. Eng. **1**(5), 0068 (2017). https://doi.org/10.1038/s41551-017-0068
2. Aguirre, J., Schwarz, M., Soliman, D., Buehler, A., Omar, M., Ntziachristos, V.: Broadband mesoscopic optoacoustic tomography reveals skin layers. Opt. Lett. **39**(21), 6297–6300 (2014). http://ol.osa.org/abstract.cfm?URI=ol-39-21-6297
3. Al Arif, S.M.M.R., Knapp, K., Slabaugh, G.: Shape-aware deep convolutional neural network for vertebrae segmentation. In: Glocker, B., Yao, J., Vrtovec, T., Frangi, A., Zheng, G. (eds.) MSKI 2017. LNCS, vol. 10734, pp. 12–24. Springer, Cham (2018). https://doi.org/10.1007/978-3-319-74113-0_2
4. Anas, E.M.A., Zhang, H.K., Kang, J., Boctor, E.M.: Towards a fast and safe LED-based photoacoustic imaging using deep convolutional neural network. In: Frangi, A.F., Schnabel, J.A., Davatzikos, C., Alberola-López, C., Fichtinger, G. (eds.) MICCAI 2018. LNCS, vol. 11073, pp. 159–167. Springer, Cham (2018). https://doi.org/10.1007/978-3-030-00937-3_19
5. BenTaieb, A., Hamarneh, G.: Topology aware fully convolutional networks for histology gland segmentation. In: Ourselin, S., Joskowicz, L., Sabuncu, M.R., Unal, G., Wells, W. (eds.) MICCAI 2016. LNCS, vol. 9901, pp. 460–468. Springer, Cham (2016). https://doi.org/10.1007/978-3-319-46723-8_53
6. Cyganek, B.: Tensor Methods in Computer Vision, Chap. 2, pp. 9–188. Wiley, Hoboken (2013)
7. Gok, A., Gologlu, C., Demirci, H., Kurt, M.: Determination of surface qualities on inclined surface machining with acoustic sound pressure. Strojniski Vestnik **58**, 587–597 (2012)
8. Hu, X., Li, F., Samaras, D., et al.: Topology-preserving deep image segmentation. In: Advances in Neural Information Processing Systems, pp. 5658–5669 (2019)
9. Kalogerakis, E., Averkiou, M., Maji, S., Chaudhuri, S.: 3D shape segmentation with projective convolutional networks. In: Proceedings of the IEEE Conference on Computer Vision and Pattern Recognition, pp. 3779–3788 (2017)
10. Kittler, H., Pehamberger, H., Wolff, K., Binder, M.: Diagnostic accuracy of dermoscopy. Lancet Oncol. **3**(3), 159–165 (2002)

11. Li, K., Wu, X., Chen, D.Z., Sonka, M.: Optimal surface segmentation in volumetric images-a graph-theoretic approach. IEEE Trans. Pattern Anal. Mach. Intell. **28**(1), 119–134 (2005)
12. Manfredi, M., Grana, C., et al.: Skin surface reconstruction and 3D vessels segmentation in speckle variance optical coherence tomography. In: VISIGRAPP (2016)
13. Moustakidis, S., Omar, M., Aguirre, J., Mohajerani, P., Ntziachristos, V.: Fully automated identification of skin morphology in raster-scan optoacoustic mesoscopy using artificial intelligence. Med. Phys. **46**(9), 4046–4056 (2019)
14. Navarro, F., et al.: Shape-aware complementary-task learning for multi-organ segmentation. In: Suk, H.-I., Liu, M., Yan, P., Lian, C. (eds.) MLMI 2019. LNCS, vol. 11861, pp. 620–627. Springer, Cham (2019). https://doi.org/10.1007/978-3-030-32692-0_71
15. Nitkunanantharajah, S., Zahnd, G., Olivo, M., Navab, N., Mohajerani, P., Ntziachristos, V.: Skin surface detection in 3D optoacoustic mesoscopy based on dynamic programming. IEEE Trans. Med. Imaging (2019)
16. Ntziachristos, V., Razansky, D.: Molecular imaging by means of multispectral optoacoustic tomography (MSOT). Chem. Rev. **110**(5), 2783–2794 (2010). https://doi.org/10.1021/cr9002566, pMID: 20387910
17. Omar, M., Aguirre, J., Ntziachristos, V.: Optoacoustic mesoscopy for biomedicine. Nat. Biomed. Eng. **3**(5), 354–370 (2019)
18. Omar, M., Schwarz, M., Soliman, D., Symvoulidis, P., Ntziachristos, V.: Pushing the optical imaging limits of cancer with multi-frequency-band raster-scan optoacoustic mesoscopy (RSOM). Neoplasia **17**(2), 208–214 (2015)
19. Paetzold, J.C., Schoppe, O., et al.: Transfer learning from synthetic data reduces need for labels to segment brain vasculature and neural pathways in 3D. In: International Conference on Medical Imaging with Deep Learning-Extended Abstract Track (2019)
20. Patiño, D., Avendaño, J., Branch, J.W.: Automatic skin lesion segmentation on dermoscopic images by the means of superpixel merging. In: Frangi, A., et al. (eds.) MICCAI 2018. LNCS, pp. 728–736. Springer, Heidelberg (2018). https://doi.org/10.1007/978-3-030-00937-3_83
21. Rajpara, S., Botello, A., Townend, J., Ormerod, A.: Systematic review of dermoscopy and digital dermoscopy/artificial intelligence for the diagnosis of melanoma. Br. J. Dermatol. **161**(3), 591–604 (2009)
22. Ravishankar, H., Venkataramani, R., Thiruvenkadam, S., Sudhakar, P., Vaidya, V.: Learning and incorporating shape models for semantic segmentation. In: Descoteaux, M., Maier-Hein, L., Franz, A., Jannin, P., Collins, D.L., Duchesne, S. (eds.) MICCAI 2017. LNCS, vol. 10433, pp. 203–211. Springer, Cham (2017). https://doi.org/10.1007/978-3-319-66182-7_24
23. Ronneberger, O., Fischer, P., Brox, T.: U-Net: Convolutional Networks for Biomedical Image Segmentation. In: Navab, N., Hornegger, J., Wells, W.M., Frangi, A.F. (eds.) MICCAI 2015. LNCS, vol. 9351, pp. 234–241. Springer, Cham (2015). https://doi.org/10.1007/978-3-319-24574-4_28
24. Schneider, M., Hirsch, S., Weber, B., Székely, G., Menze, B.H.: Joint 3-D vessel segmentation and centerline extraction using oblique hough forests with steerable filters. Med. Image Anal. **19**(1), 220–249 (2015)
25. Schneider, M., Reichold, J., Weber, B., Székely, G., Hirsch, S.: Tissue metabolism driven arterial tree generation. Med. Image Anal. **16**(7), 1397–1414 (2012)
26. Shit, S., Paetzold, J.C., et al.: clDice-a topology-preserving loss function for tubular structure segmentation. arXiv preprint arXiv:2003.07311 (2020)

27. Srivastava, R., Yow, A.P., Cheng, J., Wong, D.W., Tey, H.L.: Three-dimensional graph-based skin layer segmentation in optical coherence tomography images for roughness estimation. Biomed. Opt. Express **9**(8), 3590–3606 (2018)
28. Tetteh, G., et al.: Deepvesselnet: Vessel segmentation, centerline prediction, and bifurcation detection in 3-D angiographic volumes. arXiv preprint arXiv:1803.09340 (2018)
29. Todorov, M.I., Paetzold, J.C., et al.: Machine learning analysis of whole mouse brain vasculature. Nat. Methods **17**(4), 442–449 (2020)
30. Wiecheć, A., Nowicka, K., Błażewicz, M., Kwiatek, W.: Effect of magnetite composite on the amount of double strand breaks induced with X-rays. Acta Physica Polonica A **129**, 174–175 (2016)
31. Zhang, J., Xie, Y., Wu, Q., Xia, Y.: Skin lesion classification in dermoscopy images using synergic deep learning. In: Frangi, A.F., Schnabel, J.A., Davatzikos, C., Alberola-López, C., Fichtinger, G. (eds.) MICCAI 2018. LNCS, vol. 11071, pp. 12–20. Springer, Cham (2018). https://doi.org/10.1007/978-3-030-00934-2_2

Fairness of Classifiers Across Skin Tones in Dermatology

Newton M. Kinyanjui[1,4], Timothy Odonga[1,4], Celia Cintas[1],
Noel C. F. Codella[2], Rameswar Panda[3], Prasanna Sattigeri[2],
and Kush R. Varshney[1,2(✉)]

[1] IBM Research – Africa, Nairobi 00100, Kenya
krvarshn@us.ibm.com
[2] IBM Research – T. J. Watson Research Center, Yorktown Heights, NY 10598, USA
[3] IBM Research – Cambridge, Cambridge, MA 02142, USA
[4] Carnegie Mellon University Africa, Kigali, Rwanda

Abstract. Recent advances in computer vision have led to break-throughs in the development of automated skin image analysis. However, no attempt has been made to evaluate the consistency in performance across populations with varying skin tones. In this paper, we present an approach to estimate skin tone in skin disease benchmark datasets and investigate whether model performance is dependent on this measure. Specifically, we use individual typology angle (ITA) to approximate skin tone in dermatology datasets. We look at the distribution of ITA values to better understand skin color representation in two benchmark datasets: 1) the ISIC 2018 Challenge dataset, a collection of dermoscopic images of skin lesions for the detection of skin cancer, and 2) the SD-198 dataset, a collection of clinical images capturing a wide variety of skin diseases. To estimate ITA, we first develop segmentation models to isolate non-diseased areas of skin. We find that the majority of the data in the two datasets have ITA values between 34.5° and 48°, which are associated with lighter skin, and is consistent with under-representation of darker skinned populations in these datasets. We also find no measurable correlation between accuracy of machine learning models and ITA values, though more comprehensive data is needed for further validation.

Keywords: Algorithmic fairness · Dermatology image analysis · Medical imaging

1 Introduction

As machine learning is becoming more frequently applied to support consequential decisions, there is increasing interest in accurately measuring latent dataset characteristics and demographic representation to prevent the potential negative consequences of dataset imbalances [27], henceforth referred to as "dataset bias". Dataset bias is a critical issue because it is one of the causes of machine learning-based systems placing certain groups at a systematic disadvantage [3].

© Springer Nature Switzerland AG 2020
A. L. Martel et al. (Eds.): MICCAI 2020, LNCS 12266, pp. 320–329, 2020.
https://doi.org/10.1007/978-3-030-59725-2_31

Recognition and mitigation of unwanted bias is necessary to build machine learning systems that are trustworthy [31].

Skin diseases continue to bear significant negative impacts on human health. Skin diseases contribute 1.79% to the global burden of disease [19] and skin cancer accounts for about 7% of new cancer cases [4]. Within skin cancer, there is evidence of some outcome disparities with respect to ethnicity: although people of color are roughly 20 to 30 times less likely to develop melanoma than lighter skinned individuals, for certain melanoma sub-types they have been found to have lower [22,23,34] or higher [22] survival rates. Some studies have found that for people of color, the diagnosis of skin cancer may occur at a more advanced stage, leading to lower rates of survival and poorer outcomes [14,21]. However, increased screening also carries risks, such as unnecessary surgeries, disfigurement, disability, morbidity, and over-diagnosis [23].

Computer vision has been studied in the context of dermatology image analysis for decades [1,20,28]. The success of deep learning models has led to studies applying the technology to dermatological use cases [7,8]. Models using convolutional neural networks (CNNs) have been applied to problems such as skin cancer diagnosis and were found to outperform trained dermatologists in controlled settings and datasets [11,13,15]. However, as most of the publicly available datasets of skin images come from lighter skinned populations, due to the extreme disparities in disease prevalence, there are concerns about how to best collect data, train, and evaluate models for darker skinned populations [2,27]. Also, because of the significant risks of harm from over-diagnosis with increased screening in low-risk dark skin populations, there is a need to better discriminate between life-threatening and stable presentations of disease [23,27].

In this paper, we work towards quantifying skin tone distributions in datasets where this information is currently unavailable, and measuring downstream effects on classifier performance. Specifically, our contributions are as follows:

- We propose a pipeline to automatically estimate skin tone for images in two public benchmark skin disease datasets using the individual typology angle (ITA), which has been used previously as a measure of skin tone in absence of manually curated information [24].
- We create manually-labeled segmentation masks and automatically generated masks for non-diseased skin in both public benchmarks.
- We quantitatively confirm that the two benchmark skin disease datasets under-represent ITA values correlated with darker skin populations.
- No correlation between model performance and ITA value is measurable at this time, though more data is needed for conclusive results.

2   Related Work

Recent years have seen significant advances in automated skin lesion analysis, with hundreds of deep learning models implemented for skin cancer diagnosis. Much of this work has been enabled by the International Skin Imaging Collaboration (ISIC) [10,17,30], which has organized a public repository of annotated dermoscopic images, and hosted 4 years of public challenge benchmarks. In 2016, the first work demonstrating classification accuracy higher than the average of expert dermatologists was described [11], employing an ensemble of methods that included hand-coded feature extraction, sparse coding methods, support vector machines, CNNs for skin lesion classification, and fully convolutional networks for skin lesion segmentation. Other models have also been implemented by researchers, such as a computationally efficient skin lesion classification model that uses the MobileNet architecture implemented by [9], and an Inception architecture trained on a large dataset of over 100,000 images [13].

Outside of dermatology, there has been work on evaluating fairness in computer vision with respect to skin type. Recent studies evaluated bias in automated facial analysis models with respect to phenotypic groups [5,26]. They found poor accuracy for darker females compared to lighter females, darker males, and lighter males in gender classification systems. A related study revealed that well-performing gender classification systems are already invariant to skin type and thus the skin type by itself has a minimal effect on classification disparities [25]. Another study investigated equitable performance in state-of-the-art object detection systems on pedestrians with different skin types, finding higher precision on lighter skin than darker skin [33].

3   Datasets

Public benchmark datasets, in addition to fostering direct comparisons among various algorithms to facilitate advancement in terms of classification performance, are also capable of supporting detailed analysis of that performance with respect to various characteristics of the dataset [27]. Therefore, we focus our analysis on two of the most widely used dermatology datasets in the computer vision literature: the ISIC 2018 Challenge dataset the SD-198 dataset.

ISIC2018. This collection of dermoscopic images is separated into datasets for image segmentation (Task 1), clinical feature detection (Task 2), and disease classification (Task 3). Dermoscopic images are acquired through a digital dermatoscope, with relatively low levels of noise and consistent background illumination. The training dataset for Task 3 is the largest among the tasks and used in this work. It consists of 10,015 dermoscopic images [10,30], falling into one of 7 skin diseases: melanoma, melanocytic nevus, basal cell carcinoma, actinic keratosis, benign keratosis, dermatofibroma and vascular lesion.

SD-198. The SD-198 dataset contains 6,548 clinical images from 198 skin disease classes, varying according to scale, color, shape and structure downloaded from DermQuest [29]. Clinical images are collected via various devices, most of which are digital cameras and mobile phones [29]. Higher levels of noise and varying illumination in clinical images makes segmentation more challenging than the dermoscopic images in ISIC2018.

In this work, we preprocess the SD-198 dataset to exclude classes containing images with no observable non-diseased skin. Eventually 136 disease classes are retained, and henceforth the pre-processed dataset is referred to as "SD-136". Some of the classes excluded from the SD-136 dataset include classes of lesions inside the mouth, such as fibroma, geographic tongue, and stomatitis. Other diseases such as arsenical keratosis, pustular psoriasis, and mal perforans contain images of lesions on palms and soles of the feet from which it is difficult to determine the individual's skin tone. Other disease classes such as stasis ulcer and eccrine poroma contain images of severely scarred skin from which it is visually impossible to differentiate non-diseased and diseased skin. This preprocessing step was done manually and eventually 4,467 images were retained from the original 6,548 images.

Since there are no existing ground truth segmentation masks for the SD-136 dataset, we manually segmented a subset of 343 images. We were particularly interested in segmenting regions with non-diseased skin from other regions of the image containing diseased skin, shadows, and other artifacts. We used these ground-truth masks in training the segmentation model for the SD-136 dataset. The data is split into 90%/10% training/validation partitions.

4   Methods

Our proposed method is summarized in Fig. 1. First, we train a model to segment skin disease images to obtain the non-diseased skin in the image, this model returns a set of pairs with the image and the mask associated to it $(I_i, Mask_i)$ for all images in any of the datasets D_1 or D_2. Second, we use the provided $Mask_i$ to select and compute the metric to stratify the non-diseased skin into a skin tone category $(tone_j)$ from the categorization scheme (S_m). After that, a classification model is trained to classify skin images into one of the skin diseases in the dataset. Finally, the performance of the classifier on samples in each skin tone category is evaluated.

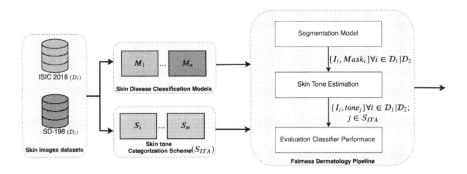

Fig. 1. Block diagram of methodology, where D_1 and D_2 correspond to the datasets ISIC 2018 and SD-198, M_i is a trained model for skin disease classification (e.g. Densenet201) over the previously mentioned datasets, S_m is a categorization scheme (e.g. ITA ranges) and $tone_j$ is a skin tone under the ITA ranges.

4.1   Quantification of Representation of Skin Tone Categories

Segmentation of the skin lesion from the non-diseased skin is done using a Mask R-CNN model [16]. Mask R-CNN was selected because it was one of the top performing skin lesion boundary segmentation models in the 2018 ISIC challenge [17] and also because it has been shown to be highly effective and efficient in performing semantic segmentation [18].

To obtain a segmentation model for the ISIC2018 dataset, a Mask R-CNN pretrained on the COCO dataset is finetuned with the lesion boundary segmentation data from ISIC 2018 Challenge Task 1. The data is split into training and validation data using 90% to 10% train validation split. The images are resized to 600×450 pixels to correspond to the size used for classification. Random horizontal flips are done on the images for data augmentation during training. The segmentation model for the ISIC2018 dataset is trained for 25 epochs. This model is used to predict segmentation masks for all the classification data. Finally, thresholding of the predicted masks is performed via contour extraction.

The segmentation model from the ISIC2018 dataset is finetuned with the 343 manually-segmented images from SD-136. All other steps are the same as for ISIC2018 except the image size is 450×450 and the number of epochs is 50.

The quality of segmentation is evaluated using accuracy and false negative rate. False negative rate is considered because it is worse to wrongly classify a diseased region as non-diseased in our analysis. To further evaluate the quality of segmentation, mean absolute error is computed between the ITA estimates from ground truth masks and ITA estimates from predicted masks.

With the segmentation masks obtained from the previous step, we obtain pixels in the non-diseased region for each image and use these pixels to categorize the skin tone. There is no universal method for characterizing skin type or skin tone among dermatologists. The Fitzpatrick skin type, used in [5,25,26], is a dermatologist's determination of a person's risk of sunburn. It is by definition, however, a subjective human determination [12]. In contrast, the melanin index is measured objectively via reflectance spectrophotometry, and has a strong correlation with the Fitzpatrick type and is useful in assigning it [32]. The metric we use to quantify skin tone in this work is the ITA (in degrees) because it has strong (anti-)correlation to the melanin index [32], and can be simply computed from images, making it a practical method for categorizing skin color [24].

The pixels from the non-diseased region are examined in CIELab-space using luminance (L) and the amount of yellow (b). To prevent the effect of outliers, we only consider L and b values within one standard deviation of their mean values in the region. The ITA value is calculated as [24]:

$$ \text{ITA} = \arctan\left(\frac{L - 50}{b}\right) \times \frac{180°}{\pi}. \tag{1} $$

We bin the mean ITA value using a scheme similar to [6], which uses 5 skin tone categories: Very Light, Light, Intermediate, Tanned, and Dark. We further subdivide the Light, Intermediate, and Tanned categories into two equal ranges, giving a total of 8 ITA categories. The scheme is summarized in Table 1.

Table 1. Skin tone categorization scheme.

ITA Range	Skin Tone Category	Abbreviation
ITA $> 55°$	Very Light	very_lt
$48° <$ ITA $\leq 55°$	Light 2	lt2
$41° <$ ITA $\leq 48°$	Light 1	lt1
$34.5° <$ ITA $\leq 41°$	Intermediate 2	int2
$28° <$ ITA $\leq 34.5°$	Intermediate 1	int1
$19° <$ ITA $\leq 28°$	Tanned 2	tan2
$10° <$ ITA $\leq 19°$	Tanned 1	tan1
ITA $\leq 10°$	Dark	dark

4.2 Evaluation of Classification Performance Across Skin Tones

A Densenet201 model pretrained on ImageNet is finetuned using our training data. The Densenet201 model is chosen because it was one of the best performing single models for lesion classification in the ISIC 2018 challenge [17]. During training, the early layers up to and including the first Dense block are frozen and all successive layers have their weights updated. Each classification model is trained for 300 epochs with a patience of 100 epochs at which early stopping would be applied to prevent overfitting.

The ISIC2018 dataset images are maintained at 600×450 pixels. Additional transformations such as random horizontal flipping are applied to augment the data. The samples in each batch are normalized using the mean and standard deviation computed on all samples in the dataset to ensure fast convergence during training. The data is split into training and validation data using an 80%/20% split. A weighted cross entropy loss function and an Adam optimizer are used for training. The weights for the loss function are obtained from the inverse of each disease class frequency. This loss function is chosen because it accounts for class imbalance.

The SD-136 dataset images are resized to 450×450 pixels and center-cropped to 360×360 pixels. Transformations including random horizontal flipping and random rotation between $-90°$ and $90°$ are applied to augment the data. All other details are the same as in ISIC2018.

5 Results

The Mask R-CNN model used for segmentation on the ISIC2018 dataset yields an accuracy of 0.956, a false negative rate of 0.024, and a mean absolute error in ITA computation of $0.428°$. The segmentation model on the SD-136 dataset yield an accuracy of 0.802, a false negative rate of 0.076, and a mean absolute error in ITA computation of $3.572°$. These are all fairly good results and sufficient for further analysis. Examples with segmented mask and ITA values for both datasets are shown in Fig. 2.

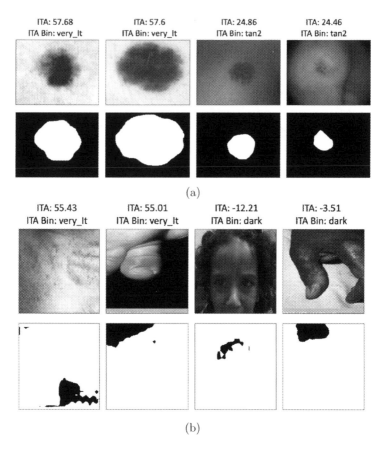

Fig. 2. Sample images (top row) and corresponding masks predicted by model (bottom row) for (a) ISIC2018 and (b) SD-136 datasets; ITA is computed on the non-diseased region which is colored black.

Figure 3 shows the distributions of the ITA values estimated from the non-diseased skin regions of the images in the entire ISIC2018 and SD-136 datasets. Both datasets are found to predominantly lie in the Light category.

On the ISIC2018 dataset, the Densenet201 model achieves an accuracy 0.869 and a balanced accuracy score of 0.814 on our internal validation partition after training approximately 140 epochs when early stopping occurred. On a separate held-out test set used for the challenge leaderboard, our model achieves a balanced accuracy score of 0.760, placing its percentile ranking around 62%. This indicates that the model scored higher balanced accuracy than 62% of the Top 200 entries in the ISIC 2018 challenge. Unfortunately, since the challenge held-out set is unavailable to us, we cannot disaggregate this result by skin tone.

The model trained on SD-136 achieves an accuracy of 0.604 and a balanced accuracy score of 0.601. The benchmark model for SD-198 achieves an accuracy 0.52, as reported in [29]. However, since we dropped the number of classes from

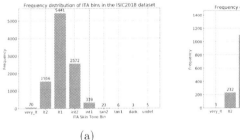
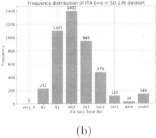

(a)   (b)

Fig. 3. Skin tone distribution for (a) ISIC2018, and (b) SD-136 entire datasets.

198 to 136, we do not have a benchmark model for comparison. Nonetheless, we are confident that we have a well-performing model.

Importantly, on evaluating the classification performance with respect to skin tone category, our results do not show a clear trend in the performance of the model. Figure 4 plots classification accuracy versus ITA for the validation set for the two datasets. The error bars indicate the standard error estimated through ten runs with random splits. The slope of the least squares line of best fit of the mean accuracy versus the midpoint ITA value of the bin for ISIC2018 is -0.000 (per degree) with a 95% confidence interval of $(-0.001, 0.001)$, whereas that for SD-136 is -0.002 (per degree) with a 95% confidence interval of $(-0.003, -0.001)$, which indicate that there are no particular trends in both datasets.

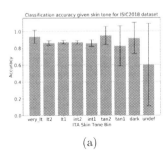
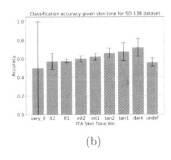

(a)   (b)

Fig. 4. Accuracy versus ITA for (a) ISIC2018, and (b) SD-136 validation sets.

6   Conclusion

In this work, we implemented an approach to measure approximate skin tone distributions in public dermatology image datasets using ITA as an estimator, and evaluated the performance of dermatology classification models with respect to the resultant ITA values. The distribution of ITA values across both ISIC2018 and SD-136 datasets are consistent with under-representation of darker skin tones. The results from the evaluation of the accuracy of the skin classification model for each skin tone category in the validation data shows that there is no observable trend in the performance of the model with respect to ITA value, which is contrary to other studies of skin color and computer vision systems. Although we have not found any evidence of model performance bias under the influence of dataset bias in this particular study, further investigation is needed on datasets with more comprehensive representation.

References

1. Abedini, M., et al.: Accurate and scalable system for automatic detection of malignant melanoma. In: Celebi, M.E., Mendonca, T., Marques, J.S. (eds.) Dermoscopy Image Analysis. CRC Press (2015)
2. Adamson, A.S., Smith, A.: Machine learning and health care disparities in dermatology. JAMA Dermatol. **154**(11), 1247–1248 (2018)
3. Barocas, S., Selbst, A.D.: Big data's disparate impact. Calif. Law Rev. **104**(3), 671–732 (2016)
4. Bray, F., Ferlay, J., Soerjomataram, I., Siegel, R.L., Torre, L.A., Jemal, A.: Global cancer statistics 2018. CA-Cancer J. Clin. **68**(6), 394–424 (2018)
5. Buolamwini, J., Gebru, T.: Gender shades: intersectional accuracy disparities in commercial gender classification. In: Proceedings of the Conference on Fairness, Accountability and Transparency, pp. 77–91 (2018)
6. Casale, G.R., Siani, A.M., Diémoz, H., Agnesod, G., Parisi, A.V., Colosimo, A.: Extreme UV index and solar exposures at Plateau Rosà (3500 m a.s.l.) in Valle d'Aosta Region, Italy. Sci. Total Environ. **512–513**, 622–630 (2015)
7. Celebi, M.E., Codella, N., Halpern, A.: Dermoscopy image analysis: overview and future directions. IEEE J. Biomed. Health **23**(2), 474–478 (2019)
8. Celebi, M.E., Codella, N., Halpern, A., Shen, D.: Guest editorial: skin lesion image analysis for melanoma detection. IEEE J. Biomed. Health **23**(2), 479–480 (2019)
9. Chaturvedi, S.S., Gupta, K., Prasad, P.: Skin lesion analyser: an efficient seven-way multi-class skin cancer classification using MobileNet. arXiv:1907.03220 (2019)
10. Codella, N., et al.: Skin lesion analysis toward melanoma detection 2018: a challenge hosted by the International Skin Imaging Collaboration (ISIC). arXiv:1902.03368 (2019)
11. Codella, N.C.F., et al.: Deep learning ensembles for melanoma recognition in dermoscopy images. IBM J. Res. Dev. **61**(4/5), 5 (2016)
12. Eilers, S., et al.: Accuracy of self-report in assessing Fitzpatrick skin phototypes I through VI. JAMA Dermatol. **149**(11), 1289–1294 (2013)
13. Esteva, A., et al.: Dermatologist-level classification of skin cancer with deep neural networks. Nature **542**(7639), 115–118 (2017)
14. Gohara, M.: Skin cancer: an African perspective. Brit. J. Dermatol. **173**(Suppl. 2), 17–21 (2015)
15. Haenssle, H.A., et al.: Man against machine: diagnostic performance of a deep learning convolutional neural network for dermoscopic melanoma recognition in comparison to 58 dermatologists. Ann. Oncol. **29**(8), 1836–1842 (2018)
16. He, K., Gkioxari, G., Dollár, P., Girshick, R.B.: Mask R-CNN. arXiv:1703.06870 (2018)
17. International Skin Imaging Collaboration: ISIC 2018: Skin lesion analysis towards melanoma detection (2018). https://challenge2018.isic-archive.com/

18. Johnson, J.W.: Automatic nucleus segmentation with mask-RCNN. In: Arai, K., Kapoor, S. (eds.) CVC 2019. AISC, vol. 944, pp. 399–407. Springer, Cham (2020). https://doi.org/10.1007/978-3-030-17798-0_32

19. Karimkhani, C., et al.: Global skin disease morbidity and mortality: an update from the global burden of disease study 2013. JAMA Dermatol. **153**(5), 406–412 (2017)

20. Korotkov, K., Garcia, R.: Computerized analysis of pigmented skin lesions: a review. Artif. Intell. Med. **56**(2), 69–90 (2012)

21. Kundu, R.V., Patterson, S.: Dermatologic conditions in skin of color: Part I. Special considerations for common skin disorders. Am. Fam. Phys. **87**(12), 850–856 (2013)

22. Mahendraraj, K., Sidhu, K., Lau, C.S.M., McRoy, G.J., Chamberlain, R.S., Smith, F.O.: Malignant melanoma in African–Americans: a population-based clinical outcomes study involving 1106 African–American patients from the surveillance, epidemiology, and end result (SEER) database (1988–2011). Medicine **96**(15), e6258 (2017)

23. Marchetti, M.A., Chung, E., Halpern, A.C.: Screening for acral lentiginous melanoma in dark-skinned individuals. JAMA Dermatol. **151**(10), 1055–1056 (2015)

24. Merler, M., Ratha, N., Feris, R.S., Smith, J.R.: Diversity in faces. arXiv:1901.10436 (2019)

25. Muthukumar, V.: Color-theoretic experiments to understand unequal gender classification accuracy from face images. In: Proceedings of the IEEE Conference on Computer Vision and Pattern Recognition Workshops (2019)

26. Raji, I.D., Buolamwini, J.: Actionable auditing: investigating the impact of publicly naming biased performance results of commercial AI products. In: Proceedings of the 2019 AAAI/ACM Conference on AI, Ethics, and Society, pp. 429–435 (2019)

27. Rotemberg, V., Halpern, A., Dusza, S.W., Codella, N.C.F.: The role of public challenges and data sets towards algorithm development, trust, and use in clinical practice. Semin. Cutan. Med. Surg. **38**(1), E38–E42 (2019)

28. Stoecker, W.V., Moss, R.H.: Editorial: digital imaging in dermatology. Comput. Med. Imag. Grap. **16**(3), 145–150 (1992)

29. Sun, X., Yang, J., Sun, M., Wang, K.: A benchmark for automatic visual classification of clinical skin disease images. In: Leibe, B., Matas, J., Sebe, N., Welling, M. (eds.) ECCV 2016. LNCS, vol. 9910, pp. 206–222. Springer, Cham (2016). https://doi.org/10.1007/978-3-319-46466-4_13

30. Tschandl, P., Rosendahl, C., Kittler, H.: Data descriptor: the HAM10000 dataset, a large collection of multi-source dermatoscopic images of common pigmented skin lesions. Sci. Data **5**, 180161 (2018)

31. Varshney, K.R.: Trustworthy machine learning and artificial intelligence. ACM XRDS **26**(3), 26–29 (2019)

32. Wilkes, M., Wright, C.Y., du Plessis, J.L., Reeder, A.: Fitzpatrick skin type, individual typology angle, and melanin index in an African population. JAMA Dermatol. **151**(8), 902–903 (2015)

33. Wilson, B., Hoffman, J., Morgenstern, J.: Predictive inequity in object detection. arXiv:1902.11097 (2019)

34. Wu, X.C., et al.: Racial and ethnic variations in incidence and survival of cutaneous melanoma in the United States, 1999–2006. J. Am. Acad. Dermatol. 65(5), S26.e1–S26.e13 (2011)

Alleviating the Incompatibility Between Cross Entropy Loss and Episode Training for Few-Shot Skin Disease Classification

Wei Zhu[1(✉)], Haofu Liao[1], Wenbin Li[2], Weijian Li[1], and Jiebo Luo[1]

[1] University of Rochester, Rochester, USA
zwvews@gmail.com
[2] Nanjing University, Nanjing, China

Abstract. Skin disease classification from images is crucial to dermatological diagnosis. However, identifying skin lesions involves a variety of aspects in terms of size, color, shape, and texture. To make matters worse, many categories only contain very few samples, posing great challenges to conventional machine learning algorithms and even human experts. Inspired by the recent success of Few-Shot Learning (FSL) in natural image classification, we propose to apply FSL to skin disease identification to address the extreme scarcity of training samples. However, directly applying FSL to this task does not work well in practice, and we find that the problem should be largely attributed to the incompatibility between Cross Entropy (CE) and episode training, which are both commonly used in FSL. Based on a detailed analysis, we propose the Query-Relative (QR) loss, which proves superior to CE under episode training and is closely related to recently proposed mutual information estimation. Moreover, we further strengthen the proposed QR loss with a novel adaptive hard margin strategy. Comprehensive experiments validate the effectiveness of the proposed FSL scheme and the possibility to diagnosis rare skin disease with a few labeled samples.

Keywords: Few-shot skin disease classification · Query-relative loss

1 Introduction

As a key step in the dermatological diagnosis, skin disease classification is quite challenging due to the extremely scarce annotations for a large number of categories. Such complexity in skin disease taxonomy requires a great deal of expertise. In addition, the diagnosis is often subjective and inaccurate even by human experts, which necessitates the research for computer-aided diagnosis [11,15]. Motivated by the unprecedented success of deep neural networks (DNNs), many researchers resort to deep learning technologies to handle this task [3,8,9]. For example, Esteva *et al.* adopt GoogleNet Inception V3 [17] to train a large-scale skin disease classification network [3]. Liao *et al.* jointly train skin lesion and

© Springer Nature Switzerland AG 2020
A. L. Martel et al. (Eds.): MICCAI 2020, LNCS 12266, pp. 330–339, 2020.
https://doi.org/10.1007/978-3-030-59725-2_32

body location classifiers using a multi-task network [9]. However, since DNN-based methods usually require a significant number of training samples for each category, categories with only a few number of samples are often discarded [8]. This reduces the applicability of DNN-based methods, especially for infrequent skin disease diagnosis.

Shi *et al.* propose to adopt active learning to reduce the annotation cost [13], but still need up to 50% of labeled samples to train their model. Alternatively, Few-Shot Learning (FSL) is usually leveraged to address such tasks with only a few training samples [2,6,7,14,16]. By assuming the availability of a large-scale auxiliary training set, one can learn generalized patterns and knowledge which facilitate the learning for unseen tasks. Formally, for each few-shot task, we are provided with a support set S, a query set Q, and an auxiliary set A, where the support set S contains C different categories and each category has K training samples, *i.e.*, C-way K-shot, and Q contains unlabeled query data. Instead of conventional minibatch training, FSL is always trained with the episode training mechanism [14]. Basically, at each training iteration, we generate an episode by drawing samples from C different categories of the auxiliary set A, with K samples in each category as support samples S_{train} and others as query samples Q_{train}. As a crucial step, we need to randomly shuffle the labels for all categories from episode to episode. Episode training mechanism benefits FSL in at least two aspects. First, it enables FSL to be trained under similar scenarios as testing tasks. Second, the labels are randomly shuffled during episode training, which enables the model to learn category-agnostic representation for a better generalization ability.

Generally, FSL employs the Cross Entropy (CE) loss as an objective for classification. Although CE is useful for conventional classification, we find that it is somewhat incompatible with the episode mechanism. Well-designed FSL methods trained with CE even perform significantly worse than the baseline methods [1]. As we will see, CE classifies the query samples individually and relies highly on well-trained category-wise representation, a.k.a. proxies in proxy-based metric learning methods which share the similar formulation as CE [10, 12]. The proxy is an category-wise *aggregation* of labeled support samples, e.g., the center used in Prototypical Network (PN). However, accurate proxies could only be obtained by a large-scale unified labeled dataset under the conventional minibatch training mechanism. This is hardily fulfilled under the episode training mechanism since we are only provided with a few training samples with randomly shuffled labels in each iteration.

We highlight our main contributions as follows:

- Upon an insightful analysis of the CE loss and episode mechanism, we propose a Query-Relative (QR) loss to better utilize the cross sample information and avoid possible sub-optimal aggregation of negative support samples, which significantly boosts the FSL performance;
- We develop an adaptive hard margin method for the QR loss to further penalize the categories with more error similarity connections;

– We evaluate our methods on a benchmark FSL suite [1], and the experiments strongly support our analysis and the proposed methods (Fig. 1).

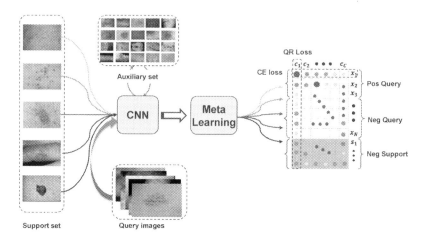

Fig. 1. Block diagram of few-shot learning-based skin disease classification and the difference between QR and CE loss. CE considers queries individually, while QR takes the relation across samples into consideration. Moreover, CE aggregates the negative support samples into proxies with possible information loss, while QR allows the model to fully exploit the information of negative support samples guided by the training objective.

2 Methodology

2.1 Discussions on FSL

Cross Entropy (CE) loss is often jointly used with episode mechanism to solve the FSL tasks. It can be generally formulated as

$$\mathcal{L}_{\text{CE}} = -\sum_i \log \frac{e^{s(c_{y_i}, x_i)}}{\sum_j e^{s(c_j, x_i)}}. \tag{1}$$

Here, $\{x_1, x_2, \ldots, x_N\} \in \mathbb{R}^{d \times N}$ are the query embeddings, and $\{c_1, c_2, \ldots, c_C\} \in \mathbb{R}^{d \times C}$ are the representations for the support categories, where N, C, and d denote the number of queries, support categories, and feature dimensions, respectively. $s(c_j, x_i)$ denotes the similarity between the support category proxy c_j and query sample x_i. Different FSL methods have different formulations of similarity measurement $s(\cdot, \cdot)$ and category proxy c aggregated by the support samples. For example, PN (Prototypical Network) uses the centers of the support samples from the j-th category as c_j and the Euclidean distance as $s(\cdot, \cdot)$ [14], Matching

Net employs an FCE (Fully Context Embedding) layer to encode the support samples and chooses cosine similarity for $s(\cdot, \cdot)$ [19], and MAML implements $s(\cdot, \cdot)$ as a Fully Connected (FC) layer where the j-th weight vector of the FC layer corresponds to c_j [4]. To unify these methods, we normalize $x_i = \frac{x_i}{\|x_i\|_2}$ and $c_j = \frac{c_i}{\|c_i\|_2}$ which leads to better performance shown in the recent literature [21]. Equation (1) can then be rewritten as $\min - \sum_i \log \frac{e^{c_{y_i}^T x_i}}{\sum_j e^{c_j^T x_i}}$.

According to Eq. (1), CE individually classifies the query samples and completely relies on the category-wise representation c_j to train the model. For the conventional classification task trained with minibatch SGD, such a mechanism could prompt c_j to learn high-level representative features of each category by exposing them to a large and balanced dataset. Unfortunately, this is not the case for FSL due to the episode training mechanism. Although episode training is important for FSL since it empowers FSL with the ability to learn generalized class agnostic representation and provides similar training scenarios as testing scenarios, it is also a double-edged sword: it makes c_j inevitably biased and inaccurate. The reasons are two aspects: first, c_j is learned from a few samples in each episode, e.g., 1 and 5 for 1-shot and 5-shot respectively, and it is difficult to learn to aggregate support samples to obtain c_j without losing useful information with so few training samples; second, the labels are randomly shuffled for each episode which limits c_j to be consistently trained across episodes. Therefore, c_j cannot be fully relied on under the episode mechanism and training the model with CE loss will eventually degrade the performance for FSL. The suboptimal performance has been observed and experimentally validated by several recent benchmark papers for natural images, where well-designed baselines could achieve similar and even better performance than CE-trained FSL counterparts [1,18]. Similar results are also disappointingly observed in the skin disease tasks according to our experiments in Sect. 3.

2.2 Query-Relative Loss

We alleviate the above problem from two aspects. First, instead of classifying the query separately, we unify all samples into a joint objective to allow them to mutually share information cross samples. Second, we avoid using negative category proxies which are aggregated by the negative support samples with a manually designed strategy (e.g., the center of support samples in PN), and the information of the support samples can then be largely preserved and extracted with the guidance of the training objective. To this end, we propose the Query-Relative (QR) loss as follows

$$\mathcal{L}_{QR} = \sum_j \log(1 + \frac{1}{2|P_j|} \sum_{x_i^+ \in P_j} e^{-s(c_j, x_i^+)} + \frac{1}{2|N_j|} \sum_{x_i^- \in N_j} e^{s(c_j, x_i^-)}), \quad (2)$$

where P_j denotes the set of positive query samples that belong to the j-th category and N_j denotes the set of *negative query and support* samples that are not from the j-th category. $|\cdot|$ denotes the number of samples in the set.

We then present an analysis on how our objective improves CE from the two aforementioned aspects. First of all, Eq. (2) implicitly utilizes the cross sample information to re-weight each sample. Specifically, taking the derivation w.r.t. $s(c_j, x_p^+)$ and $s(c_j, x_n^-)$, we have

$$\left| \frac{\partial \mathcal{L}}{\partial s(c_j, x_p^+)} \right| = \frac{\frac{1}{2|P_j|} e^{-s(c_j, x_p^+)}}{1 + \frac{1}{2|P_j|} \sum_{x_i^+ \in P_j} e^{-s(c_j, x_i^+)} + \frac{1}{2|N_j|} \sum_{x_i^- \in N_j} e^{s(c_j, x_i^-)}} \tag{3}$$

$$\left| \frac{\partial \mathcal{L}}{\partial s(c_j, x_n^-)} \right| = \frac{\frac{1}{2|N_j|} e^{s(c_j, x_n^-)}}{1 + \frac{1}{2|P_j|} \sum_{x_i^+ \in P_j} e^{-s(c_j, x_i^+)} + \frac{1}{2|N_j|} \sum_{x_i^- \in N_j} e^{s(c_j, x_i^-)}} \tag{4}$$

Here, we only focus on the absolute value of the gradient. According to Eq. (3), $s(c_j, x_p^+)$ will induce a large gradient and will be punished if (i) $s(c_j, x_p^+)$ is small; (ii) $s(c_j, x_p^+)$ is smaller than $s(c_j, x_i^+)$ where $x_i^+ \in P_j$, $p \neq i$; or (iii) $s(c_j, x_i^-)$ is small so that we could focus on intra-class relation. Moreover, a large $s(c_j, x_i^-)$ will provide $s(c_j, x_i^+)$ with tolerance to some extent, which allows our model to focus on reducing the large similarity of $s(c_j, x_i^-)$. Similar analysis can be performed with $s(c_j, x_n^-)$ based on Eq. (4), and we omit the detail here. Therefore, in contrast to CE which deals with each sample separately, QR allows the query and support samples to mutually share their information and re-weight their importance.

Second, note that the negative set N_j of each category contains not only the negative query samples but also the support samples from other categories. This avoids the information loss caused by the likely sub-optimal support sample aggregation strategy and allows the model to learn to utilize the negative support samples directly by the objective.

It turns out that the QR loss is closely related to Deep Mutual Information (MI) maximization recently proposed by [5]. Without loss of generality, following [5], the JSD-based (Jensen-Shannon Divergence) MI estimator between c_j and x can be formulated as

$$
\begin{aligned}
\mathcal{L}_{\text{JSD MI}} &= \max \frac{1}{|P_j|} \sum_{x_i^+ \in P_j} -\log(1 + e^{-s(c_j, x_i^+)}) - \frac{1}{|N_j|} \sum_{x_i^- \in N_j} \log(1 + e^{s(c_j, x_i^-)}) \\
&\geq \max -\log\left(1 + \frac{1}{2|P_j|} \sum_{x_i^+ \in P_j} e^{-s(c_j, x_i^+)} + \frac{1}{2|N_j|} \sum_{x_i^- \in N_j} e^{s(c_j, x_i^-)}\right) \\
&= \mathcal{L}_{\text{QR}}
\end{aligned}
\tag{5}
$$

Here we use the fact that $-\log(1 + x)$ is convex and the Jensen's inequality. We can thus derive that the QR loss is actually a lower bound of the JSD MI. The reason why we do not directly optimize $\mathcal{L}_{\text{JSD MI}}$ is that the re-weighting mechanism of $\mathcal{L}_{\text{JSD MI}}$ does not take both P_j and N_j into consideration for each $s(c_j, x_i)$. We experimentally verify the superiority of our formulation in Sect. 3.

2.3   Adaptive Hard Margin

The adaptive hard margin is built upon the fact that the cosine similarity between uniformly distributed normalized samples approaches $\mathcal{N}(0, \frac{1}{2d})$ [20] and is thus likely to be zero. Therefore, $s(c_j, x_i^+)$ should be at least larger than E_j^- and $s(c_j, x_i^-)$ should be at least smaller than E_j^+, where E_j^+ and E_j^- denote the average of $s(c_j, x_i^+)$ with $s(c_j, x_i^+) < 0$ and average of $s(c_j, x_i^-)$ with $s(c_j, x_i^+) > 0$, respectively. Based on this observation, we propose a QR loss with online Adaptive Hard Margin which can be written as

$$\mathcal{L}_{\mathrm{QR+margin}} = \sum_j \log\left(1 + \frac{1}{2|P_j|} \sum_{x_i^+ \in P_j} e^{-s(c_j, x_i^+) + E_j^-} + \frac{1}{2|N_j|} \sum_{x_i^- \in N_j} e^{s(c_j, x_i^-) - E_j^+}\right). \tag{6}$$

Basically, Eq. (6) imposes extra punishment on categories with more positive samples whose similarities are smaller than random or negative samples, and negative samples whose similarities are larger than random or positive samples.

3   Experiments

3.1   Datasets and Experimental Settings

We collect the dermatology images from the Dermnet atlas website[1]. To evaluate the few-shot learning methods, we discard categories with less than 10 samples, which are required for the 5-way 5-shot setting. Finally, we obtain $20,230$ images in total belonging to 334 different categories detailed in Fig. 2. The data is manually split into 186 categories for training, 74 for validation, and 74 for testing, respectively. Moreover, to better simulate the scenario of few-shot learning, we

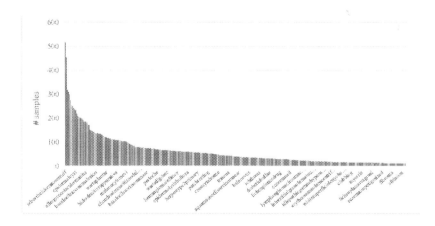

Fig. 2. The category distribution of Dermnet datasets.

---

[1] www.dermnet.com.

deliberately choose categories with more than 120 samples (38 categories in total) as the training data.

We benchmark the dataset on an FSL suite proposed by [1]. The suite contains 2 strong baseline methods (denoted as baseline and baseline++ following [1]) and 4 FSL methods including Relation Net [16], Model-Agnostic Meta-Learning (MAML) [4], Matching Net (MN) [19], and Prototypical Net (PN) [14]. The baseline methods are carefully designed and outperform FSL methods in some cases. We refer readers to [1] for details. The four FSL methods are regarded as the state-of-the-art FSL baselines in recent benchmark literature [1,18], and we train them with CE as our baselines except for the Relation Net, which is trained with Mean Square Error (MSE) Loss following the original paper. We apply the proposed QR loss to MN and PN since these two methods have proven to have superior and stable performance in natural image classification [1]. The model trained with JSD-based MI maximization Eq. (5) is denoted as JSD MI, and models trained with the proposed QR loss Eq. (2) and QR loss with adaptive hard margin Eq. (6) are denoted as QR and QR+M, respectively.

For the network structure, we follow the commonly adopted FSL settings [1,16]. The feature embedding network used in this paper is a convolutional neural network which has four convolutional blocks with each block containing a sequence of a convolutional layer with 64 filters of size 3×3, a batch normalization layer, a 2×2 max-pooling layer and a Leaky ReLU layer. For the experimental settings, the episodic training mechanism is applied to all FSL models, and $60,000$ episodes are constructed in total during training for all methods. For validation and testing, 600 episodes are randomly constructed from the validation and test set, respectively. We conduct 5-way 1-shot and 5-way 5-shot classification tasks on the collected Dermnet dataset, and 5 query samples are provided for each category within each episode for either training and testing. For optimization, we adopt the Adam algorithm with a learning rate of 0.001. Experiments are run five times and we report the performance on test set corresponding to the best validation results. The average Accuracy, Precision, and F1 score with 95% confidence interval are reported.

3.2 Result Analysis

The experimental results are reported in Table 1, and we draw several interesting points from the results as follows. First of all, the baseline methods with minibatch training and CE loss perform reasonably well in practice. The FSL methods trained with CE loss have comparable or slightly better performance. In contrast, FSL methods trained with the proposed QR loss significantly outperform the baseline methods and the FSL methods with CE. For Matching Net, our QR loss achieves 3.42 % and 5.88 % improvements compared with the CE loss in terms of accuracy for 5-way 1-shot and 5-way 5-shot tasks. Significant improvements are also observed for PN, and our QR loss outperforms CE 3.08 % and 8.32 % for 5-way 1-shot and 5-way 5-shot, respectively. The improvements are obtained by fully utilizing the cross-sample information and avoiding the information loss caused by manually designed support sample aggregation

Table 1. Experimental results on the Derment skin disease classification dataset. * denotes that the model is trained with 9 query samples per episode. - denotes that the setting is not applicable. M denotes our methods with an adaptive hard margin.

Methods	5-way 1-shot			5-way 5-shot		
	ACC%	Precision%	F1%	ACC%	Precision%	F1%
Baseline [1]	$39.89_{\pm0.89}$	$40.57_{\pm1.12}$	$37.16_{\pm0.91}$	$59.87_{\pm0.94}$	$62.37_{\pm1.10}$	$58.19_{\pm1.01}$
Baseline++ [1]	$42.47_{\pm0.94}$	$43.70_{\pm1.11}$	$40.34_{\pm0.93}$	$63.37_{\pm0.95}$	$65.80_{\pm1.07}$	$61.75_{\pm1.01}$
MAML [4]	$45.95_{\pm1.06}$	$44.82_{\pm1.29}$	$42.18_{\pm1.08}$	$66.93_{\pm0.96}$	$69.24_{\pm1.11}$	$64.92_{\pm1.05}$
Relation Net [16]	$45.50_{\pm1.07}$	$46.36_{\pm1.18}$	$44.00_{\pm1.07}$	$62.53_{\pm1.02}$	$64.90_{\pm1.11}$	$62.26_{\pm1.05}$
MN [19]	$44.59_{\pm0.97}$	$44.96_{\pm1.19}$	$41.52_{\pm1.00}$	$61.21_{\pm0.90}$	$63.15_{\pm1.13}$	$58.29_{\pm0.99}$
MN+JSD MI	$43.28_{\pm1.04}$	$43.26_{\pm1.25}$	$40.00_{\pm1.05}$	$58.99_{\pm0.94}$	$60.23_{\pm1.20}$	$55.78_{\pm1.02}$
MN+QR	$48.01_{\pm1.09}$	$48.87_{\pm1.13}$	$44.30_{\pm1.13}$	$67.09_{\pm0.97}$	$69.18_{\pm1.16}$	$64.53_{\pm1.08}$
MN+QR+M	$49.29_{\pm1.31}$	$49.95_{\pm1.05}$	$45.64_{\pm1.09}$	$66.83_{\pm0.95}$	$69.10_{\pm1.16}$	$64.25_{\pm1.05}$
MN+QR*	$48.66_{\pm1.07}$	$48.86_{\pm1.30}$	$44.98_{\pm1.11}$	-	-	-
MN+QR+M*	$49.76_{\pm1.07}$	$49.52_{\pm1.32}$	$46.01_{\pm1.13}$	-	-	-
PN [14]	$46.77_{\pm1.04}$	$46.82_{\pm1.06}$	$43.58_{\pm1.07}$	$62.06_{\pm1.02}$	$63.39_{\pm1.22}$	$59.50_{\pm1.10}$
PN+JSD MI	$47.55_{\pm1.00}$	$47.90_{\pm1.25}$	$44.33_{\pm1.05}$	$61.15_{\pm0.94}$	$61.74_{\pm1.16}$	$58.34_{\pm1.02}$
PN+QR	$49.85_{\pm1.11}$	$49.53_{\pm1.32}$	$46.34_{\pm1.14}$	$70.38_{\pm0.96}$	$72.13_{\pm1.08}$	$68.50_{\pm1.05}$
PN+QR+M	$\mathbf{52.41}_{\pm1.09}$	$\mathbf{53.21}_{\pm1.27}$	$\mathbf{49.52}_{\pm1.12}$	$\mathbf{71.99}_{\pm0.87}$	$\mathbf{74.23}_{\pm0.98}$	$\mathbf{70.30}_{\pm0.94}$
PN+QR*	$50.62_{\pm1.10}$	$50.83_{\pm1.32}$	$47.16_{\pm1.13}$	-	-	-
PN+QR+M*	$\underline{53.30}_{\pm1.11}$	$\underline{53.69}_{\pm1.35}$	$\underline{50.45}_{\pm1.17}$	-	-	-

during training. Second, we compare the QR loss with JSD MI. Although the formulations are similar, QR is significantly better than JSD MI. The reason should be attributed to the fact that JSD MI does not mutually utilize the information in P_j and N_j. Third, the adaptive hard margin consistently boosts the performance of the models trained by QR. For example, the adaptive hard margin improves PN trained with QR 2.56 % and 1.61 % for 5-way 1-shot and 5-way 5-shot, respectively. Finally, our method could be further boosted by increasing the number of queries for both training and testing. Overall, our proposed FSL methods classify skin disease with only a few available training samples and makes it possible to diagnose rare diseases using modern neural networks.

3.3   Influence of the Number of Shots and Ways

For simplicity, we only conduct experiments on PN with various ways and shots and report the accuracy. As shown in Tables 2 and 3, QR has clear advantages over CE when more samples are available per episode, suggesting that QR can better utilize the cross sample information.

Table 2. 5-way different-shot. (ACC%)

# shots	1	2	3	4	5
CE	46.77	54.04	57.15	59.65	62.06
QR	49.85	62.37	66.87	68.95	70.38

Table 3. Different-way 1-shot. (ACC%)

# ways	2	3	5	10	20
CE	69.80	59.42	46.77	35.78	24.30
QR	72.02	61.57	49.85	40.37	31.31

4 Conclusions

We propose to apply Few-Shot Learning to address the classification for rare skin diseases. We find that existing FSL methods do not perform significantly better than the baseline methods. Through careful analysis, we believe the problem should be largely attributed to the incompatibility between the episode training mechanism and cross entropy loss. Therefore, we propose a novel QR loss for FSL to make fully use of the information across samples and also allow the model to learn to extract information of the support samples guided by the training objective. With the proposed QR loss, the state-of-the-art FSL methods perform consistently better than methods training with the conventional CE loss. Our work demonstrates the promise of diagnosing rare skin diseases with one or a few labeled samples. In the future, we will investigate extensions to other medical classification problems or even natural image classification.

Acknowledgement. Research reported in this publication was supported by the National Institute Of Neurological Disorders And Stroke of the National Institutes of Health under Award Number P50NS108676. The content is solely the responsibility of the authors and does not necessarily represent the official views of the National Institutes of Health.

References

1. Chen, W.Y., Liu, Y.C., Kira, Z., Wang, Y.C., Huang, J.B.: A closer look at few-shot classification. In: International Conference on Learning Representations (2019)
2. Das, D., Lee, C.G.: A two-stage approach to few-shot learning for image recognition. IEEE Trans. Image Process. **29**, 3336–3350 (2019)
3. Esteva, A., et al.: Dermatologist-level classification of skin cancer with deep neural networks. Nature **542**(7639), 115 (2017)
4. Finn, C., Abbeel, P., Levine, S.: Model-agnostic meta-learning for fast adaptation of deep networks. In: Proceedings of the 34th International Conference on Machine Learning, vol. 70, pp. 1126–1135. JMLR. org (2017)
5. Hjelm, R.D., et al.: Learning deep representations by mutual information estimation and maximization. arXiv preprint arXiv:1808.06670 (2018)
6. Li, W., Wang, L., Xu, J., Huo, J., Gao, Y., Luo, J.: Revisiting local descriptor based image-to-class measure for few-shot learning. In: Proceedings of the IEEE Conference on Computer Vision and Pattern Recognition, pp. 7260–7268 (2019)

7. Li, W., Xu, J., Huo, J., Wang, L., Gao, Y., Luo, J.: Distribution consistency based covariance metric networks for few-shot learning. In: Proceedings of the AAAI Conference on Artificial Intelligence, vol. 33, pp. 8642–8649 (2019)

8. Liao, H., Li, Y., Luo, J.: Skin disease classification versus skin lesion characterization: achieving robust diagnosis using multi-label deep neural networks. In: 2016 23rd International Conference on Pattern Recognition (ICPR), pp. 355–360. IEEE (2016)

9. Liao, H., Luo, J.: A deep multi-task learning approach to skin lesion classification. arXiv preprint arXiv:1812.03527 (2018)

10. Movshovitz-Attias, Y., Toshev, A., Leung, T.K., Ioffe, S., Singh, S.: No fuss distance metric learning using proxies. In: Proceedings of the IEEE International Conference on Computer Vision, pp. 360–368 (2017)

11. Okuboyejo, D.A., Olugbara, O.O., Odunaike, S.A.: Automating skin disease diagnosis using image classification. In: Proceedings of the World Congress on Engineering and Computer Science, vol. 2, pp. 850–854 (2013)

12. Qian, Q., Shang, L., Sun, B., Hu, J., Li, H., Jin, R.: Softtriple loss: deep metric learning without triplet sampling. In: Proceedings of the IEEE International Conference on Computer Vision, pp. 6450–6458 (2019)

13. Shi, X., Dou, Q., Xue, C., Qin, J., Chen, H., Heng, P.-A.: An active learning approach for reducing annotation cost in skin lesion analysis. In: Suk, H.-I., Liu, M., Yan, P., Lian, C. (eds.) MLMI 2019. LNCS, vol. 11861, pp. 628–636. Springer, Cham (2019). https://doi.org/10.1007/978-3-030-32692-0_72

14. Snell, J., Swersky, K., Zemel, R.: Prototypical networks for few-shot learning. In: Advances in Neural Information Processing Systems, pp. 4077–4087 (2017)

15. Sumithra, R., Suhil, M., Guru, D.: Segmentation and classification of skin lesions for disease diagnosis. Procedia Comput. Sci. **45**, 76–85 (2015)

16. Sung, F., Yang, Y., Zhang, L., Xiang, T., Torr, P.H., Hospedales, T.M.: Learning to compare: relation network for few-shot learning. In: The IEEE Conference on Computer Vision and Pattern Recognition (CVPR), June 2018

17. Szegedy, C., Vanhoucke, V., Ioffe, S., Shlens, J., Wojna, Z.: Rethinking the inception architecture for computer vision. In: Proceedings of the IEEE Conference on Computer Vision and Pattern Recognition, pp. 2818–2826 (2016)

18. Triantafillou, E., et al.: Meta-dataset: a dataset of datasets for learning to learn from few examples. arXiv preprint arXiv:1903.03096 (2019)

19. Vinyals, O., Blundell, C., Lillicrap, T., Wierstra, D., et al.: Matching networks for one shot learning. In: Advances in Neural Information Processing Systems, pp. 3630–3638 (2016)

20. Wu, C.Y., Manmatha, R., Smola, A.J., Krahenbuhl, P.: Sampling matters in deep embedding learning. In: Proceedings of the IEEE International Conference on Computer Vision, pp. 2840–2848 (2017)

21. Ye, H.J., Chen, H.Y., Zhan, D.C., Chao, W.L.: Identifying and compensating for feature deviation in imbalanced deep learning. arXiv preprint arXiv:2001.01385 (2020)

Clinical-Inspired Network for Skin Lesion Recognition

Zihao Liu[1,2], Ruiqin Xiong[1], and Tingting Jiang[1(✉)]

[1] NELVT, Department of Computer Science, Peking University, Beijing, China
{lzh19961031,ttjiang}@pku.edu.cn
[2] Advanced Institute of Information Technology, Peking University, Hangzhou, China

Abstract. Automated skin lesion recognition methods are useful for improving the diagnostic accuracy in dermoscopy images. However, several challenges delayed the pace of the development of these methods, including limited amount of data, a lack of ability to focus on the lesion area, poor performance for distinguishing between visually-similar categories of diseases and an imbalance between different classes of training data. During practical learning and diagnosis process, doctors conduct certain strategies to tackle these challenges. Thus, it's really appealing to involve these strategies in automated skin lesion recognition method, which could be promising for a better performance. Inspired by this, we propose a new Clinical-Inspired Network (CIN) to simulate the subjective learning and diagnostic process of doctors. To mimic the diagnostic process, we design three modules, including a lesion area attention module to crop the images, a feature extraction module to extract image features and a lesion feature attention module to focus on the important lesion parts and mine the correlation between different lesion parts. To simulate the learning process, we introduce a distinguish module. The CIN is extensively tested on ISBI2016 and 2017 challenge datasets and achieves state-of-the-art performance, which demonstrates its advantages.

Keywords: Clinical-inspired · Skin lesion recognition · Neural networks

1 Introduction

Skin disease is one of the most common diseases in the world, and the number of deaths by skin diseases has increased in recent years, which aroused public attention [14,15]. Dermoscopy [2,10,16] is an essential means of improving the recognition accuracy of skin diseases for doctors. A lot of efforts are dedicated to automating the classification of dermoscopy images since the manual inspection suffers from subjective bias.

Electronic supplementary material The online version of this chapter (https:// doi.org/10.1007/978-3-030-59725-2_33) contains supplementary material, which is available to authorized users.

A. L. Martel et al. (Eds.): MICCAI 2020, LNCS 12266, pp. 340–350, 2020.
https://doi.org/10.1007/978-3-030-59725-2_33

Recently, many CNN-based methods come into being and achieve significant progress in classifying skin diseases [5,6,9,17,19–23]. One category of these methods is from the perspective of data. A synergic loss is introduced in [9,22] to tackle the inter-class similarity and intra-class variation problem in skin lesion data. To solve the imbalance problem among different classes, the authors in [19] present to dynamically sample paired data and learn domain-invariant features. The second category deliberates the design of network architecture. A deep two-stage network is proposed in [20] to acquire richer and more discriminative features for more accurate recognition. It segments the images first, then classifies them. To enhance the network's ability for discriminative representation, the attention mechanism including spatial and channel attention is integrated in [23]. A joint learning method based on both handcrafted and deep-learning features is proposed in [17] in order to construct the hybrid-prior feature representation for a better result. To take advantage of the multiple features from single image and build a global image representation, authors in [21] adopt Fisher Vector (FV) encoding to aggregate the orderless visual statistic features.

Although these methods have achieved good performance, the perspective of designing the algorithm in existing methods is similar to which for tackling natural image classification tasks, which is not fully driven by medical perspective and knowledge and ignores the medical expertise of skin disease. However, the classification task in dermoscopy images should be treated differently because it requires a lot of professional medical knowledge. Thus, having medical professionals' knowledge as much as possible involved in computer-aided methods becomes more and more appealing and significant. As a result, designing the network in a way that allows it to mimic all the processes of professionals, could let it learn and classify skin diseases more effectively.

Motivated by this, medical doctors' knowledge learning process and diagnostic process have been adopted and migrated by our method. On the side of doctors' diagnostic process, we find three steps for doctors to diagnose skin diseases. When a doctor looks at a dermoscopy image, first, in order to focus on the lesion area, he zooms in and only focuses on the lesion area, which can be called **zoom step**. Then, he observes and examines the feature of the lesion area, including shape, structure and color, which are all classical features for skin diseases [18]. This is called **observe step**. At last, the doctors will focus on the important lesion parts and mine the correlation between lesion parts before making a final diagnosis for a comprehensive judgment. This can be called **compare step**. On the side of doctors' learning process, it's really difficult to distinguish between different classes since the visual differences between images from the same kind of disease can be significantly greater than from the other kind. In order to solve this problem, doctors are devoted to comparing images which may visually look similar and decide whether they're from the same class. This is **distinguish strategy**.

In regard of all the information, we propose the Clinical-Inspired Network(CIN). **To simulate the diagnostic process, we introduce three modules.** Firstly, a lesion area attention module is adopted to mimic the zoom step, which enables the network to focus on the lesion area. To simulate the observe step, a convolutional neural network is used to extract features of input images, which forms the feature extraction module. For the compare step, the lesion feature attention module is introduced to focus on the important lesion parts and mine the correlation between different lesion parts. **To mimic the distinguish strategy in the learning process**, a distinguish module is proposed to increase the discrimination ability among different classes.

The main contributions can be summarized as follows: (1) We propose a Clinical-Inspired Network (CIN) on the perspective of simulating doctor's learning and diagnostic processes. (2) Three new modules are designed in CIN, including a lesion area attention module, a lesion feature attention module and a distinguish module, which is fully inspired by the clinical process of doctors. To sum up, our method is different from existing methods both in perspective and architecture. Our perspective is especially unique and important for medical problems since few former methods are driven by medical perspective and knowledge and systematically consider all the processes of doctors. The experimental results show that our method performs the best on both ISBI 2016 and 2017 challenge datasets.

2   Proposed Method

The whole network architecture of CIN is illustrated in Fig. 1. It consists of four modules. The first module is the Lesion Area Attention Module. It takes two original dermoscopy images as input, generates the lesion area attention map for each image and takes the smallest rectangle which includes the lesion area from the attention map to crop out the images. The following Feature Extraction Module extracts the features of the two cropped images. Next, The feature of each image is fed into the Lesion Feature Attention Module, which includes a spatial attention block and a correlation attention block. After that, the Distinguish Module takes the output features of Lesion Feature Attention Module from two branches as inputs, compares them and outputs a score reflecting whether two images are from the same class. Meanwhile, the two features are fed into two fully connected layers respectively, in order to do the classification task.

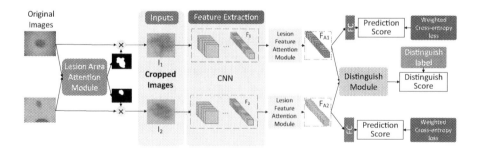

Fig. 1. An overview of the CIN.

Lesion Area Attention Module. This module is adopted to mimic the **zoom step** of diagnostic process for doctors. As illustrated in Fig. 2, we design a new encoder-decoder structure with an extra auxiliary learning part.

This module contains two parts, the attention generating part and the auxiliary learning part. The attention part uses a fully convolutional network [13] to generate the lesion area attention maps, containing a score value for each pixel, which indicates whether it refers to lesion area or normal skin area. So basically it is a binary segmentation task, which is supervised by normal cross-entropy loss, illustrated as L_{ce} in Fig. 2(b). After the attention map generated in the first part, the smallest rectangle which includes the lesion area is taken from the attention map to crop out the image. The cropped image is the output of this module, which is also the input of the next Feature Extraction Module. Meanwhile, the auxiliary learning part serves as an extra supervision for this module. The original image will be cropped by both the predicted attention map and the groundtruth map. The two cropped images are the inputs of this part and will then be fed to two convolutional blocks with same architecture, which is illustrated in Fig. 2(b). For each conv block, L_1 loss is used to compute the difference between the output features of the two branches. This loss is served as an extra regularization for the bourdaries since the mistakes may appear in these areas. To note that the auxiliary learning part will only be used at the training phase.

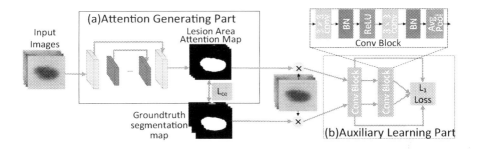

Fig. 2. The details of lesion area attention module.

Feature Extraction Module. This module is designed to simulate the **observe step** of doctors. Convolutional layers, as well as the activation layers are used in normal CNN to form this module. It takes the cropped images as the inputs and extracts the feature.

Lesion Feature Attention Module. This module is proposed to simulate the **compare step** of doctors. As illustrated in Fig. 3, we design two blocks, the spatial attention block, as to enable the network to focus on important lesion regions, and the correlation attention block, as to mine the correlation between different lesion parts. Denote the input feature as F, the shape of which is $H \times W \times C$.

For the spatial attention block, F will first go through one 1×1 conv layer and two 3×3 conv layers along with a ReLU layer. The feature now is denoted as $\hat{F}$. Then, instead of the normal softmax layer, a spatial normalization function is conducted by Eq. (1) to generate the attention map Q. Finally, an element-wise multiplication between Q and F is applied, then the result will be multiplied by a learnable weighting factor α and performed a summation to F.

$$Q = \{q | q_{ij} = \frac{\hat{F}_{ij}}{\sum_{i'=1}^{H} \sum_{j'=1}^{W} \hat{F}_{i'j'}}\} \tag{1}$$

For the correlation attention block, the feature will first go through two branches. Each branch has one 1×1 conv layer and one 3×3 conv layer. The feature now is denoted as $\overline{F}$. Then $\overline{F}$ is reshaped to $M \times C$, where M is equal to $H \times W$. After that, a matrix multiplication between $\overline{F}$ and $\overline{F}'$, the transpose of $\overline{F}$, is applied. The result is denoted as $\widetilde{F}$, which is a matrix of size $M \times M$. After that, the attention map P is generated by applying a normalization function on $\widetilde{F}$. P has a size of $M \times M$. The equation is:

$$\widetilde{F} = \overline{F} * \overline{F}', \qquad P = \{p | p_{ij} = \frac{exp(\widetilde{F}_{ij})}{\sum_{i'=1}^{M} \sum_{j'=1}^{M} exp(\widetilde{F}_{i'j'})}\} \tag{2}$$

where "*" means matrix multiplication, P is the attention map. p_{ij} represents the correlation between i^{th} position and j^{th} position in feature $\widetilde{F}$. After generating P, a matrix multiplication was applied with $\overline{F}$ and reshape the result to $H \times W \times C$. Then, the reshaped result will be multiplied by a learnable weighting factor β and applied a matrix summation to F. The final output of this module is obtained by Eq. (3):

$$F_A = \alpha \cdot (Q \cdot F) + \beta \cdot (P * \overline{F}) + F \tag{3}$$

where "$\cdot$" means element-wise multiplication, "*" means matrix multiplication, α and β are scalars which are learnable weights, F_A is the output of the lesion feature attention module.

At the same time, F_A will be fed into two fully connected layers for the classification task, and supervised by weighted cross-entropy loss:

$$L_c = -\frac{1}{N_2} \sum_{i=1}^{N_2} \sum_{j=1}^{K} w_j (y_{ij} log(x_{ij})) \tag{4}$$

where N_2 is the number of images, K is the number of classes. w_j is the weight of class j. y_{ij} denotes whether image i belongs to label j, and x_{ij} means the classification score of class j for image i.

Distinguish Module. This module is designed to mimic the **distinguish strategy** of doctors' learning process. Normally, there're three steps to distinguish whether the two images are from same class: integrate the important

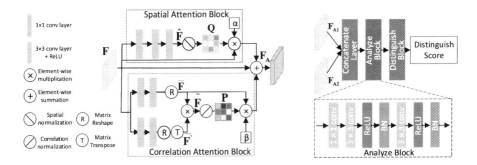

Fig. 3. Lesion feature attention module.   **Fig. 4.** Distinguish module.

features of the them, analyze them, and draw conclusions. Inspired by this, we adopt three parts in this module: a concatenate layer, an analyze block and a distinguish block, illustrated in Fig. 4.

For the two input images, the feature of two images are first fed into the concatenated layer, in which a new concatenated feature is formed. Then the analyze block, which contains a 1×1 conv layer and two 3×3 conv layers, will mine a deeper feature reflecting the relationship between two images. Finally, a distinguish score indicating whether these two images are from the same class will be obtained by the distinguish block, which contains two fully connected layer with a ReLU and dropout layer between them.

A binary cross-entropy loss is used to supervise the module, illustrated as Eq. (5):

$$L_d = -\frac{1}{N_1} \sum_{i=1}^{N_1} (y_d^i log(x_d^i) + (1 - y_d^i)log(1 - x_d^i)) \tag{5}$$

where N_1 denotes the number of image pairs, y_d^i is the distinguish label of the i_{th} image pair, and x_d^i is the distinguish score, predicted by the module.

This module is different from [22] in both perspective and architecture. The designing of the module is fully driven by the distinguish strategy in clinical process of doctors, including the three sub-steps in the strategy. However, the method in [22] is designed on the perpective of data, which is the intra-class variation and inter-class similarity in skin lesion data. Besides, the architecture is also different. The authors in [22] only adopt three single linear layers in the synergic network, in contrast we design two blocks, including several convolutional layers, activation layers and fully-connected layers, for a better non-linear learning ability of the distinguishment.

So the final loss of the whole network is:

$$L = L_c + \lambda L_d \tag{6}$$

where λ is a hyper-parameter to linearly combine two losses in order to balance them.

3 Experiments

3.1 Datasets

ISBI 2016 dataset [7] contains 900 images as training data and 379 images as testing data. The task is to classify images as either malignant (melanoma) or non-malignant (non-melanoma). ISIC 2017 dataset [3] contains 2000 training, 150 validation, and 600 test images. The dataset contains three classes, including melanoma, nevus and seborrheic keratosis. There are two binary classification sub-tasks, the first one is to distinguish between melanoma and the others (nevus and seborrheic keratosis), the second task is to distinguish between seborrheic keratosis and the others (nevus and melanoma). Since the groundtruth of the testset of ISIC2018 and 2019 dataset isn't available, we can't implement our method on these two datasets.

3.2 Implementation Details

ResNet18 are used as the backbone network for feature extraction module. At the training phase, the lesion area attention module are trained seperately first. Since in ISBI2016 Challenge Part3B, all the other methods use the data with segmentation map to aid the classification task, we use these data to supervise the lesion area attention module. After training the lesion area attention module, two original images are firstly fed into the lesion area attention module and the cropped images are the input of later architecture. The other modules are trained by Eq. (6) together. At the testing phase, a single image is fed into the lesion area attention module, the cropped image is used as the input of both branches. The probabilistic predictions obtained by each branch are summed, and the class with highest probability will be the final result. We use the same setting for three times' experiments and obtain the average result as the final result. Since ISBI2016 dataset doesn't have validation set, we randomly select 20% from the training set each time as the validation set. The images are resized to 448×448 for training and testing. The learning rate is initialized to 0.001 and decayed by 0.1 every 20 epochs. The batch size is set to 32. The image pairs are randomly chosen. In Eq. (6), λ is set to 1. Five evaluation metrics are used in the experiments, including Average Precision(AP), Accuracy(ACC), Area under the receiver operating characteristic curve(AUC), Sensitivity(SE) and Specificity(SP).

3.3 Ablation Study

We conduct the ablation study on the second sub-task of ISBI2017 dataset to investigate the impact of the proposed three modules of our method on the performance. Note that during the challenge, it is ranked only by comparing the AUC value. The specific performance is listed in Table 1. The three main modules are listed in the following order: 1. Lesion Area Attention Module($\mathbf{A}$); 2. Lesion Feature Attention Module($\mathbf{F}$); 3. Distinguish Module($\mathbf{D}$). The feature extraction module is used by default. The baseline model is trained from pretrained ResNet18, denoted as **"Baseline"**. We first compare the three modules when they are combined with the baseline respectively. The second, third, fourth rows are respectively the results of adding the lesion area attention module, as **"CIN(A)"**, the lesion feature attention module, as **"CIN(F)"** and the distinguish module, as **"CIN(D)"** to the baseline. The AUC value is improved by 1.7%, 1.3% and 1.2%, respectively. The experimental result proves that individually adopting the three modules can benefit the model, which reveals that our migration from the learning and diagnosis process to CNN is successful.

After that, we conduct the experiments with multiple modules added to the baseline. The performance of **CIN(A+F)** is better than **CIN(A)** and **CIN(F)**. The AUC value is further increased to 0.960, with a gain of 1.6% comparing to **CIN(A)**, and 2.0%, comparing to **CIN(F)**. Finally, when the Distinguish Module is introduced, it yields AUC of 0.981, which is 5.4% higher than that from the Baseline. The encouraging results show the advantage of our method. Supplementary material provides more ablation study results as well as the visualization results of Lesion Feature Attention Module.

Table 1. Ablation Study on ISIC2017 dataset sub-task2.

Method	AUC	ACC	SE	SP
Baseline	0.927	0.863	0.678	0.869
CIN(A)	0.944	0.922	0.778	0.947
CIN(F)	0.940	0.864	0.823	0.871
CIN(D)	0.939	0.905	0.711	0.939
CIN(A+F)	0.960	0.915	0.644	0.959
CIN (A+F+D)	**0.981**	**0.943**	**0.829**	**0.965**

Table 2. Performance of CIN and other methods on ISBI 2016 dataset.

Method	AP	ACC	AUC
CIN (ours)	**0.740**	**0.887**	**0.873**
DCNN-FV [21]	0.685	0.868	0.852
SDL [22]	0.664	0.858	0.818
CUMED [20]	0.637	0.855	0.804
GTDL [7]	0.619	0.813	0.802
Result2-3B [7]	0.615	0.844	0.808
USYD [7]	0.580	0.686	0.793
Mufic-IT [7]	0.534	0.760	0.685

3.4   Results

We compare our performance to DCNN-FV [21], SDL [22] and the top5 methods of the challenge in 2016. To follow the tradition in the existing methods, we evaluate the performance on AP, ACC and AUC. Note that the challenge is ranked based only on **AP**, which means that it's the most important metric. The results are shown in Table 2 and our method are shown in the second row as **"CIN(ours)"**. We achieve the best result in all the three metric, with an AP of **0.740**, which exceeds DCNN-FV [21] by 5.5% and is almost 2.6 times of the difference between the next two methods. Our ACC and AUC outperforms the second place by 1.9% and 2.1%, respectively.

Table 3. Performance of CIN and other methods on ISIC 2017 dataset.

Methods	External data	Melanoma Classification					Seborrheic Keratosis					Average AUC
		AUC	AP	ACC	SE	SP	AUC	AP	ACC	SE	SP	
CIN (ours)	0	**0.920**	**0.814**	**0.894**	0.645	0.948	**0.981**	**0.902**	**0.943**	0.829	0.965	**0.951**
RHN [17]	0	0.883	0.810	0.890	0.732	0.901	0.961	0.843	0.885	0.912	0.907	0.926
ARL-CNN [23]	1320	0.875	–	0.850	0.658	0.896	0.958	–	0.868	0.878	0.867	0.917
SDL [9]	1320	0.868	0.689	0.872	–	–	0.955	0.818	0.917	–	–	0.912
RENI [11]	1444	0.868	0.710	0.828	**0.735**	0.851	0.953	0.786	0.803	**0.978**	0.773	0.911
gpm-LSSSD [8]	900	0.856	0.747	0.823	0.103	**0.998**	0.963	0.839	0.875	0.178	**0.998**	0.910
Alea-Jacta-Est [12]	7544	0.874	0.715	0.872	0.547	0.950	0.943	0.790	0.895	0.356	0.990	0.908
EResNet [1]	1600	0.870	0.732	0.858	0.427	0.963	0.921	0.770	0.918	0.589	0.976	0.896
DLSL [4]	1341	0.836	0.665	0.845	0.350	0.965	0.935	0.808	0.913	0.556	0.976	0.886

Note that both DCNN-FV [21] and SDL [22] use ResNet50 as the backbone but we achieves better results using ResNet18 with a smaller number of parameters.

Next, we compare our performance to [17,23] and the top5 methods ISIC2017 challenge dataset. There are two sub-tasks in the challenge, to follow the tradition, we compared five metrics for each sub-task: AUC, AP, ACC, SE, and SP. Note that the challenge is ranked based only on the **Average AUC** of the two sub-tasks, which means that it's the most important metric. Table 3 shows the results. Since we use no extra images for the experiments, and couldn't find the same external data for [17,23], we compare with the result of [17] without any extra image for a fair comparison. It can be found from the table that the AUC achieved by our method of the two sub-tasks are both the highest, **0.920** and **0.981** respectively. The AUC of first sub-task improves by 3.7% and the second improves by 2.0%. The average AUC of our approach is **0.951**, which improves by 2.5% and is almost 3 times of the difference between the next two methods. Besides, the performance of AP and ACC in two sub-tasks is also the best. To sum up, our method not only achieves state-of-the-art performance among the rank metric of two challenge datasets with a significant improvement, but also gets the best results in most of other metrics, which fully demonstrates the advantages of our model. Supplymentary material provides more experiment results and visualization results.

4   Conclusion and Future Work

Motivated by the observations of learning and diagnosis of dermoscopy images from doctors, for the first time, we present a clinical-inspired network, to simulate the actual subjective process. The network includes a lesion area attention module and lesion feature attention module as to mimic the diagnosis process of doctors. Besides, the introduction of distinguish module are based on the original intention of simulating the learning process from doctors. At the same time, our method achieves encouraging performance while keeping a simple backbone network, which contains smaller parameters, and utilize no external data. The state-of-the-art experiment results on ISBI 2016 and ISIC 2017 datasets verify the effectiveness and advancement of our methods. In future works, we will try to integrate the lesion area attention module in the whole workflow.

Acknowledgement. This work was partially supported by the Natural Science Foundation of China under contracts 61572042 and 61772041. We also acknowledge the Clinical Medicine Plus X-Young Scholars Project, and High-Performance Computing Platform of Peking University for providing computational resources.

References

1. Bi, L., Kim, J., Ahn, E., Feng, D.: Automated skin lesion analysis using large-scale dermoscopy images and deep residual networks. arXiv preprint arXiv:1703.04197 (2017)
2. Binder, M., et al.: Epiluminescence microscopy: a useful tool for the diagnosis of pigmented skin lesions for formally trained dermatologists. Arch. Dermatol. **131**(3), 286–291 (1995)
3. Codella, N.C., et al.: Skin lesion analysis toward melanoma detection: a challenge at the 2017 international symposium on biomedical imaging (ISBI), hosted by the international skin imaging collaboration (ISIC). In: 2018 IEEE 15th International Symposium on Biomedical Imaging (ISBI 2018), pp. 168–172. IEEE (2018)
4. DeVries, T., Ramachandram, D.: Skin lesion classification using deep multi-scale convolutional neural networks. arXiv preprint arXiv:1703.01402 (2017)
5. Ge, Z., et al.: Exploiting local and generic features for accurate skin lesions classification using clinical and dermoscopy imaging. In: 2017 IEEE 14th International Symposium on Biomedical Imaging (ISBI 2017), pp. 986–990. IEEE (2017)
6. Ge, Z., Demyanov, S., Chakravorty, R., Bowling, A., Garnavi, R.: Skin disease recognition using deep saliency features and multimodal learning of dermoscopy and clinical images. In: Descoteaux, M., Maier-Hein, L., Franz, A., Jannin, P., Collins, D.L., Duchesne, S. (eds.) MICCAI 2017. LNCS, vol. 10435, pp. 250–258. Springer, Cham (2017). https://doi.org/10.1007/978-3-319-66179-7_29
7. Gutman, D., et al.: Skin lesion analysis toward melanoma detection: a challenge at the international symposium on biomedical imaging (ISBI) 2016, hosted by the international skin imaging collaboration (ISIC). arXiv preprint arXiv:1605.01397 (2016)
8. IV: Incorporating the knowledge of dermatologists to convolutional neural networks for the diagnosis of skin lesions. arXiv preprint arXiv:1703.09176 (2017)
9. Jianpeng, Z., Yutong, X., Qi, W., Yong, X.: Medical image classification using synergic deep learning. Med. Image Anal. **54**, 10–19 (2019)
10. Kittler, H., Pehamberger, H., Wolff, K., Binder, M.: Diagnostic accuracy of dermoscopy. Lancet Oncol. **3**(3), 159–165 (2002)
11. Matsunaga, K., Hamada, A., Minagawa, A., Koga, H.: Image classification of melanoma, nevus and seborrheic keratosis by deep neural network ensemble. arXiv preprint arXiv:1703.03108 (2017)
12. Menegola, A., Tavares, J., Fornaciali, M., Li, L.T., Avila, S., Valle, E.: RECOD titans at ISIC challenge 2017. arXiv preprint arXiv:1703.04819 (2017)
13. Ronneberger, O., Fischer, P., Brox, T.: U-Net: convolutional networks for biomedical image segmentation. In: Navab, N., Hornegger, J., Wells, W.M., Frangi, A.F. (eds.) MICCAI 2015. LNCS, vol. 9351, pp. 234–241. Springer, Cham (2015). https://doi.org/10.1007/978-3-319-24574-4_28

14. Rogers, H.W., Weinstock, M.A., Feldman, S.R., Coldiron, B.M.: Incidence estimate of nonmelanoma skin cancer (keratinocyte carcinomas) in the US population, 2012. JAMA Dermatol. **151**(10), 1081–1086 (2015)
15. Siegel, R.L., Miller, K.D., Jemal, A.: Cancer statistics, 2015. CA Cancer J. Clin. **65**(1), 5–29 (2015)
16. Silveira, M., et al.: Comparison of segmentation methods for melanoma diagnosis in dermoscopy images. IEEE J. Sel. Top. Signal Process. **3**(1), 35–45 (2009)
17. Zheng, W., Gou, C., Yan, L.: A relation hashing network embedded with prior features for skin lesion classification. In: Suk, H.-I., Liu, M., Yan, P., Lian, C. (eds.) MLMI 2019. LNCS, vol. 11861, pp. 115–123. Springer, Cham (2019). https://doi.org/10.1007/978-3-030-32692-0_14
18. Yang, J., Sun, X., Liang, J., Rosin, P.L.: Clinical skin lesion diagnosis using representations inspired by dermatologist criteria. In: Proceedings of the IEEE Conference on Computer Vision and Pattern Recognition, pp. 1258–1266 (2018)
19. Yoon, C., Hamarneh, G., Garbi, R.: Generalizable feature learning in the presence of data bias and domain class imbalance with application to skin lesion classification. In: Shen, D., et al. (eds.) MICCAI 2019. LNCS, vol. 11767, pp. 365–373. Springer, Cham (2019). https://doi.org/10.1007/978-3-030-32251-9_40
20. Yu, L., Chen, H., Dou, Q., Qin, J., Heng, P.A.: Automated melanoma recognition in dermoscopy images via very deep residual networks. IEEE Trans. Med. Imaging **36**(4), 994–1004 (2017)
21. Yu, Z., et al.: Melanoma recognition in dermoscopy images via aggregated deep convolutional features. IEEE Trans. Biomed. Eng. **66**(4), 1006–1016 (2019)
22. Zhang, J., Xie, Y., Wu, Q., Xia, Y.: Skin lesion classification in dermoscopy images using synergic deep learning. In: Frangi, A.F., Schnabel, J.A., Davatzikos, C., Alberola-López, C., Fichtinger, G. (eds.) MICCAI 2018. LNCS, vol. 11071, pp. 12–20. Springer, Cham (2018). https://doi.org/10.1007/978-3-030-00934-2_2
23. Zhang, J., Xie, Y., Xia, Y., Shen, C.: Attention residual learning for skin lesion classification. IEEE Trans. Med. Imaging **38**(9), 2092–2103 (2019)

Multi-class Skin Lesion Segmentation for Cutaneous T-cell Lymphomas on High-Resolution Clinical Images

Zihao Liu[1,2], Haihao Pan[3], Chen Gong[1], Zejia Fan[1], Yujie Wen[3],
Tingting Jiang[1(✉)], Ruiqin Xiong[1], Hang Li[3], and Yang Wang[3]

[1] NELVT, Department of Computer Science, Peking University, Beijing, China
{lzh19961031,ttjiang}@pku.edu.cn
[2] Advanced Institute of Information Technology, Peking University, Hangzhou, China
[3] Peking University First Hospital, Beijing, China

Abstract. Automated skin lesion segmentation is essential to assist doctors in diagnosis. Most methods focus on lesion segmentation of dermoscopy images, while a few focus on clinical images. Nearly all the existing methods tackle the binary segmentation problem as to distinguish lesion parts from normal skin parts, and are designed for diseases with localized solitary skin lesion. Besides, the characteristics of both the dermoscopy images and the clinical images are four-fold: (1) Only one skin lesion exists in the image. (2) The skin lesion mostly appears in the center of the image. (3) The backgrounds are similar between different images of same modality. (4) The resolution of images isn't high, with an average of about 1500×1200 in several popular datasets. In contrast, this paper focuses on a four-class segmentation task for Cutaneous T-cell lymphomas (CTCL), an extremely aggressive skin disease with three visually similar kinds of lesions. For the first time, we collect a new dataset, which only contains clinical images captured from different body areas of human. The main characteristics of these images differ from all the existing images in four aspects: (1) Multiple skin lesion parts exist in each image. (2) The skin lesion parts are widely scattered in different areas of the image. (3) The background of the images has a large variety. (4) All the images have high resolutions, with an average of 3255×2535. According to the characteristics and difficulties of CTCL, we design a new Multi Knowledge Learning Network (MKLN). The experimental results demonstrate the superiority of our method, which meet the clinical needs.

Keywords: Skin lesion segmentation · Clinical image · Neural network

1 Introduction

Skin disease is one of the most common type of disease in the world, with nearly 5 million new cases estimated every year [15,18]. In recent years, more and more

Electronic supplementary material The online version of this chapter (https://doi.org/10.1007/978-3-030-59725-2_34) contains supplementary material, which is available to authorized users.

A. L. Martel et al. (Eds.): MICCAI 2020, LNCS 12266, pp. 351–361, 2020.
https://doi.org/10.1007/978-3-030-59725-2_34

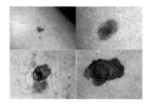 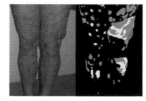 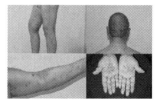

Fig. 1. Existing clinical images. **Fig. 2.** CTCL clinical image. (Color figure online) **Fig. 3.** CTCL dataset.

computer-aided methods are devoted to segmenting the lesion areas in order to assist doctors in the diagnosis process. Normally, there're two modalities of imaging: clinical images and dermoscopy images. Clinical images are captured by standard digital camera and can provide a global representation in view, angle and lighting, as for a better understanding and observation of the disease. In contrast, dermoscopy images are obtained by a microscope using incident light and oil immersion to make subsurface structures of a small skin region accessible, which permit a more detailed inspection of skin lesions. Most of the existing segmentation methods are designed for dermoscopy images [2,4,5,9,11,20,21]. In this work, we focus on clinical image segmentation.

There are also a few methods for clinical image segmentation [1,6,7,10,17]. The authors in [7] propose a texture-based method to learn the representative texture distributions, as for a better localization of lesion area. For deep-learning based methods, a FCN-based method is introduced in [17], with an extra critic module for better predictions of boundaries. To tackle the disturbing factors such as illumination variations from skin surface, the method in [1] proposes a zoom-out window to capture local and global information for accurate extraction of lesion regions. However, nearly all the above existing methods are designed for the binary segmentation task to distinguish between lesion area and skin area. Besides, the datasets mainly contain diseases with localized solitary skin lesions, which means only one lesion area appeared and is usually located in the center of the images.

In this work, we aim at a new four-class segmentation task for a supremely aggressive skin disease called Cutaneous T-cell lymphomas (CTCL), on clinical images with totally different characteristics from existing ones. CTCL is a severe group of extranodal non-Hodgkin lymphomas [8,19], which is a skin cancer with an increasing incidence rate in recent years. Survival declines dramatically as the disease progresses [16]. The unique manifestation of CTCL is three types of lesions on the skin: patches, plaques and tumors, which are increasing in severity and change at the disease progress [3,14]. Thus, for auxiliary diagnosis, four-class segmentation is needed. Meanwhile, an important clinical application called "mSWAT Score Evaluation" is needed for this disease, which is implemented by the modified Severity Weighted Assessment Tool to calculate the percentage of the area of each skin lesion type over the area of whole body to track the clinical response for the treatments [13]. However, in clinical work, due to irregular skin lesions, it's difficult for doctors to make accurate and repeatable estimates during

Table 1. Comparison between the existing skin lesion segmentation task and our task.

Attribute	Previous problem	Our problem
Number of classes	2 (Only lesion and normal skin)	4 (Three visually similar lesions)
Disease	With localized solitary skin lesion	With generalized skin lesion
Image Modality	Mostly dermoscopy, few clinical	Clinical
Characteristics of images	(1) Only one skin lesion part (2) Lesion part appears in the center (3) Similar background (4) Low to medium resolution	(1) Multiple skin lesion parts (2) Lesion parts are widely scattered (3) Large variety of background (4) High resolution

calculations. Thus, involving computers to automatically calculate the area will be very beneficial to improve the accuracy. For a better area evaluation result, segmenting all the lesion parts correctly first by the computer is important. So this four-class segmentation task is significant and the foundation of the automated mSWAT Score Evaluation.

However, there're several difficulties and uniqueness of this task, comparing to previous tasks. Firstly, the visual difference within different lesion parts or between one lesion part and a normal skin part could all be very similar. Next, multiple and connected skin lesions could appear throughout the body. Thus, for this disease, clinical, instead of dermoscopy images, need to be captured from different parts of body surface area. Besides, high-resolution images are needed for accurate estimation of lesion areas. Figure 1 contains examples of existing clinical images. Figure 2 contains an example of CTCL image and an annotation map of all the lesion areas. In the map, yellow, blue and red denote patch, plaque and tumors, respectively. As we can see, the CTCL image is different from existing clinical images no matter from the characteristics and number of lesion types. Besides, the three types of lesions could concurrently appear in a single lesion area and are visually similar. Table 1 is the comparison between our task and the previous task. To sum up, the main uniqueness and challenges for this task are: (1) It is a four-class segmentation task and the visual difference between different classes could be very similar, no matter between different lesion types within a lesion area or between lesion parts and normal skin parts. (2) The clinical images all have very high resolution and have distributed lesion parts existing in different parts, making the task more difficult. (3) The clinical images suffer from a large variety of backgrounds since they're captured in different body areas. However, to the best of our knowledge, there's no any previous work to solve the segmentation task for CTCL on clinical images. Therefore, this task is challenging yet has a great clinical application and research value.

In order to solve the above problems, for the first time, we collect a dataset, which contains 57 clinical images for CTCL, with full annotation for each lesion area and the corresponding lesion type in every image. Then, we design a novel two-branch Multi Knowledge Learning Network (MKLN), with a Lesion Area Learning Module for each branch and a Feature Co-Learning Module. One

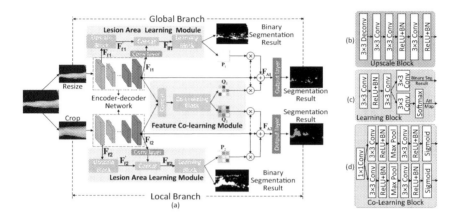

Fig. 4. The overall architecture of the MKLN is illustrated in (a). (b), (c) are the architecture of the Upscale Block and the Learning Block in Lesion Area Learning Module, (d) is the architecture of the Co-learning Block in Feature Co-learning Module.

branch resizes the image to obtain global information, called "global branch"; the other branch crops the image to obtain detailed local information, called "local branch". The Lesion Area Learning Module is equipped by each branch to tackle the two challenges: large background variety and visual similarity between lesion part and normal skin part. It utilizes multi-level feature to perform binary segmentation and provide an attention map, as for a better learning of lesion edge knowledge. The Feature Co-Learning Module is designed to deal with the other challenges, including high resolution and visual similarity between different lesion types within a lesion area. It takes the output features of the two branches, learns an attention map for each branch, and enables them to mutually facilitate and learn knowledge from each other. Our method effectively addresses the difficulties of CTCL and achieves good experimental results on our dataset, meeting clinical needs.

Our contributions are two-fold: (1) The first paper to focus on the segmentation task for CTCL and a new dataset with clinical images is collected; (2) A new method is designed based on the characteristics of CTCL and achieves state-of-the art results, meeting clinical needs.

2   Dataset

We collaborate with the hospital and collect 57 clinical images, which can be called "CTCL dataset". The images are captured by the Nikon D800 Camera. Three experienced dermatologists participated in annotating. Each lesion area and the corresponding lesion type are annotated in every image. Several examples of our dataset are showed in Fig. 3. The collected clinical images are sampled from patients with different sex and wide age range, which is from 30 to 60 years old. The images include totally 12 human body regions, containing head, neck, anterior trunk, arms, forearms, hands, posterior trunk, buttocks, thighs, legs, feet and groin. Thus the background of the images suffer from a large variety. Resolution of the images is very high, with an average of 3255×2535. The largest resolution is 6000×2921. More examples of the CTCL dataset can be found in supplementary material.

3   Method

The whole architecture of MKLN is illustrated in Fig. 4, which is a two-branch network and is primarily based on an encoder-decoder network [12] with two other modules. The global branch takes the resized image as the input and the local branch takes the randomly cropped image, which has the same size as resized image, as the input. The first module is Lesion Area Learning Module, which is equipped by each branch. It takes the feature from the first encoding block and the last decoding block as the input, generates an attention map containing the lesion edge feature and obtains a binary segmentation result which distinguishes between lesion areas and non-lesion areas. To enable the network to take the advantage of both the global and local branch, the Feature Co-learning Module takes the output feature of last decoding block from each branch and generates an attention map for each branch, containing the learned knowledge from the other branch.

Lesion Area Learning Module. As stated in Sect. 1, the background of CTCL clinical images has a large diversity since the images are captured from different body areas. Besides, the visual difference between lesion areas and normal skin areas is similar, for example between patches and normal skin in Fig. 2. Meanwhile, the lesion parts could also look similar to the background. Thus, adding additional constraint and supervision between lesion parts and other parts is really appealing and compelling for a better result. Therefore, we designed this module as an extra regularization for each branch.

For a better distinction between skin lesions and non-skin lesions, which is a binary segmentation task, edge and texture information are required. The low-level feature contains rich edge and texture information, meanwhile the high-level feature contains rich semantic information, which are all helpful for the binary segmentation. Thus, the output features of the first encoding block and the last decoding block are taken as input, denoted as F_f and F_l respectively. F_f goes through an upscale block and the upscaled feature F'_f will be obtained. Meanwhile, F_l will go through two 3×3 conv layers and concatenated with F'_f, denoted as F_{lf}. A learning block is applied to F_{lf}, which will output the binary segmentation result and an attention map P with rich edge and boundary features. The architecture of the upscale block and learning block is illustrated in Fig. 4(b) and Fig. 4(c) respectively. The prediction result of the binary segmentation is supervised through binary cross-entropy loss, as illustrated in Eq. (1):

$$L_{bce} = -\sum_i (y_i log(\hat{y}_i) + (1 - y_i)log(1 - \hat{y}_i)) \tag{1}$$

where y_i denotes the groundtruth of each pixel and $\hat{y}_i$ is the predicted results. Meanwhile, an element-wise multiplication between P and F_l is applied. The result will be multiplied by a learnable factor α and then applied a matrix sum calculation to F_l.

With the extra supervision, the learned lesion area knowledge is better able to guide the extraction of discriminative features in high-level layers to learn a better edge and texture representation and knowledge for each lesion area.

Feature Co-learning Module The global branch resizes the image, and captures more global information and comprehensive semantic information and generates a better overall segmentation result. But the loss of detailed local and texture information leads to inaccurate classification between different classes of lesions, especially when different lesion types appear in a same lesion part. In contrast, the local branch crops the image, and captures more detailed local information, including boundary and texture information. Within each lesion area, the classification between different lesion types performs better. But due to the lack of global semantic information, the performance is also limited. However, since the resolution of images in our dataset is too high, with an average of 3255×2535, neither of the single branch performs good. Therefore, we propose to combine the information of two branches and design this module as a connection between the two branches.

The module takes F_{l1} and F_{l2} as the input, which contains high-level semantic information, and concatenates them together firstly. Then, the concatenated features will go through a co-learning block, the architecture of which is illustrated in Fig. 4(d). Finally, an attention map Q for each branch will be generated. For the global branch, the generated attention map is expected to contain more detailed local information from the local branch for complementary. Similarly for the local branch, the generated attention map is expected to contain more global semantic knowledge for facilitation. Finally, an element-wise multiplication between Q and F_l is performed. The result will be multiplied by a learnable factor β and then applied a summation to F_l. The final output feature of these two modules is obtained by Eq. (2):

$$F_A = \alpha \cdot (P \cdot F_l) + \beta \cdot (Q \cdot F_l) + F_l \tag{2}$$

where "$\cdot$" means element-wise multiplication.

After that, the output feature of each branch passes a 3×3 convolutional output layer to obtain the four-class segmentation result, which is supervised by cross-entropy loss L_{ce}, as illustrated in Eq. (3):

$$L_{ce} = -\sum_i \sum_j (y_{ij} log(\hat{y}_{ij})) \tag{3}$$

where y_{ij} denotes whether pixel i belongs to label j and $\hat{y}_{ij}$ means the classification score of class j for pixel i.

This module automatically mines and exchanges the knowledge that each branch has learned, enabling them to facilitate each other. To the best of our knowledge, we are the first to take advantage of both of the two strategies and enable them to facilitate each other in the skin lesion segmentation task. To be mentioned, our method is totally different from [1] in the perspective and definition of "global" and "local", which results to a different design of the network architecture. Method in [1] just uses a slightly bigger window with the same center of local patch as global structure which still ignores the general information of the full image, and ignores to utilize the two branches to facilitate each other. But we successfully capture global information of the whole image, and directly exchange the information and knowledge from two branches to benefit each other.

A hybrid loss function is used as the final loss of each branch, which is weighted sum of two cross-entropy loss. It's illustrated in Eq. (4):

$$L = L_{bce} + \lambda L_{ce} \tag{4}$$

where λ is a hyper-parameter to linearly combine two losses in order to balance them.

4   Experiments

4.1   Implementation Details

The proposed method is evaluated on the newly introduced clinical dataset. Five-fold cross-validation is used for our experiments. Each image is resized and randomly cropped to 608×608. At the training phase, the batch size is set to 2 on two NVIDIA GTX 1080Ti GPUs. At the testing phase, after each branch gets the four-class segmentation result for the whole image, the average probabilistic score of the two branches is calculated as the final segmentation result. The learning rate is initialized to 0.001 and decayed by 0.1 every 20 epochs. In Eq. (4), λ is set to 1. We think both losses are equally important. The learnable factor α and β are trained together with the network. Four evaluation metrics are used in the experiments, including Accuracy (ACC), Dice Similarity Coefficient (DSC), Sensitivity (SE) and Specificity (SP).

4.2   Ablation Study

We conduct the ablation study on CTCL dataset to investigate the impact of the proposed two modules of our method on the performance. The specific performance is listed in Table 2. In the table, "**R**" and "**C**" mean the resize and crop strategy for the experiments respectively. "**L**" denotes Lesion Area Learning Module and "**F**" represents Feature Co-learning Module. We first conduct experiments using baseline model [12] for each branch, denoted as "**MKLN (R)**" and "**MKLN (C)**" in the first two columns. Then we add **Lesion Area Learning Module** to each branch, denoted as "**MKLN (RL)**" and

"**MKLN (CL)**". The performance of all the four metrics improves. Compared
to the baseline, **MKLN (RL)** yields an DSC of 0.695 and **MKLN (CL)** yields
an DSC of 0.636, which are 4.1% and 1.7% higher than the baseline respectively.
The Accuracy is improved by 3.1% and 2.1% respectively, which all verify the
effectiveness and necessity for the Lesion Area Learning Module.

After that, **Feature Co-learning Module** is adopted to the network, the
performance of two branch is denoted as "**MKLN (RLF)**" and "**MKLN
(CLF)**". The performance of **MKLN (RLF)** is better than **MKLN (R)** and
MKLN (RL). The DSC is further increased to 0.769, with a gain of 11.7% com-
paring to **MKLN (R)**, and 7.4%, comparing to **MKLN (RL)**. Compared to
MKLN (C) and **MKLN (CL)**, the DSC improves 5.3%and 3.6% respectively.
Besides, results of the other three metrics all have a growth. The improvement
of adopting Feature Co-learning Module is larger than the Lesion Area Learning
Module, which further proves the effectiveness and advancement of this module,
since no other methods have been utilized the knowledge from two branches.
Finally, a weighted sum is applied to combine the probabilistic score of **MKLN
(RLF)** and **MKLN (CLF)** for a final segmentation result, which is denoted as
"**MKLN (Full)**". The performance of the full model on all the four metrics is
the best, further demonstrates the effectiveness of introducing two branch and
combining both of their knowledge. Besides, global branch has a better performance than local branch, we surmise it happens because the cropped image loses
too much global semantic information, due to the high resolution of the whole
image. To sum up, the encouraging results show the advantage of our method.

4.3   Comparison with Other Methods

Since no any other related method are designed for this problem, we choose
the most related work [17] and [1] to compare the results. Both [17] and [1]
have similarities to our method in perspective of solving the segmentation task.
As introduced in Sect. 1, they are both deep-learning based method on clinical
images. Besides, [17] produces a more accurate results on lesion boundary by
introducing a critic network, and [1] also utilizes global and local information.
The source codes for these two methods are not publicly available, so we implemented them based on our best understanding and obtained the results by same
five-fold cross-validation setting.

Table 2. Four-class segmentation performance by baseline methods, our method and
other methods for comparison on the CTCL dataset.

Method	MKLN (R)	MKLN (C)	MKLN (RL)	MKLN (CL)	MKLN (RLF)	MKLN (CLF)	MKLN (Full)	Izadi et al. [17]	Jafari et al. [1]
ACC	0.643	0.656	0.678	0.667	0.715	0.683	**0.725**	0.663	0.652
DSC	0.652	0.619	0.695	0.636	0.769	0.672	**0.801**	0.716	0.709
SE	0.624	0.701	0.721	0.761	0.881	0.778	**0.884**	0.844	0.825
SP	0.593	0.598	0.645	0.640	0.736	0.707	**0.776**	0.680	0.673

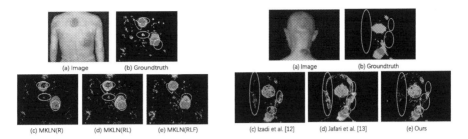

Fig. 5. Visualization results of the two modules. (Color figure online)

Fig. 6. Visualization results of three methods. (Color figure online)

The results of our method, [17] and [1] are shown in the last three columns of Table 2. The performance on ACC, DSC, SE and SP is evaluated. Our method achieve the best result in all the metrics, with an DSC of **0.801**, which exceeds [17] by 8.5% and [1] by 9.2%. Our ACC, SE and SP all perform the best.

4.4   Visualization Results

In addition to quantitative results, qualitative segmentation results are also provided. Figure 5 demonstrates the effectiveness of the two modules. The comparison illustrated by yellow circles proves that with the lesion area learning module, the details and lesion boundaries and edges are clearer. The cyan circles areas demonstrate some misclassified lesion parts are corrected classified, resulting to a better global segmentation result. The results comparison between our method and the other two methods are shown in Fig. 6. The encircled areas by yellow circles have a better boundary and edge classification result. The areas encircled by cyan circle achieve a better classification result of lesion parts. So that our results are more close to the groundtruth. Figure 7 visualizes the attention maps of the two modules. Compared to the activation of original feature map, the attention map of **L** successfully focuses on the edges of lesion parts, and the attention map of **F** further focuses on the detailed lesion parts.

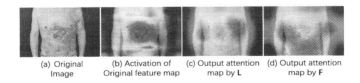

(a) Original Image   (b) Activation of Original feature map   (c) Output attention map by **L**   (d) Output attention map by **F**

Fig. 7. Visualization of attention maps of **L** and **F**.

5 Conclusion

In this paper, we focus on a new four-class segmentation problem for Cutaneous T-cell lymphomas (CTCL), an extremely severe skin disease with three types of visually similar and decentralized skin lesions. For the first time, we collect a new dataset which contains clinical images with high resolutions from different body areas. Then we propose a novel Multi Knowledge-Learning Network (MKLN), including a Lesion Area Learning Module and a Feature Co-Learning Module to address this problem, which achieves very good performance and meets the clinical needs.

Acknowledgement. This work was partially supported by the Natural Science Foundation of China under contracts 61572042, 61772041, 81922058 and National Key R&D Program of China 2019YFC0840700. We also acknowledge the Clinical Medicine Plus X-Young Scholars Project, and High-Performance Computing Platform of Peking University for providing computational resources.

References

1. Jafari, M.H., Karimi, N., Nasr-Esfahani, E., et al.: Skin lesion segmentation in clinical images using deep learning. In: International Conference on Pattern Recognition, pp. 337–342 (2016)
2. Patiño, D., Avendaño, J., Branch, J.W.: Automatic skin lesion segmentation on dermoscopic images by the means of superpixel merging. In: Frangi, A.F., Schnabel, J.A., Davatzikos, C., Alberola-López, C., Fichtinger, G. (eds.) MICCAI 2018. LNCS, vol. 11073, pp. 728–736. Springer, Cham (2018). https://doi.org/10.1007/978-3-030-00937-3_83
3. Girardi, M., Heald, P.W., Wilson, L.D.: The pathogenesis of mycosis fungoides. New Engl. J. Med. **350**(19), 1978–1988 (2004)
4. Wang, H., Wang, G., Sheng, Z., Zhang, S.: Automated segmentation of skin lesion based on pyramid attention network. In: Suk, H.-I., Liu, M., Yan, P., Lian, C. (eds.) MLMI 2019. LNCS, vol. 11861, pp. 435–443. Springer, Cham (2019). https://doi.org/10.1007/978-3-030-32692-0_50
5. Huan, W., Guotai, W., Zhihan, X., Wenhui, L., Shaoting, Z.: Star shape prior in fully convolutional networks for skin lesion segmentation. In: International Workshop on Machine Learning in Medical Imaging, pp. 611–619 (2019)
6. Filali, I., Belkadi, M.: Multi-scale contrast based skin lesion segmentation in digital images. Optik **185**, 794–811 (2019)
7. Glaister, J., Wong, A., Clausi, D.A.: A segmentation of skin lesions from digital images using joint statistical texture distinctiveness. Pattern Recogn. **61**(4), 1220–1230 (2014)
8. Korgavkar, K., Xiong, M., Weinstock, M.: Changing incidence trends of cutaneous t-cell lymphoma. JAMA Dermatol. **149**(11), 1295–1299 (2013)
9. Song, L., Lin, J., Wang, Z.J., Wang, H.: Dense-residual attention network for skin lesion segmentation. In: Suk, H.-I., Liu, M., Yan, P., Lian, C. (eds.) MLMI 2019. LNCS, vol. 11861, pp. 319–327. Springer, Cham (2019). https://doi.org/10.1007/978-3-030-32692-0_37

10. Zortea, M., Flores, E., Scharcanski, J.: A simple weighted thresholding method for the segmentation of pigmented skin lesions in macroscopic images. Pattern Recogn. **64**, 92–104 (2017)
11. Sarker, M.M.K., et al.: SLSDeep: skin lesion segmentation based on dilated residual and pyramid pooling networks. In: Frangi, A.F., Schnabel, J.A., Davatzikos, C., Alberola-López, C., Fichtinger, G. (eds.) MICCAI 2018. LNCS, vol. 11071, pp. 21–29. Springer, Cham (2018). https://doi.org/10.1007/978-3-030-00934-2_3
12. Ronneberger, O., Fischer, P., Brox, T.: U-Net: convolutional networks for biomedical image segmentation. In: Navab, N., Hornegger, J., Wells, W.M., Frangi, A.F. (eds.) MICCAI 2015. LNCS, vol. 9351, pp. 234–241. Springer, Cham (2015). https://doi.org/10.1007/978-3-319-24574-4_28
13. Olsen, E., Whittaker, S., Kim, Y., et al.: Clinical end points and response criteria in mycosis fungoides and sézary syndrome: a consensus statement of the international society for cutaneous lymphomas, the united states cutaneous lymphoma consortium, and the cutaneous lymphoma task force of the European organisation for research and treatment of cancer. J. Clin. Oncol. **29**(18), 2598 (2011)
14. Pulitzer, M.: Cutaneous t-cell lymphoma. Clin. Lab. Med. **37**(3), 527–546 (2017)
15. Rogers, H.W., Weinstock, M.A., Feldman, S.R., Coldiron, B.M.: Incidence estimate of nonmelanoma skin cancer (keratinocyte carcinomas) in the US population, 2012. JAMA Dermatol. **151**(10), 1081–1086 (2015)
16. Wilcox, R.A.: Cutaneous t-cell lymphoma: 2017 update on diagnosis, risk-stratification, and management. Am. J. Hematol. **92**(10), 1085–1102 (2017)
17. Izadi, S., Mirikharaji, Z., Kawahara, J., Hamarneh, G.: Generative adversarial networks to segment skin lesions. In: IEEE 15th International Symposium on Biomedical Imaging, pp. 881–884 (2018)
18. Siegel, R.L., Miller, K.D., Jemal, A.: Cancer statistics, 2015. CA Cancer J. Clin. **65**(1), 5–29 (2015)
19. Willemze, R., Jaffe, E.S., Burg, G., et al.: WHO-EORTC classification for cutaneous lymphomas. Blood **105**(10), 3768–3785 (2005)
20. Li, X., Yu, L., Fu, C.-W., Heng, P.-A.: Deeply supervised rotation equivariant network for lesion segmentation in dermoscopy images. In: Stoyanov, D., et al. (eds.) CARE/CLIP/OR 2.0/ISIC -2018. LNCS, vol. 11041, pp. 235–243. Springer, Cham (2018). https://doi.org/10.1007/978-3-030-01201-4_25
21. Mirikharaji, Z., Hamarneh, G.: Star shape prior in fully convolutional networks for skin lesion segmentation. In: Frangi, A.F., Schnabel, J.A., Davatzikos, C., Alberola-López, C., Fichtinger, G. (eds.) MICCAI 2018. LNCS, vol. 11073, pp. 737–745. Springer, Cham (2018). https://doi.org/10.1007/978-3-030-00937-3_84

Fetal Imaging

Deep Learning Automatic Fetal Structures Segmentation in MRI Scans with Few Annotated Datasets

Gal Dudovitch[1,5], Daphna Link-Sourani[2,5], Liat Ben Sira[3,4,5], Elka Miller[3,5], Dafna Ben Bashat[2,4,5], and Leo Joskowicz[1,5(✉)]

[1] School of Computer Science and Engineering, The Hebrew University of Jerusalem, Jerusalem, Israel
josko@cs.huji.ac.il
[2] Sagol Brain Institute, Tel Aviv Sourasky Medical Center, Tel Aviv-Yafo, Israel
[3] Division of Pediatric Radiology, Tel Aviv Sourasky Medical Center, Tel Aviv-Yafo, Israel
[4] Sackler Faculty of Medicine and Sagol School of Neuroscience, Tel Aviv University, Tel Aviv-Yafo, Israel
[5] Medical Imaging, Children's Hospital of Eastern Ontario, University of Ottawa, Ottawa, Canada

Abstract. We present a new method for end-to-end automatic volumetric segmentation of fetal structures in MRI scans with deep learning networks trained with very few annotated scans. It consists of three main stages: 1) two-step automatic structure segmentation with custom 3D U-Nets; 2) segmentation error estimation, and; 3) segmentation error correction. The automatic structure segmentation stage first computes a region of interest (ROI) on a downscaled scan and then computes a final segmentation on the cropped ROI. The segmentation error estimation stage uses prediction-time augmentations of the input scan to compute multiple segmentations and estimate the segmentation uncertainty for individual slices and for the entire scan. The segmentation error correction stage then uses these estimations to locate the most error-prone slices and to correct the segmentations in those slices based on validated adjacent slices. Experimental results of our methods on fetal body (63 cases, 9 for training, 55 for testing) and fetal brain MRI scans (35 cases, 6 for training, 29 for testing) yield a mean Dice coefficient of 0.96 for both, and a mean Average Symmetric Surface Distance of 0.74 mm and 0.19 mm, respectively, below the observer delineation variability.

Keywords: Deep learning · Fetal MRI · Segmentation · Uncertainty estimation

1 Introduction

Accurate segmentation of complex structures and pathologies in volumetric images presents a great challenge in medical image processing. Recent deep learning image

Electronic supplementary material The online version of this chapter (https://doi.org/10.1007/978-3-030-59725-2_35) contains supplementary material, which is available to authorized users.

A. L. Martel et al. (Eds.): MICCAI 2020, LNCS 12266, pp. 365–374, 2020.
https://doi.org/10.1007/978-3-030-59725-2_35

classification methods have been shown to be effective for the segmentation of a variety of structures and pathologies in CT and MRI scans [1]. State-of-the-art segmentation methods for medical images are mostly based on Convolutional Neural Networks (CNN) and their variants. These include the 2D U-Net autoencoder convolution/ deconvolution architecture with skip connections [2] and its extensions to 3D [3–5]. These networks have been demonstrated in the segmentation of complex structures, e.g. adult brain and prostate in MRI scans [5, 6] and fetal structures in MRI scans [7]. The NiftyNet platform has recently been developed for deep learning segmentation [8]. Other works rely on a developing fetal brain atlas for segmentation [9]. Its drawback is that it requires a large set of fetal brain scans to create the atlas. Fetit et al. [10] uses an algorithm for the initialization of the fetal brain segmentation, which limits its generality.

The networks performance critically depend on large, high-quality annotated data, which is seldom available, if at all. Research groups must generate their own datasets for each anatomical structure, pathology and scanning protocol. This is an expensive and time-consuming task that requires significant effort and radiological expertise. This has motivated the development of interactive machine learning [11, 12] and deep learning segmentation methods [13, 14], whose aim is to reduce the amount and complexity of the user interactions required for the necessary manual error corrections. While these methods may help to reduce user interactions, they do not yet significantly reduce the user effort and the required number of annotated training datasets.

Three key and closely related issues to the automatic segmentation of structures and their subsequent validation and correction are segmentation variability, robustness, and uncertainty. Segmentation variability and robustness estimation has been researched for deep learning classification [15, 16]. For example, Monte Carlo Dropout is a regularization technique in which random selections of active neurons is used for approximate Bayesian inference. Segmentation uncertainty estimation has been performed with an ensemble of multiple models [15]. However, this method requires a large annotated datasets to train the models. Very recently, segmentation uncertainty estimation methods based on test-time augmentation have been proposed [17, 18]. However, segmentation uncertainty estimates have not been used to prioritize manual segmentation correction and to optimize the selection of scans for manual annotation to increase the model accuracy and robustness with a small training set.

In this paper, we present an end-to-end method for volumetric segmentation of fetal structures in MRI scans with deep learning networks trained with very few annotated scans. The method relies on segmentation error estimation and correction using segmentation uncertainty measures. It increases the segmentation accuracy and robustness and optimizes the total radiologists' annotation time required for creating a dataset with validated annotations, thereby bootstrapping the automatic segmentation task with very few annotated datasets.

2   Method

Our end-to-end method consists of 3 stages: 1) automatic structure segmentation; 2) segmentation error estimation; 3) segmentation error correction. The automatic structure segmentation stage computes first a region of interest (ROI) and then computes the structure segmentation inside the ROI. The segmentation error estimation stage uses prediction-time augmentations of the input scan to compute multiple segmentations and to estimate segmentation uncertainty and error margins for individual slices and for the entire scan. The identified estimated segmentation errors are then used to prioritize the slices that require inspection and manual correction of the faulty segmentations.

The segmentation error correction stage uses individual slices corrections to automatically correct the segmentations in adjacent slices.

Custom 3D U-Net Architecture. We have developed a custom 3D U-Net architecture based on [3] and [5] for ROI localization and structure segmentation. The U-Net is an encoder/decoder architecture with residual connections whose encoding/decoding pathways classify voxels based on image patches features at different levels of abstraction. Our modifications to the standard 3D U-Net are (Fig. 1):

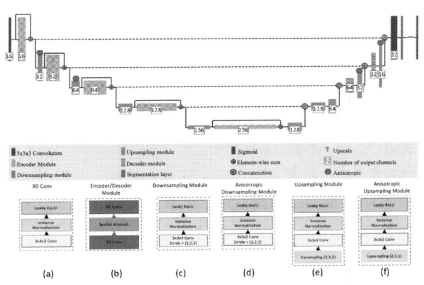

Fig. 1. Top: Architecture of our custom 3D segmentation network based on [3, 5]. The number of output channels of each unit is indicated next to it. Bottom: (a–f) network modules details.

1. Each 3D convolution layer is preceded by a leaky ReLU non-linearity activation followed by an instance normalization layer [19] (Fig. 1a). It replaces standard batch normalization and ensures classification stability for small batch sizes.
2. Each encoder layer is replaced by a residual module [20] with two 3D convolutional layers and a spatial dropout layer between them (Fig. 1b). Spatial dropout layers have been reported to yield superior results in fully convolutional networks [21].
3. Encoder modules are connected by downsampling modules (Fig. 1c) built from 3 × 3 × 3 convolutions with an input stride of 2 to reduce the feature map resolution and to incorporate additional features in the encoding pathway.

4. Up-sample modules in the decoder pathway (Fig. 1e) up-sample the low resolution feature maps with a direct upscale that repeats the feature voxels twice in each spatial dimension, followed by a $3 \times 3 \times 3$ convolution that halves the number of feature maps.
5. The up-sampled features are then recombined by concatenation with the features from the corresponding level of the encoding pathway.
6. Decoder modules (Fig. 1b) recombine the features after concatenation and reduce by half the number of features maps.

A key modification of the network is anisotropic downsampling and upsampling (Fig. 1d, 1f) [22]. Scans are usually anisotropic, e.g., the slices spacing is greater than the slice pixel resolution, resulting in a mismatch between the receptive field and the scan dimensions. Anisotropic sampling enforces this match and thus increases accuracy. The anisotropic downsampling layer performs downsampling on the slices xy plane without downsampling along the z axis. This is implemented by setting the convolution stride to 2 in the xy plane and to 1 along the z axis. The downsampling reduction proceeds anisotopically until the spatial layer dimensions are equal; it then proceeds to the next layers isotropically. The decoder pathway has matching upsampling layers.

Automatic Structure Segmentation. The two-stage segmentation method consists of ROI localization followed by structure segmentation inside the ROI. Both are performed with the custom 3D U-Net described above. Each network is trained as a supervised deep learning model with ground-truth segmentations of target structure.

The ROI localization network inputs a downscaled scan and outputs a coarse segmentation with which the ROI bounding box is computed. The structure segmentation network inputs the full resolution cropped ROI scan and outputs the structure segmentations. The networks are trained with the Dice loss function from [4]. The resulting segmentation is post-processed with Gaussian filter smoothing, connected component analysis, and binarization with a preset intensity threshold.

Spatial and intensity augmentations are used to increase the data size and variety for segmentation network training, for prediction-time augmentation, and for segmentation uncertainty estimation (Fig. 2, left). Intensity augmentations include contrast, blur by Gaussian filtering, addition of Poisson noise and additive and multiplicative Gaussian noise, and coarse dropout. Spatial augmentations include slice-wise affine and elastic deformations with a smoothed displacement field. Prediction-time augmentations yield multiple segmentations that are aligned and combined by averaging or majority voting.

Segmentation Uncertainty and Error Estimation. Neural networks trained with few annotated datasets inevitably produce segmentation errors and perform poorly on out-of-distribution inputs. In these cases, the segmentation errors should be identified and manually corrected by an expert. Currently, the manual segmentation corrections are performed by examining scans and scan slices in sequential order, which is not necessarily optimal. By estimating segmentation uncertainty and detecting possible segmentation errors, the manual segmentation correction process can be optimized to reduce radiologist time and effort. We propose to use prediction-time augmentations to estimate slice-wise and scan-wise segmentation uncertainty and errors.

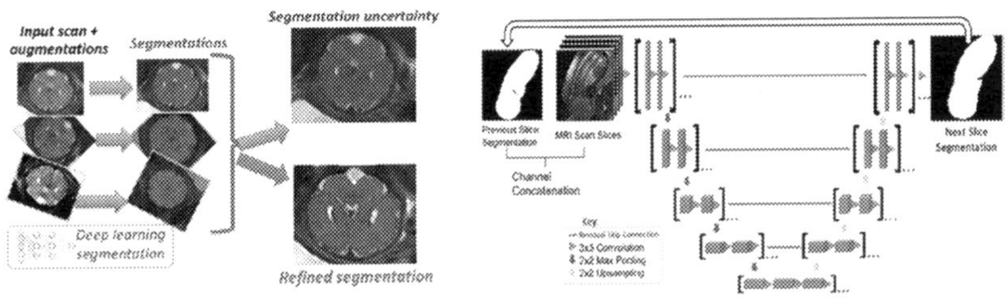

Fig. 2. Left: Prediction-time augmentations yield multiple segmentations (red). The segmentation uncertainty (blue) is then computed per voxel with the entropy of the multiple predictions (top). The final segmentation (bottom) is computed by averaging the aligned predictions. Right: slice segmentation error correction 2D U-Net architecture. The inputs are a slice, two adjacent slices above/below it, and a validated previous slice segmentation; it outputs the corrected slice segmentation. Subsequent slices can be iteratively corrected (top arrow). (Color figure online)

Slice-wise segmentation uncertainty is estimated by computing the sum of the segmentation uncertainty of each voxel v in the slice defined by the predictions binary entropy: $uncertainty(v) = P(v)\log(P(v)) + (1 - P(v))\log(1 - P(v))$ where $P(v)$ is the predicted probability of voxel v to belong to the target structure as computed by the prediction-time augmentations. The larger the voxel entropy, the higher the uncertainty value for the voxel segmentation prediction (Fig. 2 left). We use the segmentation uncertainty to distinguish between segmentation variability and segmentation error [25]. Segmentation variability is the acceptable deviation from the ground-truth: it should be removed from the segmentation uncertainty to obtain the segmentation error. We use the morphological opening operator to remove the segmentation variability around the mean segmentation contour. The estimated error value of a slice is computed by the sum of the voxels' uncertainty after filtering out the segmentation variability.

Scan-wise segmentation uncertainty measures the deviation of the individual prediction-time augmentation segmentations S_i from the mean segmentation $\bar{S}$ with an uncertainty function $u(S_i, \bar{S})$ that measures the distance of the predicted segmentation to the mean segmentation in the entire scan. Given N segmentations from the augmented scans, the mean segmentation $\bar{S} = \frac{1}{N}\sum_i^N S_i$ is computed first. Then, the uncertainty measure $u(S_i, \bar{S})$ of every segmentation S_i from the mean segmentation $\bar{S}$ is computed. Finally, the overall median uncertainty value $\bar{u} = median_i(\{u(S_i, \bar{S})\})$ is computed. The uncertainty function is computed with standard measures, e.g., Intersection-over-Union (IoU), Dice coefficient and Average Symmetric Surface Distance (ASSD).

We use the resulting value as an estimation of the network uncertainty about its segmentation and to detect possible segmentation errors. The intuition is that the confidence of a network prediction is high when it yields small variations on the segmentations resulting from the perturbations induced by the augmentation, and small otherwise. Note that the goal is compute a measure that is well correlated with the actual segmentation error and not to accurately estimate the actual segmentation error value.

Segmentation Error Correction. Slice-wise segmentation errors are corrected by using the previous slice radiologist' validated/corrected segmentation to automatically correct segmentation errors in a slice. Our method uses a 2D U-Net (Fig. 2, right) [2];

it inputs a slice, a validated previous slice segmentation, and four adjacent scan slices (two below and two above the slice); it outputs the slice's corrected structure segmentation. The corrected segmentation can then be used to correct the subsequent slices in an iterative automatic segmentation error correction process. The order in which slices are corrected can be prioritized by the estimated slice-wise segmentation error value with the largest values shown first.

3 Experimental Results

For the experimental studies, we collected two datasets of fetal brain and fetal body MRI scans from the Sourasky Medical Center acquired as part of the routine clinical fetal assessment. The fetal body dataset consists of 64 fetal body MRI coronal scans acquired on a 1.5T GE Signa Horizon Echo speed LX MRI scanner using a torso coil with the volumetric FIESTA protocol. Each scan has 50–100 slices, 256×256 pixels per slice, with resolution of $1.56 \times 1.56 \times 3.0$ mm^3. The fetal brain dataset consists of 42 fetal brain MRI coronal scans acquired on a 3T Siemens Skyra MRI scanner using a torso coil with the 3D fast imaging TrueFISP sequence. Each scan consists of 20–40 slices, 512×512 pixels per slice, with resolution of $0.74 \times 0.74 \times 3.0$–$5.0$ mm^3.

Expert-validated ground-truth fetal body and fetal brain segmentations were created for all scans as follows. Manual segmentations of the fetal body and the fetal brain were created by two expert radiologists for 13 and 8 scans (each scan requires on average 74 and 55 min to annotate). Validated segmentations for the remaining 46 and 34 scans were created by an expert radiologist by correcting the segmentations produced by the automatic segmentation method. The original and the corrected segmentations are used in Study 2 below. To quantify the manual segmentation variability, two annotators with expertise in fetal MRI performed manual segmentations: for the fetal body, 21 scans (1,741 slices) were segmented by one annotator and 10 scans were each segmented twice by both annotators; for the fetal brain, 3 scans (97 slices) were delineated twice by both annotators. Table 1 (rows 1, 2) lists the delineation observer variability results.

We conducted two studies to evaluate our methods. The results are reported for the fetal body and fetal brain on training sets of 9 and 6 scans and test sets of 55 and 29 scans, respectively. The segmentations quality is evaluated with the Dice and ASSD.

For the fetal structures segmentation networks, the 3D patch size is set to $128 \times 128 \times 32$ to ensure that the receptive field contains most of the scan. The first two encoder and the last two decoder layers perform anisotropic sampling to reach patches of size $32 \times 32 \times 32$. The remaining layers perform isotropic sampling.

Study 1: Fetal Structures Segmentation. This study compares the accuracy of various segmentation architectures and quantifies the effectiveness of the automatic fetal body and fetal brain segmentation. We performed four experiments by comparing our segmentation method to: 1) an L-Net classifier [14]; 2) a standard 2D U-Net [2]; 3) our method without and 4) with prediction-time augmentations. In all cases, the segmentation results are refined with standard post-processing techniques. Table 1 (rows 3–6) shows the results. Note that prediction-time augmentation method improves the fetal body (brain) Dice score from 0.95 to 0.96 (0.95 to 0.96) and the ASSD from 1.52

Table 1. Segmentation accuracy results for the fetal body and fetal brain. The first two rows list the inter- and intra-observer manual segmentation variability; they serve as the reference for comparing the results of the segmentation methods. The next six rows list the networks architectures and methods; the columns indicate the segmentation metric scores (mean and std).

	Method	Fetal body		Fetal brain	
		Dice	ASSD (mm)	Dice	ASSD (mm)
1	Intra-observer variability	0.94 ± 0.01	0.90 ± 0.85	0.96 ± 0.01	0.26 ± 0.15
2	Inter-observer variability	0.93 ± 0.02	0.84 ± 0.78	0.96 ± 0.01	0.22 ± 0.12
3	L-Net	0.79 ± 0.08	6.31 ± 7.51	0.84 ± 0.07	1.23 ± 0.62
4	2D U-Net	0.93 ± 0.07	1.35 ± 1.70	0.94 ± 0.06	0.75 ± 1.22
5	3D U-Net	0.93 ± 0.06	1.62 ± 1.41	0.94 ± 0.05	0.64 ± 0.48
6	Two-step segmentation	0.95 ± 0.03	1.42 ± 1.32	0.95 ± 0.03	0.21 ± 0.13
7	Two-step segmentation + prediction-time augmentation	**0.96 ± 0.02**	**0.74 ± 0.51**	**0.96 ± 0.02**	**0.19 ± 0.09**
8	Previous-slice correction	0.97 ± 0.02	0.38 ± 0.31	0.97 ± 0.02	0.13 ± 0.05

to 0.74 mm (0.21 to 0.19), both below the observer variability measures. anisotropic sampling reduces the errors of Dice and ASSD by ~5% over isotropic sampling.

Study 2: Segmentation Error Estimation and Correction Prioritization. This study evaluates the segmentation error estimation and correction methods. First, note that the relatively high std with respect to the mean (Table 1, row 6) indicates that the method fails to produce accurate segmentations for a number of slices and scans, which should be identified for correction. The correlation coefficients between the estimated and the actual segmentation Dice errors computed by linear regression are 0.94 and 0.95 for the fetal brain and body, respectively. This indicates that the segmentation error estimations are reliable and can be used to identify segmentations requiring corrections.

We investigate the use of the estimated segmentation error measure to prioritize the manual segmentation errors correction process. The goal is to optimize the radiologist time and effort by correcting first the most significant segmentation errors instead of in random or sequential order, as in the current practice. We measure the test set mean segmentation error as a function of the % of the segmentations in the individual slices and scans that were corrected by the radiologist. We evaluate scan-wise and slice-wise prioritization as follows. In scan-wise prioritization, scans are ordered and corrected in descending order of their estimated scan segmentation error value. In slice-wise prioritization, the slices are partitioned into groups of five successive slices; the slice-wise segmentation error estimations of each group is averaged, and the slices are sorted by this value. This prioritization policies are compared to random and optimal scan prioritization, in which the scan segmentations are corrected in descending order of their actual segmentation error values.

Figure 3 (left) shows the results of the prioritization on the fetal body dataset. Note that observer variability accuracy is achieved by slice-wise ordered correction of 12% of the segmentations, vs. 20% and 33% in scan-wise and random order prioritization.

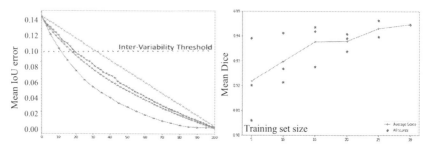

Fig. 3. Fetal body segmentation results. Left: mean IoU error (vertical axis) as a function of the # of slices corrected by the radiologist (horizontal axis). The plots show four prioritization policies: scan-wise random order (orange), scan-wise (blue) and slice-wise (red) descending order of estimated segmentation error value, and scan-wise optimal order by descending order of actual segmentation error value (green). The horizontal dotted line shows the manual segmentation observer variability. Right: mean Dice (vertical axis) as a function of the training set size (horizontal axis). Each red dot is a test result of a random training set of predefined size; blue lines show the mean test set score per training set size.

Table 1 (row 7) shows the effectiveness of using our automatic slice segmentation error correction method (Fig. 2, right), that achieves an ASSD lower by 49% and 32% from our best segmentation method for total body and brain datasets, respectively.

Study 3: Active Learning with Segmentation Error Estimation. We explore the use of segmentation error estimation for active learning [11]. The goal is to use the segmentation error estimation to select scans to augment the training set, thereby enhancing the performance of the automatic structures segmentation network.

We quantify the effect of random training sets sampling to establish a comparison baseline for training scans selection policies and to quantify the segmentation results variability of a fixed training set size. We train the fetal body segmentation networks on training sets of various sizes – 5, 10, 15, 20, 25, 30 – randomly chosen from a pool of 30 scans and tested on 30 scans. Figure 3 (right) shows the results. Note that small training sets can achieve results comparable to larger training sets when the training samples are selected differently: the best Dice (0.94) of a training set of size five is better than the worst Dice (0.93) of a training set of size 15. This suggests a potential savings in expert annotation time by judicious selection of scans added to the training set.

Finally, we quantify the addition of corrected, ground-truth annotated scans to the training dataset based on the segmentation error estimation. We test three policies for augmenting a training dataset of size 5 with 5 additional training scans based on the estimated segmentation error using the mean Dice measure of a test set of size 30. The Dice (ASSD) of the networks trained with 5 training scans is 0.92 ± 0.05 (2.1 ± 2.2). Adding 5 more scans based on their estimated segmentation errors and re-training the networks with the augmented training dataset of 10 annotated scans yields the following

Dice (ASSD) results: 0.92 ± 0.03 (2.1 ± 2.1) for the lowest estimated segmentation error, 0.93 ± 0.03 (1.7 ± 1.8) for randomly selected scans, and 0.95 ± 0.02 (1.2 ± 1.5) for the highest estimated segmentation error. These results suggest that scans with the highest estimated segmentation error should be prioritized for augmenting the training set.

4 Conclusion

We have presented a method for the end-to-end volumetric segmentation of fetal structures in MRI scan that optimizes radiologist validation and annotation time. Our method uses custom anisotropic 3D U-Net networks in a two-step process that extracts the structure ROI and computes its segmentation; the networks are trained with very few annotated scans. The segmentation error estimation stage leverages prediction-time augmentations of the input scan to compute multiple segmentations and to estimate the segmentation error for individual slices and for the entire scan based on the segmentation uncertainty estimations. These estimations are used to locate the most error prone slices and to iteratively correct the segmentations in those slices based on validated adjacent slices with a 2D U-Net slice correction network. Our method achieves state-of-the-art fetal structures segmentation results and provides effective segmentation error estimation and correction methods that enable the prioritization of the radiologist time and the effective creation of large validated datasets.

Our experimental results indicate that segmentation uncertainty and error estimation are useful for active learning and for training dataset selection and annotation optimization, thereby saving costly annotation time by utilizing the expert annotators' efforts efficiently on fewer scans. Our methods can be used to create a dataset of radiologist-validated segmentations for the accurate and robust automatic segmentation of complex structures in volumetric scans with very few annotated scans.

References

1. Litjens, G., et al.: A survey of deep learning in medical image analysis. Med. Image Anal. **42**, 60–88 (2017)
2. Ronneberger, O., Fischer, P., Brox, T.: U-Net: convolutional networks for biomedical image segmentation. In: Navab, N., Hornegger, J., Wells, W.M., Frangi, A.F. (eds.) MICCAI 2015. LNCS, vol. 9351, pp. 234–241. Springer, Cham (2015). https://doi.org/10.1007/978-3-319-24574-4_28
3. Çiçek, Ö., Abdulkadir, A., Lienkamp, S.S., Brox, T., Ronneberger, O.: 3D U-Net: learning dense volumetric segmentation from sparse annotation. In: Ourselin, S., Joskowicz, L., Sabuncu, M.R., Unal, G., Wells, W. (eds.) MICCAI 2016. LNCS, vol. 9901, pp. 424–432. Springer, Cham (2016). https://doi.org/10.1007/978-3-319-46723-8_49
4. Milletari, F., Navab, N., Ahmadi, S.A.: V-Net: fully convolutional neural networks for volumetric medical image segmentation. In: Proceedings of the 4th IEEE International Conference on 3D Vision (2016)
5. Isensee, F., Kickingereder, P., Wick, W., Bendszus, M., Maier-Hein, K.H.: Brain tumor segmentation and radiomics survival prediction: contribution to the BRATS 2017 challenge. In: Crimi, A., Bakas, S., Kuijf, H., Menze, B., Reyes, M. (eds.) BrainLes 2017. LNCS, vol. 10670, pp. 287–297. Springer, Cham (2018). https://doi.org/10.1007/978-3-319-75238-9_25

6. Karimi, D., Samei, G., Shao, Y., Salcudean, S.: A novel deep learning-based method for prostate segmentation in T2-weighted magnetic resonance imaging. arXiv:1901.09462 (2019)
7. Salehi, S.S.M., et al.: Real-time automatic fetal brain extraction in fetal MRI by deep learning. In: Proceedings of the IEEE 15th International Symposium on Biomedical Imaging – ISBI 2018, pp. 720–724 (2018)
8. Gibson, E., et al.: NiftyNet: a deep-learning platform for medical imaging. Comput. Methods Progr. Biomed. **158**, 113–122 (2018)
9. Gholipour, A., et al.: A normative spatiotemporal MRI atlas of the fetal brain for automatic segmentation and analysis of early brain growth. Sci. Rep. **7**(1), 1–13 (2017)
10. Fetit, A.E., et al.: A deep learning approach to segmentation of the developing cortex in fetal brain MRI with minimal manual labeling. In: Medical Imaging with Deep Learning (2020)
11. Veeraraghavan, H., Miller, J.V.: Active learning guided interactions for consistent image segmentation with reduced user interactions. In: Proceedings of the IEEE International Symposium on Biomed Imaging (2011)
12. Lee, N., Caban, J., Ebadollahi, S., Laine, A.: Interactive segmentation in multimodal medical imagery using a Bayesian transductive learning approach. In: Proceedings of the IEEE International Symposium on Biomedical Imaging (2009)
13. Wang, G., Li, W.: Interactive medical image segmentation using deep learning with image-specific fine tuning. IEEE Trans. Med. Imaging **37**, 1562–1573 (2018)
14. Braginsky, M., Joskowicz, L.: Interactive segmentation of structures with real-time fine-tuning of a fully convolutional neural network. M.Sc. thesis, The Hebrew University of Jerusalem (2019)
15. Lakshminarayanan, B., Pritzel, A., Blundell, C.: Simple and scalable predictive uncertainty estimation using deep ensembles. In: Advances in Neural Information Processing Systems (2017)
16. Guo, C., Pleiss, G., Sun, Y., Weinberger, K.Q.: On calibration of modern neural networks. In: Proceedings of the 34th International Conference on Machine Learning, vol. 70 (2017)
17. Wang, G., Li, W., Aertsen, M., Deprest, J., Ourselin, S., Vercauteren, T.: Aleatoric uncertainty estimation with test-time augmentation for medical image segmentation with convolutional neural networks. Neurocomputing **338**, 34–45 (2019)
18. Loquercio, A., Segu, M., Scaramuzza, D.: A general framework for uncertainty estimation in Deep Learning. arXiv preprint arXiv:1907.06890 (2019)
19. Ulyanov, D., Vedaldi, A., Lempitsky, V.: Instance normalization: the missing ingredient for fast stylization. arXiv preprint arXiv:1607.08022 (2016)
20. He, K., Zhang, X., Ren, S., Sun, J.: Identity mappings in deep residual networks. In: Leibe, B., Matas, J., Sebe, N., Welling, M. (eds.) ECCV 2016. LNCS, vol. 9908, pp. 630–645. Springer, Cham (2016). https://doi.org/10.1007/978-3-319-46493-0_38
21. Tompson, J., Goroshin, R., Jain, A., Lecun, Y., Bregler, C.: Efficient object localization using convolutional networks. In: Proceedings of the IEEE Conference on Computer Vision and Pattern Recognition (2015)
22. Lin, H., et al.: Deep learning for low-field to high-field MR: image quality transfer with probabilistic decimation simulator. In: Knoll, F., Maier, A., Rueckert, D., Ye, J.C. (eds.) MLMIR 2019. LNCS, vol. 11905, pp. 58–70. Springer, Cham (2019). https://doi.org/10.1007/978-3-030-33843-5_6

Data-Driven Multi-contrast Spectral Microstructure Imaging with InSpect

Paddy J. Slator[1]([✉]), Jana Hutter[2,3], Razvan V. Marinescu[1], Marco Palombo[1], Laurence H. Jackson[2,3], Alison Ho[4], Lucy C. Chappell[4], Mary Rutherford[2], Joseph V. Hajnal[2,3], and Daniel C. Alexander[1]

[1] Centre for Medical Image Computing, Department of Computer Science, University College London, London, UK
p.slator@ucl.ac.uk
[2] Centre for the Developing Brain, King's College London, London, UK
[3] Biomedical Engineering Department, King's College London, London, UK
[4] Women's Health Department, King's College London, London, UK

Abstract. We introduce and demonstrate an unsupervised machine learning method for spectroscopic analysis of quantitative MRI (qMRI) experiments. qMRI data can support estimation of multidimensional correlation (or single-dimensional) spectra, which allow model-free investigation of tissue properties, but this requires an ill-posed calculation. Moreover, in the vast majority of applications ground truth knowledge is unobtainable, preventing the application of supervised machine learning. Here we present a new method that addresses these limitations in a data-driven way. The algorithm simultaneously estimates a canonical basis of spectral components and voxelwise maps of their weightings, thereby pooling information across whole images to regularise the ill-posed problem. We show that our algorithm substantially outperforms current voxelwise spectral approaches. We demonstrate the method on combined diffusion-relaxometry placental MRI scans, revealing anatomically-relevant substructures, and identifying dysfunctional placentas. Our algorithm vastly reduces the data required to reliably estimate multidimensional correlation (or single-dimensional) spectra, opening up the possibility of spectroscopic imaging in a wide range of new applications.

Keywords: Quantitative MRI · Unsupervised learning · Placenta MRI

1 Introduction

Continuum modelling is an attractive method for analysing quantitative MRI (qMRI) data. The technique assumes that spins have a distribution of values

Electronic supplementary material The online version of this chapter (https://doi.org/10.1007/978-3-030-59725-2_36) contains supplementary material, which is available to authorized users.

A. L. Martel et al. (Eds.): MICCAI 2020, LNCS 12266, pp. 375–385, 2020.
https://doi.org/10.1007/978-3-030-59725-2_36

(e.g. relaxivity, diffusivity), which are quantified by a multidimensional correlation (or single-dimensional) spectrum. This approach is particularly powerful for experiments that simultaneously measure multiple MR properties, as it can resolve microstructural compartments that are indistinguishable with single contrast data. Imaging studies have recently demonstrated this in the T1-diffusion [7,8,20], T2-diffusion [12,17,25], T1-T2-diffusion [3], T2*-diffusion [24], T1-T2 [15], and T1-T2*-diffusion [11] domains.

A general continuum model gives a Fredholm integral equation on the MR signal [4], or a Laplace transform in the specific exponenetial decay case. The spectrum can be estimated using regularised inversion of this integral [9], although this is highly ill-posed. Estimating spectra in each image voxel independently therefore requires unrealistically high signal-to-noise (SNR). Moreover, to derive meaningful image maps from voxelwise spectra typically involves a procedure known as "spectral integration". In spectral integration, the user manually identifies regions of the spectrum (termed spectral regions of interest, sROI) that correspond to particular features of spectral components of interest. Scalar indices are then calculated by numerical integration of each reconstructed spectrum over these sROIs [3,12,16].

Recently, methods have been proposed for increasing the robustness of voxelwise spectral fits, utilising marginal distributions [2] or spatial regularisation [12]. These methods can improve inversion stability and give more meaningful derived spatial maps. However, inherent limitations remain. In particular, the reliance on ad-hoc choices of regularisation terms and manually defined sROIs. A recent technique automatically identifies these spectral integration regions [21], but restricts to rectangular and non-overlapping sROIs, depends on a user-defined threshold value, and still requires voxelwise spectral estimation.

Here we present a method which addresses these limitations in a data-driven way. It simultaneously estimates a canonical basis of spectral components for a whole image (or a data set comprising multiple images), and the voxelwise weighting factors of each component. We build on the discrete InSpect algorithm previously published by Slator et al. [23], adapting it for the continuous case. This allows us to capture smooth changes in parameters across the image with high resolution values, rather than forcing hard categorisation of pixels into a small set of bins, which discards subtle variation. Unlike standard inversion approaches, the InSpect method introduced here exploits the huge dependence among voxels, dramatically reducing the SNR required for stable inversion, hence enabling spectroscopic imaging in a wide variety of new situations.

2   Methods

Multidimensional Spectrum Estimation. InSpect is based on a continuum model, which assumes that single voxels contain spins with a spectrum of MR properties. For a general n-dimensional MRI experiment the voxel signal is

$$S(\boldsymbol{t}) = \int \ldots \int F(\boldsymbol{\omega}) K(\boldsymbol{t}, \boldsymbol{\omega}) \, d\omega_1 \ldots d\omega_n \tag{1}$$

where t is a vector of experimental parameters which are varied to yield contrast in intrinsic MR properties $\omega = (\omega_1, ..., \omega_n)$, via the specific form of the kernel $K(t, \omega)$. F is the n-dimensional spectrum over ω, i.e. the distribution of these values across all spins. For example, in T2*-diffusivity (or T2-diffusivity) imaging, $t = (b, T_E)$, the b-value and echo time (TE); $\omega = (ADC, T_2^*)$, the apparent diffusion coefficient and T2*; $K(t, \omega) = \exp(-T_E/T_2^*)\exp(-bADC)$ is the kernel; and F is the 2D T2*-ADC spectrum.

The standard approach for estimating the spectrum, following [9,18,22], proceeds as follows. Equation (1) is first discretised onto an n-dimensional grid, with lengths defined by the user-defined vector $N_\omega = (N_{\omega_1}, ..., N_{\omega_n})$. By choosing an ordering of the grid coordinates, the signal for all MR encodings in the experiment can thus be written in matrix form as

$$\mathbf{S} = \mathbf{KF} \tag{2}$$

where $\mathbf{S}$ is a column vector, length N_s, of the signals at each MR encoding, K is an N_s by $\prod_{m=1}^{n} N_{\omega_m}$ matrix of discretised kernel values, and F is an $\prod_{m=1}^{n} N_{\omega_m}$ length column vector of spectrum values. The spectrum F is then calculated with regularised non-negative least squares

$$\mathbf{F} = \arg\min_{\mathbf{F} \geq 0} \|\mathbf{KF} - \mathbf{S}\|_2^2 + \alpha\|\mathbf{F}\|_2^2. \tag{3}$$

By solving Equation (3) in each voxel, the spectrum can be estimated across a whole image. Volume fraction maps are then produced by numerically integrating voxelwise spectra over user-defined regions of the spectrum, e.g. [3,14,16]. However low SNR can lead to noisy spectrum estimates and hence poor mappings.

Underlying Model. Our algorithm automates spectral mapping, and undertakes data-driven regularisation of the Fredholm integral (or Laplace transform) inversion. Rather than naively fitting spectra to each voxel independently, we learn a low-dimensional representation consistent with the whole image.

The first element of the representation is a pre-specified number, M, of canonical spectral components, $\{F_1, F_2, ..., F_M\}$ (see first column of Fig. 1 for an example). Each spectral component has corresponding voxelwise weights across all N image voxels. The weighting of component m in voxel n is denoted z_{nm}, so that the full set of voxelwise weights is

$$\mathbf{z}_n = \{z_{n1}, z_{n2}, ..., z_{nM}\}_{n=1}^{N}, \text{ subject to } \sum_{m=1}^{M} z_{nm} = 1. \tag{4}$$

The second column of Fig. 1 shows example voxelwise weights. The signal from each voxel, $\mathbf{S}_n$, is described by the continuum model of Eq. (1) with the effective spectrum in each voxel n modelled as a weighted sum of the canonical spectrum components

$$F(\mathbf{z}_n) = \sum_{m=1}^{M} z_{nm} \mathbf{F}_m \tag{5}$$

where $\mathbf{z}_n = \{z_{nm}\}_{m=1}^M$ are the component weights for voxel n. The discrete model for a single voxel is therefore given by $\mathbf{S}_n = \mathbf{K}F(\mathbf{z}_n)$. We assume Gaussian noise, giving the following whole-image log-likelihood

$$\log \pi(\{\mathbf{S}_n\}_{n=1}^N | \{\mathbf{F}_m\}_{m=1}^M, \{\mathbf{z}_n, \sigma_n^2\}_{n=1}^N) = \sum_{n=1}^N \log \mathcal{N}\left(\mathbf{S}_n; \mathbf{K}F(\mathbf{z}_n), \sigma_n^2 I\right). \quad (6)$$

Note that we assume all observations in a voxel have the same variance, i.e. the covariance matrix is $\sigma_n^2 I$. For simplicity we denote the log-likelihood $\log \pi(\mathbf{D}|\theta)$, where $\mathbf{D} = \{\mathbf{S}_n\}_{n=1}^N$ is the full multi-contrast MR dataset (either a single image or set of images), and $\theta = \{\{\mathbf{F}_m\}_{m=1}^M, \{\mathbf{z}_n\}_{n=1}^N, \{\sigma_n^2\}_{n=1}^N\}$ are the model parameters.

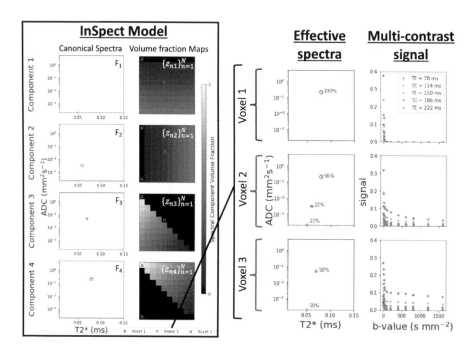

Fig. 1. The model underlying InSpect consists of a set of M components. Each component has an associated canonical spectrum and volume fraction map, which combine to give the effective voxelwise spectra.

InSpect Inference Algorithm. We seek the parameters θ that maximise $\log \pi(\mathbf{D}|\theta)$. In practice, we estimate σ_n^2 directly from the data, e.g. for a T2*-diffusion experiment we estimate by calculating the empirical variance of the volumes with $b = 0$ and the lowest TE. To derive maximisation steps for the canonical spectral components $\{F_m\}_{m=1}^M$, and voxelwise weights, $\{z_{nm}\}_{n=1}^N$ we first rearrange the log-likelihood to give

$$\log \pi(\mathbf{D}|\theta) = \sum_{n=1}^N -\frac{1}{2}\log 2\pi\sigma_n^2 - \frac{1}{2\sigma_n^2}\|\mathbf{S}_n - KF(\mathbf{z}_n)\|_2^2 . \tag{7}$$

We now derive a maximisation step for a canonical spectrum, F_j conditioned on the voxelwise weights $\{\mathbf{z}_n\}_{n=1}^N$, and the other spectra $\{F_m\}_{m\neq j}$. For the canonical spectral components, $\mathbf{F}_1, ..., \mathbf{F}_M$, we need to solve $F_k = \arg\max_{F_k \geq 0} \log \pi(\mathbf{D}|\theta)$. Taking the derivation of the function with respect to F_k (in numerator layout), setting equal to zero, and rearranging gives

$$\mathbf{F}_k = \underset{\mathbf{F}_k \geq 0}{\arg\min} \left\| \left(\sum_{n=1}^N \mathbf{K}z_{nk}\right)\mathbf{F}_k - \sum_{n=1}^N \left(\mathbf{S}_n - \mathbf{K}\left(\sum_{m\neq k} z_{nm}\mathbf{F_m}\right)\right)\right\|_2^2 . \tag{8}$$

We can solve Eq. (8) with non-negative least squares as described earlier.

To find the maximum likelihood voxelwise weightings for each of the spectral components, we first note the posterior distribution for the model, up to proportionality

$$\pi(\theta|\mathbf{D}) \propto \pi(\mathbf{D}|\theta)\pi(\theta) = \prod_{n=1}^N \mathcal{N}\left(\mathbf{S}_n; \mathbf{K}F(\mathbf{z}_n), \sigma_n^2 I\right) \tag{9}$$

where we have assumed a uniform prior on all parameters θ. The posterior distribution for each $\mathbf{z}_n$ - up to proportionality - is therefore

$$\pi(\mathbf{z}_n|\mathbf{S}_n, \theta^{(t-1)}) \propto \mathcal{N}\left(\mathbf{S}_n; \mathbf{K}F(\mathbf{z}_n), \sigma_n^2 I\right) \tag{10}$$

We therefore update $\mathbf{z}_n = \{z_{nm}\}_{m=1}^M$ by maximising this (in log-scale), subject to $\sum_{m=1}^M z_{nm} = 1$, i.e.

$$\mathbf{z}_n = \underset{\sum_{m=1}^M z_{nm}=1}{\arg\max} \log\mathcal{N}\left(\mathbf{S}_n; \mathbf{K}F(\mathbf{z}_n), \sigma_n^2 I\right) \tag{11}$$

which we solve sequentially for voxels $n = 1, ..., N$ with the interior-point algorithm. The InSpect algorithm is hence the following iterative optimisation:

1. Initialise the canonical spectral components $\{F_1, F_2, ..., F_M\}$, e.g. by assigning component to distinct elements of the whole-image spectrum.
2. Initialise the spectral weights for all voxels $\{z_{nm}\}_{n=1}^N$, given $\{F_1, F_2, ..., F_M\}$
3. Update $\mathbf{F}_m$ for some m by solving Eq. (8)
4. Update $\{z_{nm}\}_{n=1}^N$ by solving Eq. (11) for all voxels
5. Repeat steps 3 and 4 until convergence.

Application to Placenta Diffusion-Relaxometry Data. We demonstrate our InSpect algorithm on placental T2*-diffusion data acquired on 3T clinical MRI scanner using a 32-channel cardiac coil, previously published by Slator et al [24]. The protocol has 66 diffusion-weightings (ranging from $b = 5$ to $1600\,\mathrm{s\,mm^{-2}}$, including six $b = 0$ volumes) and 5 TEs (78, 114, 150, 186, 222 ms) for a total of 330 contrast-encodings. Other acquisition parameters were $\mathrm{FOV} = 300 \times 320 \times 84\,\mathrm{mm}$, $\mathrm{TR} = 7\,\mathrm{s}$, $\mathrm{SENSE} = 2.5$, halfscan $= 0.6$, resolution $= 3\,\mathrm{mm^3}$. We considered 13 scans from 12 women, of whom 9 were categorised as healthy controls, two had chronic hypertension in pregnancy, and one had pre-eclampsia (PE) with additional fetal growth restriction (FGR). One participant with chronic hypertension was scanned twice, four weeks apart, and developed superimposed PE by the second scan. The algorithm was run on all scans simultaneously, on a manually-segmented ROI comprising the whole placenta and the adjacent section of uterine wall, with $M = 4$ components specified.

Application to Simulations. We also tested on simulated diffusion-relaxometry data. Four synthetic canonical spectral components - informed by observed placental spectra [24] - were first defined (Fig. 2, first column). We next defined ground truth voxelwise weights on a 50-by-50 image (Fig. 2, third column). Given these, we simulated normalised diffusion-relaxometry scans using the same b-values and TEs as the placental data. We added Rician noise with SNR from 50-400 - this is comparable to the placenta data where we calculated SNRs ranging from 100 to 200. We applied InSpect - specifying $M = 4$ components - to these scans. We also fit voxelwise T2*-ADC spectra to all scans by solving Eq. (3), with α set to 0.01 using the L-curve method [10], and hence derived volume fraction maps with the standard spectral integration approach [3,12,16].

3 Results

Figure 2 and Supplementary Material Figures 4–5 demonstrate that our InSpect algorithm significantly outperforms the voxelwise approach on simulated data. At all noise levels, our maps more accurately recover the ground truth than voxelwise maps. We also accurately, and automatically, recover ground truth spectral components (e.g. Fig. 2 first columns) - these have to be manually identified in the standard voxelwise approach.

Figure 3 presents the joint InSpect fit to all participants' placental MR images. The four canonical spectral components (first column in top four rows) have distinct characteristics which suggests they each reflect a different microstructural environment. Although the algorithm imposes no direct anatomical analogue for any of the components, the corresponding maps identify clear anatomical structures which are consistent across control placentas, and show clear differences in dysfunctional placentas. This suggests that the canonical spectral components consistently identify distinct tissue environments, and that those tissue environments are salient to placental dysfunction.

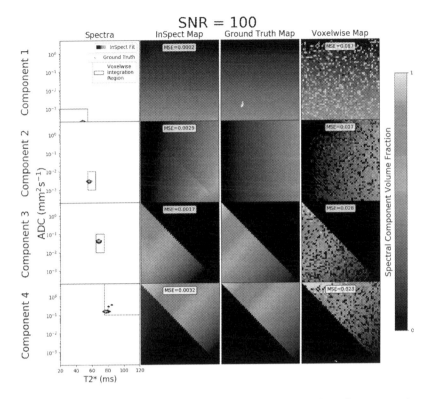

Fig. 2. InSpect algorithm applied to simulated images. The algorithm output is presented in columns 1–2: inferred spectral components (green-yellow peaks, column 1), and corresponding voxelwise weights (column 2). Column 3: ground truth voxelwise weights; ground truth spectral components (red dots in column 1) have fixed T2*-ADC values of (0.05, 0.0002), (0.06, 0.003), (0.07, 0.05) and (0.08 ms, 0.2 mm^2/s). Column 4: maps obtained by numerical integration (within blue regions of column 1) of voxelwise spectral fits. Reported mean square error (MSE) values compare each component's map and the corresponding ground truth map. See Supplementary Material for fits with other SNR values. (Color figure online)

4   Discussion and Conclusion

We introduce and demonstrate an unsupervised machine learning method for spectroscopic imaging. Our algorithm simultaneously estimates a set of canonical spectral components and their mapping across images. This offers potential advantages over typical spectrum estimation methods - such as those utilising marginal distributions [2] or spatial regularisation [12] - which can be unstable with standard MRI noise levels and require manual spectral labelling to obtain parametric maps. Our method also has advantages over blind source separation (BSS) techniques (e.g. [13] and [19]) since we incorporate a well-defined basic MRI model, allowing us to explicitly reconstruct signal components that we

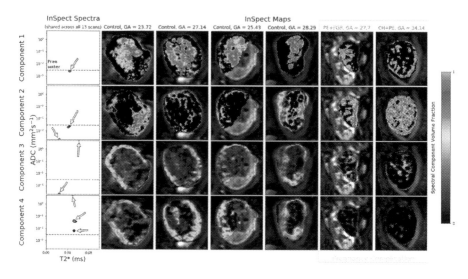

Fig. 3. Four-component InSpect model jointly fit to 13 placenta diffusion-relaxometry scans. The leftmost boxes display the InSpect spectra, which are shared across all 13 participants. The remaining boxes display the corresponding volume fraction maps for 6 of the 13 scans, with rows showing the maps for a single canonical spectral component, and columns showing the maps for a single scan. See y Material for maps from the remaining 7 scans.

can associate with distinct tissue compartments. However, BSS would be more appropriate when explicit signal models are unknown or inaccurate.

On simulated data we significantly outperform the standard voxelwise app-roach, even when the total number of voxels is relatively small compared to a typical clinical scan (Fig. 2). On placental diffusion-relaxometry MRI data InSpect maps clearly show anatomical structures (Fig. 3). This suggests that these maps provide insight into microstructure and microcirculation across the placenta.

Given the observed spatial patterns and corresponding canonical spectral component characteristics in placental data, we can make initial speculations about the tissue microenvironments associated with each component. Compo-nent one maps out lobular structures in the placenta, and consists of a single spectral peak with ADC close to free water. These observations are consistent with the characteristics we would expect of maternal blood pools within the pla-centa. Component two appears to encircle these lobules, and contains a restricted (i.e. very low ADC) spectral peak. This is consistent with this component repre-senting tissular structures, including the lobule-enclosing septa. In control par-ticipants, components three and four are prominent in the uterine wall. Both component-associated spectra contain peaks with higher ADC than free water, suggesting the presence of perfusing blood. This may be maternal blood in uter-ine wall areas and fetal blood within the placenta. Component 4 is considerably reduced in dysfunctional placentas, likely indicative of pathology.

There are limitations that motivate potential future improvements. The choice of canonical spectral components is user-defined, in future we will explore how to select this automatically, e.g. through model selection statistics, cross validation, and/or prior microstructural knowledge. Whilst the spectral components appear to show anatomical features, we could constrain the spectra to ensure this link, or add a Markov random field to make it more likely. Here we fit our InSpect model to groups of MRI scans simultaneously, which may average over some important within-individual features. In future we will compare to individual scan fits, with focus on which approach best differentiates controls from disease.

Our algorithm is widely applicable across qMRI techniques; to facilitate this our code is freely available at https://github.com/PaddySlator/inspect. Applying to single-contrast qMRI is particularly attractive, since there are numerous techniques across tissue types and imaging modalities; our approach could enable significantly improved mappings compared to standard voxelwise model fitting methods. For example, the framework is immediately applicable to spectrum-based analysis of multi-echo T2 relaxometry for myelin water imaging [1]. The method can also be applied to other multidimensional experiments beyond the diffusion-relaxometry example in this paper, such as diffusion exchange spectroscopy [5,6]. To conclude, our approach opens up spectral techniques to a wide range of situations where they are currently impossible. In particular, it paves the way to turn spectroscopic imaging into a widely used method for clinical research and practice.

References

1. Alonso-Ortiz, E., Levesque, I.R., Pike, G.B.: MRI-based myelin water imaging: a technical review. Magn. Reson. Med. **73**(1), 70–81 (2015). https://doi.org/10.1002/mrm.25198

2. Benjamini, D., Basser, P.J.: Use of marginal distributions constrained optimization (MADCO) for accelerated 2D MRI relaxometry and diffusometry. J. Magn. Reson. **271**, 40–45 (2016). https://doi.org/10.1016/j.jmr.2016.08.004

3. Benjamini, D., Basser, P.J.: Magnetic resonance microdynamic imaging reveals distinct tissue microenvironments. NeuroImage **163**, 183–196 (2017). https://doi.org/10.1016/j.neuroimage.2017.09.033

4. Benjamini, D., Basser, P.J.: Multidimensional correlation MRI. NMR Biomed. (2020). https://doi.org/10.1002/nbm.4226

5. Benjamini, D., Komlosh, M.E., Basser, P.J.: Imaging local diffusive dynamics using diffusion exchange spectroscopy MRI. Phys. Rev. Lett. **118**(15), 158003 (2017). https://doi.org/10.1103/PhysRevLett.118.158003

6. Breen-Norris, J.O., et al.: Measuring diffusion exchange across the cell membrane with DEXSY (Diffusion Exchange Spectroscopy). Magn. Reson. Med. (2019), 1–9 (2020). https://doi.org/10.1002/mrm.28207

7. De Santis, S., Assaf, Y., Jeurissen, B., Jones, D.K., Roebroeck, A.: T1 relaxometry of crossing fibres in the human brain. NeuroImage **141**, 133–142 (2016). https://doi.org/10.1016/J.NEUROIMAGE.2016.07.037

8. De Santis, S., Barazany, D., Jones, D.K., Assaf, Y.: Resolving relaxometry and diffusion properties within the same voxel in the presence of crossing fibres by combining inversion recovery and diffusion-weighted acquisitions. Magn. Reson. Med. **75**(1), 372–380 (2016). https://doi.org/10.1002/mrm.25644

9. English, A.E., Whittall, K.P., Joy, M.L., Henkelman, R.M.: Quantitative two-dimensional time correlation relaxometry. Magn. Reson. Med. **22**(2), 425–434 (1991). https://doi.org/10.1002/mrm.1910220250

10. Hansen, P.C.: Analysis of discrete Ill-posed problems by means of the L-Curve. SIAM Rev. **34**(4), 561–580 (1992). https://doi.org/10.1137/1034115

11. Hutter, J., et al.: Integrated and efficient diffusion-relaxometry using ZEBRA. Sci. Rep. **8**(1), 15138 (2018). https://doi.org/10.1038/s41598-018-33463-2

12. Kim, D., Doyle, E.K., Wisnowski, J.L., Kim, J.H., Haldar, J.P.: Diffusion-relaxation correlation spectroscopic imaging: a multidimensional approach for probing microstructure. Magn. Reson. Med. **78**(6), 2236–2249 (2017). https://doi.org/10.1002/mrm.26629

13. Kim, D., Kim, J.H., Haldar, J.P.: automatic tissue decomposition using nonnegative matrix factorization for noisy MR magnitude images. ISMRM **2015**(February), 3701 (2015)

14. Kim, D., Wisnowski, J.L., Haldar, J.P.: Improved efficiency for microstructure imaging using high-dimensional MR correlation spectroscopic imaging. In: 2017 51st Asilomar Conference on Signals, Systems, and Computers, pp. 1264–1268. IEEE, October 2017. https://doi.org/10.1109/ACSSC.2017.8335555

15. Kim, D., Wisnowski, J.L., Nguyen, C.T., Haldar, J.P.: Multidimensional correlation spectroscopic imaging of exponential decays: from theoretical principles to in vivo human applications. NMR Biomed. (2019), 1–19 (2020). https://doi.org/10.1002/nbm.4244

16. Mackay, A., Whittall, K., Adler, J., Li, D., Paty, D., Graeb, D.: In vivo visualization of myelin water in brain by magnetic resonance. Magn. Reson. Med. **31**(6), 673–677 (1994). https://doi.org/10.1002/mrm.1910310614

17. Melbourne, A., et al.: Separating fetal and maternal placenta circulations using multiparametric MRI, January 2018. https://doi.org/10.1002/mrm.27406

18. Menon, R.S., Allen, P.S.: Application of continuous relaxation time distributions to the fitting of data from model systems and excised tissue. Magn. Reson. Med. **20**(2), 214–227 (1991). https://doi.org/10.1002/mrm.1910200205

19. Molina-Romero, M., et al.: A diffusion model-free framework with echo time dependence for free-water elimination and brain tissue microstructure characterization. Magn. Reson. Med. **80**(5), 2155–2172 (2018). https://doi.org/10.1002/mrm.27181

20. Ning, L., Gagoski, B., Szczepankiewicz, F., Westin, C.F., Rathi, Y.: Joint RElaxation-diffusion imaging moments to probe neurite microstructure. IEEE Trans. Med. Imaging **39**(3), 668–677 (2020). https://doi.org/10.1109/TMI.2019.2933982

21. Pas, K., Komlosh, M.E., Perl, D.P., Basser, P.J., Benjamini, D.: Retaining information from multidimensional correlation MRI using a spectral regions of interest generator. Sci. Rep., 1–10 (2020). https://doi.org/10.1038/s41598-020-60092-5

22. Ronen, I., Moeller, S., Ugurbil, K., Kim, D.S.: Analysis of the distribution of diffusion coefficients in cat brain at 9.4 T using the inverse Laplace transformation. Magn. Reson. Imaging **24**(1), 61–68 (2006). https://doi.org/10.1016/j.mri.2005.10.023

23. Slator, P.J., et al.: InSpect: INtegrated SPECTral component estimation and mapping for multi-contrast microstructural MRI. In: Chung, A.C.S., Gee, J.C., Yushkevich, P.A., Bao, S. (eds.) IPMI 2019. LNCS, vol. 11492, pp. 755–766. Springer, Cham (2019). https://doi.org/10.1007/978-3-030-20351-1_59

24. Slator, P.J., et al.: Combined diffusion-relaxometry MRI to identify dysfunction in the human placenta. Magn. Reson. Med. (October), 1–22 (2019). https://doi.org/10.1002/mrm.27733

25. Veraart, J., Novikov, D.S., Fieremans, E.: TE dependent Diffusion Imaging (TEdDI) distinguishes between compartmental T2 relaxation times. NeuroImage **182**, 360–369 (2018). https://doi.org/10.1016/j.neuroimage.2017.09.030

Semi-supervised Learning for Fetal Brain MRI Quality Assessment with ROI Consistency

Junshen Xu[1]([✉]), Sayeri Lala[1], Borjan Gagoski[2], Esra Abaci Turk[2],
P. Ellen Grant[2,3], Polina Golland[1,4], and Elfar Adalsteinsson[1,5]

[1] Department of Electrical Engineering and Computer Science,
MIT, Cambridge, MA, USA
junshen@mit.edu

[2] Fetal-Neonatal Neuroimaging and Developmental Science Center,
Boston Children's Hospital, Boston, MA, USA

[3] Harvard Medical School, Boston, MA, USA

[4] Computer Science and Artificial Intelligence Laboratory,
MIT, Cambridge, MA, USA

[5] Institute for Medical Engineering and Science, MIT, Cambridge, MA, USA

Abstract. Fetal brain MRI is useful for diagnosing brain abnormalities but is challenged by fetal motion. The current protocol for T2-weighted fetal brain MRI is not robust to motion so image volumes are degraded by inter- and intra- slice motion artifacts. Besides, manual annotation for fetal MR image quality assessment are usually time-consuming. Therefore, in this work, a semi-supervised deep learning method that detects slices with artifacts during the brain volume scan is proposed. Our method is based on the mean teacher model, where we not only enforce consistency between student and teacher models on the whole image, but also adopt an ROI consistency loss to guide the network to focus on the brain region. The proposed method is evaluated on a fetal brain MR dataset with 11,223 labeled images and more than 200,000 unlabeled images. Results show that compared with supervised learning, the proposed method can improve model accuracy by about 6% and outperform other state-of-the-art semi-supervised learning methods. The proposed method is also implemented and evaluated on an MR scanner, which demonstrates the feasibility of online image quality assessment and image reacquisition during fetal MR scans.

Keywords: Image quality assessment · Fetal magnetic resonance imaging (MRI) · Semi-supervised learning · Convolutional neural network (CNN)

1 Introduction

Fetal brain Magnetic Resonance Imaging (MRI) is an important tool complementing Ultrasound in diagnosing fetal brain abnormalities [2,9]. While MRI

© Springer Nature Switzerland AG 2020
A. L. Martel et al. (Eds.): MICCAI 2020, LNCS 12266, pp. 386–395, 2020.
https://doi.org/10.1007/978-3-030-59725-2_37

provides higher quality tissue contrast compared to Ultrasound [7,9], it is more vulnerable to motion artifacts because data acquisition is slow relative to the motion dynamics in the body [20]. This makes it challenging to adapt MRI for fetal imaging since fetal motion is more random and larger compared to adults [9]. The current protocol for T2-weighted fetal brain MRI attempts to mitigate motion artifacts by using time-efficient (~500 ms) readouts per slice, such as the single-shot T2 weighted (SST2W) imaging acquisition. Due to safety constraints on the amount of allowable exposure to radio-frequency energy, there is a 1–2 delay between the acquisition of two consecutive slices in the stack, so obtaining an entire stack (~30 slices) takes approximately 1 min. Orthogonal stacks are acquired and used to reconstruct the fetal brain volume. However, inter-slice and even intra-slice motion artifacts occur, contaminating the volume reconstruction [17]. Therefore, in order to improve the quality, entire stacks are usually reacquired several times [6,9], which is time-consuming. Prospective detection and reacquisition of low quality slices are expected to improve both the reconstruction quality of the brain volume, as well as the efficiency of MR scans.

Several prior studies have demonstrated the potential of Convolutional Neural Networks (CNNs) for fast image quality assessment (IQA) of MRI. Esses et al. [1] trained a CNN for volume quality assessment of T2-weighted liver MRI. Sujit et al. [15] proposed an ensemble learning method for volume quality assessment of pediatric and adult brain MRI by using multiple CNNs. However, several differences exist between these problems and fetal brain MR IQA. Specifically, these works aim to evaluate the quality of the entire stack of images instead of a single slice. Furthermore, in fetal MRI, motion is a dominant source of artifacts, typically appearing as blurs and nonuniform signal voids. Although motion is also a major source of artifacts in liver, adult brain, and cardiac MRI [12], their manifestations are different, as in this applications the motion is more regular and smaller in range compared to the motion observed in the fetus [9].

Since labeling large-scale medical image datasets is usually difficult and time-consuming, numerous semi-supervised learning methods have been proposed to leverage information in unlabeled data to improve the performance and robustness of deep neural networks. One general technique of semi-supervised learning is to infer pseudo labels from partially labeled data, such as self-training [19] and label propagation methods [4]. To yield better pseudo labels, recent methods use an ensemble of multiple neural networks which is known as self-ensembling, including temporal ensembling [8] and mean teacher [16]. In temporal ensembling, for each sample, the exponential moving average of classification outputs at different training epochs are computed and used as pseudo labels. The mean squared error (MSE) between model predictions and the pseudo labels is used as a consistency loss. One drawback of the temporal ensembling method is that it needs to keep track of the pseudo labels which is memory-consuming for large datasets. To address this problem, mean teacher method is proposed, which instead of using an ensemble of network outputs, aggregates the parameters of networks at different training step to build a teacher model. The system consists of two models with the same architecture, i.e., student and teacher. The student model is updated with gradient during training, while the teacher model is the exponential moving average of the student model. The prediction of the teacher model is considered as pseudo label, and a consistency loss similar to temporal ensembling, is enforced between the predictions of student and teacher models. The consistency loss in self-ensembling method can also be interpreted as a regularization that smooths the network around unlabeled data. Following this interpretation, Miyato et al. proposed virtual adversarial training (VAT) [10] where they enforced consistency between predictions of original images and corresponding adversarial samples. These semi-supervised methods have also found their way into application of medical imaging, such as nuclei classification [14] and gastric diseases diagnosis [13].

In this work, we proposed a novel semi-supervised learning method for fetal MRI quality assessment. Our method extends the mean teacher model by introducing a region-of-interest (ROI) consistency for fetal brain, which let the network focus on the fetal brain ROI during feature extraction, and thus improves the accuracy of detecting non-diagnostic MR images. Evaluation showed that our method outperformed other state-of-the-art semi-supervised methods. We also implemented and evaluated the proposed method on a MR scanner, demonstrating the feasibility of online image quality assessment and image reacquisition during fetal MR scans.

2   Methods

2.1   Mean Teacher Model

In semi-supervised learning, let $\{x_1, x_2, ..., x_{N_l}\}$ be the labeled dataset with labels $\{y_1, y_2, ..., x_{N_l}\}$ and let $\{x_{N_l+1}, x_{N_l+2}, ..., x_N\}$ be the unlabeled dataset. The mean teacher model [16] consists of two networks with the same architecture, i.e., student network and teacher network, whose parameters are denoted as θ and θ' respectively.

During training, the student network is updated by minimizing the following loss function:

$$
\begin{aligned}
L_{\mathrm{MT}} &= L_{\mathrm{cls}} + \lambda L_{\mathrm{con}} \\
&= \frac{1}{N} \sum_{i=1}^{N_l} H(y_i, f_\theta(x_i, \eta)) + \frac{\lambda}{N} \sum_{i=1}^{N} D_{\mathrm{KL}}(f_{\theta'}(x_i, \eta') \| f_\theta(x_i, \eta))
\end{aligned}
\tag{1}
$$

The first term is the classification loss for labeled data, which is the cross entropy between student network prediction $f_\theta(x_i, \eta)$ and label y_i. The second term is the consistency loss between predictions of student and teacher networks. Inspired by VAT [10], we use Kullback–Leibler (KL) divergence to measure the distance between the student and teacher predictions, instead of MSE as used in the original mean teacher method [16], where η and η' denote the noise perturbation for the two networks and λ is the weight of consistency loss. The teacher network is updated as follows: $\theta'_{t+1} = \alpha \theta'_t + (1 - \alpha)\theta_t$, where α is the coefficient and t is training step.

2.2 Brain ROI Consistency

In fetal brain MRI, the brain occupies a small portion of the image due to imaging parameter constraints [2]. However, the fetal brain is the ROI relevant for fetal brain MRI IQA since only the artifacts occurring in the brain affect diagnostic quality of the image. Therefore, it is essential to train the model to focus on features within the brain ROI. To fulfill this goal, We propose an ROI consistency loss to regularize the network. The overall architecture of the proposed mean teacher model with brain ROI consistency is shown in Fig. 1.

First, we introduce an ROI extraction module (Fig. 1A). For each image x, it produces a brain ROI mask R. $x_R = x \odot R$ is the masked image, where $\odot$ is the Hadamard product. The implementation of ROI extraction relies on a segmentation model. We utilize a trained U-Net in [11] to segment fetal brains from MR slices. However, since the segmentation network is trained on images with different acquisition parameters, it may yield inaccurate segmentation masks and fail to detect the brain ROI for some slices in our dataset. To improve robustness of ROI detection, instead of using the output of segmentation network directly, we aggregate the masks of images belonging to the same scan to generate a single ROI mask for the whole stack of images. The proposed algorithm is described in Fig. 1C. A stack of images are fed into the pretrained network to generate raw masks. For each mask M_i in the stack, its area A_i, center q_i and radius r_i are computed. We exclude those masks with area less than a threshold $A_{\min}$, which are assumed to be inaccurate, and let $B = \{i | 1 \leq i \leq S, A_i \geq A_{\min}\}$ be the set of remaining slices. We then compute the area-weighted mean and variance of the centers over B, i.e., $q = \frac{1}{|B|} \sum_{i \in B} A_i q_i$ and $\sigma^2 = \frac{1}{|B|} \sum_{i \in B} A_i \|q_i - q\|_2^2$. The final ROI mask R is defined as the circle centered at q with radius $r = \sigma + \max_{i \in B} r_i$.

The goal of ROI consistency loss is to make the network focus on brain ROI. Let z be the output feature of the last convolution layer. $z_{\theta'}(x_i \odot R_i, \eta)$ is the feature of ROI extracted by the teacher network and $z_\theta(x_i, \eta)$ is the feature of the original image extracted by the student network. We want these features to be close to each other, so that the student can learn to detect the brain ROI from the whole image. The ROI consistency loss are defined as the MSE between these two features:

$$L_{\text{con-roi}} = \frac{1}{N} \sum_{i=1}^{N} \|z_{\theta'}(x_i \odot R_i, \eta) - z_\theta(x_i, \eta)\|_2^2 \qquad (2)$$

The ROI consistency loss use the feature of masked images extracted by the teacher network as reference. To guide the teacher network to learn meaningful features from the masked images, the classification loss for masked images in the labeled dataset is used as a regularization which is denoted as $L_{\text{cls-roi}}$.

$$L_{\text{cls-roi}} = \frac{1}{N} \sum_{i=1}^{N_l} H(y_i, f_\theta(x_i \odot R_i, \eta)) \qquad (3)$$

We also adopted conditional entropy as an additional loss:

$$L_{\text{ent}} = H(y|x) = \frac{1}{N} \sum_{i=1}^{N} H(f_\theta(x_i, \eta), f_\theta(x_i, \eta)) \tag{4}$$

which is able to exaggerate the prediction of the network on each data point [10]. Therefore, the total loss of the proposed method is as follows.

$$L = L_{\text{cls}} + L_{\text{cls-roi}} + \lambda L_{\text{con}} + \beta L_{\text{con-roi}} + \gamma L_{\text{ent}} \tag{5}$$

where λ, β and γ are weight coefficients.

At the first couple of epochs, the teacher network cannot provide a reliable guide to the student network. For this reason, we use a ramp-up function $w(t) = \exp[-5(1-\min(t,T)/T)^2]$ for coefficients λ, β and γ, where t is the current epoch and $T = 5$.

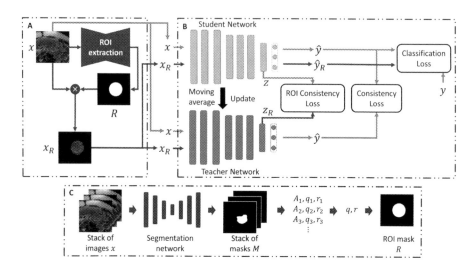

Fig. 1. Overview of the proposed method. A) Brain ROI extraction. B) Mean teacher model with ROI consistency loss. C) Details of ROI extraction algorithm.

3   Experiments and Results

3.1   Dataset

A total of 217129 images were obtained from 644 previously acquired research and clinical scans of mothers with singleton pregnancies and no pathologies, ranging in gestational age between 19 to 37 weeks. Scans were conducted at Boston Children's Hospital with Institutional Review Board approval. Scans were acquired using the SST2W sequence with median echo time $\text{TE} = 115$ ms, repetition time $\text{TR} = 1.6$ s, field of view 31 cm, and voxel size of $1.2 \times 1.2 \times 3$ mm^3.

A set of 11223 images from 42 subjects are selected as labeled set and classified into three categories: diagnostic (D), non-diagnostic (N) and images without brain region of interest (W). Diagnostic images were characterized by sharp brain boundaries while non-diagnostic images were characterized by artifacts that occlude such features (Fig. 2). Motion artifacts manifest as signal void and blurring over the brain region. Other artifacts manifest as aliasing or the fetus not being in the field of view. A research assistant trained under radiologists labeled the dataset. The labeled dataset is divided into training (7717 images), validation (1782 images), and test (1724 images) set, where the test set consists of subjects different from training and validation sets.

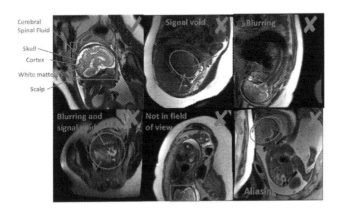

Fig. 2. Representative examples of diagnostic and nondiagnostic quality fetal brain MRI

3.2   Experiments Setup

We adopted ResNet-34 [3] as the backbone for student and teacher networks and set $\alpha = 0.994$ and $\lambda = \beta = \gamma = 1$, unless otherwise stated.

To evaluate the proposed method, we compare it with other three methods, including supervised learning, mean teacher (MT) [16] and virtual adversarial training (VAT) [10]. In addition to accuracy, we also adopted area under the ROC curves (AUC) for non-diagnostic images (N) as performance metric, since in clinical practice we are interested in detecting non-diagnostic slices and reacquiring them during MR scan.

For each method, we train the model using 1000, 2000, 4000 and all labeled data in training set (7717). For semi-supervised method, all unlabeled data are used for training. We used a batch size of 384. To balance the number of labeled and unlabeled data seen by the model, in each batch, 96 images are drawn from labeled dataset while the remains are unlabeled data. We run each experiment for 5 times and report the mean and standard deviation of evaluation metrics.

All Neural networks were implemented with PyTorch and trained on a server with an Intel Xeon E5-1650 CPU, 128 GB RAM and four NVIDIA TITAN X

GPUs. Adam [5] optimizer is used with an initial learning rate of 5×10^{-3}, and cosine learning rate decay.

3.3   Results

Results of accuracy and AUC are reported in Table 1. Results show that, the proposed method outperforms other state-of-the-art semi-supervised learning method in terms of both accuracy and AUC of non-diagnostic image. Additionally, comparing with supervised learning, the proposed approach increases accuracy and AUC by 5.82% and 0.084 respectively by learning extra information of large scale unlabeled dataset. We can also see that for smaller labeled training set (e.g., 1000 labels) the gain in accuracy from unlabeled data is higher. Besides, ablation studies were performed by setting λ, β or γ to zero to evaluate the contribution for each regularization. Results show that all the three regularization terms in our method can improve the performance of network.

Table 1. Accuracy $\pm$ std (%) and AUC for non-diagnostic image $\pm$ std over 5 runs.

Metric	Method	1000 labels	2000 labels	4000 labels	All labels
Acc.	Supervised	75.58 ± 0.93	77.40 ± 0.68	79.33 ± 0.94	79.37 ± 0.38
	VAT [10]	76.06 ± 2.18	77.51 ± 1.90	80.45 ± 2.35	81.25 ± 1.21
	MT [16]	79.27 ± 0.85	80.35 ± 0.56	81.21 ± 0.81	81.89 ± 0.63
	Proposed	**82.87 ± 0.92**	**83.73 ± 0.86**	**84.37 ± 0.37**	**85.19 ± 0.19**
	$\lambda = 0$	80.88 ± 1.07	81.38 ± 0.70	82.47 ± 0.42	82.88 ± 0.31
	$\beta = 0$	80.78 ± 0.66	82.01 ± 1.03	83.27 ± 0.34	83.81 ± 0.52
	$\gamma = 0$	80.61 ± 0.26	80.92 ± 0.61	82.68 ± 0.53	83.77 ± 0.40
AUC	Supervised	0.788 ± 0.016	0.818 ± 0.012	0.826 ± 0.008	0.815 ± 0.012
	VAT [10]	0.815 ± 0.021	0.822 ± 0.014	0.833 ± 0.017	0.844 ± 0.044
	MT [16]	0.831 ± 0.008	0.851 ± 0.005	0.856 ± 0.011	0.864 ± 0.006
	Proposed	**0.869 ± 0.008**	**0.881 ± 0.003**	**0.889 ± 0.007**	**0.899 ± 0.006**
	$\lambda = 0$	0.829 ± 0.007	0.822 ± 0.001	0.841 ± 0.011	0.854 ± 0.008
	$\beta = 0$	0.854 ± 0.006	0.872 ± 0.005	0.875 ± 0.004	0.887 ± 0.006
	$\gamma = 0$	0.855 ± 0.009	0.860 ± 0.003	0.878 ± 0.006	0.882 ± 0.005

3.4   Online Implementation

To further evaluation the proposed method and its performance in clinical practice, we developed and implemented a pipeline that runs the IQA CNN during fetal MR scans to assign a IQA score to each slice and reacquire those slices with low IQA scores. The trained CNN is deployed on a GPU (NVIDIA 1050Ti) equipped computer which is connected to the scanner's internal network. In each scan, N_{acq} slices were acquired and the IQA scores are computed as $s = 1 - P_N$, where P_N is probability of non-diagnostic image. Then the N_{re} slices with lowest IQA scores were reacquired. The proportion of re-acquisition is denoted as

$q = N_{re}/N_{acq}$. We performed a simulation study on the test set consisting of stacks of images with 20 to 40 slices where about one third of the images are of low quality in average (in the worst case, over 60% of slices in a stack are contaminated by motion artifacts). The number of missing non-diagnostic images is shown in Fig. 3a, where 'random' means random re-acquisition. The proposed method outperforms the supervised baseline and only misses one non-diagnostic slice in average when $q = 50\%$.

For in vivo study, fetal scans were performed on a 3T MR scanner with $N_{acq} = 20, q = 0.5$. Figure 3b shows 4 images from 3 separate scans, where the originally acquired slices (top row) were motion degraded, and the re-acquired ones (bottom row) were not. These results demonstrated the feasibility of online detection of non-diagnostic MR images during fetal scans using the proposed deep learning method.

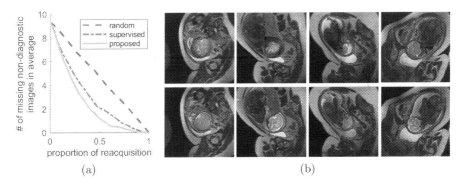

Fig. 3. (a) Number of non-diagnostic slices that are not detected by the IQA pipeline. (b) Four examples from three separate in vivo scans showing motion artifacts in the originally acquired images (top row), and much cleaner images when the same slice locations were re-acquired (bottom row).

4   Conclusions

In this paper, we proposed a novel semi-supervised learning method for fetal MRI quality assessment. Our method extend the mean teacher model by introducing a ROI consistency for fetal brain which let the network focus on brain ROI during feature extraction and therefore improve the accuracy of detecting non-diagnostic MR images. Evaluation showed that our method outperformed other state-of-the-art semi-supervised methods as well. We also implemented and evaluated the proposed method on a MR scanner, demonstrating the feasibility of online image quality assessment and image requisition during fetal MR scans, which can work in tandem with fetal motion tracking algorithm [18] to improve image quality as well as efficiency of imaging workflow.

References

1. Esses, S.J., et al.: Automated image quality evaluation of T2-weighted liver MRI utilizing deep learning architecture. J. Magn. Reson. Imaging **47**(3), 723–728 (2018)
2. Gholipour, A., et al.: Fetal MRI: a technical update with educational aspirations. Concepts Magn. Reson. Part A **43**(6), 237–266 (2014)
3. He, K., Zhang, X., Ren, S., Sun, J.: Deep residual learning for image recognition. In: The IEEE Conference on Computer Vision and Pattern Recognition (CVPR), June 2016
4. Iscen, A., Tolias, G., Avrithis, Y., Chum, O.: Label propagation for deep semi-supervised learning. In: Proceedings of the IEEE Conference on Computer Vision and Pattern Recognition, pp. 5070–5079 (2019)
5. Kingma, D.P., Ba, J.: Adam: a method for stochastic optimization. arXiv preprint arXiv:1412.6980 (2014)
6. Kline-Fath, B., Bahado-Singh, R., Bulas, D.: Fundamental and Advanced Fetal-imaging: Ultrasound and MRI. Lippincott Williams & Wilkins, Philadelphia (2014)
7. Kul, S., et al.: Contribution of MRI to ultrasound in the diagnosis of fetal anomalies. J. Magn. Reson. Imaging **35**(4), 882–890 (2012)
8. Laine, S., Aila, T.: Temporal ensembling for semi-supervised learning. arXiv preprint arXiv:1610.02242 (2016)
9. Malamateniou, C., et al.: Motion-compensation techniques in neonatal and fetal MR imaging. Am. J. Neuroradiol. **34**(6), 1124–1136 (2013)
10. Miyato, T., Maeda, S.I., Koyama, M., Ishii, S.: Virtual adversarial training: a regularization method for supervised and semi-supervised learning. IEEE Trans. Pattern Anal. Mach. Intell. **41**(8), 1979–1993 (2018)
11. Salehi, S.S.M., et al.: Real-time automatic fetal brain extraction in fetal MRI by deep learning. In: 2018 IEEE 15th International Symposium on Biomedical Imaging (ISBI 2018), pp. 720–724. IEEE (2018)
12. Schreiber-Zinaman, J., Rosenkrantz, A.B.: Frequency and reasons for extra sequences in clinical abdominal MRI examinations. Abdom. Radiol. **42**(1), 306–311 (2017)
13. Shang, H., et al.: Leveraging other datasets for medical imaging classification: evaluation of transfer, multi-task and semi-supervised learning. In: Shen, D., et al. (eds.) MICCAI 2019. LNCS, vol. 11768, pp. 431–439. Springer, Cham (2019). https://doi.org/10.1007/978-3-030-32254-0_48
14. Su, H., Shi, X., Cai, J., Yang, L.: Local and global consistency regularized mean teacher for semi-supervised nuclei classification. In: Shen, D., et al. (eds.) MICCAI 2019. LNCS, vol. 11764, pp. 559–567. Springer, Cham (2019). https://doi.org/10.1007/978-3-030-32239-7_62
15. Sujit, S.J., Coronado, I., Kamali, A., Narayana, P.A., Gabr, R.E.: Automated image quality evaluation of structural brain MRI using an ensemble of deep learning networks. J. Magn. Reson. Imaging **50**, 1260–1267 (2019)
16. Tarvainen, A., Valpola, H.: Mean teachers are better role models: weight-averaged consistency targets improve semi-supervised deep learning results. In: Advances in Neural Information Processing Systems, pp. 1195–1204 (2017)
17. Tourbier, S., Bresson, X., Hagmann, P., Thiran, J.P., Meuli, R., Cuadra, M.B.: An efficient total variation algorithm for super-resolution in fetal brain MRI with adaptive regularization. NeuroImage **118**, 584–597 (2015)

18. Xu, J., et al.: Fetal pose estimation in volumetric MRI using a 3D convolution neural network. In: Shen, D., et al. (eds.) MICCAI 2019. LNCS, vol. 11767, pp. 403–410. Springer, Cham (2019). https://doi.org/10.1007/978-3-030-32251-9_44

19. Yarowsky, D.: Unsupervised word sense disambiguation rivaling supervised methods. In: 33rd Annual Meeting of the Association for Computational Linguistics, pp. 189–196 (1995)

20. Zaitsev, M., Maclaren, J., Herbst, M.: Motion artifacts in MRI: a complex problem with many partial solutions. J. Magn. Reson. Imaging **42**(4), 887–901 (2015)

Enhanced Detection of Fetal Pose in 3D MRI by Deep Reinforcement Learning with Physical Structure Priors on Anatomy

Molin Zhang[1(✉)], Junshen Xu[1], Esra Abaci Turk[2], P. Ellen Grant[2,4], Polina Golland[1,3], and Elfar Adalsteinsson[1,5]

[1] Department of Electrical Engineering and Computer Science, MIT, Cambridge, MA, USA
molin@mit.edu
[2] Fetal-Neonatal Neuroimaging and Developmental Science Center, Boston Children's Hospital, Boston, MA, USA
[3] Computer Science and Artificial Intelligence Laboratory, MIT, Cambridge, MA, USA
[4] Harvard Medical School, Boston, MA, USA
[5] Institute for Medical Engineering and Science, MIT, Cambridge, MA, USA

Abstract. Fetal MRI is heavily constrained by unpredictable and substantial fetal motion that causes image artifacts and limits the set of viable diagnostic image contrasts. Current mitigation of motion artifacts is predominantly performed by fast, single-shot MRI and retrospective motion correction. Estimation of fetal pose in real time during MRI stands to benefit prospective methods to detect and mitigate fetal motion artifacts where inferred fetal motion is combined with online slice prescription with low-latency decision making. Current developments of deep reinforcement learning (DRL), offer a novel approach for fetal landmarks detection. In this task 15 agents are deployed to detect 15 landmarks simultaneously by DRL. The optimization is challenging, and here we propose an improved DRL that incorporates priors on physical structure of the fetal body. First, we use graph communication layers to improve the communication among agents based on a graph where each node represents a fetal-body landmark. Further, additional reward based on the distance between agents and physical structures such as the fetal limbs is used to fully exploit physical structure. Evaluation of this method on a repository of 3-mm resolution in vivo data demonstrates a mean accuracy of landmark estimation 10 mm of ground truth as 87.3%, and a mean error of 6.9 mm. The proposed DRL for fetal pose landmark search demonstrates a potential clinical utility for online detection of fetal motion that guides real-time mitigation of motion artifacts as well as health diagnosis during MRI of the pregnant mother.

M. Zhang and J. Xu—Equal contribution.

© Springer Nature Switzerland AG 2020
A. L. Martel et al. (Eds.): MICCAI 2020, LNCS 12266, pp. 396–405, 2020.
https://doi.org/10.1007/978-3-030-59725-2_38

Keywords: Multiple landmark detection · Fetal magnetic resonance imaging (MRI) · Deep reinforcement learning (DRL) · Graph communication layers · Physical structure reward

1   Introduction

Extracting localization of fetal pose plays a crucial role in fetal MRI [9]. First, current fetal MRI is heavily constrained by the non-periodic and substantial fetal motion, thus substantial efforts are taken to mitigate motion artifacts by fast, single-shot MRI as well as retrospective methods like slice intersection registration [10]. Estimation of fetal pose during scanning may benefit prospective methods to detect and mitigate fetal motion artifacts via tracking of the motion and when combined with real-time, adaptive slice prescription. Further, fetal pose and fetal body movements may prove useful for monitoring fetal growth or other antenatal surveillance [12,20].

It's time consuming to manually annotate landmarks of fetal pose, and early demonstrations by Xu et al. [19] have addressed this problem with deep learning in convolution neural networks (CNN). Deep reinforcement learning (DRL) [13] is a candidate for an alternative and powerful tool to handle this task. Previous work has been done to automatically detect single landmarks in medical imaging as a Markov Decision Process (MDP) [15] and subsequently applying DRL [5]. A multi-scale strategy [6] and hierarchical action steps [1] have been proposed to further improve the performance of this approach. For the detection of multiple landmarks, [17] proposed a collaborative DQN based on concurrent Partially Observable Markov Decision Process [7]. However, the physical structure and communications among agents are not taken into consideration during decision making. Searching all landmarks of fetal pose, e.g., 15 agents as proposed for our case, is challenging due to the many degrees of freedom in fetal gesture and position, image artifacts and intensity variations. Yet, the fetal body is characterized by robust priors on physical connections between joints, and thus we expect that incorporating realistic skeletal priors will improve detection performance of key points for pose estimation.

To exploit a prior on physical structure, the Graph Convolutional Network (GCN) [3,4,11] provides a powerful tool that represents each landmark of fetal pose as a node in a graph and combines features from both the center node and its neighbors. However, naive spatial domain GCN uses simple, fixed parameters in the adjacent matrix, which results in shared kernel weights over all edges where the relations of internal structure is not well exploited [11]. More recent efforts have been made to address this limitation by learning a semantic relationships among neighbor nodes implied in the edges [22].

Moreover, most landmark-searching DRL only adopt an immediate reward based on the distance between the agent and its target landmark [1,5,6,17]. The smallest and most peripheral joints, i.e. ankles and wrists, represent the most challenging landmark detection task due to contrast and spatial resolution limits of the EPI acquisition, and to compound the difficulty of the problem, these joints tend to be the most mobile of the set of key points that characterize fetal pose. We expect that the incorporation of the physical structure of fetal anatomy will benefit the landmark identification and provide more robust features for these challenging joints [21]. Structure-aware regression on human pose estimation has been used by reparameterizing pose representation using bones instead of joints [14, 16].

In the current work, we propose a novel end-to-end, multi-agent deep reinforcement learning network to detect fifteen landmarks of fetal pose in each frame of a time series of volumetric fetal MRI. Fifteen independent observed MRI volumes are used as input through a parameter shared convolutional network generating fifteen hidden feature presentations. Then fifteen dense layers with graph communication layers process the corresponding fifteen features for better action decisions. Further, an additional and immediate reward is designed based on the distance between the agents and physical structure connections, such as limbs that could help modify the optimal search path through a more robust structure and to improve performance. Our method achieved 87.25% accuracy 6.9 mm as the mean error of landmark detection.

2   Methods

2.1   Deep Reinforcement Learning for Landmark Detection

The task of landmark identification in fetal MRI fits a conventional MDP reinforcement learning framework. We identify five MDP components, $\mathcal{M} = (\mathcal{S}, \mathcal{A}, \mathcal{R}, \mathcal{T}, \gamma)$. $\mathcal{S}$ represents the set of states. $s_t^k \in \mathcal{S}$ is the state at step t of agent k, that is a $48 \times 48 \times 48$ MRI volume centered at the position of corresponded searching agent. $\mathcal{A}$ represents the set of actions, namely, moving *forward*, *backward*, *left*, *right*, *upward* and *downward*. $\mathcal{R}$ represents the immediate reward based on s_t^k and a_t^k. $\mathcal{T}$ is the transition function describing the sequential state under s_t^k and a_t^k, which is deterministic in our case. γ is the discount factor balancing immediate reward and future reward.

Due to the large number of dimension of $\mathcal{S}$, a single-agent RL samples states from MDP and optimizes the target function iteratively by using the experience. There are several algorithms to solve the optimization process. One of them is Q-learning [18] in which we consider the following state-action function

$$Q(s,a) = \mathbb{E}\left[\sum_{i=1}^{n} \gamma^{i-1} r_{t+i} | s, a\right] \tag{1}$$

Q-learning learns the optimal action policy by finding the highest expected future return, annotated as $Q^*(s,a)$. With the Bellman equation, it could be rewritten recursively as $Q^*(s_t, a_t) = E\left[r_t + \gamma \max_{a_{t+1}} Q^*(s_{t+1}, a_{t+1})\right]$ at step t. With the development of deep neural networks, DQN is proposed to approximate $Q(s, a; \theta) \approx Q^*(s, a)$ using a neural network [13].

2.2 Multi-agent RL with a Graph Communication Layer

In terms of multi-agent RL for landmark detection, collab-DQN [17] adopts a shared CNN to extract features from the observed environment of each agent, then estimates $Q(a, s)$ for each agent independently using separate, fully connected networks. This architecture, however, fails to share information across different agents during decision making, which is important for detecting fetal landmarks that are spatially correlated.

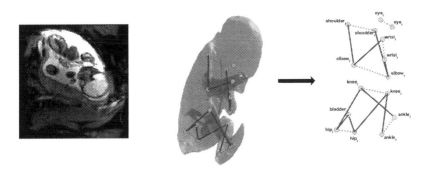

Fig. 1. Left: One slice of a 3D BOLD EPI volume. **Middle and Right**: Contour of fetus and physical structure graph based on pose landmarks. The nodes are 15 landmarks including eyeballs, bladder and joints which are annotated in the figure. There are two types of edge. The blue solid line represents edges of skeleton connection and the yellow dashed line represents edges of right and left landmark correlation. (Color figure online)

Given a graph with K nodes, Graph convolution [11] enables feature extraction from graph-structured data by combining the feature of each node with features of its neighbors, i.e., $\mathbf{H} = \sigma(\mathbf{W}\mathbf{X}\tilde{\mathbf{A}})$, where σ is the activation function, $\mathbf{H} \in \mathbb{R}^{C_o \times K}$ and $\mathbf{X} \in \mathbb{R}^{C_i \times K}$ are output and input features, $\mathbf{W} \in \mathbb{R}^{C_o \times C_i}$ is a learnable parameter matrix, and $\tilde{\mathbf{A}} \in \mathbb{R}^{K \times K}$ is symmetrically normalized from adjacency matrix $\mathbf{A}$ [2]. However, conventional GCN uses fixed parameters in the adjacent matrix which cannot model the difference in the contribution of neighbors to the target node. One way to address this problem is to make adjacent matrix trainable so that the network can learn a semantic relationships of neighbor nodes implied in the edges [22]. Another limitation of GCN is the shared kernel matrix $\mathbf{W}$ which constrains the representation ability where the fact is the correlation between each node in physical structure graph is not equal. We propose the following graph communication layer,

$$\mathbf{H} = \sigma\left(\mathop{\|}_{i=1}^{N} \left(\mathbf{W}^{(i)} x_i\right) \rho(\mathbf{M} \odot \mathbf{A})\right) \tag{2}$$

where $\mathbf{W}^{(i)}$ and x_i are the kernel matrix and input feature of node i, $\|$ represents concatenation, ρ is row-wise softmax, $\mathbf{M}$ is a trainable matrix, and $\odot$ represents element-wise operation defined in [22], resulting in a trainable adjacent matrix.

In this work, as shown in Fig. 1, we choose 15 landmarks (eyeballs, bladder, shoulders, elbows, wrists, hips, knees, and ankles) as the representation of fetal pose, as in [19]. Then we build the graph where each node represents a landmark of fetal pose and two types of edges are considered. One is the physical connection of fetal skeleton and the other is the connection of the left- and right-side landmarks.

The proposed architecture for multi-agent landmark detection is illustrated in Fig. 2. First, a shared convolution network is used to extract features from the observed environment of each agent. Then the graph-structured features are fed into a three-layer graph communication network to merge the information of correlated landmarks before producing the final estimation of Q functions. More network details are listed in the caption. The loss function is defined as,

$$L(\theta) = \sum_{k=1}^{15} \mathbb{E}\left[\left(r_t^k + \gamma \max_{a_{t+1}^k} Q_k\left(s_{t+1}, a_{t+1}^k; \theta^-\right) - Q_k(s_t, a_t^k; \theta)\right)^2\right] \quad (3)$$

where i is the index of each agent, θ is parameters of the target deep neural network and θ^- is parameters of a frozen network which will be updated by the target network every fixed iterations [13].

2.3   Physical Structure Reward

Most DRL in medical landmark searching use immediate reward design based on the distance between the location of the agent and corresponding landmark. For fetal MRI with low-resolution EPI imaging, landmarks like wrists, ankles and elbows have large location distribution and shape variations, but fetal limbs are relatively stable and present with robust signal intensity on these scans and provide a strong and clear physical connection between joint landmarks. We propose an elaboration of the reward design by incorporating the distance between agent and the corresponding limbs such that the reward is used for agent sets $\mathcal{S}_l$ (*shoulders, elbows, wrists, hips, knees and ankles*) searching landmarks on limbs. The reward for agent k at time step t is defined as below,

$$r_t^k = D_l(t, k) - D_l(t+1, k) + \beta \sum_{m \in \mathcal{N}(k)} (D_b(t, k, m) - D_b(t+1, k, m)) \quad (4)$$

where $\mathcal{N}(k)$ is the neighbor of agent k on the corresponded limb. $D_l(t, k) = \left\|p_{k,t} - p_k^{GT}\right\|_2$ is the Euclidean distance between the location of agent $p_{k,t}$ at step t and corresponded landmark p_k^{GT}. $D_b(t, k, m)$ is the Euclidean distance between the location of agent and the segment connecting landmark k and m.

$$D_b(t, k, m) = \begin{cases} \left\|p_{k,t} - p_k^{GT}\right\|_2, & \text{if } (p_{k,t} - p_k^{GT}) \cdot a_{km} > 0 \\ \left\|p_{k,t} - p_m^{GT}\right\|_2, & \text{if } (p_{k,t} - p_m^{GT}) \cdot a_{km} < 0 \\ \left\|(p_{k,t} - p_m^{GT}) \times a_{km}\right\|_2 / \|a_{km}\|_2, & \text{otherwise} \end{cases} \quad (5)$$

where $a_{k,m} = p_k^{GT} - p_m^{GT}$.

β is a parameter that needs to be adjusted to achieve the desired balance during search, where a larger β will force the agent to gravitate more aggressively towards limbs. This will help the CNN to learn more about the features of physical structures, as the optimal path would cover larger limb area, but it will also increase the difficulty of training convergence as the relative effect of reward on moving towards target landmarks is weakened.

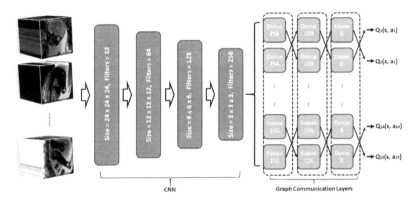

Fig. 2. The architecture of our deep reinforcement learning framework. The green box represents the convolutional block which consists two stacks, each with one 3D convolutional layer with kernel size of 3, stride step of 2, one batch-normalization layers, and one activation layers using ReLU. There are four convolutional blocks with pooling layers of stride 2 following. The channels of each blocks are 32, 64, 128, and 256. The blue dashed box represents graph communication layers (Eq. 2) where there are three graph communication layers with hidden nodes equals to 256, 128, 6. (Color figure online)

3   Experiments and Results

3.1   Dataset

The dataset consists 19,816 3D BOLD MRI volumes acquired on a 3T Skyra scanner (Siemens Healthcare, Erlangen, Germany) with multislice, single-shot, gradient echo EPI sequence. The in-plane resolution is $3 \times 3\text{mm}^2$ and slice thickness is 3 mm. The FOV captures the whole uterus and thus has a variable matrix size from subject to subject. Figure 1 shows one slice of an MRI volume. The gestational age range of the 70 fetuses ranged from 25 to 35 weeks. The mean matrix size was $120 \times 120 \times 80$; TR $= 5$–8 s, TE $= 32$–38 ms, FA $= 90°$.

All 70 fetuses, 19,816 MRI volumes were manually labeled; 49 fetuses, 12,332 volumes were used for training; 14 fetuses, 3,402 volumes were used for validation; 14 fetuses, 4,082 volumes were used for testing.

One bottleneck of DRL is the generalization. To mitigate this, all training datasets were randomly flipped and rotated, which increases the variation of spatial distribution and, further, were randomly scaled with factor in the range of 0.8–1.5.

3.2 Experiments Setup

During training, we used distributed DRL in multiple GPUs with DQN as the optimization algorithm. All experiments were performed on a server with an Intel Xeon E5-1650 CPU, 128 GB RAM and four NVIDIA TITAN X GPU.

Based on [8], our distributed DRL framework consists of N actors and M learners. Each actor fetches parameters from the global model periodically and generates experiences following the current policy, which are sent to shared memory. In each training step, each learner will copy the global model as well as a batch of experiences from shared memory. It then computes the local gradient of the DQN loss function (Eq. 3) and pushes it to a global optimizer. We use Adam as the global optimizer which collects gradients from learners and updates the global asynchronously to avoid congestion due to lock.

In the experiments, We use $N = 4$, $M = 4$, batch size of 3, action step of 1 and learning rate of 3×10^{-4}. 50 is the maximum threshold for gradient clipping.

3.3 Results

In this section, We evaluate the performance of our proposed enhanced DRL incorporating physical structure prior. We compare it with (1) conventional DRL whose frame work and architecture is similar to previous work [17], (2) DRL with graph communication layers, abbr. DRL+GC, (3) DRL with physical structure reward, abbr. DRL+PR, (4) deep learning method with Unet [19].

For fair comparison, all DRL methods are trained under same protocol described in Sect. 3.2. $\beta = 2$ is chose for the physical reward weight. The DL method is trained using the same encoder layers as the shared convolutional network in DRL. The size of the input maintains 48^3 voxels for all experiments.

The evaluation metrics are defined by a) Percentage of Correct Keypoint (PCK), the ratio of detected points if the distance between the detected and the true landmark is within a certain threshold, and b) mean error (in mm), i.e., the mean distance between detected and ground-truth landmarks. The initial location of landmark agents is randomly distributed inner 15% of the volume for both training and evaluation. All DRL evaluation experiments are repeated three times. On average, it takes 1s/volume during inference process.

Table 1 shows the results of all five experiments. Our proposed physical structure enhanced DRL achieves a significant better average detection accuracy 10 mm at 87.3% and outperforms other models in most landmarks. Also, the mean detected error of all landmarks 6.9 mm which is the smallest. Compared with Unet, the worse PCK accuracy but smaller mean error on ankle shows the fact that DRL agent detected the target ankle landmark but scattering around it due to the large size of ankle and foot, while Unet has more false detected cases. To some extent, our proposed method still behaves well.

Table 1. PCK (10 mm) performance and mean error of different models.

Metric	Method	Eye	Shoulder	Elbow	Wrist	Bladder	Hip	Knee	Ankle	All
PCK (%)	Unet	90.8	87.51	75.18	50.99	92.99	63.63	87.08	**56.01**	74.37
	DRL	94.60	96.81	78.50	45.79	96.42	74.86	80.56	31.46	73.44
	DRL+GC	91.33	92.19	86.70	68.10	97.04	83.38	92.33	42.93	80.73
	DRL+PR	94.56	97.93	81.34	55.23	96.70	86.41	86.42	35.54	78.10
	Proposed	**97.35**	**99.79**	**91.21**	**74.29**	**98.33**	**92.27**	**94.75**	55.58	**87.25**
Mean (mm)	Unet	6.33	11.61	8.29	24.87	6.99	25.74	9.60	23.67	18.63
	DRL	5.16	5.64	12.30	34.08	9.21	18.99	10.05	35.73	16.86
	DRL+GC	5.43	7.56	6.99	14.37	5.73	8.19	6.00	23.13	9.93
	DRL+PR	4.14	3.60	10.50	17.49	6.12	8.97	8.49	28.53	11.31
	Proposed	**2.94**	**2.37**	**5.70**	**10.74**	**3.93**	**4.20**	**4.26**	**19.53**	**6.90**

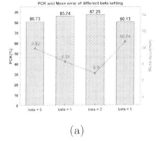

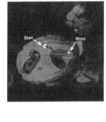

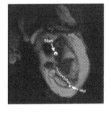

(a)   (b)

Fig. 3. (a) PCK and mean error performance of different β settings. (b) Illustration of searching paths of DRL with and without addition physical reward(PR). Blue line is the path of DRL w/ PR and yellow lines is the path of DRL w/o PR. Solid lines represent paths in current slice and dashed lines represent path in other slices. Red area represents corresponded limb in PR. The left-hand side image shows the successful case where DRL w/ PR recognizes limb structure and goes through it while DRL w/o PR fails. The right-hand side image shows the modified searching path of limb-based reward which will gravitate the agent towards the limb while another will go straight instead. (Color figure online)

In an ablation study, we compared the effects of different weights of physical reward in Eq. 4. We choose $\beta = 0, 1, 2, 5$ with graph communication layers. Seen from Fig. 3a, the model achieves best performances when $\beta = 2$. Consistent with the effect of physical reward, when β is too small, e.g. $\beta = 0$, the network intends to lack of feature extraction ability of physical structure. When β is too large, e.g. $\beta = 5$, the agent will get trapped in extracting physical structure features and moving closer instead of moving towards target landmark.

Figure 3b shows the effect of physical reward (PR). The blue and yellow paths represents DRL with and without PR respectively. On the left, both methods use the same initial location while the distance between the agent and target landmark, right wrist, is over 30 voxels which is outside the FOV of the initial location. The agent of DRL without PR converges to the wrong location, both because the target landmark is beyond the 24-voxel distance, as well as weak feature analysis ability of the surrounding structure. The agent of DRL with PR succeeds by observing the structure of the arm. The right-hand side image shows the effect of modifying searching path where the agent is likely to go through robust structures like limbs which improves accuracy of detection.

4   Conclusion

In this work we proposed the incorporation of physical fetal skeletal structure to enhance DRL to infer landmarks of fetal pose in low-resolution 3D MRI in pregnancy. The proposed method achieves an average detection accuracy of 87.3% under a 10-mm threshold 6.9 mm as the mean error.

Our proposed approach exploits graph communication layers that benefit the communication among agents, as well as rewards derived from proximity to skeletal fetal structure that enhances network performance where search paths gravitate towards skeletal voxels as they approach landmark joints. The significantly convincing performance indicates the success of our proposed method.

Overall, this work demonstrates the potential of using our proposed method to rapidly and robustly detect fetal landmarks for clinical utility that includes estimation of fetal pose, monitoring of fetal motion in health and disease, and motion tracking for real-time slice prescription to mitigating motion artifacts in diagnostic MRI in pregnancy.

Acknowledgements. This research was supported by NIH U01HD087211, NIH R01EB01733 and NIH NIBIB NAC P41EB015902.

References

1. Alansary, A., et al.: Evaluating reinforcement learning agents for anatomical landmark detection. Med. Image Anal. **53**, 156–164 (2019)
2. Biggs, N., Biggs, N.L., Norman, B.: Algebraic Graph Theory, vol. 67. Cambridge University Press, Cambridge (1993)
3. Bruna, J., Zaremba, W., Szlam, A., LeCun, Y.: Spectral networks and locally connected networks on graphs. arXiv preprint arXiv:1312.6203 (2013)
4. Defferrard, M., Bresson, X., Vandergheynst, P.: Convolutional neural networks on graphs with fast localized spectral filtering. In: Advances in Neural Information Processing Systems, pp. 3844–3852 (2016)
5. Ghesu, F.C., Georgescu, B., Mansi, T., Neumann, D., Hornegger, J., Comaniciu, D.: An artificial agent for anatomical landmark detection in medical images. In: Ourselin, S., Joskowicz, L., Sabuncu, M.R., Unal, G., Wells, W. (eds.) MICCAI 2016. LNCS, vol. 9902, pp. 229–237. Springer, Cham (2016). https://doi.org/10.1007/978-3-319-46726-9_27
6. Ghesu, F.C., et al.: Multi-scale deep reinforcement learning for real-time 3D-landmark detection in CT scans. IEEE Trans. Pattern Anal. Mach. Intell. **41**(1), 176–189 (2017)

7. Girard, J., Emami, M.R.: Concurrent Markov decision processes for robot team learning. Eng. Appl. Artif. Intell. **39**, 223–234 (2015)
8. Horgan, D., et al.: Distributed prioritized experience replay. arXiv preprint arXiv:1803.00933 (2018)
9. Jokhi, R.P., Whitby, E.H.: Magnetic resonance imaging of the fetus. Dev. Med. Child Neurol. **53**(1), 18–28 (2011)
10. Kim, K., Habas, P.A., Rousseau, F., Glenn, O.A., Barkovich, A.J., Studholme, C.: Intersection based motion correction of multislice MRI for 3-D in utero fetal brain image formation. IEEE Trans. Med. Imaging **29**(1), 146–158 (2009)
11. Kipf, T.N., Welling, M.: Semi-supervised classification with graph convolutional networks. arXiv preprint arXiv:1609.02907 (2016)
12. Lai, J., Nowlan, N.C., Vaidyanathan, R., Shaw, C.J., Lees, C.C.: Fetal movements as a predictor of health. Acta Obstet. Gynecol. Scand. **95**(9), 968–975 (2016)
13. Mnih, V., et al.: Human-level control through deep reinforcement learning. Nature **518**(7540), 529–533 (2015)
14. Sun, X., Shang, J., Liang, S., Wei, Y.: Compositional human pose regression. In: Proceedings of the IEEE International Conference on Computer Vision, pp. 2602–2611 (2017)
15. Sutton, R.S., Barto, A.G., et al.: Introduction to Reinforcement Learning, vol. 135. MIT Press, Cambridge (1998)
16. Tang, W., Yu, P., Wu, Y.: Deeply learned compositional models for human pose estimation. In: Ferrari, V., Hebert, M., Sminchisescu, C., Weiss, Y. (eds.) ECCV 2018. LNCS, vol. 11207, pp. 197–214. Springer, Cham (2018). https://doi.org/10.1007/978-3-030-01219-9_12
17. Vlontzos, A., Alansary, A., Kamnitsas, K., Rueckert, D., Kainz, B.: Multiple landmark detection using multi-agent reinforcement learning. In: Shen, D., et al. (eds.) MICCAI 2019. LNCS, vol. 11767, pp. 262–270. Springer, Cham (2019). https://doi.org/10.1007/978-3-030-32251-9_29
18. Watkins, C.J., Dayan, P.: Q-learning. Mach. Learn. **8**(3–4), 279–292 (1992). https://doi.org/10.1007/BF00992698
19. Xu, J., et al.: Fetal pose estimation in volumetric MRI using a 3D convolution neural network. In: Shen, D., et al. (eds.) MICCAI 2019. LNCS, vol. 11767, pp. 403–410. Springer, Cham (2019). https://doi.org/10.1007/978-3-030-32251-9_44
20. Yen, C.J., Mehollin-Ray, A.R., Bernardo, F., Zhang, W., Cassady, C.I.: Correlation between maternal meal and fetal motion during fetal MRI. Pediatr. Radiol. **49**(1), 46–50 (2019). https://doi.org/10.1007/s00247-018-4254-1
21. Zhang, P., Wang, F., Zheng, Y.: Deep reinforcement learning for vessel centerline tracing in multi-modality 3D volumes. In: Frangi, A.F., Schnabel, J.A., Davatzikos, C., Alberola-López, C., Fichtinger, G. (eds.) MICCAI 2018. LNCS, vol. 11073, pp. 755–763. Springer, Cham (2018). https://doi.org/10.1007/978-3-030-00937-3_86
22. Zhao, L., Peng, X., Tian, Y., Kapadia, M., Metaxas, D.N.: Semantic graph convolutional networks for 3D human pose regression. In: Proceedings of the IEEE Conference on Computer Vision and Pattern Recognition, pp. 3425–3435 (2019)

Automatic Angle of Progress Measurement of Intrapartum Transperineal Ultrasound Image with Deep Learning

Minghong Zhou, Chao Yuan, Zhaoshi Chen, Chuan Wang, and Yaosheng Lu[✉]

Department of Electronic Engineering, College of Information Science and Technology,
Jinan University, Guangzhou 510632, China
tluys@jnu.edu.cn

Abstract. Angle of progress (AOP) is an important indicator used in assessing the progress of labor during delivery. However, manually measuring AOP is time consuming and subjective. In this study, we address the challenge of automatic AOP measurement of transperineal ultrasound (TPU) to achieve accurate monitoring of maternal and infant status. We propose a multitask framework for simultaneously locating the landmark of pubic symphysis endpoints and segmenting the region of the fetal head and pubic symphysis. We then exploit the localization of the landmarks to obtain the central axis of pubic symphysis. Afterward, we calculate the tangent of fetal head as it passes through the lower endpoint of pubic symphysis. Finally, we compute AOP from the central axis and tangent. Our framework is evaluated on the basis of a TPU dataset acquired at The First Affiliated Hospital of Jinan University, which is annotated by an ultrasound physician with over 10 years of experience. Our method achieves a mean difference of 7.6° and displays promising prospects for real-time monitoring of labor progress in clinical practice. To the best of our knowledge, this study is the first to apply deep learning methods to AOP measurements.

Keywords: Multitask learning · Ultrasound image · Angle of progress

1 Introduction

Abnormal progress of labor can harm the mother and fetus and may even lead to dystocia. Therefore, monitoring of the labor process is extremely crucial during delivery. Angle of progress (AOP) describes the relationship between pubic symphysis and fetal head. AOP has been introduced as an effective indicator for monitoring the labor process on the basis of transperineal ultrasound (TPU) [1]. As shown in Fig. 1, the central axis of pubic symphysis (CAOP) connects its upper and lower endpoints. AOP is the angle between CAOP and the tangent to fetal skull contour past the lower endpoint. Manually measuring AOP requires extensive experience, is subjective, and time consuming. Thus, research on automatic AOP measurement method is important to increase the accuracy and efficiency of intrapartum diagnosis.

© Springer Nature Switzerland AG 2020
A. L. Martel et al. (Eds.): MICCAI 2020, LNCS 12266, pp. 406–414, 2020.
https://doi.org/10.1007/978-3-030-59725-2_39

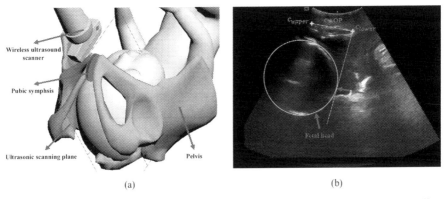

(a)   (b)

Fig. 1. The red object in (a) is the pubic symphysis, whereas the area enclosed by the red lines indicates the scanning plane of the ultrasound probe. When scanning, the blue square on the probe is at the upper end and corresponds to the small square above the image in (b), which is used to indicate the direction. (b) A transperineal ultrasound image. c_{upper} and c_{lower} represent the upper and lower endpoints of pubic symphysis, respectively. l_t denotes the tangent of fetal head passing through c_{lower}. AOP is the angle between CAOP and l_t. (Color figure online)

Few studies have attempted to tackle this challenging task. Conversano et al. [3] considered the centroid instead of the lower symphysis endpoint as the landmark for AOP measurement. They then utilized morphological filters to locate pubic symphysis and fetal head on the basis of their positions and gray levels. However, this method requires the operator to move the pubic symphysis into a fixed elliptic area at the interface during image acquisition, thereby increasing the difficulty of operation. Furthermore, the local and global contexts of the image are underutilized. Designing a robust and efficient measurement method remains a challenging task for the following reasons. (1) In addition to the target structure, other anatomical structures in the TPU image, such as the bladder, further exacerbates the effects of noise and artifacts commonly present in ultrasound images. (2) The boundaries of pubic symphysis are blurred and make the process of locating the endpoints difficult. (3) When performing an ultrasound scan, the ultrasound generally needs to pass through the pubic symphysis before it reaches the fetal head, resulting in unclear and missing fetal head contours in the image.

To address these issues, we decompose AOP measurements into several associated tasks and propose an encoder–decoder multitask network with a shared feature encoder. Our method achieves excellent performance and shows promising prospects for real-time monitoring of labor progress in clinical practice. To the best of our knowledge, this technique is the first fully automatic AOP measurement method that employs deep learning on TPU images. Moreover, this method does not require additional operations when performing ultrasound scans.

2   Methods

As shown in Fig. 2, the proposed multitask network consists of two parts: a shared feature-encoder part and a task-separated feature-decoder part. The decoder part has two branches, a segmentation branch and a localization branch. The segmentation branch is used to perform pixel-level prediction of the input image to segment the pubic symphysis and fetal head regions, whereas the localization branch aims to locate the upper and lower pubic symphysis endpoints by regressing Gaussian heatmaps [7]. Given a TPU image I, multiscale features are first extracted by the shared encoder, and then the extracted features are inputted into two task branches to generate the predicted segmentation map P and the regression Gaussian heatmap H. We treat the actual fetal head contour as an ellipse and use an elliptic curve E to fit the predicted fetal head region in the segmentation map P. We then take the pixel with the highest intensity in the Gaussian heatmap as the corresponding predicted endpoint coordinate and further connect these two endpoints as CAOP. Finally, E and CAOP are utilized to compute AOP. Each part of our methods is discussed in detail in the following subsections.

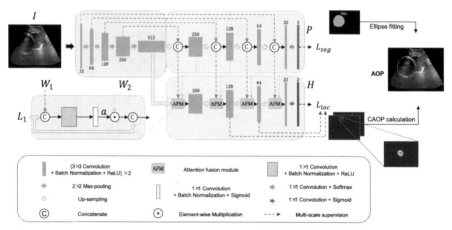

Fig. 2. Framework of the algorithm. The blue area is the shared encoder part, the green area is the segmentation task part, the red area is the localization task part, and the orange area is the internal structure of the AFM module. The network inputs a TPU image I and finally outputs the segmentation results P and predicted heatmaps H to compute AOP. a is the generated attention mask. (Color figure online)

2.1 Multitask Network

Encoder. The design of our convolutional layer structure follows the VGG network [9]. The encoder includes five convolutional blocks. Each convolution block contains two convolutional layers with a convolution kernel size of 3×3 and a 2×2 max-pooling layer to reduce image size. Each convolutional layer is followed by a batch normalization layer and a ReLU layer. In addition, we add a dropout layer with 50% dropout rate to the last convolution block to prevent overfitting.

Segmentation Branch. The segmentation branch includes four up-sampling layers and four convolutional blocks. Inspired by the success of the U-Net structure [8], we introduce skip connections to the segmentation branch to merge high-level semantic information and shallow detail information from the shared encoder. Our segmentation branch finally divides all pixels into three categories: background, pubic symphysis, and fetal head. To address the problem of imbalance in the number of pixels in the three categories, we employ dice loss [6] as the loss function of our segmentation task.

Localization Branch. Like the segmentation branch, the localization branch also adopts a fully convolutional structure and gradually increases the resolution of the feature map through the up-sampling layer. The localization branch plays an important role in our method because the predicted coordinates of the endpoints greatly affect the final measurement result, i.e., AOP. Therefore, to improve the accuracy of localization, we propose to embed an attention fusion module (AFM) to fuse segmentation task features into localization tasks inspired by [5]. Furthermore, a previous study has demonstrated that deep supervision helps improve localization accuracy [10]. Thus, we employ multiscale supervision in our localization task, as depicted in Fig. 2. The first convolution block in the localization branch does not implement this supervision operation. We observe that when the ground-truth heatmaps are reduced to the same resolution as the first convolution block, the high-intensity region generated by Gaussians almost disappears and can hardly perform any supervision.

2.2 Attention Task Feature Fusion

As mentioned earlier, we merge shallow and deep features in the segmentation task. Thus, the feature maps in this branch contain rich multilevel feature information. We propose the AFM module to fuse these multilevel features into the localization branch. Specifically, this module accepts W_1, W_2, and L_1 as inputs, which have the same resolution. W_1 represents the feature map obtained by concatenating the output of the up-sampling layer with the shallow feature map, i.e., the input of the convolution block in the segmentation branch. W_2 is the output feature map of the same convolution block with W_1 as input. L_1 denotes the output feature map of the up-sampling layer in the localization branch. First, the concatenated W_1 and L_1 are inputted into a 1×1 convolutional block to produce a feature map with the same shape as W_2. The generated feature map is then inputted into a 1×1 convolutional layer and a sigmoid activation layer to generate an attention mask a with intensity within $[0, 1]$. Finally, element-wise multiplication is applied to the attention mask a and W_2 to obtain a weighted feature map, which is then concatenated with L_1 as the output of the module. In summary, the proposed AFM weights the feature maps extracted from the segmentation task through the generated attention mask to focus on the features useful for the localization task. The weighted feature maps are merged into the localization branch by channel-dimensional concatenating, as shown in Fig. 2.

2.3 Loss and Optimization

We use dice loss in the segmentation task and mean squared error (MSE) loss in the localization task to train our network. Multiscale supervision is introduced into the localization branch to calculate the MSE loss of the middle layer and the final output.

Therefore, the localization loss is the sum of the MSE losses calculated at multiple scales:

$$L_{loc} = \sum_{k=1}^{3} \delta_k L_{MSE}^k = \sum_{k=1}^{3} \delta_k \sum_{i,j} \left\| H_k(i,j) - \hat{H}_k(i,j) \right\|^2 \tag{1}$$

where δ_k is the loss weight of L_{MSE}^k; the weights when k = 1, 2, and 3 are empirically set to be 1.0, 0.8, and 0.6, respectively; H_k and $\hat{H}_k$ represent the output heatmaps and the ground-truth heatmaps, respectively; and (i,j) correspond to pixel locations.

We simultaneously perform end-to-end training on the segmentation and localization tasks. The balance of task weights is important, but it is difficult to implement in training multitasking networks. Hence, the balance has a substantial effect on the performance of the final trained network. Therefore, we adopt a weight uncertainty method [4] to adjust the task weights automatically during training. The final loss function of the entire network is given as:

$$L_{all} = p_1 L_{dice} + p_2 L_{loc} \tag{2}$$

where p_1 and p_2 are the task weights determined via weight uncertainty method during training.

2.4   AOP Measurement

To facilitate more accurate AOP measurement, we consider the head contour as an ellipse. We first separate the head region from the predicted output and then only remain the largest connected component to extract the contour of the head region by performing contour extraction. We then use the least square method to perform ellipse fitting on the extracted fetal head contour. The contour extraction algorithm and ellipse fitting algorithm we use are both from OpenCV [2]. On the basis of the two endpoint coordinates obtained in the localization task, we can further compute the AOP as follows:

$$AOP = \cos^{-1}\left(\frac{l_1 \cdot l_t}{|l_1||l_t|} \right) \tag{3}$$

where l_1 represents CAOP, and l_t represents the fetal head tangent through the lower endpoint of pubic symphysis.

3   Experiments

3.1   Data and Implementation Detail

We first obtain a TPU video dataset consisting of 84 maternal data from The First Affiliated Hospital of Jinan University. All videos are obtained using a Youkey D8 wireless 2D ultrasound probe with its corresponding supporting software. Afterward, 313 TPU images are selected from this dataset, and the pixel values of the area where the toolbar and parameters are displayed in the image are set to zero. Each maternal datum includes 2–5 images. A physician with over 10 years of experience in gynecological and obstetric ultrasound is invited to annotate manually the fetal head region, pubic symphysis region and its endpoints, and AOP. The ground-truth heatmaps are obtained by centering Gaussians at the marked endpoints. We randomly divide all images into a training set and a testing set at a ratio of 8:2, and images of the same maternal datum is only assigned to the same set. We double the training set image by horizontal flipping and use random rotation ($-30°$, $30°$) and random scaling for data augmentation during training. Before the image is inputted to the network during training and testing, it is first resized to a size of 512×384 and normalized to $[-1, 1]$. 5-fold cross-validation is used to evaluate the performance of the method, and the reported experimental results are the average of five verification tests.

All methods in our experiments are implemented on the basis of Pytorch and run on a Nvidia Titan V GPU. During training, Adam is used to optimize the network with a learning rate of 0.0001. The network weights are initialized using the Xavier algorithm and trained for 200 epochs with a batch size of 2. Aside from the proposed multitask attention fusion network (MTAFN), we also implement several networks for comparison.

Single Task U-Net (STU). We train two independent U-Nets to perform segmentation and localization tasks as the baseline of a single task.

MTAFN Without AFM. To verify the contribution of our proposed AFM module to the overall performance, we remove the AFM module and the connection between two branches according to the proposed network structure and then add a skip connection to the localization branch to merge the shared shallow features.

Evaluation Metrics. For the segmentation task, we use four metrics to evaluate the results, including pixel accuracy (Acc), dice score of the pubic symphysis and fetal head regions ($dice_{sp}$, $dice_{head}$), and overall dice score ($dice_{all}$). $dice_{all}$ evaluates the overall similarity of the segmentation results to ground-truth except for the background. For the

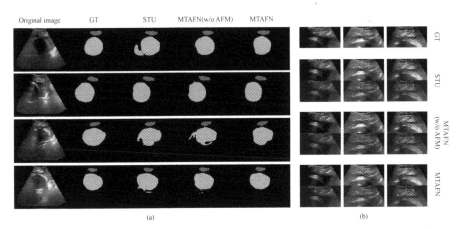

Fig. 3. (a) Visualization of segmentation results. (b) Output heatmaps and predicted coordinates of each network localization task. The cross indicates the endpoint, and the line connecting the two crosses denotes CAOP.

Table 1. Segmentation and localization results.

Method	Acc	$dice_{sp}$	$dice_{head}$	$dice_{all}$	d_{upper} (mm)		d_{lower} (mm)		$Angle_{CAOP}$ (°)	
					Median	Mean ± SD	Median	Mean ± SD	Median	Mean ± SD
STU	0.980	0.887	0.898	0.895	2.34	5.11 ± 9.44	2.76	4.14 ± 5.64	3.04	5.86 ± 11.46
MTAFN (w/o AFM)	0.982	0.886	0.903	0.903	2.03	4.04 ± 5.64	2.62	3.85 ± 5.32	2.49	4.68 ± 7.29
MTAFN	0.982	0.901	0.907	0.907	2.56	4.01 ± 4.61	2.27	3.24 ± 4.35	2.14	3.76 ± 5.56

localization task, we introduce the Euclidean distance (d_{upper}, d_{lower}) between the predicted endpoints coordinates and the ground-truth coordinates, and the included angle ($Angle_{CAOP}$) between the predicted CAOP and the ground-truth CAOP. The means and standard deviations of these two evaluation indicators are also calculated. Our ultimate goal is to measure AOP. Thus, we finally use an evaluation indicator, namely, the difference (AD_{AOP}) between the predicted angle and the true angle, to evaluate the performance of the automatic angle measurement method.

3.2   Result Analysis and Discussion

We quantitatively compare the performance of the three networks in the segmentation task, localization task, and the final AOP measurement results. As shown in Tables 1 and 2, the performance of the multitask model is better than that of training the single-task model in most metrics. Moreover, the multitask model has fewer parameters and calculations than the single-task model. Our proposed network has 12 M parameters, and the average time to infer an image in our machine is 0.29 s. Simultaneous training of two related tasks in a balanced manner can mutually promote the training process. During training, the multitask model has substantially converged faster than the single-task model. Furthermore, the performance of the proposed AFM module is also experimentally verified. In Fig. 3(b), the output heatmaps of MTAFN have higher intensity near the

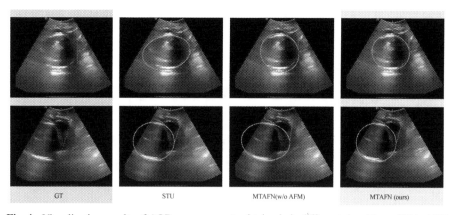

Fig. 4. Visualization results of AOP measurements obtained via different algorithms. GT is AOP manually drawn by the physician.

Table 2. Comparison of AOP measurement results.

		STU	MTAFN (w/o AFM)	MTAFN
$AD_{AOP}(°)$	Median	6.03	4.98	4.68
	Mean ± SD	9.26 ± 10.22	7.64 ± 8.29	7.6 ± 8.85

target location. The visualized heatmap shows that the AFM module can clearly focus on the target area and enhance the feature map of the area of interest. Meanwhile, it can effectively reduce the overlapping of two landmark heatmaps. Aside from improving the locating performance, the AFM module also remarkably improves $dice_{sp}$ in the segmentation task. In most cases, the AOP measured by our method is close to the label. However, some ultrasound images have missing fetal head contours, resulting in poor fetal head segmentation results and large AD_{AOP}. The method reported in [3] requires careful manual adjustment of ultrasound scanning depth and position to assist in locating pubic symphysis and fetal head. Our method can perform automatic positioning and use an elliptic curve (circular curve in [3]) to fit the contour of the fetal head. Thus, it can reduce operational complexity and is more robust and flexible in clinical applications (Fig. 4).

4   Conclusion

We propose a multitask deep learning method for fully automatic AOP measurements. By decomposing the AOP measurement task into two simple tasks and simultaneously using a multitask learning network to make inferences, the time to perform AOP measurements is substantially reduced and a promising performance is achieved. Experimental results also show that the proposed AFM module can effectively improve network performance. Our future work will further perform ablation experiments on the network and evaluate the effects of different convolution block structures, such as residual and dense blocks, on the overall network performance.

Acknowledgement. This work was supported by the National Key Research and Development Project [2019YFC0120100, 2019YFC0121907].

References

1. Barbera, A., Pombar, X., Perugino, G., Lezotte, D., Hobbins, J.: A new method to assess fetal head descent in labor with transperineal ultrasound. Ultrasound Obstet. Gynecol. 33(3), 313–319 (2009)
2. Bradski, G., Kaehler, A.: Opencv. Dr. Dobb's J. Softw. Tools 3 (2000)
3. Conversano, F., et al.: Automatic ultrasound technique to measure angle of progression during labor. Ultrasound Obstet. Gynecol. 50(6), 766–775 (2017)
4. Kendall, A., Gal, Y., Cipolla, R.: Multi-task learning using uncertainty to weigh losses for scene geometry and semantics. In: Proceedings of the IEEE Conference on Computer Vision and Pattern Recognition, pp. 7482–7491 (2018)

5. Liu, S., Johns, E., Davison, A.J.: End-to-end multi-task learning with attention. In: Proceedings of the IEEE Conference on Computer Vision and Pattern Recognition, pp. 1871–1880 (2019)
6. Milletari, F., Navab, N., Ahmadi, S.A.: V-net: Fully convolutional neural networks for volumetric medical image segmentation. In: 2016 Fourth International Conference on 3D Vision (3DV), pp. 565–571. IEEE (2016)
7. Payer, C., Štern, D., Bischof, H., Urschler, M.: Integrating spatial conguration into heatmap regression based CNNS for landmark localization. Med. Image Anal. **54**, 207–219 (2019)
8. Ronneberger, O., Fischer, P., Brox, T.: U-Net: convolutional networks for biomedical image segmentation. In: Navab, N., Hornegger, J., Wells, W.M., Frangi, A.F. (eds.) MICCAI 2015. LNCS, vol. 9351, pp. 234–241. Springer, Cham (2015). https://doi.org/10.1007/978-3-319-24574-4_28
9. Simonyan, K., Zisserman, A.: Very deep convolutional networks for large-scale image recognition. arXiv preprint arXiv:1409.1556 (2014)
10. Tompson, J.J., Jain, A., LeCun, Y., Bregler, C.: Joint training of a convolutional network and a graphical model for human pose estimation. In: Advances in Neural Information Processing Systems, pp. 1799–1807 (2014)

Joint Image Quality Assessment and Brain Extraction of Fetal MRI Using Deep Learning

Lufan Liao[1,2], Xin Zhang[1(✉)], Fenqiang Zhao[2], Tao Zhong[2], Yuchen Pei[2], Xiangmin Xu[1], Li Wang[2], He Zhang[3], Dinggang Shen[2], and Gang Li[2(✉)]

[1] School of Electronic and Information Engineering,
South China University of Technology, Guangzhou, China
`eexinzhang@scut.edu.cn`
[2] Department of Radiology and BRIC,
The University of North Carolina at Chapel Hill, Chapel Hill, NC, USA
`gang_li@med.unc.edu`
[3] Department of Radiology, Obstetrics and Gynecology Hospital,
Fudan University, Shanghai, China

Abstract. Quality assessment (QA) and brain extraction (BE) are two fundamental steps in 3D fetal brain MRI reconstruction and quantification. Conventionally, QA and BE are performed independently, ignoring the inherent relation of the two closely-related tasks. However, both of them focus on the brain region representation, so they can be jointly optimized to ensure the network to learn shared features and avoid over-fitting. To this end, we propose a novel multi-stage deep learning model for joint QA and BE of fetal MRI. The locations and orientations of fetal brains are randomly variable, and the shapes and appearances of fetal brains change remarkably across gestational ages, thus imposing great challenges to extract shared features of QA and BE. To address these problems, we firstly design a brain detector to locate the brain region. Then we introduce the deformable convolution to adaptively adjust the receptive field for dealing with variable brain shapes. Finally, a task-specific module is used for image QA and BE simultaneously. To obtain a well-trained model, we further propose a multi-step training strategy. We cross validate our method on two independent fetal MRI datasets acquired from different scanners with different imaging protocols, and achieve promising performance.

Keywords: Fetal MRI · Quality assessment · Brain extraction

1 Introduction

Fetal magnetic resonance imaging (MRI) is increasingly being used in prenatal care to better visualize the rapidly developing brain and detect abnormalities [2,16]. However, fetal MRI is susceptible to unpredictable and severe fetal movements, which negatively affect the image quality and cause motion artifacts.

© Springer Nature Switzerland AG 2020
A. L. Martel et al. (Eds.): MICCAI 2020, LNCS 12266, pp. 415–424, 2020.
https://doi.org/10.1007/978-3-030-59725-2_40

Current fetal MR protocols attempt to address this by repeating each orthogonal stack of 2D slices multiple times, but with no guarantee that the slices are of desired quality [19]. It is hence necessary to identify problematic images with poor quality and exclude them. Otherwise, the problematic images will degenerate the performance of fetal brain reconstruction and analysis. Recently, deep learning based methods have demonstrated a great potential for automated image quality assessment (QA) [8,12,14,19]. Liu et al. [12] proposes a three-stage model, with each stage responsible for slice-wise, volume-wise, and subject-wise QA, respectively. Unlike image QA methods for 3D postnatal pediatric images, our focus is on the 2D slice quality of the fetal brain region. The most related work is Lala et al. [19], which fine-tunes a 50-layer ImageNet pre-trained ResNet [5] for fetal image QA.

Fetal brain reconstruction and analysis also depend on the accurate brain extraction [1,4,6,7,9,10,13,17,18,20–22]. Manual brain extraction (BE) is cumbersome and time consuming. Therefore, automated BE in fetal MRI is highly needed to simplify the intensive manual label procedure. Salehi et al. [18] uses a 2D U-Net [15] for fetal brain extraction directly. Ebner et al. [4] proposes a two-stage convolutional neural network model for fetal brain segmentation, including a brain localization stage and a fine segmentation stage. However, these methods do not consider the fact that the shapes and appearances of fetal brains change remarkably across gestational ages.

Existing algorithms perform QA and BE independently by using different strategies, and treat them as two irrelevant tasks. However, these two tasks are closely-related as they both pay attention to the brain region representation. In this paper, we propose a novel multi-stage QA and BE model to simultaneously identify image quality and extract brain in fetal MRI. Considering the random locations and orientations of fetal brains, we design a brain detector to locate the brain region. Moreover, we introduce deformable convolution [3] to improve the network's transformation modeling capability, thus learning free deformations for various brain shapes. This multi-task architecture can be jointly optimized and ensure the network to learn shared features and avoid overfitting. To further obtain a well-trained model, we propose a multi-step training strategy.

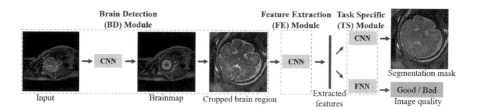

Fig. 1. The flowchart of our proposed algorithm.

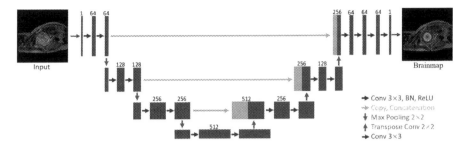

Fig. 2. Illustration of the brain detection module in Fig. 1.

2   Method

We design a three-step deep learning model for joint image QA and BE of fetal MRI. As shown in Fig. 1, our model consists of three components, i.e., 1) a brain detection (BD) module, 2) a feature extraction (FE) module, and 3) a task specific (TS) module. Fig. 2 and Fig. 3 show the detailed network architectures of three modules, which are described in details in following sections.

2.1   Brain Detection Module

Brainmap. The BD module aims to find the location of the brain region. Instead of regressing the center of the brain bounding box (CBBB) directly, we regress the heatmap of CBBB, which can improve the robustness of coordinate positioning. This is because, given an image without a brain region, the coordinate-based network still outputs an unrealistic coordinate. Herein, we use 2D probabilistic distribution to represent the probability of the CBBB. The output of the network has the same shape as the input image, called brainmap. If no brain exists, we set pixel values as 0 in the ground truth brainmap. Otherwise we place a 2D Gaussian with fixed variance at the CBBB in the ground truth brainmap.

Based on the output brainmap, we check whether the average of top k maximum pixel values is large than a threshold t, which represents the minimal probability of being a brain. If the average value is large than t, we regard the average position of top k maximum values as the CBBB. On the contrary, if there is no brain detected, we evaluate this image as poor quality and exclude it for further processing, since this type of image cannot provide effective information for fetal brain reconstruction and analysis. Once we acquire the CBBB, we can crop out the brain region with an expanded square bounding box.

Network. In this module, we train a variant of U-Net as the backbone architecture. It contains an encoder (i.e., the left part of Fig. 2) to extract spatial features at multi-resolution from the input image, and a subsequent decoder (i.e., the right part of Fig. 2) to estimate the brainmap. In this architecture, features maps at various spatial scales are fused to combine global features with local features and generate the final brainmap.

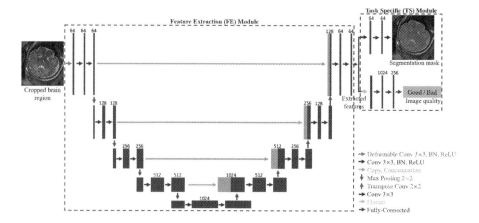

Fig. 3. Illustration of feature extraction module and task specific module in Fig. 1.

2.2   Feature Extraction Module

This sub-network takes the cropped brain region as input to learn a nonlinear mapping from an image to effective features for both QA and BE. Since the orientation of the fetal brain is randomly variable and the shape of the brain is complex, we no longer fix the regular structure of the kernel for convolution, thus the convolution can easily pay attention to interested regions, and further capture essential features of the brain. Specifically, we apply deformable convolution (i.e., the first convolution layer in Fig. 3), which adds offsets to the regular grid sampling location in the standard convolution to enable adaptive receptive field for objects with different shapes. Then, a network similar to U-Net, which can capture holistic and local information simultaneously, is leveraged to extract shared effective features for both QA and BE.

2.3   Task Specific Module

The task specific (TS) module contains two separate sub-networks for the BE task (TS_BE) and the image QA task (TS_QA), respectively. The TS_BE (i.e., the top part of the TS module in Fig. 3) is responsible for transforming the extracted features to the segmentation mask. Herein, we use two convolution layers to learn this mapping relationship. The fully-connected layer has full connections to all activations in the previous layer, which can fuse feature information globally. Hence, three fully-connected layers are used in TS_QA (i.e., the bottom part of the TS module in Fig. 3) to transfer features from the FE module to the image quality. Moreover, dropout layers are added after the first two layers to prevent the network from overfitting.

2.4   Loss Function

Given a set of image samples $\mathcal{D} = \{(I_i, H_i, Q_i, S_i)\}_{i=1,2,...,N}$, where I_i, H_i, Q_i, and S_i denote the input image, ground truth heatmap, quality category, and segmentation mask, respectively, the BD module learns the network parameters by minimizing the Mean Squared Error (MSE):

$$L_{BD} = \frac{1}{N} \sum_{i=1}^{N} ||f_{BD}(I_i) - H_i||^2, \tag{1}$$

where f_{BD} denotes the brain detection operator. After the processing of BD module, we acquire a set of cropped brain region samples $RD = \{(RI_i, RQ_i, RS_i)\}_{i=1,2,...,M}$, where RQ_i is equal to Q_i, RI_i, and RS_i are the corresponding brain region of I_i, and S_i. Note that all these M images in RD have brain regions.

The FE module learns the mapping relationship operation f_{FE} between a cropped brain region RI_i and extracted features Z_i:

$$Z_i = f_{FE}(RI_i), \tag{2}$$

For the TS_QA sub-network, we aim to minimize the discrepancy between RQ_i (i.e., also called y_i) and the predicted $R\hat{Q}_i$ (i.e., also called $\hat{y}_i$). We adopt a multi-class balanced focal loss [11] as the measurement:

$$y_i = RQ_i, \quad \hat{y}_i = R\hat{Q}_i = f_{TS\_QA}(Z_i),$$

$$L_{QA} = \frac{1}{M} \sum_{i=1}^{M} (-(1-\alpha)y_i(1-\hat{y}_i)^\lambda log(\hat{y}_i) - \alpha(1-y_i)(\hat{y}_i)^\lambda log(1-\hat{y}_i)), \tag{3}$$

where f_{TS_QA} denotes the quality assessment operator, α is used to balance the importance of positive and negative samples, and λ is a focusing parameter.

We also control the difference between the final regressed segmentation mask $f_{TS\_BE}(Z_i)$ and the corresponding ground truth RS_i by a MSE loss:

$$L_{BE} = \frac{1}{N} \sum_{i=1}^{M} ||f_{TS\_BE}(Z_i) - RS_i||^2, \tag{4}$$

where f_{TS_BE} denotes the brain extraction operator.

Actually, there is no segmentation mask when the image is of poor quality. So when we jointly train the TS module, we add an additional mask to prevent the network from backpropagating segmentation error on poor quality images. We define the joint training objective function of the TS as:

$$L_{TS} = \beta L_{QA} + \frac{1}{\gamma + \sum_{i=1}^{M} m_i} \sum_{i=1}^{M} m_i||f_{TS\_BE}(Z_i) - RS_i||^2, \tag{5}$$

where β balances the importance of the two tasks, γ is a constant value to avoid the denominator is equal to 0, and $m_i \in \{0,1\}$. When the image is of good quality, m_i is equal to 1, otherwise is equal to 0.

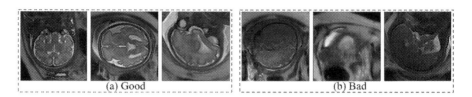

Fig. 4. Examples of fetal MR slices labeled as 'good' and 'bad'.

2.5   Training Strategy

To train the proposed three-step model, we design a four-step training strategy, including separate and joint trainings. Specifically, 1) we separately train the BD module by optimizing L_{BD} to obtain a brain detector; 2) we fix parameters in the BD module and train FE and TS_BE modules by minimizing L_{BE}; 3) we fix weights in the BD module and train FE and TS_QA modules by optimizing L_{QA}; 4) we fix weights in the BD module and jointly train FE and TS modules by minimizing L_{TS}. Here, Step 2 is before Step 3, because the segmentation task can prompt the FE module to focus on more semantic information.

3   Experiments

3.1   Datasets

To show the robustness and effectiveness of the proposed model, we perform cross-validation on two fetal datasets acquired from different scanners with different imaging protocols. Each stack is acquired through fast 2D snapshot imaging to minimize the effects of motion. And these stacks are acquired in three orthogonal views, i.e., axial, sagittal, and coronal orientations. Specifically, Dataset A contains 4152 slices from 178 stacks from 36 subjects, with the gestational age ranging from 21 weeks to 36 weeks. These stacks are acquired using a 1.5T Siemens Avanto scanner with the resolution of $0.54 \times 0.54 \times 4.4 \,\mathrm{mm}^3$. Dataset B contains 2861 slices from 123 stacks from 20 subjects. These stacks are acquired using a 1.5T Siemens Aera scanner with the resolution of $1.17 \times 1.17 \times 3.5 \,\mathrm{mm}^3$.

3.2   Data Preprocessing and Network Details

Ground truth segmentation mask and image quality of each slice in each stack are produced by manual annotation from experts. Figure 4 shows examples of image slices labeled as 'good' (minor artifacts/noises) and 'bad' (heavy artifacts/noises). The size of the expanded bounding box to crop out the brain region is 256×256. The Gaussian variance is set to 25 to get the brainmap. The threshold t, which represents the minimal probability of being a brain region, is set to 0.7. The focal loss parameters λ and α are set to 2 and 0.25, respectively. The scalar β for balancing two joint losses is set to 0.1. To improve the generalization of our model and prevent overfitting, we use two data augmentation methods. One is rotation and translation to cover the variability of the random

Table 1. Ablation study of the multi-task architecture and the joint training strategy.

Method	BE		QA		
	Dice (%)	MHD (mm)	Acc (%)	Rec (%)	F1 (%)
BE	96.66 ± 0.01	0.64 ± 0.12	–	–	–
QA	–	–	98.84	98.85	98.82
BE+QA+WOP	96.82 ± 0.02	0.59 ± 0.13	99.75	99.73	99.71
BE+QA+WP	97.21 ± 0.02	0.51 ± 0.16	99.98	99.97	99.98

locations and orientations of fetal brains. The other one is intensity transformations that simulate intensity inhomogeneity artifacts. The parameters of the network are randomly initialized using Gaussian distributions. The RMSprop optimizer is utilized for training with decay rate as 0.995. In the separate and joint training phase, we set the learning rate as 0.001 and 0.0001, respectively.

3.3 Evaluation

The BE is evaluated by means of the Dice coefficient (Dice) and modified Hausdorff distance (MHD) between the manual segmentation and automatic segmentation. These metrics are calculated in 2D, i.e., per slice, and are then averaged across all slices in the dataset. The performance of image QA is measured by the classification accuracy (Acc), Recall (Rec), and F1-score (F1).

Ablation Study. We perform some ablation experiments on the Dataset A to verify the effectiveness of heatmap for brain detection, deformable convolution, multi-task architecture and training strategy. Specifically, each experimental configuration is evaluated by 2 times of 2-fold cross-validation on Dataset A.

Ablation Study of Heatmap. To regresses the center of the brain bounding box (CBBB) directly, we add two fully-connected layers after the brain detection sub-network to generate the CBBB. According to our experiment, the coordination regression average error is 3.80 pixels and our heatmap detection average error is 2.29 pixels in the 768×768 images, which indicates that our heatmap based algorithm can improve the robustness of the CBBB position prediction.

Ablation Study of Deformable Convolution. To demonstrate the effectiveness of deformable convolution (DC), we replace DC in the FE module with regular convolution (RC). Herein, we only train FE and TS_BE module to explore the performance of different convolutions. According to our experiment, Dice and MHD of the RC based network are 95.81% and 0.72 mm, respectively and these of the DC based network are 96.66% and 0.64 mm. These results indicate the superior performance of DC. This is because the fixed receptive field in RC fails to capture the complex brain shape changing with gestational ages.

Ablation Study of Multi-task Architecture and Training Strategy. We compare results of peforming QA only, BE only, and different training strategies, with results presented in Table 1. To verify the effectiveness of the training strategy, we jointly train the TS module without pre-training (WOP) or with pre-training (WP) sub-networks. The performance is clearly improved when we do QA and BE jointly, which indicates that, by jointly performing these two tasks together, our FE module can extract more effective features. Moreover, the pre-training step makes the whole network easier to converge and allows each task specific sub-network to fully extract the features needed for its own task and finally can urge the FE module to provide more effective features.

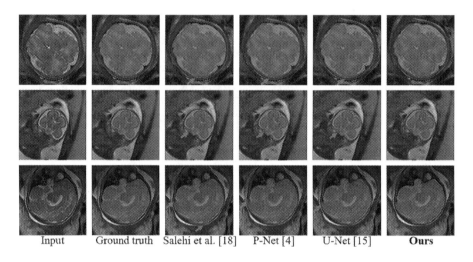

Fig. 5. Representative brain extraction results by different methods.

Comparison with Other Algorithms. In this section, we use the best network architecture and training strategy from our ablation study. We compare the proposed method with existing methods on two fetal datasets discussed above. To demonstrate the cross-source performance of our model, we use Dataset A for training and Dataset B for testing, with results shown in Table 2. Moreover, Table 3 shows the results when using Dataset B for training and Dataset A for testing. Representative qualitative results when using Dataset B for training and Dataset A for testing are provided in Fig. 5.

Table 2. Results of training on Dataset A and testing on Dataset B.

Method	BE		QA		
	Dice (%)	MHD (mm)	Acc (%)	Rec (%)	F1 (%)
U-Net [15]	94.50 ± 0.23	0.84 ± 0.13	–	–	–
Salehi et al. [18]	93.91 ± 0.28	0.83 ± 0.15	–	–	–
P-Net [4]	94.68 ± 0.24	0.79 ± 0.17	–	–	–
Ours	$\mathbf{95.45 \pm 0.22}$	$\mathbf{0.68 \pm 0.11}$	99.72	99.80	99.76

Table 3. Results of training on Dataset B and testing on Dataset A.

Method	BE		QA		
	Dice (%)	MHD (mm)	Acc (%)	Rec (%)	F1 (%)
U-Net [15]	94.26 ± 0.27	0.69 ± 0.18	–	–	–
Salehi et al. [18]	93.55 ± 0.29	0.67 ± 0.19	–	–	–
P-Net [4]	94.71 ± 0.25	0.64 ± 0.15	–	–	–
Ours	**95.79 ± 0.25**	**0.55 ± 0.13**	99.61	99.73	99.75

4   Conclusion

This paper proposes to jointly perform image quality assessment and brain extraction of fetal MRI, since both tasks focus on the brain region representation and thus can be jointly optimized to ensure the network to learn shared features and avoid overfitting. Specifically, first, we introduce heatmap to accurately detect and locate the brain region. Second, considering variable shapes of brain regions, we use the deformable convolution to learn shared feature representations for both quality assessment and brain extraction. Third, based on the learned features, we simultaneously perform quality assessment and brain extraction using their own specific sub-networks. Meanwhile, to obtain a stable well-trained model, we propose a multi-step joint training strategy. The ablation study has shown the effectiveness of every contribution, and we have achieved promising performance in cross-validation on two independent fetal brain MRI datasets.

Acknowledgements. XZ and XX are supported in part by the NSFC under grant U1801262, Guangzhou Key Laboratory of Body Data Science under grant 201605030011.

References

1. Benkarim, O.M., et al.: A novel approach to multiple anatomical shape analysis: application to fetal ventriculomegaly. Med. Image Anal. **64**, 101750 (2020)
2. Benkarim, O.M., et al.: Toward the automatic quantification of in utero brain development in 3D structural MRI: a review. Hum. Brain Mapp. **38**(5), 2772–2787 (2017)
3. Dai, J., Qi, H., Xiong, Y., et al.: Deformable convolutional networks. In: Proceedings of the IEEE International Conference on Computer Vision, pp. 764–773 (2017)
4. Ebner, M., et al.: An automated localization, segmentation and reconstruction framework for fetal brain MRI. In: Frangi, A.F., Schnabel, J.A., Davatzikos, C., Alberola-López, C., Fichtinger, G. (eds.) MICCAI 2018. LNCS, vol. 11070, pp. 313–320. Springer, Cham (2018). https://doi.org/10.1007/978-3-030-00928-1_36
5. He, K., Zhang, X., Ren, S., et al.: Deep residual learning for image recognition. In: Proceedings of the IEEE Conference on Computer Vision and Pattern Recognition, pp. 770–778 (2016)

6. Keraudren, K., Kuklisova-Murgasova, M., Kyriakopoulou, V., et al.: Automated fetal brain segmentation from 2D MRI slices for motion correction. NeuroImage **101**, 633–643 (2014)

7. Khalili, N., et al.: Automatic segmentation of the intracranial volume in fetal MR images. In: Cardoso, M.J., et al. (eds.) FIFI/OMIA -2017. LNCS, vol. 10554, pp. 42–51. Springer, Cham (2017). https://doi.org/10.1007/978-3-319-67561-9_5

8. Kim, J., Zeng, H., Ghadiyaram, D., et al.: Deep convolutional neural models for picture-quality prediction: challenges and solutions to data-driven image quality assessment. IEEE Signal Process. Mag. **34**(6), 130–141 (2017)

9. Li, J., Luo, Y., Shi, L., et al.: Automatic fetal brain extraction from 2D in utero fetal MRI slices using deep neural network. Neurocomputing **378**, 335–349 (2020)

10. Liao, L., et al.: Multi-branch deformable convolutional neural network with label distribution learning for fetal brain age prediction. In: 2020 IEEE 17th International Symposium on Biomedical Imaging (ISBI), pp. 424–427. IEEE (2020)

11. Lin, T.Y., Goyal, P., Girshick, R., et al.: Focal loss for dense object detection. In: Proceedings of the IEEE International Conference on Computer Vision, pp. 2980–2988 (2017)

12. Liu, S., Thung, K.-H., Lin, W., Yap, P.-T., Shen, D.: Multi-stage image quality assessment of diffusion MRI via semi-supervised nonlocal residual networks. In: Shen, D., et al. (eds.) MICCAI 2019. LNCS, vol. 11766, pp. 521–528. Springer, Cham (2019). https://doi.org/10.1007/978-3-030-32248-9_58

13. Lou, J., et al.: Automatic fetal brain extraction using multi-stage U-Net with deep supervision. In: Suk, H.-I., Liu, M., Yan, P., Lian, C. (eds.) MLMI 2019. LNCS, vol. 11861, pp. 592–600. Springer, Cham (2019). https://doi.org/10.1007/978-3-030-32692-0_68

14. Mostapha, M., et al.: Semi-supervised VAE-GAN for out-of-sample detection applied to MRI quality control. In: Shen, D., et al. (eds.) MICCAI 2019. LNCS, vol. 11766, pp. 127–136. Springer, Cham (2019). https://doi.org/10.1007/978-3-030-32248-9_15

15. Ronneberger, O., Fischer, P., Brox, T.: U-Net: convolutional networks for biomedical image segmentation. In: Navab, N., Hornegger, J., Wells, W.M., Frangi, A.F. (eds.) MICCAI 2015. LNCS, vol. 9351, pp. 234–241. Springer, Cham (2015). https://doi.org/10.1007/978-3-319-24574-4_28

16. Saleem, S.N.: Fetal MRI: an approach to practice: a review. J. Adv. Res. **5**(5), 507–523 (2014)

17. Salehi, S.S.M., Erdogmus, D., Gholipour, A.: Auto-context convolutional neural network (auto-net) for brain extraction in magnetic resonance imaging. IEEE Trans. Med. Imaging **36**(11), 2319–2330 (2017)

18. Salehi, S.S.M., Hashemi, S.R., Velasco-Annis, C., et al.: Real-time automatic fetal brain extraction in fetal MRI by deep learning. In: 2018 IEEE 15th International Symposium on Biomedical Imaging (ISBI 2018), pp. 720–724. IEEE (2018)

19. Sayeri, L., Nalini, S., Borjan, G., et al.: A deep learning approach for image quality assessment of fetal brain MRI. In: International Society for Magnetic Resonance in Medicine (2019)

20. Tourbier, S., Hagmann, P., Cagneaux, M., et al.: Automatic brain extraction in fetal mri using multi-atlas-based segmentation. In: Medical Imaging 2015: Image Processing, vol. 9413, p. 94130Y. International Society for Optics and Photonics (2015)

21. Tourbier, S., Velasco-Annis, C., Taimouri, V., et al.: Automated template-based brain localization and extraction for fetal brain mri reconstruction. NeuroImage **155**, 460–472 (2017)

22. Xia, J., et al.: Fetal cortical surface atlas parcellation based on growth patterns. Hum. Brain Mapp. **40**(13), 3881–3899 (2019)

Heart and Lung Imaging

Accelerated 4D Respiratory Motion-Resolved Cardiac MRI with a Model-Based Variational Network

Haikun Qi[✉], Niccolo Fuin, Thomas Kuestner, René Botnar, and Claudia Prieto

School of Biomedical Engineering and Imaging Science, King's College
London, London SE1 7EH, UK
haikun.qi@kcl.ac.uk

Abstract. Respiratory motion and long scan times remain major challenges in free-breathing 3D cardiac MRI. Respiratory motion-resolved approaches have been proposed by binning the acquired data to different respiratory motion states. After inter-bin motion estimation, motion-compensated reconstruction can be obtained. However, respiratory bins from accelerated acquisitions are highly undersampled and have different undersampling patterns depending on the subject-specific respiratory motion. Remaining undersampling artifacts in the bin images can influence the accuracy of the motion estimation. We propose a model-based variational network (VN) which reconstructs motion-resolved images jointly by exploiting shared information between respiratory bins. In each stage of VN, conjugate gradient is adopted to enforce data-consistency (CG-VN), achieving better enforcement of data consistency per stage than the classic VN with proximal gradient descent step (GD-VN), translating to faster convergence and better reconstruction performance. We compare the performance of CG-VN and GD-VN for reconstruction of respiratory motion-resolved images for two different cardiac MR sequences. Our results show that CG-VN with less stages outperforms GD-VN by achieving higher PSNR and better generalization on prospectively undersampled data. The proposed motion-resolved CG-VN provides consistently good reconstruction quality for all motion states with varying undersampling patterns by taking advantage of redundancies among motion bins.

Keywords: Motion-resolved MRI reconstruction · Model-based Deep-learning · Variational network

1 Introduction

Whole-heart MRI is an import tool for comprehensive cardiac anatomy screening. However, respiratory motion remains a major challenge for free-breathing 3D cardiac MRI. Self-navigation [1, 2] and 2D image navigation (iNAV) approaches [3, 4] have been proposed to estimate respiratory motion during scan, according to which the acquired data can be binned into different respiratory states. After inter-bin motion estimation, non-rigid motion compensated reconstruction can be obtained [5]. However, to achieve

© Springer Nature Switzerland AG 2020
A. L. Martel et al. (Eds.): MICCAI 2020, LNCS 12266, pp. 427–435, 2020.
https://doi.org/10.1007/978-3-030-59725-2_41

clinically acceptable scan time, 3D cardiac MRI is usually acquired with acceleration, leading to highly and unpredictable undersampling for each respiratory bin. Conventional iterative reconstruction with parallel imaging [6, 7] and compressed sensing [8] can be used to mitigate undersampling artifacts, relying on regularization on a priori information to solve the ill-posed inverse problem. However, traditional regularization may oversimplify MR image structure information and reconstruction quality depends on the regularization weight, which is commonly manually optimized for each specific application.

Recently, deep learning techniques have been introduced for undersampled MR image reconstruction, outperforming traditional reconstruction methods in terms of computation speed and image quality [9–12]. One category of deep learning-based approaches generalizes the concept of compressed sensing and leverages a convolutional neural network to learn the regularization term, such as the variational network (VN) [9]. VN unrolls the iterative reconstruction into several stages, with each stage corresponding to a gradient descent (GD) step. However, the number of iterations/stages is limited due to GPU memory limits. Recently, a model-based reconstruction framework is proposed to integrate a conjugate gradient (CG) optimization block into a reconstruction network [11]. The CG iterations have been shown to effectively reduce the cost per stage and thus enable faster convergence without introducing additional trainable parameters [11].

Here we aim to extend the VN approach for joint motion-resolved reconstruction by using conjugate gradient to enforce data consistency in each stage. We employ 2D VN to reduce model complexity by reconstructing the images slice-by-slice along the fully-sampled readout dimension. The performance of motion-resolved CG-VN is evaluated on both retrospectively undersampled and prospectively undersampled data for two different cardiac MRI sequences, in comparison with motion-resolved GD-VN.

2 Methods

2.1 Image Acquisition

Free-breathing ECG-triggered 3D cardiac MR images were acquired using a previously proposed iNAV-based imaging technique [13] on a 1.5T scanner (Magnetom Aera, Siemens Healthcare, Erlangen, Germany). Two different imaging sequences, T2-prepared coronary MR angiography (CMRA) and inversion-recovery magnetization transfer acquisition (MTC-BOOST) [14] were performed. CMRA was acquired on 15 healthy subjects and 2 patients with suspected cardiac diseases. Healthy subjects underwent both fully-sampled and 5-fold undersampled free-breathing CMRA scans for network training and validation, whereas patients were acquired with 5-fold acceleration. The undersampling was performed with a variable-density Cartesian acquisition with spiral-like profile order [15], featuring flexible retrospective binning and incoherent undersampling artifacts. MTC-BOOST sequence with 5-fold undersampling was acquired on 3 patients with congenital heart disease (CHD). The imaging parameters of the CMRA and MTC-BOOST acquisitions are summarized in Table 1.

Table 1. MR imaging parameters of the CMRA and MTC-BOOST techniques.

Acquisition	FOV	Spatial resolution	TR/TE	Flip angle	Preparation pulses
CMRA	$320 \times 320 \times$ $86-115$ mm^3	1.2 mm isotropic	3.7/1.6 ms	90°	T2
MTC-BOOST	$320 \times 320 \times$ $90-120$ mm^3	$1.4 \times 1.4 \times$ 2.8 mm^3	3.2/1.5 ms	90°	Magnetization transfer

2.2   Respiratory Motion and Binning

Training and Validation. Foot-head (FH) and left-right (LR) respiratory motion was estimated from 2D iNAVs acquired in each heartbeat during the 3D acquisition [13]. The fully-sampled 3D k-space data was corrected for 2D FH and LR translational motion by applying the corresponding phase shift. Then random non-rigid motion fields were simulated and used to transform the fully-sampled images, generating fully-sampled motion-resolved bin images and the corresponding k-spaces. To emulate the motion-resolved undersampling in realistic acquisitions, undersampling masks were obtained by binning the corresponding 5-fold undersampled data into four equally populated respiratory bins based on the iNAV FH motion with inter-bin soft-gating [5], resulting in undersampling factors ranging from 7 to 10 for each bin. The undersampling masks of 4 respiratory bins were applied to the simulated fully-sampled k-spaces, generating undersampled k-spaces of 4 motion states. The motion-resolved simulation and retrospectively undersampling were performed for all 15 healthy subjects with fully-sampled CMRA acquisitions. Twelve subjects were used for training whereas 3 subjects were used for validation.

Test. Similar binning process was performed on the prospectively 5-fold undersampled acquisitions of patients (CMRA and MTC-BOOST), and the k-space data within each bin was corrected for 2D FH and LR translational motion, generating undersampled respiratory motion-resolved bins with undersampling factors ranging from 7 to 10 for each bin to evaluate the reconstruction performance.

2.3   Conjugate Gradient Variational Network

The model-based reconstruction equation can be formulated as follows [11]:

$$\tilde{x} = arg\ min_x \|Ex - y\|_2^2 + \lambda \|x - D_w(x)\|^2 \tag{1}$$

where x is the image to be reconstructed, E is the forward encoding operator including sensitivity encoding, Fourier Transform and sampling mask, y is the undersampled k-space, $D_w = x - N_w$ is the VN de-aliased version of x and N_w is the learned noises/artifacts, and λ is the learnable regularization parameter. Equation (1) can be approximated by solving the following two sub-problems [11]:

$$z_{n+1} = x_n - N_w(x_n) \tag{2a}$$

$$x_n = arg \ \min_x \|Ex - y\|_2^2 + \lambda_n \|x - z_n\|^2 \tag{2b}$$

The reconstruction framework by unrolling the iterative process is shown in Fig. 1, where 3-stage VN is constructed with the initialization x_0 being the zero-filled (ZF) reconstruction. Each stage contains two blocks. The regularization block applies VN filters and activations to the motion-resolved images, which will be explained in the next section. The CG block is aimed to optimize the sub-problem (2b) numerically using several conjugate gradient (CG) steps. Compared with classic VN, we utilize several CG-SENSE iterations instead of gradient descent step in each stage, aiming to achieve more accurate enforcement of data-consistency in the forward model and to offer faster reduction of cost per stage [11], thus effectively reducing the number of unrolled iterations (stages) and model parameters. Besides, there are no trainable parameters in the CG block, and no intermediate parameters should be stored for backpropagation, so the CG iterations almost add no memory overhead during training.

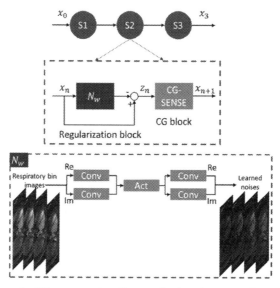

Fig. 1. The framework of the proposed motion-resolved conjugate gradient variational network (CG-VN). Each stage of CG-VN consists of two blocks: the regularization block denoises the motion-resolved images which are stacked as input channels to exploit the shared information between motion bins using VN; the CG block applies several CG-SENSE iterations to enforce data consistency.

2.4   Motion-Resolved Training

The acquired multi-channel k-space data was compressed into 8 coils using a coil compression algorithm [16]. Coil sensitivity maps were estimated from the fully-sampled central k-space region before motion binning. After respiratory binning, undersampled k-space data and undersampling masks of the four motion bins, as well as coil sensitivity maps were inputted to the network. For joint motion-resolved training, the images of different motion states were stacked as different channels of the network. As shown in Fig. 1, the VN filters mix the input channels and create a number of feature maps which enables the exchange of shared information between different motion states, and thus the redundancies along motion dimension can be exploited. Two sets of filter kernels (24 per set; filter size, 11×11) are applied respectively to the real and imaginary part of the complex images. Similar to previous VN, we use learnable activation functions defined by the Gaussian radial basis functions. After the non-linear activation, the transpose convolution which share the parameters of the convolution filters is performed to reduce channel dimension to the number of input motion states.

The network was trained on 2D complex-valued images obtained from 12 healthy subjects by slicing along the fully-sampled frequency-encoding dimension. For each volunteer, 150 central slices containing anatomical structures were extracted, resulting in 1800 motion-resolved images for training. The loss function was defined as the mean-squared error between network reconstruction and the fully-sampled reference for each bin. The network was trained using the IIPG optimizer following classic VN [9] for 1000 iterations with batch size of 8. The learning rate was set to 0.001.

2.5 Reconstruction Evaluation

From the remaining 3 healthy subjects, we generated 450 (150 per subject) random motion-resolved 2D images for validation. The reconstruction results of the proposed 3-stage CG-VN (3CG-VN) were compared with GD-VN with 3 stages (3GD-VN) and 6 stages (6GD-VN) quantitatively by calculating PSNR. The 3CG-VN and 3GD-VN have comparable trainable parameters, while the 6GD-VN has twice the number of model parameters.

For prospectively undersampled data of CMRA and CHD patients, since no fully-sampled reference is available, we compared the reconstructions of 3CG-VN, 3GD-VN and 6GD-VN qualitatively by visualizing image quality in terms of image burring and residual aliasing artifacts. In addition, a traditional method which has been proposed for Cartesian respiratory motion-resolved CMRA reconstruction using parallel imaging and total variation regularization, named XD-ORCCA [17], was performed on the prospectively undersampled acquisitions for comparison with the network reconstructions. The regularization parameter in XD-ORCCA was empirically optimized for each subject, and the total number of non-linear conjugate gradient iterations was set to 24.

3   Results

Representative CMRA images of 4 simulated motion states including the fully-sampled reference, ZF and network reconstructions of CG-VN and GD-VN are shown in Fig. 2. The proposed 3CG-VN outperforms both 6GD-VN and 3GD-VN by removing under-sampling artifacts and introducing less image blurring. The motion information across motion frames is maintained with all motion-resolved network reconstructions. The PSNR results of the images reconstructed by ZF and CG-VN, GD-VN are reported in

Fig. 3. The proposed 3CG-VN has the highest PSNR among all tested reconstruction methods.

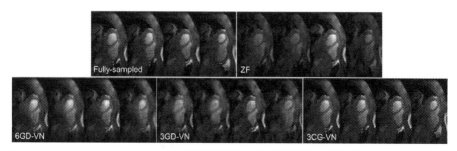

Fig. 2. Representative reconstruction results for the simulated undersampled CMRA data of 4 motion states using 6GD-VN, 3GD-VN and the proposed 3CG-VN.

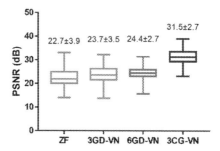

Fig. 3. PSNR values for all 450 simulated random motion-resolved CMRA images from the 3 test healthy subjects reconstructed using ZF, 6GD-VN, 3GD-VN and 3CG-VN. The PSNR range, mean and standard deviation are shown for each method.

Figure 4 shows images of four respiratory bins for one prospectively undersampled CMRA acquisition, including reconstructions of ZF, 6GD-VN, 3GD-VN, 3CG-VN and XD-ORCCA. Respiratory motion-resolved images reconstructed by the five methods for a CHD patient are compared in Fig. 5. The red dotted lines in Fig. 4 and Fig. 5 indicate respiration-induced heart displacements along the FH direction. The proposed 3CG-VN achieves better reconstruction quality than GD-VN of 3 and 6 stages in terms

Table 2. Reconstruction time of the four respiratory bins of undersampled CMRA and MTC-BOOST acquisitions.

Acquisition	Reconstruction time (seconds)			
	6GD-VN	3GD-VN	3CG-VN (proposed)	XD-ORCCA
CMRA	118	50	35	5052
MTC-BOOST	50	36	19	1546

of less residual undersampling artifacts and better keeping inter-bin motion fidelity, and outperforms XD-ORCCA by introducing less image blurring caused by total variation regularization. The reconstruction time for the 4 respiratory bins of CMRA and MTC-BOOST using VN networks and XD-ORCCA is summarized in Table 2. The network reconstruction speed is much faster than XD-ORCCA.

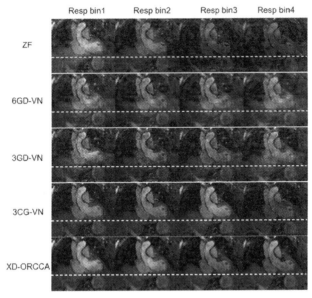

Fig. 4. Reconstruction results of the prospectively undersampled CMRA images of 4 respiratory bins using reconstruction methods of ZF, 6GD-VN, 3GD-VN, proposed 3CG-VN and XD-ORCCA. One slice of the 3D volume is shown, and the red dotted lines indicate the different heart positions in the respiratory bins. (Color figure online)

4   Discussion

The motion-resolved CG-VN and GD-VN were trained on simulated random motion-resolved images. During test, both 3CG-VN and 6GD-VN performed well on the reconstruction of simulated random motion-resolved images, efficiently removing aliasing artifacts and maintaining inter-bin motion. The 3-stage GD-VN which has the same number of model parameters to 3CG-VN, results in unacceptable reconstruction due to the insufficient gradient descent iterations. Therefore, the CG-VN achieves faster convergence than GD-VN, effectively reducing the unrolled iterations.

The gradient descent VN, whether of 6 stages or 3 stages failed to generalize to prospectively undersampled respiratory motion-resolved data, resulting in noticeable residual undersampling artifacts and blurring. On the contrast, the proposed CG-VN was able to reconstruct prospectively undersampled respiratory motion-resolved CMRA and MTC-BOOST images, which differ in image contrasts and acquisition parameters,

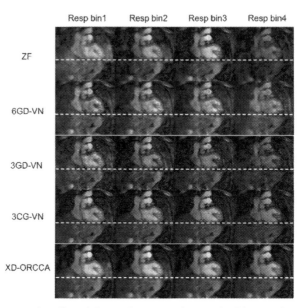

Fig. 5. One representative slice of the respiratory motion-resolved images from a CHD patient acquired with undersampled MTC-BOOST. Reconstruction results of ZF, 6GD-VN, 3GD-VN, proposed 3CG-VN and XD-ORCCA are compared. The red dotted line indicates the different heart positions in the respiratory bin images. (Color figure online)

achieving much better generalization than GD-VN. Compared with traditional XD-ORCCA, the proposed CG-VN achieves better reconstruction performance by introducing less imaging blurring benefiting from the learning-based regularizer which provides more efficient regularization than total variation [9].

The reconstructed respiratory motion-resolved images can be used to estimate non-rigid inter-bin motion to enable non-rigid motion-compensated reconstruction [5]. The motion estimation accuracy based on the CG-VN reconstructed images will be thoroughly evaluated in future studies. Furthermore, end-to-end networks enabling respiratory motion-resolved reconstruction and motion estimation will be investigated.

5   Conclusion

We proposed a CG-VN framework which reconstructs undersampled 4D respiratory motion-resolved cardiac MR images by exploiting shared information between motion phases. The proposed CG-VN is able to generalize to unpredictable undersampling of each bin due to subject-specific respiratory motion and undersampled data of CHD patients that was acquired with different image contrast and spatial resolution. Reconstruction results of prospectively undersampled acquisitions of CMRA and CHD patients show that the proposed approach outperforms GD-VN and XD-ORCCA methods. Furthermore, it results in fast reconstruction, offering easy integration into clinical workflow.

References

1. Stehning, C., Bornert, P., Nehrke, K., Eggers, H., Stuber, M.: Free-breathing whole-heart coronary MRA with 3D radial SSFP and self-navigated image reconstruction. Magn. Reson. Med. **54**, 476–480 (2005)
2. Lai, P., Bi, X., Jerecic, R., Li, D.: A respiratory self-gating technique with 3D-translation compensation for free-breathing whole-heart coronary MRA. Magn. Reson. Med. **62**, 731–738 (2009)
3. Henningsson, M., Koken, P., Stehning, C., Razavi, R., Prieto, C., Botnar, R.M.: Whole-heart coronary MR angiography with 2D self-navigated image reconstruction. Magn. Reson. Med. **67**, 437–445 (2012)
4. Wu, H.H., Gurney, P.T., Hu, B.S., Nishimura, D.G., McConnell, M.V.: Free-breathing multiphase whole-heart coronary MR angiography using image-based navigators and three-dimensional cones imaging. Magn. Reson. Med. **69**, 1083–1093 (2013)
5. Cruz, G., Atkinson, D., Henningsson, M., Botnar, R.M., Prieto, C.: Highly efficient nonrigid motion-corrected 3D whole-heart coronary vessel wall imaging. Magn. Reson. Med. **77**, 1894–1908 (2017)
6. Pruessmann, K.P., Weiger, M., Scheidegger, M.B., Boesiger, P.: SENSE: sensitivity encoding for fast MRI. Magn. Reson. Med. **42**, 952–962 (1999)
7. Griswold, M.A., et al.: Generalized autocalibrating partially parallel acquisitions (GRAPPA). Magn. Reson. Med. **47**, 1202–1210 (2002)
8. Lustig, M., Donoho, D., Pauly, J.M.: Sparse MRI: the application of compressed sensing for rapid MR imaging. Magn. Reson. Med. **58**, 1182–1195 (2007)
9. Hammernik, K., et al.: Learning a variational network for reconstruction of accelerated MRI data. Magn. Reson. Med. **79**, 3055–3071 (2018)
10. Schlemper, J., Caballero, J., Hajnal, J.V., Price, A.N., Rueckert, D.: A deep cascade of convolutional neural networks for dynamic MR image reconstruction. IEEE Trans. Med. Imaging **37**, 491–503 (2018)
11. Aggarwal, H.K., Mani, M.P., Jacob, M.: MoDL: model-based deep learning architecture for inverse problems. IEEE Trans. Med. Imaging **38**, 394–405 (2019)
12. Han, Y., Yoo, J., Kim, H.H., Shin, H.J., Sung, K., Ye, J.C.: Deep learning with domain adaptation for accelerated projection-reconstruction MR. Magn. Reson. Med. **80**, 1189–1205 (2018)
13. Bustin, A., et al.: Five-minute whole-heart coronary MRA with sub-millimeter isotropic resolution, 100% respiratory scan efficiency, and 3D-PROST reconstruction. Magn. Reson. Med. **81**, 102–115 (2019)
14. Ginami, G., et al.: Non-contrast enhanced simultaneous 3D whole-heart bright-blood pulmonary veins visualization and black-blood quantification of atrial wall thickness. Magn. Reson. Med. **81**, 1066–1079 (2019)
15. Prieto, C., et al.: Highly efficient respiratory motion compensated free-breathing coronary MRA using golden-step Cartesian acquisition. J. Magn. Reson. Imaging **41**, 738–746 (2015)
16. Zhang, T., Pauly, J.M., Vasanawala, S.S., Lustig, M.: Coil compression for accelerated imaging with Cartesian sampling. Magn. Reson. Med. **69**, 571–582 (2013)
17. Correia, T., et al.: Optimized respiratory-resolved motion-compensated 3D Cartesian coronary MR angiography. Magn. Reson. Med. **80**, 2618–2629 (2018)

Motion Pyramid Networks for Accurate and Efficient Cardiac Motion Estimation

Hanchao Yu[2], Xiao Chen[1], Humphrey Shi[3(✉)], Terrence Chen[1], Thomas S. Huang[2], and Shanhui Sun[1(✉)]

[1] United Imaging Intelligence, Cambridge, MA, USA
shanhui.sun@united-imaging.com
[2] University of Illinois at Urbana-Champaign, Urbana, IL, USA
[3] University of Oregon, Eugene, OR, USA
shihonghui3@gmail.com

Abstract. Cardiac motion estimation plays a key role in MRI cardiac feature tracking and function assessment such as myocardium strain. In this paper, we propose Motion Pyramid Networks, a novel deep learning-based approach for accurate and efficient cardiac motion estimation. We predict and fuse a pyramid of motion fields from multiple scales of feature representations to generate a more refined motion field. We then use a novel cyclic teacher-student training strategy to make the inference end-to-end and further improve the tracking performance. Our teacher model provides more accurate motion estimation as supervision through progressive motion compensations. Our student model learns from the teacher model to estimate motion in a single step while maintaining accuracy. The teacher-student knowledge distillation is performed in a cyclic way for a further performance boost. Our proposed method outperforms a strong baseline model on two public available clinical datasets significantly, evaluated by a variety of metrics and the inference time. New evaluation metrics are also proposed to represent errors in a clinically meaningful manner.

Keywords: Motion pyramid network · Motion compensation · Cyclic knowledge distillation

1 Introduction

Cardiac motion estimation in cardiac MRI (CMR) is one of the fundamental techniques for cardiac feature tracking (CMR-FT). In the feature tracking system, key points from manual annotation or automatic generation are initialized on one image and then tracked through time. Once the spatiotemporal locations of each point are known, clinical indices such as myocardium strain can be computed to assess the dynamic deformation functionality of the heart, which are

---

H. Yu—This work was carried out during the internship of the author at United Imaging Intelligence, Cambridge, MA 02140.

A. L. Martel et al. (Eds.): MICCAI 2020, LNCS 12266, pp. 436–446, 2020.
https://doi.org/10.1007/978-3-030-59725-2_42

more sensitive and earlier indicators of contractile dysfunction, compared with the frequently used ejection function (EF) [4]. Besides, motions can also help other tasks in CMR image analyses like reconstruction [5,16] and segmentation [13,24,29]. Since the ground-truth of motion is difficult to acquire, most of the deep learning-based works formulate it as an unsupervised learning problem. Under this setting, the searching space is large and the optimal is not unique due to the lack of ground truth motion field. In [13,29], motion field smoothness is used as constraints, at the cost of compromised estimation of large motions. To evaluate the tracker's performance, DICE coefficients, surface distance and endpoint error are often used in recent researches [13,22,29]. Considering the clinical applications like myocardium strain which are computed along specific directions, these metrics are not well aligned with clinical interest.

To address the cardiac feature tracking challenges, we propose a motion pyramid network, which predicts and fuses a pyramid of motion fields from multiple scales of feature representations to produce a refined motion field (Sect. 3.1). We utilize a novel cyclic teacher-student training strategy to further improve the tracking performance (Sect. 3.3). The teacher model is trained via progressive motion compensations in an iterative manner in order to handle large motion as well as minimizing smoothness constraint limitation (Sect. 3.2). The student model learns from the teacher model and matches its accuracy performing a more efficient inference (Sect. 3.3). To align with clinical interest, we propose a novel evaluation method where the error vector is decomposed into radial and circumferential directions. The extensive experiments demonstrate that our method outperforms the strong single-scale baseline model and a conventional deformable registration method on two public datasets using a variety of metrics.

2   Related Work

Many recent works show promising results in the area of deep learning-based cardiac motion estimation and myocardium tracking [8,13,28,29]. In [13], cardiac motion estimation and cardiac segmentation are formulated as a multi-task problem with shared feature encoder and independent task heads. In [29], a U-net like the apparent-flow network is proposed with a semi-supervised learning framework. These methods use the smoothness constraint to keep feasible anatomy.

There are also non-learning based methods that can be divided into two categories: optical flow-based and registration based. [22] shows the most recent gradient flow-based method. A Lagrangian displacement field based post-processing is used to reduce the end-point error, which may not be time-efficient. Image registration based methods [3,8,12,15,17,18,20,27] were applied to solve cardiac motion estimation. Rueckert et al. [15] proposed a free form deformation (FFD) method for non-rigid registration and is applied to cardiac motion estimation recently [12,17,18,20,21]. However, deformable registration methods require iterative searching in a large parameter space.

The teacher-student and knowledge distillation mechanisms are popular in computer vision and medical image analysis society [7,9]. Kong et al. [7] utilizes a L_2 loss on feature maps between the teacher and the student models to guide the training of the compressed student network.

Pyramid processing is widely accepted in the computer vision community [10, 14,19,23,26]. In PwcNet [19] the input of each level of the pyramid is the output of the last level. In contrast, our network estimates the motion independently from each level and generate a motion pyramid to perform the multi-level fusion.

3   Motion Pyramid Networks

We propose Motion Pyramid Networks (MPN) for the cardiac motion tracking problem. Our method consists of a multi-scale motion pyramid architecture for motion prediction, progressive motion compensation for post-processing and a cyclic knowledge distillation strategy to learn the motion compensation and speed up inference.

3.1   Multi-scale Motion Pyramid

Figure 1(left) illustrates the overview of the proposed motion pyramid network (MPN). The features are extracted from source and target image independently using a shared encoder. We build a feature pyramid: the different scale features from the encoder are upsampled to the same size and concatenated channel-wise. The feature pyramid is then fed to the decoder which is used to predict an initial motion field. In addition, we build a motion pyramid as following: at each feature scale-level, a corresponding scale motion field is generated. Then the motion fields are upsampled to the same size and fused with the initial motion field through a fusion module to predict the refined motion. Note that we multiply a scaling factor to each motion field in the pyramid before fusion to make the motion fields comparable across scales, following PWCnet [19]. Furthermore, we leverage a deep supervision strategy to train the motion pyramid: the source image is downsampled to corresponding scales. The downsampled images are warped with the predicted motion pyramid to generate a warped image pyramid. The total loss is composed of the MSE loss between the warped and the reference image and the second-order smoothness loss [13]. Losses at each level l are summed up as in Eq. (1).

$$L_{total} = \sum_l L_{MSE}^{(l)} + \sum_l \lambda L_{smooth}^{(l)} \tag{1}$$

3.2   Progressive Motion Compensation

Under certain smoothness constraint (L_{smooth}), it is a non-trivial task for the tracking learner to find a global optimal solution for all the cardiac motion variations even with pyramid utilization. This is because the smoothness constraint limits the searching space of the learner. Relaxing the smoothness can expand

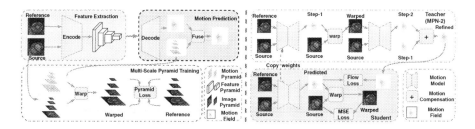

Fig. 1. Left: Overview of the proposed motion pyramid network (MPN). At each level of extracted features, the motion field is predicted. The decoder fuses the multi-scale features and generates a motion field which is further fused with the motion field pyramid to predict the refined motion field. Right: Illustration of the cyclic knowledge distillation method (MPN-C). The teacher model provides supervision through two-step compensation (MPN-2). The student model learns to estimate the motion under the guidance of the teacher as well as the same smoothness and MSE loss (MPN-S, w/o red dashed arrow). The weights of the student model are copied to the teacher (red dashed arrow) to initialize the next round of training a new student model (MPN-C).

the searching space but the tracker becomes sensitive to disturbances such as noises and abrupt intensity changes. Nevertheless, we utilize some smoothness to maintain a feasible cardiac shape. However, the tracking accuracy drops in large motion cases. To solve this problem, we utilize a progressive motion compensation approach (MPN-2) at inference stage. Consider a model M that takes frame I_A and I_B as input. Due to the aforementioned smoothness constraint problem, the warped image I_C is an intermediate result between I_A and I_B and $M(I_A, I_B)$ actually yields motion field f_{AC}. Let $M(I_C, I_B) = f_{CB}$. Suppose $\boldsymbol{x_0} = (x, y)$ is a pixel in I_A and $\boldsymbol{x_1}, \boldsymbol{x_2}$ are the corresponding pixels in I_C, I_B, respectively. We have the following equations: $\boldsymbol{x_1} = f_{AC}(\boldsymbol{x_0}) + \boldsymbol{x_0}, \boldsymbol{x_2} = f_{CB}(\boldsymbol{x_1}) + \boldsymbol{x_1}$. Replace $\boldsymbol{x_1}$ with $\boldsymbol{x_0}$ in the second equation and notice that $\boldsymbol{x_2} - \boldsymbol{x_0} = f_{AB}(\boldsymbol{x_0})$, we can get $f_{AB}(\boldsymbol{x_0}) = f_{AC}(\boldsymbol{x_0}) + f_{CB}(f_{AC}(\boldsymbol{x_0}) + \boldsymbol{x_0})$, where f_{AC}, f_{CB}, f_{AB} is step-1, step-2 and the refined motion field. The derivation is based on forward warping (use f_{AB} to warp I_A) while the same equation still holds with the backward warping (use f_{AB} to warp I_B) function. In forward warping, some pixels on the source are not on the target grids, leaving the target image with holes. In backward warping, each location in the target image is determined using backward flow to find the intensity in the source image via image interpolation. Moreover, the backward warping is differentiable and the gradients can be back-propagated, which was proved in [16]. In principle, this process could be multiple-step while in our work, we found that two-step progressive motion compensation is sufficient to solve the discussed problem.

3.3   Cyclic Knowledge Distillation

The used motion compensation method improves accuracy but increases inference time due to multi-step inferences. To solve the inference time problem, we

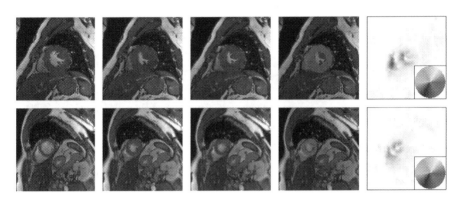

Fig. 2. Examples of tracking results using the proposed method (MPN-C). From left column to right: ED frame, ES frame, warped frame from ED, overlay of ES frame and the warped mask, and estimated motion field using HSV color coding. The color coding wheel legend indicates the motion directions. (Color figure online)

introduced another model (a student), which is a replica of the trained tracker and further learns the knowledge from the progressive motion compensation steps (a teacher). We coined this teacher-student training method as MPN-S. The model (see Sect. 3.1) used in the teacher is coined as M_t and the student M_s. M_s is initialized using M_t. Parameters of M_t are fixed in the teacher-student training step. Note that M_t is used to generate both step-1 and step-2 motion fields. The refined flow from the teacher is used to supervise training the student M_s utilizing a loss function $L_{flow} = ||f^t_{AB} - f^s_{AB}||_2$. Besides, L_{MSE} is added as a self-supervision for the student model. Thus, we have the following total loss:

$$L_{total} = L_{flow} + \mu L_{MSE} + \gamma L_{smooth}. \tag{2}$$

Through the teacher's supervision, the student model attains the teacher's inference capability (two-step motion compensation). We further improve the student's inference capability via cyclic training strategy (MPN-C): when the current teacher-student training converges, the student model takes the teacher's role and a new round of teacher-student training is started. The student's performance can be continuously improved through this self-taught learning method. The overview of the process is shown in Fig. 1(right).

4 Experiments and Results

We compared proposed methods: motion pyramid network (MPN), progressive motion compensation (MPN-2), teacher-student training (MPN-S), and cyclic knowledge distillation (MPN-C). Besides, we implemented a single-scale motion estimation network using the same network structure as MPN without the pyramids (baseline). Furthermore, we compared a conventional registration-based method: free form deformation (FFD) [15]. All models are trained and tested

Table 1. Results of compared methods in terms of Dice coefficient on the ACDC and Kaggle dataset. Mean (standard deviation) is given.

Method	ACDC			Kaggle		
	LV	RV	MYO	LV	RV	MYO
FFD	0.893(0.077)	0.850(0.124)	0.793(0.080)	0.875(0.106)	0.807(0.130)	0.793(0.113)
Baseline	0.885(0.111)	0.872(0.113)	0.830(0.057)	0.867(0.104)	0.816(0.128)	0.800(0.108)
MPN	0.901(.097)	0.880(0.108)	0.847(0.048)	0.883(0.095)	0.820(0.127)	0.820(0.102)
MPN-S	0.915(0.080)	0.883(0.099)	0.858(0.046)	0.893(0.080)	0.819(0.127)	0.833(0.084)
MPN-C	**0.918(0.073)**	**0.885(0.096)**	**0.860(0.047))**	**0.896(0.075)**	**0.820(0.127)**	**0.837(0.077)**
MPN-2	0.913(0.084)	0.887(0.102)	0.861(0.044)	0.898(0.080)	0.822(0.128)	0.842(0.083)

Table 2. Results of compared methods in terms of Hausdorff distance on the ACDC and Kaggle dataset. Mean (standard deviation) millimeter is given.

Method	ACDC			Kaggle		
	LV	RV	MYO	LV	RV	MYO
FFD	5.99(2.53)	8.19(5.06)	7.09(2.26)	6.49(4.27)	12.22(6.46)	8.35(7.00)
Baseline	5.90(2.67)	7.51(4.89)	5.96(2.23)	7.27(4.34)	12.13(6.41)	8.12(7.07)
MPN	5.51(2.49)	7.26(4.91)	5.59(2.09)	6.60(4.19)	12.04(6.47)	7.72(7.03)
MPN-S	4.93(1.88)	7.22(4.77)	5.38(1.90)	**6.28(3.95)**	**12.02(6.50)**	**7.61(6.86)**
MPN-C	**4.82(1.78)**	**7.13(4.62)**	**5.41(1.72)**	6.32(3.96)	12.20(6.46)	7.68(6.78)
MPN-2	4.95(2.03)	7.15(4.85)	5.36(1.85)	6.19(3.95)	11.99(6.52)	7.46(6.91)

on a Tesla V100 workstation. We evaluate all methods in 2 public datasets with multiple metrics. **ACDC** [2] dataset is a short-axis CMR dataset collected from real clinical exams at a hospital in France. The total number of studies is 150, where 100 studies are the training set with expert manual segmentation annotations in end-diastole (ED) and end-systole (ES) phases. **Kaggle** [6] is composed of 1140 subjects with short-axis, 4- and 2-chamber long-axis CMR scans without segmentation masks.

For the ACDC dataset, we make use of the training set, where 80 cases are used to train our model and 20 cases for testing. We first train a teacher model described in Sect. 3.1. Then we perform the cyclic training described in Sect. 3.3. Empirically we set the number of cycles to 2 since we find cycles more than 2 contribute little to the performance. For all the models, we use the same hyperparameters. We use the Kaggle dataset only for testing where we randomly pick 130 patients from short-axis studies and manually segment the myocardium on 3 image pairs at the middle slice.

Table 3. Results of compared methods in terms of average symmetrical surface distance on the ACDC and Kaggle dataset. Mean (standard deviation) millimeter is given.

Method	ACDC			Kaggle		
	LV	RV	MYO	LV	RV	MYO
FFD	1.96(1.05)	2.16(1.12)	1.81(0.50)	2.24(2.00)	2.72(1.48)	1.95(1.12)
Baseline	1.98(1.20)	1.85(1.01)	1.61(0.57)	2.44(1.84)	2.60(1.45)	1.89(1.07)
MPN	1.70(0.97)	1.71(0.93)	1.49(0.47)	2.09(1.67)	2.52(1.44)	1.76(1.04)
MPN-S	1.45(0.70)	1.65(0.84)	1.40(0.42)	1.86(1.34)	2.48(1.41)	1.70(0.98)
MPN-C	**1.38(0.59)**	**1.60(0.81)**	**1.38(0.40)**	**1.78(1.20)**	**2.47(1.40)**	**1.70(0.97)**
MPN-2	1.45(0.71)	1.57(0.82)	1.39(0.42)	1.75(1.33)	2.43(1.41)	1.62(0.98)

Table 4. Endpoint error (EPE), its decomposition (ϵ_{rr}, ϵ_{cc}) in ACDC and Kaggle dataset. The keypoint tracking error is evaluated with ACDC only. Mean (standard deviation) millimeter is given.

Method	ACDC				Kaggle		
	EPE	ϵ_{rr}	ϵ_{cc}	KPTE	EPE	ϵ_{rr}	ϵ_{cc}
FFD	2.70(1.51)	1.62(1.01)	1.44(1.14)	2.48(1.06)	1.10(0.88)	0.95(0.61)	2.17(1.70)
Baseline	1.82(1.48)	1.08(0.77)	1.17(1.15)	1.65(1.00)	0.76(0.54)	0.73(0.50)	1.38(1.08)
MPN	1.79(1.45)	0.95(0.70)	1.17(1.16)	1.60(0.95)	0.71(0.48)	0.72(0.49)	1.40(1.12)
MPN-S	1.73(1.41)	0.86(0.67)	**1.12(1.17)**	1.56(0.89)	0.65(0.44)	0.70(0.48)	1.29(1.05)
MPN-C	**1.69(1.42)**	**0.84(0.65)**	1.13(1.19)	**1.53(0.88)**	**0.64(0.42)**	**0.70(0.47)**	**1.27(1.02)**
MPN-2	1.71(1.49)	0.92(0.81)	1.11(1.25)	1.50(0.92)	0.65(0.47)	0.68(0.47)	1.33(1.16)

Since there is no ground truth cardiac motion field from real CMR cine, we evaluated methods by comparing the warped mask using the estimated motion field with the reference mask. We use standard segmentation evaluation metrics: Dice coefficients, Hausdorff distance (HD) and average symmetrical surface distance (ASSD). Similar approach is used in [8,13,28,29]. To generate more samples for evaluation, for the ACDC dataset, we manually labeled 2 extra frames: $\frac{ED+ES}{2}$ and $\frac{ED+ES}{2} + 1$, between the ED and the ES frame. For each test sample, we evaluate 3 pairs: $\{ED \rightarrow ES, ED \rightarrow \frac{ED+ES}{2}, ED \rightarrow \frac{ED+ES}{2} + 1\}$. We select the middle slice of each study to avert the impact of the out-of-plane motion. Table 1, 2, 3 show the compared results in terms of Dice, HD and ASSD in dataset ADDC and Kaggle. Results of MPN-2 are separated since it uses multi-step compensation while others are single-step. Two example results using MPN-C are depicted in Fig. 2.

4.1   Endpoint Error and Its Decomposition

In the field of motion estimation, the most commonly used evaluation metric is endpoint error (EPE) [11,19]. For 2 frames I_A and I_B, suppose the estimated flow is $\hat{f}_{AB}$ and the ground truth flow is f_{AB}, $EPE = ||\hat{f}_{AB} - f_{AB}||_2$.

However, the computation of EPE requires the known ground truth motion field, which is difficult to acquire for the cardiac motion estimation task. We use

Table 5. Performance measures for the Apparentflow and applying the proposed cyclic knowledge distillation and progressive motion compensation to it on the ACDC dataset. Mean (standard deviation) is given for the Dice coefficient.

Method	LV	RV	MYO
Apparentflow	0.880(0.111)	0.870(0.110)	0.811(0.070)
Apparentflow-C	**0.902(0.098)**	**0.883(0.095)**	**0.837(0.051)**
Apparentflow-2	0.901(0.105)	0.885(0.094)	0.841(0.049)

a different model trained with the same dataset to estimate the motion field and treat it as ground truth. Specifically, we use a single scale model similar to the motion estimation branch described in [13]. Using the generated motion field to warp the image, we can get a synthesized image pair to evaluate the EPE.

EPE measures the magnitude of the error vectors, which is sufficient for the general motion estimation tasks. The goal of cardiac motion estimation is different: motion is used to calculate clinical indices like strain along radial and circumferential directions of the myocardium, which makes it necessary to decompose the error in those directions. We propose a method to decompose the endpoint error in a clinically meaningful way. First, compute the center of the myocardium region $x_c = \frac{1}{N} \sum_{x_i \in myo}^{N} x_i$. The radial direction of every point within myocardium can be computed as $d(x_i) = x_i - x_c$ and then normalized to unit vector. Endpoint error vector e_i at x_i is $e_i = f(x_i) - \hat{f}(x_i)$ which is decomposed along radial (ϵ_{rr}) and circumferential (ϵ_{cc}) directions as: $\epsilon_{rr}^{(i)} = e_i \cdot d(x_i)$, $\epsilon_{cc}^{(i)} = e_i - \epsilon_{rr}^{(i)}$. Experiment results are presented in Table 4.

4.2 Key Point Tracking Error

Quantitative plots like the bullseye plot [25] are widely used in clinical applications, which requires an accurate partition of the myocardium. The insertion points of the left and the right ventricles are key points for the partition [1]. We propose Key Point Tracking Error (KPTE) to measure the error of these landmarks. For N predicted key points $\{\hat{x}_i\}$ and the ground truth locations $\{x_i\}$, $KPTE = \frac{1}{N} \sum_{i}^{N} ||x_i - \hat{x}_i||$. We first manually label the key points on the ED frame and these points are warped with synthesized motion as the ground truth locations on the following frames, using the same method in Sect. 4.1. Then we use the estimated motion to predict the locations of the key points and compute KPTE. The evaluation results are shown in Table 4.

4.3 Generalization Study

The proposed cyclic knowledge distillation training method with progressive motion compensation is not limited to certain neural network structure. We applied this to the recently proposed motion estimation model Apparentflow [29]. In contrast to [29], we only train their model with the self-supervision loss regardless of segmentation mask supervision. Also, we extended this work using our methods in Sect. 3.3 and 3.2. We coined them as Apparentflow-C and Apparentflow-2 respectively. The compared results are presented in Table 5.

4.4 Results Discussion

From Table 1, 2 and 3, we observe that MPN outperforms deformable registration and baseline model in terms of Dice, HD and ASSD. Moreover, the utilization of motion compensation (MPN-2) improves the MPN's performance. The teacher-student model (MPN-S) has comparable accuracy as MPN-2, which demonstrates knowledge distillation helps parameter searching during training. The cyclic training strategy (MPN-C) further pushes the performance. From Table 4, we have the same conclusion in the synthetic dataset in terms of endpoint error. Radial tracking is more accurate than circumferential tracking in light of ϵ_{rr} and ϵ_{cc} errors. This complies with our knowledge that circumferential tracking of circular shape object from 2D in-plane images is a challenging task. Results in Table 5 demonstrates our methods are not limited to our own network structure but also applicable to other motion models (i.e. Apparentflow [29]). On average, FFD takes 748 milliseconds (ms) to register one frame while our proposed model (MPN-C) needs 26 ms for tracking one frame.

5   Conclusion

In summary, we proposed a novel deep neural network model that exploits multi-scale supervision. Specifically, we utilize a progressive motion compensation strategy to overcome the limitation of motion smoothness constraints. We developed a new training strategy, the proposed cyclic knowledge distillation, which helps the learner gain inference capability of several progressive motion compensation steps. Also, we proposed and evaluated two novel evaluation metrics for CMR-FT task: error decomposition and key point tracking error in addition to the Dice coefficient and boundary distance error. In the future, we plan to apply our method to a clinical study for the short-axis CMR-FT task.

References

1. Berman, D.S., et al.: Prognostic validation of a 17-segment score derived from a 20-segment score for myocardial perfusion SPECT interpretation. J. Nucl. Cardiol. **11**(4), 414–423 (2004). https://doi.org/10.1016/j.nuclcard.2004.03.033
2. Bernard, O., et al.: Deep learning techniques for automatic MRI cardiac multi-structures segmentation and diagnosis: is the problem solved? IEEE Trans. Med. Imaging **37**(11), 2514–2525 (2018)
3. De Craene, M., et al.: Temporal diffeomorphic free-form deformation: application to motion and strain estimation from 3D echocardiography. Med. Image Anal. **16**(2), 427–450 (2012)
4. Hor, K.N., et al.: Magnetic resonance derived myocardial strain assessment using feature tracking. JoVE (J. Vis. Exp.) (48), e2356 (2011)

5. Huang, Q., Yang, D., Qu, H., Yi, J., Wu, P., Metaxas, D.N.: Dynamic MRI reconstruction with motion-guided network (2018)
6. kaggle: Data science bowl cardiac challenge data (2014). second Annual Data Science Bowl from kaggle. https://www.kaggle.com/c/second-annual-data-science-bowl/data
7. Kong, B., Sun, S., Wang, X., Song, Q., Zhang, S.: Invasive cancer detection utilizing compressed convolutional neural network and transfer learning. In: Frangi, A.F., Schnabel, J.A., Davatzikos, C., Alberola-López, C., Fichtinger, G. (eds.) MICCAI 2018. LNCS, vol. 11071, pp. 156–164. Springer, Cham (2018). https://doi.org/10.1007/978-3-030-00934-2_18
8. Krebs, J., Delingette, H., Mailhé, B., Ayache, N., Mansi, T.: Learning a probabilistic model for diffeomorphic registration. IEEE Trans. Med. Imaging 38, 2165–2176 (2019)
9. Liu, P., Lyu, M., King, I., Xu, J.: Selflow: self-supervised learning of optical flow. In: Proceedings of the IEEE Conference on Computer Vision and Pattern Recognition, pp. 4571–4580 (2019)
10. Mei, Y., et al.: Pyramid attention networks for image restoration. arXiv preprint arXiv:2004.13824 (2020)
11. Meister, S., Hur, J., Roth, S.: Unflow: unsupervised learning of optical flow with a bidirectional census loss. In: Thirty-Second AAAI Conference on Artificial Intelligence (2018)
12. Puyol-Antón, E., et al.: Fully automated myocardial strain estimation from cine MRI using convolutional neural networks. In: 2018 IEEE 15th International Symposium on Biomedical Imaging (ISBI 2018), pp. 1139–1143. IEEE (2018)
13. Qin, C., Bai, W., Schlemper, J., Petersen, S.E., Piechnik, S.K., Neubauer, S., Rueckert, D.: Joint learning of motion estimation and segmentation for cardiac MR image sequences. In: Frangi, A.F., Schnabel, J.A., Davatzikos, C., Alberola-López, C., Fichtinger, G. (eds.) MICCAI 2018. LNCS, vol. 11071, pp. 472–480. Springer, Cham (2018). https://doi.org/10.1007/978-3-030-00934-2_53
14. Ranjan, A., Black, M.J.: Optical flow estimation using a spatial pyramid network. In: Proceedings of the IEEE Conference on Computer Vision and Pattern Recognition, pp. 4161–4170 (2017)
15. Rueckert, D., Sonoda, L.I., Hayes, C., Hill, D.L., Leach, M.O., Hawkes, D.J.: Nonrigid registration using free-form deformations: application to breast MR images. IEEE Trans. Med. Imaging 18(8), 712–721 (1999)
16. Seegoolam, G., Schlemper, J., Qin, C., Price, A., Hajnal, J., Rueckert, D.: Exploiting motion for deep learning reconstruction of extremely-undersampled dynamic MRI. In: Shen, D., et al. (eds.) MICCAI 2019. LNCS, vol. 11767, pp. 704–712. Springer, Cham (2019). https://doi.org/10.1007/978-3-030-32251-9_77
17. Shen, D., Sundar, H., Xue, Z., Fan, Y., Litt, H.: Consistent estimation of cardiac motions by 4D image registration. In: Duncan, J.S., Gerig, G. (eds.) MICCAI 2005. LNCS, vol. 3750, pp. 902–910. Springer, Heidelberg (2005). https://doi.org/10.1007/11566489_111
18. Shi, W., et al.: A comprehensive cardiac motion estimation framework using both untagged and 3-d tagged MR images based on nonrigid registration. IEEE Trans. Med. Imaging 31(6), 1263–1275 (2012)
19. Sun, D., Yang, X., Liu, M.Y., Kautz, J.: PWC-Net: CNNS for optical flow using pyramid, warping, and cost volume. In: Proceedings of the IEEE Conference on Computer Vision and Pattern Recognition, pp. 8934–8943 (2018)

20. Tobon-Gomez, C., et al.: Benchmarking framework for myocardial tracking and deformation algorithms: an open access database. Med. Image Anal. **17**(6), 632–648 (2013)
21. Vigneault, D.M., Xie, W., Bluemke, D.A., Noble, J.A.: Feature tracking cardiac magnetic resonance via deep learning and spline optimization. In: Pop, M., Wright, G.A. (eds.) FIMH 2017. LNCS, vol. 10263, pp. 183–194. Springer, Cham (2017). https://doi.org/10.1007/978-3-319-59448-4_18
22. Wang, L., Clarysse, P., Liu, Z., Gao, B., Liu, W., Croisille, P., Delachartre, P.: A gradient-based optical-flow cardiac motion estimation method for cine and tagged MR images. Med. Image Anal. **57**, 136–148 (2019)
23. Xu, X., Chiu, M.T., Huang, T.S., Shi, H.: Deep affinity net: instance segmentation via affinity. arXiv preprint arXiv:2003.06849 (2020)
24. Yang, F., et al.: A deep learning segmentation approach in free-breathing real-time cardiac magnetic resonance imaging. BioMed Res. Int. **2019** (2019)
25. Young, A.A., Frangi, A.F.: Computational cardiac atlases: from patient to population and back. Exp. Physiol. **94**(5), 578–596 (2009)
26. Yu, H., et al.: Computed tomography super-resolution using convolutional neural networks. In: 2017 IEEE International Conference on Image Processing (ICIP), pp. 3944–3948. IEEE (2017)
27. Yu, H., et al.: A novel framework for 3d–2d vertebra matching. In: 2019 IEEE Conference on Multimedia Information Processing and Retrieval (MIPR), pp. 121–126. IEEE (2019)
28. Yu, H., Sun, S., Yu, H., Chen, X., Shi, H., Huang, T.S., Chen, T.: Foal: fast online adaptive learning for cardiac motion estimation. In: CVPR, pp. 4313–4323 (2020)
29. Zheng, Q., Delingette, H., Ayache, N.: Explainable cardiac pathology classification on cine MRI with motion characterization by semi-supervised learning of apparent flow. Med. Image Anal. **56**, 80–95 (2019)

ICA-UNet: ICA Inspired Statistical UNet for Real-Time 3D Cardiac Cine MRI Segmentation

Tianchen Wang[1](✉), Xiaowei Xu[2], Jinjun Xiong[3], Qianjun Jia[2], Haiyun Yuan[2], Meiping Huang[2], Jian Zhuang[2], and Yiyu Shi[1]

[1] University of Notre Dame, Notre Dame, USA
{twang9,yshi4}@nd.edu
[2] Guangdong Provincial People's Hospital, Guangzhou, China
xiao.wei.xu@foxmail.com, jiaqianjun@126.com, yhy_yun@163.com,
huangmeiping@126.com, zhuangjian5413@tom.com
[3] IBM Thomas J. Watson Research Center, Yorktown Heights, USA
jinjun@us.ibm.com

Abstract. Real-time cine magnetic resonance imaging (MRI) plays an increasingly important role in various cardiac interventions. In order to enable fast and accurate visual assistance, the temporal frames need to be segmented on-the-fly. However, state-of-the-art MRI segmentation methods are used either offline because of their high computation complexity, or in real-time but with significant accuracy loss and latency increase (causing visually noticeable lag). As such, they can hardly be adopted to assist visual guidance. In this work, inspired by a new interpretation of Independent Component Analysis (ICA) [11] for learning, we propose a novel ICA-UNet for real-time 3D cardiac cine MRI segmentation. Experiments using the MICCAI ACDC 2017 dataset show that, compared with the state-of-the-arts, ICA-UNet not only achieves higher Dice scores, but also meets the real-time requirements for both throughput and latency (up to 12.6× reduction), enabling real-time guidance for cardiac interventions without visual lag.

1 Introduction

Real-time cine Magnetic Resonance Imaging (MRI) has enabled fast and accurate visual guidance in various cardiac interventions, such as aortic valve replacement [15], cardiac electroanatomic mapping and ablation [17], electrophysiology for atrial arrhythmias [22], intracardiac catheter navigation [7], and myocardial chemoablation [18]. In these applications, it is strongly desirable to segment the

---

T. Wang and X. Xu—Both contributed equally.

Electronic supplementary material The online version of this chapter (https://doi.org/10.1007/978-3-030-59725-2_43) contains supplementary material, which is available to authorized users.

temporal frames on-the-fly, satisfying both throughput and latency requirements. The throughput should be at least above the cine MRI reconstruction rate of 22 frames per second (FPS) [13,19]. The latency should be no more than 50 ms to avoid visually noticeable lags [2]. Most of the existing segmentation methods [14,23,26–29], however, focus on accuracy. In order to handle cardiac border ambiguity and large variations among target objects from different patients, these methods come with high computation cost. Hence their inference latency and throughput are far from meeting the real-time requirements and thus can only be applied offline.

MSU-Net [24] was proposed in MICCAI'19 as the first framework achieving the real-time segmentation of 3D cardiac cine MRI. It uses a canonical form distribution to describe the multiple input frames in a snippet of cine MRI so that only a single pass through the network is needed for all the frames in the snippet. While MSU-Net increases the throughput drastically, the inference latency is also increased to well above 50 ms due to the need of input clustering, i.e., the inference is carried out only after all the frames in a snippet have arrived. When MSU-Net is applied to real-time cine MRI segmentation, such significant visual lags jeopardize the effectiveness of visual guidance in cardiac intervention.

As a popular computational method for decomposing a multivariate signal into additive independent non-Gaussian signals (bases), Independent Component Analysis (ICA) has been widely used in multiple image processing applications such as noise reduction [11], image separation in medical data [3,5] and image decomposition [20]. Through the unmixing process in ICA, any image patch out of a given image can be represented by a linear combination of a set of independent bases of the same size as the image patch. In the mixing process, the original image can be reconstructed using the bases with proper coefficients.

In this paper, based on a new interpretation of ICA for learning (Sect. 2), we propose ICA-UNet, a novel model that can not only achieve highly accurate 3D cardiac cine MRI segmentation results, but also attain both high throughput and low latency. Specifically, an input temporal frame in the cine MRI is decomposed into independent bases and a mixing tensor, composed of the coefficient tensors of all the bases, by a light-weight ICA-encoder. Such an ICA-encoder mimics the unmixing operation in ICA. A U-Net like architecture is trained to learn the transform of the mixing tensor from its original function of image reconstruction to the target function of image segmentation. As such, the transformed mixing tensors can be mixed with the bases through light-weight ICA-decoders to get the desirable features for final segmentation evaluation. Because the coefficient tensors that compose the mixing tensor are much smaller in size than the input frame and can be processed in parallel due to the independence between their corresponding bases, significant latency reduction can be achieved.

Experiment results show that, compared with the state-of-the-art real-time cardiac cine MRI segmentation method MSU-Net, ICA-UNet achieves much higher Dice scores for all cardiac classes with up to 12.6× latency reduction. More specifically, the latency of ICA-UNet is below 50 ms while its throughput is still above 22 FPS, which implies that ICA-UNet is the first method meeting

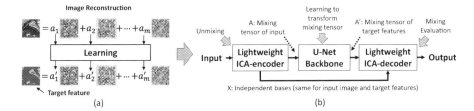

Fig. 1. (a) Conceptual illustration of the mixing tensor transform. We desire to learn the transform of the mixing tensor (formed by the coefficients a_1, a_2,...a_m), from the initial reconstruction task where it is extracted, to the target task. During the transform these coefficients can be processed in parallel due to basis independence. The bases will be saved and reused with the transformed mixing tensor to obtain target features. (b) Conceptual illustration of ICA-UNet. An encoder-backbone-decoder design where the lightweight ICA-encoder/decoder mimic the ICA unmixing/mixing operations and the U-Net learns the transform of the mixing tensor. The bases sharing between ICA-encoder/decoder also naturally forms a long-range skip connection which helps information traverse.

the real-time performance requirements in terms of both throughput and latency for MRI guided cardiac intervention with no visually noticeable lags. In fact, the accuracy achieved by ICA-UNet is on a par with state-of-the-art methods that focus on accuracy and can only run offline because of their complexity.

2   Motivation

It has long been recognized that Independent Component Analysis (ICA) can be used to extract features (bases) from images [9]. Following the similar setup as [16], we can partition an image into a set of smaller image patches such that each image patch can be represented as a linear combination of independent basis image patches along with their coefficients. Compactly put, for a set of input image patches D where each row vector of D represents an input image patch, the goal of ICA is to estimate the unmixing matrix W such that the realizations of bases $X = WD$ are as mutually independent as possible (which is called the unmixing process), while the reconstruction of input image patches, AX, is as close to D as possible (which is called the mixing process). Matrix A is called the mixing matrix, which equals the pseudo inverse of W. There are different ICA algorithms and implementations, and one popular implementation is FastICA [11]. In the rest of the paper, we extend the matrix term to tensor as mixing tensor A and basis tensor X, due to the multi-dimension nature of the input images. Each channel in X represents a basis and the number of channels m equals the basis dimension of ICA. The corresponding coefficient tensor of each basis can be extracted from the mixing tensor A. We will also use basis tensor and bases interchangeably for the simplicity of discussion.

We can have an interesting interpretation of ICA: Both the mixing tensor and the bases can be considered as some kind of feature representation of the input image, with the bases being more fundamental to the input image while the mixing tensor being more related to a particular application such as reconstruction of the input image. In other words, bases can be treated as well-behaved image features that can be reused for different applications, while the mixing tensor can be treated as weights of a simple fully connected layer used to reconstruct the input image.

With such an insight, we wonder if we can learn to transform the mixing tensor so that it can be utilized for a different set of applications (with the help of the bases) beyond the original image reconstruction. As illustrated in Fig. 1(a), the original mixing tensor for image reconstruction is transformed to that for target application, which, after mixing together with the bases, can be used to get the desirable target features for final evaluation of the target application. As the bases are shared, only the mixing tensor, which is composed of the coefficient tensors of all the bases, needs to be transformed. During this process, the coefficient tensors of different bases can be computed in parallel due to the independence between the bases, and each of which is much smaller than the original input. Thus significant latency reduction can be achieved. Since the mixing tensor still exhibits spatial patterns, a conventional image oriented deep neural networks such as U-Net can be used as the backbone to learn the transform.

In conventional ICA, the unmixing operation is lossy, which affects the downstream application (such as image reconstruction) accuracy, and runs as a separate optimization process, which can be quite time-consuming. Since the learning of the target mixing tensor is also an optimization problem, why not combining the ICA process with the learning process as a joint end-to-end training process so that we can not only mitigate the impact of lossy unmixing operation on accuracy, but also reduce one separate optimization process? Such a motivation drives us to propose a lightweight neural network based ICA encoder and decoder to mimic the unmixing and mixing operations in ICA. We further integrate them with a U-Net backbone so that they can be end-to-end trained.

3   Method

Driven by the motivation in Sect. 2, a conceptual illustration of our ICA-UNet is shown in Fig. 1(b). Its detailed architecture is shown in Fig. 2, where we use superscript to denote the time stamp of input frame (e.g. F^t). A summary of all the notations used in this section is also included in the figure. ICA-UNet is mainly made of four types of modules: the ICA-encoder, the contracting blocks ($C_k, k \in [1, n]$), the expanding blocks ($E_k, k \in [1, n]$), and the ICA-decoders. n is the number of contracting/expanding blocks acting as a hyperparameter that can affect accuracy and speed, as will be shown in the experiments.

1) ICA-encoder: The ICA-encoder extracts both the statistically independent basis tensor and the associated initial mixing tensor from the input image.

Instead of running a standard ICA process where the image needs to be explicitly partitioned and an explicit iterative optimization is needed to obtain the mixing tensor and the realization of the bases, we propose to use a neural network to obtain them as a function of the input image.

For an input frame size of (d, h, w) for depth, height, and width, respectively, we can choose the basis dimension m (in this paper $m = 32$), the size of the initial mixing tensor A_n^t as $(1, m, d/2, h/4, w/4)$, and the size of independent basis tensor X^t as $(1, u \times m, d, h/16, w/16)$, where u is the output channel width of the transposed convolution between the concatenated mixing tensor $\mathsf{A'}_{c,k}^t$ and X^t in the ICA-decoder ($u = 4$ in this paper). Each channel in X^t corresponds to an independent basis. The corresponding coefficient tensor of each basis can be obtained from the mixing tensor A_n^t, each of size $(1, 1, d/2, h/4, w/4)$. The extraction of X^t and A_n^t shares some layers for low level feature extraction before they are split channel-wise in order to reduce the computation.

After both A_n^t and X^t are obtained, A_n^t is forwarded to the following contracting block C_n, and X^t is directly forwarded to each ICA-decoder block for mixing operation.

The objective function of ICA-encoder, which is used for regulating the optimization towards sparsity, independence and accuracy, can be expressed as

$$\min \mathcal{L}_{\mathrm{ICA}} = \lambda_s ||(\mathsf{X}^t)||_1 + \lambda_i \mathrm{avg}(-\alpha \log \cosh(\mathsf{X}^t/\alpha)) + \lambda_r \parallel \mathsf{A}_n^t \otimes \mathsf{X}^t - F^t \parallel_2^2, \quad (1)$$

where λ_s, λ_i and λ_r are the weights of the loss terms which are set to 1.0 in our experiments. The first term reflects sparsity through L1 norm. The second term reflects independence through neg-entropy [10]; α is a constant number between 0.5 and 1 (we take $\alpha = 0.75$ in our experiments); avg($\cdot$) denotes element-wise average. The third term reflects reconstruction loss. We adopt transposed convolution ($\otimes$) as the mixing operation, so the L2 distance between the reconstructed frame $\mathsf{A}_n^t \otimes \mathsf{X}^t$ and the original frame F^t should be minimized.

2) Contracting Blocks: The contracting blocks of ICA-UNet are designed to further propagate the mixing tensor A_n^t, and generate the learned ones in a multi-resolution manner. As shown in Fig. 2, the contracting part is made of n contracting blocks, ranging from C_n to C_1. The contracting block C_k ($k \in [1, n]$) takes A_k^t as input, propagates it through a downsampling module (i.e., convolution with stride 2) and convolution modules (i.e., conv) sequentially, and outputs A_{k-1}^t, which is then forwarded to the next contracting block C_{k-1} as well as the corresponding expanding block E_k.

3) Expanding Blocks: The expanding blocks E_k ($k \in [1, n-1]$) are designed to process the features generated by the contracting blocks and forward the outputs to ICA-decoder blocks and the next expanding block E_{k+1} (or concatenation block T_n for E_{n-1}). An expanding block has two sub-tasks: upsampling the mixing tensor $\mathsf{A'}_k^t$ to $\mathsf{A'}_{k+1}^t$ (by upsample block U_k), and calculating dimension aligned correlation features $\mathsf{A'}_{c,k}^t$ from the neighbour frames (by concatenation block T_k). Note we use $\mathsf{A'}_k^t$ to represent the mixing tensor in the upsampling path, distinguishing it from those in the downsampling path A_k^t.

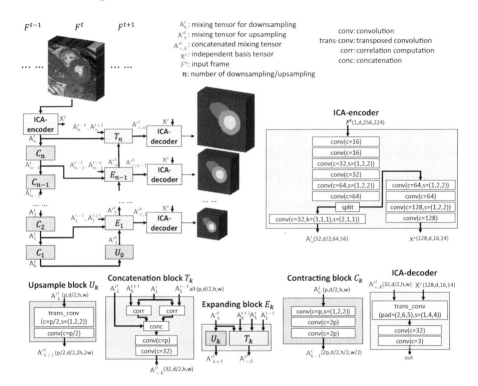

Fig. 2. The architecture of ICA-UNet. The detailed structure of ICA-UNet. p in dimension representation denotes the channel width of input features. Unless specified, the default configuration for conv block (convolution batch normalization and leaky ReLU appended) is kernel_size(k) = (1,3,3), stride(s) = (1,1,1), padding = (0,1,1) with output channel width c, and the default configuration for trans_conv block (transposed convolution) is kernel_size = (1,2,2), stride(s) = (1,2,2), padding(pad) = (0,0,0) with output channel width c.

We use transposed convolution to achieve the upsampling on A'^t_k, during which we obtain the outputs with various resolutions. Taking A^{t-1}_n, ..., A^{t-1}_1, and A^{t+1}_n, ..., A^{t+1}_1 from the outputs of contracting blocks for frames F^{t-1} and F^{t+1}, respectively, we can calculate the temporal correlation features following [12] between mixing tensors A^{t-1}_k and A^t_k, and between A^{t+1}_k and A^t_k ($k \in [1, n]$). The obtained correlation features explicitly provide matching information from the neighbouring frames for more accurate segmentation. The correlation features are then concatenated with A'^t_k, and forwarded through a convolution module with 1×1 kernel (conv-bn-leakyrelu) for dimension reduction. These computations are processed in the concatenation block T_k as shown in Fig. 2.

4) ICA-decoder: The ICA-decoder block is designed to mimic the mixing operation between the concatenated mixing tensor $A'^t_{c,k}$ ($k \in [1, n]$) and the basis X^t, as in the standard ICA, to generate the output for evaluation. As discussed earlier, transposed convolution is used as the mixing operation. The transposed convolution acts as both upsampling for evaluation and the multiplication projection field between X^t and each value in $A'^t_{c,k}$, which helps reducing both the parameter size and the computation load.

After mixing, the mixed features are propagated through a convolution module as the output for evaluation. From E_1 to E_{n-1} we can obtain a total of $n-1$ multi-resolution segmentation outputs for evaluations, denoted as y'_k, $k \in [1, n-1]$. After we get A'^t_n from E_{n-1}, we forward A'^t_n, A^{t+1}_n, A^t_n, A^{t-1}_n, and X^t to the final concatenation block T_n and its corresponding ICA-decoder, where we obtain the output with the same size as the original input. Thus, we obtain a total of n outputs in multi-resolution for evaluation.

Objective Function: We evaluate the outputs from the decoder blocks with the multi-resolutions ground-truth. The overall objective function $\mathcal{L}$ is

$$\mathcal{L} = \sum_{k=1}^{n} \alpha_k \mathcal{L}_k + \beta \mathcal{L}_{\text{ICA}}, k \in [1, n], \tag{2}$$

where $\mathcal{L}_k$ ($k \in [1, n]$) is the evaluation loss of the multi-resolution outputs in the decoder (with Cross-entropy); α_k is the corresponding loss weights, while we take $\alpha_k = 0.1$ for $k \in [1, n-1]$ and $\alpha_n = 1$; $\mathcal{L}_{\text{ICA}}$ is the loss from Equation (1), and the weight β is set to 0.2 in our experiments. Cross-entropy is used for calculating $\mathcal{L}_k$ with the rescaled versions of ground truth.

Latency Reduction Analysis. For the inference of a frame of size $(1, 1, d, h, w)$, the mixing tensor A^t_n, as the input to the backbone, is composed of the m coefficient tensors, each of size $(1, 1, d/2, h/4, w/4)$, which is $32\times$ smaller than the original frame size $(1, 1, d, h, w)$ for a regular U-Net. In addition, these coefficient tensors can be handled in parallel (i.e., a new task parallelism) due to the independence between the bases. Note that the processing of each coefficient tensor can still utilize any existing parallelization techniques such as model or operator parallelization [4,6,8,21] by applying them on the backbone. Therefore, significant latency reduction can be achieved.

4   Experiments

Experiment Setup: We evaluate our model on an extended ACDC MICCAI 2017 challenge dataset made available by MSU-Net [24] with labels on all the frames in the training data. We compare our ICA-UNet with MSU-Net, the state-of-the-art real-time cine MRI segmentation method on the same dataset [24]. We perform 5-fold cross-validation and use the average Dice score (the higher the better) to evaluate the segmentation accuracy. To further see how the accuracy of ICA-UNet compare against the state-of-the-art offline segmentation methods that achieves high accuracy but looses real-time performance, we evaluate the test data by submitting the segmentation results of ED and ES instants to ACDC online evaluation platform [1].

All the methods were implemented in PyTorch and trained from scratch with the same hyperparameters and optimizer setting. MSU-Nets were based on the implementations by [24]. All the networks are fully parallelized using CUDA/CuDNN [4]. All experiments run on a machine with 16 cores of Intel Xeon E5-2620 v4 CPU, 256G memory, and an NVIDIA Tesla P100 GPU.

Table 1. Comparison of Dice, throughput (TP) and latency (LT) between ICA-UNets and the state-of-the-art real-time 3D cardiac cine MRI segmentation method MSU-Net. For ICA-UNet, n denotes the number of contracting blocks. To satisfy real-time requirement, throughput should be above 22 FPS and the latency should be below 50 ms to avoid visually noticeable lags.

Methods	Dice score				TP	LT
	RV	MYO	LV	Average	(FPS)	(ms)
MSU-Net (span = 10)	.837 ± .034	.811 ± .049	.854 ± .040	.834 ± .020	70.2	442
MSU-Net (span = 5)	.855 ± .026	.836 ± .022	.897 ± .017	.862 ± .011	43.2	249
MSU-Net (span = 3)	.858 ± .034	.838 ± .039	.898 ± .034	.864 ± .030	29.4	169
MSU-Net (span = 2)	.860 ± .017	.837 ± .031	.901 ± .021	.867 ± .020	21.7	125
ICA-UNet (n = 3)	.900 ± .023	.869 ± .027	.934 ± .013	.901 ± .017	31.6	**35**
ICA-UNet (n = 4)	**.921 ± .017**	**.888 ± .034**	**.952 ± .015**	**.920 ± .019**	28.3	39

Table 2. The Dice scores and Hausdorff distances on ACDC test set by ICA-UNet and state-of-the-art offline segmentation methods GridNet [29], ensemble U-Net [14], Ω-Net [23]. All results are reported by ACDC evaluation platform [1]. ICA-UNet achieves comparable accuracy while satisfying real-time requirements.

Methods	Dice score			Hausdorff (mm)		
	RV	MYO	LV	RV	MYO	LV
Ensemble U-Net	0.923	0.911	0.950	11.13	8.69	7.15
Ω-Net	0.920	0.891	0.954	N/A	N/A	N/A
GridNet	0.910	0.894	0.938	11.80	9.45	7.30
ICA-UNet (n = 4)	0.920	0.890	0.940	11.91	7.93	6.85

Performance on Real-time Segmentation: The results of ACDC 3D cardiac cine MRI segmentation are shown in Table 1. We can see that ICA-UNet increases the Dice score by 0.061(RV), 0.050(MYO), 0.051(LV), and 0.053(average), respectively, compared with the best results achieved by MSU-Nets. ICA-UNets also achieve smaller Dice score variations than MSU-Nets in most cases. In terms of throughput, although both ICA-UNets and MSU-Nets can satisfy the real-time requirement of 22 FPS, only ICA-UNets can meet real-time latency requirement (below 50 ms), up to 12.6× faster than MSU-Nets. In summary,

ICA-UNet not only achieves the best Dice score, but also is the only real-time segmentation method that can simultaneously meet the real-time throughput and latency requirements for visual guidance of cardiac interventions.

From the table, we can also see that the number of convolutional decoder blocks is an effective tuning knob for Dice and speed tradeoff. A higher number of blocks result in higher Dice scores at the cost of slightly reduced throughput and increased latency. Visualization of segmentation results by ICA-UNet along with the corresponding ground truth is included in the supplementary material.

Accuracy v.s. State-of-the-art Offline Methods: To see how the accuracy of ICA-UNet compares with state-of-the-art offline segmentation methods which do not satisfy real-time requirements, we further verify our ICA-UNet on ED and ES instants of ACDC test data. The evaluation results reported by [1], in terms of both dice score and Hausdorff distance, are shown in Table 2. The results from the best approaches in the literature, including GridNet [29], Ω-Net [23], and ensemble U-Net [14], are also included for reference. With complex network structures, the latency and throughput of these methods are far from the real-time requirements, as shown in [25]. In contrast, we see that the accuracy of ICA-UNet comes very close to these state-of-the-art results while meeting the real-time throughput and latency requirements.

5   Conclusions

Inspired by ICA, ICA-UNet decomposes temporal frames in 3D cardiac cine MRI into independent bases and the corresponding coefficient tensors, which are much smaller in size and help to learn better. Experimental results show that compared with the state-of-the-arts, ICA-UNet is the only 3D cardiac cine MRI segmentation method that can satisfy both real-time throughput and latency requirements with comparable (if not better) accuracy.

References

1. Acdc challenge. https://www.creatis.insa-lyon.fr/Challenge/acdc/
2. Annett, M., Ng, A., Dietz, P., Bischof, W., Gupta, A.: How low should we go? Understanding the perception of latency while inking. In: 2014 Graphics Interface, pp. 167–174 (2014)
3. Bronstein, A.M., Bronstein, M.M., Zibulevsky, M., Zeevi, Y.Y.: Sparse ICA for blind separation of transmitted and reflected images. Int. J. Imaging Syst. Technol. **15**(1), 84–91 (2005)
4. Chetlur, S., et al.: cuDNN: efficient primitives for deep learning. arXiv preprint arXiv:1410.0759 (2014)
5. Delorme, A., Sejnowski, T., Makeig, S.: Enhanced detection of artifacts in EEG data using higher-order statistics and independent component analysis. Neuroimage **34**(4), 1443–1449 (2007)
6. Dryden, N., Maruyama, N., Benson, T., Moon, T., Snir, M., Van Essen, B.: Improving strong-scaling of CNN training by exploiting finer-grained parallelism. In: 2019 IEEE International Parallel and Distributed Processing Symposium (IPDPS), pp. 210–220. IEEE (2019)

7. Gaspar, T., Piorkowski, C., Gutberlet, M., Hindricks, G.: Three-dimensional real-time MRI-guided intracardiac catheter navigation. Eur. Heart J. **35**(9), 589–589 (2014)
8. Gholami, A., Azad, A., Jin, P., Keutzer, K., Buluc, A.: Integrated model, batch, and domain parallelism in training neural networks. In: Proceedings of the 30th on Symposium on Parallelism in Algorithms and Architectures, pp. 77–86 (2018)
9. Hoyer, P.O., Hyvärinen, A.: Independent component analysis applied to feature extraction from colour and stereo images. Netw.: Comput. Neural Syst. **11**(3), 191–210 (2000)
10. Hyvärinen, A., Karhunen, J., Oja, E.: Independent Component Analysis, vol. 46. Wiley, Hoboken (2004)
11. Hyvärinen, A., Oja, E.: Independent component analysis: algorithms and applications. Neural Netw. **13**(4–5), 411–430 (2000)
12. Ilg, E., Mayer, N., Saikia, T., Keuper, M., Dosovitskiy, A., Brox, T.: Flownet 2.0: evolution of optical flow estimation with deep networks. In: Proceedings of the IEEE Conference on Computer Vision and Pattern Recognition, pp. 2462–2470 (2017)
13. Iltis, P.W., Frahm, J., Voit, D., Joseph, A.A., Schoonderwaldt, E., Altenmüller, E.: High-speed real-time magnetic resonance imaging of fast tongue movements in elite horn players. Quant. Imaging Med. Surg. **5**(3), 374 (2015)
14. Isensee, F., Jaeger, P.F., Full, P.M., Wolf, I., Engelhardt, S., Maier-Hein, K.H.: Automatic cardiac disease assessment on cine-MRI via time-series segmentation and domain specific features. In: Pop, M., et al. (eds.) STACOM 2017. LNCS, vol. 10663, pp. 120–129. Springer, Cham (2018). https://doi.org/10.1007/978-3-319-75541-0_13
15. McVeigh, E.R., et al.: Real-time interactive MRI-guided cardiac surgery: aortic valve replacement using a direct apical approach. Magn. Reson. Med. Offi. J. Int. Soc. Magn. Reson. Med. **56**(5), 958–964 (2006)
16. Olshausen, B.A., Field, D.J.: Natural image statistics and efficient coding. Netw.: Comput. Neural Syst. **7**(2), 333–339 (1996)
17. Radau, P.E., et al.: VURTIGO: visualization platform for real-time, MRI-guided cardiac electroanatomic mapping. In: Camara, O., Konukoglu, E., Pop, M., Rhode, K., Sermesant, M., Young, A. (eds.) STACOM 2011. LNCS, vol. 7085, pp. 244–253. Springer, Heidelberg (2012). https://doi.org/10.1007/978-3-642-28326-0_25
18. Rogers, T., et al.: Transcatheter myocardial needle chemoablation during real-time magnetic resonance imaging: a new approach to ablation therapy for rhythm disorders. Circul.: Arrhythm. Electrophysiol. **9**(4), e003926 (2016)
19. Schaetz, S., Voit, D., Frahm, J., Uecker, M.: Accelerated computing in magnetic resonance imaging: real-time imaging using nonlinear inverse reconstruction. Comput. Math. Methods Med. **2017** (2017)
20. Starck, J.L., Elad, M., Donoho, D.L.: Image decomposition via the combination of sparse representations and a variational approach. IEEE Trans. Image Process. **14**(10), 1570–1582 (2005)
21. Vasudevan, A., Anderson, A., Gregg, D.: Parallel multi channel convolution using general matrix multiplication. In: 2017 IEEE 28th International Conference on Application-specific Systems, Architectures and Processors (ASAP), pp. 19–24. IEEE (2017)
22. Vergara, G.R., et al.: Real-time magnetic resonance imaging-guided radiofrequency atrial ablation and visualization of lesion formation at 3 tesla. Heart Rhythm **8**(2), 295–303 (2011)

23. Vigneault, D.M., Xie, W., Ho, C.Y., Bluemke, D.A., Noble, J.A.: ω-net (omega-net): fully automatic, multi-view cardiac MR detection, orientation, and segmentation with deep neural networks. Med. Image Anal. **48**, 95–106 (2018)
24. Wang, T., et al.: MSU-Net: multiscale statistical U-Net for real-time 3D cardiac MRI video segmentation. In: Shen, D., et al. (eds.) MICCAI 2019. LNCS, vol. 11765, pp. 614–622. Springer, Cham (2019). https://doi.org/10.1007/978-3-030-32245-8_68
25. Wang, T., Xiong, J., Xu, X., Shi, Y.: SCNN: a general distribution based statistical convolutional neural network with application to video object detection. arXiv preprint arXiv:1903.07663 (2019)
26. Xu, X., et al.: Quantization of fully convolutional networks for accurate biomedical image segmentation. In: Proceedings of the IEEE Conference on Computer Vision and Pattern Recognition, pp. 8300–8308 (2018)
27. Xu, X., et al.: Whole heart and great vessel segmentation in congenital heart disease using deep neural networks and graph matching. In: Shen, D., et al. (eds.) MICCAI 2019. LNCS, vol. 11765, pp. 477–485. Springer, Cham (2019). https://doi.org/10.1007/978-3-030-32245-8_53
28. Yan, W., Wang, Y., Li, Z., van der Geest, R.J., Tao, Q.: Left ventricle segmentation via optical-flow-net from short-axis Cine MRI: preserving the temporal coherence of cardiac motion. In: Frangi, A.F., Schnabel, J.A., Davatzikos, C., Alberola-López, C., Fichtinger, G. (eds.) MICCAI 2018. LNCS, vol. 11073, pp. 613–621. Springer, Cham (2018). https://doi.org/10.1007/978-3-030-00937-3_70
29. Zotti, C., Luo, Z., Lalande, A., Jodoin, P.M.: Convolutional neural network with shape prior applied to cardiac MRI segmentation. IEEE J. Biomed. Health Inform. **23**, 1119–1128 (2018)

A Bottom-Up Approach for Real-Time Mitral Valve Annulus Modeling on 3D Echo Images

Yue Zhang[1], Abdoul-aziz Amadou[1], Ingmar Voigt[2], Viorel Mihalef[1], Helene Houle[3], Matthias John[4], Tommaso Mansi[1], and Rui Liao[1(✉)]

[1] Digital Technology and Innovation, Siemens Healthineers, Princeton, NJ, USA
rui.liao@siemens-healthineers.com
[2] Digital Technology and Innovation, Siemens Healthineers, Erlangen, Germany
[3] Siemens Healthineers, Ultrasound, Issaquah, WA, USA
[4] Advanced Therapies, Siemens Healthineers, Forchheim, Germany

Abstract. 3D+t Transesophageal Echocardiography (TEE) performs 4D scans of mitral valve (MV) morphology at frame rate providing real-time guidance for catheter-based interventions for MV repair and replacement. A key anatomical structure is the MV annulus, and live quantification of the dynamic annulus at acquisition rates 15 fps or higher have proven to be technically challenging. In this paper, we propose a bottom-up approach inspired by clinicians' manual workflow for MV annulus modeling on 3D+t TEE images in real time. Specifically, we first detect annulus landmarks with clear 3D anatomical features via agents trained using Deep Reinforcement Learning. Leveraging the circular structure of the annulus, cross-annular planes are extracted and additional landmarks are then detected through 2D image-to-image networks on the 2D cutting planes. The complete 3D annulus is finally fitted through all detected landmarks using Splines. We validate the proposed approach on 795 3D+t TEE sequences with 1906 annotated frames, and achieve a speed 20 fps with a median accuracy 2.74 mm curve-to-curve error. Furthermore, device simulation is utilized to augment the training data that results in promising accuracy improvement on challenging echos with visible devices and warrants further investigation.

Keywords: Live cardiac analytics · Deep Reinforcement Learning · Mitral valve annulus · 3D ultrasound imaging

1 Introduction

Heart valve disease constitutes an important subgroup of cardiovascular disease (CVD), the leading cause of death worldwide, with the mitral valve (MV) frequently being affected by various dysfunctions [1,2]. Inspired by the remarkable success of transcatheter aortic valve implantation, device manufacturers are developing numerous catheter-based strategies for MV repair and replacement

© Springer Nature Switzerland AG 2020
A. L. Martel et al. (Eds.): MICCAI 2020, LNCS 12266, pp. 458–467, 2020.
https://doi.org/10.1007/978-3-030-59725-2_44

that are becoming more complex to perform and typically require additional real-time guidance provided by 3D+t transesophageal echocardiography (TEE) [3]. 4D scans of valve morphology and function are generated over the complete cardiac cycle at average frame rates of 15 to 20 fps. For many mitral devices a key structure is the mitral annulus, which separates the left atrium and left ventricle and is attached by the MV leaflets. Live visualization of the annulus in the ultrasound images or overlaid onto X-ray could help the user with anatomy orientation and instrument guidance.

Various approaches for MV annulus modeling have been proposed in the literature and either take seconds [4–6] to minutes [7] to process on a single frame, or semi-automatic [8], or require a complex multithreading scheme, which still needs seconds to initialize [9]. Overall real-time annulus modeling has been rarely addressed. Recently, Andreassen et al. [10] leverages a 2D network model to segment annulus in 2D slices from the 3D TEE images. This approach, however, assumes the volume centerline coincides with the centerline of the left ventricle via standardized image acquisition, which does not hold for a significant portion our experimental data, where 64% of the data shows angles that are greater than 15° between the two centerlines and for 10% of the data the volume centerline does not even pass through the annulus ring. Further, due to the lack of explicit landmark/orientation detection, clinically important measurement such as septolateral diameter could not be derived. In addition, surgical devices presented in the Field-of-View (FOV) during interventional procedures could pose additional challenges for annulus modeling due to device/anatomy interaction and/or shadowing effect, but are rarely handled in the reported methods. Last, reported methods were only tested on a small number of patients (19 in [10], 65 in [4], 18 in [8]). In this work, we provide a systematic approach for real-time annulus modeling with extensive studies on over 600 patients.

In this paper, we present a highly efficient approach for MV annulus detection and modeling in 3D TEE images. Instead of exhaustive global search of the candidate locations as used in [4,5,9], our proposed approach uses a bottom-up scheme and first traces in real-time the clinically well-defined 3D anatomical landmarks, called trigones of the annulus, with an optimal pathway provided by an agent trained using Deep Reinforcement Learning (DRL). Leveraging the circular structure of the annulus, the detected trigone points are then used as clues to extract the 2D septo-lateral plane, and a third landmark, namely mid-point of the posterior annulus, is then detected through a 2D Image-to-Image (I2I) network. Similarly, fast and parallel 2D search for additional landmarks on the annulus are performed, and the 3D annulus is finally constructed from all the detected points using a cubic spline fitting. An overview of the proposed workflow as well as the spatial relation of all related landmarks are summarized in Fig. 1. Furthermore, to tackle the challenge of interventional data with visible devices, we augment the training data using simulated images that are generated from the fusion of diagnostic TEEs with simulated mitral-clip devices.

The proposed approach leverages clinical knowledge of the anatomical structure and is inspired by the way that clinical experts navigate through the data

to find the MV annulus in practice. At the same time it is a generic and efficient bottom-up scheme, which could be applied for detecting any circular structures in 3D images without prior knowledge of the location or the orientation of
the annulus. The resulting modeling speed 20 fps is unprecedented in reported
state-of-art methods and could potentially enable a wide range of applications
requiring live quantification and workflow optimization.

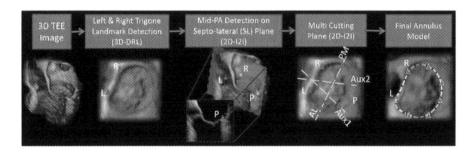

Fig. 1. Overview of proposed model workflow. Primary landmarks include left and
right trigones (L/R), antero-lateral and postero-medial landmarks (AL/PM), auxillary
points for smooth construction (Aux1/Aux2), middle point on posterior annulus (P).

The contributions of our work are: 1) We propose a novel bottom-up pipeline
for real-time annulus modeling in 3D TEE images, leveraging optimal 3D landmark search using deep reinforcement learning together with 2D image-to-image
networks for efficient 2D landmark localization with a large field of view. 2)
We introduce simulated interventional TEEs into the training pipeline to tackle
challenging data with devices in the FOV, which are typically not collected
during the interventional procedure and therefore scarce in the database for
training purposes. 3) We demonstrate the efficiency and effectiveness of the proposed workflow for two potential clinical applications, namely *(i)* automatic view
alignment for optimal TEE imaging as well as *(ii)* real-time annulus modeling
for intraoperative guidance, and test our method on a large database acquired
both before and during MV interventional procedures.

2   Methodology

2.1   Modeling Rationale

The proposed bottom-up workflow consists of three steps shown in Fig. 1. Firstly,
we train artificial agents to identify anatomical landmarks in the 3D image,
namely left (L) and right (R) trigone landmarks on the annulus, that have clear
anatomical features and are commonly used for annulus quantification. Secondly,
we construct the septo-lateral (SL) plane that passes through the middle of the
line segment between the detected L, R trigones and is perpendicular to it. A 2D

Image-to-image network (I2I) is trained to detect (the projected) middle point of posterior annulus (P). With detected L, R and P, the location as well as orientation of the annulus can be estimated, which could be used to define the optimal view orientation for TEE imaging of the MV. Thirdly, we construct additional cross-annular planes to detect additional landmarks on the annulus, and the complete annulus ring can be constructed with spline interpolation over all the detected points for the purpose of full annulus modeling and size estimation.

The benefits of this bottom-up approach, that essentially translates 3D structure detection into 2D landmark localization, are threefold: (1) There are no well-defined 3D landmarks with clear anatomical features on posterior annulus, which hamper 3D modeling strategies. In contrast, with a proper cutting strategy, the annulus is a unique landmark and can be easily identified in the cross-annular planes, as depicted in Fig. 1. (2) Prediction with 2D networks can be very efficiently executed in parallel, and hence is well-suited for real-time applications, as a 2D I2I predicts a single landmark at a speed of 125 fps, c.f. Sect. 3. (3) The approach is highly flexible in terms of number of landmarks to be detected, which allows balancing desired accuracy vs. runtime.

2.2   Trigone Landmark Detection with Deep Q-Learning

We first train artificial agents via deep Q-learning to detect the trigone landmarks in the 3D TEE images [11]. The trained agent is able to learn the anatomical structures presented in the 3D image volume and move towards the target landmark. In the setting of deep reinforcement learning, we consider a setting of Markov decision process defined by the tuple $\{S, \mathcal{A}, \tau, r\}$, where

- S stands for the state of the agent, a $19 \times 19 \times 19$ sub-volume of the 3D TEE image centered at the current location of agent.
- $\mathcal{A}$ represents a finite set of actions executed by the agent. In this case, $\mathcal{A}$ is an action space of displacements along each axis $\{x, y, z\}$. For example, ± 1 step towards the positive/negative direction along with the axis.
- $\tau : S \times \mathcal{A} \times S \rightarrow [0; 1]$ is a stochastic transition function, describing the probability of arriving in a certain state with a given action.
- r is the rewards collected by the agent by interacting with the environment. Here $r = \|p^c - p^{gt}\|_2^2 - \|p^n - p^{gt}\|_2^2$, where p^c represents the current location of the agent, p^{gt} is the ground truth location of the landmark it is pursuing, p^n is the next location of the agent after it executes an action. In other words, the agent collects a positive reward if moves closer to the target and a negative reward otherwise.

We use the standard deep Q learning strategy to train the agent [12], where the loss is defined as

$$\mathcal{L}_{DQL}(\theta) = \mathbb{E}_{s,a,r,s'} \left[(r + \gamma \max_{a' \in \mathcal{A}} Q(s', a'; \theta) - Q(s, a; \theta^-))^2 \right].$$

Here the Q-function is parametrized by a six-layer fully convolutional network with $3 \times 3 \times 3$ kernels cross all layers. Given any 3D local patch, it outputs a

6-dimensional vector, corresponding to the Q-value evaluation of the actions in space $\mathcal{A}$. The agent is trained to collect the maximum amount of rewards. The two agents are trained independently for L and R accordingly. Starting from any location, a well-trained agent moves closer to the target landmark iteratively. In other words, instead of searching through the entire image, the agent only needs to process a collection of small sub-volumes (the states) along its trajectory to the target landmark. Furthermore, due to its small size, the Q-network predicts with high speed at testing time.

2.3   Landmark Detection on 2D Cross-Annular Planes

Once the trigone landmarks (L, R) are detected, we extract the SL plane from the 3D TEE image. The SL plane takes the L-R connecting line as plane normal and passes through its middle point. The cross sections between this plane and the annulus contains the middle point on posterior annulus (P), where the posterior leaflet is connected to the wall (Fig. 1). The middle point (P) is not a well-defined 3D anatomical landmark as it does not manifest distinguishing 3D features, but rather a clinically defined one, used for annulus quantification, i.e. for measuring anterior-posterior distance, and brings important insights on the location and the orientation of the annulus in 3D space. Detection of such 2D landmarks can thus benefit from a global structural understanding of the 2D image. We therefore train a 2D deep image-to-image network over the entire 2D cross-sected image for the landmark detection.

The specific network used in this work is a UNet with densely connected blocks [13]. It is a 14-layer fully convolutional network and employs an symmetric encoder-decoder UNet structure with 4 dense blocks of growth rate 16. Filter size is set to be 3×3 for all layers. We convert the ground truth binary mask to a continuous 2D Gaussian map G to stabilize the training process. Specifically, given the ground truth landmark location as (x^{gt}, y^{gt}) on the SL plane, the value of the (i, j)th entry in the converted mask is

$$G(i, j) = \exp(-\frac{(i - x^{gt})^2 + (j - y^{gt})^2}{\sigma^2}),$$

where σ is defined empirically to achieve best detection performance. A standard L^2 loss is used between the predicted probability map and ground truth map.

Given the detected point P, additional cross-annular planes can be computed in a similar fashion to detect other landmarks such as the antero-lateral (AL) and postero-medial (PM) landmarks, which are generally used for width estimation of the annulus. To obtain a smooth and accurate curve construction, we extract two additional planes with angles of $\pm 30°$ to the SL plane. Furthermore, detection all these additional landmarks is highly efficient since performed in parallel.

2.4   Simulation of Interventional TEEs

Interventional TEEs may include devices during different stages of an intervention. In contrast to diagnostic TEE examinations, they are generally not collected. This creates a bottleneck for training deep models. To tackle this challenge, we generate simulated interventional TEEs by implanting virtual MitraClip devices into real volumes (without devices), see Fig. 2. The virtual implantation is performed by first generating synthetic ultrasound images of both anatomy and device, where a scattering response of virtual ultrasound waves is simulated based on both the original TEE scattering intensities and device intensities. The extra scattering response due to the presence of the virtual device is then used to augment the original no-device TEE.

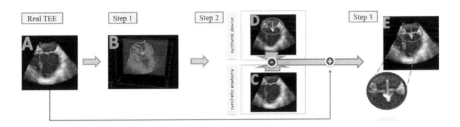

Fig. 2. Augmentation of real data with simulated devices (details in the text). The final intensity of the augmented volume E is computed as $I_E = I_A + \mathcal{N}(I_D, I_C)$, where a simple option for the device response $\mathcal{N}(I_D, I_C) = I_D - I_C$ is illustrated above.

In a preparatory stage, several mesh models of the device, corresponding to different stages of its deployment, are created based on the manufacturer specifications [14]. A variable length deployment catheter mesh is connected to each device mesh and includes a bend, which is directed toward the insertion point located on the inter-atrial septum.

Step 1. *Automatic positioning of the device and catheter mesh within the TEE volume*: The device mesh is placed in various locations ±1 cm away from the valve center, corresponding to its deployment position and aligned along a direction normal to MV annulus ground truth curve (obtained from the original volume, depicted in Fig. 2 A). The septal wall insertion point of the catheter is set 4–6 above annulus plane, according to atrial size.

Step 2. *Virtual TEE generation*: two virtual TEE volumes are created, the first one approximating only the TEE intensity due to original tissue response $I_C = I_A + \epsilon$ (with ϵ being the model approximation error), and the second one (I_D) approximating the TEE intensity in presence of the device. The first volume includes only tissue information (Fig. 2 C), encoded as a 3D point-cloud (scatterers) sampling the US cone, with power-adjusted scatterer intensities [15]. The second volume additionally includes the device and catheter (Fig. 2 D), as 3D point-clouds with intensities set as constants – determined from real interventional TEEs. From each 3D scatterer set a TEE volume is simulated using the fast simulation algorithm COLE [16]. The same scan parameters (angles, dimensions, frequencies etc.) as in the original TEE are used.

Step 3. *Final TEE generation from virtual images*: We define the final intensity as $I_E = I_A + \mathcal{N}(I_D, I_C)$, where $\mathcal{N}$ is an operator that, given one image with device and one without, computes the extra response induced by the device. We consider a simple version, $\mathcal{N}(I_D, I_C) = I_D - I_C$, which, in the context of the COLE model, gives $\mathcal{N}(I_D, I_C) = H \otimes T_{Tissue+Device} - H \otimes T_{Tissue} = H \otimes T_{Device}$, where H is the point spread function (PSF) and T_m denotes the echographic response of the material m. This shows that our final intensity is in essence the original intensity, augmented with a model device response. Finally, to avoid intensity range overshoot we also apply a range-limiting filter to I_E.

3   Experiments

We first validate the proposed approach on a collection of datasets collected from six clinical sites and do not include any devices. There are 659 patients included in this study. Most patients suffered from secondary mitral regurgitation (MR) with a smaller subset with primary MR. This collection consists of 795 3D+t TEE sequences[1] with a total number of 1906 annotated 3D frames, and 1474 frames are used for training and 432 frames for testing, without overlapping patients between the two groups. The annotation procedure follows a guideline where two annotations per sequence were typically performed by clinical experts, one for diastolic frame and the other for systolic frame, to cover maximum motion. The images cover a wide range of imaging settings and FOV, to exclusively cover the mitral valve but also other anatomies like ventricles and other valves. All volumes are resampled 1.0 mm isotropic resolution. The frame rate ranges from 15 to 20 fps. All models are implemented with Pytorch and trained on 12 GB NVIDIA TITAN X GPUs.

We evaluate the results with various metrics that are relevant for clinical practice (Table 1). The accuracy of 3D landmark detection (L, R) is measured by Euclidean distance. For landmarks without clear anatomical features (P, AL, PM, Aux1, Aux2), accuracy is measured by their projection distance to the annotated annulus. An important application of the proposed method is automated selection of surgical view that is defined by a 2D plane computed from L, R and P. The median angle error between the predicted view plane normal and ground truth is 7.2°, which, according to our collaborating physicians, is clinically acceptable to provide optimal TEE volume visualization for qualitative assessment of valve morphology and for guiding device deployment. The accuracy of final annulus model is measured by average curve-to-curve distance between the detection result (D) and the ground truth (G),

$$d(D, G) = (\mathbb{E}_{d \in D} \inf_{g \in G} \|d - g\|_2 + \mathbb{E}_{g \in G} \inf_{d \in D} \|g - d\|_2)/2.$$

The resulting accuracy 2.74 mm median annulus error is also considered to be clinically relevant for triage patient selection for device therapy. Accuracy on

---

[1] Each sequence or case refers to a TEE recording over time. It consists of 10 to 60 frames. Each frame is a 3D TEE image.

septo-lateral diameter has also been evaluated, with a mean error 3.52 mm and standard deviation of 3.95 mm. Accuracy on perimeter measurements that determines the implant size in valvular surgeries, is presented via relative error in percentage, and the median relative error is 6%. To better interpret this result, three experts annotated the annulus over a randomly selected 20 3D+t TEE sequences, and maximum disagreement on the annulus perimeter for each sequence is calculated. Summarizing over all 20 sequences, the median and maximum annulus perimeter inter-observer difference are 4.8% and 15.5% accordingly. Examples of predicted annulus on the surgical view, as well as on randomly selected cross-annular planes are shown in Fig. 3[2].

 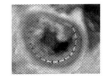 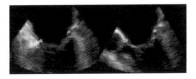

Fig. 3. Visualization of predicted annulus (yellow) and ground truth annotation (red). Left: predicted annulus curve and ground truth in 3D coordinates. Middle: detected surgical view of TEE volume overlaid with predicted annulus. Right: cross-section of prediction and ground truth on randomly selected cross-annular planes. (Color figure online)

Table 1. Quantitative results of landmark detection, view angle prediction, and annulus modeling on data without (D1) and with devices (D2). Various percentiles, mean and standard deviation of the errors are presented.

Exp Data	Acc\Objs	L (mm)	R (mm)	P (mm)	View Angle (degree)	Aux Lmks (mm)	Annulus Curve (mm)	Perim (%)
D1	25%	2.15	2.13	1.56	4.00	2.42	2.25	3
	50%	3.15	3.20	2.52	7.21	2.97	2.74	6
	80%	4.87	5.77	4.63	12.50	4.37	4.19	12
	mean	3.94	4.24	3.84	9.62	3.57	3.49	10
	std	3.67	3.37	5.25	10.46	2.04	2.21	16
D2	Vanilla (50%)	4.21	6.64	5.30	14.00	6.09	5.89	10
	Augmented (50%)	3.74	4.53	4.96	12.36	5.27	5.11	9

In terms of processing time, the detection speed of individual trigone landmark in 3D and additional landmarks in 2D 30.5 fps 125 fps accordingly. We point out that both the detection of L/R as well as auxiliary landmarks are parallelizable. As a result, the proposed approach is able to achieve a mean accuracy 3.49 mm at a processing speed of 20.4 fps, which significantly outperforms related

---

[2] Courtesy of Dr Mani Vannan, Piedmont Heart Institute, Atlanta, GA.

approaches in terms of accuracy and speed (4.04 mm 12.5 fps in [9] with seconds of re-initialization, and 0.58fps in [10][3]).

Furthermore, we validate our proposed approach on a collection of data with devices to demonstrate improved performance in presence of devices by training with simulated data. This dataset consists of 48 cases with 95 annotated frames. All these cases are collected during or after MitraClip deployment, with the mitral-clip devices visible. The entire dataset is used for testing only. For training, we randomly select a set of 46 cases from the dataset without devices and a total number of 138 simulated frames with devices are generated. We add these simulated frames to the training set mentioned in the previous experiment and fine-tune the training weights for all networks. Here we compare two setups: 1) 'vanilla' refers to directly application of the networks trained in the previous experiment; 2) 'Augmented' refers to results from network fine-tuned with the simulated data. As is shown in Table 1, adding simulated data in training results in clearly improved performance in all metrics compared to the vanilla model. While the result is preliminary given the limited size of the testing data, it warrants further investigation using simulated data for general training purpose when real clinical data is difficult to collect.

4   Discussion and Conclusion

In this paper, we presented a systematic approach for real-time MV annulus modeling on 3D+t TEE images. We leveraged reinforcement learning for efficient searching of 3D landmarks and used them as clues for additional landmark detection on cross-annular planes. The proposed approach is generic and can be easily adapted to annulus construction of other valves (e.g. aortic valve and tricuspid valve) and in different image modalities. Furthermore, it also provides possibility to additional interventional applications. Positioning the device correctly relative to the anatomy is generally challenging due to the limited FOV and the image artifacts created by the device. Live annulus modeling in the presence of the device therefore potentially can help position the MitraClip device properly. In addition, having the mitral annulus detected in real-time will allow quantification during the procedure, updated continuously. This may help complex procedures like Cardioband and other annulo-plasty like interventions. The achieved accuracy in this work is promising to facilitate real-time mitral valve viewing orientation, live monitoring of annulus dimensions/size, and real-time Fluoro-echo overlay of the annulus for instrument guidance. The field of interventional planning and real time guidance is bright with endless possibilities in our not so distant future, and we leave this development for future direction.

Disclaimer. The concepts and information presented in this paper are based on research results that are not commercially available. Due to regulatory reasons its future availability cannot be guaranteed.

---

[3] [10] presents accuracy over 2D cutting-planes with weighted mean error as 2.0 mm, which the same metric in this work is 1.57 mm.

References

1. Benjamin, E.J., Muntner, P., Bittencourt, M.S.: Heart disease and stroke statistics-2019 update: a report from the american heart association. Circulation **139**(10), e56–e528 (2019)
2. El Sabbagh, A., Reddy, Y.N., Nishimura, R.A.: Mitral valve regurgitation in the contemporary era. JACC: Cardiovasc. Imaging **11**(4), 628–643 (2018)
3. Bax, J.J., et al.: Transcatheter interventions for mitral regurgitation. JACC: Cardiovasc. Imaging **12**(10), 2029–2048 (2019)
4. Ionasec, R.I., et al.: Patient-specific modeling and quantification of the aortic and mitral valves from 4-d cardiac CT and tee. IEEE Trans. Med. Imaging **29**(9), 1636–1651 (2010)
5. Voigt, I., et al.: Robust physically-constrained modeling of the mitral valve and subvalvular apparatus. In: Fichtinger, G., Martel, A., Peters, T. (eds.) MICCAI 2011. LNCS, vol. 6893, pp. 504–511. Springer, Heidelberg (2011). https://doi.org/10.1007/978-3-642-23626-6_62
6. Schneider, R.J., Perrin, D.P., Vasilyev, N.V., Marx, G.R., Pedro, J., Howe, R.D.: Mitral annulus segmentation from four-dimensional ultrasound using a valve state predictor and constrained optical flow. Med. Image Anal. **16**(2), 497–504 (2012)
7. Pouch, A.M., et al.: Modeling the myxomatous mitral valve with three-dimensional echocardiography. Ann. Thorac. Surg. **102**(3), 703–710 (2016)
8. Graser, B., et al.: Using a shape prior for robust modeling of the mitral annulus on 4D ultrasound data. Int. J. Comput. Assist. Radiol. Surg. **9**(4), 635–644 (2014). https://doi.org/10.1007/s11548-013-0942-3
9. Voigt, I., et al.: Robust live tracking of mitral valve annulus for minimally-invasive intervention guidance. In: Navab, N., Hornegger, J., Wells, W.M., Frangi, A.F. (eds.) MICCAI 2015. LNCS, vol. 9349, pp. 439–446. Springer, Cham (2015). https://doi.org/10.1007/978-3-319-24553-9_54
10. Andreassen, B.S., Veronesi, F., Gerard, O., Solberg, A.H.S., Samset, E.: Mitral annulus segmentation using deep learning in 3D transesophageal echocardiography. IEEE J. Biomed. Health Inform. **24**, 994–1003 (2019)
11. Ghesu, F.C., et al.: Multi-scale deep reinforcement learning for real-time 3d-landmark detection in CT scans. IEEE Trans. Pattern Anal. Mach. Intell. **41**(1), 176–189 (2017)
12. Mnih, V., et al.: Human-level control through deep reinforcement learning. Nature **518**(7540), 529 (2015)
13. Jégou, S., Drozdzal, M., Vazquez, D., Romero, A., Bengio, Y.: The one hundred layers tiramisu: Fully convolutional densenets for semantic segmentation. In: Proceedings of the IEEE Conference on Computer Vision and Pattern Recognition Workshops, pp. 11–19 (2017)
14. Abbott Laboratories: Mitraclip clip delivery system: instructions for use. https://www.accessdata.fda.gov/cdrh_docs/pdf10/P100009c.pdf (2013). Accessed 16 Mar 2020
15. Alessandrini, M., et al.: A pipeline for the generation of realistic 3D synthetic echocardiographic sequences: methodology and open-access database. IEEE Trans. Med. Imaging **34**, 1436–1451 (2015)
16. Gao, H., et al.: A fast convolution-based methodology to simulate 2-d/3-d cardiac ultrasound images. IEEE Trans. Ultrasonics Ferroelectr. Freq. Control **56**, 404–409 (2009)

A Semi-supervised Joint Network for Simultaneous Left Ventricular Motion Tracking and Segmentation in 4D Echocardiography

Kevinminh Ta[1]([✉]), Shawn S. Ahn[1], John C. Stendahl[2], Albert J. Sinusas[2,4], and James S. Duncan[1,3,4]

[1] Department of Biomedical Engineering, Yale University, New Haven, CT, USA
kevinminh.ta@yale.edu
[2] Department of Internal Medicine, Yale University, New Haven, CT, USA
[3] Department of Electrical Engineering, Yale University, New Haven, CT, USA
[4] Department of Radiology and Biomedical Imaging, Yale University, New Haven, CT, USA

Abstract. This work presents a novel deep learning method to combine segmentation and motion tracking in 4D echocardiography. The network iteratively trains a motion branch and a segmentation branch. The motion branch is initially trained entirely unsupervised and learns to roughly map the displacements between a source and a target frame. The estimated displacement maps are then used to generate pseudo-ground truth labels to train the segmentation branch. The labels predicted by the trained segmentation branch are fed back into the motion branch and act as landmarks to help retrain the branch to produce smoother displacement estimations. These smoothed out displacements are then used to obtain smoother pseudo-labels to retrain the segmentation branch. Additionally, a biomechanically-inspired incompressibility constraint is implemented in order to encourage more realistic cardiac motion. The proposed method is evaluated against other approaches using synthetic and in-vivo canine studies. Both the segmentation and motion tracking results of our model perform favorably against competing methods.

Keywords: Echocardiography · Motion tracking · Segmentation

1 Introduction

Echocardiography is a non-invasive and cost-efficient tool that allows clinicians to visually evaluate the left ventricular (LV) wall and detect any motion or structural abnormalities in order to evaluate cardiovascular health and diagnose cardiovascular diseases (CVD). However, qualitative assessment is prone to inter-observer variability and cannot completely characterize the severity of

© Springer Nature Switzerland AG 2020
A. L. Martel et al. (Eds.): MICCAI 2020, LNCS 12266, pp. 468–477, 2020.
https://doi.org/10.1007/978-3-030-59725-2_45

the abnormality. As a result, many efforts have been made to develop objective, quantitative methods for assessing cardiovascular health through the use of echocardiography.

Motion tracking and segmentation both play crucial roles in the detection and quantification of myocardial dysfunction and can help in the diagnosis of CVD. Traditionally, however, these tasks are treated uniquely and solved as separate steps. Often times, motion tracking algorithms will use segmentations as an anatomical guide to sample points and regions of interest used to generate displacement fields [8,11,12,17]. If initial segmentations are poorly done, errors in the segmentation will propagate and lead to inaccurate displacement fields, which can further propagate to inaccurate clinical measurements. This is problematic as the task of segmentation is nontrivial, especially in echocardiography where the low signal-to-noise ratio (SNR) inherent in ultrasound results in poorly delineated LV borders. Additionally, there is limited ground truth segmentations available for clinical images due to the impracticality of having an expert manually annotate complete volumetric echocardiographic sequences. Often, only the end-diastolic or end-systolic frames are segmented. This makes it difficult to train and implement automatic segmentation models that rely on supervised learning techniques [15,23].

Recent works in the computer vision and magnetic resonance (MR) image processing fields suggests that the tasks of motion tracking and segmentation are closely related and information used to complete one task may complement and improve the overall performance of the other. In particular, Tsai et al. proposed ObjectFlow, an algorithm that iteratively optimizes segmentation and optical flow in a multi-scale framework until both tasks reach convergence [21]. Building on this, Chen et al. proposed SegFlow, a deep learning approach that combines segmentation and optical flow in an end-to-end unified network that simultaneously trains both tasks. The net exploits the commonality of these two tasks through bi-directional feature sharing [4]. However, these approaches have practical limitations. ObjectFlow is optimized online and, therefore, is computationally intensive and time-consuming [21]. SegFlow is trained in a supervised manner and requires ground truth segmentation and flow fields [4]. Qin et al. successfully implements the idea of combining motion and segmentation on 2D cardiac MR sequences by developing a dual Siamese style recurrent spatial transformer network and fully convolutional segmentation network to simultaneously estimate motion and generate segmentation masks. Features are shared between both branches [13,14]. However, this work is limited to MR images, which have higher SNR than echocardiographic images and, therefore, more clearly delineated LV walls which makes it challenging to directly apply to echocardiography. Furthermore, similar works in echocardiography are limited to 2D images [1,20]. Because of this, out of plane motion cannot be accurately captured, which provides valuable clinical information for cardiac deformation analysis.

This paper proposes a 4D (3D+t) semi-supervised joint network to simultaneously track LV motion while segmenting the LV wall. The network is trained in an iterative manner where results from one branch influences and regularizes the other. Displacement fields are further regularized by a biomechanically-inspired incompressibility constraint that enforces realistic cardiac motion behavior. The proposed model is different from other models in that it expands the network to 4D in order to capture out of plane motion. Furthermore, it addresses the issue of limited ground truth in clinical datasets by employing a training framework that only requires a single segmented frame per sequence and no ground truth displacement fields. To the knowledge of the authors, this work is the first to successfully combine segmentation and motion tracking simultaneously on volumetric echocardiographic sequences.

2   Method

The architecture of the proposed model is illustrated in Fig. 1. The objective is to simultaneously generate displacement fields and LV masks in 4D echocardiography by taking advantage of the complementary nature between the tasks of segmentation and motion tracking with the assistance of a biomechanical incompressibility constraint.

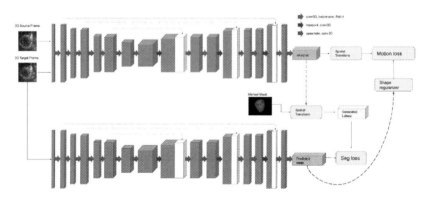

Fig. 1. Architecture of our proposed joint network: The motion branch (top) and the segmentation branch (bottom).

2.1   Motion Network (Unsupervised)

Large amounts of ground truth clinical data is often difficult to obtain. A 3D U-Net inspired architecture is designed to input an image pair. This pair is comprised of two volumetric images (a source frame and a target frame, stacked as a 2 channel single input) from a single sequence. The network consists of a downsampling analysis path followed by an upsampling synthesis path with skip connections that concatenate features learned in the analysis path with features learned in the synthesis path [23]. The output of the network is a 3 channel volumetric displacement map, corresponding to displacements in the x-y-z directions. In order for the network to train without the usage of ground truth, a VoxelMorph inspired training framework is implemented [3]. The output x-y-z displacement field is used to transform the input source frame via trilinear interpolation. Network weights are trained by minimizing the mean square difference between the transformed source frame and the target frame, effectively encoding the displacement field between the two frames. The loss function can be described as follows:

$$L_{motion} = \lambda_{motion} \frac{1}{N} \sum_{i=1}^{N} (I_{i,t} - F(I_{i,s}, U_i))^2 \tag{1}$$

where $I_{i,s}$ and $I_{i,t}$ are the source and target images, respectively of the i-th image pair, U_i is the predicted displacement field that maps the source and target images, $F = (I_{i,s}, U_i))$ is a spatial transforming operator that morphs $I_{i,s}$ to $I_{i,t}$ using U_i, and λ_{motion} is a weighting term.

2.2   Segmentation Network (Weakly-supervised)

The segmentation branch of the proposed model follows generally the same 3D U-Net inspired architecture as the motion branch [23]. The primary difference being that the input of the segmentation branch is a single volumetric image (the same target frame used to generate the displacement field of the motion network), and the output is a single volmetric segmented LV mask. The displacement field generated by the motion network is used to transform a manually segmented source frame (corresponding to the inputted target frame) in a similar Voxelmorph-inspired framework as the motion branch [6]. This transformed segmentation acts as a pseudo-ground truth label for training the segmentation branch. The network seeks to optimize a combined binary cross entropy and dice score between the propagated source segmentation and the predicted target segmentation. The loss function can be described as follows:

$$L_{dice} = \frac{1}{N} \sum_{i=1}^{N} (1 - \frac{|M_{i,t} \cap F(Y_{i,s}, U_i)|}{|M_{i,t}| + |F(Y_{i,s}, U_i)|}) \tag{2}$$

$$L_{bce} = \frac{1}{N} \sum_{i=1}^{N} (-y_i(log(p_i) + (1 - y_i)(log(1 - p_i)) \tag{3}$$

$$L_{seg} = \lambda_{dice} L_{dice} + \lambda_{bce} L_{bce} \tag{4}$$

where $Y_{i,s}$ is the manually segmented mask of the source image, $M_{i,t}$ is the predicted mask of the target image, y is a binary indicator for if a voxel is correctly labeled, and p is the predicted probability a voxel is part of the LV segmentation, and λ_{dice} and λ_{bce} are weighting terms. All other terms are as previously defined.

2.3   Combining Networks (Joint Learning)

Each network is optimized separately with their respective loss functions. An iterative training framework is designed such that the results and training of one network can positively influence the other in order to create a connection between the two branches. Initially, the motion tracking network is trained in a completely unsupervised manner, as described in Sect. 2.1. This generates a rough 3D displacement field that effectively maps the source frame to the target frame. Using this displacement field, the corresponding source frame segmentation is propagated to obtain a rough target frame segmentation. These rough target frame segmentations are used to retrain the motion tracking branch and act as an additional shape regularization term to guide the network to produce smoother displacement estimations. This regularization term is added to L_{motion} and can be described as follows:

$$L_{shape} = \lambda_{shape} \frac{1}{N} \sum_{i=1}^{N} (G_{i,t} - F(Y_{i,s}, U_i))^2 \qquad (5)$$

where $G_{i,t}$ is the pseudo-ground truth label and λ_{shape} is a weighting term. All other terms are as previously defined.

These shape regularized displacement estimations are used to generate new, smoother pseudo-ground truth labels, which are then used to retrain the segmentation network to produce more accurate segmentations.

2.4   Incompressibility Constraint

To ensure spatial smoothness and encourage more realistic cardiac motion patterns, flow incompressibility is enforced by penalizing divergence as seen in [9,12,18]. In real cardiac motion, tissue trajectories cannot collapse to a single point nor can a single point generate multiple tissue trajectories. To discourage this unrealistic behavior, sources or sinks in the motion field are penalized. This term is added to L_{motion} can be described as follows:

$$L_{inc} = \lambda_{inc} \frac{1}{N} \sum_{i=1}^{N} \|\nabla U_i\| \qquad (6)$$

where λ_{inc} is a weighting term. All other terms are as previously defined.

3   Experiments and Results

The general framework is qualitatively evaluated on a synthetic dataset with ground truth displacement fields and segmentations and the joint model is quantitatively evaluated on an in-vivo canine dataset with implanted sonomicrometer crystals for motion detection [19] and manual segmentations. Images are resampled and resized to $[64 \times 64 \times 64]$ for computational purposes. Experiments and processing were performed using MATLAB and Python. The network was built using PyTorch and trained on a GTX 1080 Ti GPU in batch sizes of 1 for 200 epochs with a learning rate of 1e-4 using Adam optimizer. Online data augmentation included random rotations, flips, and shears. Model hyperparameters were fine-tuned to each dataset.

3.1 Evaluation Using Synthetic Data

An open access dataset, 3D Strain Assessment in Ultrasound (STRAUS) [2], was used. The dataset contained 8 different volumetric sequences with different physiological conditions: 2 left anterior descending artery (LAD) occlusions in the proximal and distal arteries, 1 left circumflex artery occlusion, 1 right circumflex artery occlusion, 2 left bundle branch blocks, a sychronous sequence, and a normal (healthy) sequence. 1 sequence is left out for each testing and validation and 6 sequences are used for training. In total, the model is trained on 204 pairs, validated on 32 pairs, and tested on 32 pairs.

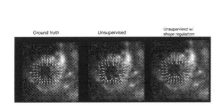

Fig. 2. A short-axis view of the displacement vectors for a normal (healthy) synthetic sequence using different methods

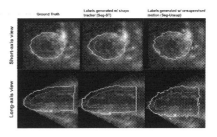

Fig. 3. Epicardium (green) and endocardium (red) segmentations for a normal (healthy) synthetic sequence using different methods (Color figure online)

As a proof-of-concept, the effect of a shape regularization term on unsupervised motion tracking and the feasibility of training a segmentation network in a weakly-supervised manner using propagated pseudo-ground truth labels is qualitatively evaluated. For the motion tracking branch, the performance of the network after implementing the shape regularization term using ground truth segmentations is compared to the network trained in a completely unsupervised manner. For the segmentation branch, the network is trained on weak labels generated by propagating an initial manual label using motion fields generated via a shape-tracking algorithm (which originally tracked ground truth labels) and an unsupervised motion network. Figures 2, 3 show improved results after including the shape regularization term and feasible segmentation predictions when trained in a weakly-supervised manner.

3.2   Evaluation Using Animal Study

In vivo animal studies were done on 8 anesthetized open-chest canines, and images were captured using a Philips iE33 scanner and a X7-2 probe. Each study was conducted under five physiological conditions: baseline, mild LAD stenosis, moderate LAD stenosis, mild LAD stenosis with low-dose dobutamine (5µg/kg/min), and moderate LAD stenosis with low-dose dobumatine. 1 full study is used each for testing and validation and 6 studies are used for training. In total, the model is trained on 745 pairs, validated on 133 pairs, and tested on 126 pairs. All procedures were approved under Institutional Animal Care and Use Committee policies.

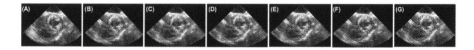

Fig. 4. A short-axis view of the displacement vectors for a normal (healthy) in vivo sequence using different methods: A) crystal derived displacement, B) nonrigid registration (NRR), C) Lucas-Kanade Optical Flow (LK), D) Shape Tracking (ST), E) Unsupervised w/ shape regularizer (Unsup+Shape), F) Unsupervised G) Proposed Model

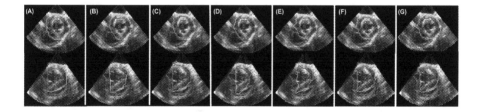

Fig. 5. Epicardium (green) and endocardium (red) segmentations for a normal (healthy) in vivo sequence using different methods: A) manual, B) dynamic appearance model (DAM), C) trained with crystal-generated labels (Seg-CD) (LK), D) nonrigid registration-generated labels (Seg-NRR), E) Lucas-Kanade generated labels (Seg-LK), F) Unsupervised motion generated labels (Seg-Unsup) G) Proposed Model (Color figure online)

Each task of the joint model is evaluated separately. The displacement predictions are compared against displacements derived from an implanted array of sonomicrometers as previously reported [19]. It is important to note that dense displacement fields from the sonomicrometer crystals are generated through RBF based interpolation [5] and cannot be considered absolute ground truth, but act as a useful validation metric. The root mean squared error (RMSE) of the displacement fields generated by the joint model are compared against a non-rigid registration algorithm with b-spline parameterization (NRR) [16], and Lucas-Kanade optical flow (LK) algorithm [10], a shape-tracking algorithm (ST) [12], and the unsupervised single motion tracking branch without (Usup) and with (Usup+Shape) manually segmented shape regularization. According to Table 1 and Fig. 4, the joint model performs comparably to Unsup+Shape and favorably against all other methods in all metrics. For segmentation results, label predictions are evaluated against manually traced segmentations [22]. The Dice score and Hausdorff distance (HD) of the predicted endocardium and epicardium borders of the joint model are compared to a dictionary learning-based dynamic appearance model (DAM) [7], and weakly supervised versions of the joint model using crystal derived displacements (Seg-CD), nonrigid registration (Seg-NRR), optical flow (Seg-LK), and unsupervised motion (Seg-Unsup) to generate pseudo-ground truth labels. According to Table 2 and Fig. 5, the joint model performs comparably to Seg-CD and favorably against all other methods in all metrics.

Table 1. Root mean squared error (RMSE) in the x-y-z direction. Lower RMSE means better performance

Method	Ux (mm)	Uy (mm)	Uz (mm)
NRR	0.95 ± 0.38	1.06 ± 0.34	0.62 ± 0.17
LK	0.85 ± 0.37	0.91 ± 0.38	0.58 ± 0.17
ST	0.81 ± 0.32	0.72 ± 0.40	0.63 ± 0.21
Unsup+Shape	0.80 ± 0.33	0.70 ± 0.33	0.60 ± 0.18
Unsup	1.07 ± 0.42	1.31 ± 0.41	0.74 ± 0.17
Proposed model	0.79 ± 0.33	0.70 ± 0.36	0.62 ± 0.20

Table 2. Dice and Hausdorff Distance (HD) for the endo- and epi- cardium. Higher Dice score and lower HD means better performance

Methods	Endocardium		Epicardium	
	Dice	HD (mm)	Dice	HD (mm)
DAM	0.81 ± 0.05	3.02 ± 0.29	0.92 ± 0.04	3.12 ± 0.24
Seg-CD	0.84 ± 0.07	2.63 ± 0.20	0.96 ± 0.01	2.75 ± 0.08
Seg-NRR	0.80 ± 0.08	2.80 ± 0.28	0.88 ± 0.06	3.22 ± 0.43
Seg-LK	0.83 ± 0.05	2.79 ± 0.24	0.93 ± 0.03	3.16 ± 0.31
Seg-Unsup	0.84 ± 0.06	2.91 ± 0.26	0.94 ± 0.02	3.11 ± 0.16
Proposed model	0.88 ± 0.04	2.70 ± 0.17	0.95 ± 0.02	2.97 ± 0.19

4   Conclusions

This paper proposes a novel joint learning network for simultaneous LV segmentation and motion tracking in 4D echocardiography. Motion tracking and segmentation branches are trained iteratively such that the results of one branch positively influences the other. Motion is trained in an unsupervised manner and the resulting displacement fields are used to create pseudo-ground truth labels by propagating a single manually segmented time frame. Predicted labels are then used as landmarks to smooth the displacement fields. An incompressibility constraint is added to enforce spatially realistic LV motion patterns. Experimental results show our proposed model performs favorably against competing methods. Future work includes further validation on larger datasets and exploring temporal regularization.

Acknowledgement. The authors are thankful for the technical assistance provided by the staff of the Yale Translational Research Imaging Center and Drs. Nabil Boutagy, Imran Alkhalil, Melissa Eberle, and Zhao Liu for their assistance with the in vivo canine imaging studies.

References

1. Ahn, S.S., Ta, K., Lu, A., Stendahl, J.C., Sinusas, A.J., Duncan, J.S.: Unsupervised motion tracking of left ventricle in echocardiography. In: Medical Imaging 2020: Ultrasonic Imaging and Tomography, p. 113190Z (2020)
2. Alessandrini, M., et al.: A pipeline for the generation of realistic 3D synthetic echocardiographic sequences: Methodology and open-access database. IEEE Trans. Med. Imaging **34**, 1436–1451 (2015)
3. Balakrishnan, G., et al.: An unsupervised learning model for deformable medical image registration. In: The IEEE Conference on Computer Vision and Pattern Recognition (CVPR) (2018)
4. Cheng, J., et al.: Segflow: joint learning for video object segmentation and optical flow. In: IEEE International Conference on Computer Vision (ICCV) (2017)
5. Compas, C., et al.: Radial basis functions for combining shape and speckle tracking in 4D echocardiography. IEEE Trans. Med. Imaging **33**, 1275–1289 (2014)
6. Dalca, A.V., Yu, E., Golland, P., Fischl, B., Sabuncu, M.R., Eugenio Iglesias, J.: Unsupervised deep learning for Bayesian brain MRI segmentation. In: Shen, D., et al. (eds.) MICCAI 2019. LNCS, vol. 11766, pp. 356–365. Springer, Cham (2019). https://doi.org/10.1007/978-3-030-32248-9_40
7. Huang, X., et al.: Contour tracking in echocardiographic sequences via sparse representation and dictionary learning. Med. Image Anal. **18**, 253–271 (2014)
8. Lin, N., et al.: Generalized robust point matching using an extended free-form deformation model: application to cardiac images. In: IEEE International Symposium on Biomedical Imaging: Nano to Macro. IEEE (2004)
9. Lu, A., et al.: Learning-based regularization for cardiac strain analysis with ability for domain adaptation. CoRR (2018). http://arxiv.org/abs/1807.04807
10. Lucas, B.D., Kanade, T.: An iterative image registration technique with an application to stereo vision (darpa). In: Proceedings of the 1981 DARPA Image Understanding Workshop, pp. 121–130 (1981)

11. Papademetris, X., et al.: Estimation of 3-D left ventricular deformation from medical images using biomechanical models. IEEE Trans. Med. Imaging **21**, 786–800 (2002)
12. Parajuli, N., et al.: Flow network tracking for spatiotemporal and periodic point matching: Applied to cardiac motion analysis. Med. Image Anal. (2019). https://doi.org/10.1016/j.media.2019.04.007. http://www.sciencedirect.com/science/article/pii/S1361841518304559
13. Qin, C., et al.: Joint learning of motion estimation and segmentation for cardiac MR image sequences. In: Frangi, A.F., Schnabel, J.A., Davatzikos, C., Alberola-López, C., Fichtinger, G. (eds.) MICCAI 2018. LNCS, vol. 11071, pp. 472–480. Springer, Cham (2018). https://doi.org/10.1007/978-3-030-00934-2_53
14. Qin, C., et al.: Joint motion estimation and segmentation from undersampled cardiac MR image. In: Knoll, F., Maier, A., Rueckert, D. (eds.) MLMIR 2018. LNCS, vol. 11074, pp. 55–63. Springer, Cham (2018). https://doi.org/10.1007/978-3-030-00129-2_7
15. Ronneberger, O., Fischer, P., Brox, T.: U-Net: convolutional networks for biomedical image segmentation. In: Navab, N., Hornegger, J., Wells, W.M., Frangi, A.F. (eds.) MICCAI 2015. LNCS, vol. 9351, pp. 234–241. Springer, Cham (2015). https://doi.org/10.1007/978-3-319-24574-4_28. http://lmb.informatik.uni-freiburg.de/Publications/2015/RFB15a
16. Rueckert, D., et al.: Nonrigid registration using free-form deformations: application to breast MR images. IEEE Trans. Med. Imaging **18**, 712–721 (1999)
17. Shi, P., et al.: Point-tracked quantitative analysis of left ventricular surface motion from 3-D image sequences. IEEE Trans. Med. Imaging **19**, 36–50 (2000)
18. Song, S., Leahy, R.: Computation of 3-D velocity fields from 3-D cine CT images of a human heart. IEEE Trans. Med. Imaging **10**(3), 295–306 (1991)
19. Stendahl, J.C., et al.: Regional myocardial strain analysis via 2D speckle tracking echocardiography: validation with sonomicrometry and correlation with regional blood flow in the presence of graded coronary stenoses and dobutamine stress. Cardiovasc. Ultrasound **18**, 2 (2020). https://doi.org/10.1186/s12947-019-0183-x
20. Ta, K., Ahn, S.S., Lu, A., Stendahl, J.C., Sinusas, A.J., Duncan, J.S.: A semi-supervised joint learning approach to left ventricular segmentation and motion tracking in echocardiography. In: 2020 IEEE 17th International Symposium on Biomedical Imaging (ISBI), pp. 1734–1737 (2020)
21. Tsai, Y., et al.: Video segmentation via object flow. In: 2016 IEEE Conference on Computer Vision and Pattern Recognition (CVPR), pp. 3899–3908 (2016). https://doi.org/10.1109/CVPR.2016.423
22. Yushkevich, P.A., et al.: User-guided 3D active contour segmentation of anatomical structures: significantly improved efficiency and reliability. Neuroimage **31**(3), 1116–1128 (2006)
23. Çiçek, Ö., Abdulkadir, A., Lienkamp, S.S., Brox, T., Ronneberger, O.: 3D U-Net: learning dense volumetric segmentation from sparse annotation. In: Ourselin, S., Joskowicz, L., Sabuncu, M.R., Unal, G., Wells, W. (eds.) MICCAI 2016. LNCS, vol. 9901, pp. 424–432. Springer, Cham (2016). https://doi.org/10.1007/978-3-319-46723-8_49

Joint Data Imputation and Mechanistic Modelling for Simulating Heart-Brain Interactions in Incomplete Datasets

Jaume Banus[1(✉)], Maxime Sermesant[1], Oscar Camara[2], and Marco Lorenzi[1]

[1] Inria, Epione Team, Université Côte d'Azur, Sophia Antipolis, France
jaume.banus-cobo@inria.fr
[2] PhySense, Department of Information and Communication Technologies,
Universitat Pompeu Fabra, Barcelona, Spain

Abstract. The use of mechanistic models in clinical studies is limited by the lack of multi-modal patients data representing different anatomical and physiological processes. For example, neuroimaging datasets do not provide a sufficient representation of heart features for the modeling of cardiovascular factors in brain disorders. To tackle this problem we introduce a probabilistic framework for joint cardiac data imputation and personalisation of cardiovascular mechanistic models, with application to brain studies with incomplete heart data. Our approach is based on a variational framework for the joint inference of an imputation model of cardiac information from the available features, along with a Gaussian Process emulator that can faithfully reproduce personalised cardiovascular dynamics. Experimental results on UK Biobank show that our model allows accurate imputation of missing cardiac features in datasets containing minimal heart information, e.g. systolic and diastolic blood pressures only, while jointly estimating the emulated parameters of the lumped model. This allows a novel exploration of the heart-brain joint relationship through simulation of realistic cardiac dynamics corresponding to different conditions of brain anatomy.

Keywords: Gaussian Process · Variational inference · Lumped model · Missing features · Biomechanical simulation

1 Introduction

Heart and brain are characterized by several common physiological and pathophysiological mechanisms [6]. The study of this multi-organ relationship is of great interest, in particular to better understand neurological diseases such as vascular dementia or Alzheimer's disease. The development of computational models simulating heart and brain dynamics is currently limited by the lack of

Electronic supplementary material The online version of this chapter (https://doi.org/10.1007/978-3-030-59725-2_46) contains supplementary material, which is available to authorized users.

databases containing information for both organs. Neuroimaging datasets often provide a limited number of cardiac-related measurements, usually restricted to brachial diastolic and systolic blood pressure (DBP, SBP) [7]. These quantities provide a limited assessment of the cardiac function, thus compromising the possibility of further analysis of cardiovascular factors in brain disorders.

On the contrary, the availability of a rich set of cardiac information in heart studies allows the use of cardiovascular models to estimate descriptors of the cardiac function that are not possible to measure in-vivo, such as contractility or stiffness of the heart fibers. These models optimize the parameters through data-assimilation procedures to reproduce the observed clinical measurements [11], but usually do not include neurological factors. The ability to jointly account for cardiovascular descriptors and brain information is key to gather novel insights about the relationship between heart dynamics and brain conditions.

Most of current studies relating heart and brain are based on statistical association models, such as based on multivariate regression [5]. While this kind of analysis allows to easily formulate and test association hypotheses, it usually offers a limited interpretation of the complex relationship between organs. This issue is generally addressed by mechanistic modeling, allowing deeper insights on physiological and biomechanical aspects. These models allow for example to describe the brain vasculature, and to quantify physiological aspects such as blood flow auto-regulation effects in the brain [1], up to the simulation of the whole-body circulation with detailed compartmental components [3]. Although these approaches offer a high level of interpretability, they are usually severely ill-posed and require large data samples and arrays of measurements to opportunely tune their parameters.

To bridge the gap between data-driven and mechanistic approaches to heart-brain analysis, in this work we propose to learn cardiovascular dynamics from brain imaging and clinical data by leveraging on large-scale datasets with missing cardiac information. This is achieved through an inference framework composed of two nested models accounting respectively for the imputation of missing cardiac information conditioned on the available cardiac and brain features, and for a Gaussian Process emulator that mimics the behavior of a lumped cardiovascular model. This setting allow us to formulate a probabilistic end-to-end generative model enabling imputation of missing measurements and estimation of cardiovascular parameters given a subset of observed heart and brain features.

Results on real data from the UK Biobank show that our framework can be used to reliably estimate and simulate cardiac function from datasets in which we have minimal cardiovascular information, such as SBP and DBP only. Moreover, the proposed framework allows novel exploration of the joint heart-brain relationship through the simulation of realistic cardiac dynamics corresponding to different scenarios of brain anatomy and damage.

2   Methods

2.1   Problem Statement

We denote by ν the vector representing brain image-derived phenotypes (IDPs) and clinical information such as age or body surface area (BSA), and by x the vector of cardiac IDPs and blood pressure measurements. The vector x can be represented as $x = \{\hat{x}, x_{obs}\}$ where $\hat{x}$ represents the unobserved information we wish to impute and x_{obs} the observed one. Moreover, we assume that for each observation x a corresponding set of parameters y of the associated lumped model is available. We would like to learn a generative model in which we assume that the unobserved measurements are generated by a latent random variable z conditioned on the variables ν. Hence, our generative process can be seen as sampling from a distribution $p(z|\nu)$ and then obtaining $\hat{x}$ with probability $p(\hat{x}|z,\nu)$. Due to the association between cardiac IDPs, x, and lumped model parameters y, we also assume that the latter are dependent from ν through z. Our graphical model is shown in Fig. 1, while the evidence lower bound (ELBO) of the joint data marginal $p(y, \hat{x}|x_{obs}, \nu)$ writes as:

$$
\begin{aligned}
\log p(y, \hat{x}|x_{obs}, \nu) &= \int \log p(y, \hat{x}|x_{obs}, z, \nu)p(z|\nu)dz \\
&\geq \mathbb{E}_{q_\phi(z|x_{obs},\nu)} \log p(y, \hat{x}|x_{obs}, z, \nu) \\
&\quad - KL(q_\phi(z|x_{obs},\nu)||p_\theta(z|\nu)) \\
&= \mathbb{E}_{q_\phi(z|x_{obs},\nu)} \log p_\omega(y|x_{obs}, z) \\
&\quad + \mathbb{E}_{q_\phi(z|x_{obs},\nu)} \log p_\theta(\hat{x}|\nu, z) \\
&\quad - KL(q_\phi(z|x_{obs},\nu)||p_\theta(z|\nu)) \equiv \mathcal{L}(\theta, \phi, \omega; x_{obs}, \nu)
\end{aligned}
\tag{1}
$$

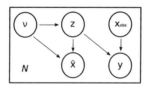

Fig. 1. Graphical model of our framework. From a latent variable z we generate the unobserved features $\hat{x}$ conditioned on the variables ν and we estimate y via Gaussian process regression. During inference x_{obs} and ν are used to estimate the approximated posterior $q_\phi(z|x_{obs}, \nu)$.

The approximation of the posterior distribution by $q_\phi(z|x_{obs}, \nu)$ defines the optimization of the ELBO through variational inference. The variational distributions $q_\phi(z|x_{obs}, \nu)$ and $p_\theta(\hat{x}|\nu, z)$ are parametrized respectively with parameters ϕ and θ. The term $\log p_\omega(y|x_{obs}, z)$ in the ELBO denotes the emulator which approximates the mechanistic behavior of the lumped model via Gaussian Process regression parametrized by ω. The choice of a GP as emulator is motivated by the uncertainty of the data, hence it is desirable to obtain a distribution of interpolating functions rather than a single deterministic function. Moreover, GPs have already been proved to be valid emulators of 1D mechanistic vascular models [10]. The GP allow us to sample functions $f(x)$ from a given prior parameterized by a mean $\mu(x)$ and covariance $\Sigma(x)$, i.e: $f(x) \sim \mathcal{N}(\mu(x), \Sigma(x))$ to obtain the marginal $y \sim \mathcal{N}(\mu(x), \Sigma(x) + \sigma^2 I)$. The prior mean $\mu(x)$ is here set to 0 and we use a radial-basis function (RBF) kernel for the covariance:

$$k^j(x_i, x_i') = \alpha_j^2 exp\left(-\frac{(x_i - x_i')^2}{2\beta_i^2}\right), \tag{2}$$

Since our data is multi-dimensional, $k^j(x_i, x_i')$ is the kernel for the jth target and the ith predictor. The hyper-parameters of the kernel $\omega = \{\alpha, \beta\}$ represent the output amplitude α and length scale β of the sampled functions. The goal during training is to learn the hyper-parameters that maximize the marginal likelihood of the observed data y.

The second term of the ELBO is related to the imputation of $\hat{x}$. The term $\log p_\theta(\hat{x}|\nu, z)$ denotes the log-likelihood of the imputed features, which can be seen as the reconstruction error. The last term $KL(q_\phi(z|x_{obs}, \nu)||p_\theta(z|\nu))$ is the Kullback–Leibler divergence between variational approximation and prior for z, that can be expressed in a closed form given that both distributions are Gaussians. The imputation scheme is equivalent to a conditional variational autoencoder (CVAE) which has become a popular approach to feature imputation [8]. Fast and efficient optimization in our model is possible by means of stochastic gradient descent thanks to the closed form for data fit and KL terms, the use of the reparametrisation trick and Monte Carlo sampling.

2.2 Data Processing and Cardiovascular Model

From UK Biobank we selected a subset of 3445 subjects for which T1, T2 FLAIR magnetic resonance images (MRI) and several brain and cardiac IDPs were available. Among the brain IDPs we used total grey matter (GM), total white matter (WM) and ventricles volumes. We used T1 and T2 FLAIR images to obtain the number of white matter hyper-intensities (WMHs) and their total volume relying on the lesion prediction algorithm (LPA), available from the lesion segmentation toolbox (LST) [14] of SPM[1]. WMHs are a common indicator of brain damage of presumably vascular origin [15]. We combined WM and GM volumes into a single measurement that we denoted as brain volume. The WMHs total volume and number of lesions presented a skewed distribution, and were Box-Cox transformed prior to the analysis. Regarding the cardiac IDPs we selected stroke volume (SV), ejection fraction (EF) and end-diastolic volume (EDV) for the left ventricle. All brain-related volumes were normalized by head size. Besides

---

[1] https://www.fil.ion.ucl.ac.uk/spm.

IDPs, we had access to blood pressure measurements (DBP, SBP) and socio-demographic features such as age and body surface area (BSA). We used DBP and SBP to compute MBP as $MBP = DP + (SP - DP)/3$. Next, we used the lumped cardiovascular model derived in [4] to obtain additional indicators of the cardiac function. In particular we estimated the contractility of the main systemic arteries (τ), peripheral resistance (R_p), the radius of the left ventricle (R_0), contractility of the cardiac fibers (σ_0) and their stiffness (C_1). The parameters of the model were selected based on the available clinical data. To obtain the target values for the emulator, the data-assimilation procedure was carried out according to the approach presented in [2,12].

2.3   Experiments

The data was split in two sets: one containing the full-information (2309 subjects), and one in which cardiac IDPs ($\hat{x}$) and the estimated model parameters (y) were removed (1136 subjects, Table 1). The quality of imputation was compared to conventional methods such as mean, median and k-nearest neighbors (KNN) imputation. We subsequently assessed the relationship learnt by our model between brain features and cardiovascular parameters through simulation. Starting from the mean values of each parameter we sampled along the dimension of the different conditional variables ν that we used to parameterize the prior. This procedure allowed us to assess their influence in the inferred simulation parameters.

Table 1. Variables used in our framework. Mean blood pressure (MBP), diastolic blood pressure (DBP), stroke volume (SV), end-diastolic volume (EDV), ejection fraction (EF), heart fibers contractility (σ_0), ventricle size (R_0), heart fibers stiffness (c_1), peripheral resistance (R_p) and aortic compliance (τ)

Input (x_{obs})	Condition (ν)	Imputed ($\hat{x}$)	Predicted (y)
MBP	Brain volume	SV	σ_0
DBP	Ventricles volume	EDV	R_0
	WMHs volume	EF	c_1
	Num. WMHs		R_p
	BSA		τ
	Age		

3   Results

Data Imputation and Regression. We assessed the performance of our model by measuring the mean squared error (MSE) on the testing data, for both the imputation of missing cardiac information and the emulation of the lumped model parameters. The results in Fig. 2a show that our method gives significantly

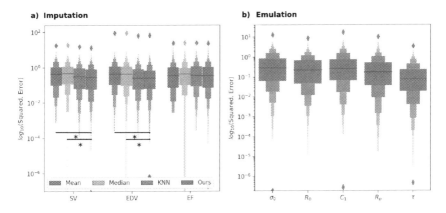

Fig. 2. Mean squared error (MSE) a) of the imputation of missing cardiac measurements ($\hat{x}$) b) of the estimated parameters of the emulated cardiovascular lumped model (y). * denotes that the MSE distributions are significantly different with respect to our method according to the Wilcoxon rank-sum test using a significance level of $a = 0.05$, Bonferroni corrected by multiple comparison.

better estimates than mean and median imputation, and comparable predictions to KNN, for which the optimal number of neighbors was optimized through 10-fold cross-validation and corresponds to $K = 10$. At the same time the emulator consistently gives low errors for the parameters' estimation (Fig. 2b). In supp. Fig. 3 we present the most relevant predictors for each emulated feature based on their β values, while a qualitative comparison between the distributions of imputed and emulated features compared to the ground truth data is available in supp. Fig. 4.

Cardiovascular Dynamics Simulation and Model Plausibility. In Fig. 3a we observe the change in the predicted parameters of the cardiovascular model as we sample along the range of values of the different conditional variables ν. Figure 3b shows the pressure-volume (PV) loops generated by the lumped model using the inferred parameters. The simulated PV loops highlight meaningful relationships:

- An increase in the volume of WMHs is associated to decreased SV and EDV, together with a smaller reduction of ESV, leading to a decrease of EF which is related to reduced contractility.
- Similar dynamics are associated to brain volume loss.
- The number of WMHs exhibits different dynamics than WMHs volume, associated to the increase in afterload and the increase of EDV.

From the plots showing the estimation of the parameters we can observe the ones driving the observed dynamics. For example, WMHs and brain volumes changes are mainly driven by the joint evolution of peripheral resistance (R_p),

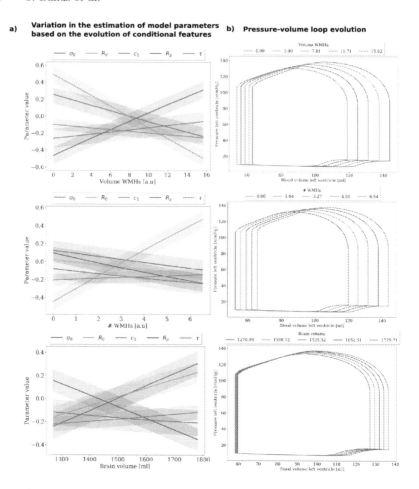

Fig. 3. a) Inferred model parameters and their respective confidence interval as we sample along the dimension of the different conditional variables ν while keeping the other elements of the generative framework constant. b) Pressure-volume loops generated by the cardiovascular lumped model given the mean inferred parameters

contractility (σ_0) and size of the left ventricle (R_0). The changes in the number of WMHs are related to heart-remodelling changes, driven by R_0 and by a decrease of σ_0 and the ventricular stiffness c_1. In supp. Fig. 2 we provide analysis and discussion of the remaining conditional features (age, BSA and brain ventricles volume). Overall, in the simulated dynamics we can identify several physiological responses in line with the clinical literature. Moreover, our model may be useful to give insights in currently controversial topics such as concerning the pathogenesis of WMHs. Our results suggest that the evolution of cardiac function with respect to brain and WMHs volumes is similar to the one due to aging (see supp. Fig. 2), while the effect induced by the number of WMHs is similar to the one related to ventricles enlargement. In both cases the changes in R_0, σ_0 and C_1 suggest that the increase in the number of WMHs and the enlargement of the brain ventricles are related to heart-remodelling processes, such as loss of contractility or decrease in compliance. These findings are in line with clinical observations [9] relating lower cardiac output with higher burden of WMHs and reduced brain volume.

4   Conclusion

We presented a generative model that enables the analysis of complex physiological relationships between heart and brain in datasets where we have minimal available features. The framework allow us to emulate a lumped cardiovascular model through data-driven inference of mechanistic parameters, and provides us a generative model to explore hypothetical scenarios of heart and brain relationships. In the future, the model will allow to potentially transfer the knowledge learnt in UK Biobank to datasets where we have minimal cardiac information, to explore the relationship between brain conditions and cardiovascular factors in specific clinical contexts, such as in neurodegeneration. Our approach could also be extended to account for deep learning architectures, and the framework could be further improved by jointly accounting for multiple outputs, which are currently modelled independently, or by including spatial information from imaging data, beyond the modelling of scalar volumetric features. Furthermore, while the cardiovascular features considered in this study are rather general, more complex features will allow to study more realistic cardiovascular models. For example, while the mechanistic model used in this study does not simulate cerebral blood flow, previous studies suggested that WMHs may be due to local vascular impairment [13]. Hence, by selecting appropriate clinical features we could constrain the imputation by means of any biophysical model representing the desired aspect of systems biology. This could represent an innovative tool in real world scenarios, for which multi-modal patient data is often limited or not available.

Acknowledgments. This work has been supported by the French government, through the 3IA Côte d'Azur Investments in the Future project managed by the National Research Agency (ANR) with the reference number ANR-19-P3IA-0002, and by the ANR JCJC project Fed-BioMed 19-CE45-0006-01. The project was also supported by the Inria Sophia Antipolis - Méditerranée, "NEF" computation cluster and by the Spanish Ministry of Science, Innovation and Universities under the Retos I+D Programme (RTI2018-101193-B-I00) and the Maria de Maeztu Units of Excellence Programme (MDM-2015-0502). This research has been conducted using the UK Biobank Resource under Application Number 20576 (PI Nicholas Ayache). Additional information can be found at: https://www.ukbiobank.ac.uk.

References

1. Acosta, S., et al.: An effective model of cerebrovascular pressure reactivity and blood flow autoregulation. Microvasc. Res. **115**(November 2016), 34–43 (2018)

2. Banus, J., et al.: Large scale cardiovascular model personalisation for mechanistic analysis of heart and brain. Interactions **3504**, 285–293 (2019)
3. Blanco, P., et al.: An anatomically detailed arterial network model for one-dimensional computational hemodynamics. T-BME **62**(2), 736–753 (2015)
4. Caruel, M., Chabiniok, R., Moireau, P., Lecarpentier, Y., Chapelle, D.: Dimensional reductions of a cardiac model for effective validation and calibration. Biomech. Model. Mechanobiol. **13**(4), 897–914 (2013). https://doi.org/10.1007/s10237-013-0544-6
5. Cox, S.R., et al.: Associations between vascular risk factors and brain MRI indices in UK Biobank. Eur. Heart J. **44**, 1–11 (2019)
6. Doehner, W., et al.: Heart and brain interaction in patients with heart failure: overview and proposal for a taxonomy. Eur. J. Heart Fail. **20**(2), 199–215 (2018)
7. Epstein, N.U., et al.: Cognitive dysfunction and greater visit-to-visit systolic blood pressure variability. J. Am. Geriatr. Soc. **61**(12), 2168–2173 (2013)
8. Ivanov, O., et al.: Variational autoencoder with arbitrary conditioning. In: 7th International Conference on Learning Representations, ICLR 2019, pp. 1–25 (2019)
9. Jefferson, A.L., et al.: Lower cardiac output is associated with greater white matter hyperintensities in older adults with cardiovascular disease. J. Am. Geriatr. Soc. **55**, 1044–1048 (2009)
10. Melis, A., et al.: Bayesian sensitivity analysis of a 1D vascular model with Gaussian process emulators. Int. J. Numer. Method Biomed. Eng. **33**(12), 1–11 (2017)
11. Molléro, R., Pennec, X., Delingette, H., Garny, A., Ayache, N., Sermesant, M.: Multifidelity-CMA: a multifidelity approach for efficient personalisation of 3D cardiac electromechanical models. Biomech. Model. Mechanobiol. **17**(1), 285–300 (2017)
12. Molléro, R., et al.: Population-based priors in cardiac model personalisation for consistent parameter estimation in heterogeneous databases. Int. J. Numer. Method Biomed. Eng. **35**, e3158 (2018)
13. Müller, L.O., Toro, E.F.: Enhanced global mathematical model for studying cerebral venous blood flow. J. Biomech. **47**(13), 3361–3372 (2014)
14. Schmidt, P.: Bayesian inference for structured additive regression models for large-scale problems with applications to medical imaging. Ph.D., January 2017
15. Wardlaw, J.M., Valdés Hernández, M.C., Muñoz-Maniega, S.: What are white matter hyperintensities made of? Relevance to vascular cognitive impairment. J. Am. Heart Assoc. **4**(6), 001140 (2015)

Learning Geometry-Dependent and Physics-Based Inverse Image Reconstruction

Xiajun Jiang[(⊠)], Sandesh Ghimire, Jwala Dhamala, Zhiyuan Li, Prashnna Kumar Gyawali, and Linwei Wang

Rochester Institute of Technology, Rochester, NY 14623, USA
xj7056@rit.edu

Abstract. Deep neural networks have shown great potential in image reconstruction problems in Euclidean space. However, many reconstruction problems involve imaging physics that are dependent on the underlying non-Euclidean geometry. In this paper, we present a new approach to learn inverse imaging that exploit the underlying geometry and physics. We first introduce a non-Euclidean encoding-decoding network that allows us to describe the unknown and measurement variables over their respective geometrical domains. We then learn the geometry-dependent physics in between the two domains by explicitly modeling it via a bipartite graph over the graphical embedding of the two geometry. We applied the presented network to reconstructing electrical activity on the heart surface from body-surface potential. In a series of generalization tasks with increasing difficulty, we demonstrated the improved ability of the presented network to generalize across geometrical changes underlying the data in comparison to its Euclidean alternatives.

Keywords: Geometric deep learning · Physics-based · Inverse problems

1 Introduction

Deep learning has shown state-of-the-art performance in image reconstruction tasks across a variety of medical modalities [1,10,15,18,20]. These approaches typically formulate the problems in standard Euclidean image grids. However, in many problems, the unknown variables of interests and the corresponding measurements are defined over non-Euclidean geometrical domains: their physics-based relationship, both forward and inverse, is largely reliant on the underlying geometry. Examples include electrical activity in the heart and the potential it generates on the body surface [3,9,12], or electrical activity in the brain and its potential measurements on the skull surface [16]. Standard Euclidean deep learning neglecting the underlying geometry not only ignores the geometry-dependent imaging physics, but also has difficulty in generalizing over different geometry.

© Springer Nature Switzerland AG 2020
A. L. Martel et al. (Eds.): MICCAI 2020, LNCS 12266, pp. 487–496, 2020.
https://doi.org/10.1007/978-3-030-59725-2_47

To design inverse imaging (image reconstruction) networks that can generalize across geometry, there are two general approaches. One is to make the network invariant to geometry by, for instance, an information bottleneck that removes geometrical information from the input data [11]. While demonstrating improved generalization to geometrical changes [11], the treatment of non-Euclidean data as Euclidean data ties the network to the training mesh and prevents its direct application to unseen meshes from new patients. Alternatively, one can make the network equivariant to the geometry. In [3], for instance, the reconstruction of electrical activity in the heart is formulated and conditioned on 2D image scans of the heart [3]. Rather than explicitly describing the geometry, this approach defines non-Euclidean variables at a small region of interest within the Euclidean image grid. How to extend it to consider the geometry of both the unknown (*e.g.*, the heart) and the measurement (*e.g.*, the body), and to explicitly consider the geometry-dependent physics in between, is not clear.

Graph convolutional neural networks (GCNN) provide an appealing alternative to solving inverse imaging between non-Euclidean variables defined over geometrical domains [4]. Significant efforts in GCNN have been made for node- and graph-level classifications, graph embedding, and graph generation [19]. However, no existing work has considered learning geometry-dependent relationship between signals defined on two separate graphs, which is a critical component of achieving physics-based inverse imaging.

In this paper, we present a non-Euclidean inverse imaging (image reconstruction) network that 1) directly models the unknown and its measurement over their geometrical domains, and 2) models and learns their inverse relationship – as informed by the physics – as a function of the geometry. It consists of two novel contributions. First, to describe the spatiotemporal variables (unknowns and measurements) over their respective geometrical domain, we introduce an encoding-decoding architecture composed of spatial-temporal graph convolutional neural networks (ST-GCNN) defined separately for each domain. Second, to learn the geometry-dependent physics in between, we model it with a bipartite graph between the graphical embedding of the two geometrical domains. We applied the presented method for reconstructing spatiotemporal electrical potential on the ventricular surface from body-surface potential. In synthetic and real-data experiments, we tested the presented network in a series of generalization tasks with increasing difficulty, and compared it to Euclidean baselines without and with a geometry-invariant bottleneck [11]. By learning inverse imaging in a geometry-dependent and physics-informed fashion, the presented network showed an improved generalization to geometrical changes in the data.

2   Methodology

Cardiac electrical excitation produces time-varying voltage signals on the body surface, following quasi-static approximation of the electromagnetism [17]. Given a pair of heart and torso geometry, the governing physics can be numerically approximated to relate signals in the heart $\mathbf{X}_t$ to those on the body surface $\mathbf{Y}_t$:

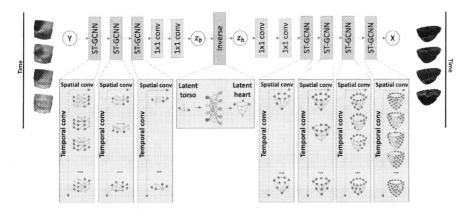

Fig. 1. Illustration of the presented non-Euclidean inverse imaging network.

$$\mathbf{Y}_t = \mathbf{H}\mathbf{X}_t \quad \forall t \in \{1, ..., T\}. \tag{1}$$

Note that $\mathbf{X}_t$ and $\mathbf{Y}_t$ live on the 3D geometry of the heart and torso surface, respectively. The forward operator $\mathbf{H}$ defines the physics of their relationship and is highly dependent on the given heart-torso geometry. Traditional approaches to reconstructing $\mathbf{X}_t$ from $\mathbf{Y}_t$ starts with this forward model, exploiting the geometry and physics behind the inverse relationship. When using Euclidean deep learning for direct inference of $\mathbf{X}_t$ from $\mathbf{Y}_t$, the network becomes solely reliant on labeled data pairs, incorporating neither the physics nor the geometry underlying the problem. The proposed method is set to bridge these gaps by 1) allowing the description of $\mathbf{X}_t$ and $\mathbf{Y}_t$ in their geometrical domains, and 2) explicitly modeling their physics relationship as a function of the geometry.

As summarized in Fig. 1, we present an encoder-decoder architecture with ST-GCNNs to embed/generate $\mathbf{Y}_t$ and $\mathbf{X}_t$ over their respective geometry. The geometry-dependent relationship between the latent variables of $\mathbf{Y}_t$ and $\mathbf{X}_t$ is learned via a bipartite graph over the graph embedding of the two geometry.

2.1   Encoding-Decoding with ST-GCNNs

As $\mathbf{X}_t$ and $\mathbf{Y}_t$ are temporal sequences living on 3D geometry, we describe their generation/embedding with ST-GCNNs that consist of interlaced graph convolution in space and regular convolution in time. As illustrated in Fig. 1, both spatial and temporal dimensions are reduced/expanded during encoding/decoding.

Geometrical Representation in Graphs: We represent triangular meshes of the heart and torso as two separate undirected graphs: $\mathcal{G} = (\mathcal{V}, \mathcal{E}, \mathbf{U}, \mathbf{F})$, where vertices $\mathcal{V}$ consist of all V mesh nodes and edges $\mathcal{E}$ describe the vertex connection as defined by the triangular mesh. $\mathbf{U} \in [0, 1]^{V \times V \times 3}$ consists of edge attributes $\mathbf{u}(i, j)$ between vertex i and j as normalized differences in their 3D coordinates $((x_i - x_j)/s, (y_i - y_j)/s, (z_i - z_j)/s)$ if an edge exists, and 0 otherwise, where

$s = \sqrt{(x_i - x_j)^2 + (y_i - y_j)^2 + (z_i - z_j)^2}$. $\mathbf{F} \in \mathrm{R}^{V \times M \times T}$ represents the time sequences of node features across all vertices.

Spatial Graph Convolution: A continuous spline kernel for spatial convolution is used such that it can be applied across graphs [8]. Given graph node features $\mathbf{f} \in \mathrm{R}^{V \times M}$ at each time instant, the convolution kernel is defined as:

$$g_l(\mathbf{u}) = \sum_{\mathbf{p} \in \mathcal{P}} w_{\mathbf{p},l} B_{\mathbf{p}}(\mathbf{u}), \tag{2}$$

where $1 \leq l \leq M$, the spline basis $B_{\mathbf{p}}(\mathbf{u}) = \prod_{r=1}^{d} N_{r,p_r}^{m}(\mathbf{u})$ with N_{r,p_r}^{m} denoting d open B-spline basis of degree m based on equidistant knot vectors, $\mathcal{P} = (N_{1,r}^{m})_r \times \ldots \times (N_{d,r}^{m})_r$ is the Cartesian product of the B-spline bases, and $w_{\mathbf{p},l}$ are trainable parameters. Given kernel $\mathbf{g} = (g_1, ..., g_M)$, spatial convolution for vertex $i \in \mathcal{V}$ with its neighborhood $N(i)$ is defined as

$$(f_l * g_l)(i) = \sum_{j \in N(i), \mathbf{p} \in \mathcal{P}(\mathbf{u}(i,j))} f_l(j) \cdot g_l(\mathbf{u}(i,j)). \tag{3}$$

Since the B-spline basis in Eq. (2) is conditioned on local geometry, the learned kernel can be applied across graphs and the convolution incorporates geometrical information within the graph. This spatial convolution is independently applied to each time frame of the signal sequence in parallel.

To make the network deeper and more expressive, we introduce residual blocks here to pass the input of spatial convolution through a skip connection with 1D convolution before adding it to the output of the spatial convolution.

Temporal Modeling: After spatial convolution, temporal convolution using standard 1D convolution is applied to the time sequence for each node and feature. The number of filters is set to compresses the time sequence in dimension in the encoder, while expanding in the decoder. The geometry graph remains the same for the complete temporal sequences.

Hierarchical Graph Composition: To allow pooling and unpooling in space, we further introduce a hierarchical graph representation of the two geometry. While various graph clustering [7] and pooling methods [19] exist, a unique constraint needs to be met here due to the underlying physics: the topology of the geometry must be preserved in its hierarchical representations to prevent non-physical spatial propagation of signals. Here, we obtain hierarchical geometry representations by specialized mesh coarsening method in CGAL [5,14].

The hierarchical graph representation is predefined and stored in matrices to allow efficient matrix multiplications for pooling/unpooling [6]. If $\mathcal{G}_o$ is a graph with N_1 vertices and $\mathcal{G}_c$ is its coarsened graph with N_2 vertices, we use a binary matrix $\mathbf{P} \in \mathrm{R}^{N_1 \times N_2}$, where $\mathbf{P}_{ij} = 1$ if vertex i in $\mathcal{G}_o$ is grouped to vertex j in $\mathcal{G}_c$, and $\mathbf{P}_{ij} = 0$ otherwise. Given feature map $\mathbf{f}_o \in \mathrm{R}^{N_1 \times M}$ over $\mathcal{G}_o$ and

$\mathbf{f}_c \in \mathrm{R}^{N_2 \times M}$ over $\mathcal{G}_c$, the pooling operation is defined by $\mathbf{f}_c = \mathbf{P}_n^T \mathbf{f}_o$ and the unpooling operation is defined by $\mathbf{f}_o = \mathbf{P}\mathbf{f}_c$, where $\mathbf{P}_n^T$ is column normalized from $\mathbf{P}$.

Summary: As summarized in Fig. 1, each ST-GCNN block consists of spatial graph convolution, temporal convolution, and spatial pooling/unpooling as described above. Using these building blocks, we obtain an encoder that embeds body-surface signal $\mathbf{Y}_t$ over its torso geometry, and a decoder that generates heart-surface potential $\mathbf{X}_t$ over its heart geometry. Next, we learn the physics-based relationship between the two latent space as a function of their geometry.

2.2 Learning Geometry-Dependent Physics in Latent Space

As explained earlier, the physics between $\mathbf{X}_t$ and $\mathbf{Y}_t$ is heavily reliant on the underlying heart-torso geometry: according to Eq. (1), the potential on one torso node can be represented as a linear combination of the potential from all heart nodes, where the coefficients are determined by the relative position between each pair of torso-heart nodes. We assume the linearity to hold between the heart and torso signals in the latent space during inverse imaging, and explicitly model it as a function of the relative position between embedded heart and torso geometry, where a quadratic function exists between the coefficients of the linear function and the geometry.

To do so, we construct a bipartite graph where the edge exists between each pair of heart and torso vertices from their respective graph embedding: the edge attribute $\mathbf{u}(i, j)$ between torso vertex i and heart vertex j thus describes their relative geometrical relationship. The bipartite graph is also learned using the complete temporal sequences. For latent representation $\mathbf{z}_h(i)$ on vertex i of the latent heart mesh, we define it as a linear combination of latent representation $\mathbf{z}_b(j)$ across all vertices j of the latent torso mesh:

$$\mathbf{z}_h(i) = \sum_j \mathbf{z}_b(j) \cdot \hat{\mathbf{h}}(\mathbf{u}(i, j)), \tag{4}$$

where the coefficients $\hat{\mathbf{h}}(\mathbf{u}(i, j))$ are dependant on the relative position $\mathbf{u}(i, j)$ between the two graphs. Aside from being a physics-informed function, this geometric parameterization allows the learned function to generalize across different torso-heart geometry. None of this would be achievable by, for instance, using fully connected layers between $\mathbf{z}_b$ and $\mathbf{z}_h$. Exploiting the similarity between Eq. (4) and Eq. (3), we recast linear relationship in Eq. (4) using spline convolution, with the geometry-dependent coefficients $\hat{\mathbf{h}}$ learned as the spline convolution kernel.

2.3 Loss Function

Denoting the encoder as $\mathbf{z}_b = E_\theta(\mathbf{Y})$, the geometry-dependent inverse function as $\mathbf{z}_h = h_\rho(\mathbf{z}_b)$, and the decoder as $\mathbf{X} = D_\phi(\mathbf{z}_h)$, parameters θ, ρ and ϕ of the network are optimized by minimizing the mean square error between the reconstructed $\hat{\mathbf{X}}_i$ on given pairs of training data $\{\mathbf{X}_i, \mathbf{Y}_i\}_{i=1}^{N}$:

$$\mathcal{L} = \sum_i ||\mathbf{X}_i - D_\phi \left(h_\rho \left(E_\theta \left(\mathbf{Y}_i \right) \right) \right) ||_2^2. \tag{5}$$

3   Experiments

We design a series of generalization tasks of increasing difficulty. In specific, we trained the network using synthetic data simulated on a specific pair of heart-torso geometry, including geometrical variations introduced by rotating the heart along the longitudinal axis (z-axis) for a predefined range. We then tested the trained network regarding generalization to: 1) synthetic data simulated on the same heart-torso geometry but with z-axis heart rotations beyond the training range, 2) synthetic data simulated on the same heart-torso geometry but with novel heart rotations along frontal axis (x-axis) and sagittal axis (y-axis), 3) synthetic data simulated on new heart-torso geometry from new patients, and 4) real data on different heart-torso geometry.

The first two tests considered comparisons to Euclidean encoding-decoding networks [11], both in a deterministic formulation and in a stochastic formulation with improved invariance to input geometry. These Euclidean networks will not apply without re-training on the new geometry in the last two tests.

Models and Training: In all experiments, the presented network consists of three ST-GCNN blocks and two standard convolutional layers in the encoder, one spline convolutional layer in the inverse block, and four ST-GCNN blocks and two standard convolutional layers in the decoder. We used ELU activation, ADAM optimizer [13], and a learning rate of 5×10^{-4}. The Euclidean baselines followed the architectures presented in [11], which consist of cascaded LSTMs and fully connected layers in the encoder and decoder.

For training, we generated pairs of simulated potential data on a specific heart-torso mesh. On the heart, we simulated spatiotemporal propagation sequence of action potential by the Aliev-Panfilov (AP) model [2], considering a combination of 38 different origins of activation and 16 spatial distribution of scar tissue in the heart. We then rotated the heart by $-2°$ to $2°$ around the z-axis, obtaining approximately 2700 sets of different body-surface potential embodying changes of heart orientations in the data. All body-surface potential were corrupted with 20 dB Gaussian noises for inverse imaging. Using NVIDIA Tesla T4 with 16 GB memory, the geometric model took 3 days for training.

Synthetic data for testing were generated in a similar fashion, with additional geometry changes as detailed in later sections. The reconstruction accuracy was measured by the mean square error (MSE) and correlation coefficient (CC) between the reconstructed and actual potential sequence on the heart surface.

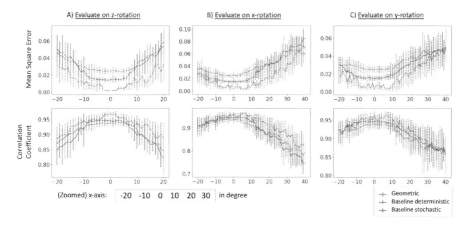

Fig. 2. Comparison of reconstruction accuracy among the three comparison models in test data with A) heart rotations outside training range, and B)–C) novel rotations not seen in training. X-axis represents the degree of rotation relative to training. (Color figure online)

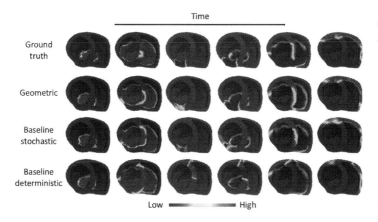

Fig. 3. Reconstructed electrical activity by three comparison models when $z = -19°$.

Generalization to Rotations Outside Training Range: We first applied the trained models to body-surface potential data generated when the heart was rotated by $-20°$ to $20°$ around the z-axis, a range far outside that considered in training. Figure 2A summarizes the quantitative metrics of the three models on approximately 22,000 test cases, against the change in heart rotations from training data. As shown, the presented method (red) outperformed the deterministic (green) and stochastic (blue) Euclidean baseline in all metrics for all heart rotations. The standard deviation of the geometric method lies in between that of deterministic and stochastic baseline. Figure 3 provides visual examples of the reconstructed image sequence of the three models.

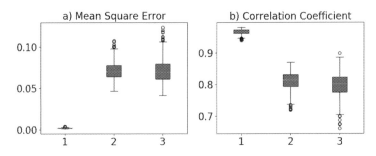

Fig. 4. Accuracy of reconstruction on training geometry (1) and new geometry (2/3).

Generalization to Novel Rotations: We then tested the trained models on approximately 66,000 body-surface data generated from novel heart rotations around the x-axis ($-20°$ to $+40°$) and y-axis ($-20°$ to $+40°$). As summarized in Fig. 2B–C, the presented model (red) significantly outperformed the two Euclidean models in all metrics. Furthermore, we observe in Fig. 2B that the geometric method performs better as the test set deviates more from the training set (up to $40°$ of rotation). This supports that, as test data move further away from training, the gain in the generalization ability of the presented method would become more significant in comparison to its Euclidean alternatives. The standard deviations of the three models are comparable.

Generalization to New Geometry: We then moved to apply the trained network to simulated data generated on two new heart-torso meshes. This represents a realistic scenario where the network trained on a group of patients will be applied to new patients. Figure 4 provides box plots of the two metrics obtained on the two new geometry over, respectively, 491 and 444 test data. Despite a drop in performance in comparison to the earlier results on the training geometry, reasonable accuracy was achieved considering the difficulty of the generalization task. Note that Euclidean networks will not be applicable here unless being re-trained on data generated on the new geometry [11].

Generalization to Real Data: Finally, we tested the presented network on *in-vivo* 120-lead body-surface potential data obtained on two patients with scar-related ventricular arrhythmia. Since the heart-torso geometry of patient ♯1 was used in training, we were able to apply the Euclidean baselines for comparison purpose. From each reconstructed potential sequence on each patient, we identified the region of scar tissue by nodes whose activation was shorter than a predefined duration. The results summarized in Fig. 5 demonstrated the ability of the presented network to not only generalize across geometry but across the shifts between simulated and real data, approximating the location of scar tissue with evident visual improvement over its Euclidean alternatives.

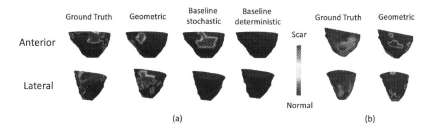

Fig. 5. Region of scar identified from reconstructed potential sequence on a) the training patient and b) a new patient. The ground truth was from *in-vivo* voltage mapping.

4   Conclusion

In this work, we present a novel non-Euclidean network for learning geometry-dependent and physics-based inverse imaging between spatiotemporal variables living on 3D geometrical domains. In generalization tests with increased difficulty, we demonstrated the ability of the presented network to better generalize to unseen geometrical variations in comparison to its Euclidean alternatives, and to directly apply to new geometry which is not possible with Euclidean approaches. An immediate future work is to explore the use of fine-tuning with a small number of labeled data in order to improve the performance of the network when applying it to new patients. To our knowledge, this is the first geometry-dependent inverse imaging network over non-Euclidean domains and its application to reconstructing cardiac electrical activity from surface potential.

Acknowledgement. This work is supported by the National Heart, Lung and Blood Institute of the National Institutes of Health under Award Number R01HL145590, and the National Science Foundation under CAREER Award Number ACI-1350374.

References

1. Adler, J., Oktem, O.: Learned primal-dual reconstruction. IEEE Trans. Med. Imaging **37**(6), 1322–1332 (2018)
2. Aliev, R.R., Panfilov, A.V.: A simple two-variable model of cardiac excitation. Chaos Solitons Fractals **7**(3), 293–301 (1996)
3. Bacoyannis, T., Krebs, J., Cedilnik, N., Cochet, H., Sermesant, M.: Deep learning formulation of ECGI for data-driven integration of spatiotemporal correlations and imaging information. In: Coudière, Y., Ozenne, V., Vigmond, E., Zemzemi, N. (eds.) FIMH 2019. LNCS, vol. 11504, pp. 20–28. Springer, Cham (2019). https://doi.org/10.1007/978-3-030-21949-9_3
4. Bronstein, M.M., Bruna, J., LeCun, Y., Szlam, A., Vandergheynst, P.: Geometric deep learning: going beyond Euclidean data. IEEE Signal Process. Mag. **34**(4), 18–42 (2017)
5. Cacciola, F.: Triangulated surface mesh simplification. In: CGAL User and Reference Manual. CGAL Editorial Board, 5.0.2 edn. (2020). https://doc.cgal.org/5.0.2/Manual/packages.html#PkgSurfaceMeshSimplification

6. Dhamala, J., Ghimire, S., Sapp, J.L., Horáček, B.M., Wang, L.: Bayesian optimization on large graphs via a graph convolutional generative model: application in cardiac model personalization. In: Shen, D., et al. (eds.) MICCAI 2019. LNCS, vol. 11765, pp. 458–467. Springer, Cham (2019). https://doi.org/10.1007/978-3-030-32245-8_51

7. Dhillon, I.S., Guan, Y., Kulis, B.: Weighted graph cuts without eigenvectors a multilevel approach. IEEE Trans. Pattern Anal. Mach. Intell. **29**(11), 1944–1957 (2007)

8. Fey, M., Eric Lenssen, J., Weichert, F., Müller, H.: SplineCNN: fast geometric deep learning with continuous B-spline kernels. In: The IEEE Conference on Computer Vision and Pattern Recognition (CVPR), pp. 869–877 (2018)

9. Ghimire, S., et al.: Overcoming barriers to quantification and comparison of electrocardiographic imaging methods: a community-based approach. In: 2017 Computing in Cardiology (CinC), pp. 1–4. IEEE (2017)

10. Ghimire, S., Dhamala, J., Gyawali, P.K., Sapp, J.L., Horacek, M., Wang, L.: Generative modeling and inverse imaging of cardiac transmembrane potential. In: Frangi, A.F., Schnabel, J.A., Davatzikos, C., Alberola-López, C., Fichtinger, G. (eds.) MICCAI 2018. LNCS, vol. 11071, pp. 508–516. Springer, Cham (2018). https://doi.org/10.1007/978-3-030-00934-2_57

11. Ghimire, S., Gyawali, P.K., Dhamala, J., Sapp, J.L., Horacek, M., Wang, L.: Improving generalization of deep networks for inverse reconstruction of image sequences. In: Chung, A.C.S., Gee, J.C., Yushkevich, P.A., Bao, S. (eds.) IPMI 2019. LNCS, vol. 11492, pp. 153–166. Springer, Cham (2019). https://doi.org/10.1007/978-3-030-20351-1_12

12. Ghimire, S., Sapp, J.L., Horáček, B.M., Wang, L.: Noninvasive reconstruction of transmural transmembrane potential with simultaneous estimation of prior model error. IEEE Trans. Med. Imaging **38**(11), 2582–2595 (2019)

13. Kingma, D.P., Ba, J.: Adam: a method for stochastic optimization. arXiv preprint arXiv:1412.6980 (2014)

14. Lindstrom, P., Turk, G.: Fast and memory efficient polygonal simplification. In: Proceedings Visualization 1998 (Cat. No. 98CB36276), pp. 279–286. IEEE (1998)

15. Lucas, A., Iliadis, M., Molina, R., Katsaggelos, A.K.: Using deep neural networks for inverse problems in imaging: beyond analytical methods. IEEE Signal Process. Mag. **35**(1), 20–36 (2018)

16. Michel, C.M., Murray, M.M.: Towards the utilization of EEG as a brain imaging tool. NeuroImage **61**(2), 371–385 (2012)

17. Plonsey, R.: Bioelectric Phenomena. Wiley Encyclopedia of Electrical and Electronics Engineering. Wiley, New York (2001)

18. Sun, J., Li, H., Xu, Z., et al.: Deep ADMM-net for compressive sensing MRI (2016)

19. Wu, Z., Pan, S., Chen, F., Long, G., Zhang, C., Yu, P.S.: A comprehensive survey on graph neural networks. IEEE Trans. Neural Netw. Learn. Syst. https://doi.org/10.1109/TNNLS.2020.2978386

20. Zhu, B., Liu, J.Z., Cauley, S.F., Rosen, B.R., Rosen, M.S.: Image reconstruction by domain-transform manifold learning. Nature **555**(7697), 487 (2018)

Hierarchical Classification of Pulmonary Lesions: A Large-Scale Radio-Pathomics Study

Jiancheng Yang[1,2,3], Mingze Gao[3], Kaiming Kuang[3], Bingbing Ni[1,2,4(✉)],
Yunlang She[5], Dong Xie[5], and Chang Chen[5]

[1] Shanghai Jiao Tong University, Shanghai, China
nibingbing@sjtu.edu.cn
[2] MoE Key Lab of Artificial Intelligence, AI Institute,
Shanghai Jiao Tong University, Shanghai, China
[3] Dianei Technology, Shanghai, China
[4] Huawei Hisilicon, Shanghai, China
[5] Shanghai Pulmonary Hospital, Tongji University, Shanghai, China

Abstract. Diagnosis of pulmonary lesions from computed tomography (CT) is important but challenging for clinical decision making in lung cancer related diseases. Deep learning has achieved great success in computer aided diagnosis (CADx) area for lung cancer, whereas it suffers from label ambiguity due to the difficulty in the radiological diagnosis. Considering that invasive pathological analysis serves as the clinical golden standard of lung cancer diagnosis, in this study, we solve the label ambiguity issue via a large-scale radio-pathomics dataset containing 5,134 radiological CT images with pathologically confirmed labels, including cancers (*e.g.*, invasive/non-invasive adenocarcinoma, squamous carcinoma) and non-cancer diseases (*e.g.*, tuberculosis, hamartoma). This retrospective dataset, named Pulmonary-RadPath, enables development and validation of accurate deep learning systems to predict invasive pathological labels with a non-invasive procedure, *i.e.*, radiological CT scans. A three-level hierarchical classification system for pulmonary lesions is developed, which covers most diseases in cancer-related diagnosis. We explore several techniques for hierarchical classification on this dataset, and propose a Leaky Dense Hierarchy approach with proven effectiveness in experiments. Our study significantly outperforms prior arts in terms of data scales (6× larger), disease comprehensiveness and hierarchies. The promising results suggest the potentials to facilitate precision medicine.

Keywords: Pulmonary lesion · Hierarchical classification · Radio-pathomics

---

J. Yang and M. Gao—These authors have contributed equally.

© Springer Nature Switzerland AG 2020
A. L. Martel et al. (Eds.): MICCAI 2020, LNCS 12266, pp. 497–507, 2020.
https://doi.org/10.1007/978-3-030-59725-2_48

1   Introduction

Lung cancer is the most commonly diagnosed cancer worldwide, which accounts for 18.4% of global cancer-related mortality in 2018 [2]. Remarkable success has been achieved in deep learning for lung nodule detection [4,19,23] and diagnosis [13,22,25], thanks to medical datasets, *e.g.*, LIDC-IDRI [1]. However, the radiological diagnosis of lung cancer suffers from ambiguous labels [24]. For instance, 4 expert annotators make diverse diagnosis in LIDC-IDRI dataset [1]. A possible solution to reduce label ambiguity is to leverage clinical golden standard. For lung cancer diagnosis, invasive pathological analysis serves as the golden standard. To this end, we build a large-scale **radio-pathomics** dataset, consisting of 5,134 **radiological** CT images with **pathologically** confirmed labels. Once a deep learning system is well trained on this dataset, it could predict invasive pathological labels via non-invasive procedures, *i.e.*, radiological CT scans. The in-house dataset, named Pulmonary-RadPath, is collected retrospectively from a single clinical center. It covers most diseases in lung cancer-related diagnosis, including cancer (*e.g.*, invasive/non-invaisve adenocarcinoma, squamous carcinoma) and non-cancer diseases (*e.g.*, tuberculosis, hamartoma) (Fig. 1).

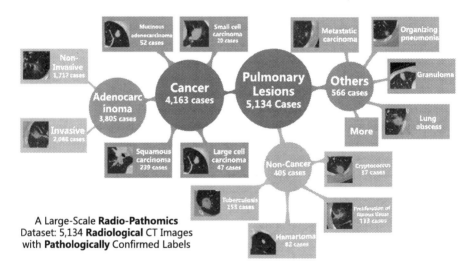

A Large-Scale **Radio-Pathomics** Dataset: 5,134 **Radiological** CT Images with **Pathologically** Confirmed Labels

Fig. 1. We develop a hierarchical multi-disease classification system of pulmonary lesions based on an in-house large-scale Pulmonary-RadPath dataset, which consists of 5,134 radiological CT cases with pathologically confirmed labels, *i.e.*, the labels are retrospectively collected via invasive pathological analysis. By utilizing this dataset, our hierarchical classification system could predict the **invasive** pathological labels from **non-invasive** radiological CT images. Notably, it significantly outperforms prior arts in terms of data scale (6× larger), disease comprehensiveness and hierarchies.

Considering the heterogeneity of included lesions, we develop a three-level hierarchical classification deep learning system of pulmonary lesions with the help of Pulmonary-RadPath dataset. Several hierarchical classification strategies are explored on this real-world clinical dataset, including Leaf-Node baseline approach, Flattened Hierarchy and Leaky Node (details in Sect. 2.2). We further propose (Leaky) Dense Hierarchy approach to encourage hierarchical feature reuse, and prove its effectiveness in our experiments. These strategies improve the three-level hierarchical classification performance compared with naive baseline (Leaf-Node).

Contributions. We develop a three-level hierarchical classification system for pulmonary lesions, via an in-house large-scale radio-pathomics dataset, where 5,134 radiological CT images with pathologically confirmed labels (clinical golden standard) are collected retrospectively. We explore several hierarchical classification strategies and propose a Leaky Dense Hierarchy approach with proven effectiveness in experiments. Our study significantly outperforms previous studies in terms of data scales ($6\times$ larger), disease comprehensiveness and hierarchies.

2   Materials and Methods

2.1   Pulmonary-RadPath Dataset

Dataset Overview. Classification of lung cancers is exceedingly complicated due to its complex and diverse nature. According to WHO lung tumor classification guide, there are 77 subtypes of lung tumors [20]. In this study, we compile a large dataset named Pulmonary-RadPath that contains 5,134 cases with CT scans and invasive pathological diagnosis. All data is collected from a single clinical center (Shanghai Pulmonary Hospital, Tongji University, Shanghai, China). Thicknesses of these scans range 0.2 mm and 1.0 mm, and the average long diameter of all lesions is 1.9 cm. More than 80% of the cases are diagnosed as malignant. Classes are unbalanced in our dataset, including some diseases with less than 10 cases present. This results from the uneven distribution of lung diseases.

Since that our research is based on CT scans rather than pathological images, we select some categories as our classification target according to their quantities and visibility in CT volumes. We categorize all individual diseases into three major types: cancer (*e.g.*, adenocarcinoma and squamous carcinoma), non-cancer (*e.g.*, tuberculosis and hamartoma) and others (long tail diseases, *e.g.*, metastatic carcinoma and granuloma). All classes of interest, their tags and corresponding sample sizes in each subset are summarized in Table 1. Note that in our study, "invasive" denotes minimally invasive adenocarcinoma (MIA) and invasive pulmonary adenocarcinoma (IA), while "non-invasive" denotes atypical adenomatous hyperplasia (AAH) and adenocarcinomas in situ (AIS).

Annotation and Pretreatment. The hierarchical pathological label and mass center of each lesion is manually labelled by a junior thoracic radiologist, according to corresponding pathological reports. These annotations are then confirmed

Table 1. Overview of Pulmonary-RadPath Dataset. There are 5,134 radiological CT cases with pathologically confirmed labels in total, where 80% are regarded as Train-Dev, and 20% are regarded as Test. Classification categories are structured into 3 hierarchical levels. Tags are named as "HeadID+ClassID" (Fig. 2).

Level	Class	Tag	Train-Dev	Test	Total
0	ALL	H0	4,109	1,025	5,134
1	Cancer	H1a	3,330	833	4,163
2	Adenocarcinoma	H2a	3,042	763	3,805
3	Invasive	H4a	1,672	416	2,088
3	Non-invasive	H4b	1,370	3,47	1,717
2	Squamous carcinoma	H2b	192	47	239
2	Large cell carcinoma	H2c	38	9	47
2	Small cell carcinoma	H2d	16	4	20
2	Mucinous adenocarcinoma	H2e	42	10	52
1	Non-Cancer	H1b	326	79	405
2	Tuberculosis	H3a	123	30	153
2	Hamartoma	H3b	66	16	82
2	Proliferation of fibrous tissue	H3c	107	26	133
2	Cryptococcus	H3d	30	7	37
1	Others	H1c	453	113	566

by a senior radiologist with 15 years of experience in chest CT. Decisions on CT findings are made by consensus. The data is anonymized so that patient identities cannot be traced. We preprocess the data following a common practice [28]: (1) Resample CT volumes to $1\,mm \times 1\,mm \times 1\,mm$. (2) Normalize Hounsfield Units to $[-1,1]$. (3) Crop a $48 \times 48 \times 48$ volume centered at the centroid of the each lesion. The dataset is randomly split into 5 subsets of same size, numbered as 0 to 4. Each subset retains roughly the same proportion for each disease as the entire dataset. We use subset 0–3 for training and hyperparameter validation, and hold out subset 4 only for testing.

2.2   Approaches for Hierarchical Classification

Model Overview. We use a custom 3D DenseNet [11] as the backbone. DenseNet has shown compelling accuracies on natural image recognizing tasks with efficient use of parameters. To leverage the power of dense connectivities, we transform the 2D DenseNet into its 3D variant. The 3D DenseNet comprises of stacked densely-connected blocks (*i.e.*, DenseBlock), each of which consists of several convolutional modules. In each convolutional module, a $1 \times 1 \times 1$ convolution layer with 64 filters is followed by a $3 \times 3 \times 3$ convolution layer with 16 filters, Dense connections enable efficient 3D representation learning. In this study, we use a 3D DenseNet that has a growth rate of 16 and three DenseBlocks with depths of 8, 4 and 4, respectively. The features after global average pooling layer are then processed by either of the following hierarchical classification heads, *i.e.*, (b) (c) (d) (e) (f) in Fig. 2.

We arrange the hierarchical system of disease categories in a tree structure, as shown in Fig. 2(a). Three root nodes, H1a, H1b and H1c, represent general disease categories: cancer, non-cancer and others. Individual diseases are initialized as leaf nodes. The entire tree is then developed in a bottom-up fashion. Nodes under the same pathological class are merged as internal nodes, until all classes and sub-classes are included in the tree.

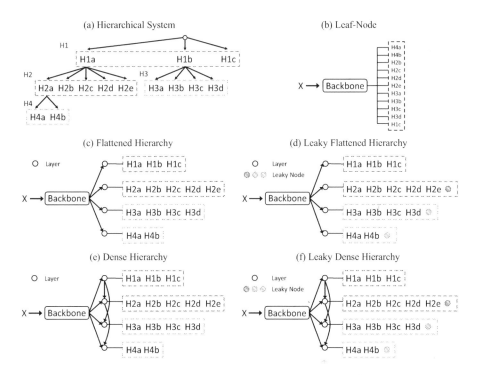

Fig. 2. Approaches for hierarchical classification in this study. (a) The hierarchical classification system for pulmonary lesions. The tags in the tree follow the names from Table 1. We use a same 3D DenseNet backbone [11] for all our experiments. (b) Leaf-Node approach. The leaf nodes are ordered from left to right. (c) Flattened Hierarchy approach. (d) Leaky Flattened Hierarchy approach. (e) Dense Hierarchy approach. (f) Leaky Dense Hierarchy approach.

Leaf-Node Approach. We first introduce the naive baseline Leaf-Node approach [5] for hierarchical classification. Leaf-Node approach treats hierarchical classification as a simple multi-class problem, where each leaf node in the hierarchical system is a class. The model outputs the probability for each individual diseases. The probabilities of internal or root nodes equal to the sum of those of all the children nodes.

Flattened Hierarchy Approach. Inspired by the multiple softmax heads for individual annotators [9], we propose the Flattened Hierarchy approach with multiple heads for different hierarchies, *i.e.*, H1, H2, H3 and H4 shown in

Fig. 2(a). Instead of structuring it as a plain multi-class problem, the Flattened Hierarchy approach includes each training sample's hierarchical information, *i.e.*, all categories on the path from itself to the top node. The problem is then decomposed into several multi-class classifications within each hierarchy. Probabilities of each leaf node is calculated by multiplying those of nodes on the path from the leaf node to the root. Compared with the Leaf-Node approach, Flattened Hierarchy enables the model to learn the hierarchical relations between classes. Loss is ignored for a certain hierarchy if the target class is not in the hierarchy.

Dense Hierarchy Approach. To reuse features in different hierarchies, we propose the Dense Hierarchy approach as shown in Fig. 2(e). Directed connections are made between any parent and child hierarchy, where features from the parent hierarchy are passed down to the next level and concatenated with features of child hierarchy. This design is similar to DenseNet [11], in that features of the higher hierarchy are reused at a lower level through jump connections. Loss is ignored for a certain hierarchy if the target class is not in the hierarchy.

Leaky Node as a Virtual Class. We observe better performances compared with the Leaf-Node approach when adopting strategies above. Moreover, we introduce a virtual node inspired by hierarchical novelty detection [17], named Leaky Node, to further improve the performance in hierarchical classification settings. This technique could be easily applied to both Flattened and Dense Hierarchy approaches as a plug-and-play unit, and it stably delivers better performances in our experiments. The Leaky Node is a fall-back class when a certain training sample does not belong to any nodes in this hierarchy. In such cases, the leaky node label is set positive while all others are negative, and its loss is back-propagated along with those of other nodes in the same hierarchy. The leaky nodes make the full data distributions accessible to the hierarchical head classifiers; as the softmax classification with cross-entropy training loss is naturally proper scoring rules [7], the leaky node approach encourages calibration [10] of predictive uncertainty in each hierarchy classifier, which leads to better classification performance in our experiments (Sect. 3.2).

2.3   Model Training and Inference

During training, we use cross entropy loss with class-imbalance weights allocated for each head. Online data augmentations, including random rotation, flipping and translation are applied at each volume. We use Adam optimizer [16] to train all models end-to-end for 200 epochs. We set the initial learning rate at 0.01, and decay it by a factor of 1/3 every 20 epochs. The batch size is 16 for training. We use synchronized batch normalization [14,26] for multiple GPUs. All experiments are implemented with PyTorch 1.2 [18] and 4 NVIDIA 1080Ti GPUs.

For the Leaf-Node approach, we calculate an internal or root node's probability by summing up those of all its children. In terms of Hierarchy approach, the probability of leaf or internal node is calculated by multiplying those of nodes on the path from itself to the root.

3   Experiments

3.1   Evaluation

We calculate the Area Under the Receiver Operating Characteristic Curve (AUROC) for each class to assess the agreement between the ground truth and the model output. Within each head, the sample's class is labelled as 1 and the others are set to 0. We calculate each head's mAUC to evaluate each hierarchical level performance. The mAUC is mean AUC weighted by number of samples in each class. The mAUC@L is calculated by averaging AUCs of all leaf nodes, weighted by number of samples at each leaf node. Similarly, the mAUC@H1-4 is the average of AUCs at all child nodes within the hierarchy.

Table 2. Mean AUC (mAUC) within each head (H1/2/3/4) and leaf nodes (L). Performance of each node is depicted in Table 3.

Methods	mAUC@H1	mAUC@H2	mAUC@H3	mAUC@H4	mAUC@L
Leaf-Node	68.9	86.9	76.5	43.8	80.7
Flattened hierarchy	64.3	85.5	76.2	92.2	80.4
Leaky flattened hierarchy	74.1	87.3	70.0	92.4	81.9
Dense hierarchy	66.1	87.3	**80.0**	93.0	81.1
Leaky dense hierarchy	**75.9**	**89.1**	79.9	**93.8**	**84.1**

Table 3. Model performance in AUC of each node class within H1/2/3/4. AUC is computed between each node class and every other classes in a same node.

Methods	H1a	H1b	H1c	H2a	H2b	H2c	H2d	H2e	H3a	H3b	H3c	H3d	H4a	H4b
Leaf-Node	69.3	67.8	66.6	87.1	86.8	86.2	76.4	74.6	70.6	71.6	88.0	**69.8**	55.3	31.0
Flattened hierarchy	64.4	62.9	64.3	85.7	87.2	84.8	78.3	68.2	76.2	61.9	88.2	63.9		92.2
Leaky flattened hier	74.6	72.9	71.1	87.4	86.2	**92.8**	80.3	**81.9**	76.6	**78.3**	88.8	54.2		92.4
Dense hierarchy	66.2	67.2	64.6	87.5	87.2	89.0	**83.8**	75.0	77.8	69.9	**91.5**	69.6		93.0
Leaky dense hierarchy	**76.4**	**76.9**	**71.5**	**89.2**	**90.4**	86.9	**83.8**	75.9	**79.5**	71.1	90.2	63.3		**93.8**

3.2   Results

Table 2 shows the mAUC of each head and the mAUC over all leaf nodes. Table 3 gives the details of each node's AUC. At each level, at least one of our hierarchical methods achieves better AUC score than the Leaf-Node approach. Leaf-Node does not even delivers the best performance on mAUC@L, which only includes AUC scores on leaf nodes. This is because the Leaf-Node approach does not include the hierarchical relation and the differences in the level of difficulty between each classification task. For instance, Leaf-Node gets the lowest mAUC in H4 even though this hierarchy makes up more than 70% of total data.

It is worth noting that Dense Hierarchy approaches reach higher mAUC than its Flattened counterparts in all hierarchies, as they share features through jump connections between levels. Besides, leaky nodes consistently improve the network's performance, which demonstrates the usefulness of learning calibrated classification as real-world or full data distribution.

4   Discussion

Compared with Previous Studies. In this study, we develop a radio-pathomics hierarchical classification system of pulmonary lesions based on 3D deep learning. The labels are retrospectively collected via invasive pathological analysis. Although the labels are collected invasively, once the deep models are trained on this dataset, we could predict the invasive labels via a non-invasive procedure (*i.e.*, radiological CT scans). Note that the clinical golden standard pathological labels avoid the annotation ambiguity [24] from human annotators. Even for experts, the diagnosis could be diverse for controversial cases (refer to the diverse annotations from 4 experts in LIDC-IDRI dataset [1]). *DenseSharp* [28] introduces deep learning methodology in the radio-pathomics research for lung cancer. A dataset of 653 adenocarcinoma CT scans is built to develop and validate the deep learning system. It achieves a promising AUC performance of 78.8 in differentiating invasive and non-invasive adenocarcinoma, which outperforms radiologists in the observer study. This study focuses on adenocarcinoma; though clinically important, it is only a possible diagnosis in the pulmonary lesion diagnosis system, which hinders the clinical usage in real world. Several studies [8,15] follows the same setting as *DenseSharp* [28] with 828 and 525 cases, which results in AUC of 92.0 and 92.1, respectively. As a comparison, the classification system of our study is more comprehensive in terms of disease coverage and hierarchical levels. We do not only include adenocarcinoma, but also squamous carcinoma and other cancer and non-cancer diseases. On our dataset, we achieve a AUC performance of 93.8 in differentiating invasive and non-invasive adenocarcinoma[1]. We believe the performance comes from the overwhelming superiority over previous studies on data scale, disease comprehensiveness and hierarchies.

Data Bias and Long-Tailed Issues. A limitation of this study is the data bias and long-tailed issues for subjects with pathological analysis. Pathological analysis as a invasive or minimally invasive examination, is generally performed on subjects regarded as high-risk by clinicians. It is interesting to consider how to calibrate the trained models with pathologically confirmed labels to real-world distribution [10]. Besides, cancers (*e.g.*, squamous carcinoma) not suitable for surgery are also lack of pathological analysis. On the other hand, certain types of cancer or diseases are naturally very long-tailed, *e.g.*, small cell carcinoma. Techniques for long-tailed classification [3] or even few-shot classification [6] are

---

[1] Note that the AUC metrics could not directly compare with each other since the datasets, inclusion criteria, and experiment settings are different.

worth explorating the hierarchical settings. It is also beneficial to integrate model uncertainty [12] into the deep hierarchical classification system and improve model generalization [29] upon diverse imaging protocols.

5   Conclusion

In this study, we develop a hierarchical multi-disease classification system of pulmonary lesions with on an in-house large-scale radio-pathomics dataset, named Pulmonary-RadPath. Several hierarchical classification strategies are explored and developed on our dataset. Our study significantly outperforms prior arts in terms of data scale, disease comprehensiveness and hierarchies.

In future studies, we will continuously complement the Pulmonary-RadPath data with more classes and more samples (especially the highly imbalanced classes). Research is urged on the data bias and long-tailed issues. It is also interesting to integrate radio-genomics (*e.g.*, predicting EGFR-mutation [21,27]) into our radio-pathomics hierarchical classification system.

Acknowledgment. This work was supported by National Science Foundation of China (61976137, U1611461). This work was also supported by Shanghai Municipal Health Commission (2018ZHYL0102, 2019SY072, 201940018). Authors appreciate the Student Innovation Center of SJTU for providing GPUs.

References

1. Armato, S.G., et al.: The Lung Image Database Consortium (LIDC) and Image Database Resource Initiative (IDRI): a completed reference database of lung nodules on CT scans. Med. Phys. **38**(2), 915–931 (2011)
2. Bray, F., Ferlay, J., Soerjomataram, I., Siegel, R.L., Torre, L.A., Jemal, A.: Global cancer statistics 2018: GLOBOCAN estimates of incidence and mortality worldwide for 36 cancers in 185 countries. CA Cancer J. Clin. **68**(6), 394–424 (2018)
3. Cui, Y., Jia, M., Lin, T.Y., Song, Y., Belongie, S.J.: Class-balanced loss based on effective number of samples. In: CVPR, pp. 9260–9269 (2019)
4. Dou, Q., Chen, H., Jin, Y., Lin, H., Qin, J., Heng, P.-A.: Automated pulmonary nodule detection via 3D ConvNets with online sample filtering and hybrid-loss residual learning. In: Descoteaux, M., Maier-Hein, L., Franz, A., Jannin, P., Collins, D.L., Duchesne, S. (eds.) MICCAI 2017. LNCS, vol. 10435, pp. 630–638. Springer, Cham (2017). https://doi.org/10.1007/978-3-319-66179-7_72
5. Esteva, A., et al.: Dermatologist-level classification of skin cancer with deep neural networks. Nature **542**, 115–118 (2017)
6. Finn, C., Abbeel, P., Levine, S.: Model-agnostic meta-learning for fast adaptation of deep networks. In: ICML (2017)
7. Gneiting, T., Raftery, A.E.: Strictly proper scoring rules, prediction, and estimation. J. Am. Stat. Assoc. **102**, 359–378 (2007)
8. Gong, J., et al.: A deep residual learning network for predicting lung adenocarcinoma manifesting as ground-glass nodule on CT images. Eur. Radiol. **30**, 1847–1855 (2019). https://doi.org/10.1007/s00330-019-06533-w

9. Guan, M.Y., Gulshan, V., Dai, A.M., Hinton, G.E.: Who said what: modeling individual labelers improves classification. In: AAAI (2017)
10. Guo, C., Pleiss, G., Sun, Y., Weinberger, K.Q.: On calibration of modern neural networks. In: ICML (2017)
11. Huang, G., Liu, Z., Van Der Maaten, L., Weinberger, K.Q.: Densely connected convolutional networks. In: CVPR, vol. 1, p. 3 (2017)
12. Huang, X., Yang, J., et al.: Evaluating and boosting uncertainty quantification in classification. arXiv preprint arXiv:1909.06030 (2019)
13. Hussein, S., Cao, K., Song, Q., Bagci, U.: Risk stratification of lung nodules using 3D CNN-based multi-task learning. In: Niethammer, M., et al. (eds.) IPMI 2017. LNCS, vol. 10265, pp. 249–260. Springer, Cham (2017). https://doi.org/10.1007/978-3-319-59050-9_20
14. Ioffe, S., Szegedy, C.: Batch normalization: accelerating deep network training by reducing internal covariate shift. In: ICML (2015)
15. Kim, H., et al.: CT-based deep learning model to differentiate invasive pulmonary adenocarcinomas appearing as subsolid nodules among surgical candidates: comparison of the diagnostic performance with a size-based logistic model and radiologists. Eur. Radiol. 30(6), 3295–3305 (2020). https://doi.org/10.1007/s00330-019-06628-4
16. Kingma, D.P., Ba, J.: Adam: a method for stochastic optimization. arXiv preprint arXiv:1412.6980 (2014)
17. Lee, K., Lee, K., Min, K., Zhang, Y., Shin, J., Lee, H.: Hierarchical novelty detection for visual object recognition. In: CVPR, pp. 1034–1042 (2018)
18. Paszke, A., et al.: Automatic differentiation in PyTorch (2017)
19. Tang, H., Zhang, C., Xie, X.: NoduleNet: decoupled false positive reduction for pulmonary nodule detection and segmentation. In: Shen, D., et al. (eds.) MICCAI 2019. LNCS, vol. 11769, pp. 266–274. Springer, Cham (2019). https://doi.org/10.1007/978-3-030-32226-7_30
20. Travis, W.D., et al.: The 2015 world health organization classification of lung tumors: impact of genetic, clinical and radiologic advances since the 2004 classification. J. Thorac. Oncol. 10(9), 1243–1260 (2015)
21. Wang, S., et al.: Predicting EGFR mutation status in lung adenocarcinoma on computed tomography image using deep learning. Eur. Respir. J. 53, 1–11 (2019)
22. Xie, Y., Xia, Y., Zhang, J., Feng, D.D., Fulham, M., Cai, W.: Transferable multi-model ensemble for benign-malignant lung nodule classification on chest CT. In: Descoteaux, M., Maier-Hein, L., Franz, A., Jannin, P., Collins, D.L., Duchesne, S. (eds.) MICCAI 2017. LNCS, vol. 10435, pp. 656–664. Springer, Cham (2017). https://doi.org/10.1007/978-3-319-66179-7_75
23. Yang, J., Deng, H., Huang, X., Ni, B., Xu, Y.: Relational learning between multiple pulmonary nodules via deep set attention transformers. In: ISBI (2020)
24. Yang, J., Fang, R., Ni, B., Li, Y., Xu, Y., Li, L.: Probabilistic radiomics: ambiguous diagnosis with controllable shape analysis. In: Shen, D., et al. (eds.) MICCAI 2019. LNCS, vol. 11769, pp. 658–666. Springer, Cham (2019). https://doi.org/10.1007/978-3-030-32226-7_73
25. Yang, J., Huang, X., Ni, B., Xu, J., Yang, C., Xu, G.: Reinventing 2D convolutions for 3D medical images. arXiv preprint arXiv:1911.10477 (2019)
26. Zhang, H., et al.: Context encoding for semantic segmentation. In: CVPR, June 2018
27. Zhao, W., et al.: Toward automatic prediction of EGFR mutation status in pulmonary adenocarcinoma with 3D deep learning. Cancer Med. 8, 3532–3543 (2019)

28. Zhao, W., et al.: 3D deep learning from CT scans predicts tumor invasiveness of subcentimeter pulmonary adenocarcinomas. Cancer Res. **78**(24), 6881–6889 (2018)
29. Zhao, W., et al.: Convolution kernel and iterative reconstruction affect the diagnostic performance of radiomics and deep learning in lung adenocarcinoma pathological subtypes. Thorac. Cancer **10**(10), 1893–1903 (2019)

Learning Tumor Growth via Follow-Up Volume Prediction for Lung Nodules

Yamin Li[1,2,3], Jiancheng Yang[1,2,3], Yi Xu[1,2,3(✉)], Jingwei Xu[1,2,3],
Xiaodan Ye[1,4], Guangyu Tao[1,4], Xueqian Xie[1,5], and Guixue Liu[1,5]

[1] Shanghai Jiao Tong University, Shanghai, China
xuyi@sjtu.edu.cn
[2] Shanghai Institute for Advanced Communication and Data Science,
Shanghai, China
[3] MoE Key Lab of Artificial Intelligence, AI Institute,
Shanghai Jiao Tong University, Shanghai, China
[4] Shanghai Chest Hospital, Shanghai, China
[5] Shanghai General Hospital, Shanghai, China

Abstract. Follow-up serves an important role in the management of pulmonary nodules for lung cancer. Imaging diagnostic guidelines with expert consensus have been made to help radiologists make clinical decision for each patient. However, tumor growth is such a complicated process that it is difficult to stratify high-risk nodules from low-risk ones based on morphologic characteristics. On the other hand, recent deep learning studies using convolutional neural networks (CNNs) to predict the malignancy score of nodules, only provides clinicians with black-box predictions. To this end, we propose a unified framework, named Nodule Follow-Up Prediction Network (*NoFoNet*), which predicts the growth of pulmonary nodules with high-quality visual appearances and accurate quantitative results, given any time interval from baseline observations. It is achieved by predicting future displacement field of each voxel with a WarpNet. A TextureNet is further developed to refine textural details of WarpNet outputs. We also introduce techniques including Temporal Encoding Module and Warp Segmentation Loss to encourage time-aware and shape-aware representation learning. We build an in-house follow-up dataset from two medical centers to validate the effectiveness of the proposed method. *NoFoNet* significantly outperforms direct prediction by a U-Net in terms of visual quality; more importantly, it demonstrates accurate differentiating performance between high- and low-risk nodules. Our promising results suggest the potentials in computer aided intervention for lung nodule management.

Keywords: Lung nodule · Follow-up · Tumor growth prediction

Y. Li and J. Yang—These authors have contributed equally.

Electronic supplementary material The online version of this chapter (https://doi.org/10.1007/978-3-030-59725-2_49) contains supplementary material, which is available to authorized users.

A. L. Martel et al. (Eds.): MICCAI 2020, LNCS 12266, pp. 508–517, 2020.
https://doi.org/10.1007/978-3-030-59725-2_49

1   Introduction

Pulmonary nodule management strategy influences the cost-effectiveness of a lung cancer screening program [3]. It remains difficult to differentiate high-risk nodules from low-risk ones based on morphologic characteristics [13]. In order to help radiologists and clinicians to make precise clinical decision for each patient, researchers have made several categorical management recommendation and scoring systems according to morphology, diameters or volume in recent years, *e.g.*, NCCN [16], Fleischner [9], Lung-RADS [12]. However, tumor growth is such a complicated progress that more advanced strategies are worth exploring to facilitate precision medicine. Emerging deep learning technology suggests a potential alternative to develop end-to-end lung nodule management system in a data-driven fashion. Although numerous studies have explored end-to-end approaches to predict malignancy scores [5,17,20] or categories [19,22,23] of lung nodules, while only a few studies [1,4] address the lung nodule follow-up problem. Nevertheless, these studies only provide black-box predictions without intuitive explanations. There is also study [11] on predicting tumor growth with a model-free appearance modeling approach using a probabilistic U-Net [14], however it could not provide any quantitative assessment on the risk of tumors.

In this study, we aim at a unified approach to predict growth of lung nodules, with both high-quality visual appearances and accurate quantitative results. The core of our approach is based on a WarpNet, predicting displacement field **u** (or motion [6]) on a future volume from a baseline volume. With the field **u**, we could obtain not only the predicted **visual appearance** of the future volume by warping the baseline, but also the feature segmentation mask from the baseline mask, which could be used for **quantitative assessment** of tumor growth. This approach is inspired from VoxelMorph [14], where the displacement field for registration is conditional on both the baseline and future volumes; instead, our predictive displacement field is conditional only on the baseline volume and could be dynamically estimated. Moreover, a TextureNet is designed to refine textural details of the outputs from WarpNet. We introduce techniques including Temporal Encoding Module and Warp Segmentation Loss to encourage time-aware and shape-aware representation learning. The whole network, named Nodule Follow-Up Prediction Network (*NoFoNet*), establishes a unified framework to produce both high-quality visual appearances and accurate quantitative assessment for lung nodule follow-up. Our in-house follow-up dataset from two medical centers validates the effectiveness of *NoFoNet*.

2   Materials and Methods

2.1   Task Formalization and Dataset

We aim at a unified framework to predict future volume of a lung nodule, given any time interval and a baseline volume. An in-house dataset is collected, containing 622 LDCT scans from 246 patients (114 males and 132 females) with a total of 315 long-standing pulmonary nodules. Each patient has at least two time points of thin layer LDCT (slice thickness ≤ 1.25 mm), with the time interval of 30–1351, 136 days (min-max, median). We select nodules at every two time points as a sample (for example if a nodule has 3 follow-up scans at time points $t_1 t_2 t_3$, we choose time points $t_1 \& t_2$, $t_1 \& t_3$, $t_2 \& t_3$ as 3 samples), resulting in 731 pairs. The age of the patients at first examination is 23–97, 62 years. The segmentation VOI of each selected nodule (diameter 3 mm to 30 mm) is delineated by an expert radiologist and checked by another.

We pre-process the data as follows [20]: CT scans are resampled isotropically into 1 mm × 1 mm × 1 mm. The voxel intensity is normalized to $[-1, 1)$ from the Hounsfield unit (HU), using the mapping function $I = \lfloor \frac{I_{HU} + 1024}{400 + 1024} \times 255 \rfloor / 128 - 1$. Each data sample is a cubic volume image with the size of $48 \times 48 \times 48$, which covers the size of all nodules in our study.

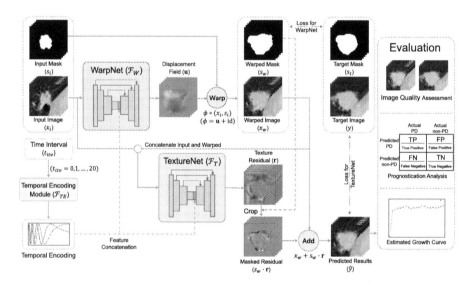

Fig. 1. Overview of the proposed *NoFoNet* architecture. *NoFoNet* consists of a Warp-Net and a TextureNet, each with a Temporal Encoding Module (TEM), which encodes follow-up time interval into the nodule representation. WarpNet and TextureNet are two 3D CNNs based on U-Net [14], modeling the spatial and texture transformation for nodule growth respectively. During evaluation, we perform the image quality assessment, prognostication analysis and growth estimation. Note PD means progressive disease, i.e., significant growth of nodule in our study. The red point in the growth curve represents actual volume of the nodule in the future time. (Color figure online)

2.2   *NoFoNet*: Nodule Follow-Up Prediction Network

To model the growth of nodules, we develop a Nodule Follow-Up Prediction Network (*NoFoNet*, see Fig. 1) consisting of a WarpNet $\mathcal{F}_W$ and a TextureNet

$\mathcal{F}_T$ for spatial and texture (intensity) transformations [21] respectively, where an integrated temporal encoding module (TEM) $\mathcal{F}_{TE}$ is addressed to encode different follow-up time interval information into the lesion representation. As we will show later, the WarpNet and TextureNet are able to model the shape and texture variation of nodule growth well.

Given a pair of follow-up input and target images[1] with time interval t_{itv}, each of which has corresponding nodule segmentation map $\{x_i, s_i\}$ and $\{y, s_t\}$, the WarpNet $\mathcal{F}_W$ with parameter θ_w first predicts a smooth voxel-wise displacement field $\mathbf{u} = \mathcal{F}_W(x_i, \mathcal{F}_{TE}(t_{itv}); \theta_w)$ for spatial transformation. Following the registration literature [2], we have the warp function $\phi = \mathbf{u} + \mathrm{id}$, where id is identity function. We apply the warp function ϕ to x_i to get the warped image x_w, and denote this as $x_w = \phi \circ x_i$. Similarly, the warped segmentation map $s_w = \phi \circ s_i$. The TextureNet $\mathcal{F}_T$ with parameter θ_t takes the concatenation of x_i and x_w as inputs and generates a voxel-wise residual $\mathbf{r} = \mathcal{F}_T(x_i, x_w, \mathcal{F}_{TE}(t_{itv}); \theta_t)$. Then we get the results $\hat{y} = x_w + s_w \cdot \mathbf{r}$, where $s_w \cdot \mathbf{r}$ denotes the residual cropped by warped segmentation. The overall formulation of our *NoFoNet* is as follows:

$$\begin{aligned}
\mathbf{u} &= \mathcal{F}_W(x_i, \mathcal{F}_{TE}(t_{itv}); \theta_w), \quad \phi = \mathbf{u} + \mathrm{id}, \quad x_w = \phi \circ x_i, \quad s_w = \phi \circ s_i; \\
\mathbf{r} &= \mathcal{F}_T(x_i, x_w, \mathcal{F}_{TE}(t_{itv}); \theta_t), \quad \hat{y} = x_w + s_w \cdot \mathbf{r}.
\end{aligned} \tag{1}$$

2.3 Temporal Encoding Module (TEM)

Since the time interval between two follow-up scans can be rather different, inspired by positional encoding [15] we develop a Temporal Encoding Module (TEM) to embed time interval information into the prediction model. Due to the limitation of dataset size, we discretize the interval using time mapping function $t_{itv} = \lceil t_{day}/30 \rceil$ with an upper cut-off value 20, for most of the intervals are less than 600 days. Sine and cosine functions with different frequencies are used in the TEM to generate values of different dimensions of the encoded temporal feature vector:

$$\begin{aligned}
\mathcal{F}_{TE}(t_{itv}, 2i) &= \sin(t_{itv}/100^{2i/d_{fm}}), \\
\mathcal{F}_{TE}(t_{itv}, 2i + 1) &= \cos(t_{itv}/100^{2i/d_{fm}}),
\end{aligned} \tag{2}$$

where t_{itv} is the discretized time interval, d_{fm} is the total number of channels of the encoded feature vector and i is the dimension. That is, the even/odd dimensions of the temporal encoding are generated by sin/cos function with different wavelengths (2π to $100 \times 2\pi$), which makes the relative time information encoded in a redundant way. Besides, the value range of the encoding result is within a certain numerical interval due to the boundedness of sinusoid. These two points ensure that the temporal encoding method can generate a more meaningful high-dimensional representation space. Then the feature vector is expanded repetitively and concatenated with the bottom feature map of WarpNet and TextureNet.

---

[1] If no otherwise specified, image mentioned here and later in this article refers to $48 \times 48 \times 48$ cubic volume image with a nodule in the center.

2.4   WarpNet for Spatial Transformation

As the core of our method, WarpNet predicts a displacement field $\mathbf{u}$ to model the shape variation of nodule growth, which is similar to the motion prediction in video tasks [6,8,18]. The architecture of WarpNet is based on a CNN similar to U-Net [14] with skip connections, and the temporal encoding from TEM is connected to the bottom of WarpNet.

The loss function for training WarpNet θ_w has four terms: similarity loss $\mathcal{L}_{sim}$ between warped images x_w and target images y, segmentation loss $\mathcal{L}_{seg}$ between warped segmentation maps s_i and target maps s_t, smoothness loss $\mathcal{L}_{smooth}$ for the deformation field and regularization loss $\mathcal{L}_{reg}$ for the output of WarpNet when $t_{itv} = 0$. In summary, the learning of WarpNet is formulated as:

$$\hat{\theta}_w = \underset{\theta_w}{\operatorname{argmin}} \{ \mathcal{L}_{sim}(x_w, y) + \lambda_1 \mathcal{L}_{seg}(s_w, s_t) + \lambda_2 \mathcal{L}_{smooth}(\mathbf{u}) + \lambda_3 \mathcal{L}_{reg}(\phi_0) \} \quad (3)$$

with weights $\lambda_1, \lambda_2, \lambda_3 > 0$, where ϕ_0 is the predicted spatial warp function when time interval $t_{itv} = 0$. All loss functions are designed as follows:

a) similarity loss and regularization loss: In our experiments we find that for spatial transformation normalized cross correlation (NCC) loss leads to more reasonable and robust results than MSE loss. The NCC loss between warped image x_w/target image y and the regularization loss for ϕ_0 is defined as:

$$\mathcal{L}_{sim}(x_w, y) = 1 - NCC(x_w, y) = 1 - NCC(\phi \circ x_i, y),$$
$$\mathcal{L}_{reg}(\phi_0) = \mathcal{L}_{sim}(\phi_0 \circ x_i, x_i) + \mathcal{L}_{sim}(\phi_0 \circ y, y). \quad (4)$$

b) segmentation loss: We use Dice loss to constrain the similarity between warped segmentation mask s_w and target mask s_t:

$$\mathcal{L}_{seg}(s_w, s_t) = 1 - \frac{2 \cdot \sum_{p \in \Omega} s_w(p) s_t(p)}{\sum_{p \in \Omega} s_w(p) + \sum_{p \in \Omega} s_t(p)}. \quad (5)$$

c) smoothness loss: Considering that the contour of nodule changes continuously as it grows, we use a diffusion regularization loss to encourage the smoothness of displacement field $\mathbf{u}$:

$$\mathcal{L}_{smooth}(\mathbf{u}) = \frac{1}{|\Omega|} \sum_{p \in \Omega} ||\nabla \mathbf{u}(p)||^2. \quad (6)$$

where finite differences between neighboring voxels are used to approximate the spatial gradients $\nabla \mathbf{u}(p)$ (for x, y, z 3 dimensions).

2.5   TextureNet for Texture Transformation

In addition to the shape variation, there is also a texture variation in nodule growth caused by the change of CT value distribution of nodules. So a TextureNet is needed to estimate the residual between warped image x_w and target

Table 1. Quantitative results of multiple models. We choose U-Net w/ or w/o TEM and WarpNet w/ or w/o segmentation loss as comparisons of *NoFoNet*. The performance is estimated by PSNR, PSNR* (PSNR in the nodule parts), dice coefficient between warped/target segmentation maps and sensitivity/specificity/G-mean for PD/non-PD classification. We evaluate the performance of our models on our in-house dataset (see Sect. 2.1) with 5-fold cross validation.

Method	PSNR	PSNR*	Dice	Sensitivity	Specificity	G-mean
Baseline (U-Net)	4.1213	29.7490	–	–	–	–
+TEM	6.0380	31.8821	–	–	–	–
WarpNet	18.0915	43.1140	0.6301	0.7656	**0.9083**	0.8339
+Warp Seg Loss	18.1952	43.2464	0.6474	0.8594	0.8805	0.8699
+TextureNet	**18.2089**	**43.4904**	**0.6474**	**0.8594**	0.8805	**0.8699**

image y. TextureNet follows the architecture of WarpNet. To train TextureNet we need an intensity similarity loss $\mathcal{L}'_{sim}$ between textured images $\hat{y}$ (see Eq. 1) and target images y, and a regularization loss $\mathcal{L}'_{reg}$ for the predicted residual $\mathbf{r_0}$ when $t_{itv} = 0$. So the texture transformation learning is formulated as:

$$\hat{\theta}_t = \underset{\theta_t}{\mathrm{argmin}}\{\mathcal{L}'_{sim}(\hat{y}, y) + \lambda'_1 \mathcal{L}'_{reg}(\mathbf{r_0})\} \tag{7}$$

with weight $\lambda'_1 > 0$. We choose MSE loss to encourage maximal intensity similarity. The loss functions of TextureNet are defined as:

$$\mathcal{L}'_{sim}(\hat{y}, y) = \frac{1}{|\Omega|} \sum_{p \in \Omega} [\hat{y}(p) - y(p)]^2,$$
$$\mathcal{L}'_{reg}(\mathbf{r_0}) = \frac{1}{|\Omega|} \sum_{p \in \Omega} |\mathbf{r_0}(p)|^2. \tag{8}$$

2.6   Implementation Details

NoFoNet can use any CNN architecture for WarpNet and TextureNet, and we use the network design of Appendix Fig. A.1 in this work. All of the experiments in this study are implemented on an NVIDA Titan X GPU and an Intel i7-6700 CPU. Our codes are based on Python 3.7.3 and PyTorch-1.2.0 [10]. We use $\lambda_1 = 0.5, \lambda_2 = 10$ and $\lambda_3 = \lambda'_1 = 1$ for the loss weights in Eq. 3 and Eq. 7. Online data augmentation methods, including rotation and flipping along a random axis, are applied on the input images. Each part of *NoFoNet* is trained using Adam optimizer [7] with an initial learning rate of 0.001 for 200 epochs. Specifically, we emphasize the similarity loss inside the segmentation map to put more attention on the nodule.

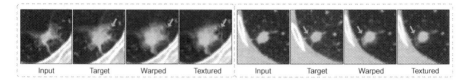

| Input | Target | Warped | Textured | Input | Target | Warped | Textured |

Fig. 2. Comparison of warped images and textured images for a PD case (left) and a non-PD case (right). Areas where intensity is changed significantly by TextureNet are indicated by red arrows. (Color figure online)

3   Experiments

3.1   Evaluation Protocol

Our *NoFoNet* is trained to predict what the nodule may be visually like after a certain time interval, then we can determine whether it is a PD (progressive disease, i.e., significant growth of nodule in size) case. Since some nodules in our dataset have multiple follow-ups, we stipulate that a nodule is judged as a PD case as long as one of its follow-up pairs (see Sect. 2.1) meets specific criterion, which is determined with the help of two senior radiologists.

Define V_1, V_2 (mm^3) as the two nodule volumes of a follow-up pair with time interval T (d), the criterion is as follows: (1) Considering that the fast-growing nodules have higher risks, we calculate the average volume growth rate AVGR $= (V_1 - V_2)/T$, and set a threshold of 1; (2) Some cases may have AVGR less than 1 but eventually grow significantly in size, we set a threshold of 200 for volume difference VD $= V_1 - V_2$ and a threshold of 50% for relative volume difference RVD $= (V_1 - V_2)/V_1$. In summary, a nodule is classified as PD case if one of its observed AVGR $\geq 1\,\mathrm{mm}^3$/d, or VD $\geq 200\,\mathrm{mm}^3$ and RVD $\geq 50\%$.

The 315 nodules are divided into two parts according to the aforementioned criterion, resulting in 64 positive cases (PD, significant growth) and 251 negative cases (non-PD, stable or shrinking). We split our dataset randomly into 5 groups based on patients (i.e., all nodules of one patient must be in same subset) and perform 5-fold cross validation to evaluate our models.

3.2   Performance Analysis

In this section we will present some quantitative results and qualitative results. Table 1 shows the performance of our models and baselines using 5-fold cross validation method. Note that U-Net w/ or w/o TEM predicts output images directly so it only has PSNR and PSNR* (PSNR in the nodule parts) for output/target images. As is shown in Appendix Fig. A.2, U-Net baselines generate predicted images with low visual quality. It is noticeable that when added segmentation loss for warped/target images, WarpNet predicts more accurate displacement fields, resulting in higher dice coefficient between warped/target images and better performance for PD/non-PD classification than WarpNet without segmentation loss. We use the geometrical mean (G-mean) of sensitivity (TP/TP+FN) and

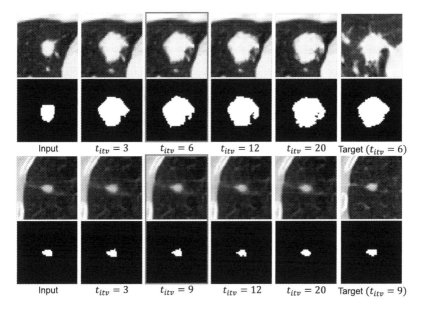

Fig. 3. Continuous prediction results of a PD case (top) and a non-PD case (bottom) by WarpNet. The first and last columns are the input image/segmentation and the target image/segmentation, and columns in the middle are the warped images/segmentations by WarpNets with different temporal encodings. Warped results that have the same time interval as the targets are highlighted in red. (Color figure online)

specificity (TN/TN+FP) as main evaluation index for the unbalanced dataset. The TextureNet in *NoFoNet* improves the visual quality of the warped images and achieves higher PSNR/PSNR* scores, as visually shown in Fig. 2.

Figure 2 shows the results of spatial transformation for input images by WarpNet and voxel-wise texture addition for warped images by TextureNet. We select a PD case and a non-PD case to demonstrate the performance of *NoFoNet* on different types of nodules. It can be seen that TextureNet is able to refine the warped images from WarpNet and increase the intensity similarity between the predicted and target nodules. Please refer to Appendix Fig. A.2 for more comparison results (including results from U-Net).

Figure 3 illustrates the continuous prediction results of two nodules using WarpNet. Note that results with the same follow-up time interval as the targets are highlighted in red. We choose a PD (progressive disease) case and a non-PD case for contrast to show that our WarpNet can represent both significant growth and stabilization of nodules in size well. For PD case it can also be seen that the model is able to generate reasonable nodules as time interval changes and the variation tendency is plausible, indicating the effectiveness of TEM.

4   Conclusion

We develop the *NoFoNet*, a unified network to predict the tumor growth for lung nodules. By explicitly learning spatial transformation and texture transformation, it yields high-quality visual appearances and accurate quantitative results, with validated effectiveness on an in-house dataset from two clinical centers. To the best of our knowledge, this is one of the first study to predict nodule growth quantitatively (size) and visually (appearance) given any time interval during.

A limitation of this study is that we only model the tumor growth as the indicator of nodule risk. However, according to TNM tumor staging system, tumor size (T), lymph node (N) and metastasis (M) are considered in tumor prognosis assessment. In future studies, we will address the N and M information to develop a more advanced risk stratification system for lung nodule follow-up. Besides, we will expand the dataset from cooperative hospitals and explore more effective architectures and temporal encoding methods for our framework.

Acknowledgment. This work was supported in part by National Natural Science Foundation of China (61671298), 111 project (BP0719010), Shanghai Science and Technology Committee (18DZ2270700) and Shanghai Jiao Tong University Science and Technology Innovation Special Fund (ZH2018ZDA17).

References

1. Ardila, D., et al.: End-to-end lung cancer screening with three-dimensional deep learning on low-dose chest computed tomography. Nat. Med. **25**, 954–961 (2019)
2. Balakrishnan, G., Zhao, A., Sabuncu, M.R., Guttag, J., Dalca, A.V.: VoxelMorph: a learning framework for deformable medical image registration. IEEE Trans. Med. Imaging **38**(8), 1788–1800 (2019)
3. Cressman, S., et al.: Resource utilization and costs during the initial years of lung cancer screening with computed tomography in Canada. J. Thorac. Oncol. **9**, 1449–1458 (2014)
4. Huang, P., et al.: Prediction of lung cancer risk at follow-up screening with low-dose CT: a training and validation study of a deep learning method. Lancet Digit. Health **1**, e353–e362 (2019)
5. Hussein, S., Cao, K., Song, Q., Bagci, U.: Risk stratification of lung nodules using 3D CNN-based multi-task learning. In: Niethammer, M., et al. (eds.) IPMI 2017. LNCS, vol. 10265, pp. 249–260. Springer, Cham (2017). https://doi.org/10.1007/978-3-319-59050-9_20
6. Jin, X., et al.: Predicting scene parsing and motion dynamics in the future. In: NIPS, pp. 6915–6924 (2017)
7. Kingma, D.P., Ba, J.: Adam: a method for stochastic optimization. arXiv preprint arXiv:1412.6980 (2014)
8. Luc, P., Neverova, N., Couprie, C., Verbeek, J., LeCun, Y.: Predicting deeper into the future of semantic segmentation. In: ICCV, pp. 648–657 (2017)
9. MacMahon, H., et al.: Guidelines for management of incidental pulmonary nodules detected on CT images: from the Fleischner Society 2017. Radiology **284**(1), 228–243 (2017)
10. Paszke, A., et al.: Automatic differentiation in PyTorch (2017)

11. Petersen, J., et al.: Deep probabilistic modeling of glioma growth. In: Shen, D., et al. (eds.) MICCAI 2019. LNCS, vol. 11765, pp. 806–814. Springer, Cham (2019). https://doi.org/10.1007/978-3-030-32245-8_89
12. Pinsky, P.F., et al.: Performance of lung-RADS in the national lung screening trial. Ann. Intern. Med. **162**, 485–491 (2015)
13. Pinsky, P.F., Gierada, D.S., Nath, P., Kazerooni, E.A., Amorosa, J.: National lung screening trial: variability in nodule detection rates in chest CT studies. Radiology **268**(3), 865–73 (2013)
14. Ronneberger, O., Fischer, P., Brox, T.: U-Net: convolutional networks for biomedical image segmentation. In: Navab, N., Hornegger, J., Wells, W.M., Frangi, A.F. (eds.) MICCAI 2015. LNCS, vol. 9351, pp. 234–241. Springer, Cham (2015). https://doi.org/10.1007/978-3-319-24574-4_28
15. Vaswani, A., et al.: Attention is all you need. In: NIPS, pp. 5998–6008 (2017)
16. Wood, D.E.: National Comprehensive Cancer Network (NCCN) clinical practice guidelines for lung cancer screening. Thorac. Surg. Clin. **25**(2), 185–97 (2015)
17. Xie, Y., Xia, Y., Zhang, J., Feng, D.D., Fulham, M., Cai, W.: Transferable multimodel ensemble for benign-malignant lung nodule classification on chest CT. In: Descoteaux, M., Maier-Hein, L., Franz, A., Jannin, P., Collins, D.L., Duchesne, S. (eds.) MICCAI 2017. LNCS, vol. 10435, pp. 656–664. Springer, Cham (2017). https://doi.org/10.1007/978-3-319-66179-7_75
18. Xu, J., Ni, B., Yang, X.: Video prediction via selective sampling. In: Advances in Neural Information Processing Systems, pp. 1705–1715 (2018)
19. Yang, J., Deng, H., Huang, X., Ni, B., Xu, Y.: Relational learning between multiple pulmonary nodules via deep set attention transformers. In: 2020 IEEE 17th International Symposium on Biomedical Imaging (ISBI), pp. 1875–1878. IEEE (2020)
20. Yang, J., Fang, R., Ni, B., Li, Y., Xu, Y., Li, L.: Probabilistic radiomics: ambiguous diagnosis with controllable shape analysis. In: Shen, D., et al. (eds.) MICCAI 2019. LNCS, vol. 11769, pp. 658–666. Springer, Cham (2019). https://doi.org/10.1007/978-3-030-32226-7_73
21. Zhao, A., Balakrishnan, G., Durand, F., Guttag, J.V., Dalca, A.V.: Data augmentation using learned transformations for one-shot medical image segmentation. In: CVPR, pp. 8543–8553 (2019)
22. Zhao, W., et al.: Toward automatic prediction of EGFR mutation status in pulmonary adenocarcinoma with 3D deep learning. Cancer Med. **8**(7), 3532–3543 (2019)
23. Zhao, W., et al.: 3D deep learning from CT scans predicts tumor invasiveness of subcentimeter pulmonary adenocarcinomas. Cancer Res. **78**(24), 6881–6889 (2018)

Multi-stream Progressive Up-Sampling Network for Dense CT Image Reconstruction

Qiuyue Liu[1,2], Zhen Zhou[7(✉)], Feng Liu[3], Xiangming Fang[4], Yizhou Yu[3,5(✉)], and Yizhou Wang[2,6,7]

[1] Center for Data Science, Peking University, Beijing, China
[2] Advanced Institute of Information Technology, Peking University, Hangzhou, China
[3] Deepwise AI Lab, Beijing, China
yizhouy@acm.org
[4] Department of Radiology, Wuxi People's Hospital, Nanjing Medical University, Wuxi, China
[5] The University of Hong Kong, Pokfulam, Hong Kong, China
[6] Center on Frontiers of Computing Studies, Peking University, Beijing, China
[7] Computer Science Department, Peking University, Beijing, China
z.zhou@pku.edu.cn

Abstract. Pulmonary computerized tomography (CT) images with small slice thickness (thin) is very helpful in clinical practice due to its high resolution for precise diagnosis. However, there are still a lot of CT images with large slice thickness (thick) because of the benefits of storage-saving and short taking time. Therefore, it is necessary to build a pipeline to leverage advantages from both thin and thick slices. In this paper, we try to generate thin slices from the thick ones, in order to obtain high quality images with a low storage requirement. Our method is implemented in an encoder-decoder manner with a proposed progressive up-sampling module to exploit enough information for reconstruction. To further lower the difficulty of the task, a multi-stream architecture is established to separately learn the inner- and outer-lung regions. During training, a contrast-aware loss and feature matching loss are designed to capture the appearance of lung markings and reduce the influence of noise. To verify the performance of the proposed method, a total of 880 pairs of CT images with both thin and thick slices are collected. Ablation study demonstrates the effectiveness of each component of our method and higher performance is obtained compared with previous work. Furthermore, three radiologists are required to detect pulmonary nodules in raw thick slices and the generated thin slices independently, the improvement in both *sensitivity* and *precision* shows the potential value of the proposed method in clinical applications.

Keywords: Deep learning · Image reconstruction · Pulmonary computed tomography

Q. Liu and Z. Zhou—Equal contribution.

© Springer Nature Switzerland AG 2020
A. L. Martel et al. (Eds.): MICCAI 2020, LNCS 12266, pp. 518–528, 2020.
https://doi.org/10.1007/978-3-030-59725-2_50

1   Introduction

Pulmonary CTs are an important tool in disease screening and diagnosis. Thus, high quality of CT images is required in clinical scenarios, which is controlled by various parameters, e.g., the manufacturer, radiation dose, reconstructed slice thickness, etc. Among them, slice thickness is one of the most crucial factors, which affects the spatial resolution of the CT images significantly and further influences the appearance of potential lesions. Therefore, recent studies and guidelines [12,17] suggest that reconstructed slice thickness used for screening [5] should be lower than or at least equal 1.5 mm to obtain enough clear observation of CT images.

However, CT images with large slice thickness (e.g., 5 mm) are still widely used in practice. The reason lies in three folds. Firstly, it is quick to obtain the thick slices after taking the CT scan, which is necessary in emergency scenarios. Secondly, it is storage-saving and fast-passing through internet. Thirdly, it takes low cost, which is especially important in wide health examinations. Nevertheless, the drawback of thick slices is obvious. The spatial resolution is not enough, resulting in the difficulty of observing small lesions and the details of the lesions (vessel segments and small nodules), thus leading to misdiagnosis.

In this paper, we aim to reconstruct thin slices from the thick ones, in the hope that the ambiguous lesions in thick slices could be diagnosed in generated thin slices with increased details. The proposed method is an encoder-decoder architecture, the backbone of which is a modified 3D U-Net [14] with the proposed progressive up-sampling module. Observing the huge difference of the HU distribution between regions inside the lung and regions outside the lung, a multi-stream approach is built, where the voxels of inner-lung and outer-lung regions are learned separately by different branches. Besides, a contrast-aware loss is exploited to pay attention to the generation of low contrast tissues inside the lung. Furthermore, a feature matching loss is designed to reduce the influence of noise, which is implemented by comparing the predicted thin slices and ground truth slices in the feature space. In the experiments, 880 pairs of CT images with both thin and thick slices are collected. We firstly compare our method with previous work to demonstrate its effectiveness and then verify the contribution of each component of our model. Finally, a clinical comparison is taken to show that the generated thin slices are helpful in real applications.

2   Related Work

CT slice reconstruction can be understood as an image super resolution (SR) problem, where the resolution of the depth dimension is increased. In natural image, Dong et al. [6] propose the first deep learning based SR model, and it is further improved in [7,10,15] by using fully convolutional neural networks [16], and generative adversarial networks (GAN) [9]. These advances have also been applied in medical images [2,3,18].

For the problem of thin slice reconstruction with varied input and output shapes, Bae et al. [1] propose a 2D based approach by performing SR on coronal

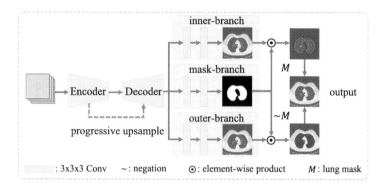

Fig. 1. The overall architecture of the proposed MPU-Net.

or sagittal planes and then compose them to 3D output. As 3D context cannot be modeled in 2D networks, 3D GAN [14] is utilized by firstly resizing the input to the shape of output. Instead of directly resizing the input, Ge et al. [8] adopt a subpixel convnet [21] based approach to encode the depth in channels such that a V-Net [19] like architecture can be used, and they additionally investigated stereo-correlation and coupled discriminators for GAN model training.

Our work is distinct from theirs in that we propose a progressive up-sampling network that can capture more information to reconstruct images and that inner- and outer-lung are learned separately in our method via the multi-stream architecture. Moreover, we deal with the recovery of the lung markings by the contrast-aware loss and the influence of noise by the feature matching loss, which are rarely studied in previous works.

3   Methodology

In this section, we introduce our multi-stream progressive up-sampling network (MPU-Net) for thin slices reconstruction. The overall architecture is shown in Fig. 1: Consecutive slices from thick images are firstly fed into the backbone network to extract multi-scale features, in which with gradually up-sampling, the depth of input data reaches the same as the target thin images. The upsampled feature maps are sent into different branches to generate the masks of lung regions, inner- and outer-lung voxels, and the final thin images are composed from the output of these three branches.

3.1   Progressive Up-Sampling Backbone

We consider the slice thickness and slice intervals of CT images to be equal, and the slice thickness of thick images is 5 mm, and that of thin images is 1 mm. So we set the number of input thick slices to 8 and output thin slices to 36. This is to ensure that the first and last slices in both inputs and outputs are aligned in world coordinates.

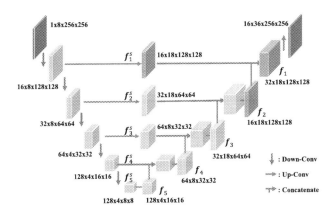

Fig. 2. The architecture of the progressive up-sampling backbone.

The architecture of our backbone is illustrated in Fig. 2. Each 3D patch of size $8 \times 256 \times 256$ from thick slice CT will be transformed to a size of $36 \times 256 \times 256$. The input patch is firstly down-sampled five times to learn high-level features and modeling 3D context. Unlike existing networks where the down-sampling rate is the same for all dimensions, we down-sample 32 times on transverse plane but only down-sample twice along depth, as we want to retrain more spatial information as long as the receptive field is large enough to cover the whole depth dimension. During up-sampling, we design progressive up-sampling operator g to aggregate features from short-cut connection $\boldsymbol{f}_i^s$ and previous level $\boldsymbol{f}_{i+1}$: the feature maps $\boldsymbol{f}_i^s$ are firstly up-sampled to match the shape of $\boldsymbol{f}_{i+1}$, and then they are concatenated and passed to another deconvolution layer for further up-sampling as in Eq. 1:

$$g(\boldsymbol{f}_i^s, \boldsymbol{f}_{i+1}) = Deconv(Deconv(\boldsymbol{f}_i^s) \oplus \boldsymbol{f}_{i+1}), \tag{1}$$

where $Deconv$ is deconvolution operation and $\oplus$ is concatenation. As more up-sampling is conducted along depth than down-sampling, the final depth is increased. Compared with Subpixel Convnet [21], our method aggregates the features of different levels gradually, and exploits the spatial relation between different dimensions with 3D convolution. Note that this design is easy to be adapted to other slice thickness/intervals setting by changing the down/up-sampling rate and the kernel size/stride of deconvolution layers.

3.2   Multi-stream Architecture

Due to having different organs and tissues, there is a large discrepancy between inner-lung and outer-lung HU distribution. Compared with the outer part, the lung area has smaller HU values and is more condensed with low contrast structures. Inspired by this statistic, we propose a multi-stream architecture to reconstruct inner- and outer-lung regions independently. It contains three branches corresponding to the prediction of lung mask M, inner-lung voxels X_i, and outer-lung voxels X_o. We feed the thick images to the progressive backbone obtaining the depth up-sampled features and use 3D convolution layers transforming the features into target output. The final output X is composed by combining the inner- and outer-lung predictions following the predicted lung mask: $M \odot X_i + (1 - M) \odot X_o$, where $\odot$ is element-wise product. To save the computation cost, we share the backbone network for all branches and a detailed paradigm is shown in Fig. 1.

3.3 Loss Function

We designed multiple loss functions for effective training: cross entropy loss for lung segmentation, contrast-aware loss for the inner-lung reconstruction, L1 loss for the outer-lung reconstruction, and feature matching loss to deal with noise.

Segmentation Loss. A cross entropy loss $\mathcal{L}_{Mask}$ is applied to the mask prediction branch for supervising the learning of high quality segmentation. It is used to compose the final thin slices and meanwhile introduce additional anatomical priors as an auxiliary training signal.

Contrast-Aware Loss. The contrast-aware loss $\mathcal{L}_{contrast}$ put more weights on hard samples (data points) aiming to enhance the contrast of low attenuation tissues. The hardness of a sample is defined as the reconstruction error compared with its true value as shown in Eq. 2:

$$\mathcal{L}_{contrast} = \frac{1}{MND} \sum_{i,j,k} \left| \hat{X}_{i,j,k} - X_{i,j,k} \right| \cdot w_{i,j,k},$$

$$w_{i,j,k} = \left(\frac{|\hat{X}_{i,j,k} - X_{i,j,k}|}{X_{i,j,k} + \alpha} \right)^\gamma \cdot \beta \tag{2}$$

where $\hat{X}$ and X are samples from reconstructed and target thin slices; $M \times N \times D$ is the shape of the target output; α, γ, β are hyper-parameters that control the hardness weight for each sample, α and γ separately control the highlighted HU range and the sparseness of weights, β is a weight balancing different losses, a reasonable setting of these hyper-parameters can be easily obtained by visualizing the volume of weights. In our study, we set α, β, γ to 0.5, 5, 0.5. These weights can further be processed by a Gaussian filter to smooth and expand the scope of attention. This formulation enforces samples of smaller HU value to weight more than those of high value given same amount of error. It is desirable as small vessels or nodules which are hard to learn are usually of low attenuation and contain only few data points.

Feature Matching Loss. Considering the higher level noises within thin slice CTs, we additionally compare the reconstructed and real thin images in the feature space of a CNN model to minimize the influence of noise. As the lack of pretrained CNN models on large scale CT dataset, we resort to the CNN model pretrained on natural images, e.g., ImageNet [20]. To make CT image more similar to natural image, we extract a single thin slice $\boldsymbol{X}_{..k}$ each time and replicate the channels to a 3-channel gray scale image. We presume with millions of data and the harsh data augmentation, the ImageNet pretrained model is capable of representing CT slices and being insensitive to noise. The feature matching loss [13] is formulated as:

$$\mathcal{L}_{feat} = \frac{1}{C_i M_i N_i D} \sum_k \|\phi_i(\hat{\boldsymbol{X}}_{..k}) - \phi_i(\boldsymbol{X}_{..k})\|_2^2, \tag{3}$$

where $\phi_i(\boldsymbol{X})$ is the mapping from X to the i-th layer of a pretrained VGG-16 network; $\hat{\boldsymbol{X}}_{..k}$, $\boldsymbol{X}_{..k}$ are the k-th thin slice in the output and targets respectively; $C_i \times M_i \times N_i$ is the size of the i-th feature map, and D is the depth of the thin slices.

4 Experiment

In this section, we firstly present the comparison between our model and other previous works, and then show the ablation results. Finally, we conduct a lung nodule screening experiment to investigate the clinical usage of this technique and show some reconstructed pathological cases for reasonable generalization ability.

4.1 Datasets and Settings

All data used in this paper are first collected by using Siemens machines from 880 patients, and then further reconstructed to the thin slices (1 mm) and thick slices (5 mm) with LUNG reconstruction (I50f). This study has obtained the local committee approval, and all data has been anonymized to ensure the privacy of patients before study. We randomly sample 80% of the data for training and validation, and the rest 20% for testing, resulting in 600 training, 104 validation, and 176 testing CT pairs. We rescale the input images of HU range [-1024, 3071] to [0, 1] as input to the model. During training, we randomly crop 8 × 256 × 256 voxels from the thick images as input data and select corresponding 36 × 256 × 256 voxels from the thin images as ground-truth. During inference we feed 8 consecutive thick slices to the model in a sliding window manner. The stride of sliding window is set to 1, and we average the thin slices if there are multiple reconstructions at the same world coordinate. For evaluation, we use mean absolute error (MAE), root mean square error (RMSE) and peak signal-to-noise ratio (PSNR) to measure direct pixel errors in HU; meanwhile, structural similarity (SSIM) is also used to serve as approximation of human subjective evaluation for image quality [22].

4.2 Validation of MPU-Net

Results. We compare our method with the following baselines. *Bicubic* is to directly upsample the thick slices to the shape of thin slices using bicubic interpolation. *SRCNN* is to train a 2D super resolution network on coronal or sagittal

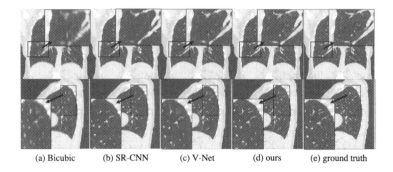

(a) Bicubic   (b) SR-CNN   (c) V-Net   (d) ours   (e) ground truth

Fig. 3. Visualization of the reconstructed thin slice CTs of different models.

Table 1. Results of model comparison and ablation study, where prog-up, feature, contrast, stream is abbreviation of progressive up-sampling, feature matching loss, contrast-aware loss, multi-stream architecture respectively.

Models	prog-up	feature	contrast	stream	MAE(↓)	RMSE(↓)	PSNR(↑)	SSIM(↑)
Bicubic	–	–	–	–	33.31	69.63	35.41	0.917
SRCNN [1]	–	–	–	–	30.82	61.98	36.88	0.920
VNet [19]	–	–	–	–	29.55	54.54	37.53	0.932
Ours					25.34	46.25	38.96	0.948
	✓				24.37	42.28	39.75	0.950
	✓	✓			24.99	41.95	39.81	0.948
	✓		✓		23.54	40.24	40.17	0.953
	✓	✓	✓		23.60	40.14	40.19	0.952
	✓	✓	✓	✓	**23.37**	**38.79**	**40.50**	**0.955**
Ours+W-GAN		✓	✓		24.72	42.31	39.60	0.950

planes, and we follow the same setting as [1]. For *VNet*, we use a similar architecture as [14], where the input is firstly upsampled using trilinear interpolation and then a VNet-like [19] model is used as generator. The results of all models is shown in Table 1. Our model performs significantly better than others on all metrics. Meanwhile, a visualization from coronal and sagittal views of the reconstructed thin slice CT is shown in Fig. 3, where our model indeed recovers richer details and looks more realistic. It indicates the necessity of properly modeling 3D context and setting effective training objectives.

Ablation Study. To verify the contribution of each components of network design, we compare our model with several stripped-down variants. Our basis is a subpixel convnet as used in [8], then we compared it with the proposed progressive up-sampling backbone, multi-stream architecture, and different loss combinations. The results are shown in Table 1, and we can make the following observations: 1) reconstruction model can get sustainable improvement by using progressive up-sampling compared with post up-sampling like subpixel convnet; 2) using feature matching loss could greatly improve the quality of reconstructed thin slices as more subtle anatomical structures can be recovered (Fig. 4), even

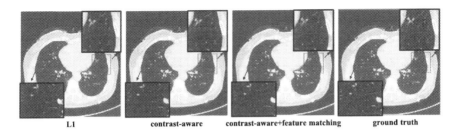

L1   contrast-aware   contrast-aware+feature matching   ground truth

Fig. 4. The effectiveness of each proposed loss in our method.

though the numerical results does not change much; 3) the proposed contrast-aware loss together with the multi-stream architecture gives the largest improvement, indicating the effectiveness of modeling inner- and outer-lung separately and applying appropriate loss functions. We have also tried using adversarial loss as in [8,14], such as W-GAN [11], but we did not observe any quantitative or qualitative improvements. This could due to the noise distribution in real thin slice CTs from which the discriminator is easy to distinguish real and fake thin slices. However, such noise is hard and pointless to learn for a generative model.

4.3 Applications in Lung Nodule Detection

To further investigate the practical value of reconstructed thin slices in clinical scenarios, we conduct an experiment on CT images of 50 patients in terms of lung nodule detection. This experiment tries to verify that the generated thin slices can help radiologists find more nodules compared with the corresponding thick slices. Following *Lung-Rads* [4], three radiologists are required to independently annotate nodules larger or equal 4 mm on the real thin slices. Then a senior radiologist makes the final decision providing all annotations of the three radiologists, the result of which is referred as *golden standard*. A total of 146 pulmonary nodules are obtained, 45 of which are larger or equal than 6 mm. Another three radiologists are then asked to independently label pulmonary nodules on the corresponding thick slices. The labelling of the remaining thin slices generated from thick scans are conducted 30 days later. The detection performance of the three radiologists in both thick slices and the generated thin slices can be calculated by comparing with the golden standard. Figure 5 shows the result of the radiologists on two nodule size settings. The results clearly demonstrate that both the sensitivity and precision of all three radiologists on generated thin slices are boosted, indicating the usefulness of our method in clinical practice.

4.4 Reconstruction Examples of Pathological Lung

Figure 6 shows the reconstructed thin images of pathological lung compared with the real ones. As a prospective study, our original dataset does not cover all disease types, but our model still shows reasonable generalization ability and good reconstruction performance.

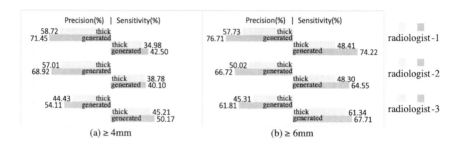

Fig. 5. Results of radiologists on lung nodule detection. Each color represents a radiologist, and darker ones are based on generated thin slices.

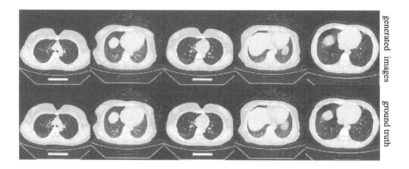

Fig. 6. Reconstruction examples of pathological lung.

5   Conclusions

In this paper, we propose a novel multi-stream progressive up-sampling network for dense CT reconstruction, with the incorporation of a feature matching loss and contrast-aware loss to learn stable and detailed structures. Experimental results demonstrate that our proposed method can achieve a superior performance in both quantitative level and image perceptual level. Clinical experiment further implies its potential in practice for radiologists.

Acknowledgment. This work was supported in part by National Key Research and Development Program of China (MOST-2018AAA0102004, 2019YFC0118100), Beijing Municipal Science and Technology Planning Project (Grant No. Z201100005620008), National Natural Science Foundation of China (NSFC-61625201), and Hong Kong Research Grants Council through Research Impact Fund under Grant R-5001-18.

References

1. Bae, W., Lee, S., Park, G., Park, H., Jung, K.H.: Residual CNN-based image super-resolution for CT slice thickness reduction using paired CT scans: preliminary validation study (2018)

2. Chen, Y., Shi, F., Christodoulou, A.G., Xie, Y., Zhou, Z., Li, D.: Efficient and accurate MRI super-resolution using a generative adversarial network and 3D multi-level densely connected network. In: Frangi, A.F., Schnabel, J.A., Davatzikos, C., Alberola-López, C., Fichtinger, G. (eds.) MICCAI 2018. LNCS, vol. 11070, pp. 91–99. Springer, Cham (2018). https://doi.org/10.1007/978-3-030-00928-1_11

3. Chen, Y., Xie, Y., Zhou, Z., Shi, F., Christodoulou, A.G., Li, D.: Brain MRI super resolution using 3D deep densely connected neural networks. In: 2018 IEEE 15th International Symposium on Biomedical Imaging (ISBI 2018), pp. 739–742. IEEE (2018)

4. Clark, T.J., Flood, T.F., Maximin, S.T., Sachs, P.B.: Lung CT screening reporting and data system speed and accuracy are increased with the use of a semiautomated computer application. J. Am. Coll. Radiol. **12**(12), 1301–1306 (2015)

5. Cristiano, R., Daniela, O., Massimo, B.: Low-dose CT: technique, reading methods and image interpretation. Cancer Imaging Official Publ. Int. Cancer Imaging Soc. **12**(3), 548–556 (2012)

6. Dong, C., Loy, C.C., He, K., Tang, X.: Learning a deep convolutional network for image super-resolution. In: Fleet, D., Pajdla, T., Schiele, B., Tuytelaars, T. (eds.) ECCV 2014. LNCS, vol. 8692, pp. 184–199. Springer, Cham (2014). https://doi.org/10.1007/978-3-319-10593-2_13

7. Dong, C., Loy, C.C., Tang, X.: Accelerating the super-resolution convolutional neural network. In: Leibe, B., Matas, J., Sebe, N., Welling, M. (eds.) ECCV 2016. LNCS, vol. 9906, pp. 391–407. Springer, Cham (2016). https://doi.org/10.1007/978-3-319-46475-6_25

8. Ge, R., Yang, G., Xu, C., Chen, Y., Luo, L., Li, S.: Stereo-correlation and noise-distribution aware ResVoxGAN for dense slices reconstruction and noise reduction in thick low-dose CT. In: Shen, D., et al. (eds.) MICCAI 2019. LNCS, vol. 11769, pp. 328–338. Springer, Cham (2019). https://doi.org/10.1007/978-3-030-32226-7_37

9. Goodfellow, I., et al.: Generative adversarial nets. In: Advances in Neural Information Processing Systems, pp. 2672–2680 (2014)

10. Gu, J., Lu, H., Zuo, W., Dong, C.: Blind super-resolution with iterative kernel correction. In: Proceedings of the IEEE Conference on Computer Vision and Pattern Recognition, pp. 1604–1613 (2019)

11. Gulrajani, I., Ahmed, F., Arjovsky, M., Dumoulin, V., Courville, A.C.: Improved training of Wasserstein GANs. In: Advances in Neural Information Processing Systems, pp. 5767–5777 (2017)

12. Iwano, S., Makino, N., Ikeda, M., Itoh, S., Ishigaki, T.: Solitary pulmonary nodules: optimal slice thickness of high-resolution CT in differentiating malignant from benign. Clin. Imaging **28**(5), 322–328 (2004)

13. Johnson, J., Alahi, A., Fei-Fei, L.: Perceptual losses for real-time style transfer and super-resolution. In: Leibe, B., Matas, J., Sebe, N., Welling, M. (eds.) ECCV 2016. LNCS, vol. 9906, pp. 694–711. Springer, Cham (2016). https://doi.org/10.1007/978-3-319-46475-6_43

14. Kudo, A., Kitamura, Y., Li, Y., Iizuka, S., Simo-Serra, E.: Virtual thin slice: 3D conditional GAN-based super-resolution for CT slice interval. In: Knoll, F., Maier, A., Rueckert, D., Ye, J.C. (eds.) MLMIR 2019. LNCS, vol. 11905, pp. 91–100. Springer, Cham (2019). https://doi.org/10.1007/978-3-030-33843-5_9

15. Ledig, C., et al.: Photo-realistic single image super-resolution using a generative adversarial network. In: Proceedings of the IEEE Conference on Computer Vision and Pattern Recognition, pp. 4681–4690 (2017)

16. Long, J., Shelhamer, E., Darrell, T.: Fully convolutional networks for semantic segmentation. In: Proceedings of the IEEE Conference on Computer Vision and Pattern Recognition, pp. 3431–3440 (2015)
17. Macmahon, H., et al.: Guidelines for management of incidental pulmonary nodules detected on CT images: from the fleischner society 2017. Radiology **284**, 228–243 (2017). 161659
18. Mahapatra, D., Bozorgtabar, B., Garnavi, R.: Image super-resolution using progressive generative adversarial networks for medical image analysis. Comput. Med. Imaging Graph. **71**, 30–39 (2019)
19. Milletari, F., Navab, N., Ahmadi, S.A.: V-Net: fully convolutional neural networks for volumetric medical image segmentation. In: 2016 Fourth International Conference on 3D Vision (3DV), pp. 565–571. IEEE (2016)
20. Russakovsky, O., et al.: ImageNet large scale visual recognition challenge. Int. J. Comput. Vis. **115**(3), 211–252 (2015)
21. Shi, W., et al.: Real-time single image and video super-resolution using an efficient sub-pixel convolutional neural network. In: Proceedings of the IEEE Conference on Computer Vision and Pattern Recognition, pp. 1874–1883 (2016)
22. Silpa, K., Mastani, S.A.: Comparison of image quality metrics. Int. J. Eng. Res. Technol. (IJERT) **1**(4), 5 (2012)

Abnormality Detection in Chest X-Ray Images Using Uncertainty Prediction Autoencoders

Yifan Mao[1], Fei-Fei Xue[1], Ruixuan Wang[1,2(✉)], Jianguo Zhang[3(✉)],
Wei-Shi Zheng[1,2,4], and Hongmei Liu[1,5]

[1] School of Data and Computer Science, Sun Yat-sen University, Guangzhou, China
wangruix5@mail.sysu.edu.cn
[2] Key Laboratory of Machine Intelligence and Advanced Computing, MOE,
Guangzhou, China
[3] Department of Computer Science and Engineering,
Southern University of Science and Technology, Shenzhen, China
zhangjg@sustech.edu.cn
[4] Pazhou Lab, Guangzhou, China
[5] Guangdong Province Key Laboratory of Information Security Technology,
Guangzhou, China

Abstract. Chest radiography is widely used in annual medical screening to check whether lungs are healthy or not. Therefore it would be desirable to develop an intelligent system to help clinicians automatically detect potential abnormalities in chest X-ray images. Here with only healthy X-ray images, we propose a new abnormality detection approach based on an autoencoder which outputs not only the reconstructed normal version of the input image but also a pixel-wise uncertainty prediction. Higher uncertainty often appears at normal region boundaries with relatively larger reconstruction errors, but not at potential abnormal regions in the lung area. Therefore the normalized reconstruction error by the uncertainty provides a natural measurement for abnormality detection in images. Experiments on two chest X-ray datasets show the state-of-the-art performance by the proposed approach.

Keywords: Abnormality detection · Uncertainty prediction · Anomaly detection · Chest X-ray

1 Introduction

Chest X-ray has been widely adopted for annual medical screening, where the main purpose is to check whether the lung is healthy or not. Considering the huge amount of regular medical tests worldwide, it would be desirable if there exists an intelligent system helping clinicians automatically detect potential abnormality in chest X-ray images. Here we consider such a specific task of abnormality detection, for which there is only *normal* (i.e., healthy) data available during model

© Springer Nature Switzerland AG 2020
A. L. Martel et al. (Eds.): MICCAI 2020, LNCS 12266, pp. 529–538, 2020.
https://doi.org/10.1007/978-3-030-59725-2_51

training. Compared to diagnosis with supervised learning, the key challenge of the task is the lack of abnormal data for training an abnormality detector.

For medical image analysis, the approaches thus far proposed for abnormality detection include parametric and non-parametric statistical models, one-class SVM, and deep learning models like generative adversarial networks (GANs). Parametric models usually refer to Gaussian and Gaussian mixture models, which estimate the density distribution of normal data from training set to predict the abnormality of a test sample [17]. Parametric models often assume that the normal data distribution is a Gaussian or a mixture of Gaussian distributions. In comparison, *non*-parametric statistical models, such as Gaussian process, are more capable of modelling complex distributions but have more computational loads [20]. Both parametric and non-parametric models are bottom-up generative approaches. In contrast, one-class SVM is a top-down classification-based method for abnormality detection, which constructs a hyperplane as a decision boundary that best separates normal data and the origin point, and meanwhile maximises the distance between the origin and the hyperplane [15]. It has been applied to abnormal detection based on fMRI and retinal OCT images [11,16]. While the above conventional approaches have been widely used in the medical domain, there is one serious drawback to restrict their performance, i.e., the feature representation of images need to be manually designed in advance. Without the need to extract hand-crafted features, generative adversarial networks (GANs) [8] and autoencoders are recently becoming popular for medical abnormality detection due to their capability of implicitly modelling more complex data distribution than the conventional approaches. The early GAN-based approach for anomaly detection, called AnoGAN, was proposed for abnormality detection in retinal OCT images [14]. The basic idea is to train a generator in the AnoGAN which can generate only normal image patches, such that any abnormal patch would not be well reconstructed by the generator. A fast version of the AnoGAN called f-AnoGAN [13] was recently proposed with an additional encoder included to make the generator become an autoencoder. More autoencoder models which are often combined with GANs have also been recently developed for abnormality detection in medical image analysis [1–4,18] and natural image analysis [7,12]. One issue in most GAN and autoencoder models is about the relative large reconstruction errors particularly at region boundaries although the regions are normal, which would cause false detection of abnormality in normal images.

This paper for the first time applies an autoencoder model to not only reconstruct the corresponding normal version of any input image, but also estimate the uncertainty of reconstruction at each pixel [5,6] to enhance the performance of anomaly detection. Higher uncertainty often appears at normal region boundaries with relatively larger reconstruction errors, but not at potential abnormal regions in the lung area. As a result, the normalized reconstruction error by the uncertainty can then be used to better detect potential abnormality. Our approach obtains state-of-the-art performance on two chest X-ray datasets.

2   Method

The problem of interest is to automatically determine whether any new chest X-ray image is abnormal ('unhealthy') or not, only based on a collection of normal ('healthy') images. Since abnormality in X-ray images could be due to small area of lesions or unexpected change in subtle contrast between local regions, extracting an image-level feature representation may suppress such small-scale features, while extracting features for each local image patch may fail to detect the contrast-based abnormalities, both resulting in the failing of abnormality detection. In comparison, reconstruction error based on pixel-level differences between the original image and its reconstructed version by an autoencoder model may be a more appropriate measure to detection abnormality in X-ray images, because both local and global features have been implicitly considered to reconstruct each pixel by the autoencoder. However, it has been observed that there often exists relatively large reconstruction errors around the boundaries between different regions (e.g., lung vs. the others, foreground vs. background, Fig. 2) even in normal images. Such large errors could result in false positive detection, i.e., considering a normal image as abnormal. Therefore, it would be desirable to automatically suppress the contribution of such reconstruction errors in anomaly detection. Simply detecting edges and removing their contributions in reconstruction error may not work well due to the difficulty in detecting low-contrast boundaries in X-ray images and due to possibly larger reconstruction errors close to region boundaries. In this paper, we applied a probabilistic approach to automatically downgrade the contribution of normal regions with larger reconstruction errors. The basic idea is to train an autoencoder to simultaneously reconstruct the input image and estimate the pixel-wise uncertainty in reconstruction (Fig. 1), where larger uncertainties often appear at normal regions with larger reconstruction errors. On the other hand, there are often relatively large reconstruction errors with small reconstruction uncertainties at abnormal regions in the lung area. All together, normal images would be more easily separated from abnormal images based on the uncertainty-weighted reconstruction errors.

Fig. 1. Autoencoder with both reconstruction $\mu(\mathbf{x})$ and predicted pixel-wise uncertainty $\sigma^2(\mathbf{x})$ as outputs.

2.1   Autoencoder with Pixel-Wise Uncertainty Prediction

In order to reconstruct the input image and estimate pixel-wise uncertainty for the reconstruction, the autoencoder needs to somehow automatically learn to find where the reconstruction is more uncertain without ground-truth uncertainty available. As in the related work for estimation of uncertainty [5,6,9,10], here we formulate the reconstruction uncertainty prediction problem by a probabilistic model, with the special (unusual) property that each variance element in the model is not fixed but varies depending on input data. Formally, given a collection of N normal images $\{\mathbf{x}_i, i = 1, \ldots, N\}$, where $\mathbf{x}_i \in \mathbb{R}^D$ is the vectorized representation of the corresponding i-th original image, an autoencoder can be trained to make each reconstructed image $\boldsymbol{\mu}(\mathbf{x}_i)$ as similar to the corresponding input image $\mathbf{x}_i$ as possible. In general, there are always more or less pixel-wise differences between the autoencoder's expected output $\mathbf{y}_i$ (i.e., same as the input $\mathbf{x}_i$) and the real output $\boldsymbol{\mu}(\mathbf{x}_i)$. Suppose such differences are noise sampled from an *input-dependent* (note traditionally noise is assumed input-independent) multivariate Gaussian distribution $\mathcal{N}(\mathbf{0}, \boldsymbol{\Sigma}(\mathbf{x}_i))$, i.e., $\mathbf{y}_i = \boldsymbol{\mu}(\mathbf{x}_i) + \boldsymbol{\epsilon}(\mathbf{x}_i)$, where $\boldsymbol{\epsilon}(\mathbf{x}_i) \sim \mathcal{N}(\mathbf{0}, \boldsymbol{\Sigma}(\mathbf{x}_i))$. Then the conditional probability density of the ideal output $\mathbf{y}_i$ (same as the input $\mathbf{x}_i$) given the input to the autoencoder is

$$p(\mathbf{y}_i|\mathbf{x}_i, \boldsymbol{\theta}) = \frac{1}{(2\pi)^{\frac{D}{2}}|\boldsymbol{\Sigma}(\mathbf{x}_i)|^{\frac{1}{2}}} \exp\left\{-\frac{1}{2}(\mathbf{y}_i - \boldsymbol{\mu}(\mathbf{x}_i))^\mathsf{T}\boldsymbol{\Sigma}^{-1}(\mathbf{x}_i)(\mathbf{y}_i - \boldsymbol{\mu}(\mathbf{x}_i))\right\}, \tag{1}$$

where $\boldsymbol{\theta}$ denotes the parameters of the model which can output both the reconstructed image $\boldsymbol{\mu}(\mathbf{x}_i)$ and the covariance matrix $\boldsymbol{\Sigma}(\mathbf{x}_i)$. By simplfying $\boldsymbol{\Sigma}(\mathbf{x}_i)$ to a diagonal matrix $\boldsymbol{\Sigma}(\mathbf{x}_i) = \mathrm{diag}(\sigma_1^2(\mathbf{x}_i), \sigma_2^2(\mathbf{x}_i), ..., \sigma_D^2(\mathbf{x}_i))$, the negative logarithm of Eq. (1) gives

$$-\log p(\mathbf{y}_i|\mathbf{x}_i, \boldsymbol{\theta}) = \frac{1}{D}\sum_{k=1}^{D}\left\{\frac{(x_{i,k} - \mu_k(\mathbf{x}_i))^2}{\sigma_k^2(\mathbf{x}_i)} + \log\sigma_k^2(\mathbf{x}_i)\right\} + \frac{D}{2}\log(2\pi), \tag{2}$$

where $x_{i,k}$ is the k-th element of the expected output $\mathbf{y}_i$ (i.e., the input $\mathbf{x}_i$), and $\mu_k(\mathbf{x}_i)$ is the k-th element of the real output $\boldsymbol{\mu}(\mathbf{x}_i)$. Then the autoencoder can be optimized by maximizing the log-likelihood over all the normal (training) images, i.e., by minimizing the negative log-likelihood function $\mathcal{L}(\boldsymbol{\theta})$,

$$\mathcal{L}(\boldsymbol{\theta}) = \frac{1}{ND}\sum_{i=1}^{N}\sum_{k=1}^{D}\left\{\frac{(x_{i,k} - \mu_k(\mathbf{x}_i))^2}{\sigma_k^2(\mathbf{x}_i)} + \log\sigma_k^2(\mathbf{x}_i)\right\}. \tag{3}$$

Equation (3) would be simplified to the mean squared error (MSE) loss based on either Mahalanobis distance or Euclidean distance, when the variance elements $\sigma_k^2(\mathbf{x}_i)$'s are fixed and not dependent on the input $\mathbf{x}_i$ or when they are not only fixed but also equivalent.

Note that for each input image $\mathbf{x}_i$, the model generates two outputs, the reconstruction $\boldsymbol{\mu}(\mathbf{x}_i)$ and the noise variance $\boldsymbol{\sigma}^2(\mathbf{x}_i) = (\sigma_1^2(\mathbf{x}_i), \sigma_2^2(\mathbf{x}_i), ..., \sigma_D^2(\mathbf{x}_i))^\mathsf{T}$ (Fig. 1). Interestingly, while $\boldsymbol{\mu}(\mathbf{x}_i)$ is supervised to approach to $\mathbf{x}_i$,

$\boldsymbol{\sigma}(\mathbf{x}_i)$ is totally unsupervised during model training, only based on minimization of the objective function $\mathcal{L}(\boldsymbol{\theta})$. From the definition of the noise variance (above Eq. (1)), each element $\sigma_k^2(\mathbf{x}_i)$ of the noise variance represents *not the reconstruction error* but the degree of uncertainty for the i-th element of the reconstruction $\boldsymbol{\mu}(\mathbf{x}_i)$. This uncertainty is used to naturally normalize the reconstruction error for the i-th element of the reconstruction (first loss term in Eq. (3)). During model training, the first loss term discourages the autoencoder from predicting very small uncertainty values for those pixels with higher reconstruction errors, because smaller $\sigma_k^2(\mathbf{x}_i)$ will enlarge the contribution of the already large reconstruction errors by the first loss term. Therefore, the autoencoder will automatically learn to generate relatively larger uncertainties for those pixels (e.g., around region boundaries) with relatively larger reconstruction errors in normal images. On the other hand, the second loss term $\log \sigma_k^2(\mathbf{x}_i)$ in Eq. (3) will prevent the autoencoder from predicting larger uncertainty for all reconstructed pixels. Therefore, the two loss terms together will help train an autoencoder such that the predicted uncertainty will be smaller at those regions where the model can reconstruct well and relatively larger otherwise in normal images.

It is worth noting that the positive correlation between the uncertainty prediction and the reconstruction error may hold mainly for normal image pixels or regions. For anomaly in the lung area which has not been seen during model training, the uncertainty prediction is often small (see Sect. 3.2), probably because the model has learned to reconstruct well (with smaller uncertainty) inside the lung area during model training and therefore often predicts low uncertainty for lung area for any new image, no matter whether there exists anomaly in the area or not. On the other hand, the reconstruction errors at abnormal regions in the lung area are often relatively large because the well-trained autoencoder learns to just reconstruct normal lung by removing any potential noise or abnormal signals in this area. As a result, anomaly with larger reconstruction errors and small uncertainty would become distinctive from normal regions which have positive correlation between reconstruction errors and predicted uncertainties.

2.2   Abnormality Detection

Based on the above analysis, for any new image $\mathbf{x}$, it is natural to use the *pixel-wise* normalized reconstruction error (as first term in Eq. (3)) to represents the degree of abnormality for each pixel x_k, and the average of such errors over all pixels for the abnormality $\mathcal{A}(\mathbf{x})$ of the image, i.e.,

$$\mathcal{A}(\mathbf{x}) = \frac{1}{D} \sum_{k=1}^{D} \frac{(x_k - \mu_k(\mathbf{x}))^2}{\sigma_k^2(\mathbf{x})}. \qquad (4)$$

Since the pixel-wise uncertainties $\sigma_k^2(\mathbf{x})$ depend on the input $\mathbf{x}$, it is not as easily estimated as for fixed variance. As far as we know, it is the first time to apply such pixel-wise input-dependent uncertainty to estimate of abnormality. If the image $\mathbf{x}$ is normal, pixels or regions with larger reconstruction errors are often accompanied with larger uncertainties, therefore often resulting in the overall smaller abnormality score $\mathcal{A}(\mathbf{x})$. In contrast, if there is certain anomaly in the image, the relatively larger reconstruction errors still with small uncertainties at the abnormal region would lead to a relatively larger abnormality score $\mathcal{A}(\mathbf{x})$.

3 Experiments

3.1 Experimental Setup

Datasets. Our method is tested on two publicly available chest X-ray datasets: 1) RSNA Pneumonia Detection Challenge dataset[1] and 2) pediatric chest X-ray dataset[2]. The RSNA dataset is a subset of ChestXray14 [19]; it contains 26,684 X-rays with 8,851 normal, 11,821 no lung opacity/not normal and 6,012 lung opacity. The pediatric dataset consists of 5,856 X-rays from normal children and patients with pneumonia.

Protocol. For the RSNA dataset, we used 6,851 normal images for training, 1,000 normal and 1,000 abnormal images for testing. On this dataset, our method was tested on three different settings: 1) *normal* vs. *lung opacity*; 2) *normal* vs. *not normal* and 3) *normal* vs. *all* (lung opacity and not normal). For the pediatric dataset, 1,249 normal images were used for training, and the original author-provided test set was used to evaluate the performance. The test set contains 234 normal images and 390 abnormal images. All images were resized to 64 × 64 pixels and pixel values of each image were normalized to [-1,1]. The area under the ROC curve (AUC) is used to evaluate the performance, together with equal error rate (EER), F1-score (at EER) reported.

Implementation. The backbone of our method is a convolutional autoencoder. The network is symmetric containing an encoder and a decoder. The encoder contains four layers (each with one 4 × 4 convolution with a stride 2), which is then followed by two fully connected layers whose output sizes are 2048 and 16 respectively. The decoder is connected by two fully connected layers and four transposed convolutions, which constitute the encoder. The channel sizes are 16-32-64-64 for encoder and 64-64-32-16 for decoder. All convolutions and transposed convolutions are followed by batch normalization and ReLU nonlinearity except for the last output layer. We trained our model for 250 epochs. The optimization was done using the Adam optimizer with a learning rate 0.0005. For numerical stability we did not directly predict σ^2 in Eq. (3). Instead, the uncertainty output by the model is the log variance (i.e., $\log \sigma^2$).

3.2 Evaluations

Baselines. Our method is compared with three baselines as well as state-of-the-art methods for anomaly detection. Below summarizes the methods compared.

---

[1] https://www.kaggle.com/c/rsna-pneumonia-detection-challenge.
[2] https://doi.org/10.17632/rscbjbr9sj.3.

Table 1. Comparison with others with different metrics. Bold face indicates the best, and italic face for the second best.

Method	RSNA Setting-1			RSNA Setting-2			RSNA Setting-3			Pediatric		
	EER↓	F1↑	AUC↑	EER↓	F1↑	AUC↑	EER↓	F1↑	AUC↑	EER↓	F1↑	AUC↑
AE	0.36	0.64	0.68	0.40	0.60	0.63	0.38	0.62	0.65	0.41	0.65	0.64
OC-SVM-1	0.41	0.59	0.63	0.45	0.54	0.57	0.42	0.58	0.60	0.38	0.67	0.67
OC-SVM-2	0.31	0.69	0.74	0.40	0.60	0.64	0.46	0.64	0.69	0.39	0.66	0.68
f-AnoGAN	*0.21*	*0.79*	*0.84*	*0.31*	*0.68*	*0.73*	*0.27*	*0.73*	*0.79*	*0.33*	*0.72*	*0.71*
Ours	**0.18**	**0.81**	**0.89**	**0.28**	**0.72**	**0.78**	**0.22**	**0.77**	**0.83**	**0.29**	**0.75**	**0.78**

– *Autoencoder (AE).* A vanilla autoencoder is the most relevant baseline. For a fair comparison, the backbone of the vanilla AE is designed exactly the same as ours. We use the L_2 reconstruction error as anomaly score for this method.
– *OC-SVM.* The one-class support vector machine (OC-SVM)[15] is a traditional model for one-class learning. For OC-SVM, we use the feature representations (i.e., the output of the encoder) learned from a vanilla AE and ours as the input to SVM respectively, resulting in two versions OC-SVM-1 and OC-SVM-2.
– *f-AnoGAN.* It is a state-of-the-art anomaly detection method in medical imaging [13]. During inference in this model, we fed an image into the encoder-generator to acquire an reconstructed image. A hybrid score combining pixel-level and feature reconstruction error is used to measure abnormality.

Comparison and analysis. The abnormality detection performance with different methods was summarized in Table 1. The state-of-the-art method f-AnoGAN clearly outperforms the other baselines, but performs worse than ours. OC-SVM-2 (with our encoder) is consistently better than OC-SVM-1, suggesting that the encoder in our approach may have mapped normal data into a more compact region in the latent feature space, which can be easily learned by one-class SVM. The superior performance of our method is probably due to the suppression of larger reconstruction error at normal region boundaries by the predicted pixel-wise uncertainties. As Fig. 2 (columns 3, 5, 7) demonstrated, while the reconstruction errors are relatively large at some normal region boundaries for all methods, only our method can estimate the pixel-wise uncertainty (column 8), by which the pixel-wise normalized reconstruction errors at normal region boundaries has been largely reduced (column 9). On the other hand, larger reconstruction errors in abnormal regions in the lung area often do not correspond to larger uncertainties.

As a result, the uncertainty normalized abnormality score can help separate abnormal images from normal ones, as confirmed in Fig. 3 (right). In comparison, the two histograms are largely overlapped when using the vanilla reconstruction error (Fig. 3, left). In addition, it is worth noting that, as in other autoencoder and GAN based image reconstruction methods, our method can also provide the pixel-level localization of potential abnormalities (Fig. 2, last column), which could be helpful for clinicians to check and analyze the abnormality details in practice.

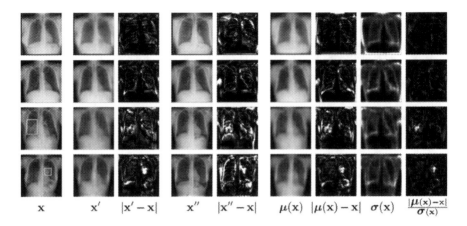

$$\mathbf{x} \qquad \mathbf{x}' \qquad |\mathbf{x}' - \mathbf{x}| \qquad \mathbf{x}'' \qquad |\mathbf{x}'' - \mathbf{x}| \qquad \boldsymbol{\mu}(\mathbf{x}) \quad |\boldsymbol{\mu}(\mathbf{x}) - \mathbf{x}| \quad \boldsymbol{\sigma}(\mathbf{x}) \quad \frac{|\boldsymbol{\mu}(\mathbf{x}) - \mathbf{x}|}{\boldsymbol{\sigma}(\mathbf{x})}$$

Fig. 2. Exemplar reconstructions of normal (rows 1–2) and abnormal (rows 3–4) test images. $\mathbf{x}$ is input; $\mathbf{x}'$, $\mathbf{x}''$, and $\boldsymbol{\mu}(\mathbf{x})$ are reconstructions from AE, f-AnoGAN, and our method; operators are pixel-wise. Green bounding boxes for abnormal regions.

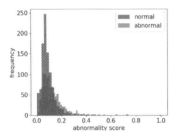 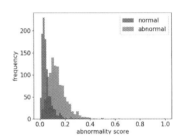

Fig. 3. Histograms of abnormality score for normal (blue) and abnormal (red) images in the test set (RSNA Setting-1). Left: without uncertainty normalization. Right: with uncertainty normalization. Scores are normalized to $[0, 1]$ in each subfigure. (Color figure online)

Ablation Study. Table 2 shows that only incorporating uncertainty loss with autoencoder (i.e., without uncertainty normalization) doesn't improve the performance (Table 2, 'without-**U**', AUC $= 0.68$ which is similar to that of vanilla AE). In contrast, uncertainty normalized abnormality score ('with-**U**') largely improves the performance. Interestingly, adding skip connections downgraded

Table 2. Ablation study on RSNA Setting-1. '**U**' denotes uncertainty output. '0'–'4': number of skip connections between encoder and decoder convolutional layers, with '1' for the connection between encoder's last and decoder's first convolutional layers.

Skip connections	0	1	2	3	4
with-**U**	**0.89**	0.62	0.50	0.44	0.38
without-**U**	**0.68**	0.43	0.38	0.33	0.33

performance. This is probably because skip connections prevents the encoder learning the true low-dimensional distribution of normal data.

4 Conclusion

We proposed an uncertainty normalized abnormality detection method which is capable of reconstructing the image with the pixel-wise prediction uncertainty. Experiments on two chest X-ray datasets shows that the uncertainty can well suppress the adversarial effect of larger reconstruction errors around normal region boundaries, and consequently state-of-the-art performance was obtained.

Acknowledgement. This work is supported in part by the National Key Research and Development Program (grant No. 2018YFC1315402), the Guangdong Key Research and Development Program (grant No. 2019B020228001), the National Natural Science Foundation of China (grant No. U1811461), the Guangzhou Science and Technology Program (grant No. 201904010260) and the National Key R&D Program of China (grant No. 2017YFB0802500).

References

1. Alaverdyan, Z., Jung, J., Bouet, R., Lartizien, C.: Regularized siamese neural network for unsupervised outlier detection on brain multiparametric magnetic resonance imaging: application to epilepsy lesion screening. Med. Image Anal. **60**, 101618 (2020)
2. Baur, C., Wiestler, B., Albarqouni, S., Navab, N.: Fusing unsupervised and supervised deep learning for white matter lesion segmentation. In: International Conference on Medical Imaging with Deep Learning, pp. 63–72 (2019)
3. Chen, X., Konukoglu, E.: Unsupervised detection of lesions in brain MRI using constrained adversarial auto-encoders. In: International Conference on Medical Imaging with Deep Learning (2018)
4. Chen, X., Pawlowski, N., Glocker, B., Konukoglu, E.: Unsupervised lesion detection with locally Gaussian approximation. In: Suk, H.-I., Liu, M., Yan, P., Lian, C. (eds.) MLMI 2019. LNCS, vol. 11861, pp. 355–363. Springer, Cham (2019). https://doi.org/10.1007/978-3-030-32692-0_41
5. Diederik, P.K., Welling, M., et al.: Auto-encoding variational bayes. In: Proceedings of the International Conference on Learning Representations, vol. 1 (2014)
6. Dorta, G., Vicente, S., Agapito, L., Campbell, N.D., Simpson, I.: Structured uncertainty prediction networks. In: Proceedings of the IEEE Conference on Computer Vision and Pattern Recognition, pp. 5477–5485 (2018)
7. Gong, D., et al.: Memorizing normality to detect anomaly: memory-augmented deep autoencoder for unsupervised anomaly detection. In: Proceedings of the IEEE International Conference on Computer Vision, pp. 1705–1714 (2019)
8. Goodfellow, I., et al.: Generative adversarial nets. In: Advances in Neural Information Processing Systems, pp. 2672–2680 (2014)
9. He, Y., Zhu, C., Wang, J., Savvides, M., Zhang, X.: Bounding box regression with uncertainty for accurate object detection. In: Proceedings of the IEEE Conference on Computer Vision and Pattern Recognition, pp. 2888–2897 (2019)

10. Kendall, A., Gal, Y.: What uncertainties do we need in Bayesian deep learning for computer vision? In: Advances in Neural Information Processing Systems, pp. 5574–5584 (2017)
11. Mourão-Miranda, J., et al.: Patient classification as an outlier detection problem: an application of the one-class support vector machine. NeuroImage **58**(3), 793–804 (2011)
12. Sabokrou, M., Khalooei, M., Fathy, M., Adeli, E.: Adversarially learned one-class classifier for novelty detection. In: Proceedings of the IEEE Conference on Computer Vision and Pattern Recognition, pp. 3379–3388 (2018)
13. Schlegl, T., Seeböck, P., Waldstein, S.M., Langs, G., Schmidt-Erfurth, U.: f-AnoGAN: fast unsupervised anomaly detection with generative adversarial networks. Med. Image Anal. **54**, 30–44 (2019)
14. Schlegl, T., Seeböck, P., Waldstein, S.M., Schmidt-Erfurth, U., Langs, G.: Unsupervised anomaly detection with generative adversarial networks to guide marker discovery. In: Niethammer, M., et al. (eds.) IPMI 2017. LNCS, vol. 10265, pp. 146–157. Springer, Cham (2017). https://doi.org/10.1007/978-3-319-59050-9_12
15. Schölkopf, B., Williamson, R.C., Smola, A.J., Shawe-Taylor, J., Platt, J.C.: Support vector method for novelty detection. In: Advances in Neural Information Processing Systems, pp. 582–588 (2000)
16. Seeböck, P., et al.: Unsupervised identification of disease marker candidates in retinal OCT imaging data. IEEE Trans. Med. Imaging **38**(4), 1037–1047 (2018)
17. Sidibe, D., et al.: An anomaly detection approach for the identification of DME patients using spectral domain optical coherence tomography images. Comput. Methods Programs Biomed. **139**, 109–117 (2017)
18. Tang, Y.X., Tang, Y.B., Han, M., Xiao, J., Summers, R.M.: Abnormal chest x-ray identification with generative adversarial one-class classifier. In: IEEE International Symposium on Biomedical Imaging, pp. 1358–1361 (2019)
19. Wang, X., Peng, Y., Lu, L., Lu, Z., Bagheri, M., Summers, R.M.: ChestX-ray8: hospital-scale chest x-ray database and benchmarks on weakly-supervised classification and localization of common thorax diseases. In: Proceedings of the IEEE Conference on Computer Vision and Pattern Recognition, pp. 2097–2106 (2017)
20. Ziegler, G., Ridgway, G.R., Dahnke, R., Gaser, C., Initiative, A.D.N., et al.: Individualized Gaussian process-based prediction and detection of local and global gray matter abnormalities in elderly subjects. NeuroImage **97**, 333–348 (2014)

Region Proposals for Saliency Map Refinement for Weakly-Supervised Disease Localisation and Classification

Renato Hermoza[1(✉)], Gabriel Maicas[1], Jacinto C. Nascimento[2], and Gustavo Carneiro[1]

[1] Australian Institute for Machine Learning, The University of Adelaide, Adelaide, Australia
renato.hermozaaragones@adelaide.edu.au
[2] Institute for Systems and Robotics, Instituto Superior Tecnico, Lisbon, Portugal

Abstract. The deployment of automated systems to diagnose diseases from medical images is challenged by the requirement to localise the diagnosed diseases to justify or explain the classification decision. This requirement is hard to fulfil because most of the training sets available to develop these systems only contain global annotations, making the localisation of diseases a weakly supervised approach. The main methods designed for weakly supervised disease classification and localisation rely on saliency or attention maps that are not specifically trained for localisation, or on region proposals that can not be refined to produce accurate detections. In this paper, we introduce a new model that combines region proposal and saliency detection to overcome both limitations for weakly supervised disease classification and localisation. Using the ChestX-ray14 data set, we show that our proposed model establishes the new state-of-the-art for weakly-supervised disease diagnosis and localisation. We make our code available at https://github.com/renato145/ RpSalWeaklyDet.

Keywords: Weakly supervised learning · Object localisation · Gumbel softmax · Region proposal · Saliency maps · Attention maps · ChestXray14

1 Introduction

An important way to explain a disease classification made by a medical image computing (MIC) system relies on showing the image region(s) associated with

Supported by Australian Research Council through grant DP180103232.

Electronic supplementary material The online version of this chapter (https:// doi.org/10.1007/978-3-030-59725-2_52) contains supplementary material, which is available to authorized users.

A. L. Martel et al. (Eds.): MICCAI 2020, LNCS 12266, pp. 539–549, 2020.
https://doi.org/10.1007/978-3-030-59725-2_52

the classification. Even though object detection is a classic MIC problem, it usually relies on the availability of fully annotated data sets that contain not only the disease classification, but also the localisation of image regions associated with the classification [15]. Unfortunately, such data sets tend to be expensive to acquire and small, which is challenging for the training of classification and detection models, particularly the ones based on deep learning. Such issues motivated the community to consider data sets that are larger but weakly supervised [27]. Since the deployability of disease diagnosing systems partly depends on the localisation of image regions associated with the image classification, the medical image analysis community is increasingly developing systems that can classify and localise diseases from weakly annotated training sets [5,24,27,30].

Currently, weakly-supervised disease detection and classification methods merge the classification model with saliency maps [27,30], region proposal [13, 28], or attention maps [5,16,25]. Methods based on saliency maps [27,30] represent the most common approach in the field. Saliency map methods produce classification results based on a pooling operation from the model's last layer that contains an activation pattern that is likely to highlight image regions that are active for the classification result. However, there is no penalisation when the saliency map highlights image regions that are not associated with the classification of a particular disease. Region proposal methods explicitly encourage regions to be associated with correct classification labels and penalise regions associated with incorrect classification labels [13]. However, they do not have a way to differentiabily extract particular crops (or regions), so they aggregate the information from all extracted crops [1,13]. Consequently, they are unable to refine the initially proposed regions as done by supervised methods [8,22]. Attention maps extend saliency maps by enforcing the classification of a highlighted image region [5,16,25] and penalising regions associated with incorrect classification [19]. However, comparison with these methods is difficult given that they use quantitative evaluation measures that cannot be used for a fair comparison with other approaches [25] and they also use unpublished data set splits [5,16] for ChestX-ray 14 data set [27].

In this paper, we introduce a novel model that jointly produces disease diagnosis and localisation of relevant image regions associated with the diagnosis, where the localisation process relies on combining the results from the region proposal and saliency map detectors. Such combination is enabled by the use of the Gumbel softmax function [11] to differentiablily sample discrete regions from a set of region proposals, which allows us to refine the proposed regions and potentially increase the weakly-supervised detection precision. We test our new approach on the ChestX-ray14 data set [27] using the published train-test split and widely used quantitative evaluation measures. Results show that we establish a new state-of-the-art for both classification, with 0.82 average area under the receiver operating characteristic curve (AUC), weakly supervised localisation results with 0.29 average intersection over union (IoU) and 0.37 average continuous Dice (cDice).

2   Related Works

Weakly supervised disease classification and localisation is gaining increasing attention by the MIC community. This is partly due to the availability of relatively large chest X-ray data sets designed to be used in the development of models that can address weakly supervised disease classification and localisation [10,27]. Chest X-ray imaging is one of the most widely available modalities for screening and diagnosis. However, automatic disease classification and localisation from chest X-ray images is recognised as being technically challenging [3]. This is primarily due to (i) the large diversity in the appearance, size and location of the visual patterns of different types of thoracic diseases, and (ii) the relatively scarcity of high quality disease annotations. The initial models proposed for this problem were based on standard deep learning models [21,27] that produced relatively accurate classification, but poor detection results. More recent approaches improve classification accuracy by handling label noise [10], incorporating information from the associated radiology reports into the training process [2,25], or including strong annotations to add extra supervision [6,13,16].

Research on diagnosing and localising diseases from chest x-rays (especially based on ChestX-ray14) [27] has been hindered by the following two factors: 1) the data set split proposed by [27] for the evaluation is not often used, and 2) the localisation results are reported using different evaluation measures. These issues make a fair comparison between different approaches challenging. Regarding data set splits, some approaches [4,6,17,25–27,30] use the published split [27], while others [5,13,16,21,29] use a random split, which is not appropriate because it leads to unfair comparisons, as mentioned above – for example, it is possible that images from the same patient can be present in both training and testing sets in these random splits. This is discussed by Wang et al. [26] by showing that results on the suggested split [27] can be worse than in random splits by more than 10% average AUC. Regarding localisation measures, most methods [13,16,27] use IoU (but Li et al. [13] uses a few bounding box annotations for training and Liu et al. [16] relies on an unpublished train/test split and also uses a few bounding box annotations for training), while [30] uses cDice. While IoU is a standard measure for object detection, it is sensitive to the threshold applied to binarise the saliency map [30]. This issue is alleviated by the cDice measure that does not binarise the detection map – instead it is based on the continuous values of the saliency map. In this paper, we use the experimental setup proposed by Wang et al. [27] to ensure that our results are fairly compared with previous methods and can be used as baseline for future approaches.

3   Method

The proposed weakly supervised disease classification and detection consists of a joint classification and detection approaches, where the detection combines the results from saliency map and region proposal, as shown in Fig. 1. We first explain the training and testing sets used, followed by an explanation of the model inference and training approaches.

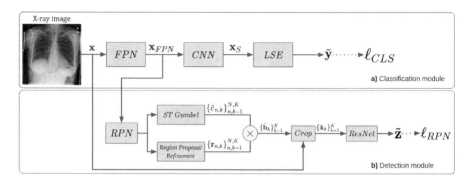

Fig. 1. The architecture of the proposed model consists of two modules: a) a classification module that produces saliency map $\mathbf{x}_S$ (via class activation maps) and classification $\tilde{\mathbf{y}}$ (trained with loss ℓ_{CLS} in Eq. 1); and b) a detection module that produces a set of regions $\{\mathbf{r}_{n,k}\}_{n,k=1}^{N,K}$ with class confidences $\{c_{n,k}\}_{n,k=1}^{N,K}$ and region proposal classification $\tilde{\mathbf{z}}$ (trained with loss ℓ_{RPN} in Eq. 2).

3.1   Data Set

The training set is defined by $\mathcal{D} = \{(\mathbf{x}, \mathbf{y})_i\}_{i=1}^{|\mathcal{D}|}$ and the testing set is formed by $\mathcal{T} = \{(\mathbf{x}, \mathbf{y}, \{\mathbf{b}_k\}_{k=1}^{K})_i\}_{i=1}^{|\mathcal{T}|}$, where $\mathbf{x} : \Omega \to \mathbb{R}$ denotes an X-ray image, with Ω being the image lattice, $\mathbf{y} \in \{0, 1\}^K$ indicates the presence or absence of K pathologies, and $\mathbf{b}_k \in \mathbb{R}^4$ indicates the bounding box (center coordinates, width and height) localising each of the K pathologies (note that if the kth pathology is not present in a test image or if it was not annotated, the kth element of the set $\{\mathbf{b}_k\}_{k=1}^{K}$ contains a token indicating that the annotation is not available).

3.2   Weakly Supervised Disease Classification and Detection

The system integrates a classification and a detection modules – see Fig. 1. The classification module follows a fully convolutional model [20], consisting of a feature pyramid network (FPN) [14] that extracts $\mathbf{x}_{FPN} \in \mathbb{R}^{Q \times (H_x/4) \times (W_x/4)}$, which contains Q feature maps of size $(H_x/4) \times (W_x/4)$ (with $H_x \times W_x$ representing the height and width of image $\mathbf{x}$), followed by a convolutional neural network (CNN) that produces K saliency maps of size $(H_x/4) \times (W_x/4)$ (one map for each of the K classes), denoted by $\mathbf{x}_S \in \mathbb{R}^{K \times (H_x/4) \times (W_x/4)}$, and the final classification logit $\tilde{\mathbf{y}} \in \mathbb{R}^K$ is produced by pooling the results from each of the saliency maps using the log-sum-exp (LSE) function.

The detection module takes $\mathbf{x}_{FPN}$ and uses *RoiAlign* [8] to extract the features of N pre-defined region proposals for each of the K classes, where each region proposal is defined by a 4-dimensional bounding box vector. The features from these $N \times K$ region proposals are used by two regressors: one to predict the class confidence for each region proposal $\{c_{n,k}\}_{n,k=1}^{N,K}$, and another to predict the refined region proposal bounding box vector $\{\mathbf{r}_{n,k}\}_{n,k=1}^{N,K}$, with $\mathbf{r}_{n,k} \in \mathbb{R}^4$.

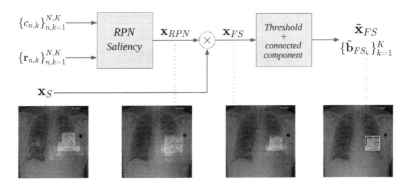

Fig. 2. Inference procedure to obtain the final bounding boxes from saliency maps. Sources of saliency maps from left to right: 1) classification module $\mathbf{x}_S$; 2) ROI detection scores $\mathbf{x}_{RPN}$; 3) saliency map combination $\mathbf{x}_{FS} = \mathbf{x}_S \odot \mathbf{x}_{RPN}$ 4) binarised saliency $\tilde{\mathbf{x}}_{FS}$ with bounding box $\tilde{\mathbf{b}}_{FS_k}$ (predicted bounding box is shown in yellow and ground truth in green). (Color figure online)

Instead of aggregating all region proposals that can produce inaccurate detection results because of the large value of N, we use the differentiable operator Straight-Through (ST) Gumbel-Softmax estimator [11] to sample a single region proposal per class based on the confidence value – effectively, the result of this operator forms a binarised $c_{n,k}$, denoted by $\tilde{c}_{n,k} \in \{0,1\}$, where $\sum_{n=1}^{N} \tilde{c}_{n,k} = 1$ for each class k. By selecting the region proposals $\mathbf{r}_{n,k}$ for which $\tilde{c}_{n,k} = 1$ we build the bounding box set $\{\tilde{\mathbf{b}}_k\}_{k=1}^{K}$ (with $\tilde{\mathbf{b}}_k \in \mathbb{R}^4$), which is used to crop the input image $\mathbf{x}$ to produce K feature maps $\{\tilde{\mathbf{x}}_k\}_{k=1}^{K}$ (where $\tilde{\mathbf{x}}_k \in \mathbb{R}^{3 \times H_f \times W_f}$, with $H_f \times W_f$ denoting the height and width of the crop operation) that are used by a ResNet [7] to produce a region proposal classification denoted by the logit $\tilde{\mathbf{z}} \in \mathbb{R}^K$.

The training procedure minimises the binary cross entropy loss for each class k for the classification and the detection modules. Due to the high number of cases with no pathologies, we adopt a balancing strategy for the labels using positive and negatives weight factors $\beta_P, \beta_N \in \mathbb{R}^K$. The loss for the classification module for each sample i is:

$$\ell_{CLS}(\tilde{\mathbf{y}}_i, \mathbf{y}_i) = -\beta_P \mathbf{y}_i \log(\tilde{\mathbf{y}}_i) - \beta_N(1 - \mathbf{y}_i) \log(1 - \tilde{\mathbf{y}}_i) \tag{1}$$

where $\beta_P(k) = 1 - \frac{P_k}{|\mathcal{D}|}$, $\beta_N(k) = \frac{P_k}{|\mathcal{D}|}$, where P_k is the total number of positive cases for class k and $|\mathcal{D}|$ is the training set size. In a similar way, we define the following loss for the detection module:

$$\ell_{RPN}(\tilde{\mathbf{z}}_i, \mathbf{y}_i) = -\beta_P \mathbf{y}_i \log(\tilde{\mathbf{z}}_i) - \beta_N(1 - \mathbf{y}_i) \log(1 - \tilde{\mathbf{z}}_i). \tag{2}$$

The training of the detection module is sensitive because region proposals are unstable at the beginning of the optimisation. To address this issue, we divide the training process into 3 stages: 1) training of the classification module with

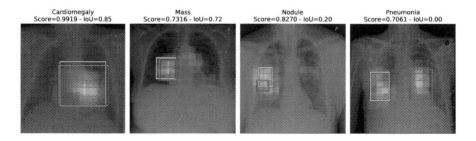

| Cardiomegaly | Mass | Nodule | Pneumonia |
| Score=0.9919 - IoU=0.85 | Score=0.7316 - IoU=0.72 | Score=0.8270 - IoU=0.20 | Score=0.7061 - IoU=0.00 |

Fig. 3. Examples of classification and detection results. Predicted bounding boxes are shown in yellow and ground truth in green. (Color figure online)

ℓ_{CLS} in (1), 2) training of the detection module with ℓ_{RPN} in (2), and 3) joint training of the classification and detection modules with loss $\ell_{CLS} + \ell_{RPN}$.

The inference procedure combines the saliency map and region proposal detection results, by: 1) computing the average of all refined region proposals weighted by their confidence, denoted by $\mathbf{x}_{RPN} \in \mathbb{R}^{K \times (H_x/4) \times (W_x/4)}$; 2) followed by an element-wise multiplication with the saliency map from the classification module, defined by $\mathbf{x}_{FS} = \mathbf{x}_S \odot \mathbf{x}_{RPN}$; 3) thresholding followed by the selection of the largest connected component which produces the final binary segmentation map $\tilde{\mathbf{x}}_{FS} \in \{0,1\}^{K \times (H_x/4) \times (W_x/4)}$; and 4) obtaining the parameters of a bounding box $\{\tilde{\mathbf{b}}_{FS_k}\}_{k=1}^{K}$ (with $\tilde{\mathbf{b}}_{FS_k} \in \mathbb{R}^4$) as the smallest rectangle able to cover the pixels $\omega \in \Omega$ where $\tilde{\mathbf{x}}_{FS}(\omega) = 1$. The inference procedure is shown in Fig. 2.

4   Experiments

4.1   Data Set

We conduct the experiments in this paper on the data set ChestX-ray14 [27], which contains 112,120 frontal-view chest X-ray images that are weakly labelled for 14 pathologies (this is a multi-label problem, where each image can have between 0 and 14 annotated pathologies) and bounding box annotations for 880 images relative to 8 different diseases. We use the published training and test split provided with the data set [27], and we use part of the training set as validation for model selection (i.e., hyper-parameter estimation). Note that the bounding box annotations are only present in some of the testing set images, i.e. they are not used for training.

4.2   Experimental Set up

We train the model in three stages (as described in Sect. 3.2), where in stages 1 and 2 we use a learning rate of 0.001 and for stage 3 a learning rate of 0.0003. We use Adam [12] with a momentum of 0.9, weight decay of 0.001 and a mini-batch size of 8. Images are down-sampled to 512×512 and normalised using

ImageNet [23] mean and standard deviation. While training, we apply random data augmentation operations such as: zoom between 0 and 0.1, translation in the four directions between -50 and 50 pixels, rotation between -10 and 10 degrees and random horizontal flipping. We use ImageNet [23] pre-trained Densenet-121 [9] and ResNet-34 [7] models to initialise CNN and $ResNet$ respectively (see Fig. 1), where we replaced all ReLU activations with Leaky ReLU [18], using a 0.1 negative slope. The initial region proposals in RPN uses regions of size 64, 128 and 256, all with ratio 1 and strides of 8, 16 and 32, to produce a total of $N = 395$ region proposals. The ST Gumbel-Softmax estimator temperature τ is exponentially annealed from 1 to 0.001 on stages 2 of 3 of the training procedure. The thresholds used to obtain the predicted bounding boxes are empirically selected for each class using the ground truth annotations[1].

To evaluate our model, we use the AUC for each pathology and the average over pathologies. For detection accuracy, we evaluate in terms of average IoU and cDice. We compared our method against baselines methods using saliency maps [27,30], region proposals [13] and attention maps [4,17]. Note that to allow a fair comparison, we only include methods that used the published train-test split [27] and the widely used detection measures IoU and cDice.

4.3 Results

Table 1 compares the AUC classification results between our approach (labelled as 'Ours') and several baselines [4,13,17,27]. We show an ablation study for the detection results of our method in Table 2 using different saliency maps to select the final bounding box: the classification module saliency map $\mathbf{x}_S$ (denoted by Sal), the region proposal map $\mathbf{x}_{RPN}$ (denoted by Det), and the combined saliency and region proposal maps $\tilde{\mathbf{x}}_{FS}$ (denoted by Mix). As activation maps tends to highlight more regions, we can observe that saliency map Sal performs well on Cardiomegaly (the largest pathology on the dataset) while saliency map Det performs better on the rest of labels. On Table 3 we show T(IoU), which measures the proportion of test images with IoU $\geq \kappa$, with $\kappa \in \{0.3, 0.5, 0.6\}$. Figure 3 shows visual examples of classification and detection results produced by our approach.

5 Discussion and Conclusion

Experimental results in Tables 1 and 3 show that our proposed method establishes the new state-of-the-art performance results for the problem of disease classification and weakly supervised localisation in the ChestX-ray14 data set. Compared to previous methods, our classification model benefits from using a feature pyramid network [14] pipeline that considers features maps at several scales to account for the size variation of the pathologies. In terms of localisation results, our method shows superior results compared with previous approaches

---

[1] This practice follows the protocol of other methods [5,27] in the field.

Table 1. Comparison on classification results of state-of-the-art methods on ChestX-ray14.

Label	Wang *et al.* [27]	Li *et al.* [13]	CRAL [4]	Ma *et al.* [17]	Ours
Atelectasis	0.700	0.729	**0.781**	0.777	0.775
Cardiomegaly	0.810	0.846	0.883	**0.894**	0.881
Effusion	0.759	0.781	**0.831**	0.829	**0.831**
Infiltration	0.661	0.673	**0.697**	0.696	0.695
Mass	0.693	0.743	0.830	**0.838**	0.826
Nodule	0.669	0.758	0.764	0.771	**0.789**
Pneumonia	0.658	0.633	0.725	0.722	**0.741**
Pneumothorax	0.799	0.793	0.866	0.862	**0.879**
Consolidation	0.703	0.720	**0.758**	0.750	0.747
Edema	0.805	0.710	**0.853**	0.846	0.846
Emphysema	0.833	0.751	0.911	0.908	**0.936**
Fibrosis	0.786	0.761	0.826	0.827	**0.833**
Pleural thickening	0.684	0.730	0.780	0.779	**0.793**
Hernia	0.872	0.668	0.918	**0.934**	0.917
Mean	0.745	0.739	0.816	0.817	**0.821**

Table 2. Comparison of localisation measures: average IoU and cDice [30] (Pne1 represents Pneumonia and Pne2 Pneumothorax). Our methods Sal, Det and Mix are describe as follows: 1) Sal, uses saliency map x_S; 2) Det, uses saliency map x_{RPN}; and 3) Mix, uses saliency map $\tilde{x}_{FS}$. We show in bold the best results within a 0.005 confidence.

Metric	Method	Atel	Card	Effu	Infi	Mass	Nodu	Pne1	Pne2	Mean
IoU	Sal	0.201	0.558	0.208	0.300	0.230	0.080	0.335	**0.144**	0.257
	Det	0.216	**0.663**	**0.221**	0.322	0.216	0.081	0.327	0.128	0.272
	Mix	**0.240**	0.662	**0.226**	**0.343**	**0.240**	**0.092**	**0.346**	0.133	**0.285**
cDice	[30]	0.204	0.180	0.293	0.325	0.202	**0.295**	0.112	0.039	0.206
	Sal	0.296	**0.737**	0.333	0.374	0.294	0.059	0.424	**0.222**	0.342
	Det	0.376	0.590	**0.363**	**0.449**	0.361	0.143	**0.494**	0.200	**0.372**
	Mix	**0.403**	0.500	0.355	0.431	**0.403**	0.181	**0.493**	0.190	**0.370**

Table 3. Comparison of localisation accuracy using IoU (Pne1 represents Pneumonia and Pne2 Pneumothorax).

T(IoU)	Model	Atel	Card	Effu	Infi	Mass	Nodu	Pne1	Pne2	Mean
0.3	[27]	0.24	0.46	0.30	0.28	0.15	**0.04**	0.17	0.13	0.22
	Ours	**0.37**	**0.99**	**0.37**	**0.54**	**0.35**	**0.04**	**0.60**	**0.21**	**0.43**
0.5	[27]	0.05	0.18	**0.11**	0.06	0.01	**0.01**	0.03	0.03	0.06
	Ours	**0.11**	**0.92**	0.05	**0.30**	**0.13**	0.00	**0.27**	**0.06**	**0.23**
0.6	[27]	0.02	0.08	**0.05**	0.02	0.00	**0.01**	0.02	**0.03**	0.03
	Ours	**0.04**	**0.73**	0.01	**0.20**	**0.05**	0.00	**0.18**	0.01	**0.15**

– we argue that this happens because methods based on saliency maps [27,30] suffer from the low resolution of intermediate feature maps. Similarly to classification, we alleviate this problem using FPN [14] that produces the initial saliency map at higher resolution. We also observe that the saliency map tends to include areas bigger than the actual targeted area. Thus, we believe that saliency maps alone are not suited to obtain good localisation predictions. By including individual region proposals during training, our method is able to focus on different regions of the input separately and effectively refine localisation results.

In this paper, we proposed a new model for disease classification and weakly supervised localisation from chest X-ray images. Our model produces the disease classification using a saliency map that indicates the relevant regions for the classification. This localisation information is then refined using the straight-through Gumbel-Softmax estimator to discretely sample region proposals, allowing the method to refine initially proposed regions in an end-to-end training set-up. Future work will focus on simultaneously improving the classification performance and refining detection results by modelling interactions between potential multiple regions of interest from the same image.

References

1. Bilen, H., Vedaldi, A.: Weakly supervised deep detection networks. In: Proceedings of the IEEE Conference on Computer Vision and Pattern Recognition, pp. 2846–2854 (2016)
2. Chen, D., et al.: Deep learning and alternative learning strategies for retrospective real-world clinical data. NPJ Digit. Med. **2**(1), 1–5 (2019). Number: 1 Publisher: Nature Publishing Group
3. Folio, L.R.: Chest Imaging: An Algorithmic Approach to Learning. Springer, Heidelberg (2012). https://doi.org/10.1007/978-1-4614-1317-2
4. Guan, Q., Huang, Y.: Multi-label chest X-ray image classification via category-wise residual attention learning. Pattern Recogn. Lett. **130**, 259–266 (2018)
5. Guan, Q., Huang, Y., Zhong, Z., Zheng, Z., Zheng, L., Yang, Y.: Thorax disease classification with attention guided convolutional neural network. Pattern Recogn. Lett. **131**, 38–45 (2020)
6. Gündel, S., Grbic, S., Georgescu, B., Liu, S., Maier, A., Comaniciu, D.: Learning to recognize abnormalities in chest X-rays with location-aware dense networks. In: Vera-Rodriguez, R., Fierrez, J., Morales, A. (eds.) CIARP 2018. LNCS, vol. 11401, pp. 757–765. Springer, Cham (2019). https://doi.org/10.1007/978-3-030-13469-3_88
7. He, K., Zhang, X., Ren, S., Sun, J.: Deep residual learning for image recognition. In: 2016 IEEE Conference on Computer Vision and Pattern Recognition (CVPR), pp. 770–778 (2016). https://doi.org/10.1109/CVPR.2016.90
8. He, K., Gkioxari, G., Dollár, P., Girshick, R.B.: Mask R-CNN. In: 2017 IEEE International Conference on Computer Vision (ICCV), pp. 2980–2988 (2017)
9. Huang, G., Liu, Z., van der Maaten, L., Weinberger, K.Q.: Densely connected convolutional networks. In: 2017 IEEE Conference on Computer Vision and Pattern Recognition (CVPR), pp. 2261–2269 (2017). https://doi.org/10.1109/CVPR.2017. 243

10. Irvin, J., et al.: CheXpert: a large chest radiograph dataset with uncertainty labels and expert comparison. In: Proceedings of the AAAI Conference on Artificial Intelligence, vol. 33, pp. 590–597 (2019)

11. Jang, E., Gu, S., Poole, B.: Categorical Reparameterization with Gumbel-Softmax (2017)

12. Kingma, D.P., Ba, J.: Adam: a method for stochastic optimization. In: International Conference on Learning Representations (ICLR) (2015)

13. Li, Z., et al.: Thoracic disease identification and localization with limited supervision. In: Proceedings of the IEEE Conference on Computer Vision and Pattern Recognition, pp. 8290–8299 (2018)

14. Lin, T.Y., Dollár, P., Girshick, R., He, K., Hariharan, B., Belongie, S.: Feature pyramid networks for object detection. In: Proceedings of the IEEE Conference on Computer Vision and Pattern Recognition, pp. 2117–2125 (2017)

15. Litjens, G., et al.: A survey on deep learning in medical image analysis. Med. Image Anal. **42**, 60–88 (2017)

16. Liu, J., Zhao, G., Fei, Y., Zhang, M., Wang, Y., Yu, Y.: Align, attend and locate: chest X-ray diagnosis via contrast induced attention network with limited supervision. In: Proceedings of the IEEE International Conference on Computer Vision, pp. 10632–10641 (2019)

17. Ma, C., Wang, H., Hoi, S.C.H.: Multi-label thoracic disease image classification with cross-attention networks. In: Shen, D., et al. (eds.) MICCAI 2019. LNCS, vol. 11769, pp. 730–738. Springer, Cham (2019). https://doi.org/10.1007/978-3-030-32226-7_81

18. Maas, A.L., Hannun, A.Y., Ng, A.Y.: Rectifier nonlinearities improve neural network acoustic models. In: Proceedings of the ICML, vol. 30, issue 1, p. 3 (2013)

19. Maicas, G., Snaauw, G., Bradley, A.P., Reid, I., Carneiro, G.: Model agnostic saliency for weakly supervised lesion detection from breast DCE-MRI. In: 2019 IEEE 16th International Symposium on Biomedical Imaging (ISBI 2019), pp. 1057–1060 (2019)

20. Oquab, M., Bottou, L., Laptev, I., Sivic, J.: Is object localization for free? - weakly-supervised learning with convolutional neural networks. In: 2015 IEEE Conference on Computer Vision and Pattern Recognition (CVPR), pp. 685–694 (2015). https://doi.org/10.1109/CVPR.2015.7298668. ISSN 1063-6919, 1063-6919

21. Rajpurkar, P., et al.: CheXnet: radiologist-level pneumonia detection on chest x-rays with deep learning. arXiv preprint arXiv:1711.05225 (2017)

22. Ren, S., He, K., Girshick, R.B., Sun, J.: Faster R-CNN: towards real-time object detection with region proposal networks. IEEE Trans. Pattern Anal. Mach. Intell. **39**, 1137–1149 (2015)

23. Russakovsky, O., et al.: ImageNet large scale visual recognition challenge. Int. J. Comput. Vis. **115**(3), 211–252 (2015). https://doi.org/10.1007/s11263-015-0816-y

24. Tang, P., et al.: Weakly supervised region proposal network and object detection. In: Ferrari, V., Hebert, M., Sminchisescu, C., Weiss, Y. (eds.) ECCV 2018. LNCS, vol. 11215, pp. 370–386. Springer, Cham (2018). https://doi.org/10.1007/978-3-030-01252-6_22

25. Tang, Y., Wang, X., Harrison, A.P., Lu, L., Xiao, J., Summers, R.M.: Attention-guided curriculum learning for weakly supervised classification and localization of thoracic diseases on chest radiographs. In: Shi, Y., Suk, H.-I., Liu, M. (eds.) MLMI 2018. LNCS, vol. 11046, pp. 249–258. Springer, Cham (2018). https://doi.org/10.1007/978-3-030-00919-9_29

26. Wang, H., Jia, H., Lu, L., Xia, Y.: Thorax-net: an attention regularized deep neural network for classification of thoracic diseases on chest radiography. IEEE J. Biomed. Health Inform. **24**(2), 475–485 (2019)

27. Wang, X., Peng, Y., Lu, L., Lu, Z., Bagheri, M., Summers, R.M.: ChestX-ray8: hospital-scale chest X-ray database and benchmarks on weakly-supervised classification and localization of common thorax diseases. In: Proceedings of the IEEE Conference on Computer Vision and Pattern Recognition, pp. 2097–2106 (2017)

28. Wang, Y., et al.: Weakly supervised universal fracture detection in pelvic X-rays. In: Shen, D., et al. (eds.) MICCAI 2019. LNCS, vol. 11769, pp. 459–467. Springer, Cham (2019). https://doi.org/10.1007/978-3-030-32226-7_51

29. Yao, L., Poblenz, E., Dagunts, D., Covington, B., Bernard, D., Lyman, K.: Learning to diagnose from scratch by exploiting dependencies among labels. arXiv preprint arXiv:1710.10501 (2017)

30. Yao, L., Prosky, J., Poblenz, E., Covington, B., Lyman, K.: Weakly supervised medical diagnosis and localization from multiple resolutions. arXiv:1803.07703 [cs] (2018)

CPM-Net: A 3D Center-Points Matching Network for Pulmonary Nodule Detection in CT Scans

Tao Song[1,2], Jieneng Chen[2,5], Xiangde Luo[1], Yechong Huang[2], Xinglong Liu[2], Ning Huang[2], Yinan Chen[2], Zhaoxiang Ye[3], Huaqiang Sheng[4], Shaoting Zhang[1,2], and Guotai Wang[1(✉)]

[1] School of Mechanical and Electrical Engineering,
University of Electronic Science and Technology of China, Chengdu, China
guotai.wang@uestc.edu.cn
[2] SenseTime Research, Shanghai, China
[3] Department of Radiology,
Tianjin Medical University Cancer Institute and Hospital, Tianjin, China
[4] Department of Radiology,
The First Affiliated Hospital of Shandong First Medical University, Jinan, China
[5] College of Electronics and Information Technology,
Tongji University, Shanghai, China

Abstract. Automatic and accurate lung nodule detection from Computed Tomography (CT) scans plays a vital role in efficient lung cancer screening. Despite the state-of-the-art performance obtained by recent anchor-based detectors using Convolutional Neural Networks (CNNs) for this task, they require pre-determined anchor parameters such as the size, number and aspect ratio of anchors, and have limited robustness when dealing with lung nodules with a massive variety of sizes. To overcome this problem, we propose a 3D center-points matching detection network (CPM-Net) that is anchor-free and automatically predicts the position, size and aspect ratio of nodules without manual design of nodule/anchor parameters. The CPM-Net uses center-points matching strategy to find center-points, and then uses features of these points correspondingly to regress the size of the bounding box of nodule and local offset of the center points. To better capture spatial information and 3D context for the detection, we propose to fuse multi-level spatial coordinate maps with the feature extractor and combine it with 3D squeeze-and-excitation attention modules. To deal with the enormous imbalance between the number of positive and negative samples during center points matching, we propose a hybrid method of adaptive points mining and re-focal loss. Experimental results on LUNA16 dataset showed that our proposed CPM-Net achieved superior performance for lung nodule detection compared with state-of-the-art anchor-based methods.

T. Song and J. Chen—Equal contribution.

© Springer Nature Switzerland AG 2020
A. L. Martel et al. (Eds.): MICCAI 2020, LNCS 12266, pp. 550–559, 2020.
https://doi.org/10.1007/978-3-030-59725-2_53

1   Introduction

Lung cancer is one of the leading life-threatening cancer around the world, and the diagnosis at an early stage is crucial for the best prognosis [11]. As one of the essential computer-aided diagnosis technologies, lung nodule detection from medical images such as Computed Tomography (CT) has been increasingly studied for automatic screening and diagnosis of the lung cancer. Detecting pulmonary nodules is very challenging due to the large variation of nodule size, location, and appearance. With the success of deep Convolutional Neural Network (CNN) for object detection in natural images, CNN-based algorithms are widely used for detecting pulmonary nodules from CT scans [1,2,12].

Current automatic CNN-based lung nodule detection approaches mainly include one-stage anchor-based detection frameworks [8,9] and two-stage methods with multiple modules [1,2]. Anchor-based methods such as Faster RCNN [9] and RetinaNet [8] are originally designed to deal with 2D images and trained to distinguish whether each pre-determined anchor box sufficiently overlaps with a certain nodule. However, they are faced with several limitations when applied to 3D medical images. First, they typically need a very large set of anchor boxes, e.g. more than 100k in 2D RetinaNet [8], resulting in huge computational and memory consumption for 3D medical images. Second, the use of anchor boxes introduces many hyper-parameters and design choices, including the number, size and aspect ratio of the anchor boxes. Manual design of these hyper-parameters is not only time-consuming but also subject to human experience, which may limit the detection performance. It was indicated in [1] that detection of small objects is sensitive to the manual design of anchors. Moreover, it has been shown that the default anchor configuration is ineffective for detecting lesions with a small size and large aspect ratio [13], let alone the more complex pulmonary nodules, of which the size can vary by as much as ten times. The two-stage frameworks [1,2] use multiple modules, where a detector in the first stage obtains a set of candidates with many false positives and a False Positive Reduction (FPR) module is used in the second stage to get higher accuracy. However, these modules at different stages are independently trained and cannot be optimized jointly, which may limit the detection performance.

To overcome the limitation of existing anchor-base methods, we propose a novel 3D Center-Points Matching Detection Network (CPM-Net) for pulmonary nodule detection in CT scans. The CPM-Net is a one-stage anchor-free detection method, which predicts the probability of a pixel i in an $S \times S \times S$ grid being around the center of a nodule, and simultaneously regresses the bounding box size and relative offset from pixel i to the real nodule center. It is worth mentioning that rare works focus on 3D anchor-free detection in medical images, and unlike recent anchor-free methods [3,6] for 2D natural images, we don't need to use a key points estimation network to generate a heatmap. Moreover, our CPM-Net is a one-stage end-to-end network without any false positive reduction module adopted in [1,2], so that it has a higher efficiency and lower memory cost.

There are three key contributions in the proposed CPM-Net. First, we mitigate the ineffectiveness of current anchor-based detectors [8,9] by discarding pre-determined anchor boxes, as we instead predict a center-point map directly via a points matching strategy. Second, to effectively capture the 3D context, we propose an attentive module as a strong feature extractor that takes advantage of multi-level spatial coordinate maps and Squeeze-and-Excitation (SE) [5] for better performance. Third, to solve the enormous imbalance between positive and negative samples, we propose a hybrid method of adaptive points mining and re-focal loss. We evaluated our approach on LUNA16 dataset [10], and experimental results showed that the proposed CPM-Net achieved superior performance compared with anchor-based methods for lung nodule detection from 3D CT scans.

2   Method

The overall structure of the proposed CPM-Net is illustrated in Fig. 2. It consists of a feature extractor that takes advantages of multi-level spatial coordinate maps and SE-based channel attention for better feature extraction, and a head that predicts the existence of a nodule and its size and offset simultaneously at two resolution levels. To assist the training process, we propose a center points matching strategy to predict K points that are nearest to the center of a nodule.

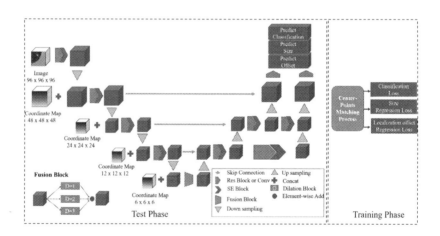

Fig. 1. Overall network structure of our CPM-Net

Let the 3D path image as $I \in R^{D \times H \times W}$, where D, H and W denote depth, height and width of image, respectively. The predicted feature map with the size of $\frac{D}{R} \times \frac{H}{R} \times \frac{W}{R} \times C$ can be obtained by forward process of the CPM-Net, where R denotes down-sampling ratio and C denotes channel. Let Map_C represent the classification map with 1 channel indicating the probability of each pixel, called point in this paper, being the center of a target, Map_S represent the size map with 3 channels that gives the 3D sizes of the target, and Map_O represent the offset map with 3 channels indicating offsets in three directions.

2.1   Architecture of the CPM-Net

The architecture of our detection backbone follows an encoder-decoder flowchart as show in Fig. 1. It has a series of convolutional layers to learn 3D patterns from training data. Each convolution uses $3 \times 3 \times 3$ kernels with 1 as stride, each pooling $2 \times 2 \times 2$ with 2 as stride, and each down-sampling and up-sampling are implemented by max pooling and deconvolutional layers, respectively. Skip connections are used to link low levels and high levels of the network.

Each layer of the last two layers in our CPM-Net decodes three prediction maps, i.e. classification, size and offset predictions, respectively, similarity as feature pyramid networks [7]. To be clear, the classification prediction is a pixel-wise binary classification, where 0 stands for non-center-points, and 1 stands for predicted center-points.

In testing phase, a bounding boxes of detection result can be constructed by the position of predicted center-points, size and offset (See Postprocessing in Sect. 2.2 for details). In training phase, we have an significant center-points matching procedure as mentioned in Sect. 2.2.

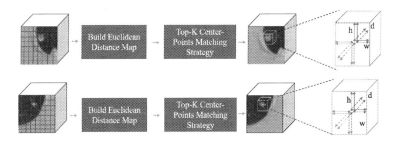

Fig. 2. Illustration of center-points matching used for training. K (i.e., 4) nearest points to the center of ground truth nodule is first obtained, and then their corresponding features at size map are used to predict the size of the nodule.

Attentive Modules. In particular, we introduce multi-path normalized coordinate map fusion blocks to generate attentive features for being aware of spatial position. The 3-channel interpolated coordinate maps are concatenated with the feature maps via four paths in total with corresponding sizes 48^3, 24^3, 12^3 and 6^3 respectively. Following [5], SE-blocks are adopted to learn spatial and channel-wise attention to further obtain discriminative feature. In the deepest layer, we utilize a dilated fusion block, consisting of three dilated convolutions with dilation of 1, 2, 3 respectively.

2.2   Center-Points Matching Detection

Center-Points Matching Strategy. In anchor-based methods, they assign positive and negative anchor according to the threshold of overlap between anchor boxes and ground-truth for training. In contrast, without any anchor prior, a novel center-points matching strategy for training is proposed in our CPM-Net to assign positive, negative, and ignored labels to make the network predict whether a point belongs to center-points. In short, this center-points matching process is to online generate classification labels in the training phase.

The whole strategy can be divided into the following four steps: 1) calculate the distance map serving as a prior to assign positive, negative and ignored points; 2) select the top-K nearest points as positive points $\in \mathbb{P}$ 3) to reduce the false positive, a certain amount of points around positive points will be assigned to ignored points $\in \mathbb{I}_{\mathbb{P}}$. Till now, the rest points $\notin \mathbb{P} + \mathbb{I}_{\mathbb{P}}$ make up a temporary negative points-set $\mathbb{N}_{\mathbb{P}}$; 4) use adaptive points mining (APM) to further sample hard negative points as point-set $\mathbb{N} \subset \mathbb{N}_{\mathbb{P}}$, while those points $\in \mathbb{N}_{\mathbb{P}} - \mathbb{N}$ are reset to the ignored points, and therefore we can obtain an ignored point-set $\mathbb{I}$ equal to $\mathbb{I}_{\mathbb{P}} + \mathbb{N}_{\mathbb{P}} - \mathbb{N}$. The center-point matching process is illustrated in Fig. 2. Leveraging each ground-truth, the distance map can be calculated by Eq. (1),

$$D_{map} = \sum_{d=0}^{\frac{D}{R}} \sum_{h=0}^{\frac{H}{R}} \sum_{w=0}^{\frac{W}{R}} \left[\left(x_{d,h,w} - \frac{x'}{R} \right)^2 + \left(y_{d,h,w} - \frac{y'}{R} \right)^2 + \left(z_{d,h,w} - \frac{z'}{R} \right)^2 \right]$$

(1)

here, (z', y', x') denotes the coordinate of the center point of ground truth.

We loop all the annotations in the patch images using the above steps. Note that (i) the top-K positive points are selected according to the lowest distance values sorted in distance map (ii) ignored points do not participate in training.

Loss Function. The imbalance of positive and negative samples in 3D classification is more severe than that in 2D classification. A hybrid method of adaptive points mining (APM) and re-focal loss is used in our CPM-Net to deal with such a huge imbalance. In APM, we first randomly choose N points from all negative points to reduce the effects of noise annotation, then we sort them using the highest confidence loss for each negative and pick the top ones so that the ratio between the negatives and positives is at most 100:1. In the end, these negative points which were not selected are reset as ignored points. Note that negative samples containing no annotation can only sample M negative points without any positive points. A re-focal loss is shown as Eq. (2), which can improve the sensitivity and further balance the gradient of the positives and the negatives.

$$L_{cls} = \sum_{j=0}^{J} -w_j \alpha (1 - p_j)^\gamma \log(p_j) \begin{cases} w_j = 1 \text{ if } j \in \mathbb{P} \text{ or } \mathbb{N} \\ w_j = 0 \text{ if } j \in \mathbb{I} \\ w_j = 4 \text{ if } j \in \mathbb{P} \text{ and } p_j < t \end{cases}$$

(2)

where p_j denotes the probability of the class at point j, $J = \frac{D}{R} \times \frac{H}{R} \times \frac{W}{R}$ is total number of points, and t is a threshold to filter unqualified points, setting to 0.9. A smooth L1 loss [9] in Eq. (3) is also used to regress the normalized size of bounding boxes.

$$L_{size}(r, r^*) = \sum_{r \in \{d,h,w\}} \begin{cases} 0.5(r - r^*)^2 / \beta & \text{if } |(r - r^*)| < \beta \\ |(r - r^*)| - 0.5 * \beta & \text{otherwise} \end{cases}$$

(3)

where r denotes predicted size of ground-truth by network, and r^* denotes the target size. To obtain more accurate locations of small objects, we use L1 loss to regresses the offset of between the center points of ground truth and integer localization of positive points. We take L1 loss as the follow form:

$$L_{offset}(f, f^*) = \sum_{f \in \{x,y,z\}} |f - f^*| \tag{4}$$

where f denotes predicted offset in x, y and z directions, and f^* denotes target offsets to be regressed. The total loss consisting of re-focal loss, smooth L1 loss and L1 loss in our CPM-Net can be expressed by the following equation:

$$L_{total} = \lambda_{cls} L_{cls} + L_{size} + L_{offset} \tag{5}$$

Note, only points $\in \mathbb{P}$ are calculated in L_{size} and L_{offset}. In our experiment, k, M, N, λ_{cls}, β, α and γ are set to 7, 100, 10000, 2, $\frac{1}{9}$, 0.75 and 2, respectively. Due to sphere characteristics of LUNA16 annotation, the channel of size regression map is 1 to regress the radius of pulmonary nodule.

Post-processing. In inference phase, we firstly pick up top-n candidate predicted points $\hat{P}_{\{(\hat{z}_j, \hat{y}_j, \hat{x}_j)\}_{j=1}^n}$ in a classification map with a size of $D \times H \times W \times 1$ by sorting the probabilities. Each predicted candidate point have a probability $\hat{P}_{(\hat{z}_j, \hat{y}_j, \hat{x}_j)}$ in an integer $(\hat{z}_j, \hat{y}_j, \hat{x}_j)$ location. Also in the same integer location, we can get offsets prediction $\hat{O}_{\{(\hat{z}_j, \hat{y}_j, \hat{x}_j)\}_{j=1}^n}$ in x, y, z direction and radius prediction $\hat{R}_{\{(\hat{z}_j, \hat{y}_j, \hat{x}_j)\}_{j=1}^n}$, respectively. Then the top-n detected bounding boxes can be written as

$$(\hat{z}_j + \hat{O}\hat{z}_j - \hat{R}_j/2, \hat{y}_j + \hat{O}\hat{y}_j - \hat{R}_j/2, \hat{x}_j + \hat{O}\hat{x}_j - \hat{R}_j/2,$$
$$\hat{z}_j + \hat{O}\hat{z}_j + \hat{R}_j/2, \hat{y}_j + \hat{O}\hat{y}_j + \hat{R}_j/2, \hat{x}_j + \hat{O}\hat{x}_j + \hat{R}_j/2).$$

All detected bounding boxes will pass a 3D IoU-based non-maxima suppression (NMS) to filter overlapping bounding boxes.

3   Experimental Results

In this section, we validated the proposed framework on the large-scale public challenge dataset of LUNA16 [10], which contains 888 low-dose CT scans with the location centroids and diameters of the pulmonary nodules annotated. In the LUNA16 dataset, performances of detection systems are evaluated using the Free-Response Receiver Operating Characteristic (FROC) [10].

Implementation Details. In the training phase, for each model, we use 150 epochs in total with stochastic gradient descent optimization, momentum as 0.9, and weight decay as 1e-4. The batch size is set to 24. The initial learning rate is

0.01, 0.001 after 80 epochs, and 0.0001 after 150 epochs. The parameter n in post-processing is set as 100. All experiments were done in SenseCare Platform [4].

Lung Nodule Detection Results and Comparison. In this paper, the evaluation metrics are sensitivity and the average number of false positives per scan (FPs/scan). We draw the FROC curve in Fig. 3(a) demonstrating the sensitivities at seven predefined FPs/Scan rates: 1/8, 1/4, 1/2, 1, 2, 4, 8. As the radiologists can tolerate 1 to 2 FPs/Scan, we report the performance of our method and that of other methods at 1,2 and 8 FPs/Scan in Table 1, here subsets 0–8 are used for training and subset 9 for testing. Our CPM-Net achieve sensitivities of 91.2% at 1 FPs/Scan and 92.4% at 2 FPs/Scan, which demonstrate promising results for further clinical use. We compare our method with 3D CenterNet, Deeplung, 3D RetinaNet, and 3D RetinaNet++, where we implemented the 3D version of CenterNet [3] and RetinaNet [8]. It is to be noted that 3D RetinaNet++ is an enhanced version of 3D RetinaNet, adopting the same APM and re-focal loss with our CPM-Net and using an online anchors matching strategy instead of offline pre-determining strategy in [8]. The reason why the 3D CenterNet obtains horrible results is that the anchor-free-based CenterNet only care single center-point obtained from a heatmap for one object but the nodules are much smaller than objects in natural images, resulting in location failure and bringing many false negatives. In contrary to CenterNet, our CPM-Net matching multi-points to reduce the false negatives, and thus increases the sensitivity.

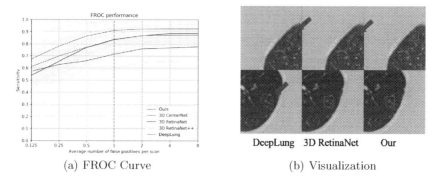

(a) FROC Curve   (b) Visualization

Fig. 3. (a): Comparison of FROC curves using different network configurations, with shaded areas presenting the 95% confidence interval. (b): Visualizations of detection results in different methods, where dotted circles represent missed cases and solid boxes represent correct detection.

Ablation Study. As center-points matching strategy is the main contribution in this paper, we want to evaluate the performance of our network at different K for selecting the top-K positive center-points, as mentioned in Sect. 2.2. When K is greater or smaller than 7, the sensitivity goes down significantly as shown

Table 1. Comparison with different detector on LUNA16 dataset.

Method	Sensitivity		
	FPs/Scan=1	FPs/Scan=2	FPs/Scan=8
3D CenterNet	71.4%	75.8%	77.5%
Deeplung	85.8%	86.9%	87.5%
3D RetinaNet	83.6%	86.8%	88.7%
3D RetinaNet++	88.8%	91.3%	91.4%
Our CPM-Net	**91.2%**	**92.4%**	**92.5%**

in Table 2. This is because too many positive points during training will bring in false positives while few positive points weaken the model's ability to locate the nodules. Furthermore, the attentive module with coordinate attention and SE attention is a crucial component of our CPM-Net. To understand its contribution to performance, we train a network without attentive modules and another network without coordinate attention but with SE attention. Table 3 shows that adding an attentive module gives significant improvement: 4.4% sensitivity at 1 FPs/Scan and 3% sensitivity at 2 FPs/Scan. We also see that coordinate attention further improves the detection results.

Table 2. Analysis of performance at different top-K.

K	Sensitivity(FPs/Scan)		
	1	2	8
5	85.5%	87.5%	91.3%
6	89.4%	89.8%	90.4%
7	**91.2%**	**92.4%**	**92.5%**
8	85.6%	88.5%	92.5%
9	87.7%	90.6%	90.6%

Table 3. Ablation study of CPM-Net.

	Sensitivity(FPs/Scan)		
	1	2	8
w/o $\mathbb{A}$	86.8%	89.4%	90.4%
w/o $\mathbb{C}$	90.5%	90.8%	91.5%
Ours	**91.2%**	**92.4%**	**92.5%**

* $\mathbb{A}$ represents Attentive Modules
** $\mathbb{C}$ represents Coordinate Map

We also compare our method with several state-of-the-art methods in 10-fold cross validation on LUNA dataset as show in Table 4. Note that Zhu et al. [12] employed three-stage framework consisting of candidate generation, feature extraction, and classification while both Dou et al. [2] and Ding et al. [1] used two-stage framework with a anchor-based detector and a false positive reduction module (FPR). In contrast, our CPM-Net show the advantage over other methods as it's an end-to-end single-stage detection network.

Table 4. Comparison with state-of-the-art methods on LUNA16 dataset.

Method	Sensitivity			FPR	Stages
	FPs/Scan=1	FPs/Scan=2	FPs/Scan=8		
Zhu et al. [12]	86.0%	90.0%	92.3%		3
Dou et al. [2]	86.5%	90.6%	94.6%	✓	2
Ding et al. [1]	92.8%	93.1%	94.6%	✓	2
Ours	91.1%	92.8%	94.8%		**1**

4   Conclusion

In this paper, we propose a novel 3D center-points matching detection network (CPM-Net) for pulmonary nodule detection. In addition to center-points matching strategy, we use an attentive module consisting of coordinate attention and squeeze-and-excitation attention to capture spatial position. Besides, we adopt a hybrid method of adaptive points mining and re-focal loss to solve the imbalance between positive points and negative points. Experimental results on the LUNA16 Nodule Detection Dataset show that the proposed CPM-Net achieve competitive performance compared with several state-of-the-art approaches. The high sensitivities in single stage inference demonstrate promising potential for further clinical use.

References

1. Ding, J., Li, A., Hu, Z., Wang, L.: Accurate pulmonary nodule detection in computed tomography images using deep convolutional neural networks. In: Descoteaux, M., Maier-Hein, L., Franz, A., Jannin, P., Collins, D.L., Duchesne, S. (eds.) MICCAI 2017. LNCS, vol. 10435, pp. 559–567. Springer, Cham (2017). https://doi.org/10.1007/978-3-319-66179-7_64
2. Dou, Q., Chen, H., Jin, Y., Lin, H., Qin, J., Heng, P.-A.: Automated pulmonary nodule detection via 3D ConvNets with online sample filtering and hybrid-loss residual learning. In: Descoteaux, M., Maier-Hein, L., Franz, A., Jannin, P., Collins, D.L., Duchesne, S. (eds.) MICCAI 2017. LNCS, vol. 10435, pp. 630–638. Springer, Cham (2017). https://doi.org/10.1007/978-3-319-66179-7_72
3. Duan, K., Bai, S., Xie, L., Qi, H., Huang, Q., Tian, Q.: CenterNet: keypoint triplets for object detection. In: ICCV, pp. 6569–6578 (2019)
4. Duan, Q., et al.: SenseCare: a research platform for medical image informatics and interactive 3D visualization. arXiv preprint arXiv:2004.07031 (2020)
5. Hu, J., Shen, L., Sun, G.: Squeeze-and-excitation networks. In: CVPR, pp. 7132–7141 (2018)
6. Law, H., Deng, J.: CornerNet: detecting objects as paired keypoints. In: ECCV, pp. 734–750 (2018)
7. Lin, T.Y., Dollár, P., Girshick, R., He, K., Hariharan, B., Belongie, S.: Feature pyramid networks for object detection (2017)
8. Lin, T.Y., Goyal, P., Girshick, R., He, K., Dollár, P.: Focal loss for dense object detection. In: ICCV, pp. 2980–2988 (2017)

9. Ren, S., He, K., Girshick, R., Sun, J.: Faster R-CNN: towards real-time object detection with region proposal networks. In: NIPS, pp. 91–99 (2015)
10. Setio, A.A.A., et al.: Validation, comparison, and combination of algorithms for automatic detection of pulmonary nodules in computed tomography images: the luna16 challenge. Med. Image Anal. **42**, 1–13 (2017)
11. Siegel, R.L., Miller, K.D., Jemal, A.: Cancer statistics, 2015. CA Cancer J. Clin. **65**(1), 5 (2015)
12. Zhu, W., Liu, C., Fan, W., Xie, X.: DeepLung: deep 3D dual path nets for automated pulmonary nodule detection and classification. In: WACV, pp. 673–681 (2018)
13. Zlocha, M., Dou, Q., Glocker, B.: Improving RetinaNet for CT lesion detection with dense masks from weak RECIST labels. In: Shen, D., et al. (eds.) MICCAI 2019. LNCS, vol. 11769, pp. 402–410. Springer, Cham (2019). https://doi.org/10.1007/978-3-030-32226-7_45

Interpretable Identification of Interstitial Lung Disease (ILD) Associated Findings from CT

Yifan Wu[✉], Jiancong Wang, William D. Lindsay, Tarmily Wen, Jianbo Shi, and James C. Gee

University of Pennsylvania, Philadelphia, PA, USA
yfwu@seas.upenn.edu

Abstract. In this study, we present a method to identify radiologic findings associated with interstitial lung diseases (ILD), a heterogeneous collection of progressive lung diseases, from thoracic CT scans. Prior studies have relied on densely supervised methods using 2D slices or small 3D patches as input, requiring significant manual labor to create dense labels. This limits the amount of data available for algorithm development and thus hinders generalization performance. To harness available large, but sparsely labeled datasets, we present a weakly supervised method to identify imaging findings associated with ILD. We test this framework to classify and roughly localize 14 radiologic findings on the LTRC dataset of 3380 thoracic CT scans. We conduct 5-fold cross-validation and achieve 0.8 mean AUC scores on 5 out of 14 findings classification. We visualize attention energy maps which demonstrate that our classifier is able to learn representative features with meaningful differences between radiologic findings, and is capable of approximately localizing the findings of interest, thereby adding interpretability of our model (This work was supported by the USPHS under NIH grant R01-HL133889).

1 Introduction

Interstitial Lung Disease (ILD), a heterogeneous group of more than 200 chronic lung diseases, is characterized by progressive pulmonary inflammation leading to fibrosis and permanent respiratory impairment. Accurate diagnosis of ILD is a major clinical challenge requiring collaboration between radiologists, pathologists, and pulmonologists. A recent nationwide survey of ILD patients revealed that 55% of patients are misdiagnosed at the initial clinical visit with 38% of patients being misdiagnosed more than once [5].

High-resolution CT is an essential tool to characterize ILD and monitor its longitudinal progression through identification of discriminative imaging findings. However, consistently diagnosing various ILDs remains challenging since the intricate visual patterns of ILDs can vary widely in extent, appearance, and

---

Y. Wu and J. Wang—Equal contribution.

© Springer Nature Switzerland AG 2020
A. L. Martel et al. (Eds.): MICCAI 2020, LNCS 12266, pp. 560–569, 2020.
https://doi.org/10.1007/978-3-030-59725-2_54

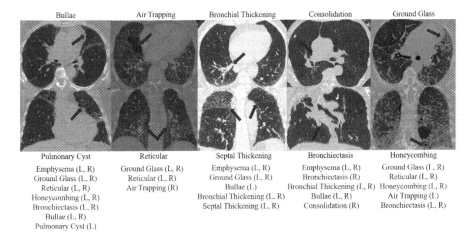

Fig. 1. Examples of ILD findings. Each column shows an axial slice and a coronal slice from a CT scan of a single subject. The arrows indicate the rough location of the finding with corresponding finding in red text above or below. The black text below each column shows all present CT findings with L indicating the left lung, R indicating the right lung. (Color figure online)

location. Differences in physicians' skill and experience may lead to different interpretations of dominant findings [2]. An automated method to identify and localize imaging findings would aid physicians in providing accurate diagnoses.

The recent emergence of multi-institute, large-scale, openly available chest X-ray datasets [10,11,14], has accelerated development of deep learning algorithms to classify and localize findings on 2D images, and enabled automatic radiology report generation [15,17]. In contrast, development of these methods for CT imaging, particularly for ILD, has lagged. This can be attributed to the heterogeneity of visual patterns, lack of large, suitable datasets, and limitation of computing power. Recent studies focused on identifying visual findings have used either 2D slices or small 3D patches as input and have trained on datasets consisting of approximately 200 CT scans [1,3,7,13]. This exacerbates the risk of overfitting, hindering the generalization ability of their models. Furthermore, fully supervised frameworks require expensive human labor to label imaging findings on each slice or patch, and labeling of ILD findings requires sub-speciality clinical expertise, making dense labeling highly expensive for large ILD datasets.

In this work, we present a multi-label classification and approximate localization framework, capable of simultaneously identifying 14 findings associated with ILD from CT imaging. We note three contributions of this work. First, we develop a findings identification algorithm capable of identifying fine-grained findings associated with ILD. Although ubiquitous findings such as Emphysema, Ground Glass Opacities, Reticular Infiltrates, and Honeycombing have been explored in prior work, we expand this range to include classification of Air Trapping, Bronchial Thickening, Bronchiectasis, Bullae, Consolidation,

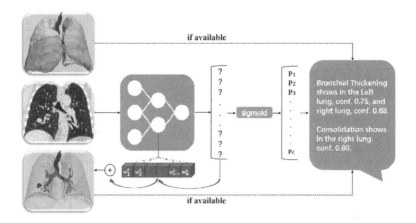

Fig. 2. Schematic overview for ILD finding identification.

Micronodules, Mosaic Attenuation, Pulmonary Cysts, Septal Thickening and Tree in Bud Pattern. Accurate classification of these findings is challenging because they are less common and more fine-grained, thus more likely to have intra-class variations and inter-class similarities, requiring greater physician expertise in real-world diagnostic scenarios. Moreover, less common findings result in very unbalanced data distributions, with 8 out of 14 findings possessing less than 10% positive samples in our experiment. Second, we developed a method capable of using the entire CT volume as input requiring only positive/negative binary labels at the image level. This allows us to take advantage of large-scale datasets of CT and radiology report pairs that are routinely available and which can be mined automatically from hospital databases. In our experiments, we use the Lung Tissue Research Consortium (LTRC) dataset, which includes 3380 lung CT scans. To the best of our knowledge, we are the first to explore findings classification of ILD findings on a dataset of this magnitude. Third, we achieve an AUC greater than 0.8 for 5 out of 14 findings, and use an attention mechanism based on Grad-CAM [12] to localize presence of specific findings. By combining our attention energy maps with detailed lung lobe masks, there is the potential for this framework to further describe imaging findings by their specific lung location, which will be crucial for accurate clinical diagnosis of certain ILD subtypes.

2   Methods

Given a pulmonary CT, we first classify 14 binary findings using a single convolution neural network, then we locate each finding of interest by a 3D extension to Grad-CAM [12]. To this end, we aim to output a list of confidence scores, and rough location of the findings. When additional lobe masks are available, we could further describe the lesion's location as the central or peripheral and generate radiological reports automatically. Figure 2 shows our schematic overview.

2.1 Multi-label Classification on 3D Volume

For each input volume from either left or right lung, we define a 14-dimensional label vector $\mathbf{y} = [y_1, \ldots, y_c, \ldots, y_C], y_c \in \{0, 1\}, C = 14$, where y_c indicates the presence of corresponding finding while a all-zero vector represents "Normal". It should be noted that each lung could have different $\mathbf{y}'s$ since some radiologic findings might only be present in one side of the lung. We use the standard Binary Cross Entropy loss for the multi-label binary classification loss:

$$\mathcal{L} = \mathbb{E}_{x,y \in V}[(1 - y)\log(1 - D(x|\theta)) - y\log(D(x|\theta))], \qquad (1)$$

where V is the given dataset. x is the input volume and y is the corresponding multi-label vector. D is the network and θ is the parameters of the network that we are optimizing for. For the network architecture, we refer to [4] and adapt it to 3D volume. Detailed architecture configuration is shown in Table 1.

2.2 Attention Map Localization

We adapt the gradient-weighted class activation mapping (Grad-CAM) [12] to 3D for finding localization within lung volume. Given the finding class c, we calculate the gradient of the class logit, i.e., the quantity before sigmoid activation with respect to the final convolution activation feature A. We apply spatial pooling over 3 dimensions to derive the channel importance weights α_m^c. Formally,

$$\alpha_m^c = \frac{1}{Z}\sum_i\sum_j\sum_k\frac{\partial \hat{y}^c}{\partial A_{ijk}^m}, \qquad (2)$$

where $\hat{y}^c$ is network prediction, i, j, k are spatial indices, m is channel index and Z is total number of elements in A. We then weight the feature map A channel wise by α_m^c and sum them together:

$$Att_{Grad-CAM}^c = ReLU(\sum_m\alpha_m^c A^m), \qquad (3)$$

which produces single channel attention with the same spatial size as A. We denote this map as $Att_{Grad-CAM}^c$. For visualization we resize $Att_{Grad-CAM}^c$ to the size of input image and overlay them together (Fig. 3).

Table 1. Model architecture. K, S, P refers to kernal size stride and padding.

Layer	Inout → Output Shape	Layer Configuration
Input Layer	$[1, 98, 98, 128] \rightarrow [32, 46, 46, 62]$	Conv3d(K7, S2, P1), LeakyRelu
Hidden Layer	$[32, 46, 46, 62] \rightarrow [64, 23, 23, 31]$	Conv3d(K4, S2, P1), LeakyRelu
Hidden Layer	$[64, 23, 23, 31] \rightarrow [128, 11, 11, 15]$	Conv3d(K4, S2, P1), LeakyRelu
Hidden Layer	$[128, 11, 11, 15] \rightarrow [256, 5, 5, 7]$	Conv3d(K4, S2, P1), LeakyRelu
Hidden Layer	$[256, 5, 5, 7] \rightarrow [512, 2, 2, 3]$	Conv3d(K4, S2, P1), LeakyRelu
Hidden Layer	$[512, 2, 2, 3] \rightarrow [1024, 1, 1, 1]$	Conv3d(K4, S2, P1), LeakyRelu
Output Layer	$[2048, 1, 1, 1] \rightarrow [C, 1, 1, 1]$	Conv3d(K1, S1, P0)

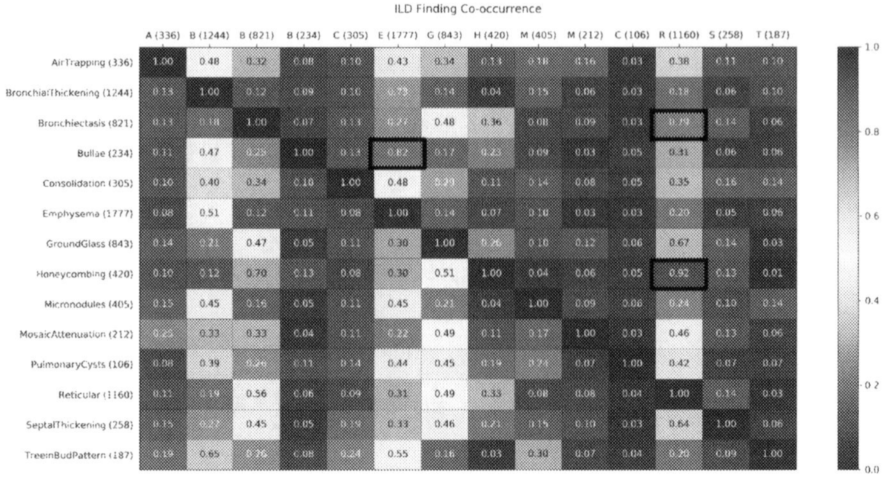

ILD Finding Co-occurrence

Fig. 3. Finding co-occurrence. While we regress each individual feature as independent separate binary labels for formulation simplicity, we also visualize their co-occurrence. Top 3 co-occurring findings are 1). Bullae refers to a permanent sub type of emphysema. 2). Reticulation and honey combing both refer to a porous pattern of different densities. 3). Bronchiactasis occurs largely on smoking population who also tend to carry reticulation.

3   Experiments

3.1   Experimental Setup

Data. We use the LTRC dataset of 3380 lung CT scans after removing data without labels or lungs mask. We extract a single side, i.e., left and right lung as independent samples using lung masks provided in the dataset. We perform random left/right and anterior/posterior flip during training. In order to adapt this framework to new lung samples, we test the open-source lung mask extraction network, PHNN [8], which achieves a dice score of 0.987. We crop the lungs according to the mask then resize to $196 \times 196 \times 256$. After augmentation we have 6760 samples for training and testing. Two sides from the same lung are in the same split so there is no patient overlap between training and testing.

Each sample has 15 findings labels in the dataset. One finding has less than 1% positive samples and we remove it from the experiment. We conduct binary classification on the remaining 14 findings in our experiments (presence/absence of finding). Figure 1 shows sample data and ground truth organization. We observe that the visual findings classes have large intra-class variations and inter-class similarities. We present each finding's prevalence within the LTRC dataset in Table 2, 8 out of 14 class have less than <10% positive samples, which makes this multi-class classification with sparse labels more challenging.

Experiments. We conduct a 5-fold cross-validation. We normalize input volumes to roughly $[-1, 1]$ by mean/std normalization. Since the findings only occur

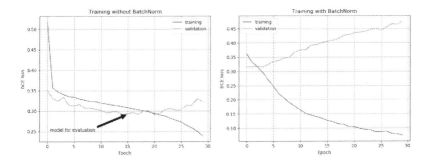

Fig. 4. Learning loss curves for w and w/o BatchNorm.

within the lung, and since other contents within the CT (rib cage, stomach, etc) may interfere the learning, we optionally apply the lung mask to the input volume for normalization then set region outside the lung to 0 as a separate experiment. We compare these 2 settings in the following quantitative analysis section. During training, we use SGD optimizer with a learning rate of 0.001. We set batch size as 4 full lungs, which totals 8 volumes including left/right sides. We train our model for 30 epochs and found that the model achieves the best validation results at around 20 epochs.

3.2 Ablation Study for Architecture Selection

To improve the model's performance, we attempt to increase model complexity and learn better representative feature by incorporating attention modules such as squeeze and excitation [9], CBAM [16], and WFildCAT [6]. We also add a batch normalization to each convolution layer in the network. However, despite accelerating training convergence significantly, all of these lead to over-fitting and reduce testing accuracy. We plot the training and validation loss curves in Fig. 4. Hence we propose a simple model without batch normalization layers or any attention modules. We argue that for this specific task, as the data is highly un-balanced and labelling is sparse, avoiding over-fitting is more important than adding model complexity.

3.3 Quantitative Analysis

We conduct a 5 fold cross validation and calculate the mean area-under-the-curve (AUC) for classification for 14 findings. They are reported in Table 2. Among the 14 findings, the mean AUC scores for Emphysema, Ground Glass, Honeycombing, Reticular, and Bronchial Thickening reach 0.8. The mean AUC for all findings are higher than 0.55. We compare the performance with and without lung masks and report the mean and standard deviation of AUC in Fig. 5. From the figure we can see that the two experimental settings have similar performances and neither is consistently better than the other on all findings.

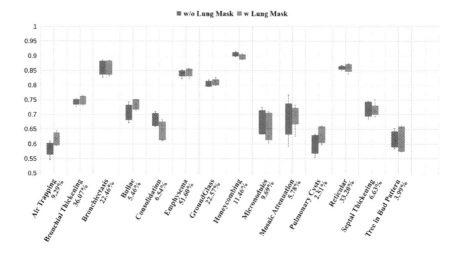

Fig. 5. Testing Classification performance. The prevalence along with the mean 5-fold cross validation AUC for each finding and finding label is shown along the x-axis. Mean AUC (indicated by x), mean quartiles (indicated by small cirles), median quartiles (indicated by box), and range are shown in the y-axis for training with and without a lung mask.

We observe that the mean is higher and standard deviation is consistently lower when the findings have more balanced positive/negative samples. Pulmonary Cysts and Tree in Bud Pattern are challenging because of low prevalence, but achieve reasonable results with only <5% positive samples. Air Trapping has the potential to be confused with Mosaic Attenuation, and shows a relatively weak performance.

To the best of our knowledge, we are the first to conduct findings classification using full lung 3D input over a dataset of thousands of CT samples. All previous works either perform patch wise classification [1,3], which effectively requires

Table 2. Prevalence and AUC scores for each finding within the LTRC dataset. The first row shows the positive rates for the corresponding finding, the second row shows mean AUC scores of 5-folds cross validation. The result without/with mask is are the left/right respectively.

Air Trapping	Bronchial Thickening	Bronchiectasis	Bullae	Consolidation
9.29%	36.07%	22.46%	5.46%	6.54%
0.5845 | 0.6175	0.7435 | 0.7476	0.8585 | 0.8573	0.7045 | 0.7372	0.6841 | 0.6454
Emphysema	Ground Glass	Honeycombing	Micronodules	Mosaic Attenuation
51.60%	22.57%	11.46%	9.69%	5.78%
0.8419 | 0.8455	0.8048 | 0.8093	0.9053 | 0.8979	0.6671 | 0.6586	0.6892 | 0.6975
Pulmonary Cysts	Reticular	Septal Thickening	Tree in Bud Pattern	
2.51%	33.20%	6.63%	3.99%	
0.6037 | 0.6293	0.8600 | 0.8606	0.7177 | 0.7141	0.6136 | 0.6123	

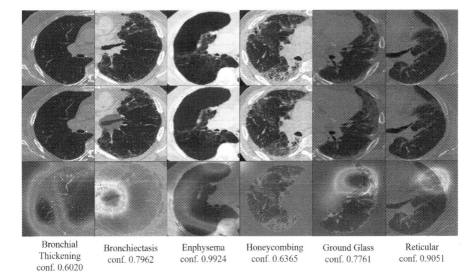

| Bronchial Thickening conf. 0.6020 | Bronchiectasis conf. 0.7962 | Enphysema conf. 0.9924 | Honeycombing conf. 0.6365 | Ground Glass conf. 0.7761 | Reticular conf. 0.9051 |

Fig. 6. Testing Localization examples. Columns correspond to imaging findings. First row show unlabeled axial slide. Second row shows manual voxel-wise labeling of the finding of interest. Third row shows results of generated attention map localization.

segmentation labeling, or classification on 2D slices, which requires dense segmentation [7] or low-level feature engineering [13]. Our work is the first to utilize image-level labels, capable of being derived from radiological reports, without further supervision.

3.4 Qualitative Analysis

We visualize sample cases from the testing data of the first fold for disease localization in Fig. 6. The Grad-CAM localization (third-row) agrees well with the lesion segmented by a human reader (second row). Within these cases our model demonstrates good localization capability for the underlying lesion despite not being trained on image segmentation or dense labels.

4 Conclusion

In this paper we present a weakly supervised learning method for identifying radiologic findings associated with ILD. Our network receives a full lung CT study as input and directly outputs a multi-label classification, applying Grad-CAM attention mapping to localize findings. To our knowledge, we are the first to demonstrate the feasibility of weakly supervised learning on full 3D CT volumes, for not only classifying but approximately localizing a group of findings with heterogeneous appearance, locus, and spatial extent. Our work does not require expensive dense labels, and can scale to very large datasets with findings labels mined automatically from electronic medical records or radiology reports.

References

1. Anthimopoulos, M., Christodoulidis, S., Ebner, L., Christe, A., Mougiakakou, S.: Lung pattern classification for interstitial lung diseases using a deep convolutional neural network. IEEE Trans. Med. Imaging **35**(5), 1207–1216 (2016)
2. Bartholmai, B.J., et al.: Quantitative CT imaging of interstitial lung diseases. J. Thoracic Imaging **28**(5) (2013)
3. Bermejo-Peláez, D., Ash, S.Y., Washko, G.R., Estépar, R.S.J., Ledesma-Carbayo, M.J.: Classification of interstitial lung abnormality patterns with an ensemble of deep convolutional neural networks. Sci. Rep. **10**(1), 1–15 (2020)
4. Choi, Y., Choi, M., Kim, M., Ha, J.W., Kim, S., Choo, J.: StarGAN: unified generative adversarial networks for multi-domain image-to-image translation. In: Proceedings of the IEEE Conference on Computer Vision and Pattern Recognition, pp. 8789–8797 (2018)
5. Cosgrove, G.P., Bianchi, P., Danese, S., Lederer, D.J.: Barriers to timely diagnosis of interstitial lung disease in the real world: the intensity survey. BMC Pulm. Med. **18**(1), 9 (2018)
6. Durand, T., Mordan, T., Thome, N., Cord, M.: WILDCAT: weakly supervised learning of deep convnets for image classification, pointwise localization and segmentation. In: Proceedings of the IEEE Conference on Computer Vision and Pattern Recognition, pp. 642–651 (2017)
7. Gao, M., et al.: Holistic classification of CT attenuation patterns for interstitial lung diseases via deep convolutional neural networks. Comput. Methods Biomech. Biomed. Eng. Imaging Vis. **6**(1), 1–6 (2018)
8. Harrison, A.P., Xu, Z., George, K., Lu, L., Summers, R.M., Mollura, D.J.: Progressive and multi-path holistically nested neural networks for pathological lung segmentation from CT images. In: Descoteaux, M., Maier-Hein, L., Franz, A., Jannin, P., Collins, D.L., Duchesne, S. (eds.) MICCAI 2017. LNCS, vol. 10435, pp. 621–629. Springer, Cham (2017). https://doi.org/10.1007/978-3-319-66179-7_71
9. Hu, J., Shen, L., Sun, G.: Squeeze-and-excitation networks. In: Proceedings of the IEEE Conference on Computer Vision and Pattern Recognition, pp. 7132–7141 (2018)
10. Irvin, J., et al.: CheXpert: a large chest radiograph dataset with uncertainty labels and expert comparison. In: Proceedings of the AAAI Conference on Artificial Intelligence, vol. 33, pp. 590–597 (2019)
11. Johnson, A.E., et al.: Mimic-CXR: a large publicly available database of labeled chest radiographs. arXiv preprint arXiv:1901.07042 1(2) (2019)
12. Selvaraju, R.R., Cogswell, M., Das, A., Vedantam, R., Parikh, D., Batra, D.: Grad-CAM: visual explanations from deep networks via gradient-based localization. In: Proceedings of the IEEE International Conference on Computer Vision, pp. 618–626 (2017)
13. Wang, C., et al.: Weakly-supervised deep learning of interstitial lung disease types on CT images. In: Medical Imaging 2019: Computer-Aided Diagnosis, vol. 10950, p. 109501H. International Society for Optics and Photonics (2019)
14. Wang, X., Peng, Y., Lu, L., Lu, Z., Bagheri, M., Summers, R.M.: ChestX-ray8: hospital-scale chest X-ray database and benchmarks on weakly-supervised classification and localization of common thorax diseases. In: Proceedings of the IEEE Conference on Computer Vision and Pattern Recognition, pp. 2097–2106 (2017)

15. Wang, X., Peng, Y., Lu, L., Lu, Z., Summers, R.M.: TieNet: text-image embedding network for common thorax disease classification and reporting in chest X-rays. In: Proceedings of the IEEE Conference on Computer Vision and Pattern Recognition, pp. 9049–9058 (2018)
16. Woo, S., Park, J., Lee, J.-Y., Kweon, I.S.: CBAM: convolutional block attention module. In: Ferrari, V., Hebert, M., Sminchisescu, C., Weiss, Y. (eds.) ECCV 2018. LNCS, vol. 11211, pp. 3–19. Springer, Cham (2018). https://doi.org/10.1007/978-3-030-01234-2_1
17. Yuan, J., Liao, H., Luo, R., Luo, J.: Automatic radiology report generation based on multi-view image fusion and medical concept enrichment. In: Shen, D., et al. (eds.) MICCAI 2019. LNCS, vol. 11769, pp. 721–729. Springer, Cham (2019). https://doi.org/10.1007/978-3-030-32226-7_80

Learning with Sure Data
for Nodule-Level Lung Cancer Prediction

Hanxiao Zhang[1(✉)], Yun Gu[1], Yulei Qin[1], Feng Yao[2], and Guang-Zhong Yang[1]

[1] Institute of Medical Robotics, Shanghai Jiao Tong University, Shanghai, China
hanxiao.zhang@sjtu.edu.cn
[2] Shanghai Chest Hospital, Shanghai, China

Abstract. Recent evolution in image-based disease prediction based on deep learning has significantly extended the clinical capabilities of these systems. However, in certain cases (e.g. lung nodule prediction), ground truth labels manually annotated by radiologists (unsure data) are often based on subjective assessment, which lack pathological-proven benchmarks (sure data) at the nodule-level. To address this issue, we build a small yet definite CT dataset (171 patients) called SCH-LND focusing on solid lung nodules (90 benign/90 malignant cases). Under the supervision of SCH-LND dataset, many hidden drawbacks of unsure data (484 solid nodules selected from LIDC-IDRI dataset) served for malignancy prediction are objectively revealed. Explanations to this phenomenon are inferred in this paper from the view of model training and data annotation bias. Although learning from scratch over sure data with commonly used model can surpass the performance of unsure data in large scales, we additionally propose two frameworks to make the best use of these cross-domain resources, among which, transfer learning is verified as an effective approach for LIDC-IDRI knowledge adaptation. Results show that the proposed method can achieve good performance for nodule-level malignancy prediction with a small SCH-LND dataset.

Keywords: Lung nodule · Malignancy prediction · Deep learning · Transfer learning

1 Introduction

Data-driven techniques are advantageous for fast and robust computer-aided diagnosis. In practice, however, it is difficult to define definite annotations due to insufficient patient information. For instance, in Lung Image Database Consortium and Image Database Resource Initiative (LIDC-IDRI) [1], all the nodule characteristics were independently assessed by multiple radiologists, in which the rating of "likelihood of malignancy" can be especially subjective.

This is defined as "unsure data" by its nature of uncertainty at the early stage of disease [12]. It is hard to assign a binary label for unsure cases (e.g. radiologists' rating on a five-point scale for malignancy in LIDC-IDRI). To avoid misdiagnosis, further examinations are often required.

© Springer Nature Switzerland AG 2020
A. L. Martel et al. (Eds.): MICCAI 2020, LNCS 12266, pp. 570–578, 2020.
https://doi.org/10.1007/978-3-030-59725-2_55

In common practice, different radiologists' decisions based on selected inter-section nodules are averaged to achieve more stable and confident labels, as described in [2]. To date, most of the relevant literatures adopt such a prepara-tion method for nodule malignancy annotation. Performance can be improved by using additional semantic information (e.g. high-level nodule attributes [5], nodule segmentation results [13], discovery radiomics [7]), fusing predictions from multi-scale nodule patches [11], or employing pre-trained model from other domains [4]. However, without the verification of pathological-proven labels, the accuracy of these methods can be quite questionable.

LIDC-IDRI dataset also contains a set of cases (157 patients) with diagnosis data at the patient-level [8] where four ratings (0: unknown, 1: non-malignant dis-ease, 2: primary lung cancer, 3: metastatic lesion) were recorded along with five diagnosis methods. Based on this small diagnostic dataset, Shen et al. [10] devel-oped a framework for patient-level lung cancer prediction using multiple instance learning (MIL), indicating the potential of using definite pathological-proven CT data for lung cancer diagnosis. However, the limitations of LIDC-IDRI diagnosis dataset (e.g. lack of nodule detection, trouble of multi-classification, inter-class imbalance problem) hinder end-to-end individual nodule prediction, which is still regarded as unsure data at the nodule-level.

To alleviate the limitations of LIDC-IDRI dataset mentioned above, we pro-pose a small yet definite dataset called Shanghai Chest Hospital Lung Nod-ule Dataset (SCH-LND), with details described in Sect. 4.1. Given the accurate benchmark to judge the nodule malignancy at the nodule-level, we seek to inter-pret the questions as follows: What's the effect of unsure data (LIDC-IDRI) on target model training? Whether the nodule-level sure data SCH-LND is infor-mative for model learning? How to make the best use of SCH-LND dataset to yield better performance in nodule classification?

We answer the first two questions with detailed experiments in Sect. 2. Although good performance can be achieved by a commonly used model trained and tested only based on the LIDC-IDRI dataset, certain drawbacks are evident as this model is revalidated under the reference of SCH-LND dataset. We also find that with single sure data one can hardly achieve better performance due to the small quantity of the data and the difficulty to learn discriminative features, our challenge turns to the third question which is looking for suitable approaches of integrating all the nodule resources.

In this study, we firstly formulate the problem as a multi-confidence learning problem and use an end-to-end network to model the mixed data with multi-class annotations. However, its performance is found to be limited by the data source bias towards the majority classes. Recently, transfer learning has shown to be generalizable across different medical image domains [15]. Inspired by this work, we put forward a CNN framework that is pre-trained on the large unsure dataset (484 nodules) to obtain low-level radiologists' knowledge and fine-tuned by the small sure dataset (180 nodules), aiming to enhance the deep feature extraction at the nodule-level.

Overall, our main contributions in this paper can be summarized as follows: (1) We build a lung nodule dataset with definite and balanced malignancy labels at the nodule-level. (2) Verified by the SCH-LND dataset, we objectively assess the practicability of LIDC-IDRI for malignancy prediction and give explanation for each anomaly occurring with empirical evidence. (3) Further experiments are conducted to identify the problems of unsure data and feasibilities when integrating the cross-domain datasets for better malignancy determination. (4) We demonstrate how transfer learning is used to effectively adapt the knowledge from LIDC-IDRI to SCH-LND to enhance nodule prediction using limited sure data.

2   Preliminary Work

In the majority of related works, people often use Accuracy and AUC (Area Under the Curve) for evaluating the malignancy predictions. Wu et al. [12] argued that the benign and positive samples should be treated differently based on the conservative/aggressive strategy that the former should be assessed with Precision in benign class (Precision$_b$ treats benign as positive sample) and the latter should be assessed with Recall in malignant class. We additionally introduce Precision and Specificity in malignant class to observe the performance in detail.

Table 1. Performances of different experiments using 3D ResNet-18. Set P = 0.5 as a threshold (P < 0.5: benign, P geq 0.5: malignant). Precision$_b$ represents Precision in benign class.

	Training	Testing	Accuracy	Precision	Specificity	AUC	Recall	Precision$_b$
1	LIDC	LIDC	**0.8889**	**0.9043**	**0.8929**	**0.9536**	0.8854	0.8721
2	LIDC	SCH	0.6056	0.5605	0.2333	0.6333	**0.9778**	**0.9130**
3	SCH	SCH	0.6167	0.6129	0.6000	0.6552	0.6333	0.6207

The first experiment is made based on LIDC-IDRI dataset using a 3D ResNet-18 [3] with 304 nodule volumes randomly selected to form the training set, from which 10% of the training set is used for validation, and the remaining 180 nodule volumes selected for testing. Data information and other training details are described in Sect. 4. The result (1st row in Table 1) shows that good performance can be easily achieved by using a commonly used model trained simply on unsure data. However, the drawbacks of this unsure dataset begin to reveal in the second experiment where the same model is trained over the whole LIDC-IDRI dataset and tested on SCH-LND dataset. As shown in the 2nd row of Table 1, the result falls sharply in terms of Accuracy, Precision, Specificity and AUC, but abnormal growth happens in both Recall and Precision$_b$. The reasons for causing these phenomena are analyzed as follows:

Explain the Good Performance in the First Experiment: Due to lack of suitable reference to assess the likelihood of malignancy, low-level and visible features (e.g. size, shape, brightness) are likely to be regarded as scoring criteria by radiologists' observation. Built on consensus agreement within multiple radiologists, apparent features of these nodules can be easily extracted and classified by a commonly used model, whose power can successfully emulate the radiologist's one. However, without the sure data, this model will always take the human ability as golden standard rather than real malignancy labels. This is our first motive to build SCH-LND dataset.

Explain the Poor Result in the Second Experiment: Although this result satisfies the requirement of conservative and aggressive strategy, other evaluation metrics fail to meet our expectation. According to the high value in Recall and Precision$_b$ but the low value in Precision and Specificity, we can conclude that the model decisions contain too many false-positive predictions. Consequently, inference can be given that, in LIDC-IDRI, the radiologists have misdiagnosed plenty of benign nodules to high scores of malignancy, which yield inaccurate features during model training. It is reasonable that in real life people often assume the worst to prevent worse, but this situation should not happen for prediction models. Thus, we want to collect a certain number of benign nodules with definite labels and locations to correct this bias at the nodule-level. This is our second motive to build SCH-LND dataset.

To validate the effectiveness of SCH-LND dataset for nodule prediction, the third experiment adopts 5-fold cross-validation over SCH-LND dataset with the same model as the second experiment. The results are shown in the 3rd row of Table 1. Comparing the second experiment on metrics, growth has been witnessed in Accuracy, Precision, Specificity and AUC at the cost of the drops in Recall and Precision$_b$. By training with sure data, this model effectively reduces the false-positive predictions but more false-negative predictions appear due to limited model capacity and small dataset.

The current challenge remains as how can we develop the framework to make the best use of the integrated cross-domain resources to yield a comprehensive high performance in nodule prediction. Two independent approaches are put forward in the following section.

3   Methods

3.1   Multi-confidence Learning with Mixed Labels

Given the fact that the sure data has the priority to judge the nodule malignancy, we hypothesize that allocating various confidence levels to data could not only help fuse two datasets for richer feature representation, but also increase the weight of sure data during training. Thus, in this approach, we apply the multi-confidence learning for nodule prediction by adding the factor of uncertainty during annotation. We change the binary-classification label method of LIDC-IDRI dataset that described in Sect. 4.1 and re-assign new labels to each nodule,

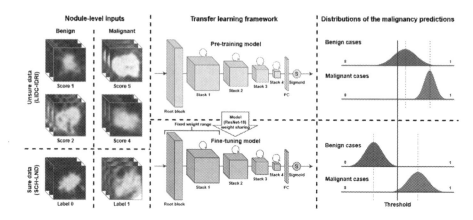

Fig. 1. Illustration of the framework of our proposed methods for nodule cancer prediction. Left: the 3D nodule-level inputs of unsure (LIDC-IDRI) and sure data (SCH-LND) are cropped from CT scans of each dataset, whose annotations can be given according to the malignancy scores or pathological-proven labels. Middle: a commonly used model (3D ResNet-18) is firstly trained on the unsure data with/without multi-confidence learning. The weights of the pre-trained network are shared for fine-tuning by domain transfer to model sure data. To enhance the training efficiency, we can fix the weights shared in root block and stack1 and fine-tune the remaining layers. Right: the model decisions contain too many false-positive predictions if trained on LIDC-IDRI, which can be corrected by fine-tuning on sure data.

where benign nodules with the highest confidence level (average score 1) are annotated with 0.2 and the second (average score 2) are annotated with 0.4; malignant nodule with the highest confidence level (average score 5) is annotated with 0.8 and the second (average score 4) is annotated with 0.6. No label change happens for SCH-LND. Then we evaluate this approach trained on mixed data with 6 classes (both LIDC-IDRI and SCH-LND are provided), and tested on SCH-LND dataset with 5-fold cross-validation.

3.2 Transfer Learning

We perform separate training rather than training with mixed data in this approach. Given the preliminary experiment report in Sect. 2, we note that features defined by radiologists and pathological examinations are different. Hence, it is reasonable to assume that the knowledge learned from radiologists' experience is a useful resource, but currently lacks a strong supervision signal with definite information. Since learning from scratch is difficult for sure data to determine latent features efficiently, we employ the low-level features as a stepping stone and bridge over the knowledge from unsure to sure data.

Based on this inference, we propose a framework (Fig. 1) with two modes for domain transfer from big unsure data to small sure data for nodule malignancy prediction. **For Mode 1,** we pre-train the model over LIDC-IDRI with another multi-confidence label fashion (score 1-label 0, score 2-label 0.25, score 4-label 0.75, score 5-label 1). The purpose of learning from scratch on unsure data is to generate a common visual representation close to the ability of the radiologists. The weights of this model are then mapped to the other domain with SCH-LND dataset for the follow-up fine-tuning step. **For Mode 2,** we replace the pre-trained model with binary-class model that we used in the second experiment in Sect. 2.

Considering the first few layers may preserve the common features, we make trials in Sect. 4 to fix the weights shared in different preceding modules (root block, stack1) while the weights in the remaining part can be optimized during fine-tuning under the supervision of sure data, which may enhance feature propagation and increase the learning efficiency.

4   Experiments and Results

4.1   Dataset

Unsure Data. The unsure dataset used in this work comes from LIDC-IDRI database, which consists of 1018 CT scans, where nine characteristics of the nodules were given by multiple radiologists. We excluded scans with slice thickness greater than 3 mm and sampled nodules accepted by at least three radiologists as in [9]. In this study, only solid nodules (texture $= 5$) include because their malignancy is much more challenging to annotate than ground-glass opacity (GGO). We followed the nodule relabel and extraction method in [2] that choose average score 3 as a division point (benign < 3: label 0; malignant > 3: label 1). Our final unsure dataset consists of 223 benign nodules and 261 malignant nodules.

Sure Data. The sure dataset called SCH-LND consists of 180 solid nodules (90 benign/90 malignant) collected from 171 patients' CT scans with ethical approval. All the cases were annotated to single definite class (benign or malignant) diagnosed by pathological-proven examination via biopsy.

Due to varied slice thicknesses of CT scans, both the CT scans from sure and unsure datasets were resampled to 1 mm/pixel along z axis if their slice thickness is above 1 mm/pixel. We fixed the x and y axes to 512×512 pixels rather than making resampling to prevent distortion. Nodule volumes were cropped from the resampled CT scans according to their spatial coordinates and radii.

4.2   Implementation

We use 3D ResNet-18 [3] throughout this paper where the output (512 channels) of convolutional network is down-sampled by adaptive average pooling to the size of $1 \times 1 \times 1$. Finally, a fully connected layer followed by a Sigmoid function is applied to produce the predicted score. Due to various sizes of nodule inputs that share the same CNN, the batch size is set to 1 and group normalization [14] is used after each convolution operation. The loss function is chosen using L2 loss between the predicted score and malignancy label.

All the experiments were implemented in PyTorch with a single NVIDIA GeForce GTX 1080 Ti GPU and trained using the Adam optimizer [6] with the learning rate of 1e−3 for pre-training (100 epochs) and that of 1e−4 for fine-tuning (50 epochs). No augmentation method is applied to any dataset in this paper. We did not implement any reference models from other literatures due to different research objectives. Others mainly focus on how to achieve better performance within the same unsure domain, while we intend to explore the new domain with sure data for training and testing and the way of the best use of it.

Table 2. Performances of models trained or pre-trained using multi-confidence data by 5-fold cross-validation. mc, pre and fx denote multi-confidence learning, pre-training and fixed weights in module x, respectively.

	Training	Testing	Accuracy	Precision	Specificity	AUC	Recall	Precision$_b$
1	LIDC+SCH	SCH	0.5944	0.5578	0.2778	**0.6564**	0.9111	0.7576
2	LIDC(mc)+SCH	SCH	0.5889	0.5615	0.3667	0.6409	0.8111	0.6600
3	LIDC(mc)	SCH	**0.6222**	0.5724	0.2778	0.6551	**0.9667**	**0.8929**
4	LIDC(mc, pre)+SCH	SCH	0.5722	0.5631	**0.5000**	0.6328	0.6444	0.5844
5	LIDC(mc, pre)+SCH(f1)	SCH	0.5889	0.5678	0.4333	0.6249	0.7444	0.6290
6	LIDC(mc, pre)+SCH(f12)	SCH	0.6056	**0.5748**	0.4000	0.6336	0.8111	0.6792

Table 3. Performances of transfer learning by 5-fold cross-validation.

	Training	Testing	Accuracy	Precision	Specificity	AUC	Recall	Precision$_b$
1	LIDC(pre)+SCH	SCH	0.6833	0.6634	0.6222	**0.7486**	0.7444	0.7089
2	LIDC(pre)+SCH(f1)	SCH	0.7056	0.6762	0.6222	0.7321	**0.7889**	**0.7467**
3	LIDC(pre)+SCH(f12)	SCH	**0.7111**	**0.6939**	**0.6667**	0.7401	0.7556	0.7317

4.3   Quantitative Evaluation

Compared with the second and third experiments in Sect. 2, the performances of the first two experiments in Table 2 with model trained by mixed data do not improve the situation that using single training data (2nd and 3rd row in Table 1). Because the unsure data constitutes a major part of mixed data training, the performance is found limited by the data source bias towards the unsure data class. Comparing within the mixed data training experiments (1st and 2nd row in Table 2), multi-confidence learning seems to have a worse overall result. We also test the multi-confidence learning in transfer learning Mode 1. Although learning from scratch with multi-confidence data (3rd row in Table 2) outperforms the corresponding binary-class training model (2nd row in Table 1), the fine-tuning results (4th to 6th row in Table 2) even get worse due to the negative impact of the pre-trained model with multi-confidence learning. According to these undesirable

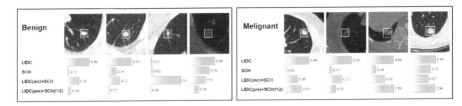

Fig. 2. Comparisons between different methods in terms of malignancy predicted scores below each nodule marked by yellow box. (Color figure online)

performances, we re-affirm the existence of some hidden inter-class problems of LIDC-IDRI dataset that deserve further investigation.

Since using binary classification can avoid some trouble, we apply this method in transfer learning with Mode 2. The results in Table 3 and Fig. 2 show that Mode 2 can achieve a remarkable improvement over training from scratch either with LIDC-IDRI or SCH-LND dataset in Table 1, increasing the AUC, Accuracy and Precision by at least 8 points, 6 points and 5 points, respectively. Our extensive experiments with fixed weights modules demonstrate that significant performance gain can be achieved in this way to reconcile the trade-off between the demand of conservative/aggressive strategy and other metrics in evaluation system.

5   Conclusion and Future Work

Focusing on the malignancy prediction of solid lung nodules, we first challenged in this paper about the model ability powered by unsure data of LIDC-IDRI used in common practice and verified its hidden weaknesses by comprehensive assessment based on a small newly built dataset (SCH-LND) with definite pathological-proven labels. Furthermore, we investigated the significance of SCH-LND dataset as the dual role of golden standard and strong supervision signal at the nodule-level. Two individual approaches were put forward to integrate cross-domain resources and came to some conclusions: (1) multi-confidence learning helps to confirm the subjective bias problem of unsure data; (2) transfer learning is proven as an effective way for knowledge adaptation where the features defined by radiologists can be learned by the pre-trained model and then transferred to the domain of sure data. The empirical results support the evidence that with limited demand for sure data, transfer learning can achieve remarkable performance which is beneficial for computer-aided diagnosis.

As an extension to this study, our future work will investigate the label assignment strategy of unsure data malignancy scores based on the advice from sure data supervision feedback. Under such strategy and the reference of inter-observer variability, nodules with average malignancy score equalling to 3 in LIDC-IDRI can also be included for training. Moreover, an observer study with a group of radiologists could also be conducted afterwards to provide convincing judgments of our research.

References

1. Armato III, S.G.: The lung image database consortium (LIDC) and image database resource initiative (IDRI): a completed reference database of lung nodules on CT scans. Med. Phys. **38**(2), 915–931 (2011)
2. Han, F., et al.: A texture feature analysis for diagnosis of pulmonary nodules using LIDC-IDRI database. In: 2013 IEEE International Conference on Medical Imaging Physics and Engineering, pp. 14–18. IEEE (2013)
3. He, K., Zhang, X., Ren, S., Sun, J.: Deep residual learning for image recognition. In: Proceedings of the IEEE Conference on Computer Vision and Pattern Recognition, pp. 770–778 (2016)
4. Hussein, S., Cao, K., Song, Q., Bagci, U.: Risk stratification of lung nodules using 3D CNN-based multi-task learning. In: Niethammer, M., et al. (eds.) IPMI 2017. LNCS, vol. 10265, pp. 249–260. Springer, Cham (2017). https://doi.org/10.1007/978-3-319-59050-9_20
5. Hussein, S., Gillies, R., Cao, K., Song, Q., Bagci, U.: TumorNet: lung nodule characterization using multi-view convolutional neural network with Gaussian process. In: 2017 IEEE 14th International Symposium on Biomedical Imaging (ISBI 2017), pp. 1007–1010. IEEE (2017)
6. Kingma, D.P., Ba, J.: Adam: a method for stochastic optimization. arXiv preprint arXiv:1412.6980 (2014)
7. Kumar, D., Chung, A.G., Shaifee, M.J., Khalvati, F., Haider, M.A., Wong, A.: Discovery radiomics for pathologically-proven computed tomography lung cancer prediction. In: Karray, F., Campilho, A., Cheriet, F. (eds.) ICIAR 2017. LNCS, vol. 10317, pp. 54–62. Springer, Cham (2017). https://doi.org/10.1007/978-3-319-59876-5_7
8. McNitt-Gray, M.F., et al.: The lung image database consortium (LIDC) data collection process for nodule detection and annotation. Acad. Radiol. **14**(12), 1464–1474 (2007)
9. Setio, A.A.A., et al.: Validation, comparison, and combination of algorithms for automatic detection of pulmonary nodules in computed tomography images: the LUNA16 challenge. Med. Image Anal. **42**, 1–13 (2017)
10. Shen, W., et al.: Learning from experts: developing transferable deep features for patient-level lung cancer prediction. In: Ourselin, S., Joskowicz, L., Sabuncu, M.R., Unal, G., Wells, W. (eds.) MICCAI 2016. LNCS, vol. 9901, pp. 124–131. Springer, Cham (2016). https://doi.org/10.1007/978-3-319-46723-8_15
11. Shen, W., Zhou, M., Yang, F., Yang, C., Tian, J.: Multi-scale convolutional neural networks for lung nodule classification. In: Ourselin, S., Alexander, D.C., Westin, C.-F., Cardoso, M.J. (eds.) IPMI 2015. LNCS, vol. 9123, pp. 588–599. Springer, Cham (2015). https://doi.org/10.1007/978-3-319-19992-4_46
12. Wu, B., Sun, X., Hu, L., Wang, Y.: Learning with unsure data for medical image diagnosis. In: Proceedings of the IEEE International Conference on Computer Vision, pp. 10590–10599 (2019)
13. Wu, B., Zhou, Z., Wang, J., Wang, Y.: Joint learning for pulmonary nodule segmentation, attributes and malignancy prediction. In: 2018 IEEE 15th International Symposium on Biomedical Imaging (ISBI 2018), pp. 1109–1113. IEEE (2018)
14. Wu, Y., He, K.: Group normalization. In: Proceedings of the European Conference on Computer Vision (ECCV), pp. 3–19 (2018)
15. Zhou, Z., et al.: Models genesis: generic autodidactic models for 3D medical image analysis. In: Shen, D., et al. (eds.) MICCAI 2019. LNCS, vol. 11767, pp. 384–393. Springer, Cham (2019). https://doi.org/10.1007/978-3-030-32251-9_42

Cascaded Robust Learning at Imperfect Labels for Chest X-ray Segmentation

Cheng Xue[1(✉)], Qiao Deng[1], Xiaomeng Li[1,2], Qi Dou[1], and Pheng-Ann Heng[1,3]

[1] Department of Computer Science and Engineering,
The Chinese University of Hong Kong, Sha Tin, Hong Kong
cxue@cse.cuhk.edu.hk
[2] Department of Radiation Oncology, Stanford University, Stanford, USA
[3] Guangdong Provincial Key Laboratory of Computer Vision and Virtual Reality
Technology, Shenzhen Institutes of Advanced Technology,
Chinese Academy of Sciences, Shenzhen, China

Abstract. The superior performance of CNN on medical image analysis heavily depends on the annotation quality, such as the number of labeled images, the source of images, and the expert experience. The annotation requires great expertise and labor. To deal with the high inter-rater variability, the study of the imperfect label has great significance in medical image segmentation tasks. In this paper, we present a novel cascaded robust learning framework for chest X-ray segmentation with imperfect annotation at the boundary. Our model consists of three independent networks, which can effectively learn useful information from peer networks. The framework includes two stages. In the first stage, we select the clean annotated samples via a model committee setting, the networks are trained by minimizing a segmentation loss using the selected clean samples. In the second stage, we design a joint optimization framework with label correction to gradually correct the wrong annotation and improve the network performance. We conduct experiments on the public chest X-ray image datasets collected by Shenzhen Hospital. The results show that our methods could achieve a significant improvement on the accuracy in segmentation tasks compared to the previous methods.

Keywords: Robust learning · Imperfect label · Lung segmentation

1 Introduction

Deep neural networks (DNNs) have achieved human-level performance on many medical image analysis tasks, such as melanoma diagnosis [5], pulmonary nodules detection [14], retinal disease [4], and lumpy node metastases detection [1]. These outstanding performances heavily rely on massive training data with high-quality annotations. Annotation of medical images, especially for pixel-level annotation for segmentation tasks, is costly and time-consuming. The process is experience-prone, while the annotations from different clinical experts may have disagreements that are usually inevitable for the blurred boundary of lesions and organs.

© Springer Nature Switzerland AG 2020
A. L. Martel et al. (Eds.): MICCAI 2020, LNCS 12266, pp. 579–588, 2020.
https://doi.org/10.1007/978-3-030-59725-2_56

Previous studies show that the DNNs trained by noisy labeled datasets can cause performance degradation. That is because the huge memory capacity and strong learning ability of DNNs can remember the noisy labels and easily overfit to them [15,18,19]. Tackling the issue of annotation noises is a complicated and challenging topic. Manually reducing the presence of incorrect labels, for example by requiring a stronger committee of expert clinicians to do labelling, has to be expensive, time-consuming and impractical. In this paper, we address this problem in the insight of robust learning with the noisy labelled data inherent in the training procedure. Many studies have addressed the issue of the noisy label in medical analysis community. Goldberger et al. [6] added an additional softmax layer to estimate the correct labels. Xue et al. [17] proposed to consider the noisy sample and hard sample by an on-line sample selection module and re-weighting module. Zhu et al. [19] proposed the automatic quality evaluation module and overfitting control module to update the network parameters. Shu et al. [15] presented an LVC-Net losses function by combining noisy labels with image local visual cues to generate better semantic segmentation. Most of the approaches adopted the strategy of selecting samples for training [17,19], exhibited their feasibility in robust learning. However, these methods exist a strong accumulated error caused by sample selection bias. The wrongly selected samples will influence the network performance and further decrease the quality of selected samples. Le et al. [10] addressed the sample selection bias issue by utilized a small set of clean training samples to assign weights to training samples. The main drawback of this approach was the extra clean labels were usually unavailable in the real-world scenarios.

To tackle the challenging problem of noisy labeled segmentation masks, we present a cascaded learning framework for lung segmentation using the X-ray images with imperfectly annotated ground truth. In the first stage, our framework selects clean annotated samples according to the prediction confidence and uncertainty of samples, which is inspired by the ideas of Co-teaching [7]. Specifically, our model consists of three independent networks being trained simultaneously, each network is real-time updated according to the prediction results of the other two networks. For a clean annotated sample, the three networks tend to produce high confidence prediction with smaller inter-rater variance. Thus, the samples with close prediction and high confidence are selected as the high-quality sample, which will be used to contribute to the weight backpropagation process. Since the selection stage leads to a low utilization efficiency of the valuable training data, we propose a label correction module in the second stage, which can correct the imperfect label. Furthermore, a joint optimization scheme is designed to cooperatively supervise the three networks with the original label and the corrected one. Our method was extensively evaluated on the Shenzhen chest x-ray dataset [3,8,16]. The results demonstrate a good capability of our method to the issue of the noisy labeled boundary, that the cascaded robust learning framework can more accurately perform the lung segmentation comparing to other methods.

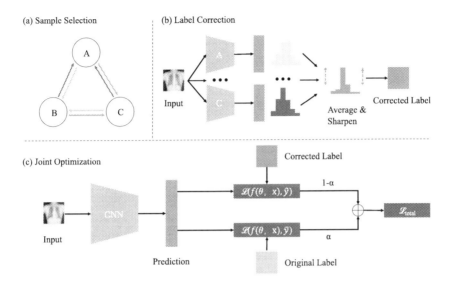

Fig. 1. Illustration of the pipeline of our cascaded robust learning framework. (a) shows the first sample selection stage, where three networks trained independently, but updated according to the prediction of the other two peer networks. (b) and (c) are the second stage. (b) shows our proposed label correction module, using the average prediction of three networks followed by a sharpening function to produce the corrected label $\bar{y}$. (c) shows the joint optimization scheme, the network is supervised by the original label $\hat{y}$ and the corrected label $\bar{y}$. The final output is given by the average of the three networks.

2   Method

Figure 1 illustrates the framework of our cascaded robust learning method. In the first stage, we design a three-networks sample selection module. The module filters the clean samples and updates the three networks with the selected clean samples. In the second stage, our method starts to correct the imperfect labels, then use the corrected label and original label to jointly optimize the three networks.

2.1   Sample Selection Stage

We study the task of chest x-ray segmentation, where the training set contains images x and noisy labeled ground truth $\hat{y}$, while the clean ground truth y is unknown. The goal for this fully supervised segmentation task is to minimize the following object function:

$$\min_{\theta} \sum_{i=1}^{N} \mathcal{L}(f(x_i; \theta), \hat{y}_i) \tag{1}$$

where $\mathcal{L}$ denotes the loss function (e.g., cross-entropy loss) to evaluate the quality of the network output on inputs. $f(\theta)$ denotes the segmentation neural network with weights θ.

Recent studies show that by updating the network with high confidence samples can improve the robustness to noisy labels [7,9,12]. Therefore, we propose a novel sample selection framework (SS) to select high confidence samples as the useful training instances. Our framework consisted of three independent networks, where they have identical architecture. We adopt the vanilla U-Net [13] as the classifier in our experiment. In the training process, we select the high confidence samples with small uncertainty to update each network, because those samples are more likely to be clean labeled instances. In our experiment, we empirically select half batch data as useful information. Concretely, the three networks feed forward and predict the same mini-batch of data. Then for each network, the useful samples for weight updating are obtained by the other two networks as shown in Fig. 1(a). Taking network A as an example, the useful sample for network A is obtained from network B and C, where we first filter out the high uncertainty (μ) samples by excluding the ones showing disagreed prediction, then among the low uncertainty samples, the small loss samples was further selected as useful samples for network A. We employ the agreement between two models as uncertainty of each samples and calculate the uncertainty according to Eq. 2.

$$\mu = |\mathcal{L}(f_B(x_i; \theta_B), \hat{y}_i) - \mathcal{L}(f_C(x_i; \theta_C), \hat{y}_i)| \qquad (2)$$

where $\mathcal{L}$ denotes the cross-entropy loss. f_B and f_C denote the network B and network C. θ_B and θ_C represent the weight of network B and network C. Note that the three networks has different training parameters as they are updated by different selected samples in each mini-batch, they did not learn the bias in the noisy labels at the same speed, and μ is not close to 0.

2.2 Joint Optimization with Label Correction

In the stage of sample selection, only partial samples can be used for training, where it does not take full advantage of the imperfect training data. Therefore, we design a joint optimization (JO) framework to train the network with the original label and corrected label, so that the utilization efficiency of training data can be maintained. In order to correct the noisy label, we design a label correction module to work together with the joint optimization scheme.

Label Correction. The sample selection stage first trains an initial network by using image x with noisy label $\hat{y}$. Then we proceed to the label correction phase, as shown in Fig. 1 (b). We compute the average of three model' prediction in each iteration, that is followed by an entropy minimization step widely adopted in semi-supervised learning [2,11]. Specifically, for the average prediction of the three models, we apply a sharpening function to reduce the entropy of the per pixel label distribution through adjusting the temperature:

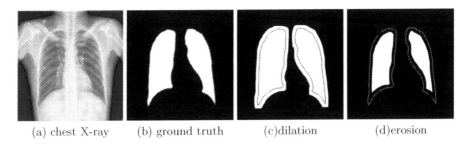

| (a) chest X-ray | (b) ground truth | (c)dilation | (d)erosion |

Fig. 2. Examples of the noisy annotations. Red line indicates the clean ground truth. (Color figure online)

$$q = \frac{1}{3}(f_A((x_i;\theta_A),\hat{y}_i) + f_B((x_i;\theta_B),\hat{y}_i) + f_C((x_i;\theta_C),\hat{y}_i))$$

$$\bar{y} = sharpen(q,T)_i = q_i^{\frac{1}{T}} / \sum_{j=1}^{L} q_j^{\frac{1}{T}} \tag{3}$$

where q is the average prediction feature map over three models, T is a hyperparameter that adjusts the temperature. As T closes to zero, the output of $Sharpen(q,T)$ will approach a one-hot distribution. Since we will use $\bar{y} = Sharpen(q,T)$ as a corrected target for the model's prediction later, following the setting of [2], $T = 0.5$ is chosen to encourage the model to produce lower-entropy prediction.

Joint Optimization. We start the joint optimization stage after k epochs of sample selection. For each uncertain sample, we produce a corrected label for the imperfect input by the label correction module. The corrected label is used in the training process together with the original label as a complementary supervision to jointly supervise the network:

$$\mathcal{L}_{total} = \alpha \times \mathcal{L}(f(x_i;\theta),\hat{y}_i) + (1-\alpha) \times \mathcal{L}(f(x_i;\theta),\bar{y}_i) \tag{4}$$

where $\mathcal{L}$ is the cross entropy loss, $\hat{y}$ is the original noisy label, and $\bar{y}$ is the corrected label produced by the label correction phase. The weight factor α controls the weights of the two terms. In our study, we set $\alpha = 0.5$ that gives the best results.

3   Experiments

3.1   Dataset and Pre-processing

We evaluated our method on the public Shenzhen chest x-ray dataset [3,8,16], the segmentation masks were prepared manually by Computer Engineering Department, Faculty of Informatics and Computer Engineering, National Technical University of Ukraine. The dataset contains 566 chest x-ray images and

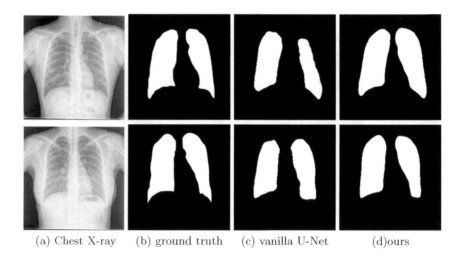

(a) Chest X-ray   (b) ground truth   (c) vanilla U-Net   (d)ours

Fig. 3. Two examples of the segmentation results in the test data by different methods. (a) is the input image, (b) is the ground truth. (c) and (d) shows the results of U-Net and our method under 75% noise ratio.

each image has the left and the right lungs. We split the 566 chest x-ray images into 396 images for training and 170 for evaluation. All the images were resized to 256×256, and normalized as zero mean and unit variance.

3.2   Implementation

The framework was implemented in PyTorch, using a TITAN Xp GPU. We used the Stochastic Gradient Descent optimizer to update the network parameters with weight decay of 0.001 and a momentum of 0.9. We adopt an exponential learning rate with an initial learning rate set as 0.001. We totally trained 100 epochs, the batch size was 32. We adopted the data augmentation including random rotation and random horizontal flipping. In order to produce noisy labels for the training data, we simulate imperfect annotation with noisy boundary in real word scenarios. We randomly selected (noise ratio) 25%, 50%, 75% samples from the training set to erode or dilate with the number of iterations (noise level) n between 5–15 ($5 \leq n \leq 15$). We adopted the dice coefficient as evaluation criteria for segmentation accuracy evaluation. Figure 2 shows the example of some noisy annotation of the segmentation mask.

3.3   Quantitative Evaluation

The experiments were conducted on the Chest X-ray dataset. We trained the network on the samples with different ratio of noisy labels and tested it by the clean labels. Table 1 presents the segmentation performance of vanilla U-Net (baseline) and our cascaded robust learning framework that were all trained by

Table 1. Comparison between our method and various methods.

Noise ratio	Noise level	Strategy	Dice (%)	k		
				20	50	80
No noise	–	Vanilla U-Net	89.89	–	–	–
No noise	–	Co-teaching [7]	91.46	–	–	–
No noise	–	Ours	–	**92.52**	92.50	92.36
25%	$5 \leq n \leq 15$	Vanilla U-Net	87.58	–	–	–
25%	$5 \leq n \leq 15$	Co-teaching [7]	89.06	–	–	–
25%	$5 \leq n \leq 15$	Ours-SS	91.42	–	–	–
25%	$5 \leq n \leq 15$	Ours	–	92.11	92.81	**93.06**
50%	$5 \leq n \leq 15$	Vanilla U-Net	86.65	–	–	–
50%	$5 \leq n \leq 15$	Co-teaching [7]	88.56	–	–	–
50%	$5 \leq n \leq 15$	Ours-SS	88.87	–	–	–
50%	$5 \leq n \leq 15$	Ours	–	**90.14**	90.05	89.56
75%	$5 \leq n \leq 15$	Vanilla U-Net	84.96	–	–	–
75%	$5 \leq n \leq 15$	Co-teaching [7]	90.23	–	–	–
75%	$5 \leq n \leq 15$	Ours-SS	90.41	–	–	–
75%	$5 \leq n \leq 15$	Ours	–	91.07	90.19	**91.17**

noisy labels. We first trained the fully supervised vanilla U-Net with the noisy ratio set to zero, which can be regarded as the upper-line performance. Compared with the vanilla U-Net, our framework improves the segmentation performance and achieves an average Dice of 0.925 on the clean annotated dataset, indicating that the sample selection stage and joint-optimization stage can encourage the model to learn more distinguishing features.

For the training dataset with different ratio of noisy labels, we observed that as the noise ratio increases, the segmentation performance of the vanilla U-Net decreases dramatically. Compared with vanilla U-Net, the sample selection stage (SS) can consistently improve the performance by encouraging the model to be trained by the selected data. Through the joint optimization (JO) stage supervised by the corrected label and original ones, the segmentation accuracy is further improved, suggesting that our method can effectively eliminate the effect of the noise and gain performance by producing the correct label. In Fig. 3, we show some segmentation results under 75% noise, in which our results have higher Dice score than the vanilla U-Net. At all the noise ratio, we compared our method with the state-of-the-art noise robust method [7], which select the small loss samples according to the prediction of peer network. The results show that our method outperforms the state-of-the-art method in all the noise ratio setting.

In our experiment, we also investigated the impact of the starting epoch k on the performance of our method. As shown in Table 1, the joint optimization

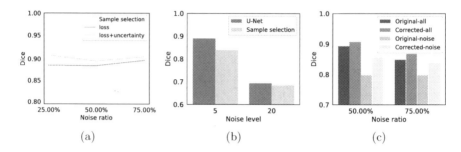

Fig. 4. (a) The segmentation accuracy of different sample selection criteria. (b) The segmentation accuracy of the U-Net and U-Net with only sample selection stage on 100% noise ratio setting. (c) The label accuracy of labels in the original dataset, and labels corrected by the model at the end of training.

(JO) with label correction stage is started at 20, 50, and 80 epochs, respectively. The experimental results show that the joint optimization stage can consistently produce good results with different starting epoch k.

3.4   Analysis of Our Method

Sample Selection. Compared with the vanilla U-Net, our sample selection stage (SS) shows higher segmentation accuracy under different noisy ratio, as shown in Table 1. To validate the criteria of our sample selection, we conducted another experiment that only selected the small loss sample. Figure 4(a) shows the test accuracy with different sample selection criteria. It reveals that the test accuracy significantly improved when considering the uncertainty in the selection stage. To further validate the effectiveness of our method at the sample selection stage, we applied our method on training dataset with 100% noise and noise level $n = 5, 20$. As shown in Fig. 4(b), under this setting, the sample selection stage shows worse segmentation accuracy than vanilla U-Net, because no clean sample can be selected. The results decreased due to the low sample utilization efficiency.

Joint Optimization. To analyze the contribution of the joint optimization stage, we explore the label accuracy with and without the stage of joint optimization and label correction. We calculated the Dice coefficient of the initial noisy label ($\hat{y}$) and the corrected label ($\bar{y}$) of the final model at the end of the training. Figure 4(c) shows the overall accuracy for severe noise situation (50%, 75%), where the Dice coefficient for all the original (Original-all) and corrected label (Corrected-all), and the Dice coefficient only for the original noise label (Original-noise) and corrected noise label (Corrected-noise) are presented. We see that the label quality is improved by the scheme of joint optimization and label correction, especially for those original noise labels.

4   Conclusion

In this paper, we present a novel Cascaded Robust Learning framework for the segmentation of noisy labeled chest x-ray images. Our method consists of two stages: sample selection stage, and the stage of joint optimization with label correction. In the first stage, the clean annotated samples are selected for network updating, so that the influence of noisy sample can be interactively eliminated in the three networks. In the second stage, the label correction module works together with the joint optimization scheme to revise the imperfect labels. Thus the training of the whole network is supervised by the corrected labels and the original ones. Compared with other state-of-the-art models, our cascaded robust learning framework keeps high robustness when the training data contains imperfect annotate boundaries. Experimental results on the benchmark dataset demonstrate that our network outperforms other methods on segmentation tasks and achieves very competitive results on the noisy labeled dataset.

Acknowledgments. This work is supported by Hong Kong Innovation and Technology Commission (Project No. ITS/311/18FP), Shenzhen Science and Technology Program (JCYJ20170413162256793) and a CUHK Direct Grant for Research.

References

1. Bejnordi, B.E., et al.: Diagnostic assessment of deep learning algorithms for detection of lymph node metastases in women with breast cancer. JAMA **318**(22), 2199–2210 (2017)
2. Berthelot, D., Carlini, N., Goodfellow, I., Papernot, N., Oliver, A., Raffel, C.A.: MixMatch: a holistic approach to semi-supervised learning. In: Advances in Neural Information Processing Systems, pp. 5050–5060 (2019)
3. Candemir, S., et al.: Lung segmentation in chest radiographs using anatomical atlases with nonrigid registration. IEEE Trans. Med. Imaging **33**(2), 577–590 (2013)
4. De Fauw, J., et al.: Clinically applicable deep learning for diagnosis and referral in retinal disease. Nat. Med. **24**(9), 1342–1350 (2018)
5. Esteva, A., et al.: Dermatologist-level classification of skin cancer with deep neural networks. Nature **542**(7639), 115 (2017)
6. Goldberger, J., Ben-Reuven, E.: Training deep neural-networks using a noise adaptation layer. In: ICLR (2017)
7. Han, B.,et al.: Co-teaching: robust training of deep neural networks with extremely noisy labels. In: Advances in Neural Information Processing Systems, pp. 8527–8537 (2018)
8. Jaeger, S., et al.: Automatic tuberculosis screening using chest radiographs. IEEE Trans. Med. Imaging **33**(2), 233–245 (2013)
9. Jiang, L., Zhou, Z., Leung, T., Li, L.J., Fei-Fei, L.: MentorNet: regularizing very deep neural networks on corrupted labels. In: ICML (2018)
10. Le, H., Samaras, D., Kurc, T., Gupta, R., Shroyer, K., Saltz, J.: Pancreatic cancer detection in whole slide images using noisy label annotations. In: Shen, D., et al. (eds.) MICCAI 2019. LNCS, vol. 11764, pp. 541–549. Springer, Cham (2019). https://doi.org/10.1007/978-3-030-32239-7_60

11. Li, X., Yu, L., Chen, H., Fu, C.W., Xing, L., Heng, P.A.: Transformation-consistent self-ensembling model for semisupervised medical image segmentation. IEEE Trans. Neural Netw. Learn. Syst. (2020)
12. Ren, M., Zeng, W., Yang, B., Urtasun, R.: Learning to reweight examples for robust deep learning. In: ICML (2018)
13. Ronneberger, O., Fischer, P., Brox, T.: U-Net: convolutional networks for biomedical image segmentation. In: Navab, N., Hornegger, J., Wells, W.M., Frangi, A.F. (eds.) MICCAI 2015. LNCS, vol. 9351, pp. 234–241. Springer, Cham (2015). https://doi.org/10.1007/978-3-319-24574-4_28
14. Setio, A.A.A., et al.: Validation, comparison, and combination of algorithms for automatic detection of pulmonary nodules in computed tomography images: the LUNA16 challenge. Med. Image Anal. **42**, 1–13 (2017)
15. Shu, Y., Wu, X., Li, W.: LVC-Net: medical image segmentation with noisy label based on local visual cues. In: Shen, D., et al. (eds.) MICCAI 2019. LNCS, vol. 11769, pp. 558–566. Springer, Cham (2019). https://doi.org/10.1007/978-3-030-32226-7_62
16. Stirenko, S., et al.: Chest X-ray analysis of tuberculosis by deep learning with segmentation and augmentation. In: 2018 IEEE 38th International Conference on Electronics and Nanotechnology (ELNANO), pp. 422–428. IEEE (2018)
17. Xue, C., Dou, Q., Shi, X., Chen, H., Heng, P.A.: Robust learning at noisy labeled medical images: applied to skin lesion classification. In: ISBI (2019)
18. Zhang, C., Bengio, S., Hardt, M., Recht, B., Vinyals, O.: Understanding deep learning requires rethinking generalization. In: ICLR (2017)
19. Zhu, H., Shi, J., Wu, J.: Pick-and-learn: automatic quality evaluation for noisy-labeled image segmentation. arXiv preprint arXiv:1907.11835 (2019)

Class-Aware Multi-window Adversarial Lung Nodule Synthesis Conditioned on Semantic Features

Qiuli Wang, Xingpeng Zhang, Wei Chen, Kun Wang, and Xiaohong Zhang[✉]

School of Big Data and Software Engineering, Chongqing University,
Chongqing 400030, China
{wangqiuli,xpzhang,wchen,kun.wang,xhongz}@cqu.edu.cn

Abstract. Nodule CT image synthesis is effective as a data augmentation method for deep learning tasks about lung nodules. To advance the realistic malignant/benign lung nodule synthesis, the conditional Generative Adversarial Networks have been widely adopted. In this paper, we argue about an issue in the existing technique for class-aware nodule synthesis: the class-aware controllability of semantic features. To address this issue, we propose a adversarial lung nodule synthesis framework based on conditional Generative Adversarial Networks and class-aware multi-window semantic feature learning. By learning semantic features from multi-window CT images, our framework can generate realistic nodule CT images, and has better controllability of class-aware nodule features. Our framework provides a new perspective for nodule CT image synthesis that has never been noticed before. We train our framework on the public dataset LIDC-IDRI. Our framework improves the malignancy prediction F1 score by more than 3% and shows promising results as a solution for lung nodule augmentation. The source code can be found at https://github.com/qiuliwang/CA-MW-Adversarial-Synthesis.

Keywords: Lung nodule synthesis · Multi-window · Generative Adversarial Networks · Computed Tomography

1 Introduction

Deep learning methods have been proved to be sufficient for lung nodule analysis in many ways [11,15,16]. Existing methods heavily rely on the quality and the diversity of the labeled Computed Tomography (CT) datasets. As a result, these methods suffer a lot when dealing with imbalanced distribution in the real world medical data. It is thus meaningful to improve the data quality and balance the distribution of lung nodule images. Recently, several methods based

This research was supported in part by the National Natural Science Foundation of China under Grant 61772093, in part by the National Key R&D Project of China under Grant 2018YFB2101200, and in part by the Chongqing Major Theme Projects under Grant cstc2018jszx-cyztzxX0017.

© Springer Nature Switzerland AG 2020
A. L. Martel et al. (Eds.): MICCAI 2020, LNCS 12266, pp. 589–598, 2020.
https://doi.org/10.1007/978-3-030-59725-2_57

on Generative Adversarial Networks (GAN) have been proposed to overcome this challenge, and demonstrated superior performances as data augmentation methods for deep learning tasks of lung nodules, such as image segmentation, feature extraction, and malignancy prediction [4,17,18].

Although shown to be promising, current GAN-based methods still suffer from limited practical use: the poor controllability of semantic features for class-aware nodule synthesis, which aims to synthesize malignant/benign nodules according to the requirements. To address this problem, one possible solution is to follow the clinical practice and seek for more visual features from the multi-window CT images. According to the Fleischner recommendation [3], the diagnosis of malignant/benign nodules is closely related to the semantic feature observation, and these features can only be clearly shown with the appropriate CT image window settings. Malignant nodules tend to have spiculation signs and great density changes. The spiculation sign is a low-dense feature that needs to be observed with the wide (lung) window settings, on the other hand, the density changes of nodules need to be observed with the soft-tissue (mediastinal) windows [10]. Figure 1 shows three benign nodules and three malignant nodules from LIDC-IDRI, and demonstrate how the window settings affect the nodule observation.

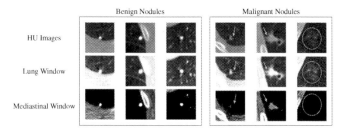

Fig. 1. Nodule observation with different image window settings. We show three benign nodules and three malignant nodules with three image windows: HU, lung window, and mediastinal window. As can be seen, malignant nodules show different appearances with different image windows (pointed by red arrows and circles). (Color figure online)

So we summarize the nodule diagnosis as two tasks: (1) Observing semantic features with appropriate window settings; (2) Making diagnosis decisions according to the semantic features. And it can be easily understood that (1) correct diagnosis relies on the accurate extraction of semantic features; (2) accurate extraction of semantic features relies on the appropriate window settings. We further propose two tasks for class-aware lung nodule synthesis: (1) Accurate semantic feature learning through multi-window observation; (2) Malignant/benign nodule synthesis.

Based on these findings, we propose a class-aware lung nodule synthesis framework for nodule CT conditioned on semantic features, which is shown in Fig. 2. Our framework contains two sub-networks: (1) *HU synthesis sub-network,* and (2) *class-aware multi-window synthesis sub-network*. Based on mask-guided cGAN [7], the *HU synthesis sub-network* learns the mappings between masks and real nodule CT HU images and gives out the synthesized HU images. The *class-aware multi-window synthesis sub-network* transforms the synthesized HU images into images with two window settings (lung window, mediastinal window). Two image windows contribute to a better synthesis of delicate semantic features, improving the controllability of malignant/benign nodule synthesis, and learning the relationships between semantic features and malignancy levels automatically.

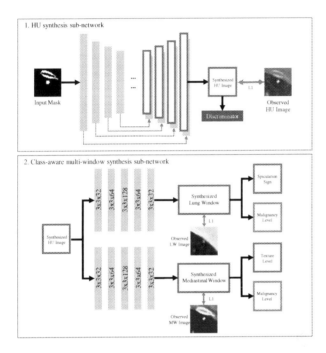

Fig. 2. Architecture of the proposed framework. The *HU synthesis sub-network* uses a U-Net structure as its backbone, and contains one discriminator to discriminate synthesized nodule CT HU images. The *class-aware multi-window synthesis sub-network* takes synthesized HU image as its input, and generates two windows using two branches of CNN.

The main contributions of this paper are: (1) We propose a class-aware multi-window lung nodule synthesis framework conditioned on semantic features. Our framework can synthesize realistic malignant/benign nodules conditioned on semantic features with different CT image windows. (2) Our framework improves malignancy prediction F1 score by more than 3%, and shows promising performances as a solution for data augmentation.

2   Related Work

2.1   Generative Adversarial Network

Generative Adversarial Network (GAN) was first proposed by Goodfellow *et al.* [6] in 2014, and has received significant attention in the field of computer vision [12,14,19]. To achieve more controllability of the adversarial learning process, Mehdi Mirza and Simon Osindero proposed conditional GAN (cGAN) [9]. Later in 2017, Isola *et al.* [8] applied cGAN as a solution to image-to-image translation problems. The proposed framework could learn the mapping between the input images and output images. Based on this research, Gu *et al.* [7] further developed a framework for mask-guided portrait editing. The proposed framework could synthesize and edit portraits from face masks and control the synthesized face photos by editing the face masks.

Invoked by the study [7], our framework also uses masks as the guidance of nodule synthesis. The masks we use are composed of nodule masks and background masks, which can help to 'edit' the synthesized images. The backbone of *HU synthesis sub-network* is similar to the study [8]. Moreover, we attach two CNN branches for multi-window transforming.

2.2   Lung Nodule Synthesis

It has been proved that synthesizing medical images from annotated datasets can help to improve the performances of the supervised learning [5]. In the field of nodule image analyzing, the applications of GAN have drawn much attention due to their advantages. In 2018, Chuquicusma *et al.* [4] first attempted to synthesize lung nodule CT images with deep convolutional GAN. In 2019, Jie Yang *et al.* [18] proposed an in-painting based framework for synthesizing lung nodule images with class-aware manipulations. Also, in 2019, Ziyue Xu *et al.* [17] proposed a tunable CT lung nodule synthesis based on conditional adversarial learning strategy with nodule masks.

As mentioned before, we summarize two tasks for class-aware lung nodule synthesis. Existing studies usually handle these two tasks separately or fuse all semantic features without manual design. Our framework achieves two tasks at a time by combining semantic feature learning and malignancy level prediction.

3   Methods

3.1   Architecture

Our framework contains two sub-networks: (1) *HU synthesis sub-network*, and (2) *class-aware multi-window synthesis sub-network*. The inputs of the whole framework contain (a) masks, (b) the observed HU images, (c) the observed lung window and mediastinal window images, (d) the labels for spiculation signs, texture, and malignancy levels. The outputs of the framework are synthesized

HU images, lung window images, and mediastinal window images conditioned on spiculation sign, texture, and malignancy levels.

HU synthesis sub-network uses a modified U-Net structure [13] as its backbone. The U-Net contains eight down-sampling processes and eight up-sampling processes. We use the L1 loss to learn the mappings between masks and nodule HU images and encourage less blurring [8]. A discriminator based on conventional CNN is used to discriminate the input tuples: {synthesized nodule, mask} and {real nodule, mask}. This structure is similar to the study of [8]. HU value in CT volume is the Hounsfield Unit value, which has a range from -1000 (air) to more than $+2000$ (dense bone). The details of discriminator can be found on our website.

Class-aware multi-window synthesis sub-network takes the synthesized HU image, observed lung window images, and observed mediastinal window images as its inputs. We design this sub-network for capturing semantic features and their relationships between malignancy levels automatically. Two branches of CNN transform the synthesized HU image into two image windows. The lung window ranges from -1250HU to 50HU, the mediastinal window ranges from -160HU to 240HU. Two L1 loss functions are adopted to synthesized realistic lung window and mediastinal window images. With the branch for the lung window, classifiers need to capture spiculation sign levels and malignancy levels. With the branch for mediastinal window, classifiers need to capture texture levels and malignancy levels. By doing so, our framework not only synthesizes malignancy nodules but also controls the semantic features which are closely related to the malignancy levels. All classifiers are constructed by conventional CNN, and the details of each classifier are listed on our website.

3.2   Loss Functions

Multi-window Synthesis. Following the study [8], we use three L1 loss functions between {input HU image, lung window (LW) image, mediastinal window (MW) image}, and {observed HU image, observed LW image, observed MW image}:

$$\mathcal{L}_{L1}(G) = \mathcal{L}_{HU} + \mathcal{L}_{LW} + \mathcal{L}_{MW} \tag{1}$$

where G represents the multi-window learning processes, $\mathcal{L}_{L1}(G)$ represents the overall loss for synthesizing. $\mathcal{L}_{L1}(G)$ involves three parts: $\mathcal{L}_{HU}$, $\mathcal{L}_{LW}$, and $\mathcal{L}_{MW}$, which are calculated as follows:

$$\begin{aligned}
\mathcal{L}_{HU} &= \mathbb{E}_{X,M}[|||X - G_{HU}(M)||_1] \\
\mathcal{L}_{LW} &= \mathbb{E}_{X_{LW},Y}[|||X_{LW} - G_{LW}(Y)||_1] \\
\mathcal{L}_{MW} &= \mathbb{E}_{X_{MW},Y}[|||X_{MW} - G_{MW}(Y)||_1]
\end{aligned} \tag{2}$$

where $\mathcal{L}_{HU}$, $\mathcal{L}_{LW}$, and $\mathcal{L}_{MW}$ are three L1 losses. The framework learns a transformation from observed X and corresponding mask M to Y. Then Y will be transformed into LW image and MW image. X_{LW} and X_{MW} are observed LW and MW images. G_{HU}, G_{LW}, and G_{MW} are three generative processes. The inputs for two CNN branches are synthesized HU images, and two branches are aimed at transforming HU images to a specific image window rather than learning from scratch.

Adversarial Loss. The loss of the discriminator for HU images is defined as:

$$\mathbb{E}_{X,Y}[logD(X,Y)] + \mathbb{E}_{X,M}[log(1 - D(X, G_{HU}(M)))] \tag{3}$$

The discriminator is trained to do as well as possible at discriminating two input tuples: {real nodule, mask} and {synthesized nodule, mask}.

The classifiers for synthesized LW images and MW images are aimed at capturing spiculation sign levels, texture levels, and malignancy levels with different image windows. Each classifier loss $C(Y, L)$ is calculated by cross-entropy loss. Then we have Eq. 4 for the adversarial learning process:

$$
\begin{aligned}
\mathcal{L}_{GAN}(G, D, C) = {} & \mathbb{E}_{X,Y}[logD(X,Y)] \\
& + \mathbb{E}_{X,M}[log(1 - D(X, G_{HU}(M)))] \\
& + C_{spic}(Y_{LW_{spic}}, L_{spic}) + C_{text}(Y_{MW_{text}}, L_{text}) \\
& + C_{LW_{mali}}(Y_{LW_{mali}}, L_{mali}) + C_{MW_{mali}}(Y_{MW_{mali}}, L_{mali})
\end{aligned} \tag{4}
$$

where C_{spic} and C_{text} are the loss functions for semantic feature prediction, $C_{LW_{mali}}$ and $C_{MW_{mali}}$ are the loss functions for malignancy prediction with different image windows. L_{mali}, L_{spic} and L_{text} are labels for malignancy level, spiculation sign level, and texture level of the observed input nodule X. $Y_{LW_{spic}}$, $Y_{MW_{text}}$ are prediction results for spiculation sign and texture. $Y_{LW_{mali}}$ and $Y_{MW_{mali}}$ are malignancy level predictions with different window settings.

Overall Loss. The final loss is a combination of Eq. 1 and Eq. 4:

$$G* = \arg\min_{G,C} \max_{D} \mathcal{L}_{GAN}(G, D, C) + \lambda \mathcal{L}_{L1}(G) \tag{5}$$

We manually set λ to 100 for a better image quality. Other trainable parameters of the whole framework are updated by the back-propagation algorithm during training.

4 Experiments and Results

4.1 Dataset and Experimental Setup

We extract 2635 annotated lung nodules from LIDC-IDRI [2]. The malignancy levels, spiculation sign levels, and texture levels range from 1 (highly unlikely) to 5 (highly suspicious). We keep the original HU values as the target of synthesized HU images. Our HU ranges are manually selected for better training. We use segmentation maps provided by LIDC-IDRI as the nodule masks, and extract HU values higher than 400 as background masks. The tissues that have HU values higher than 400 are mostly bones so that they can help to synthesize realistic backgrounds. Each nodule mask has a corresponding background mask. And during the testing, each nodule mask can be combined with a random background mask for more synthesized data. The original nodule images are 56×56, and we resize them into 128×128. The learning rate is 2×10^{-4}. The whole framework is built on the Tensorflow framework [1].

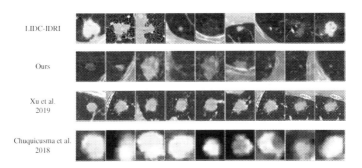

Fig. 3. Comparisons between LIDC-IDRI, [4,17] and ours. All images from LIDC-IDRI are HU images, so as ours. As can be seen, our framework can synthesize nodules with density changes and delicate details.

4.2   Realistic Nodule Synthesis

Figure 3 shows the direct comparisons between other studies and ours. As can be seen, the quality of our synthesized nodules are similar to those from LIDC-IDRI. Compared to [4,17], our synthesized nodules have smoother edges, and more detailed semantic features like spiculation signs, changes in densities. It is because two branches of CNN learn different appearances of semantic features with different image window and force the backbone to synthesize realistic CT images through the back-propagation algorithm. This design allows more density and semantic changes in synthesized nodules.

4.3   Control of Malignancy Levels and Semantic Features

We summarize the class-aware nodule synthesis as two tasks: (1) *Accurate semantic feature learning through multi-window observation*; (2) *Malignant/benign nodule synthesis.* In this section, we demonstrate the performances of our framework for these two tasks.

(1) *Accurate semantic feature learning through multi-window observation.* We show the multi-window control of semantic feature appearances in Fig. 4. As can be seen, our synthesized nodules (blue block) have similar appearances with nodules from LIDC-IDRI (red block), and show different appearances

with different windows as our expectation. Synthesized nodules tend to have 'fuzzy' edges with lung window, and show the density changes with mediastinal window. Moreover, low-dense nodules disappear with the mediastinal window, which is consistent of clinical observation. In Fig. 5 we show the semantic controllability of our framework. Each line of Fig. 5 contains four malignant nodules synthesized with the same masks, and these four nodules are conditioned on different semantic features. As we can see, the low texture levels lead to synthesizing low-dense nodules, and high spiculation signs lead to synthesizing 'fuzzy' nodules.

(2) *Malignant/benign nodule synthesis.* In Fig. 6 we demonstrate the malignant/benign nodule synthesis. In this figure, we show six pairs of malignant/benign nodules, and each pair is conditioned on the same semantic features (high spiculation sign, and high texture level). As can be seen, the synthesized malignant nodules tend to be more 'fuzzy', and have more connections with lung tissues. It means our framework learns the knowledge that malignant nodules are more likely to have spiculation signs and density changes, which indicates the effectiveness of our framework. We believe the second sub-network successfully helps the framework 'pay attention' to the semantic features with different window settings.

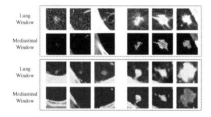

Fig. 4. Multi-window control of nodule synthesis. (Color figure online)

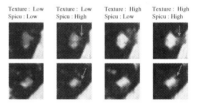

Fig. 5. Semantic control of nodule synthesis. (Color figure online)

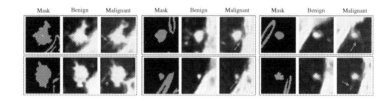

Fig. 6. Malignant/benign nodule synthesis conditioned on the same semantic feature settings.

To verify the effectiveness of our framework as a data augmentation method, we synthesize some malignant nodules to improve the malignancy prediction.

The experimental settings are similar to the study of [18], and are listed on our website. The experimental results are shown in Table 1. The results show that our methods help the malignancy prediction task significantly improve the F1 score to 0.8903, and have the best accuracy. Experiments demonstrate that our framework is an effective data augmentation framework.

Table 1. Experimental results

Method	Acc	F1
Yang *et al.* [18]	0.8216	0.7608
VGG16 (LIDC)	0.8934	0.8588
VGG16 (LIDC+resampling)	0.8804	0.8556
VGG16 (LIDC+synthesized)	**0.9051**	**0.8903**

'resampling' indicates re-sampled data.

5   Conclusions

In this paper, we propose a class-aware multi-window adversarial lung nodule synthesis framework conditioned on semantic features. The proposed framework simulates the clinical practice of multi-window lung observation and captures the appearances of semantic features with different CT image windows. This practice leads to a better quality of synthesized semantic features and better control of class-aware lung nodule synthesis. We train our framework on LIDC-IDRI, and show promising results as a data augmentation framework.

References

1. Abadi, M., Agarwal, A., Barham, P., et al.: TensorFlow: large-scale machine learning on heterogeneous distributed systems. arXiv preprint arXiv:1603.04467 (2016)
2. Armato III, S.G., McLennan, G., Bidaut, L., et al.: The lung image database consortium (LIDC) and image database resource initiative (IDRI): a completed reference database of lung nodules on CT scans. Med. Phys. **38**(2), 915–931 (2011)
3. Bankier, A.A., MacMahon, H., Goo, J.M., et al.: Recommendations for measuring pulmonary nodules at CT: a statement from the Fleischner Society. Radiology **285**(2), 584–600 (2017)
4. Chuquicusma, M.J., Hussein, S., Burt, J., et al.: How to fool radiologists with generative adversarial networks? A visual turing test for lung cancer diagnosis. In: ISBI 2018, pp. 240–244. IEEE (2018)
5. Costa, P., Galdran, A., Meyer, M.I., et al.: End-to-end adversarial retinal image synthesis. IEEE T. Med. Imaging **37**(3), 781–791 (2017)
6. Goodfellow, I., Pouget-Abadie, J., Mirza, M., et al.: Generative adversarial nets. In: NeurIPS 2014, pp. 2672–2680 (2014)
7. Gu, S., Bao, J., Yang, H., et al.: Mask-guided portrait editing with conditional GANs. In: CVPR 2019, pp. 3436–3445. IEEE (2019)

8. Isola, P., Zhu, J.Y., Zhou, T., et al.: Image-to-image translation with conditional adversarial networks. In: CVPR 2017, pp. 1125–1134. IEEE (2017)

9. Mirza, M., Osindero, S.: Conditional generative adversarial nets. arXiv preprint arXiv:1411.1784 (2014)

10. Okada, M., Nishio, W., Sakamoto, T., et al.: Correlation between computed tomographic findings, bronchioloalveolar carcinoma component, and biologic behavior of small-sized lung adenocarcinomas. J. Thorac. Cardiovasc. Surg. **127**(3), 857–861 (2004)

11. Qi, D., Hao, C., Yu, L., et al.: Multilevel contextual 3-D CNNs for false positive reduction in pulmonary nodule detection. IEEE Trans. Bio-Med. Eng. **64**(7), 1558–1567 (2016)

12. Radford, A., Metz, L., Chintala, S.: Unsupervised representation learning with deep convolutional generative adversarial networks. arXiv preprint arXiv:1511.06434 (2015)

13. Ronneberger, O., Fischer, P., Brox, T.: U-Net: convolutional networks for biomedical image segmentation. In: Navab, N., Hornegger, J., Wells, W.M., Frangi, A.F. (eds.) MICCAI 2015. LNCS, vol. 9351, pp. 234–241. Springer, Cham (2015). https://doi.org/10.1007/978-3-319-24574-4_28

14. Shaham, T.R., Dekel, T., Michaeli, T.: SinGAN: learning a generative model from a single natural image. In: ICCV 2019, pp. 4570–4580. IEEE (2019)

15. Shen, S., Han, S.X., Aberle, D.R., et al.: An interpretable deep hierarchical semantic convolutional neural network for lung nodule malignancy classification. Expert Syst. Appl. **128**, 84–95 (2019)

16. Wu, B., Zhou, Z., Wang, J., et al.: Joint learning for pulmonary nodule segmentation, attributes and malignancy prediction. In: ISBI 2018, pp. 1109–1113. IEEE (2018)

17. Xu, Z.: Tunable CT lung nodule synthesis conditioned on background image and semantic features. In: Burgos, N., Gooya, A., Svoboda, D. (eds.) SASHIMI 2019. LNCS, vol. 11827, pp. 62–70. Springer, Cham (2019). https://doi.org/10.1007/978-3-030-32778-1_7

18. Yang, J., Liu, S., Grbic, S., et al.: Class-aware adversarial lung nodule synthesis in CT images. In: ISBI 2019, pp. 1348–1352. IEEE (2019)

19. Yi, X., Walia, E., Babyn, P.: Generative adversarial network in medical imaging: a review. Med. Image Anal. **58**, 101552 (2019)

Nodule2vec: A 3D Deep Learning System for Pulmonary Nodule Retrieval Using Semantic Representation

Ilia Kravets[1]([envelope]), Tal Heletz[1], and Hayit Greenspan[2]

[1] Y-Data, Yandex School of Data Analysis, Tel Aviv, Israel
ilia.kravets@gmail.com, talheletz123@gmail.com
[2] Tel Aviv University, Tel Aviv, Israel
hayit@eng.tau.ac.il

Abstract. Content-based retrieval supports a radiologist decision making process by presenting the doctor the most similar cases from the database containing both historical diagnosis and further disease development history. We present a deep learning system that transforms a 3D image of a pulmonary nodule from a CT scan into a low-dimensional embedding vector. We demonstrate that such a vector representation preserves semantic information about the nodule and offers a viable approach for content-based image retrieval (CBIR). We discuss the theoretical limitations of the available datasets and overcome them by applying transfer learning of the state-of-the-art lung nodule detection model. We evaluate the system using the LIDC-IDRI dataset of thoracic CT scans. We devise a similarity score and show that it can be utilized to measure similarity 1) between annotations of the same nodule by different radiologists and 2) between the query nodule and the top four CBIR results. A comparison between doctors and algorithm scores suggests that the benefit provided by the system to the radiologist end-user is comparable to obtaining a second radiologist's opinion.

Keywords: CBIR · Pulmonary nodules · Deep learning · Image retrieval

1 Introduction

1.1 Motivation and Background

Lung cancer is a leading cause of cancer mortality in both men and women, accounting for nearly 25% of all cancer deaths [16]. The chances of treating lung cancer successfully are much higher if the treatment starts at an early stage.

I. Kravets and T. Heletz—Equal contribution.

© Springer Nature Switzerland AG 2020
A. L. Martel et al. (Eds.): MICCAI 2020, LNCS 12266, pp. 599–608, 2020.
https://doi.org/10.1007/978-3-030-59725-2_58

Currently, Low-Dose Computed Tomography (LDCT) screening is the most effective way for pulmonary nodules detection and diagnosis, and its usage has increased dramatically over the last two decades. However, scan examination and diagnosis is a very time-consuming task that requires a lot of invaluable radiologist time.

To assist radiologists to quickly and effectively diagnose tumors, it is important to present them with similar historical cases. Examination of similar cases can be beneficial in two aspects. First, the radiologist can have access to the labeling information that other doctors gave in similar cases and thus can deduce the status of the current case. Second, the radiologist can examine the related case development history past the similarly looking LDCT scan, as if peeking at a possible future of the current case and infer a more accurate prognosis. For example, a similar case biopsy outcome can suggest whether it is advisable to perform a biopsy in the current case.

Many contemporary works that apply machine learning techniques to medical imaging aim to replace the doctors and directly produce a diagnosis. The usual downsides of this approach are the lack of the output interpretability and uncertain robustness guarantees in the light of potential bugs, input variation, or even adversarial attacks. Moreover, they are usually lacking in the amount of diagnostic data they can provide, often limited by a handful of bits of information, like benign-vs-malignant binary classification.

The solution we describe is a content-based image retrieval (CBIR) system. Given a pulmonary nodule, our system retrieves several similar nodules from the historical database, potentially enriched with the relevant clinical records, to aid the doctor in the diagnostic process. Our main contributions include:

- We develop a system to provide the radiologist semantically-meaningful decision support in nodule analysis, in contrast to providing an automated nodule diagnosis.
- We identify architectural constraints to our deep learning system and provide theoretical justification for utilizing transfer learning.
- We define a proxy task to learn the transformation of a 3D nodule to a latent vector based on semantic features defined by medical experts.
- We study semantic information preservation by our system, devise an evaluation technique that considers the lack of consensus between the doctors, and show that our method retrieves highly relevant results.

1.2   Related Work

Most of the previous research uses classic computer vision methods for feature extraction, using a 2D representation of the nodules. This approach fails to capture the full spatial nodule information. A significant challenge in CBIR is the definition of the distance between two entities so that they will be considered semantically similar in addition to being visually similar. Often, the evaluation of the models is hindered by the lack of consensus between human annotators.

A few works in the field include the following: Lam et al. [6] performed CBIR on 2D slices of the 3D nodules using classical image descriptors. Dhara et al. [3] extended this approach with manually defined volumetric features. Pan et al. [13] used spectral clustering to transfer a 3D nodule to hash code used to retrieve similar nodules. Wei et al. [18] proposed a learned distance metric. Finally, Loyman and Greenspan [8] study included LIDC-IDRI rating regression from 2D slices to obtain embeddings using deep learning.

2 Methods

2.1 Data

LIDC-IDRI. The Lung Image Database Consortium and Image Database Resource Initiative (LIDC-IDRI) image collection consists of diagnostic and lung cancer screening thoracic computed tomography (CT) scans with marked-up annotated lesions [1,2]. A panel of four experienced radiologists performed independent segmentation and initial categorization. Lesions categorized as nodules larger than 3mm in diameter were further assessed for nine subjective characteristics: subtlety, internal structure, calcification, sphericity, margin, lobulation, spiculation, radiographic solidity, and malignancy [10]. Each characteristic consists of either a discrete category set or an integer rating on a five-point scale. We use pylidc software [4] to access radiologist annotations.

Analysis of nodule characteristic distribution revealed four characteristics with very low variability between nodules, which we decided to omit from further processing. Adopting the LUNA16 [15] approach we also limited the analysis to nodules accepted by at least three out of four radiologists. This resulted in 1186 nodules, each with three or four sets of segmentation and five-dimensional rating characteristic (subtlety, sphericity, margin, lobulation, and malignancy). We normalize all rating values to $[0,1]$ range to aid the implementation.

2.2 Methodology

CBIR Using Embeddings. Our approach to CBIR is partially inspired by the natural language processing technique called word2vec [11]. We learn a function $f : V \to S$ from a high dimensional space of CT voxels V to a much lower dimensional space S, which we call "a semantic space". A desirable property of S is to capture nodule similarity as perceived by radiology experts, that is, for some distance function $d : S^2 \to \mathbb{R}$ we expect two vectors $s_1, s_2 \in S$ to be relatively close (have small $d(s_1, s_2)$) if the corresponding nodules would be considered similar by the doctors and vice versa. We call vectors in S "embeddings". Not necessarily interpretable per se, we would like an embedding to incorporate both characteristical information as defined by radiologists as well as some visual information about the nodule. A traditional approach to learn embeddings is to define a proxy task, such that training a deep learning model to solve this task would produce the embeddings as a byproduct.

Theoretical Architecture Constraints. According to PAC theory (e.g. see [12, ch. 2]) bound of the generalization error over hypothesis space $\mathcal{H}$ and number of training samples N is:

$$R(h) \leq \hat{R}_S(h) + O\left(\sqrt{\frac{\log |\mathcal{H}|}{N}}\right) \tag{1}$$

Our dataset has a very limited number of nodules, while the deep learning network capable of extracting the information from the 3D space of voxels can have many millions of parameters, that is, $N \ll \log |\mathcal{H}|$. While some researchers question the tightness of such theoretical generalization error bounds, we still wary of the model overfitting in our setting. Therefore, we decide that our proxy task may not be sufficient to train a robust feature extraction.

Feature Extraction. Fortunately, thanks to LUNA16 [15] and Kaggle Data Science Bowl (DSB) 2017 [5] competitions there are a lot of previous works tackling pulmonary nodule analysis. In this work we apply a transfer learning technique to the winning DSB solution [7] which is based on the U-net architecture [14]. It has initially been trained on a dataset including lots of scans not found in LIDC-IDRI and optimized for different tasks, namely, pulmonary nodule detection and binary whole-scan classification. For our feature extraction, we reuse the pre-trained U-net backbone of the nodule detector and drop the region proposal head comprising of the last two convolutional layers.

Rating Regression. Here we define a proxy task as a radiologist rating regression. We extend the feature extraction network with three fully connected layers with the output being a five-dimensional vector (Fig. 1). A target is defined

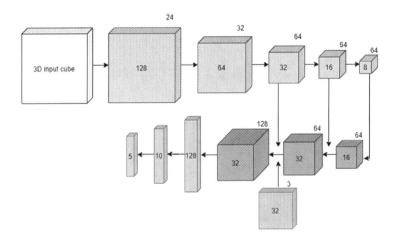

Fig. 1. Regression network for embeddings learning (only feature map is shown). The trainable head comprises of the last three fully connected layers. See [7] for a detailed explanation about the backbone.

as a mean of radiologist ratings applied to the nodule at hand. We use MSE loss, training only the head while keeping the backbone weights fixed. A ten-dimensional vector from the second-to-last layer is used as an embedding.

3   Results and Discussion

3.1   Semantic Information Preservation

In this section we study whether the embeddings produced by our method preserve semantic meaning, that is, nodules with similar characteristics produce similar embeddings.

t-SNE. We run t-SNE [9] dimensionality reduction over the random sample of the embeddings space (Fig. 2a). For simplicity, the coloring here only reflects malignancy: samples with high malignancy (mean rating > 3 out of 5) are colored red and others blue. The separation of red and blue is not ideal because embedding vectors preserve more information than just a malignancy level. However, we observe that the malignant nodules are clustered together, unlike the benign nodules. The more detailed analysis of malignant embeddings reveals that the distance from the center of mass of the red cluster roughly corresponds to the inverse malignancy rating of the nodule (not shown).

Hierarchical Clustering. We also run a hierarchical clustering of a random subset of embeddings using Ward's minimum variance method [17] (Fig. 2b). We analyze the top three splits and observe that the mean malignancy rating of nodules in the leftmost group (green) is much higher than in other groups, that is, the first split happens to cluster malignant vs benign nodules. Similarly, the second split isolates low-margin benign nodules (red) from other benign ones, and finally, the third split differentiates based on subtlety rating (among benign nodules with high margin score).

3.2   Usability: Top-k Evaluation

The purpose of our system is to aid a radiologist by presenting top-k cases similar to the query nodule. A radiologist can then inspect each presented case, assessing morphological similarity and reviewing historical diagnosis or even broader clinical records of the patients with presumably similar cases. Aiming for interface simplicity and maximum user productivity we prefer a minimum k which still provides enough information to support a doctor's decision. We study the semantic space distance between the retrieved samples and a query and decide that beyond $k = 4$ CBIR results provide diminishing returns.

 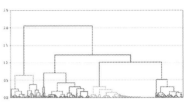

(a) t-SNE visualization of embedding space. Embeddings of the malignant nodules (red) are clustered together, in contrast to the benign nodules (blue)

(b) Hierarchical clustering of embeddings with top three splits corresponding to malignancy (green), margin (red) and subtlety (cyan, magenta) ratings

Fig. 2. Semantic information preservation in embedding space (Color figure online)

3.3 Qualitative Assessment: Example Output

Figure 3 shows example results of our system. The query nodule is presented on the left side (gray border). Only a single CT slice is shown, with doctors segmentation marks superimposed. A 3D mesh of consensus segmentation is presented below the slice. Query ratings (before normalization) together with the mean are provided on the left. The Top-4 CBIR results are shown on the right (green border) together with their mean ratings. Notice the shape and rating similarity of the CBIR results to that of the query. A non-match nodule is displayed (red border) for comparison. We remind the reader that in contrast to Fig. 3 visualization the CBIR query consists of a 3D CT patch only.

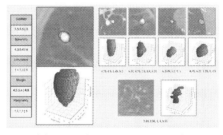

(a) Low malignancy query example

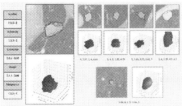 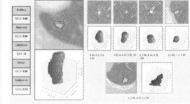

(b) High malignancy query example (c) Uncertain malignancy query example

Fig. 3. CBIR evaluation examples. See Subsect. 3.3 text for the description. (Color figure online)

3.4   Quantitative Assessment

To quantify the CBIR performance we conduct two evaluations: First, we compare the CBIR-based algorithm to medical experts. Then we compare to a recent work in the field [8].

Comparison to Human Experts. We define a nodule rating consensus as a mean vector of ratings assigned to the same nodule by several radiologists. A *dissent score* of the specific radiologist ratings is defined as an RMSE between it and the consensus among other radiologists. Similarly, a *dissent score* of an algorithm is the RMSE between ratings predicted by the algorithm and the consensus of all radiologists (using normalized rating range $[0, 1]$).

We use a rating consensus of k top CBIR results as a naïve algorithm prediction. We also provide, for comparison, a dissent score of a *random algorithm* that predicts one of the existing ratings randomly regardless of the input.

Figure 4 shows a distribution of dissent scores, while Table 1 reports their mean and standard deviation (measured over five-fold cross-validation). We can see that the naïve CBIR-based algorithm for $k = 4$ improves on the doctors' diagnosis on average as much as the latter improves on a random guess (1.5 times lower mean dissent score). That is, a distance between the query and CBIR results is at least comparable to the ranking uncertainty of the query itself, which is a strong indicator of the embedding space quality.

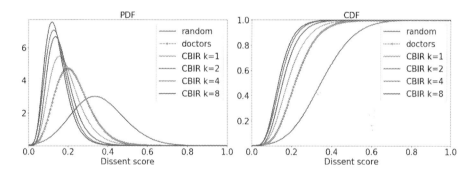

Fig. 4. PDF and CDF of the dissent score of random predictions, doctor ratings, and naïve CBIR-based algorithm with $k \in \{1, 2, 4, 8\}$. For clarity, all graphs show MLE fit of the log-normal distribution. The lower dissent score is better.

Comparison to Recent Results. We extend Table 1 with a comparison to [8]. Following the methodology defined in [8], we compute a prediction RMSE and STD over the dataset for each rating component separately (using rating range $[1, 5]$) and compare to the rating regression results from [8]. We also compute a CBIR malignancy retrieval precision, that is: what portion of the retrieved k results has a correct binary malignancy class (benign vs malignant). Since [8]

Table 1. Quantitative comparison of methods: Random guess, human expert, presented CBIR solution ("CBIR"), and Loyman et al. [8]. For the RMSE-based metrics, as defined here and in [8], lower is better. For Retrieval precision, higher is better. The best results are in bold.

Method	Dissent score mean	STD	Rating RMSE subtlety	sphericity	lobulation	margin	malignancy	Rating STD subtlety	sphericity	lobulation	margin	malignancy	Precision
random	0.35	0.13	1.43	1.13	1.27	1.44	1.47	2.81	1.59	2.64	3.08	2.86	
doctors	0.23	0.09	0.72	0.69	0.74	0.73	0.75	0.83	0.64	0.98	0.95	0.85	
CBIR k=1	0.19	0.08	0.89	0.79	0.84	0.94	0.87	0.59	0.48	0.58	0.61	0.56	**0.81**
CBIR k=2	0.17	0.07	0.76	0.70	0.74	0.81	0.73	0.49	0.41	0.48	0.52	0.47	0.80
CBIR k=4	0.15	0.06	0.70	0.65	0.68	0.74	0.68	0.45	0.38	0.44	0.46	0.44	0.80
CBIR k=8	**0.14**	**0.06**	**0.66**	**0.63**	**0.64**	**0.70**	**0.64**	**0.43**	**0.37**	0.42	0.44	**0.41**	0.79
CBIR mean													**0.79**
Loyman et al.			0.93	0.83	0.89	0.94	0.68	0.84	0.47	**0.27**	**0.37**	0.84	0.75

only presents mean precision for $k \in \{1, 3, 5, 7, 9, 11, 13, 15\}$ we compute it as well for a meaningful comparison ("CBIR mean" in the Table). From the results shown in the Table, we can see that our CBIR-based rating prediction improves over [8] by 0.20 for RMSE and by 0.14 for STD on average. It also compares favorably to human experts. The retrieval precision is improved by 4%.

4   Conclusion

In this work we prototype a CBIR system for pulmonary nodules. We develop a methodology to learn low-dimensional embeddings and present a theoretical justification for the architectural design selected. Our methodology facilitates learning of high-quality embeddings preserving both spatial and semantic information, all this despite very high data dimensionality and sample scarcity. We determine optimal usability settings and perform a qualitative analysis of CBIR results. Finally, we conduct a quantitative study demonstrating state-of-the-art results. While we believe that in reality, the diagnostic process is more complex than a 5-tuple rating can convey, we conclude that the CBIR output is highly relevant to the query, such that the benefit provided by the system to the radiologist end-user is comparable to obtaining a second radiologist's opinion.

Acknowledgments. Part of the work presented in this paper was done by the first two authors in the course of Y-Data program by Yandex School of Data Analysis. The authors would like to thank Kostya Kilimnik and Shlomo Kashani for the organization and support of the project initiation.

References

1. Armato, S.G., et al.: The lung image database consortium (LIDC) and image database resource initiative (IDRI): a completed reference database of lung nodules on CT scans. Med. Phys. **38**(2), 915–931 (2011). https://doi.org/10.1118/1.3528204

2. Armato III, S.G., et al.: Data from LIDC-IDRI (2015). https://doi.org/10.7937/K9/TCIA.2015.LO9QL9SX. https://wiki.cancerimagingarchive.net/x/rgAe

3. Dhara, A.K., Mukhopadhyay, S., Dutta, A., Garg, M., Khandelwal, N.: Content-based image retrieval system for pulmonary nodules: assisting radiologists in self-learning and diagnosis of lung cancer. J. Digit. Imaging **30**(1), 63–77 (2016). https://doi.org/10.1007/s10278-016-9904-y

4. Hancock, M.C., Magnan, J.F.: Lung nodule malignancy classification using only radiologist-quantified image features as inputs to statistical learning algorithms: probing the lung image database consortium dataset with two statistical learning methods. J. Med. Imaging **3**(4), 044504 (2016). https://doi.org/10.1117/1.jmi.3.4.044504

5. Kaggle data science bowl 2017 (2017). https://www.kaggle.com/c/data-science-bowl-2017. Accessed January 2020

6. Lam, M.O., Disney, T., Raicu, D.S., Furst, J., Channin, D.S.: BRISC—an open source pulmonary nodule image retrieval framework. J. Digit. Imaging **20**(S1), 63–71 (2007). https://doi.org/10.1007/s10278-007-9059-y

7. Liao, F., Liang, M., Li, Z., Hu, X., Song, S.: Evaluate the malignancy of pulmonary nodules using the 3-D deep leaky noisy-OR network. IEEE Trans. Neural Netw. Learn. Syst. **30**(11), 3484–3495 (2019). https://doi.org/10.1109/tnnls.2019.2892409

8. Loyman, M., Greenspan, H.: Lung nodule retrieval using semantic similarity estimates. In: Hahn, H.K., Mori, K. (eds.) Medical Imaging 2019: Computer-Aided Diagnosis. SPIE, March 2019. https://doi.org/10.1117/12.2512115

9. van der Maaten, L., Hinton, G.: Visualizing data using t-SNE. J. Mach. Learn. Res. **9**, 2579–2605 (2008)

10. McNitt-Gray, M.F., et al.: The lung image database consortium (LIDC) data collection process for nodule detection and annotation. Acad. Radiol. **14**(12), 1464–1474 (2007). https://doi.org/10.1016/j.acra.2007.07.021

11. Mikolov, T., Chen, K., Corrado, G., Dean, J.: Efficient estimation of word representations in vector space (2013)

12. Mohri, M., Rostamizadeh, A., Talwalkar, A.: Foundations of Machine Learning, 2nd edn. The MIT Press, Cambridge (2018)

13. Pan, L., Qiang, Y., Yuan, J., Wu, L.: Rapid retrieval of lung nodule CT images based on hashing and pruning methods. BioMed Res. Int. **2016**, 1–10 (2016). https://doi.org/10.1155/2016/3162649

14. Ronneberger, O., Fischer, P., Brox, T.: U-Net: convolutional networks for biomedical image segmentation. In: Navab, N., Hornegger, J., Wells, W.M., Frangi, A.F. (eds.) MICCAI 2015. LNCS, vol. 9351, pp. 234–241. Springer, Cham (2015). https://doi.org/10.1007/978-3-319-24574-4_28

15. Setio, A.A.A., et al.: Validation, comparison, and combination of algorithms for automatic detection of pulmonary nodules in computed tomography images: The LUNA16 challenge. Med. Image Anal. **42**, 1–13 (2017). https://doi.org/10.1016/j.media.2017.06.015

16. Siegel, R.L., Miller, K.D., Jemal, A.: Cancer statistics, 2020. CA: A Cancer J. Clin. **70**(1), 7–30 (2020). https://doi.org/10.3322/caac.21590
17. Ward, J.H.: Hierarchical grouping to optimize an objective function. J. Am. Stat. Assoc. **58**(301), 236–244 (1963). https://doi.org/10.1080/01621459.1963.10500845
18. Wei, G., Ma, H., Qian, W., Jiang, H., Zhao, X.: Content-based retrieval for lung nodule diagnosis using learned distance metric. In: 2017 39th Annual International Conference of the IEEE Engineering in Medicine and Biology Society (EMBC). IEEE, July 2017. https://doi.org/10.1109/embc.2017.8037711

Deep Active Learning for Effective Pulmonary Nodule Detection

Jingya Liu[1], Liangliang Cao[2,3], and Yingli Tian[1(✉)]

[1] The City College of New York, New York, NY 10031, USA
ytian@ccny.cuny.edu
[2] UMass CICS, Amherst, MA 01002, USA
[3] Google Inc., New York, NY 10011, USA

Abstract. Expensive and time-consuming medical imaging annotation is one of the big challenges for the deep learning-based computer-aided diagnosis (CAD) on the low-dose computed tomography (CT). To address this problem, we propose a novel active learning approach to improve the training efficiency for a deep network-based lung nodule detection framework as well as reduce the annotation cost. The informative CT scans, such as the samples that inconspicuous or likely to produce high false positives, are selected and further annotated for the nodule detector network training. A simple yet effective schema suggests the samples by ranking the uncertainty loss predicted by multi-layer feature maps and the Region of Interests (RoIs). The proposed framework is evaluated on a public dataset DeepLesion and achieves results that surpass the active learning baseline schema at all the training cycles.

Keywords: Lung nodule detection · Active learning · Low-dose CT · Deep learning

1 Introduction

Lung cancer is the leading cause of cancer death in the United States [12]. To better assist the clinic diagnosis, the low-dose computed tomography (CT) screening is recommended as the most effective lung nodule diagnostic tool. The automatic computer-aided diagnosis (CAD) systems based on CT scans have been widely exploited for automatic lung nodule classification, segmentation, and detection.

In recent years, many researchers have devoted to applying deep learning-based frameworks to medical imaging analysis [15]. Since deep learning algorithms are data-hungry, they may suffer from the expensive data acquisition and annotation of medical images. To label the medical images, the oracle requires clinical and biomedical background knowledge while the nature scene labeling needs mainly common sense. Time-consuming pulmonary nodule annotation brings difficulty to acquire large annotated datasets. With the limited data, the deep learning-based models are prone to overfitting. Recently, semi-supervised

© Springer Nature Switzerland AG 2020
A. L. Martel et al. (Eds.): MICCAI 2020, LNCS 12266, pp. 609–618, 2020.
https://doi.org/10.1007/978-3-030-59725-2_59

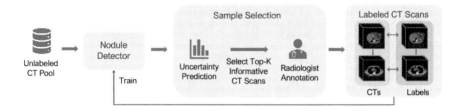

Fig. 1. The flowchart of the proposed active learning nodule detection framework. A group of randomly sampled data is selected to train the nodule detection network. In each of the training cycle, the trained network evaluates the remaining unlabeled CT scans in the unlabeled CT pool by predicting the uncertainty for each CT scan. Based on the uncertainty scores, the top-k informative CT scans are selected in the current training cycle considered as the most informative samples for labeling. The newly annotated CTs are further joined the labeled CTs for the detector training. The steps are iterated until the network achieves good performance.

and unsupervised learning-based methods utilize partial or unlabeled data to avoid the annotation cost [5,6]. These methods mainly focus on extracting good features, however, the data annotations are still required for the downstream tasks, such as classification, segmentation, and detection.

Existing approaches consider active learning methods to reduce data annotation cost by selecting valuable representatives from unlabeled data [2,4,9,10,13]. For medical imaging, the active learning strategy has shown high potential to reduce the annotation cost [1], such as in biomedical image segmentation [3,14,18] and pathology image classification [11,13]. Most of the active learning methods for object detection tasks are based on the uncertainty scores predicted by image-level representation extracted from backbone network to select informative samples. Recently, Yoo *et al.* [19] proposed a state-of-the-art learning-loss schema to predict the loss value of unlabeled data based on the target network, which can be applied to the object detection task. Multi-level features from the target model are fused to map the scalar value for loss prediction. Since the loss prediction network only considers the model loss and despite the task, the loss prediction strategy is robust to various tasks and has high potential to apply to nodule detection tasks. However, the performance of object detection task is not guaranteed with small objects or 3D volumetric data. The image-level distribution is not sufficient and not detailed for pulmonary nodule detection.

There are two major challenges in implementing the existing active learning methods for pulmonary nodule detection. First, unlike natural scene images, CT scans contain relatively small volume of lung nodules relative to the whole 3D CT. Second, the feature similarity between normal tissues and nodules may mislead the sample selection. The image-level uncertainty prediction cannot be directly applied to the 3D features. Although the uncertainty of the whole CT scan can be learned, the informative features that distinguish nodules and tissues are tended to be ignored.

To address the above challenges, this paper makes the following contributions: 1) A deep active learning-based nodule detection framework is proposed to achieve comparable performance with less data and annotations, as shown in Fig. 1. The unlabeled CT scans are assessed by a trained model with predicted uncertainties. By sorting the uncertainty scores of CTs, the top-K informative candidates are selected, further annotated as the training data. 2) A simple yet effective uncertainty selection method predicts the loss through the features extracted from the basic blocks of the backbone network which evaluates the learning-loss uncertainty, and the features from region proposals for an RoI-level prediction. 3) The proposed framework is conducted on the large public DeepLesion dataset [17] and the proposed method surpasses the active learning baseline [19] at all the training cycles.

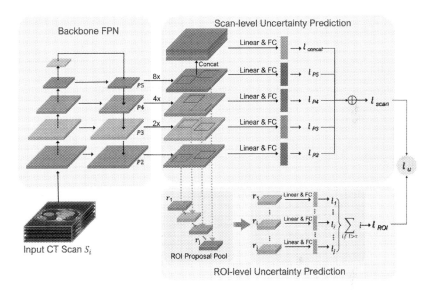

Fig. 2. The proposed active learning schema for pulmonary nodule detection. A trained 3DFPN detector takes the input of 3D CT volume and obtains the learning-loss features maps through multiple layers $[P2, P3, P4, P5]$. The learning-loss uncertainty is computed by aggregating the loss of four-layer feature maps and a global loss predicted by concatenated four-layer features. The RoI-level uncertainty prediction considers the local features of the region proposals prediction and is obtained by summing predicted scores greater than the threshold τ. The uncertainty of sample selection combines learning-loss with RoI-level loss for each unlabeled CT scan.

2   Method

The informative CT scans with higher uncertainty values are defined as hard samples, while the non-informative CT scans with lower scores are defined as easy samples. The key insight of our method is to learn from the hard samples of CT scans for pulmonary nodule detection. More specifically, we consider the following two scenarios. 1) The CT scans contain normal tissues with similar features of nodules. The high probability to be detected as nodules leads to a high false-positive rate. 2) The CT scans include the nodules that are difficult to detect, with a small size or low intensity. In this section, we introduce a simple yet effective active learning scheme for pulmonary nodule detection network.

2.1 Nodule Detection Framework

By appending the low-level texture features with higher-level strong-semantic features, the feature pyramid networks (FPNs) [7] achieved good performance on detecting small objects by combining both local and global features through the multi-layered feature extraction. Since the original FPN is designed to process 2D images, considering the spatial information of nodules in consecutive CT slices, we follow the nodule detection network of 3DFPN [8] by applying 3D ResNet-18 as the backbone network and detecting nodule locations in 3D CT volumes. The feature pyramid network consists of four layers of $[P2, P3, P4, P5]$, which integrates low-level features through a top-down pathway by lateral connections. The feature maps are further applied to evaluate the uncertainty. Currently, two-stage object detection methods such as RCNN, are widely used in small object detection combined with the FPN backbone framework. Yan *et al.* [16] proposed a 3D Context Enhanced (3DCE) RCNN model for lesion detection, however, the spatial information is not guaranteed by the three-channel images. In this paper, we employ the 3D volume CT scan as input to extract multi-layer features from 3D Feature Pyramid ConvNet ($3DFPN$) with the region of interest (RoI) proposal selection. The prediction includes a confidence score of each nodule candidate, as well as the corresponding position and nodule size, as $[x, y, z, d]$, where $[x, y]$ are the spatial coordinates of the candidate, z is the CT slice index number, d is the diameter of the nodule.

2.2 Active Learning with Nodule Detection

This section introduces the proposed active learning framework for lung nodule detection. During the initial network training, a group of CT scans is randomly selected from all unlabeled CT pool for annotation and defined as S_{train}. In the first learning cycle, a deep learning framework is trained by S_{train} and predicts uncertainties l for each unlabeled CT in the unlabeled pool $S_{unlabel}$. As the CT scans are not equally contributed to the performance of the model, the higher predicted uncertainty loss indicates the greater difficulty for the nodule prediction, with the high false-negative rate for the missing nodules or high false-positive prediction for tissues detected as nodules. We define informative CT scans with high uncertainty value as hard samples and non-informative CT scans with lower scores as easy samples. The hard samples aim to efficiently improve the performance of the nodule detection network. By sorting the predicted loss of l in descending order, top-k CT scans are selected as the hard samples, which would be annotated by radiologists and aggregated to the training data S_{label}. The data augmentation methods of the random flip, crop, and rotate are applied to the labeled k CT scans to avoid overfitting. The augmented data are joined as the input to train the lung nodule detector. The trained model is applied to predict the loss for unlabeled data to select another set of top-k CT scans for annotation. Repeat the active learning schema until the nodule detector achieves good performance.

As illustrated in Fig. 2, an active learning-based sample selection approach is proposed to predict loss through global features and RoIs. 3DFPN is applied as the backbone network by the multi-layer feature maps extracting by a top-down path. The feature maps of $\{P2, P3, P4, P5\}$ extracted from the last four convolution layers represent the global features, which are normalized by 3D global average pooling and a fully connected (FC) layer. Following the learning-loss schema [19], a scan-level prediction consists of multi-layer uncertainty losses of the four feature scalars and a loss predicted by the concatenated four feature maps. The objective function for the learning-loss uncertainty is obtained by the five uncertainty scores, shown as Eq. 1:

$$l_{scan} = l_{P2} + l_{P3} + l_{P4} + l_{P5} + l_{P_{concat}}, \tag{1}$$

where i indicates the current feature layers, P_{concat} concatenates the feature maps of $\{P2, P3, P4, P5\}$.

To obtain the detailed local features for CT scans contained nodules with the small scale and tissues with the similar feature as nodules, the learning-loss schema is not sufficient to select the most valuable CT scans by predicting the loss through the feature maps, where the informative features of the true negative and false positives candidates are not statistically significant shown. As the region proposal network predicts nodule candidate location and the region of interest (RoI) cropped from the multi-layer feature maps, in particular, we introduce a simple yet effective sample selection method based on uncertainty prediction by statistically selecting by RoI level uncertainty prediction, as shown in Fig. 2. We aim to select CT scans with a large number of high uncertainty RoI regions as additional criteria for CT scan sample selection.

In order to select the CT scans with the most false-positive samples, a set of uncertainty scores is predicted for each RoI region of the entire 3D CT scan. For each CT scan s_i, a set of RoI regions r_{s_i} is obtained. RoI-level loss prediction L_{RoI} sums the scores of the region proposal l_i for the value greater than the threshold τ, shown as Eq. 2:

$$L_{RoI} = \begin{cases} \sum\limits^{i} l_i & if \ \ l_i > \tau \\ 0 & otherwise. \end{cases} \tag{2}$$

Therefore, the objective function is combined with learning-loss, RoI-level loss, and detector loss as Eq. 3:

$$l_{final} = l_{target}(\hat{y}, y) + \lambda_1 \cdot L_{scan}\{\hat{l}, l\} + \lambda_2 \cdot L_{RoI}\{\hat{l}, l\}, \tag{3}$$

where λ_1, λ_2 are the weights of learning-loss and RoI-level prediction respectively. $(\hat{l}, l)$ are the two samples selected from different batches followed the batch strategy of [19] by comparing two samples from different batches.

3   Experiments

3.1   Experimental Settings

Dataset and Evaluation. The NIH DeepLesion dataset [17] contains $32,000$ annotated lesions on the CT scans acquired from $4,400$ patients. In this paper, $1,281$ CT scans with $2,592$ lung nodules are conducted in the evaluation. The $1,000$ CT scans are for training and 281 CT scans from the official split are for testing. To assess the proposed framework, we assume all the CT scans are unlabeled and in each active learning cycle, 10% of the unlabeled dataset is selected and annotated. We follow the evaluation method of the baseline detector 3DCE [16] and employ the sensitivity at certain false positives per image similar to the Free-Response Receiver Operating Characteristic (FROC). In this paper, we present the results for sensitivity at 4 false positives per image for a fair comparison with the baseline detector 3DCE [16] for the performance of lung nodule detection.

Training. In the first active learning cycle, the detector is trained by a randomly selected 10% CT scans from the unlabeled data pool with annotations (100 CT scans). The trained model is applied to predict the loss of the remaining 90% unlabeled data. By sorting the loss prediction, the top 100 CTs, which is 10% of the initial unlabeled CT pool, are annotated and added to the training data to finetune the trained model. A total of 30 epochs are applied for each training cycle. The learning rate is initialized as 0.01 and decreased by 1/10 at the 280 epoch. 300 epochs are conducted for the framework. The 18-layer residual network (ResNet-18) is applied as the backbone network of the 3DFPN detector. Following the same implementation of Yoo *et al.* [19], loss prediction module of learning-loss schema is conducted for scoring the four feature maps of the backbone network. The RoIs with the prediction score greater than 0.5 are applied to obtain the RoI-level loss. To avoid overfitting in the training process, the random flip, rotation and crop are also applied for the data augmentation.

Table 1. Comparison between 3DCE [16] and our proposed active learning-based nodule detection network. With the 80% annotated training data, the proposed method achieves a comparable result with the baseline 3DCE network.

Methods	3DCE [16]	Ours-60%	Ours-70%	Ours-80%	Ours-90%	Ours-100%
Sensitivity	0.910	0.875	0.896	0.908	0.915	0.921

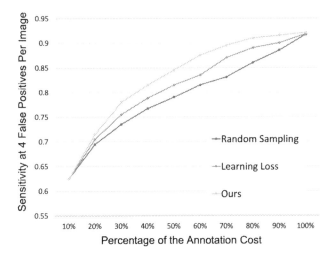

Fig. 3. The sensitivities at 4 false positives per image for the proposed active learning-based framework compared with random sampling and learning-loss sample selection.

3.2 Results

Performance Evaluation. Table 1 shows the results compared to the baseline detector. By using 80% of all the training data, the sensitivity at 4 false positives per image of our proposed framework is comparable to the baseline 3DCE model with the full training data. By only trained with 60% annotated data, the result of our model is approaching to the sensitivity of 3DCE trained by the full training data with only 3.5% less. With all the training data annotated, our model surpasses the baseline detector by 1.1% sensitivity.

In addition, we compare the sensitivity of the proposed active learning method with random sampling and learning-loss sampling schema [19]. The sensitivities of our proposed method and the baseline active learning methods at 4 false positives per image are shown in Fig. 3. Our active learning strategy surpasses learning-loss prediction [19] and random sampling at all active learning cycles. The learning-loss features may fail to capture the features of the nodule location, while the RoI feature region prediction may lose the global information. As shown in Fig. 3, by training with 60% annotated data, the proposed method obtains an 87.5% sensitivity at 4 false positives per image which is 6% higher than the random sampling baseline, and 4% higher than the learning-loss sample selection strategy. The performance of the nodule detector is greatly improved by combining the learning-loss and RoI-level prediction shown the effectiveness of the proposed active learning sample selection strategy.

Selected Sample Analysis. The samples selected by the proposed method, learning-loss prediction, and random sampling within 20%, 40%, 60%, 80% active learning cycles are compared in Fig. 4. X-axes and Y-axes indicate the nodule

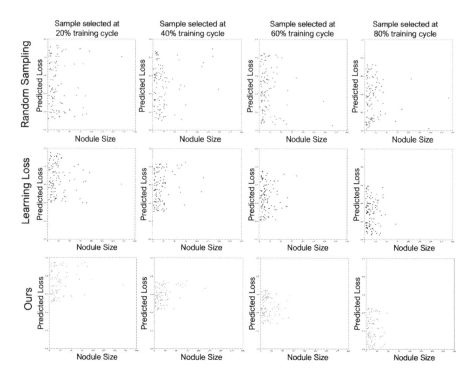

Fig. 4. Visualization of the selected nodule samples on 20%, 40%, 60%, 80% active learning cycle respectively. The x-axis of the scatter map indicates the nodule diameter and y-axis correspond to the uncertainty loss prediction. We compare the samples for the random sampling (marked as blue dots), and learning-loss sampling (marked with black dots) with the proposed active learning schema (marked as cyan dots). (Color figure online)

diameter and the predicted loss value for the corresponding CT scan. As the RoI regions selected by the proposed detector are the potential candidate regions, by selecting the candidates containing RoI regions with higher loss in the object function, the algorithm aims at finding the candidates with the highest potential to be predicted as miss-detected candidates (negative candidates) and the false detected candidates (positive candidates). For the samples selected from the 20% training cycle, we can observe that most of the predicted losses from the proposed active learning strategy (marked as cyan dots) are higher than other methods especially random sampling (marked as blue). With the learning cycle increasing from 40% to 60%, the proposed method leans to focus on the nodules with the small size, while the random sampling still selects the large nodule with low loss values. The learning-loss (marked as black) predicts higher and clusters are more sparse. When the training cycle increases to 80%, it is worth noting that the predicted loss values of the active learning-based sample selection strategies gradually decreases due to the priority to select informative samples, while the remaining 20% of the unlabeled data contains majority easy samples.

4 Conclusion

To eliminate expensive annotation costs and effort to acquire large datasets, we have proposed an active learning schema to select the valuable CT scans to train a deep learning-based pulmonary nodule detector. A simple yet effective active learning schema predicts loss from multi-layer features and RoIs for sample selection. The experimental results have demonstrated that the proposed framework has great potential in accelerating clinic diagnosis.

Acknowledgement. This material is based upon work supported by the National Science Foundation under award number IIS-1400802.

References

1. Budd, S., Robinson, E.C., Kainz, B.: A survey on active learning and human-in-the-loop deep learning for medical image analysis. arXiv preprint arXiv:1910.02923 (2019)
2. Budd, S., et al.: Confident head circumference measurement from ultrasound with real-time feedback for sonographers. In: Shen, D., et al. (eds.) MICCAI 2019. LNCS, vol. 11767, pp. 683–691. Springer, Cham (2019). https://doi.org/10.1007/978-3-030-32251-9_75
3. Chen, X., Williams, B.M., Vallabhaneni, S.R., Czanner, G., Williams, R., Zheng, Y.: Learning active contour models for medical image segmentation. In: Proceedings of the IEEE Conference on Computer Vision and Pattern Recognition, pp. 11632–11640 (2019)
4. Gal, Y., Islam, R., Ghahramani, Z.: Deep Bayesian active learning with image data. In: Proceedings of the 34th International Conference on Machine Learning, vol. 70, pp. 1183–1192. JMLR.org (2017)
5. Jing, L., Tian, Y.: Self-supervised visual feature learning with deep neural networks: a survey. arXiv preprint arXiv:1902.06162 (2019)
6. Károly, A.I., Fullér, R., Galambos, P.: Unsupervised clustering for deep learning: a tutorial survey. Acta Polytech. Hung. **15**(8), 29–53 (2018)
7. Lin, T.Y., Dollár, P., Girshick, R., He, K., Hariharan, B., Belongie, S.: Feature pyramid networks for object detection. In: Proceedings of the IEEE Conference on Computer Vision and Pattern Recognition, pp. 2117–2125 (2017)
8. Liu, J., Cao, L., Akin, O., Tian, Y.: 3DFPN-HS^2: 3D feature pyramid network based high sensitivity and specificity pulmonary nodule detection. In: Shen, D., et al. (eds.) MICCAI 2019. LNCS, vol. 11769, pp. 513–521. Springer, Cham (2019). https://doi.org/10.1007/978-3-030-32226-7_57
9. Lowell, D., Lipton, Z.C., Wallace, B.C.: Practical obstacles to deploying active learning. In: Proceedings of the 2019 Conference on Empirical Methods in Natural Language Processing and the 9th International Joint Conference on Natural Language Processing (EMNLP-IJCNLP), pp. 21–30 (2019)
10. Mahapatra, D., Bozorgtabar, B., Thiran, J.-P., Reyes, M.: Efficient active learning for image classification and segmentation using a sample selection and conditional generative adversarial network. In: Frangi, A.F., Schnabel, J.A., Davatzikos, C., Alberola-López, C., Fichtinger, G. (eds.) MICCAI 2018. LNCS, vol. 11071, pp. 580–588. Springer, Cham (2018). https://doi.org/10.1007/978-3-030-00934-2_65

11. Nalisnik, M., Gutman, D.A., Kong, J., Cooper, L.A.: An interactive learning framework for scalable classification of pathology images. In: 2015 IEEE International Conference on Big Data (Big Data), pp. 928–935. IEEE (2015)

12. Siegel, R.L., Miller, K.D., Jemal, A.: Cancer statistics, 2020. CA: A Cancer J. Clin. **70**(1), 7–30 (2020)

13. Wang, K., Zhang, D., Li, Y., Zhang, R., Lin, L.: Cost-effective active learning for deep image classification. IEEE Trans. Circ. Syst. Video Technol. **27**(12), 2591–2600 (2016)

14. Wen, S., et al.: Comparison of different classifiers with active learning to support quality control in nucleus segmentation in pathology images. AMIA Summits on Transl. Sci. Proc. **2018**, 227 (2018)

15. Wu, J., Qian, T.: A survey of pulmonary nodule detection, segmentation and classification in computed tomography with deep learning techniques. J. Med. Artif. Intell. **2**, 2–8 (2019)

16. Yan, K., Bagheri, M., Summers, R.M.: 3D context enhanced region-based convolutional neural network for end-to-end lesion detection. In: Frangi, A.F., Schnabel, J.A., Davatzikos, C., Alberola-López, C., Fichtinger, G. (eds.) MICCAI 2018. LNCS, vol. 11070, pp. 511–519. Springer, Cham (2018). https://doi.org/10.1007/978-3-030-00928-1_58

17. Yan, K., Wang, X., Lu, L., Summers, R.M.: DeepLesion: automated mining of large-scale lesion annotations and universal lesion detection with deep learning. J. Med. Imaging **5**(3), 036501 (2018)

18. Yang, L., Zhang, Y., Chen, J., Zhang, S., Chen, D.Z.: Suggestive annotation: a deep active learning framework for biomedical image segmentation. In: Descoteaux, M., Maier-Hein, L., Franz, A., Jannin, P., Collins, D.L., Duchesne, S. (eds.) MICCAI 2017. LNCS, vol. 10435, pp. 399–407. Springer, Cham (2017). https://doi.org/10.1007/978-3-319-66179-7_46

19. Yoo, D., Kweon, I.S.: Learning loss for active learning. In: Proceedings of the IEEE Conference on Computer Vision and Pattern Recognition, pp. 93–102 (2019)

Musculoskeletal Imaging

Towards Robust Bone Age Assessment: Rethinking Label Noise and Ambiguity

Ping Gong[1], Zihao Yin[1,2], Yizhou Wang[1,2,3,4], and Yizhou Yu[1,5(✉)]

[1] Deepwise AI Lab, Beijing, China
yizhouy@acm.org
[2] Department of Computer Science, Peking University, Beijing, China
[3] Center on Frontiers of Computing Studies, Peking University, Beijing, China
[4] Advanced Institute of Information Technology, Peking University, Hangzhou, China
[5] The University of Hong Kong, Pokfulam, Hong Kong

Abstract. The effects of label noise and ambiguity are widespread, especially for subjective tasks such as bone age assessment (BAA). However, most existing BAA algorithms ignore these issues. We propose a robust framework for BAA supporting Tanner & Whitehouse 3 (TW3) method, which is clinically more objective and reproducible than Greulich & Pyle (GP) method, but has received less attention from the research community. Since the publicly available RSNA BAA dataset was annotated using GP method, we contribute additional TW3 annotations. We formulate TW3 BAA as an ordinal regression problem, and address both label noise and ambiguity with a two stage deep learning framework. The first stage focuses on correcting erroneous labels with ambiguity tolerated, while the latter stage introduces a module called Residual Context Graph (RCG) to conquer label ambiguity. Inspired by the way human experts handle ambiguity, we combine fine-grained local features with a graph based context. Experiments show the proposed framework outperforms previously reported TW3-based BAA systems by large margins. TW3 annotations of bone maturity levels for a portion of the RSNA BAA dataset will be made publicly available.

Keywords: Bone age assessment · Noisy label · Graph convolutional network

1 Introduction

Pediatric Bone Age Assessment (BAA) is a radiological measure of skeletal maturity widely used in the diagnosis of growth disorders and the prediction of adult height. Common BAA standards are Greulich & Pyle (GP) [6] and Tanner & Whitehouse (TW2 [15], TW3 [14]). Compared with GP method, TW methods are more precise and reproducible, which is crucial for monitoring therapy outcome with endocrine, metabolic, and genetic diseases.

P. Gong and Z. Yin—Contributed equally.

A. L. Martel et al. (Eds.): MICCAI 2020, LNCS 12266, pp. 621–630, 2020.
https://doi.org/10.1007/978-3-030-59725-2_60

This paper focuses on the most useful TW3-RUS method, which require individual assessment for Radius, Ulna, and 11 Short bones (RUS) to determine their maturity level, as illustrated in Fig. 1(a). Each maturity level, denoted from A to I, is assigned a predefined maturity score. The summed maturity score (SMS) is then mapped to TW-RUS BA according to the subject's gender.

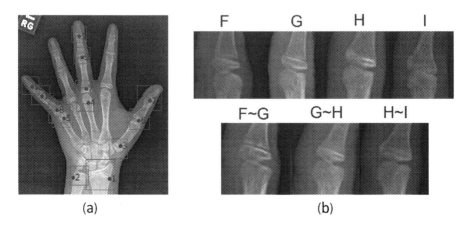

(a)   (b)

Fig. 1. (a) Regions of interest associated with the bones considered in the TW3-RUS standard. (b) The 1st Distal Phalanx ROIs and their maturity levels(denoted F to I): the first row shows examples matching with TW3-RUS maturity indicators. The second row shows ambiguous cases that 'fall between' two adjacent maturity levels.

One challenge of TW3-RUS BAA is the ambiguity of maturity levels ratings, as shown in Fig. 1(b). Many cases do not match well the standard TW3-RUS descriptions, forcing physicians to rely on their own understandings. To make the ambiguity even worse, X ray imaging defects, rotation or occlusion of hand bones are not unusual in clinical practices.

Another problem is label noise, due to the tedious nature of manual maturity level rating. One could suppress label noise by employ multiple annotators per sample, however, annotation cost would be tremendous. We seek an approach to work with only one annotation per training sample.

We attack the above challenges with a two stage framework. For the first stage, we propose a validation guided label correction scheme. Inspired by [1,13], we prevent model from memorizing noise with large initial learning rate, and then progressively optimize pseudo labels by utilizing self-supervision signals from best predictions of learning history. The second stage tackles ambiguous cases whose maturity level can not be determined within local ROI. We introduce a Residual Context Graph (RCG) module to exploit hand bones' overall growth pattern. With combined local features and RCG context features, our model achieves Mean Absolute Error (MAE) of 0.216 year on a RSNA TW3 test set, significantly outperforms previously reported TW3-based BAA systems [11,16].

In summary, this paper has the following contributions:

1. We propose a novel label noise correction scheme called Validation Guided Label Correction. It fixes obviously erroneous labels and allows the subsequent stage to focus on harder, ambiguous cases.
2. We propose a graph network module, called Residual Context Graph, to effectively resolve ambiguities in maturity level rating by exploiting the hand bones' overall growth pattern.
3. We contribute TW3-RUS annotations of maturity levels for a selected portion of the RSNA BAA dataset to encourage future research. To the best of our knowledge, it is the first large scale publicly available TW3 annotations.

2   Related Work

Before deep learning era, the BoneXpert method [16] extracted shape, intensity and texture features to derive 'intrinsic bone age', which was then calibrated to GP or TW Methods. They reported TW2-RUS BAA performance of 0.8 year standard deviation on 84 clinical images.

Recently, deep learning approaches achieved impressive performance [3, 4, 8]. PRSNet [8] achieved GP BAA MAE of 4.49 months on RSNA test set by mining the relationship among different parts and select the top ranking parts for BAA. BoNet [4] used local information embedded in keypoint heatmap to help learn fine-grained features. They reported single model GP MAE of 4.37 months on RSNA test set. Son [11] proposed an end-to-end TW3-based BAA system and achieved MAE of 0.46 years. Average voting ensembles were used for TW3 maturity level rating. None of these methods addressed the label noise and ambiguity issues.

3   Methodology

Our pipeline starts with detecting hand and 13 RUS anchor points. As shown in Fig. 1(a), TW3-RUS associated ROIs can be deduced from those anchor points through anatomical relationships. ROIs are then aligned with finger directions and fed to the Maturity Level Rating Framework. Finally, the estimated maturity levels are mapped to TW3-RUS BA according to standard rules.

Our Maturity Level Rating framework has two stages: correction of label noise and learning the Residual Context Graph. In the following sections, we describe each stage in detail.

3.1   Validation Guided Noisy Label Correction

In this paper, label noise refers to obvious annotator mistakes, while label ambiguity refers to the hardness or impossibility to uniquely determine ground truth labels from images, as discussed in the introduction section. The challenge is, how to identify and clean noisy labels, while preserving normal and ambiguous labels?

Tanaka et al. empirically found [13] that under a high learning rate DNNs do not memorize noise and maintain high performance for clean data. We confirmed this observation on our TW3 BAA dataset.

In addition, we have two other observations under high learning rate: firstly, validation loss fluctuates violently from epoch to epoch. Secondly, even for epochs with low validation loss, unreasonable predictions occur randomly.

Based on those observations, we design a simple yet effective noisy label correction algorithm. Firstly, the original one hot label Y_{noisy} is converted to discrete Gaussian distribution $\widetilde{Y}_{noisy}$. Then, under a large learning rate, the model is trained for $N + M$ epochs ($N \geq 10$) as illustrated in Algorithm 1. For the first N epochs, the model is trained with fixed noisy labels $\widetilde{Y}_{noisy}$ to exploit ambiguity between adjacent maturity levels. For the next M epochs, the model and Y_{pseudo_label} are jointly optimized. We use the MAE loss of bone maturity levels on validation set, which was independently annotated by two experienced physicians, to rank model performance. The training set Y_{pseudo_label} is updated as the mean of 10 best model's predictions $\hat{Y}$. The overall loss function for training set label correction is shown in Eq. 1. Y is $\widetilde{Y}_{noisy}$ for the first N epochs and Y_{pseudo_label} afterwards.

$$L_{correction} = L_{KL}(Y, \hat{Y}) + \alpha \cdot L_{mean}(Y, \hat{Y}) + \beta \cdot L_{var}(Y, \hat{Y}) \tag{1}$$

We perform distribution learning [5] by penalizing KL divergence between model output $\hat{Y}$ and label distribution Y. Mean loss and variance loss from [10] are added to further penalize the deviation of $E(\hat{Y})$ from $E(Y)$ and avoid over smoothing predictions.

Algorithm 1. Validation Guided Label Correction

---
  for $t = 1, \ldots, N$ **do**
    update $\theta^{(t+1)}$ by Optimize on $L_{correction}(\theta^{(t)}, \widetilde{Y}_{noisy}|X_{train})$
    if validation set MAE loss $< max$(best 10 validation set MAE losses) **then**
      update best 10 validation set MAE queue and associated model predictions
    end if
  end for
  $Y_{pseudo\_label} \leftarrow Mean$(best 10 models' predictions)
  for $t = N + 1, \ldots, N + M$ **do**
    upddate $\theta^{(t+1)}$ by Optimize on $L_{correction}(\theta^{(t)}, Y_{pseudo\_label}|X_{train})$
    if validation set MAE loss $< max$(best 10 validation set MAE losses) **then**
      update best 10 validation set MAE queue and associated model predictions
      $Y_{pseudo\_label} \leftarrow Mean$(best 10 models' predictions)
    end if
  end for

Our scheme is inspired by [13] but quite different. First, they use the running average of the last ten epochs' model predictions as pseudo labels, which means the model is prone to confirm its mistakes; in contrast, we persist with the ten best models' predictions to ensure supervision signals get better, not worse. Second, to avoid trivial solution, they add a regularization loss term by assuming the prior distribution of all classes is known and can be approximated by each mini-batch, which is not practical for imbalanced bone maturity levels. Our method, guided by the validation set, completely gets rid of this regularization loss. To the best of our knowledge, we are the first to solve the noisy ordinal regression problem by combining distribution learning and label correction, while previous noisy label works are focused on classification.

3.2   Learning Residual Context Graph

According to the feedback of annotators, there are some hard cases caused by irregular hand poses and abnormal growth, where epiphyses may rotate, deform or even be partially obscured. Under these circumstances, experienced experts attempt to take other bones with similar growth patterns for references. Inspired by this observation, we design a novel module named **Residual Context Graph** (RCG) which combines local features with relevant context and enables flexible information flow among RUS ROIs, for robust bone age assessment.

Preliminary Knowledge Recap

Graph Convolution Network. (GCN) was introduced in [9] to perform semi-supervised classification. The GCN architecture has been proven to generalize traditional convolution operator to non-Euclidean data quite successfully.

The function form of GCN can be simply written as follows:

$$X^{l+1} = f(\widehat{A}X^l\Theta) \tag{2}$$

which takes both $X^l \in \mathbb{R}^{N \times C}$ and normalized adjacency matrix $\widehat{A}$ as inputs. $\Theta \in \mathbb{R}^{C \times F}$ denotes a matrix of filter parameters which GCN needs to learn. f denotes a non-linear operator which activates the convolved feature matrix and yields the next layer output $X^{l+1} \in \mathbb{R}^{N \times F}$. For numerical stability of repeated applications of this operator, the original graph matrix A needs to be renormalized. $\widetilde{A} = A + I_N$ add self-connections and $\widetilde{D}_{ii} = \sum_j \widetilde{A}_{ij}$ is the corresponding degree matrix. Then the renormalization trick is defined as:

$$\widehat{A} = \widetilde{D}^{-\frac{1}{2}}\widetilde{A}\widetilde{D}^{-\frac{1}{2}} \tag{3}$$

where $\widehat{A}$ enables information exchange among graph nodes. The linear formulation here is both effective and efficient. By stacking multiple such layers, our model can learn a rich class of inter-relationships of nodes.

GCN for Learning Context Graph. In our work, each node of the graph represents a single RUS ROI and utilizes local features extracted by backbone as initial information. Then we want GCN to learn a proper information exchange pattern between these 13 RUS ROIs. We anticipate output features after a stack of GCN layers include relevant context information for robust maturity level ratings. For adjacency matrix, we adopt the classical design of fully connected graph. The overall architecture of stage 2 is shown in Fig. 2.

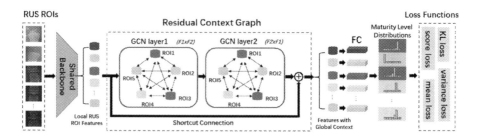

Fig. 2. Overall architecture of stage 2. Stack of two GCN layers enables information exchange among RUS ROIs. GCN context features are combined with local ROI features by shortcut connections. Design of RCG follows bottleneck fashion with $F_1 > F_2$. Considering label ambiguity, we apply mean loss, variance loss and KL loss to each RUS ROI individually and propose a novel total score loss for the whole hand, to achieve robust and unbiased distribution learning of maturity level estimation.

Resdiual Learning. The popular residual learning scheme is applied to the design of RCG. We stack two GCN layers and the first GCN layer takes extracted features as inputs. Features output by the second GCN layer are added together with the initial input features. We also follow the design of bottleneck in [7] to make our RCG module light-weighted. For input $X \in \mathbb{R}^{N \times F_1}$, we let first GCN layer to shrink the feature dimension to a relatively small number F_2 ($F_2 < F_1$) and then let second GCN layer to expand feature dimension to F_1.

Score Loss. Since the level-to-score mapping under TW3 standard is only for discrete integer levels, we utilize both predicted distribution output by our model and target distribution generated by DLDL method to compute two expectations of maturity scores for each RUS ROI. By summing maturity scores for RUS ROI together, we get total scores for the whole hand. We propose a new loss function named **score loss** to penalize L1 distance between predicted scores and target scores as follows:

$$\mathcal{L}_{score} = \mathcal{L}_1(sum_{pred}, sum_{target}), sum = \sum_{i=1}^{N} \sum_{j=1}^{C_i} p_{ij} * S_{ij} \quad (4)$$

where N is 13, C_i is the number of maturity levels for bone i, p_{ij} denotes probability for bone i with maturity level j and S_{ij} denotes the corresponding scores under TW3 standard. The introduction of score loss forces maturity level error for each RUS ROI to randomly deviate, but not systematically larger or smaller.

4 Experiments

4.1 Dataset

All experiments are conducted on the public RSNA BAA dataset, for which we contribute TW3-RUS annotations. All recruited annotators were radiologists or pediatricians experienced in TW3 BAA. The RSNA test set and validation set were fully annotated. Two thousand RSNA training samples were selected towards balanced GP BA distribution. Six senior physicians independently annotated the test set, to give reliable test results. The validation set was annotated twice by different physicians. The training set was annotated once per sample.

4.2 Implementation Details

We adopt MMDetection [2] for hand detection and HRnet [12] for RUS keypoint detection. Random vertical flip and random rotation of $\pm 180°$ are employed for robustness. To train maturity level rating models, cropped RUS ROIs are augmented by random rotation of $\pm 10°$, random translation of $\pm 10\%$ and random scaling from 0.9 to 1.1. All ROIs are resized to 256×256.

For maturity level rating, Resnet34 is used as backbone with batch size 32. Adam optimizer with weight decay 0.001 is used. For stage 1, $lr = 0.001, N = 40, M = 40$. For stage 2 training, lr is set to 0.001 and halves every 40 epochs. For RCG module, $F1 = 512, F2 = 64$. LeakyReLU with negative slope 0.2 is used as activation function. Score loss weight is 0.1.

4.3 Performance

Because there are few TW3-RUS algorithms[11, 16] in the literature, we build a strong baseline using a combined DLDL [5] and MVL [10] method in the same way as Eq. 1, with α set to 0.5, β set to 0.05. Table 1 shows our method significantly improves the baseline, with TW3-RUS maturity level MAE reduced by 23.6% and BA MAE reduced by 25.5%. Recently, Son [11] developed a TW3-based deep learning system using private data set of 3,344 images. They reported TW3-RUS BA MAE of 0.46 years but did not report maturity level MAE. The Performance for each RUS bone is shown in Table 2. Clear improvements are achieved for all bones except middle phalanx III.

4.4 Ablation Studies

Ablation studies are performed to demonstrate contributions from label correction (VGLC) module, the RCG module and the Score Loss. Without VGLC,

Table 1. Results on RSNA test set

Method	TW3-RUS method	
	Maturity level MAE	Bone age MAE (year)
DLDL+MVL	0.178	0.271
OURS	0.144	0.216
son [11]	N.A	0.46

Table 2. TW3-RUS maturity level rating errors

Method	Ulna	Radius	Metacarpal			Proximal			Middle		Distal		
			I	III	V	I	III	V	III	V	I	III	V
DLDL+MVL	0.218	0.267	0.176	0.231	0.248	0.131	0.152	0.135	0.118	0.146	0.188	0.156	0.152
OURS	0.152	0.221	0.128	0.199	0.208	0.103	0.129	0.119	0.125	0.124	0.131	0.112	0.126

maturity level ratings MAE increases by 20.1% and TW3-RUS BA MAE increases by 18.4%. With VGLC module, the RCG module can further improve both metrics by 7.6% and 7.4%. Score loss reduces MAE of BAA by 8.8% with almost no increase of maturity level MAE (Table 3).

Table 3. Ablation studies

Method	TW3-RUS method	
	Maturity level MAE	BA MAE (year)
Ours (VGLC+RCG+ScoreLoss)	0.144	0.216
No VGLC	0.173 (+20.1%)	0.255 (+18.4%)
VGLC: no KL loss	0.168	0.267
VGLC: no mean and variance loss	0.148	0.245
No RCG	0.155 (+7.6%)	0.232 (+7.4%)
No Score Loss	0.144 (+0.0%)	0.235 (+8.8%)

Ablation studies for adjacency matrix design are conducted in Table 4. The fully connected graph achieves better performance. Our second design, which is based on prior knowledge and data analysis, connects bones with shared maturity level descriptions or have large correlation coefficients of normalized maturity levels.

Table 4. Adjacency matrix comparison

Method	TW3-RUS method	
	Maturity level MAE	Bone age MAE
Full graph	0.144	0.216
Prior+Corr	0.144	0.229

5   Conclusions and Future Work

Robust bone age assessment is important for the diagnosis and treatment of growth-related diseases of children. In this paper, a Validation Guided Label Correction scheme and a Residual Context Graph module are presented to address label noise and ambiguity. Experiments on RSNA test set show the proposed framework outperforms state-of-the-art TW3 BAA systems by large margins.

Future work will include (1) Unifying TW3 method with GP method in a single model; (2) Improving model generalization abilities on different data sets.

Acknowledgments. We thank the anonymous reviewers for their constructive comments. This work was supported by Capital's Funds for Health Improvement and Research (2020-2-2104), MOST-2018AAA0102004 and NSFC-61625201.

References

1. Arpit, D., et al.: A closer look at memorization in deep networks. In: Proceedings of the 34th International Conference on Machine Learning, vol. 70, pp. 233–242. JMLR.org (2017)
2. Chen, K., et al.: MMDetection: Open MMLab detection toolbox and benchmark. arXiv preprint arXiv:1906.07155 (2019)
3. Cicero, M., Bilbily, A.: Machine learning and the future of radiology: how we won the 2017 RSNA ML challenge (2017). https://www.16bit.ai/blog/ml-and-future-of-radiology
4. Escobar, M., González, C., Torres, F., Daza, L., Triana, G., Arbeláez, P.: Hand pose estimation for pediatric bone age assessment. In: Shen, D., et al. (eds.) MICCAI 2019. LNCS, vol. 11769, pp. 531–539. Springer, Cham (2019). https://doi.org/10.1007/978-3-030-32226-7_59
5. Gao, B.B., Xing, C., Xie, C.W., Wu, J., Geng, X.: Deep label distribution learning with label ambiguity. IEEE Trans. Image Process. **26**(6), 2825–2838 (2017)
6. Greulich, W.W., Pyle, S.I.: Radiographic Atlas of Skeletal Development of the Hand and Wrist, 2nd edn. Stanford University Press, Standord (1959)
7. He, K., Zhang, X., Ren, S., Sun, J.: Deep residual learning for image recognition. In: Proceedings of the IEEE Conference on Computer Vision and Pattern Recognition, pp. 770–778 (2016)
8. Ji, Y., Chen, H., Lin, D., Wu, X., Lin, D.: PRSNet: part relation and selection network for bone age assessment. In: Shen, D., et al. (eds.) MICCAI 2019. LNCS, vol. 11769, pp. 413–421. Springer, Cham (2019). https://doi.org/10.1007/978-3-030-32226-7_46

 9. Kipf, T.N., Welling, M.: Semi-supervised classification with graph convolutional networks. arXiv preprint arXiv:1609.02907 (2016)
10. Pan, H., Han, H., Shan, S., Chen, X.: Mean-variance loss for deep age estimation from a face. In: Proceedings of the IEEE Conference on Computer Vision and Pattern Recognition, pp. 5285–5294 (2018)
11. Son, S., et al.: TW3-based fully automated bone age assessment system using deep neural networks. IEEE Access **7**, 33346–33358 (2019)
12. Sun, K., Xiao, B., Liu, D., Wang, J.: Deep high-resolution representation learning for human pose estimation. In: Proceedings of the IEEE Conference on Computer Vision and Pattern Recognition, pp. 5693–5703 (2019)
13. Tanaka, D., Ikami, D., Yamasaki, T., Aizawa, K.: Joint optimization framework for learning with noisy labels. In: Proceedings of the IEEE Conference on Computer Vision and Pattern Recognition, pp. 5552–5560 (2018)
14. Tanner, J.M., Healy, M.J.R., Goldstein, H., Cameron, N.: Assessment of Skeletal Maturity and Prediction of Adult Height (TW3 Method). Saunders, London (2001)
15. Tanner, J.M., Whitehouse, R., Cameron, N., Marshall, W., Healy, M.J.R., Goldstein, H.: Assessment of Skeletal Maturity and Prediction of Adult Height (TW2 Method). Academic press, London (1976)
16. Thodberg, H.H., Kreiborg, S., Juul, A., Pedersen, K.D.: The BoneXpert method for automated determination of skeletal maturity. IEEE Trans. Med. Imaging **28**(1), 52–66 (2009)

Improve Bone Age Assessment by Learning from Anatomical Local Regions

Dong Wang[1]([✉]), Kexin Zhang[2], Jia Ding[3], and Liwei Wang[1,2]

[1] Center for Data Science, Peking University, Beijing, China
{wangdongcis,wanglw}@pku.edu.cn
[2] Key Laboratory of Machine Perception, MOE, School of EECS, Peking University, Beijing, China
zhangkexin@pku.edu.cn
[3] Yizhun Medical AI Co., Ltd., Beijing, China
jia.ding@yizhun-ai.com

Abstract. Skeletal bone age assessment (BAA), as an essential imaging examination, aims at evaluating the biological and structural maturation of human bones. In the clinical practice, Tanner and Whitehouse (TW2) method is a widely-used method for radiologists to perform BAA. The TW2 method splits the hands into Region Of Interests (ROI) and analyzes each of the anatomical ROI separately to estimate the bone age. Because of considering the analysis of local information, the TW2 method shows accurate results in practice. Following the spirit of TW2, we propose a novel model called Anatomical Local-Aware Network (ALA-Net) for automatic bone age assessment. In ALA-Net, anatomical local extraction module is introduced to learn the hand structure and extract local information. Moreover, we design an anatomical patch training strategy to provide extra regularization during the training process. Our model can detect the anatomical ROIs and estimate bone age jointly in an end-to-end manner. The experimental results show that our ALA-Net achieves a new state-of-the-art single model performance of 3.91 mean absolute error (MAE) on the public available RSNA dataset. Since the design of our model is well consistent with the well recognized TW2 method, it is interpretable and reliable for clinical usage.

Keywords: Bone age assessment · Medical imaging · Anatomical information

1 Introduction

Skeletal bone age assessment (BAA) aims at evaluating the biological and structural maturation of human bones. As an essential imaging examination of clinical diagnosis, bone age assessment is acting in plenty of scenarios, such as identifying the growth disorder of the body or investigating endocrinology problems. In clinical practice, radiologists perform bone age assessment by analyzing the

© Springer Nature Switzerland AG 2020
A. L. Martel et al. (Eds.): MICCAI 2020, LNCS 12266, pp. 631–640, 2020.
https://doi.org/10.1007/978-3-030-59725-2_61

ossification patterns visually of the non-dominant hands in X-Ray images. The two most popular methods adopted by radiologists are Greulich and Pyle (G&P) [4] and Tanner and Whitehouse (TW2) [18]. The G&P method assesses the entire hands as a whole, while the TW2 splits the hands into 20 Region Of Interests (ROI) and analyzes each of them separately to estimate the bone age. Since the TW2 method considers the analysis of local information, it shows more accurate results than the G&P method [9]. Nevertheless, these methods heavily rely on the expertise, and may be affected by subjective factors during observations and are considerably time-consuming, which are limitations for clinical applications.

Recently, accompanied by the development of deep learning and computer vision, the automatic BAA methods have shown great potential. Thanks to the public BAA dataset released by the Radiological Society of North America (RSNA) [6], a series of deep learning methods are proposed and achieve impressive performance based on the dataset. Since the RSNA dataset only provides image-level annotations, most of these works adopt end-to-end training approaches and are not built to explicitly exploit the information of hand structure and local information. To make better use of anatomical local information, BoNet [2] proposes to perform BAA by using hand keypoints heatmap as the input of network along with the hand image, to leverage the information of anatomical ROIs. The BoNet raises a useful viewpoint for using local information, while there are further limitations. The BoNet requires hand keypoints as input of the network during not only training but also inference, which should be provided by another network or human annotations in the paper.

In this paper, to make better use of anatomical information and facilitate clinical usage, we propose a novel Anatomical Local-Aware Network (ALA-Net) for bone age assessment. We list the advantages of our method below:

1. We propose an anatomical local extraction module to learn the hand structure and extract local information for BAA. Unlike BoNet, our ALA-Net can extract the anatomical ROIs (Fig. 1) and evaluate the bone age jointly in an end-to-end manner. Benefiting from multi-task learning, the performances of anatomical region detection and BAA are significantly improved.
2. Inspired by the TW2 method, we further propose an anatomical patch training strategy, which can be regarded as an effective regularization technique during the training process.
3. Our approach is well consistent with the well recognized clinical assessment method (i.e., TW2) in BAA, which means the model structure is reliable and interpretable for clinical usage.

Experiments show that ALA-Net outperforms other state-of-the-art single-model methods, and achieves 3.91 Mean Absolute Error (MAE) on the RSNA dataset. Meanwhile, the effectiveness of each component of the network design is validated through the ablation study.

2   Related Work

Earlier deep learning based methods for bone age assessment adopt the end-to-end deep neural models, which take the whole hand image as input and make

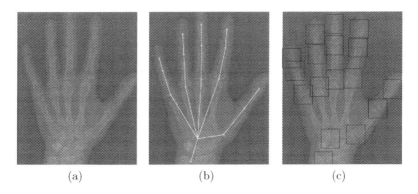

(a)   (b)   (c)

Fig. 1. (a) Original image. (b) Hand keypoint annotation used in BoNet [2]. (c) We convert the keypoints to bounding boxes which are used in our method to learn hand structure and extract local information.

prediction for bone age. Larson *et al.* [10] use a deep residual network with 50 layers to predict probability scores for each month. Spampinato *et al.* [15] validate the effectiveness of deep CNNs pre-trained on general imagery in the bone age regression model. Torres *et al.* [19] introduce a carefully tuned architecture called GPNet for BAA. Even though achieving promising results, these models do not consider the local information of different bones. Accordingly, the models are lack of interpretability and have limited performance.

More recently, with the rapid progress of applying deep learning in medical image [1,3,5,20], researchers have paid more attention to excavate the anatomical information for BAA or tried to incorporate human prior knowledge in the task. PRSNet [8] uses a part selection module to select the most helpful hand parts for BAA and uses part relation module to model the multi-scale context information. Liu *et al.* [11] propose to use the attention agent to discover the discriminative bone parts and extract features from these parts. Although the importance of bone part information is emphasized in their methods, there are also certain limitations. Due to the lack of prior knowledge of hand structure, the selected parts are more likely to the central regions of the hand, which can not capture the anatomical information of hand. To solve this problem, Escobar *et al.* [2] proposes BoNet, which uses hand pose estimation as the new task to extract local information. Specifically, BoNet uses the heatmap generated from hand keypoints as the input of the network along with the image. The manually annotated keypoints or pre-computed keypoints are required in both the training and inference stage, which leads to limited usability. Unlike BoNet, our ALA-Net learns the hand structure and anatomical ROIs itself, which boosts the performance through multi-task learning. Meanwhile, our end-to-end model is more convenient for practical usage since we do not need keypoints input during inference. Besides, there are also some works trying to estimate ages from MR images by deep learning methods [16,17].

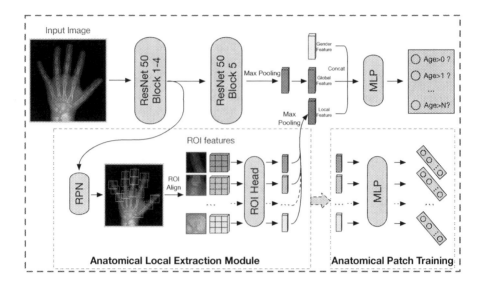

Fig. 2. Overview of our proposed anatomical local-aware network.

3   Methodology

Overall Framework: The overview of our method is illustrated in Fig. 2. Taking the hand image as input, the network combines global feature, local feature, and gender feature to conduct bone age assessment. The global feature is generated by the ResNet-50, while the local feature is the output of our proposed local extraction module. We use the same gender feature as in [8]. We formulate the bone age assessment problem as an ordinal regression problem, and the final prediction is made by the multi-layer perceptron. Meanwhile, we use the anatomical patch training strategy to regularize the network during training. In the following sections, we describe the design of each part in detail.

Anatomical Local Extraction Module: The anatomical structure provides rich information for the assessment of bone age. It has been proven that the local analysis of hand is essential in BAA in the TW2 method [9]. Furthermore, from the perspective of multi-task training, the understanding of hand structure can be regarded as an additional task to promote the training of the network. Thus, we propose an anatomical local extraction module to perform local analysis and learn the hand structure at the same time.

As shown in Fig. 2, the anatomical local extraction module takes the output feature of the 4th block of ResNet as input, and a region proposal network (RPN) is applied to generate ROI proposals. We only leave the top 17 ROIs according to the definition of hand keypoints. Then, the ROI align module extracts local features for the ROIs. We use a shared ROI head with four convolution layers and one fully-connection layer to transform the ROI features. Finally, we apply max pooling across the ROI features to aggregate the transformed ROI features and output the local feature for further analysis of BAA.

The original annotations of hand pose are a series of points, and we need to use boxes for local extraction. Hence, during the training time, we replace the point annotations by boxes centered at the points (Fig. 1). We use the same loss function as in [13] to train RPN, and we denote the loss function of RPN as L_{RPN}.

Ordinal Regression: The commonly used L_2 loss varies widely in scale during the training process, which causes difficulty for convergence in multi-task training. To tackle this problem, we use ordinal regression [12] to predict bone age. In the ordinal regression method, the regression problem of K ranks is transformed into a $K - 1$ binary classification problem. For each rank $r \in \{0, 1, ..., K - 2\}$ and an example with rank y, a binary classifier is adopted to classify whether the y is larger than r. We implement the prediction by using $K - 1$ dimension logits followed by the sigmoid activation function, and it is intuitive to employ the binary cross-entropy loss as loss function,

$$L_{ord}(o, y) = \frac{1}{K - 1} \sum_{k=0}^{K-2} BCE(o_k, 1\{y > k\}), \tag{1}$$

where o is the $K - 2$ dimension output probabilities, y is the target rank, and BCE denotes binary cross entropy loss. During the inference time, we could simply add the probabilities of all ranks to output the predicted bone age.

Anatomical Patch Training: Crop augmentation is a frequently-used technique for training neural networks. The augmentation method can enrich the dataset and make the network more robust to the small variations of the input images. For general images, it is a common practice to crop the images in a random region. While in our task, to integrate anatomical knowledge into our network, we propose to train BAA on the detected anatomical ROIs as a data augmentation method. As shown in the bottom right part in Fig. 2, during the training stage, we use a multi-layer perceptron to predict bone ages on the features of each ROI patch and train the model by ordinal regression losses described above. The 17 ROIs detected by the local extraction module are used in this step.

The anatomical patch training method is inspired by the clinical practice of the TW2 method. The TW2 method performs intensive local analysis on different bone parts and combine the results to make a final decision. Following this spirit, our anatomical patch training strategy aims at teaching the network to evaluate the growth of each bone, which promotes the performance of BAA.

In consequence, the anatomical patch training method is well consistent with the motivation of the TW2 method.

Loss Function: Our network can be trained end-to-end using a compound loss function:

$$L_{total} = L_{ord} + L_{ord}^{patch} + L_{RPN}, \tag{2}$$

where L_{ord} and L_{RPN} are introduced in previous subsections, and L_{ord}^{patch} is the ordinal regression loss for anatomical patch training.

4   Experiments

4.1   Experimental Setup

Dataset and Metric: We conduct all experiments on the RSNA Pediatric Bone Age Challenge dataset [6]. The RSNA dataset consists of 12611 images for training, 1425 images for validation, and 200 images for testing. We use the hand keypoints annotations provided by [2]. We report all the results on the test set. The Mean Absolute Error (MAE) between ground-truth age and predicted age is reported to evaluate the model performance, and the unit of measurement is month.

Network: We implement the network using PyTorch. We resize the long side of the input images to 512 pixels and keep the original aspect ratios. The box size of anatomical ROI is set to be 64 × 64. During the training stage, we use Adam optimizer to optimize the network. The learning rate is set to be 0.001 initially and reduced by a factor of 10 at 30k and 40k iteration, respectively. We train the network for 50k iterations and use a batch size of 32. Apart from the anatomical patch training method, we use horizontal flipping and random scaling as data augmentation. The total training process costs about 5 hours with 2 TITAN RTX GPU cards, and the inference time is 0.036 s per image.

Table 1. Ablation study of bone age assessment.

ResNet-50	Local extraction	Patch training	MAE
✓			4.92
✓	✓		4.53
✓		✓	4.49
✓	✓	✓	**3.91**

4.2   Ablation Study

Bone Age Assessment: We first study the effect of each component in the ALA-Net, i.e., anatomical local extraction module and anatomical patch training. The experimental results are shown in Table 1. Without both of these components, the ALA-Net degrades to the ResNet-50, which produces the MAE score of 4.92. When the anatomical local extraction module is applied to this model, the MAE score becomes 4.53, which is an improvement of 0.39 in MAE. The improvement confirms the necessity of learning anatomical structures and extracting local features. We also test the performance of ResNet-50 with the anatomical patch training strategy alone. In this circumstance, the network can not generate ROI proposals itself, hence we use the annotated ROIs during the training stage to replace the proposals predicted by the region proposal network. The anatomical patch training strategy performs a strong regularization for the network, and improves the MAE score to 4.49. Combining these two components, our ALA-Net achieves 3.91 MAE.

Table 2. Ablation study of ROI detection.

Method	AP (%)	AP50 (%)	AP75 (%)
RPN	86.2	97.9	96.2
ALA-Net (our)	89.2	98.0	98.0

Anatomical ROI Detection: Our model detects anatomical ROIs and performs bone age assessment jointly. We observe that the ROI detection also benefits from the multi-task learning experimentally. We use mean Average Precision (AP) at Intersection over Union (IoU) at $[0.5 : 0.05 : 0.95]$ to measure the performance. The results are shown in Table 2. We use a vanilla RPN as the baseline, and our model outperforms the RPN significantly, which demonstrates that our model has a better understanding of the hand structure.

4.3   Comparison with State-of-the-Arts

In this subsection, we compare our network with previous state-of-the-art single-model methods. The results are shown in Table 3. For each method, we show whether it uses global or local information for BAA. The human performance of bone age assessment is 7.32 MAE [10]. The approaches only using global feature have a slightly better performance than humans. Our ALA-Net yields better performance since we consider both global and local features. Compared with the approaches using both global and local features, our model still has significant advantages. It is attributed to not only the learning of the anatomical

Table 3. Comparison with state-of-the-art methods on the RSNA dataset. Notations: "Global" – using global information; "Local" – using local information.

Method	Global	Local	MAE
Human [10]			7.32
Larson [10]	✓		6.24
Ren [14]	✓		5.20
Iglovikov [7]	✓	✓	4.97
PRSNet [8]	✓	✓	4.49
AR-CNN [11]	✓	✓	4.38
BoNet [2]	✓	✓	4.14
ALA-Net (Our)	✓	✓	**3.91**

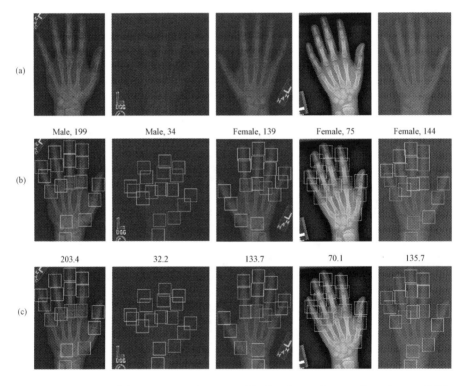

Fig. 3. Case study. (a) The original image. (b) Gender and ground truth age, ROIs. (c) Predicted age and ROIs.

structure of hand but also the effectiveness of anatomical local information. As a result, our method achieves a performance gain of 0.23 MAE over the previous SOTA [2], which is a relative improvement of 5.6%.

For a better understanding of our method, we show some randomly sampled examples in Fig. 3. We can observe that the detected anatomical ROIs are close to the annotations, which leads to accurate BAA results.

5   Conclusion

In this work, we propose a novel model called ALA-Net for bone age assessment. The anatomical local extraction model is introduced to learn the structure of hands and extract local information. The anatomical patch training strategy further provides a regularization term to enhance the performance of bone age assessment. Experimental results on the RSNA dataset demonstrates that our model achieves a new state-of-the-art result. Moreover, since the design of our model takes human priors into account, it is well interpretable and reliable for medical practice. Our work emphasizes the importance of medical prior knowledge for model design. We encourage further exploration of how to integrate medical knowledge into medical image analysis.

Acknowledgement. This work was supported by National Key R&D Program of China (2018YFB1402600), BJNSF (L172037), Key-Area Research and Development Program of Guangdong Province (No. 2019B121204008) and Beijing Academy of Artificial Intelligence.

References

1. Ding, J., Li, A., Hu, Z., Wang, L.: Accurate pulmonary nodule detection in computed tomography images using deep convolutional neural networks. In: Descoteaux, M., Maier-Hein, L., Franz, A., Jannin, P., Collins, D.L., Duchesne, S. (eds.) MICCAI 2017. LNCS, vol. 10435, pp. 559–567. Springer, Cham (2017). https://doi.org/10.1007/978-3-319-66179-7_64
2. Escobar, M., González, C., Torres, F., Daza, L., Triana, G., Arbeláez, P.: Hand pose estimation for pediatric bone age assessment. In: Shen, D., Liu, T., Peters, T.M., Staib, L.H., Essert, C., Zhou, S., Yap, P.-T., Khan, A. (eds.) MICCAI 2019. LNCS, vol. 11769, pp. 531–539. Springer, Cham (2019). https://doi.org/10.1007/978-3-030-32226-7_59
3. Esteva, A., et al.: Dermatologist-level classification of skin cancer with deep neural networks. Nature **542**(7639), 115 (2017)
4. Greulich, W.W., Pyle, S.I.: Radiographic Atlas of Skeletal Development of the Hand and Wrist. Stanford University Press, Palo Alto (1959)
5. Gulshan, V., et al.: Development and validation of a deep learning algorithm for detection of diabetic retinopathy in retinal fundus photographs. JAMA **316**(22), 2402–2410 (2016)
6. Halabi, S.S., et al.: The RSNA pediatric bone age machine learning challenge. Radiology **290**(2), 498–503 (2019)
7. Iglovikov, V.I., Rakhlin, A., Kalinin, A.A., Shvets, A.A.: Paediatric bone age assessment using deep convolutional neural networks. In: Stoyanov, D., et al. (eds.) DLMIA/ML-CDS -2018. LNCS, vol. 11045, pp. 300–308. Springer, Cham (2018). https://doi.org/10.1007/978-3-030-00889-5_34

8. Ji, Y., Chen, H., Lin, D., Wu, X., Lin, D.: PRSNet: part relation and selection network for bone age assessment. In: Shen, D., et al. (eds.) MICCAI 2019. LNCS, vol. 11769, pp. 413–421. Springer, Cham (2019). https://doi.org/10.1007/978-3-030-32226-7_46

9. King, D., et al.: Reproducibility of bone ages when performed by radiology registrars: an audit of Tanner and Whitehouse ii versus Greulich and Pyle methods. Br. J. Radiol. **67**(801), 848–851 (1994)

10. Larson, D.B., Chen, M.C., Lungren, M.P., Halabi, S., Stence, N.V., Langlotz, C.P.: Performance of a deep-learning neural network model in assessing skeletal maturity on pediatric hand radiographs. Radiology **287**(1), 313–322 (2017)

11. Liu, C., Xie, H., Liu, Y., Zha, Z., Lin, F., Zhang, Y.: Extract bone parts without human prior: end-to-end convolutional neural network for pediatric bone age assessment. In: Shen, D., et al. (eds.) MICCAI 2019. LNCS, vol. 11769, pp. 667–675. Springer, Cham (2019). https://doi.org/10.1007/978-3-030-32226-7_74

12. Niu, Z., Zhou, M., Wang, L., Gao, X., Hua, G.: Ordinal regression with multiple output CNN for age estimation. In: Proceedings of the IEEE Conference on Computer Vision and Pattern Recognition, pp. 4920–4928 (2016)

13. Ren, S., He, K., Girshick, R., Sun, J.: Faster R-CNN: towards real-time object detection with region proposal networks. In: Advances in Neural Information Processing Systems, pp. 91–99 (2015)

14. Ren, X., et al.: Regression convolutional neural network for automated pediatric bone age assessment from hand radiograph. IEEE J. Biomed. Health Inform. **23**(5), 2030–2038 (2019)

15. Spampinato, C., Palazzo, S., Giordano, D., Aldinucci, M., Leonardi, R.: Deep learning for automated skeletal bone age assessment in x-ray images. Med. Image Anal. **36**, 41–51 (2017). https://doi.org/10.1016/j.media.2016.10.010

16. Štern, D., Payer, C., Lepetit, V., Urschler, M.: Automated age estimation from hand MRI volumes using deep learning. In: Ourselin, S., Joskowicz, L., Sabuncu, M.R., Unal, G., Wells, W. (eds.) MICCAI 2016. LNCS, vol. 9901, pp. 194–202. Springer, Cham (2016). https://doi.org/10.1007/978-3-319-46723-8_23

17. Štern, D., Payer, C., Urschler, M.: Automated age estimation from MRI volumes of the hand. Med. Image Anal. **58**, 101538 (2019)

18. Tanner, J., Whitehouse, R., Marshall, W., Carter, B.: Prediction of adult height from height, bone age, and occurrence of menarche, at ages 4 to 16 with allowance for midparent height. Arch. Dis. Child. **50**(1), 14–26 (1975)

19. Torres, F., González, C., Escobar, M.C., Daza, L., Triana, G., Arbeláez, P.: An empirical study on global bone age assessment. In: 15th International Symposium on Medical Information Processing and Analysis, vol. 11330, p. 113300E. International Society for Optics and Photonics (2020)

20. Wang, D., Zhang, Y., Zhang, K., Wang, L.: FocalMIX: semi-supervised learning for 3D medical image detection. In: IEEE/CVF Conference on Computer Vision and Pattern Recognition (CVPR), June 2020

An Analysis by Synthesis Method that Allows Accurate Spatial Modeling of Thickness of Cortical Bone from Clinical QCT

Stefan Reinhold[1]([✉])[iD], Timo Damm[3][iD], Sebastian Büsse[2][iD],
Stanislav Gorb[2][iD], Claus-C. Glüer[3][iD], and Reinhard Koch[1][iD]

[1] Department of Computer Science, Kiel University, Kiel, Germany
{sre,rk}@informatik.uni-kiel.de
[2] Functional Morphology and Biomechanics, Institute of Zoology, Kiel University,
Kiel, Germany
{sbuesse,sgorb}@zoologie.uni-kiel.de
[3] Section Biomedical Imaging, Molecular Imaging North Competence Center
(MOIN CC), Department of Radiology and Neuroradiology,
University Medical Center Schleswig-Holstein (UKSH), Kiel University,
Kiel, Germany
{timo.damm,glueer}@rad.uni-kiel.de

Abstract. Osteoporosis is a skeletal disorder that leads to increased fracture risk due to decreased strength of cortical and trabecular bone. Even with state-of-the-art non-invasive assessment methods there is still a high underdiagnosis rate. Quantitative computed tomography (QCT) permits the selective analysis of cortical bone, however the low spatial resolution of clinical QCT leads to an overestimation of the thickness of cortical bone (Ct.Th) and bone strength.

We propose a novel, model based, fully automatic image analysis method that allows accurate spatial modeling of the thickness distribution of cortical bone from clinical QCT. In an analysis-by-synthesis (AbS) fashion a stochastic scan is synthesized from a probabilistic bone model, the optimal model parameters are estimated using a maximum a-posteriori approach. By exploiting the different characteristics of in-plane and out-of-plane point spread functions of CT scanners the proposed method is able assess the spatial distribution of cortical thickness.

The method was evaluated on eleven cadaveric human vertebrae, scanned by clinical QCT and analyzed using standard methods and AbS, both compared to high resolution peripheral QCT (HR-pQCT) as gold standard. While standard QCT based measurements overestimated Ct.Th. by 560% and did not show significant correlation with the gold standard ($r^2 = 0.20$, $p = 0.169$) the proposed method eliminated the overestimation and showed a significant tight correlation with the gold standard ($r^2 = 0.98$, $p < 0.0001$) a root mean square error below 10%.

Electronic supplementary material The online version of this chapter (https://doi.org/10.1007/978-3-030-59725-2_62) contains supplementary material, which is available to authorized users.

A. L. Martel et al. (Eds.): MICCAI 2020, LNCS 12266, pp. 641–651, 2020.
https://doi.org/10.1007/978-3-030-59725-2_62

Keywords: Quantitative computed tomography · Cortical thickness · Analysis by synthesis

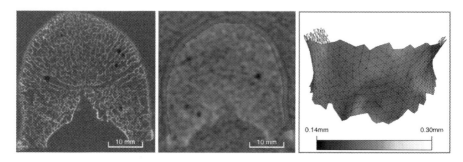

Fig. 1. Comparison of HR-pQCT (left) and standard QCT (center): axial slice (due to different orientation of the specimen the two slices appear flipped) of a vertebral body. The apparent cortical bone is highlighted in red. The sponge like structure of trabecular bone is clearly visible in the left image. Right: vertical cortex mesh with color coded thickness estimated by our AbS algorithm. (Color figure online)

1   Introduction

The world health organization (WHO) estimates the life time risk of a osteo-porotic fracture at 30–40% [15]. Even though non-invasive assessment methods are widely available, there is still a very high underdiagnosis rate [21]. Current osteoporosis diagnosis is based on the assessment of trabecular bone mineral density (BMD). However, recent studies [5, 25] show that the thin cortical shell contributes at least 50% to overall bone strength and that there is a strong cor-relation between the thickness of vertebral cortical bone (Ct.Th) and failure load estimated by finite element analysis (FEA). While quantitative computed tomog-raphy (QCT) permits the selective analysis of cortical bone, the limited spatial resolution (pixel size 300–500 μm, slice width 1–3 mm) of clinical QCT prevents accurate assessment. The thickness of the thin cortical shell (150–400 μm) is highly overestimated by current methods. Figure 1 shows a comparison of a scan taken with high resolution peripheral QCT (HR-pQCT) and clinical QCT pro-tocols. The amount of overestimation in the clinical QCT scan is clearly visible.

Related Work. BMD is defined as the volumetric density of calcium hydroxya-patite (CaHA). However, typical BMD measurements also include surrounding soft tissue and marrow. *Tissue mineral density* (TMD) is the density of the calcified bone, excluding soft tissue and lies around $1200\,\mathrm{mg\,cm^{-3}}$ for fully min-eralized bone [7, 8, 10, 11]. The standard method to assess Ct.Th is to apply a maximum sphere approach to a segmented scan [6]. Since it solely depends on the cortex as apparent in the reconstructed volume, the thickness measurements from this standard method is denoted as *apparent Ct.Th* (aCt.Th) throughout this article. As aCt.Th is highly affected by the spatial resolution of the reconstructed volume Ct.Th is tremendously overestimated in clinical QCT while cortical BMD (Ct.BMD) underestimates cortical TMD. One way to account for the overestimation is to combine thickness with density measurements into *density weighted Ct.Th* (wCt.Th) by multiplying aCt.Th with Ct.BMD normalized by a TMD of $1200 \mathrm{mg\,cm^{-3}}$, however this method is still limited by the spatial resolution and the quality of the segmentation.

Prevrhal et al. [16,17] introduced a method for improved assessment of Ct.Th by using analytical models to describe the blurring of the imaging system. However the improvement vanishes for thin cortices. Hangartner et al. proposed an iterative mathematical model to correct aCt.Th and Ct.BMD measurements based on per scanner density-versus-width curves for peripheral CT, yielding low error rates in phantom experiments for cortices thicker than 0.5 mm. Other improvements based on star-line tracing [14] or fuzzy distance transform [13] exist for HR-pQCT of tibia. However, those method are not applicable to vertebrae where HR-pQCT is not an option in a clinical setting. Treece et al. [22–24] describe a method to correct thickness measurement from clinical QCT of the femoral cortex. Damm et al. [3] propose an *Iterative Convolution OptimizatioN* (ICON) method. They showed that *deconvolved Ct.Th* (dcCt.Th) can reduce the overestimation of Ct.Th to 20% in high resolution QCT (HR-QCT) and increase the correlation between clinical QCT and HR-pQCT thickness measurements.

Recently, Reinhold et al. [19] proposed a method to identify the center of the vertebral cortical bone with sub-voxel accuracy. They used an analysis-by-synthesis (AbS) approach to fit a geometrical model of the cortical shell to the input scan.

Our Contribution. We propose a fully automatic AbS based algorithm that allows the accurate spatial modeling of the thickness distribution of cortical bone from clinical QCT. Starting from an approximate surface along the center of the cortical bone, the bone thicknesses and densities are modeled as latent variables of a stochastic measurement process. We exploit the different characteristics of the in-plane and out-out-plane point spread functions (PSF) of CT scanners in our optimization procedure. For that we propose a novel analytical approximation of the in-plane PSF that meets the required accuracy. To estimate the a-posteriori distributions of the latent model, a Monte Carlo expectation maximization (MCEM) scheme, tailored to our requirements is proposed. The result is an accurate spatial model of the cortical bone permitting the analysis of intra bone variations of Ct.Th which has the potential in improving future FEA and osteoporosis diagnosis in general.

In Sect. 2 an overview of the proposed algorithm is given. It is evaluated in an ex-vivo experiment, described in Sect. 3; we compare our approach with standard methods. After presenting and discussing the results in Sect. 4, we conclude this articles in Sect. 5.

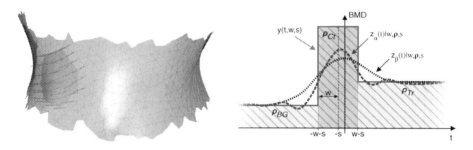

Fig. 2. Left: Cortex surface mesh with local patch (green); profiles (red lines) used to sample the input volume. Right: Schematic bone model: the cortex (green) center is slightly shifted. Two possible measurements of the same underlying signal under different angles are depicted (dashed/dotted). (Color figure online)

2   Algorithm Overview

The aim of our algorithm is to estimate the parameters of a three dimensional cortex model from a clinical QCT scan (Fig. 1 right). Because of the low spatial resolution the imaging process cannot be reversed. Instead, a synthetic version of a reconstructed volume is generated from the model; the synthesis results are compared with the input volume until the best model parameters are found. To make this AbS process feasible, simplifications and approximations must be made. First of all, it is assumed that the cortical bone is a dense plate with varying thickness and density, modeled as a triangle mesh along the cortex center with per-vertex thickness and density properties. Although the trabecular region consists of trabecles surrounded by bone marrow (Fig. 1 left), it is modeled as an area with homogeneous density. This choice is justified by the fact that the synthesis results for discretely modeled trabecles are indistinguishable from a homogeneous density distribution for low resolutions (Fig. 1 center). Since simulating a complete CT reconstruction in each synthesis step is computationally expensive, the imaging system is approximated by a blurring with an in-plane and an out-out-plane PSF. Furthermore, the complete synthesis of a volume is avoided by using sparse synthesis where only one dimensional profiles perpendicular through the cortex are synthesized.

Bone Model. The bone model is an extension of the one used in [19]. Following their notation, the density of the background (e.g. surrounding soft tissue), the cortical and the trabecular bone is modeled as a multivariate gaussian latent random vector $\boldsymbol{\rho} = (\rho_{BG}, \rho_{Ct}, \rho_{Tr})^{\mathrm{T}} \sim \mathcal{N}(\boldsymbol{\mu}_{\rho}, \Sigma_{\rho})$ and the half cortex width w

is modeled as a log normally distributed latent random variable with $\log w \sim \mathcal{N}(\mu_w, \sigma_w^2)$. Due to the shape constraints in the cortex identification process, there might be a small offset from any point on the estimated surface to the real cortex center. To account for this small but unknown offset $s \sim \mathcal{N}(\mu_s, \sigma_s^2)$, we add it as another latent variable to the profile process:

$$y(t, w, s) = \Phi(t - s, w) \cdot \boldsymbol{\rho}, \tag{1}$$

where $\Phi(t, w) = [1 - \mathrm{H}(t + w), \mathrm{H}(t + w) - \mathrm{H}(t - w), \mathrm{H}(t - w)]$, t is the position along the profile and H is the Heaviside step function (see Fig. 2 (right) for an example). To simplify the model, we assume that all latent variables $\boldsymbol{\rho}, w, s$ are independent and Σ_ρ is diagonal. From literature and μCT experiments we have a basic understanding of the distribution of those parameters. Therefore we define a joint weakly informative Normal-Inverse-χ^2 (NIχ^2) prior on the parameter vector $(\boldsymbol{\mu}_\rho^{\mathrm{T}}, \mu_w, \mu_s, \mathrm{diag}(\Sigma_\rho)^{\mathrm{T}}, \sigma_w^2, \sigma_s^2)^{\mathrm{T}}$.

Measurement Model. As in [19], the imaging process is approximated by a blurring with an in-plane and out-of-plane PSF. Both PSFs can be combined into a one dimensional, angle dependent PSF g_α along a profile. The influence of each PSF varies with the angle of the profile with the z-axis. Profiles from regions with equal latent distributions but different angles will therefore have slightly different synthesis results (Fig. 2 right). This is the key observation that leads to our algorithm. However, in contrast to previous works [3, 19, 22–24], a convenient gaussian approximation for the in-plane PSF is not sufficient anymore. Using a normalized symmetric sum of gaussian in the Fourier domain, the PSF can be approximated with arbitrary precision. After inverse Fourier transform the in-plane PSF states:

$$g_{\mathrm{ip}}(t) = \left(2 \sum_{k=1}^{N_{\mathrm{ip}}} a_k \exp\left(-\frac{b_k^2}{2c_k^2}\right) \right)^{-1} \sum_{k=1}^{N_{\mathrm{ip}}} a_k \xi_k(t), \tag{2}$$

where $\xi_k(t) = \sqrt{2\pi}c_k \left[\exp\left(-2\pi t(\pi c_k^2 t - \imath b_k)\right) + \exp\left(-2\pi t(\pi c_k^2 + \imath b_k)\right) \right]$ and $a_k, b_k, c_k \in \mathbb{R}$, $c_k > 0$ are obtained by fitting the positive part of $\mathcal{F}\{g_{\mathrm{ip}}\}$ to an empirically measured MTF. Given the combined PSF g_α the stochastic measurement process can be formulated:

$$z_\alpha(t) | \boldsymbol{\rho}, w, s = \Psi_\alpha(t, w, s) \cdot \boldsymbol{\rho} + g_\alpha(t) * \epsilon(t) + \xi(t), \tag{3}$$

where $\Psi_\alpha(t, w, s) = \Phi(t - s, w) * g_\alpha(t)$, $\epsilon(t) \sim \mathcal{N}(0, \sigma_\epsilon^2)$ is a gaussian white noise process simulating the measurement noise, $\xi(t) \sim \mathcal{N}(0, \sigma_\xi^2)$ is a gaussian noise process accounting for model errors and the notation $z_\alpha(t) | \boldsymbol{\rho}, w, s$ denotes that the stochastic process $z_\alpha(t)$ is conditioned on the random variables $\boldsymbol{\rho}, w$ and s.

Optimization. Let Z be a set of profiles sampled from the input volume, perpendicular to the cortex surface (cf. Fig. 2 left). Each profile in Z is assumed to be a realization of the stochastic process $z_\alpha(t)|\boldsymbol{x},\boldsymbol{\theta}$, where $\boldsymbol{x} = (\ln w, \boldsymbol{\rho}^{\mathrm{T}}, \boldsymbol{s}^{\mathrm{T}})^{\mathrm{T}}$ is the coalesced random vector and $\boldsymbol{\theta} = (\mu_w, \sigma_w, \boldsymbol{\mu}_\rho^{\mathrm{T}}, \boldsymbol{\sigma}_\rho^{\mathrm{T}}, \boldsymbol{\mu}_s^{\mathrm{T}}, \boldsymbol{\sigma}_s^{\mathrm{T}})^{\mathrm{T}} \sim \mathrm{NI}\chi^2(\boldsymbol{\theta}_0)$ is the vector of its distribution parameters. Our goal is to find the parameters $\boldsymbol{\theta}^\star$ that maximize the posterior density of $\boldsymbol{\theta}$ given Z. However, the posterior density contains an intractable integral over $\mathbb{R}^{N+4}$. Using an expectation maximization (EM) scheme, only the following lower bound needs to be maximized:

$$\ln p(\boldsymbol{\theta}|Z, \boldsymbol{\theta}_0) \geq \int p(\boldsymbol{x}|Z, \boldsymbol{\theta}^{[i]}) \ln\left(p(\boldsymbol{x}|\boldsymbol{\theta})p(\boldsymbol{\theta}|\boldsymbol{\theta}_0)\right) d\boldsymbol{x}. \tag{4}$$

Using Monte Carlo integration, (4) can be approximated as

$$\int p(\boldsymbol{x}|Z, \boldsymbol{\theta}^{[i]}) \ln\left(p(\boldsymbol{x}|\boldsymbol{\theta})p(\boldsymbol{\theta}|\boldsymbol{\theta}_0)\right) d\boldsymbol{x} \approx \frac{1}{K}\sum_{k=1}^{K} \gamma_i \ln p(\boldsymbol{\theta}|\boldsymbol{x}_k, \boldsymbol{\theta}_0), \tag{5}$$

with $\boldsymbol{x}_k \sim q$, $\gamma_i = p(\boldsymbol{x}_k|Z, \boldsymbol{\theta}^{[i]})(q(\boldsymbol{x}_k))^{-1}$. Since the posterior $p(\boldsymbol{\theta}|\boldsymbol{x}_k, \boldsymbol{\theta}_0)$ is $\mathrm{NI}\chi^2$, Eq. (5) can be maximized in closed form [12], yielding the estimate $\boldsymbol{\theta}^{[i+1]}$ for the next iteration. As $p(\boldsymbol{x}_k|Z, \boldsymbol{\theta}^{[i]})$ is intractable, we use an iterative adaptive multiple importance sampling scheme based on the work of El-Laham et al. [4] to approximate it. To improve the convergence properties of the algorithm, an ascend-based MCEM scheme [2] is utilized. To avoid local maxima, we start with a small sample size K, increase it dependent on the amount of improvement that was made in the last few iterations ensuring convergence. The algorithm stops when the estimated upper bound of the log likelihood improvement falls below a given threshold.

3   Materials and Methods

Eleven excised human vertebrae, obtained from the anatomical institute of our institution were examined. Ethics were approved by the responsible ethics review committee. The vertebrae were embedded into a body phantom and scanned[1] on a CT scanner with a clinical QCT (resolution $0.234 \times 0.234 \times 1\,\mathrm{mm}$) and a HR-pQCT protocol (isotropic resolution $0.082\,\mathrm{mm}$). Hounsfield Units (HU) were calibrated to mg CaHA cm^{-3} using a QRM CT calibration phantom. Several slices of cortical bone were acquired from a non-embedded excised vertebra for detailed analysis in a high-resolution μCT scanner[2] (isotropic resolution $1.73\,\mu\mathrm{m}$).

The cortices of the clinical QCT scans were identified using the method of Reinhold et al. [19]. The resulting triangle meshes were used to generate a voxel based segmentation and as starting point for the proposed AbS method. The HR-pQCT scans were segmented using the dual thresholding method of Buie et al. [1]. An established in-house software tool was used to assess standard aCt.Th and wCt.Th for the vertical cortex region of the segmented scans, QCT and HR-pQCT. The vertical cortex region of the surface mesh was divided into 48

---

[1] A complete list of all parameters can be found in the supplement.

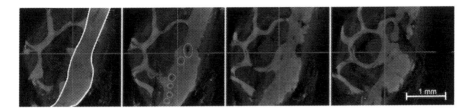

Fig. 3. High-resolution µCT scans of four different cortical bone slices. The cortex has been outlined in green on the first slice to highlight the variability of the cortical thickness. Pores have been marked in red on the second slice. (Color figure online)

Table 1. Mean deviation from and correlation with gold standard by method.

	Mean±SD [mm]	Mean±SD deviation [mm]	[%]	Correlation [r^2]
QCT				
Proposed[†]	0.29 ±0.08	-0.05 ±0.03	-15 ±8	0.98***
Standard (aCt.Th)	2.24 ±0.13	1.90 ±0.13	560 ±37	0.20[n.s.]
Density Weighted (wCt.Th)	0.62 ±0.11	0.28 ±0.05	83 ±14	0.82***
HR-pQCT				
Standard (aCt.Th)	0.69 ±0.17	0.35 ±0.06	103 ±18	0.99***
Density Weighted (wCt.Th)	0.34 ±0.11	–	–	–

[n.s.] $p \geq 0.05$; *** $p < 0.001$; [†] Evaluated on 48 patches per specimen

quasi randomly placed, partially overlapping patches[2]. For each patch between 11 and 51 profiles were sampled from the QCT scans, excluding profiles outside the target VOI. Those profiles were used as the input for the proposed method and cortical thickness distributions were estimated for each scan and patch using our AbS method (See footnote 1). For comparison with other methods a per-specimen average was computed by merging all patch distributions into a single distribution, adjusting for overlaps[3]. To gather per patch reference data, the statistics of HR-pQCT wCt.Th in cylindrical regions approximately reflecting the corresponding surface patches of the QCT meshes were computed.

All statistical calculations were performed using the R [18] programming language. Levels of probability $p < 0.05$ were considered significant.

4   Results and Discussion

The aim of our method is to spatially model the real thickness distribution of the cortical bone. But even the resolution of HR-pQCT is not sufficient to

---

[2] With 48 patches the vertical cortex was completely covered with minimal overlap.

[3] A single big patch is not feasible since the dimension of the latent space linearly increases with the patch size requiring an exponentially increasing number of samples.

assess Ct.Th directly. As can be seen in the second last row of Table 1, the average HR-pQCT based aCt.Th is 0.69 ± 0.17 mm which is clearly above the thickness range reported in literature [20]. For HR-pQCT, wCt.Th should be a good approximation of the real Ct.Th. Figure 3 shows four representative slices of cortical bones obtained by high-resolution μCT. The cortical bone is visible as a compact structure with only few small pores. Its thickness varies between locations; the span is similar to the one reported in literature [20]. We therefore choose HR-pQCT wCt.Th as the gold standard for the following evaluations.

Table 2. Intra-bone correlation between the proposed method and gold standard.

		V1	V2	V3	V4	V5	V6	V7	V8	V9	V10	V11
Correlation	$[r^2]$	0.69	0.74	0.76	0.42	0.75	0.49	0.54	0.89	0.60	0.83	0.40
RMSE	[mm]	0.02	0.03	0.07	0.06	0.06	0.04	0.07	0.02	0.02	0.02	0.05
RMSE	[%]	12	13	23	22	15	11	29	11	14	6	12

Average Cortical Thickness. Table 1 compares the mean deviation from gold standard for the proposed method, QCT based aCt.Th and wCt.Th. Our AbS method completely eliminates the overestimation introduced by all other methods and tightly correlates with the gold standard; it is able to explain 98% of the variance. A Bland-Altman analysis [9], showed that a slight proportional error might be introduced but there is still a good agreement with the gold standard. We would like to remark here that the proposed method was only evaluated at 48 patches not on the complete vertical cortex and the error might by induced by the different samples[4]. Nevertheless, our method reduces the mean deviation to less than 25% of the in-plane pixel size and is able to tightly correlate Ct.Th estimated from clinical QCT scans to wCt.Th measurements from HR-pQCT.

Intra-bone Correlation Analysis. Unlike other methods, our AbS method is able to model the spatial thickness distribution. Figure 1 (right) shows the resulting distribution for a representative vertebra. Visualizations of the spatial distribution of the differences between our method and the gold standard can be found in Fig. 4 in the supplement. Table 2 depicts the correlation of the proposed method with the gold standard per specimen. For all specimens there is a significant ($p < 0.001$) correlation able to explain 40%–89% of the variance; the RMSE ranges from 6% to 29% but does not exceed 70 μm which is below 30% of the in-plane and below 7% of the out-of-plane resolution of the input scan. Our method permits to estimate a complete three-dimensional model of the cortical bone: the initial surface mesh is augmented by local per vertex thickness estimates and can be transferred into a volumetric tetrahedral mesh. Starting from

---

[4] Additional supporting tables and figures can be found in the supplement.

there, a ready to use finite element model could be generated that accurately models spatial thickness variations. Such a model should permit more accurate finite element analysis.

5   Conclusion

We propose a fully automatic, AbS based method that permits the accurate spatial modeling of the thickness distribution of cortical bone from clinical QCT. From a probabilistic bone model, stochastic measurement processes are synthesized. The maximum a-posteriori model parameters, including cortical thickness, are estimated in an optimized MCEM process exploiting the different characteristics of the in-plane and out-of-plane PSF of clinical CT scanners.

We showed that our method is in tight agreement with the gold standard and completely eliminates the overestimation induced by other methods. Besides from giving accurate thickness estimates per specimen, it permits the assessment of intra-bone thickness variations. Because of its high accuracy our AbS method has the potential in improving estimation of bone strength. The resulting spatial model of the cortical bone also has the potential in increasing the accuracy of FEA and osteoporosis diagnosis and monitoring in general.

There is of course a limitation: the method was only validated on a few specimens, ex-vivo. However, in in-vivo there is no gold standard available so accuracy cannot be analyzed directly. The in-vivo validation could only be performed indirectly, e.g. by checking if fracture risk prediction or the distinction of treatment effects caused by different drugs can be improved. This will be future work.

Acknowledgments. This work was supported by the German Research Foundation, DFG, No. KO2044/9-1, and the Bundesministerium für Bildung (BMBF), Förderkennzeichen 01EC1005 (Diagnostik Bilanz Study, BioAsset Project).

References

1. Buie, H.R., Campbell, G.M., Klinck, R.J., MacNeil, J.A., Boyd, S.K.: Automatic segmentation of cortical and trabecular compartments based on a dual threshold technique for in vivo micro-CT bone analysis. Bone **41**(4), 505–515 (2007). https://doi.org/10.1016/j.bone.2007.07.007
2. Caffo, B.S., Jank, W., Jones, G.L.: Ascent-based Monte Carlo expectation- maximization. J. R. Stat. Soc.: Ser. B (Stat. Methodol.) **67**(2), 235–251 (2005). https://doi.org/10.1111/j.1467-9868.2005.00499.x
3. Damm, T., Peña, J.A., Campbell, G.M., Bastgen, J., Barkmann, R., Glüer, C.C.: Improved accuracy in the assessment of vertebral cortical thickness by quantitative computed tomography using the iterative convolution optimization (icon) method. Bone **120**, 194–203 (2019). https://doi.org/10.1016/j.bone.2018.08.024
4. El-Laham, Y., Elvira, V., Bugallo, M.F.: Robust covariance adaptation in adaptive importance sampling. IEEE Signal Process. Lett. **25**(7), 1049–1053 (2018). https://doi.org/10.1109/LSP.2018.2841641

5. Eswaran, S.K., et al.: The micro-mechanics of cortical shell removal in the human vertebral body. Comput. Methods Appl. Mech. Eng. **196**(31), 3025–3032 (2007). https://doi.org/10.1016/j.cma.2006.06.017

6. Hildebrand, T., Rüegsegger, P.: A new method for the model-independent assessment of thickness in three-dimensional images. J. Microsc. **185**(1), 67–75 (1997). https://doi.org/10.1046/j.1365-2818.1997.1340694.x

7. Kazakia, G.J., Burghardt, A.J., Cheung, S., Majumdar, S.: Assessment of bone tissue mineralization by conventional x-ray microcomputed tomography: comparison with synchrotron radiation microcomputed tomography and ash measurements. Med. Phys. **35**(7), 3170–3179 (2008). https://doi.org/10.1118/1.2924210

8. Kazakia, G.J., Burghardt, A.J., Link, T.M., Majumdar, S.: Variations in morphological and biomechanical indices at the distal radius in subjects with identical BMD. J. Biomech. **44**(2), 257–266 (2011). https://doi.org/10.1016/j.jbiomech.2010.10.010

9. Krouwer, J.S.: Why Bland-Altman plots should use x, not (y+x)/2 when x is a reference method. Stat. Med. **27**(5), 778–780 (2008). https://doi.org/10.1002/sim.3086

10. Laib, A., Häuselmann, H.J., Rüegsegger, P.: In vivo high resolution 3D-QCT of the human forearm. Technol. Health Care **6**(5–6), 329–337 (1998). https://doi.org/10.3233/THC-1998-65-606

11. Laval-Jeantet, A.M., Bergot, C., Carroll, R., Garcia-Schaefer, F.: Cortical bone senescence and mineral bone density of the humerus. Calcif. Tissue Int. **35**(1), 268–272 (1983). https://doi.org/10.1007/BF02405044

12. Lee, P.M.: Bayesian Statistics. An Introduction, 4th edn. Wiley, Chichester (2012)

13. Li, C., et al.: Automated cortical bone segmentation for multirow-detector CT imaging with validation and application to human studies. Med. Phys. **42**(8), 4553–4565 (2015). https://doi.org/10.1118/1.4923753

14. Liu, Y., et al.: A robust algorithm for thickness computation at low resolution and its application to in vivo trabecular bone CT imaging. IEEE Trans. Biomed. Eng. **61**(7), 2057–2069 (2014). https://doi.org/10.1109/TBME.2014.2313564

15. Nojiri, S., Burge, R.T., Flynn, J.A., Foster, S.A., Sowa, H.: Who scientific group on the assessment of osteoporosis at primary health care level: summary meeting report. Brussels, Belgium (2004). https://www.who.int/chp/topics/Osteoporosis.pdf

16. Prevrhal, S., Engelke, K., Kalender, W.A.: Accuracy limits for the determination of cortical width and density: the influence of object size and CT imaging parameters. Phys. Med. Biol. **44**(3), 751–764 (1999). https://doi.org/10.1088/0031-9155/44/3/017

17. Prevrhal, S., Fox, J.C., Shepherd, J.A., Genant, H.K.: Accuracy of CT-based thickness measurement of thin structures: modeling of limited spatial resolution in all three dimensions. Med. Phys. **30**(1), 1–8 (2003). https://doi.org/10.1118/1.1521940

18. R Core Team: R: a language and environment for statistical computing. R Foundation for Statistical Computing, Vienna, Austria (2013). http://www.R-project.org/

19. Reinhold, S., et al.: An analysis by synthesis approach for automatic vertebral shape identification in clinical QCT. In: Brox, T., Bruhn, A., Fritz, M. (eds.) GCPR 2018. LNCS, vol. 11269, pp. 73–88. Springer, Cham (2019). https://doi.org/10.1007/978-3-030-12939-2_6

20. Ritzel, H., Amling, M., Pösl, M., Hahn, M., Delling, G.: The thickness of human vertebral cortical bone and its changes in aging and osteoporosis: a histomorphometric analysis of the complete spinal column from thirty-seven autopsy specimens. J. Bone Miner. Res. **12**(1), 89–95 (1997). https://doi.org/10.1359/jbmr.1997.12.1. 89

21. Smith, M., Dunkow, P., Lang, D.: Treatment of osteoporosis: missed opportunities in the hospital fracture clinic. Ann. R. Coll. Surg. Engl. **86**(5), 344 (2004). https://doi.org/10.2298/VSP1205420D

22. Treece, G., Gee, A.: Independent measurement of femoral cortical thickness and cortical bone density using clinical CT. Med. Image Anal. **20**(1), 249–264 (2015). https://doi.org/10.1016/j.media.2014.11.012

23. Treece, G., Gee, A., Mayhew, P., Poole, K.: High resolution cortical bone thickness measurement from clinical CT data. Med. Image Anal. **14**(3), 276–290 (2010). https://doi.org/10.1016/j.media.2010.01.003

24. Treece, G., Poole, K., Gee, A.: Imaging the femoral cortex: thickness, density and mass from clinical CT. Med. Image Anal. **16**(5), 952–965 (2012). https://doi.org/10.1016/j.media.2012.02.008

25. Yamada, S., et al.: Correlation between vertebral bone microstructure and estimated strength in elderly women: an ex-vivo HR-pQCT study of cadaveric spine. Bone **120**, 459–464 (2019). https://doi.org/10.1016/j.bone.2018.12.005

Segmentation of Paraspinal Muscles at Varied Lumbar Spinal Levels by Explicit Saliency-Aware Learning

Jiawei Huang[1], Haotian Shen[1], Bo Chen[2], Yue Wang[1(✉)], and Shuo Li[2]

[1] Spine Lab, Department of Orthopedic Surgery, The First Affiliated Hospital, Zhejiang University School of Medicine, Hangzhou, China
wangyuespine@zju.edu.cn
[2] Department of Medical Imaging and Medical Biophysics, Western University, London, ON, Canada
slishuo@gmail.com

Abstract. Automated segmentation for paraspinal muscles on axial lumbar MRIs of varied spinal levels is clinically demanded. However, it is challenging and there is no reported success due to the large inter- and intra-organ variations, unclear muscle boundaries and unpredictable muscle degeneration patterns. In this paper, we propose a novel explicit saliency-aware learning framework (BS-ESNet) for fine segmentation of multiple paraspinal muscles and other major components at varied spinal levels across the full lumbar spine. BS-ESNet is designed to first detect the location of each organ in forms of bounding box (b-box); then performs accurately segmentation which utilizes detected b-boxes to enable spatial saliency awareness. BS-ESNet creatively conducts detection upon a preliminary segmentation mask instead of input MRI, which eliminates the influence of inter-organ variations and is robust against unclear muscle boundaries. Such segment-then-detect workflow also provides a paradigm to formulate multi-organ detection in an end-to-end trainable process. Our framework also embeds an elaborate spatial attention gate which adopts detection b-boxes to obtain a saliency activation map in an explicitly supervised manner. Acquired salient attention map can automatically correct and enhance segmentation features, and further guides the adaptation of variable precise anatomical structures. The method is validated on a challenging dataset of 320 MRIs. Evaluation results demonstrate that our BS-ESNet achieves high segmentation performance with mean Dice score of 0.94 and outperforms other state-of-the-art frameworks.

1   Introduction

Paraspinal muscles segmentation on axial spine magnetic resonance images (MRIs) targets at bilateral multifidus, erector spinae, and psoas major, as well as the disc, canal, and lamina (9 organs in total) simultaneously. Accurate

© Springer Nature Switzerland AG 2020
A. L. Martel et al. (Eds.): MICCAI 2020, LNCS 12266, pp. 652–661, 2020.
https://doi.org/10.1007/978-3-030-59725-2_63

paraspinal muscles segmentation is important in spine research, and is also clinically demanded. For instance, the cross-sectional area and fat percentage (pixels exceeding a threshold which are regarded as fat tissues, relative to the total pixels within a muscle) are crucial indicators to muscle degeneration [12], which can further assist clinicians in medical and surgical treatment guidance, and pre-operative assessment of various lumbar spinal diseases [9].

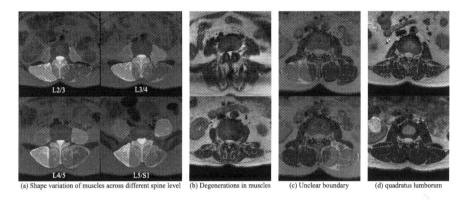

(a) Shape variation of muscles across different spine level   (b) Degenerations in muscles   (c) Unclear boundary   (d) quadratus lumborum

Fig. 1. Challenges of paraspinal muscles segmentation on axial lumbar spine MRI. (a) Demonstrates variations in shape and size among 9 targeted organs, as well as variations across different spinal level. (b) Shows muscle degeneration (atrophy and fatty infiltration) introduces remarkable variations. (c) Shows the unclear boundary between muscles. (d) Illustrates the uninterested quadratus lumborum.

However, segmentation of paraspinal muscles at varied spinal levels simultaneously is challenging because: (1) The 9 targeted organs vary remarkably with each other in the shape and size. Morphology of each paraspinal muscle also changes considerably across different spinal levels (Fig. 1a); (2) Variations in muscle degeneration may affect the morphology and signal intensity of a muscle in an unpredictable manner (Fig. 1b). (3) A clear boundary between the multifidus and erector spinae is often absent (Fig. 1c); (4) Quadratus lumborum, a small muscle immediately adjacent to erector spinae, is not a research interest but may present on some MRIs at the upper lumbar spinal levels, leading to anatomical variations (Fig. 1d).

While automatic analysis on sagittal spine MRI has achieved vertebrae and disc segmentation [5,16] and quantification [4,8], as well as spinal disease identification [3], axial spine MRI is less studied. There is no reported success to achieve paraspinal muscles segmentation at varied spine levels, although some similar methods [1,6,15] have been proposed to segment an individual muscle or at fixed spine level. While Engstrom et al. [1] used statistical shape model to segment only the quadratus lumborum on axial MRI at fixed L3/4 intervertebral spine level. Xia et al. [15] used U-Net and conditional random fields to segment paraspinal muscles at merely lower lumbar region. However, these methods cannot be directly generalized to segment multiple muscles at varied spine levels, as they pay equal attention to each pixel during the pixel-wise dense classification, resulting in a poor discriminative ability against large inter-organ variations. Roth et al. [10] proposed aggregation of two holistically-nested convolutional networks (HNNs) for pancreas localization and segmentation, with the first HNN conducts bounding box (b-box) localization and the second HNN performs segmentation inside the b-box, thus could restrict segmentation error in the b-box. However, such framework was based on segmentation for one single organ, generalization of such method to multiple targeted muscles on axial spine MRI may lead to excessive computational expenses.

Attention mechanism has great potential since it can guide the learning architecture focusing on target regions, which has been well recognized to promote generalization ability against large inter-organ variations [11,13,14]. Wang et al. [14] proposed nonlocal, a self-attention operation to capture internal dependencies of features in a convolutional neural network (CNN). Schlemper et al. [11] proposed attention gate, through which low-level localized features are filtered by semantic features to leverage salient regions. However, such self-attention methods are not explicitly trained and supervised to focus on target regions, which limits their effect in segmentation and often overconsume computational expenses.

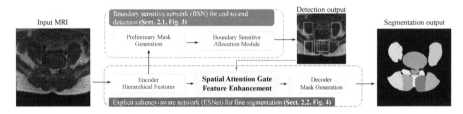

Fig. 2. The framework of BS-ESNet. It consists of a Boundary Sensitive Network (BSN) for end-to-end multi-organ detection, and an Explicit Saliency-Aware Network (ESNet) for fine segmentation, which embeds an elaborately designed spatial attention gate to enhance segmentation features by detections from BSN.

In this paper, we propose an explicit saliency-aware learning framework (BS-ESNet, Fig. 2) for segmentation of paraspinal muscles at varied spine levels. We first develop a boundary sensitive network (BSN) to detect paraspinal muscles and other spinal components. BSN creatively performs detection upon a preliminary segmentation mask instead of original input MRI, which eliminates the influence of inter- and intra-organ variations and deals with the unclear boundary and quadratus lumborum challenge. Then, explicit saliency-aware network (ESNet) is proposed to segment fine boundaries for targeted organs. ESNet embeds an elaborate spatial attention gate (SAG) to utilize BSN detected b-boxes as the attention query to obtain a spatial saliency activation map, which guides ESNet focusing on organ relevant regions and deals with the morphology and degeneration variation challenge.

Our work has the following contributions: (1) For the first time, segmentation of the multiple paraspinal muscles and other major spinal components on axial lumbar MRIs simultaneously at both upper and lower spinal levels is achieved; (2) Our boundary sensitive network provides a novel segment-then-detect work-flow, which is robust to unclear organ boundaries and can further simplify multi-organ detection as an end-to-end trainable process. (3) Our explicit saliency-aware network provides an elaborately designed architecture which can utilize detection b-boxes to automatically correct and enhance segmentation features in an explicitly supervised manner, and facilitates the adaptation of variable precise anatomical structures.

2 Methodology

BS-ESNet (Fig. 2) is composed of two parallel networks: (1) Boundary sensitive network (BSN, Sect. 2.1, Fig. 3) performs multi-organ detection upon preliminary segmentation masks, which enable robustness against unclear organ boundaries and simplifies detection task as an end-to-end trainable process. BSN separately allocates each organ's mask to a corresponding detector to promote detection accuracy and avoid the false positive and false negative in anchor-proposal based detection methods. (2) Explicit saliency-aware network (ESNet, Sect. 2.2, Fig. 4) is a segmentation network and embeds our newly proposed spatial attention gate (SAG), which adopts detected b-boxes from BSN to automatically correct and enhance segmentation features and thus, guide the adaptation of variable anatomical structures. Altogether, our proposed BS-ESNet performs end-to-end multi-organ detection and fine segmentation in a one-shot workflow.

2.1 Boundary Sensitive Network (BSN) for End-to-end Detection

BSN (Fig. 3) performs multi-organ detection in an end-to-end trainable process and handles the unclear boundaries and the presence of quadratus lumborum challenge through two carefully designed cascade modules:

The Preliminary Mask Generator. (Fig. 3a) shares encoder with ESNet (Sect. 2.2) and is trained to generate a down-sampled preliminary segmentation mask. As the spatial size of preliminary mask would determine lower bound of subsequent detection error, the generator is designed to output $\frac{1}{4}$ down-sampled mask with respect to the original size of MRI, to ensure a considerable detection accuracy and take less computational resources as well.

The Boundary Sensitive Allocation Module. (Fig. 3b) splits the prelim-inary mask to N (number of targeted organs) binary masks, each mask corre-sponds to a target organ and is then allocated to a separate boundary sensitive detector. Each detector is a lightweight CNN, with only two convolutional layers and two fully connected layers to directly regress a 4-dimension b-box. Normal-ized pixel coordinates features are also fed to each detector to enhance its spatial

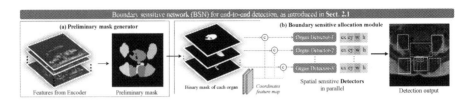

Fig. 3. Boundary sensitive network (BSN). (a) is trained to generate a preliminary segmentation mask, upon which (b) performs multi-organ detection by allocating each organ's mask to the corresponding detector, thus avoids false positive and false negative as in anchor proposal methods. Coordinates feature map is concatenated into each detector to promote spatial sensitivity. BSN provides a novel segment-then-detect workflow, which is robust against unclear organ boundaries and can simplify multi-organ detection as an end-to-end trainable process.

sensitivity, as inspired by the CoordConv operator [7]. N parallel organ detectors together achieves detection for all targeted organs.

Summarized Advantages: (1) Our BSN provides a novel segment-then-detect paradigm to formulate multi-organ detection as an end-to-end trainable process, which gets rid of the complicated anchor-proposal workflow. (2) As BSN embeds parallel detectors and each is responsible for detection of one particular organ, such delicate design avoids false positive and false negative as in anchor proposal based methods.

2.2   Explicit Saliency-Aware Network (ESNet) for Fine Segmentation

ESNet (Fig. 4) performs fine segmentation through U-Net-like encoder-decoder architecture and skip connection (Fig. 4a (1–2)), and our elaborate spatial attention gate (SAG), which utilizes detected b-boxes from BSN (Sect. 2.1) to guide the adaptation of variable precise anatomical structures. A delicate multi-task cross strategy is proposed to obtain higher detection and segmentation accuracy.

Spatial Attention Gate (SAG). SAG (Fig. 4b) propagates BSN detected b-boxes into a set of binary feature maps Q^N aligned to each organ, and takes Q^N as the explicit attention query. Meanwhile, SAG introduces an additional supervision to generate another preliminary mask $x_{h \times w}^N$ from the semantic features $x_{h \times w}^C$, and takes $x_{h \times w}^N$ as the aligned attention key. We designed a cascade of element-wise multiplication and addition to filter out wrong semantics and enhance organ relevant regions, then a 1×1 convolution layer to compress its channel dimension, and finally a Sigmoid function to obtain the salient activation map α, which guides the decoder to focus on organ relevant salient regions.

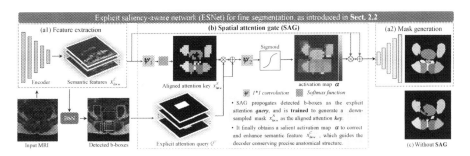

Fig. 4. The explicit saliency-aware network (ESNet). It adopts U-Net-like architecture (a1–a2, the skip connection is omitted in this figure for visual simplicity), and embeds our elaborate SAG (b). Element-wise multiply and addition between the attention query and key are calculated to filter out wrong semantics and enhance organ relevant regions. Then a Sigmoid function as activation gate is applied to obtain the spatial activation map to guide the decoder focusing on salient regions and acquiring precise segmentation boundaries (a2), as compared with (c) Which is from the same encoder-decoder without SAG module.

Multi-task Asynchronous Training Strategy is adopted to ensure SAG being robust to b-boxes with various deviations and simultaneously balance different convergence rate between detection and segmentation tasks. During the training process, BSN and ESNet are simultaneously optimized for the first 10 epochs. Then BSN is trained for 1 epoch each time after ESNet have been trained for 9 epochs. Such training cycle is repeated till a fine segmentation result is acquired. Finally, BSN is independently optimized to obtain better detection performance. We also introduce deep-supervision in the decoder to compel features to be semantically discriminative at each level.

Summarized Advantages: (1) Our elaborate designed SAG module provides a paradigm which can utilize detection b-boxes to automatically correct and enhance segmentation features in an explicitly supervised manner. (2) A delicate multi-task asynchronous training strategy is proposed to balance convergence rate between detection and segmentation task, and leads to stronger robustness and higher generalization ability.

3   Experiments and Results

Experiment Setup. Lumbar spine MRIs of 150 subjects are randomly selected from a population-based database. The dataset studied includes 320 axial lumbar MRIs at both upper and lower lumbar levels. Specifically, there are 80 T2W axial MRIs at L2/3 level, 60 L3/4 MRIs, 100 L4/5 MRIs, and 80 L5/S1 MRIs. Each MRI was manually segmented for the 9 targeted organs by an orthopedic resident and then adjusted by a senior spine surgeon. Detection labels were directly generated from each segmentation mask. BS-ESNet takes 2D single slice MRI at arbitrary spine level as input and generates corresponding detection and segmentation mask. The framework was trained on single 11 Gb GPU for 200 epochs. Pixel-wise cross-entropy loss is adopted to train ESNet and the preliminary segmentation in BSN, and the combination of smooth L1 loss and mean square error loss is used to optimize bbox regression in BSN. Standard five-fold cross-validation was performed for evaluation. Metrics of Intersection-over-union (IoU) was used to evaluate detection accuracy and dice coefficient (Dice) for the evaluation of segmentation performance.

High Performance with Respect to the Ground Truth. The last column in Fig. 5 visually presents segmentation results of each targeted muscle and spinal component from BS-ESNet. The 5th and the last row in Table 1 demonstrates that BS-ESNet achieves high segmentation and detection accuracy for each targeted muscle and spinal component, with a mean DICE of 94.4% and mean IoU of 77.8%. The detection accuracy of BSN is limited by the preliminary segmentation errors, however, such detection is effective for promoting final segmentation performance as demonstrated in the ablation experiments. In addition, our delicate design in BSN avoids false positive and false negative as in anchor proposal detection methods, and achieves 100% detection precision and recall.

Table 1. Quantitative evaluations of the BS-ESNet framework

DICE (%)	Disc	Canal	Lamina	Psoas		Miltifidus		Erector	
				Left	Right	Left	Right	Left	Right
U-Net (baseline)	96.7	91.7	89.7	93.3	93.9	93.8	94.3	93.5	93.5
	±4.8	±8.9	±5.5	±8.0	±7.6	±4.2	±3.7	±4.7	±4.5
Without SAG	97.0	91.2	89.7	93.9	93.9	94.3	94.6	93.6	93.5
	±3.5	±9.4	±5.3	±6.4	±6.9	±3.5	±3.3	±4.2	±4.5
Without deep-supervision	97.3	93.1	90.1	93.6	94.6	94.2	**95.0**	94.0	93.9
	±2.5	±5.6	±4.6	±7.3	±4.2	±3.9	**±2.6**	±3.9	±4.2
BS-ESNet (proposed)	**97.5**	**93.8**	**90.4**	**94.8**	**95.2**	94.4	94.9	**94.4**	**94.4**
	±1.8	**±3.5**	**±4.6**	**±4.0**	**±3.2**	**±3.5**	±3.0	**±3.4**	**±3.4**
Mask RCNN	94.3	79.0	80.4	89.8	88.5	90.7	91.4	91.8	91.6
	±2.2	±4.5	±4.8	±3.5	±12	±4.5	±3.0	±3.6	±3.7
Non-local U-Net	96.9	92.1	89.0	93.5	93.9	92.8	93.4	92.9	93.0
	±2.0	±6.5	±5.1	±5.2	±4.3	±4.6	±4.0	±4.2	±3.9
Attention U-Net	96.8	92.6	90.4	94.0	93.6	93.9	94.5	93.7	93.6
	±5.3	±7.1	±5.2	±6.8	±7.0	±4.7	±4.0	±4.4	±4.5
Detection IoU (%)	82.5	73.1	81.7	74.8	76.7	75.3	78.2	77.5	80.0

Ablation Experiments was conducted to validate the effectiveness of SAG module by comparing BS-ESNet with (1) baseline U-Net, (2) BS-U-Net (BS-ESNet without SAG module), and (3) BS-ESNet without deep-supervision. Segmentation improvement of BS-ESNet comes from the elaborate SAG module, as BS-ESNet without SAG merely obtains a similar result with baseline U-Net, which is inferior to BS-ESNet without deep-supervision (row 2–5, Table 1). Deep-supervision further promotes segmentation results(row 4–5, Table 1).

Comparison with the State-of-the-Art. Our approach is compared with two self-attention based methods in [11,14], and an instance segmentation methods, Mask-RCNN [2]. Our BS-ESNet achieves the best segmentation performance, as compared with other methods (Table 1).

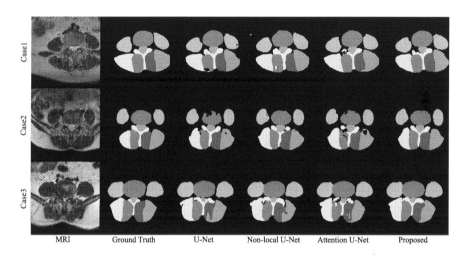

Fig. 5. Visual superiority of the proposed BS-ESNet in segmentation of the paraspinal muscles on T2W axial lumbar spine MRIs.

Visual Superiority. Figure 5 demonstrates the visual advantages of segmentation for the BS-ESNet. Case 1 is an axial L3/4 MRI on which the boundaries between bilateral multifidus and erector spinae are difficult to identify. With the help of BSN and SAG module, our method can better depict the appearance of each muscle. Case 2 is the L5/S1 MRI of a 70-years-old woman and case 3 is the L4/L5 MRI of a 72-years-old man. Severe muscle atrophy and fatty infiltration present on both images which introduce remarkable variations in the shape and signal of paraspinal muscles, and puzzled other methods. As our SAG module could guide the mask decoder focusing on organ relevant salient regions, the achieved performance is as good as that of manual segmentation.

4 Conclusion

For the first time, this paper proposed an explicit saliency-aware learning framework for segmentation of the paraspinal muscles on axial lumbar MRIs at multiple lumbar spinal levels. The proposed BS-ESNet consists of two novel modules: a lightweight subnetwork (BSN) for multi-organ detection which is end-to-end trainable, and a segmentation network (ESNet) which embeds detections from BSN to correct and enhance segmentation features and further guide the adaptation of precise anatomical structures. The BSN is highly robust to unclear muscle boundaries, and acquired detections can further be utilized to promote segmentation results. The performance and effectiveness are validated in extensive experiments. The results revealed that our proposed BS-ESNet is reliable and accurate, and outperforms other state-of-the-art methods, suggesting it can be generalized in epidemiological studies of the lumbar spine.

Acknowledgements. This study was supported by National Key R&D Program of China (No. 2018YFB1105600), National Natural Science Foundation of China (NSFC, No. 81772382), and Innovation research grant program for 8-year-system medical students at Zhejiang University (No. 119000-5405A1).

References

1. Engstrom, C.M., Fripp, J., Jurcak, V., Walker, D.G., Salvado, O., Crozier, S.: Segmentation of the quadratus lumborum muscle using statistical shape modeling. J. Magn. Reson. Imaging **33**(6), 1422–1429 (2011). https://doi.org/10.1002/jmri. 22188
2. He, K., Gkioxari, G., Dollár, P., Girshick, R.: Mask r-cnn. IEEE Trans. Pattern Anal. Mach. Intell. **42**(2), 386–397 (2020). https://doi.org/10.1109/TPAMI.2018. 2844175
3. Hong, Y., Wei, B., Han, Z., Li, X., Zheng, Y., Li, S.: Mmcl-net: spinal disease diagnosis in global mode using progressive multi-task joint learning. Neurocomputing **399**, 307–316 (2020). https://doi.org/10.1016/j.neucom.2020.01.112
4. Huang, J., et al.: Spine explorer: a deep learning based fully automated program for efficient and reliable quantifications of the vertebrae and discs on sagittal lumbar spine mr images. Spine J. **20**(4), 590–599 (2020). https://doi.org/10.1016/j.spinee. 2019.11.010
5. Li, S., Zhang, R., Xiao, X., Liu, Z., Li, Y.: Multi-task relational learning network for MRI vertebral localization, identification and segmentation. IEEE J. Biomed. Health Inform. 1 (2020). https://doi.org/10.1109/JBHI.2020.2969084
6. Lin, L., et al.: Multiple axial spine indices estimation via dense enhancing network with cross-space distance-preserving regularization. IEEE J. Biomed. Health Inform. 1 (2020). https://doi.org/10.1109/JBHI.2020.2977224
7. Liu, R., et al.: An intriguing failing of convolutional neural networks and the coordconv solution (2018)
8. Pang, S., et al.: Direct automated quantitative measurement of spine by cascade amplifier regression network with manifold regularization. Med. Image Anal. **55**, 103–115 (2019). https://doi.org/10.1016/j.media.2019.04.012

9. Ranger, T.A., et al.: Are the size and composition of the paraspinal muscles associated with low back pain? A systematic review. Spine J. **17**(11), 1729–1748 (2017). https://doi.org/10.1016/j.spinee.2017.07.002

10. Roth, H.R., et al.: Spatial aggregation of holistically-nested convolutional neural networks for automated pancreas localization and segmentation. Med. Image Anal. **45**, 94–107 (2018). https://doi.org/10.1016/j.media.2018.01.006

11. Schlemper, J., et al.: Attention gated networks: learning to leverage salient regions in medical images. Med. Image Anal. **53**, 197–207 (2019). https://doi.org/10.1016/j.media.2019.01.012

12. Tosato, M., et al.: Measurement of muscle mass in sarcopenia: from imaging to biochemical markers. Aging Clin. Exp. Res. **29**(1), 19–27 (2017). https://doi.org/10.1007/s40520-016-0717-0

13. Wang, G., et al.: Automatic segmentation of vestibular schwannoma from t2-weighted MRI by deep spatial attention with hardness-weighted loss. In: Shen, D., et al. (eds.) Medical Image Computing and Computer Assisted Intervention - MICCAI 2019. pp. 264–272. Springer International Publishing, Cham (2019). https://doi.org/10.1007/978-3-030-32245-8_30

14. Wang, X., Girshick, R., Gupta, A., He, K.: Non-local neural networks. In: 2018 IEEE/CVF Conference on Computer Vision and Pattern Recognition, pp. 7794–7803, June 2018. https://doi.org/10.1109/CVPR.2018.00813

15. Xia, W., et al.: Automatic paraspinal muscle segmentation in patients with lumbar pathology using deep convolutional Neural Network. In: Shen, D., et al. (eds.) MICCAI 2019. LNCS, vol. 11765, pp. 318–325. Springer, Cham (2019). https://doi.org/10.1007/978-3-030-32245-8_36

16. Zheng, G., et al.: Evaluation and comparison of 3D intervertebral disc localization and segmentation methods for 3D T2 MR data: A grand challenge. Med. Image Anal. **35**, 327–344 (2017). https://doi.org/10.1016/j.media.2016.08.005

Manifold Ordinal-Mixup for Ordered Classes in TW3-Based Bone Age Assessment

Byeonguk Bae, Jaewon Lee, Seo Taek Kong, Jinkyeong Sung,
and Kyu-Hwan Jung[✉]

VUNO Inc., Seoul, South Korea
{byuk.bae,jaewon1209,stkong,jinkyeong.sung,khwan.jung}@vuno.co

Abstract. Bone age assessment (BAA) is vital to detecting abnormal growth in children and can be used to investigate its cause. Automating assessments could benefit radiologists by reducing reader variability and reading time. Recently, deep learning (DL) algorithms have been devised to automate BAA using hand X-ray images mostly based on GP-based methods. In contrast to GP-based methods where radiologists compare the whole hand's X-ray image with standard images in the GP-atlas, TW3 methods operate by analyzing major bones in the hand image to estimate the subject's bone age. It is thus more attractive to automate TW3 methods for their lower reader variability and higher accuracy; however, the inaccessibility of bone maturity stages inhibited wide-spread application of DL in automating TW3 systems. In this work, we propose an unprecedented DL-based TW3 system by training deep neural networks (DNNs) to extract region of interest (RoI) patches in hand images for all 13 major bones and estimate the bone's maturity stage which in turn can be used to estimate the bone age. For this purpose, we designed a novel loss function which considers ordinal relations among classes corresponding to maturity stages, and show that DNNs trained using our loss not only attains lower mean absolute error, but also learns a path-connected latent space illuminating the inherent ordinal relations among classes. Our experiments show that DNNs trained using the proposed loss outperform other DL algorithms, known to excel in other tasks, in estimating maturity stage and bone age.

Keywords: Bone age assessment · TW3 · Ordinal learning · Manifold mixup · Deep learning

1 Introduction

An adolescent's bone maturity is mostly affected by genetic, hormonal, and nutritional factors, and can significantly differ from the subject's chronological age.

Electronic supplementary material The online version of this chapter (https://doi.org/10.1007/978-3-030-59725-2_64) contains supplementary material, which is available to authorized users.

© Springer Nature Switzerland AG 2020
A. L. Martel et al. (Eds.): MICCAI 2020, LNCS 12266, pp. 662–670, 2020.
https://doi.org/10.1007/978-3-030-59725-2_64

Detecting abnormal growth early-on is thus essential for proper treatment of its causes including nutrition deficiency, obesity, or precocious puberty. Traditional approaches to bone age assessment (BAA) using X-ray images include Greulich-Pyle (GP) [5] and Tanner-Whitehouse (TW) [18] methods. When using GP-based BAA, radiologists compare the hand's image with the GP atlas to find the most similar atlas based on visual features, and can be used to quickly evaluate the bone age; however, due to its subjective nature, there is a large variation among radiologists' interpretations. TW methods assess bone age by collectively analyzing major bones in a hand X-ray, with its most recent version TW3 using scores obtained from 13 major bones shown in Fig. 1, and converting their sum to predict bone age. While the bone-specific assessments allow for more accurate BAA, the reading time associated with TW3 is considerably longer than with GP. Nonetheless, both methods require extensive reading time in assessing bone age and their accuracy highly varies with the reader's experience.

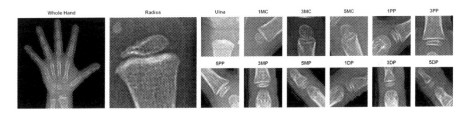

Fig. 1. Bone patches used to assess bone age using TW3. MC, PP, MP, and DP stand for metacarpal, proximal phalanx, middle phalanx, and distal phalanx.

To reduce the reading time and estimation variations among radiologists for BAA, automated systems based on computer vision and machine learning have been proposed in [19]. More recently, algorithms using convolutional neural networks (CNNs) achieved leading performances in the Radiological Society of North America (RSNA) BAA competition [3,6]. Modern deep learning (DL) algorithms applied to BAA can be primarily categorized in three directions: using a strong pre-processing algorithm to handle diverse bone shapes and sizes [3,6], modifying deep neural network (DNN) architectures, or adjusting optimization objectives for better feature representation [1,7]. However, all aforementioned DL algorithms evaluate bone age using full hand images instead of analyzing the maturity stage of each bone as there is no public dataset with annotations apt for automating TW3. Similarly, a few works [6,9] attempt at solving BAA while extracting region of interest (ROI) patches in an unsupervised manner, but the ROI patches do not focus on major bones in the hand X-ray due to its unsupervised nature. A review of existing BAA algorithms is provided in [2].

In this work, we developed a fully automated TW3 method using an in-house dataset with annotated ROI patches and bone maturity stages, unlike previous approaches to BAA using full hand X-ray images. A key-point detection network is trained to extract all 13 major bone patches used for TW3, and individual local classifiers estimate the maturity stage in each patch which can in turn be used to compute the Radius, Ulna, and short bone (RUS) scores using a standard closed-form expression. The sum of these RUS scores can then be used to calculate the TW3 bone age using a gender-dependent table. When trained on a novel loss function which considers non-integral values by imputing continuous ages as interpolations of pairs of integral values, e.g. 4.5 years of age on top of 4 or 5, our model is able to learn a smooth decision boundary and outperform all considered algorithms in automating TW3 methods.

2   Methods

Our BAA system uses a region of interest (ROI) extractor to first extract 13 major bone patches relevant to TW3 from a hand X-ray image, and uses local classifier models to predict their corresponding Radius, Ulna, and short bone (RUS) scores. As done in TW3, the sum of these RUS scores are then converted to predict the bone age. The overall framework is illustrated in Fig. 2.

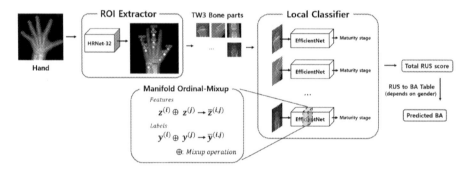

Fig. 2. Proposed automated BAA diagram. An ROI extractor retrieves major bone patches from a hand image and local classifiers estimate the maturity stage from each ROI. The sum of the 13 RUS scores converted from maturity stages are then used to predict bone age using a gender dependent bone age (BA) table.

2.1   BAA System

An ROI Extractor. $\mathcal{R} : \mathcal{X} \to \otimes_{r=1}^{R} \mathcal{X}_r$ is designed to extract $R = 13$ bone ROIs in a hand X-ray image $x \in \mathcal{X}$. The ROI extractor design is mainly motivated by a pose estimation network proposed by Escobar *et al.* [3], where they use hand detection and hand pose estimation to assess bone age using diverse hand positions. Our model is trained to localize key points indicating the centers of ROI patches from hand images using patch annotations as illustrated in Fig. 1. Each ROI output $\mathcal{R}_r : x \mapsto W_r \in \mathcal{X}_r$ is a spatial map with the model's confidence describing the likelihood $W_r(p)$ of a key point being located at each pixel p. After training, estimated key points $\hat{p}_r = \arg\max_{p \in \mathcal{X}_r} W_r(p)$ are obtained from each heatmap, and an ROI crop is obtained using a fixed window centered at $\hat{p}_r$.

Table 1. Percentage correct keypoint (PCK@0.1) [12] in extracting major bones.

Bone	R	U	1MC	3MC	5MC	1PP	3PP	5PP	3MP	5MP	1DP	3DP	5DP	Avg.
%	98.6	99.7	99.7	99.8	99.8	99.9	99.8	99.6	99.6	99.5	99.8	99.8	99.3	99.6

Local Classifiers. $f = (f_1, \ldots, f_R)$ are then trained to estimate the maturity stage which can be mapped to compute the RUS scores using a closed form expression. Each bone r is associated with an ordered set of maturity stages $\mathcal{Y}_r$, and each local classifier $f_r : \mathcal{X}_r \to \Delta(\mathcal{Y}_r)$ is a distribution over the possible maturity scores. The models are obtained by minimizing a loss function $\mathcal{L}(p_r, f_r(w))$ with target one-hot vector (or dirac-delta distribution) $p_r \in \Delta(\mathcal{Y}_r)$ indicating the maturity stage y_r of each bone r. In our experiments, we used the state-of-the art classification network EfficientNet-B4 [17] for each local classifier f_r by modifying the number of output nodes to match $|\mathcal{Y}_r|$. The estimated maturity stage is then mapped to its corresponding RUS score, and the sum of the scores is converted to the predicted bone age using a table depending on the subject's gender.

2.2 Ordinal Mixup for Ordered Classes

For this section, we drop the subscript r for brevity of notation. Mixup [21] is a modification of empirical loss minimization, which computes loss functions at samples in the training set, i.e. $\mathcal{L}(p, f(x))$, to computing the loss at a convex combination between samples i, j, i.e. $\mathcal{L}\left(M_\lambda\left(p^{(i)}, p^{(j)}\right), f\left(M_\lambda(w^{(i)}, w^{(j)})\right)\right)$ where $M_\lambda(a, b) = \lambda a + (1 - \lambda)b$ is the line segment between a and b controlled by $\lambda \in [0, 1]$. We combine this with label distribution learning (LDL) to design ordinal-Mixup to suit ordinal classification. LDL is often used in face age estimation [4,13,14], where a target class y is smoothed using a Gaussian kernel with parameters (y, σ) to produce a smoothed target p. Under this context, LDL uses an objective loss function $d(p, f(w))$ with some divergence or integral probability metric d to minimize a proximity measure between distributions instead of targeting a one-hot vector.

Inspired by the above algorithms, we propose a Manifold ordinal-Mixup strategy applicable when classes have ordinal relations. For simplicity, we first describe an input-output ordinal-Mixup in analogy to input-output Mixup. Again, consider two samples $\{(w^{(i)}, y^{(i)}), (w^{(j)}, y^{(j)})\}$, but instead of using a convex combination between the distributions $p^{(i)}, p^{(j)}$ as in Mixup, ordinal-Mixup uses a convex combination between ordinal classes $y^{(i)}, y^{(j)} : \bar{y}^{(ij)} = M_\lambda(y^{(i)}, y^{(j)}) = \lambda y^{(i)} + (1 - \lambda)y^{(j)}$. We now borrow from LDL to produce another target $p^{(ij)}$ by first smoothing $\bar{y}^{(ij)}$ with a Gaussian kernel to produce a temporary distribution $q^{(ij)}$ i.e. $q^{(ij)} = \mathcal{N}(\bar{y}^{(ij)}, \sigma)$. The Gaussian kernel $q^{(ij)}$ is then quantized by letting $p^{(ij)}(y) = q^{(ij)}(y), \forall y \in \mathcal{Y}$ and normalizing $p^{(ij)}$ over its support $\mathcal{Y}$. Now denote the local classifier's prediction $\hat{p}^{(ij)} = f\left(M_\lambda(w^{(i)}, w^{(j)})\right) \in \Delta(\mathcal{Y})$ on a mixed input $M_\lambda(w^{(i)}, w^{(j)})$. The input-output ordinal-Mixup is defined as:

$$\mathcal{L}_{OM}\left(x^{(i)}, x^{(j)}, y^{(i)}, y^{(j)}\right) = D_{KL}\left(p^{(ij)}||\hat{p}^{(ij)}\right) + \left|\bar{y}^{(ij)} - \mathbb{E}_{Y \sim \hat{p}^{(ij)}}[Y]\right|, \quad (1)$$

where D_{KL} is the Kullback-Leibler (KL) divergence.

Since its proposal, Mixup has been modified to a variation known as manifold Mixup [20]. We now go on to describe Manifold ordinal-Mixup which 'mixes' vectors at the latent space instead of the input space. Denote $f_{1:l}(w)$ the lth layer output of an L-layer DNN given patch w, and $f_{l:L}(w)$ a mapping from the latent space induced by $f_{1:l}$ to the output space $\mathcal{Y}$. The only difference between Manifold ordinal-Mixup and input-output ordinal-Mixup is in generating $\hat{p}^{(ij)}$, where instead of using mixed inputs $\bar{w}^{(ij)} = M_\lambda(w^{(i)}, w^{(j)})$, the latent vectors are mixed: $\bar{z}^{(ij)} = M_\lambda\left(f_{1:l}(w^{(i)}), f_{1:l}(w^{(j)})\right)$. Thus, the predictions on mixed latent vectors $\hat{p}^{(ij)} = f_{l:L}\left(\bar{z}^{(ij)}\right)$ are used to compute ordinal-Mixup (Eq. 1).

We have thus far described how ordinal-Mixup is computed when two random samples indexed by i, j are given. In typical DL settings, DNNs are trained using sample batches. Given a batch $\mathcal{B} = \{(x^{(i)}, y^{(i)})\}_{i=1}^{B}$, we assert the target class used to mix with sample i is not too far by finding candidate $Q(y^{(i)}) = \{y \in \mathcal{B} : |y - y^{(i)}| \le C\}$, where $C \ge 1$ is a hyper-parameter controlling the size of the candidates. Then, for each sample $(x^{(i)}, y^{(i)})$, we match a sample $(x^{(j)}, y^{(j)})$ such that $y^{(j)} \in Q(y^{(i)})$ for each $i \in \{1, \dots, B\}$. In our experiments, we use $\lambda \sim Beta(\alpha, \alpha)$ with $\alpha = 0.5$ to control the line segment for Mixup.

3　Experimental Results

We first train an ROI model using the deep high-resolution network (HR-Net) [16] with $R = 13$ output channels on the binary cross entropy loss to extract major bone ROI patches and fix its weights. Then, local classification models are trained using the extracted ROIs on the cross entropy loss (baseline), Manifold Mixup, label distribution learning (LDL), and Manifold ordinal-Mixup (proposed).

3.1　Dataset

Because there is no publicly available dataset with hand X-ray images annotated for TW3 methods, we used a subset of the publicly available hand images from

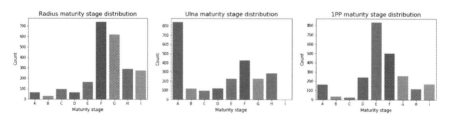

Fig. 3. Maturity stage distributions for Radius, Ulna, and 1PP. Notice the high class imbalance, with a few maturity stages nearly absent, e.g. 'B' and 'C' in 1PP.

the well-known public dataset Radiological Society of North America (RSNA) BAA [8] with 17 key point annotations provided by [3] indicating the joints and bones, removing all but the major bones necessary for TW3. A musculoskeletal radiologist with more than 10 years of clinical practice selected 2,342 clean images from the dataset and annotated the additional four bones (Radius, Ulna, 3MC, 5MC), as well as the maturity stage ('A'-'I') of all 13 bones as in Fig. 1 in accordance to the TW3 guidelines. The resulting maturity stage distribution for Radius, Ulna, and 1PP are shown in Fig. 3. This histogram shows the high class imbalance in bone maturity in each bone, with nearly-absent samples for 'B' and 'C' in 1PP. The ROI extraction network is obtained using a random 80%–20% train/validation split and fixed. All algorithms used to train local classifiers use the same ROI extraction network and are evaluated using 5-fold validation, and their performances averaged across the 5 trials are reported.

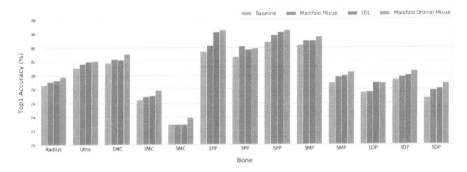

Fig. 4. Performance of all local classifiers using different algorithms. We compare the proposed algorithm (Manifold ordinal-Mixup) with a baseline (using neither Mixup nor LDL), Mixup, and LDL.

3.2 Training Details

We remove extensive noise present in X-ray images by pre-processing as in [6]. All input hand images are resized to 512×512 for the ROI extractor model, but the patches are cropped from the original image's resolution to maintain their high-resolutions. The key point annotations used to train the ROI extractor are sets of coordinates $t = (t_x, t_y)$ smoothened using a Gaussian filter with $\sigma = 2$. ROI patches corresponding to different bones are then extracted by cropping a window centered at the estimated key point. Local classifiers are initialized using ImageNet-pretrained weights and trained using random augmentation (flip, rotation, shift, size scale) and RAdam optimizer [10] with batch size 72. The Gaussian kernel's variance parameter for LDL and ordinal-Mixup was set as 1, and the candidate size parameter (C) for ordinal-Mixup was set as 1. Each algorithm's hyperparameter configuration was selected based on its highest average performance across 5-fold validation found over a search space guided by the choices reported in their respective references. All algorithms were allocated the same computational power for tuning.

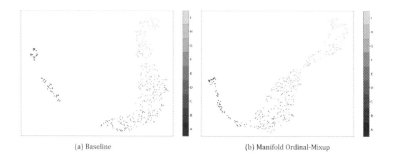

(a) Baseline   (b) Manifold Ordinal-Mixup

Fig. 5. Visualization of the features extracted at the local classifier's penultimate layer when trained using (left) standard classification and (right) Manifold ordinal-Mixup using UMAP [11]. Since bone age takes on continuous values, it is expected that a well-trained classifier has a path-connected latent space. Notice how the disconnected paths in baseline are rectified when using ordinal-Mixup.

3.3 Ordinal Learning Results

Top-1 accuracies of the local classifiers for each bone are shown in Fig. 4. Models trained using the proposed Manifold ordinal-Mixup loss achieve the highest performance across all bones but 1DP, with a small margin. This shows that by exploiting ordinal relations among bone ages, the proposed loss effectively benefits local classifiers across all bones. While the baseline's features do not form a path-connected latent space in Fig. 5 (a), using Manifold ordinal-Mixup induces a path-connected latent space, showing how the local classifiers learn continuous values as opposed to integral values, which is desired as the progression of bone maturity is continuous by nature. In turn, the visualization of the features extracted at the local classifiers' penultimate layers show how ordinal-Mixup yields smooth decision boundaries.

3.4 BAA Results

We lastly evaluate the automated BAA systems using different algorithms averaged over 5 fold cross-validation in Table 2. Manifold ordinal-Mixup outperforms all considered algorithms in both genders and overall, estimating the bone age within 4.04 months of the ground-truths on average. This performance is comparable to the state-of-the-art GP-based DL algorithms [3,7,9,15] although the comparison is quite difficult due to the different annotations used. The Bland-Altman plot illustrating the age differences between those predicted by our model and ground truth labels is shown in Fig. 6.

Table 2. Mean average error (MAE) of BAA using different algorithms. We report the average ± std across 5-fold cross-validation.

	MAE (months)		
	Male	Female	All
Baseline	4.63 ± 0.22	3.83 ± 0.37	4.24 ± 0.22
Manifold Mixup	4.45 ± 0.17	3.92 ± 0.26	4.19 ± 0.21
LDL	4.54 ± 0.17	3.74 ± 0.26	4.15 ± 0.22
Manifold ordinal-Mixup	**4.31 ± 0.16**	**3.65 ± 0.25**	**4.04 ± 0.21**

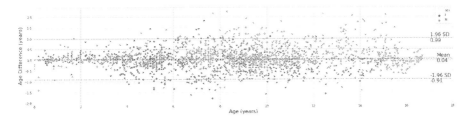

Fig. 6. Bland-Altman plot of the proposed model's BAA predictions and the ground-truth bone ages. 'M' denotes male and 'F' denotes female.

4   Conclusions

We proposed an automated TW3 algorithm that can extract the major bones used in TW3 from hand X-ray images and infer the maturity stage to estimate the bone age. By using a loss function which incorporates ordinal relations between classes, local classifier models are able to learn intermediate features corresponding to imputed ages using an interpolation of the available data samples, hence alleviating the lack of data corresponding to certain maturity stages. The resulting model outperforms all other algorithms considered in our experiments which are otherwise widely used for other tasks due to their prominent success. A visualization of the learnt representations elicit how the proposed training algorithm allows learning smooth decision boundaries and path-connected feature spaces which can thus be used to reliably assess bone age from previously unseen hand X-ray images.

References

1. Bae, B.U., Bae, W., Jung, K.H.: Improved deep learning model for bone age assessment using triplet ranking loss (2018)
2. Dallora, A.L., Anderberg, P., Kvist, O., Mendes, E., Diaz Ruiz, S., Sanmartin Berglund, J.: Bone age assessment with various machine learning techniques: a systematic literature review and meta-analysis. PloS One **14**(7), e0220242 (2019)

3. Escobar, M., González, C., Torres, F., Daza, L., Triana, G., Arbeláez, P.: Hand pose estimation for pediatric bone age assessment. In: Shen, D., et al. (eds.) MICCAI 2019. LNCS, vol. 11769, pp. 531–539. Springer, Cham (2019). https://doi.org/10.1007/978-3-030-32226-7_59

4. Gao, B.B., Zhou, H.Y., Wu, J., Geng, X.: Age estimation using expectation of label distribution learning. In: IJCAI, pp. 712–718 (2018)

5. Greulich, W.W., Pyle, S.I.: Radiographic Atlas of Skeletal Development of the Hand and Wrist. Stanford University Press, Palo Alto (1959)

6. Iglovikov, V., et al.: Pediatric bone age assessment using deep convolutional neural networks. arXiv preprint arXiv:1712.05053 (2017)

7. Ji, Y., Chen, H., Lin, D., Wu, X., Lin, D.: PRSNet: part relation and selection network for bone age assessment. In: Shen, D., et al. (eds.) MICCAI 2019. LNCS, vol. 11769, pp. 413–421. Springer, Cham (2019). https://doi.org/10.1007/978-3-030-32226-7_46

8. Larson, D.B., Chen, M.C., Lungren, M.P., Halabi, S.S., Stence, N.V., Langlotz, C.P.: Performance of a deep-learning neural network model in assessing skeletal maturity on pediatric hand radiographs. Radiology **287**(1), 313–322 (2018)

9. Liu, C., Xie, H., Liu, Y., Zha, Z., Lin, F., Zhang, Y.: Extract bone parts without human prior: end-to-end convolutional neural network for pediatric bone age assessment. In: Shen, D., et al. (eds.) MICCAI 2019. LNCS, vol. 11769, pp. 667–675. Springer, Cham (2019). https://doi.org/10.1007/978-3-030-32226-7_74

10. Liu, L., et al.: On the variance of the adaptive learning rate and beyond. arXiv preprint arXiv:1908.03265 (2019)

11. McInnes, L., Healy, J., Melville, J.: Umap: uniform manifold approximation and projection for dimension reduction. arXiv preprint arXiv:1802.03426 (2018)

12. Newell, A., Yang, K., Deng, J.: Stacked hourglass networks for human pose estimation. In: Leibe, B., Matas, J., Sebe, N., Welling, M. (eds.) ECCV 2016. LNCS, vol. 9912, pp. 483–499. Springer, Cham (2016). https://doi.org/10.1007/978-3-319-46484-8_29

13. Pan, H., Han, H., Shan, S., Chen, X.: Mean-variance loss for deep age estimation from a face. In: Proceedings of the IEEE Conference on Computer Vision and Pattern Recognition, pp. 5285–5294 (2018)

14. Rothe, R., Timofte, R., Van Gool, L.: Dex: deep expectation of apparent age from a single image. In: Proceedings of the IEEE International Conference on Computer Vision Workshops, pp. 10–15 (2015)

15. Son, S.J., et al.: TW3-based fully automated bone age assessment system using deep neural networks. IEEE Access **7**, 33346–33358 (2019)

16. Sun, K., Xiao, B., Liu, D., Wang, J.: Deep high-resolution representation learning for human pose estimation. In: Proceedings of the IEEE Conference on Computer Vision and Pattern Recognition, pp. 5693–5703 (2019)

17. Tan, M., Le, Q.V.: Efficientnet: rethinking model scaling for convolutional neural networks. arXiv preprint arXiv:1905.11946 (2019)

18. Tanner, J.M., et al.: Assessment of Skeletal Maturity and Prediction of Adult Height (TW2 Method). Saunders, London (2001)

19. Thodberg, H.H., van Rijn, R.R., Jenni, O.G., Martin, D.D.: Automated determination of bone age from hand x-rays at the end of puberty and its applicability for age estimation. Int. J. Legal Med. **131**(3), 771–780 (2017)

20. Verma, V., et al.: Manifold mixup: better representations by interpolating hidden states. arXiv preprint arXiv:1806.05236 (2018)

21. Zhang, H., Cisse, M., Dauphin, Y.N., Lopez-Paz, D.: mixup: Beyond empirical risk minimization. arXiv preprint arXiv:1710.09412 (2017)

Contour-Based Bone Axis Detection for X-Ray Guided Surgery on the Knee

Florian Kordon[1,2,3]([✉])([iD]), Andreas Maier[1,2]([iD]), Benedict Swartman[4],
Maxim Privalov[4], Jan Siad El Barbari[4], and Holger Kunze[3]([iD])

[1] Pattern Recognition Lab, Friedrich-Alexander-Universität
Erlangen-Nürnberg (FAU), Erlangen, Germany
florian.kordon@fau.de
[2] Erlangen Graduate School in Advanced Optical Technologies (SAOT),
Friedrich-Alexander-Universität Erlangen-Nürnberg (FAU), Erlangen, Germany
[3] Siemens Healthcare GmbH, Forchheim, Germany
[4] Department for Trauma and Orthopaedic Surgery,
BG Trauma Center Ludwigshafen, Ludwigshafen, Germany

Abstract. The anatomical axis of long bones is an important reference line for guiding fracture reduction and assisting in the correct placement of guide pins, screws, and implants in orthopedics and trauma surgery. This study investigates an automatic approach for detection of such axes on X-ray images based on the segmentation contour of the bone. For this purpose, we use the medically established two-line method and translate it into a learning-based approach. The proposed method is evaluated on 38 clinical test images of the femoral and tibial bone and achieves a median angulation error of 0.19° and 0.33° respectively. An inter-rater study with three trauma surgery experts confirms reliability of the method and recommends further clinical application.

Keywords: Surgical planning · Orthopedics · Bone axis detection · X-ray imaging · Intra-operative guidance

1 Introduction

The reconstruction of anatomical joint surface and angular relationships is a paramount aspect in surgical management of fractures or ligament injuries. Intra-operative fluoroscopic guidance, 3D imaging, or navigation is typically used to ensure anatomically and mechanically correct reduction, so that irregular joint loading and complications caused by aberrant biomechanics can be alleviated or avoided. Moreover, for technically demanding procedures, a pre-operative planning sketch is obligatory and helps the surgeon to achieve operational safety [4]. In many of these planning and verification steps, the bone axis serves as

The authors gratefully acknowledge funding of the Erlangen Graduate School in Advanced Optical Technologies (SAOT) by the Bavarian State Ministry for Science and Art.

an important reference line (Fig. 1). While planning such axes can be easily done on pre-operative static data, doing so consistently on live images during surgery is inherently more complex due to motion and a limited field of view. In addition, non-sterile interaction with a planning software is unwanted. For this reason, axial alignment is typically verified by visual inspection and use of hardware-based solutions such as the cable method, alignment rods, goniometers, or optical navigation amongst others [12,13,20]. However, these methods either increase task complexity, are inherently imprecise, or require an open reduction or additional incisions regardless of the surgical technique used. To this end, several methods were proposed to automate detection of the bone axis on image data. Tian et al. [18] compute the femoral shaft axis by using a combination of contour extraction and analysis of intersecting line normals to the shaft contour. They recover the contour by using Canny edge detection and identify the relevant straight line sections with Hough transformation and active contour mode via Gradient Vector Flow. While this approach can deal with truncated bones, it prerequisites the bone to be oriented in an upward position on the X-ray image to isolate the relevant intersection points. Donnelley et al. [3] use a scale-space approach and approximate line straight parameters via Hough transformation. To deal with ambiguous peak spread in the dual space encountered in real-world radiographs, this methods relies on prior spread quantification which falls short in the case of truncated bones. Subburaj et al. [17] use a 3D-reconstructed bone model from pre-operative CT scans. They combine geometrically detected landmarks and maximal inscribed sphere fitting to detect the medial axis, which is then used for identification of anatomical and mechanical axes. Although very accurate results can be achieved, such 3D information is oftentimes not available and requires registration with the intra-operative 2D image.

To circumvent these limitations, we propose a simple and clinically motivated image-guided approach for detection of the anatomical axis of long bones on 2D X-ray images. We translate the established two-line/two-circle manual method [6–8,10] to a learning based extraction of anatomical features and subsequent geometric construction based on segmentation of the bone cortex outline. With reference to [9], region of interest (ROI) encoding of the relevant contour sections is used to cope with variability in image truncation and arbitrary image rotation. Moreover, the segmentation results can directly be used for registration of the detected axis on fluoroscopic live images. The method is evaluated for the femur and tibia in the knee joint, which are amongst the most prominent anatomies treated in trauma surgery. The reliability of the proposed method is evaluated and confirmed in an inter-rater study with three expert trauma surgeons.

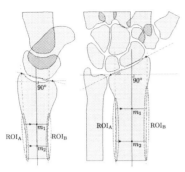
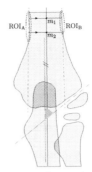
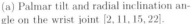

(a) Palmar tilt and radial inclination angle on the wrist joint [2, 11, 15, 22].

(b) Baumann angle on frontal radiograph of the elbow [16, 23].

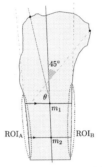
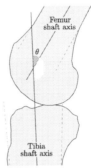

(c) Angles for tibial intramedullary nail
insertion (solid) [5] and transtibial tunnel drilling (dashed) [6, 8].

(d) Approximation of knee flexion angle
based on femoral and tibial bone axes.

Fig. 1. Examples for using the shaft axis of long bones as reference line.

2   Methods

The anatomical axis of long bones in a 2D image plane can be described
by two auxiliary lines that follow the orientation of the anterior/posterior or
medial/lateral contour of the bone shaft. In contrast to conventional radiographs
with rather standardized imaging, this shaft area is usually truncated on intra-
operative images due to a limited field of view and a joint-centered acquisition
protocol. Furthermore, the largely linear shaft contour can suffer from structural
changes due to e.g. bony proliferation. To this end, first the relevant contour sections are estimated and extracted from the image. Subsequently, these sections
are masked based on positional probability and smoothed to reduce the influence
of outliers. Lastly, the clinically motivated two-line method is used to calculate
the bone axis.

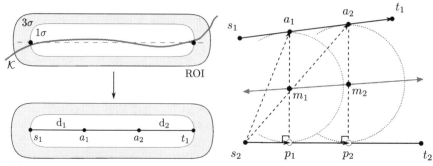

(a) Masking of segmentation contour $\mathcal{K}$ by evaluation the predicted likelihood in ROI and parametrization of the line segment.

(b) Vector projection of intermediary line points onto the opposing segment and geometric construction of the axis points m_1 and m_2.

Fig. 2. Implementation of the two line method for bone axis estimation based on the extracted segmentation contour.

2.1 Likelihood Encoding of Relevant Contour Regions

Given a binary bone segmentation mask S we extract the complete cortex contour $\mathcal{K}$ by using a morphological erosion operation. With a cross-shaped 3×3 structuring element $X = \{(-1,0),(0,-1),(0,0),(0,1),(1,0)\}$ this equates to

$$\mathcal{K} = \text{XOR}\left(S, \text{erode}(S, X)\right). \tag{1}$$

To constrain the relevant contour section, a ROI similar to [9] is constructed (Fig. 1). Its bounds are defined by the start and end points of an additional line segment. Positional variance both in the parallel as well as in the orthogonal direction to this line segment is encoded by a 2D Gaussian distribution with a standard deviation of $\sigma = 6\,\text{px}$ and truncation bounds at 3σ. This gives us a symmetrical fall-off in probability orthogonal to the line within a margin of $37\,\text{px}$. This spatial likelihood distribution is used to decide whether a contour point should be considered part of the relevant contour region. Since we can assume a mainly linear contour, we argue that using a threshold at 1σ retains the most probable points while eliminating most outliers (Fig. 2a).

2.2 Axis Construction with Two-Line Method

The auxiliary contour extension lines are obtained by fitting two linear functions to the pair of relevant contour regions. Since we cannot assume a designated dependent variable due to unknown image rotation, major axis regression[1] is used [21]. Given these two lines, we can now perform a geometric construction

---

[1] Ordinary least squares with dependent variable of highest variance is also possible.

of the in-between axis based on the midpoints of two parallel and intersecting line segments. This method is known as the two-line method and is a clinically known and trusted procedure especially in pre-operative manual planning [6–8, 10]. First, a line segment is parametrized for each contour line which is bounded by the relevant contour region. One of these segments is subdivided by two points at distance d_1 and d_2 from the respective start and end points (Fig. 2a). The actual distances can be selected depending on the target anatomical structure to facilitate easier correction by the user. In a second step, these intermediary points are then projected onto the opposing line segment (Fig. 2b). This procedure allows for different orientation and length of each segment and is close to clinical practice.

2.3 Neural Network Architecture

The proposed construction relies on a segmentation mask of the target long bone and ROI encodings for both relevant contour regions. For combined prediction we use a multi-task variant of the hourglass network architecture by Newell et al. [9, 14]. This network architecture allows to optimize a joint representation of both tasks and benefits execution time and computational footprint upon inference. We separate segmentation and prediction of ROIs into two tasks. The segmentation task is trained with binary cross entropy to delineate the target bone (foreground) and all other image content (background). The ROIs are optimized by direct matching of the pixel intensity values with a mean squared error loss. In addition, we employ gradient normalization [1, 9] to cope with different loss function characteristics and task difficulties. To limit the hardware requirements in consideration of the intra-operative application area, we refrain from a stacked network variant.

2.4 Data and Evaluation

Network training and geometric construction were evaluated for the femur and tibia on a dataset of 221 clinical X-ray images of the knee joint. Each image was acquired as a lateral standard projection where the outlines of both femoral condyles are aligned. The ground truth segmentation masks and line segments representing the ROIs were annotated by a medically-trained engineer with the *labelme* annotation tool [19]. Our experiment and evaluation setup on this dataset was two-fold.

1. Training of the network in three configurations (a) femur only, (b) tibia only, (c) joint training of femur and tibia, followed by a quantitative evaluation of the performance. For variant (c) the number of output channels for each task head of the network was increased accordingly.
2. Assessment of clinical reliability of the automatic axis detection in an inter-rater study. To this end, three expert trauma surgeons (one site) were asked to annotate the femoral and tibial axes on all 38 evaluation images via two axis control points.

For both experiment series a hold-out test set of 38 images with a 3 mm calibration sphere was defined. Representative variability in bone truncation and
absolute joint rotation was confirmed. The remaining data was split into training
and validation subsets of 167/16 images respectively. The data was split up in
such a way that disjoint patient groups are ensured in the training/validation
and test datasets. Optimization for the first experiment step was performed
using Stochastic Gradient Descent (SGD) with a batch size of 2 over 300 epochs
on a NVIDIA TITAN RTX graphics card in the PyTorch (v.1.2) Deep Learning framework. We used a learning rate of $2.5e-4$ which we halved every 50
epochs. To aid generalization and to prevent early overfitting, we applied L2
weight decay with a factor of $5e-5$ and a basic online augmentation sequence
during training. This sequence comprised affine transformations (scaling, rotation, shearing, horizontal flipping) and margin crops of random strength. Upon
propagation in the network, min-max normalization to the interval of $[0, 1]$ was
applied and the image resolution was standardized to 256×256 px by resizing
and a subsequent center-crop. All reported results are based on the respective
model parameters for which the minimum combined task error on the validation
split was observed.

3   Results and Discussion

Bone Segmentation. The results for bone segmentation by the multi-task
neural network variants are given in Table 1. In general we observe segmentation results which closely resemble the annotated ground truth. Despite missing
annotations of other bony structures in the knee joint, the single-anatomy model
is capable of delineating the target bone from other structures, even in ambiguous overlap areas. On the other hand, prediction quality of the combined variant
does not suffer from a doubling of inference tasks which benefits fast execution
time and a smaller computational footprint. A very low contour error indicates
that the networks do not only learn the global shape but also successfully capture small details which are often caused by bony erosion and proliferation.
This allows for marginal error propagation into geometric axis construction.

Table 1. Evaluation of segmentation performance for the femur and tibia (DICE =
Sørensen–Dice coefficient; ASD = Average Surface Distance; HD = Hausdorff Distance).

Bone	Network	DICE Mean ± Std	ASD (mm) Mean ± Std	HD (mm) Mean ± Std
Femur	Single	0.99 ± 0.003	0.57 ± 0.45	7.91 ± 13.86
	Comb.	0.99 ± 0.004	0.57 ± 0.58	11.38 ± 22.17
Tibia	Single	0.99 ± 0.005	0.62 ± 0.53	7.07 ± 7.08
	Comb.	0.99 ± 0.003	0.51 ± 0.16	4.23 ± 1.95

Table 2. Angulation and displacement error for the femur and tibia in single/combined anatomy training. Results are reported for the anterior/posterior auxiliary lines and bone shaft axis. The displacement error is constructed as the mean orthogonal point-to-line distance of predicted points s_1/t_1, s_2/t_2, m_1/m_2 onto the respective ground truth axis and combines translation and angulation error components. The best results for each axis are marked in bold ($\text{CI}_{95} = 95\%$ confidence interval).

		Angulation (deg)			Displacement (mm)		
Axis	Netw.	Mean ± Std	Median & CI_{95}	Max	Mean ± Std	Median & CI_{95}	Max
Ant.	Single	0.55 ± 0.98	**0.31** [0.18, 0.57]	6.17	0.57 ± 1.25	0.34 [0.25, 0.49]	8.09
	Comb.	**0.48** ± 0.64	0.34 [0.24, 0.52]	**3.89**	**0.47** ± 0.79	**0.31** [0.19, 0.44]	**5.10**
Post.	Single	**0.56** ± 0.38	**0.49** [0.36, 0.68]	**1.50**	**0.54** ± 0.33	**0.43** [0.36, 0.55]	**1.53**
	Comb.	0.64 ± 0.44	0.54 [0.38, 0.83]	1.94	0.59 ± 0.47	0.45 [0.36, 0.56]	2.71
Shaft	Single	0.35 ± 0.42	0.21 [0.14, 0.36]	2.54	0.18 ± 0.22	**0.13** [0.09, 0.16]	4.11
	Comb.	**0.28** ± 0.24	**0.19** [0.11, 0.23]	**0.98**	**0.15** ± 0.13	**0.13** [0.07, 0.15]	**2.65**

(a) Femoral axes detection.

		Angulation (deg)			Displacement (mm)		
Axis	Netw.	Mean ± Std	Median & CI_{95}	Max	Mean ± Std	Median & CI_{95}	Max
Ant.	Single	1.59 ± 1.99	**0.64** [0.43, 1.59]	7.37	0.86 ± 0.95	**0.48** [0.35, 0.92]	5.26
	Comb.	**1.43** ± 1.62	0.81 [0.67, 1.18]	**6.98**	**0.75** ± 0.50	0.62 [0.52, 0.74]	**1.93**
Post.	Single	**0.66** ± 0.55	**0.51** [0.32, 0.68]	**1.98**	**0.38** ± 0.21	0.36 [0.25, 0.45]	**0.95**
	Comb.	0.83 ± 0.84	0.52 [0.31, 0.81]	3.37	0.43 ± 0.28	**0.33** [0.26, 0.46]	1.30
Shaft	Single	0.78 ± 0.95	0.48 [0.32, 0.78]	4.11	0.21 ± 0.19	**0.16** [0.11, 0.25]	0.96
	Comb.	**0.62** ± 0.66	**0.33** [0.23, 0.69]	**2.65**	**0.17** ± 0.11	0.17 [0.11, 0.21]	**0.38**

(b) Tibial axes detection.

Segmentation outliers indicated by higher Hausdorff error points are exclusively caused by the inserted measuring spheres which are not represented in the training data.

Axes Detection. The performance of the proposed geometric axis construction is presented in Table 2. We observe an average angulation error of less than $0.65°$ for the anterior and posterior auxiliary lines on both bones and only minor differences between single and combined training. This indicates that the predicted ROIs can provide masking of relevant contour sections on a sufficiently fine scale. We can also qualitatively confirm that the likelihood distribution follows along the actual anatomical contour, albeit this area is only approximated by a straight line in the ground truth annotations. These observations strengthen our assumption that we can retain all relevant contour points by masking at a likelihood threshold of 1σ. In addition, low values for the displacement error (Table 2) indicate minor deviation of the line's shift off the ground truth bone contour. The constructed bone axes generally benefit from the combined train-

Table 3. Comparison of automatically detected shaft axis (Auto) to the annotation of three expert readers (E-1, E-2, E-3) and assessment of inter-rater variability. Due to missing midpoints m_1 and m_2 in the expert reader annotations, the respective displacement error is based on the two annotated control points. Here, $\mapsto$ denotes a mapping of the 1st rater's control points on the predicted axis of the 2nd rater. $\leftarrowtail$ marks a mapping in reverse order.

1st rater	2nd rater	Femur Angulation (deg)	Femur Displacement (mm)	Tibia Angulation (deg)	Tibia Displacement (mm)
Auto	E-1	0.61 [0.38, 1.01]	0.91 [0.57, 1.16]	1.48 [1.12, 2.23]	1.90 [1.36, 2.62]
Auto	E-2	1.12 [0.63, 1.82]	0.58 [0.39, 0.72]	2.36 [1.68, 3.07]	1.68 [1.22, 2.13]
Auto	E-3	0.49 [0.29, 0.68]	0.74 [0.43, 1.17]	5.76 [4.87, 6.44]	3.89 [3.38, 4.56]
E-1	E-2	0.93 [0.39, 1.32]	$\mapsto$ 1.85 [1.51, 2.21] $\leftarrowtail$ 1.68 [1.52, 1.93]	1.75 [1.01, 2.21]	$\mapsto$ 1.93 [1.60, 2.58] $\leftarrowtail$ 2.02 [1.72, 2.61]
E-1	E-3	0.70 [0.49, 1.06]	$\mapsto$ 1.38 [1.20, 1.77] $\leftarrowtail$ 1.82 [1.35, 2.15]	4.53 [3.47, 5.25]	$\mapsto$ 4.64 [3.72, 5.22] $\leftarrowtail$ 4.09 [3.27, 4.81]
E-2	E-3	1.02 [0.69, 1.29]	$\mapsto$ 1.41 [1.15, 1.74] $\leftarrowtail$ 1.38 [1.20, 1.77]	3.22 [2.40, 4.24]	$\mapsto$ 4.71 [3.33, 5.47] $\leftarrowtail$ 3.71 [2.85, 4.25]

ing variant and exhibit a comparatively lower maximum error bound (Table 2). Furthermore, it can be observed that by training both anatomies together, the respective confidence intervals taper off and follow the downward shift of the position measure. Based on these results, we chose the combined network for evaluation in the inter-rater study.

Inter-rater Comparison. The reliability of our method in comparison to expert rater annotations is analyzed in Table 3. For the femur, low angulation and displacement errors indicate reliable axis estimates which are independent of the amount of truncation and rotation present in the image data. A significantly higher angular deviation of the tibial axis can be explained by comparatively more divergent contour lines. Together with structural variation of the anterior tibia (tibial tuberosity), this leads to higher complexity and differences in the individual approach to manual annotation. This reasoning is strengthened by comparison with rater E-3 for whom a systematically more posterior position and orientation can be observed. If compared to the differences between expert raters (Table 3), the automatic approach yields very comparable performance and achieves axis predictions that lie within the inter-rater error bounds. It should be noted that agreement between raters could be further increased if a dedicated tool for semi-automatic two-line planning is used.

4   Conclusion

This study investigated a method for automatic detection of the shaft axis on long bone X-rays. The experiments reveal encouraging results which match expert rater performance. A major strength of the proposed method is the flexibility of ROI masking which we use to select relevant sections of the bone contour without strong prerequisites on image truncation and rotation. We see limitations in that no evaluation was performed for bones that suffer from increased antecurvation/recurvation (e.g. due to natural deformity or increased weight bearing) or major occlusion of the contour by surgical implants. In addition, future work should analyze potential extensions to our method to promote axis estimation in cases of multi-fragment fractures.

Disclaimer. The methods and information presented here are based on research and are not commercially available.

References

1. Chen, Z., Badrinarayanan, V., Lee, C.Y., Rabinovich, A.: Gradnorm: gradient normalization for adaptive loss balancing in deep multitask networks. In: International Conference on Machine Learning, pp. 794–803 (2018)
2. Dée, W., Klein, W., Rieger, H.: Reduction techniques in distal radius fractures. Injury **31**, 48–55 (2000). https://doi.org/10.1016/s0020-1383(99)00263-6
3. Donnelley, M., Knowles, G., Hearn, T.: A CAD system for long-bone segmentation and fracture detection. In: Elmoataz, A., Lezoray, O., Nouboud, F., Mammass, D. (eds.) ICISP 2008. LNCS, vol. 5099, pp. 153–162. Springer, Heidelberg (2008). https://doi.org/10.1007/978-3-540-69905-7_18
4. Ewerbeck, V., et al. (eds.): Standardverfahren in der operativen Orthopädie und Unfallchirurgie, 4 edn. Thieme, Stuttgart (2014)
5. Franke, J., et al.: Infrapatellar vs. suprapatellar approach to obtain an optimal insertion angle for intramedullary nailing of tibial fractures. Eur. J. Trauma Emerg. Surg. **44**(6), 927–938 (2017). https://doi.org/10.1007/s00068-017-0881-8
6. Hiesterman, T.G., Shafiq, B.X., Cole, P.A.: Intramedullary nailing of extra-articular proximal tibia fractures. J. Am. Acad. Orthop. Surg. **19**(11), 690–700 (2011). https://doi.org/10.5435/00124635-201111000-00005
7. James, E.W., LaPrade, C.M., Ellman, M.B., Wijdicks, C.A., Engebretsen, L., LaPrade, R.F.: Radiographic identification of the anterior and posterior root attachments of the medial and lateral menisci. Am. J. Sports Med. **42**(11), 2707–2714 (2014). https://doi.org/10.1177/0363546514545863
8. Johannsen, A.M., Anderson, C.J., Wijdicks, C.A., Engebretsen, L., LaPrade, R.F.: Radiographic landmarks for tunnel positioning in posterior cruciate ligament reconstructions. Am. J. Sports Med. **41**(1), 35–42 (2013). https://doi.org/10.1177/0363546512465072
9. Kordon, F., et al.: Multi-task localization and segmentation for x-ray guided planning in knee surgery. In: Shen, D., et al. (eds.) MICCAI 2019. LNCS, vol. 11769, pp. 622–630. Springer, Cham (2019). https://doi.org/10.1007/978-3-030-32226-7_69

10. Kostogiannis, I., Swärd, P., Neuman, P., Fridén, T., Roos, H.: The influence of posterior-inferior tibial slope in ACL injury. Knee Surg. Sports Traumatol. Arthrosc. **19**(4), 592–597 (2011). https://doi.org/10.1007/s00167-010-1295-x. official journal of the ESSKA

11. Kreder, H.J., Hanel, D.P., McKee, M., Jupiter, J., McGillivary, G., Swiontkowski, M.F.: X-ray film measurements for healed distal radius fractures. J. Hand Surg. **21**(1), 31–39 (1996). https://doi.org/10.1016/S0363-5023(96)80151-1

12. Krettek, C., Miclau, T., Grün, O., Schandelmaier, P., Tscherne, H.: Intraoperative control of axes, rotation and length in femoral and tibial fractures technical note. Injury **29**, 29–39 (1998). https://doi.org/10.1016/S0020-1383(98)95006-9

13. Lee, Y.K., Kim, J.W., Kim, T.Y., Ha, Y.C., Koo, K.H.: Validity of the intra-operative measurement of stem anteversion and factors for the erroneous estimation in cementless total hip arthroplasty using postero-lateral approach. Orthop. Traumatol. Surg. Res. OTSR **104**(3), 341–346 (2018). https://doi.org/10.1016/j.otsr.2017.11.023

14. Newell, A., Yang, K., Deng, J.: Stacked hourglass networks for human pose estimation. In: Leibe, B., Matas, J., Sebe, N., Welling, M. (eds.) ECCV 2016. LNCS, vol. 9912, pp. 483–499. Springer, Cham (2016). https://doi.org/10.1007/978-3-319-46484-8_29

15. Schmitt, R., Lanz, U. (eds.): Bildgebende Diagnostik der Hand. Georg Thieme Verlag, Stuttgart (2015)

16. Silva, M., et al.: Inter- and intra-observer reliability of the Baumann angle of the humerus in children with supracondylar humeral fractures. Int. Orthop. **34**(4), 553–557 (2010). https://doi.org/10.1007/s00264-009-0787-0

17. Subburaj, K., Ravi, B., Agarwal, M.: Computer-aided methods for assessing lower limb deformities in orthopaedic surgery planning. Comput. Med. Imaging Graph. **34**(4), 277–288 (2010). https://doi.org/10.1016/j.compmedimag.2009.11.003

18. Tian, T.P., Chen, Y., Leow, W.K., Hsu, W., Howe, T.S., Png, M.A.: Computing neck-shaft angle of femur for x-ray fracture detection. In: Petkov, N., Westenberg, M.A. (eds.) CAIP 2003. LNCS, vol. 2756, pp. 82–89. Springer, Heidelberg (2003). https://doi.org/10.1007/978-3-540-45179-2_11

19. Wada, K.: labelme: image polygonal annotation with python (2016). https://github.com/wkentaro/labelme

20. Waelkens, P., van Oosterom, M.N., van den Berg, N.S., Navab, N., van Leeuwen, F.W.B.: Surgical navigation: an overview of the state-of-the-art clinical applications. In: Herrmann, K., Nieweg, O.E., Povoski, S.P. (eds.) Radioguided Surgery. LNCS, pp. 57–73. Springer, Cham (2016). https://doi.org/10.1007/978-3-319-26051-8_4

21. Warton, D.I., Wright, I.J., Falster, D.S., Westoby, M.: Bivariate line-fitting methods for allometry. Biol. Rev. Camb. Philos. Soc. **81**(2), 259–291 (2006). https://doi.org/10.1017/S1464793106007007

22. Watson, N.J., Asadollahi, S., Parrish, F., Ridgway, J., Tran, P., Keating, J.L.: Reliability of radiographic measurements for acute distal radius fractures. BMC Med. Imaging **16**(1), 44 (2016). https://doi.org/10.1186/s12880-016-0147-7

23. Williamson, D.M., Coates, C.J., Miller, R.K., Cole, W.G.: Normal characteristics of the Baumann (humerocapitellar) angle: an aid in assessment of supracondylar fractures. J. Pediatr. Orthop. **12**(5), 636–639 (1992)

Automatic Segmentation, Localization, and Identification of Vertebrae in 3D CT Images Using Cascaded Convolutional Neural Networks

Naoto Masuzawa[1]([✉]), Yoshiro Kitamura[1], Keigo Nakamura[1], Satoshi Iizuka[2], and Edgar Simo-Serra[3]

[1] Imaging Technology Center, Fujifilm Corporation, Tokyo, Minato, Japan
naoto.masuzawa@fujifilm.com
[2] Center for Artificial Intelligence Research, University of Tsukuba, Tsukuba, Ibaraki, Japan
[3] Department of Computer Science and Engineering, Waseda University, Tokyo, Shinjuku, Japan

Abstract. This paper presents a method for automatic segmentation, localization, and identification of vertebrae in arbitrary 3D CT images. Many previous works do not perform the three tasks simultaneously even though requiring a priori knowledge of which part of the anatomy is visible in the 3D CT images. Our method tackles all these tasks in a single multi-stage framework without any assumptions. In the first stage, we train a 3D Fully Convolutional Networks to find the bounding boxes of the cervical, thoracic, and lumbar vertebrae. In the second stage, we train an iterative 3D Fully Convolutional Networks to segment individual vertebrae in the bounding box. The input to the second networks have an auxiliary channel in addition to the 3D CT images. Given the segmented vertebra regions in the auxiliary channel, the networks output the next vertebra. The proposed method is evaluated in terms of segmentation, localization, and identification accuracy with two public datasets of 15 3D CT images from the MICCAI CSI 2014 workshop challenge and 302 3D CT images with various pathologies introduced in [1]. Our method achieved a mean Dice score of 96%, a mean localization error of 8.3 mm, and a mean identification rate of 84%. In summary, our method achieved better performance than all existing works in all the three metrics.

Keywords: Vertebrae · Segmentation · Localization · Identification · Convolutional neural networks

1   Introduction

Automatic segmentation, localization, and identification of individual vertebrae from 3D CT (Computed Tomography) images play an important role in a pre-processing step of automatic analysis of the spine. However, many previous works

© Springer Nature Switzerland AG 2020
A. L. Martel et al. (Eds.): MICCAI 2020, LNCS 12266, pp. 681–690, 2020.
https://doi.org/10.1007/978-3-030-59725-2_66

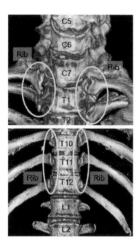

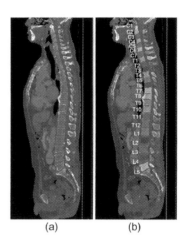

Fig. 1. Differences in anatomy between cervical and thoracic vertebrae, and thoracic and lumbar vertebrae.

Fig. 2. a) A sagittal slice of 3D CT images which includes cervical (C1–C7), thoracic (T1–T12), and lumbar (L1–L5) vertebrae. b) Segmentation and identification of the individual vertebrae.

are not able to perform segmentation, localization, and identification simultaneously and require a priori knowledge of which part of the anatomy is visible in the 3D CT images.

We overcome those drawbacks with a single multi-stage framework. More specifically, in the first stage, we train a 3D Fully Convolutional Networks (we call it "Semantic Segmentation Net"), which segments cervical, thoracic, and lumbar vertebrae so as to find the bounding boxes. As shown in Fig. 1, thoracic vertebrae are distinguished from the cervical and lumbar vertebrae by whether they connect to their ribs and therefore it appears that the Semantic Segmentation Net performs well even if the field-of-view (FOV) is limited. In the second stage, we train an iterative 3D Fully Convolutional Networks (we call it "Iterative Instance Segmentation Net"), which segments (i.e., predicts the labels of all voxels in the 3D CT images), localizes (i.e., finds the centroids of all vertebrae), and identifies (i.e., assigns the anatomical labels) the vertebrae in the bounding box one-by-one. Figure 2 shows an example input image and the corresponding image synthesized by the proposed method. In summary, our contribution is as follows. 1) A two-stage coarse-to-fine approach for vertebrae segmentation, localization, and identification. 2) In-depth experiments and comparisons with existing approaches.

2   Related Work

The challenges associated with automatic segmentation, localization, and identification of individual vertebrae are due to the following three points. 1) High similarity in appearance of the vertebrae. 2) The various pathologies such as the abnormal spine curvature and vertebral fractures. 3) The variability of input 3D

CT images such as FOV, resolution, and image artifacts. To address these challenges, many methods have been proposed. Traditionally, vertebral segmentation has used mathematical methods such as atlas-based segmentation or deformable models [5,8,9]. Speaking of localization and identification, Glocker et al. [1,2] proposed a method based on regression forests with a challenging dataset. They introduced 302 3D CT images with various pathologies, the narrow FOV, and metal artifacts. Recently, deep learning has been employed in the applications of vertebral segmentation, localization, and identification. Yang et al. [13] proposed a deep image-to-image network (DI2IN) to predict centroid coordinates of vertebrae. On the other hand, the common way to segment vertebrae using deep learning is to use semantic segmentation to predict the labels of all voxels in input 3D CT images. For example, Janssens et al. [4] proposed a 3D fully convolutional neural networks (FCN) to segment lumbar vertebrae. However, the way based on the semantic segmentation can segment vertebrae such as lumbar only when whole of the vertebrae is visible in 3D CT images. This motivated Lessmann et al. [10] to consider vertebral segmentation as an instance segmentation problem. The networks introduced by Lessman et al. [10] have an auxiliary channel in addition to the input. Given the segmented vertebra regions in the auxiliary channel, the networks output the next vertebra. Thus, the method proposed by Lessmann et al. [10] is able to perform vertebral segmentation even though whole of the vertebrae is not visible in 3D CT images and the number of vertebra is not known a priori.

Although the method by Lessmann et al. [10] achieves high segmentation accuracy, it does not predict anatomical labels (i.e., cervical C1–C7, thoracic T1–T12, lumbar L1–L5) for each vertebra and it does not handle general 3D CT images where it is not known in advance which part of the anatomy is visible. In fact, their method requires a priori knowledge of anatomy, such as lumbar 5. On the other hand, our approach is able to predict anatomical labels and handle general 3D CT images.

3   Proposed Method

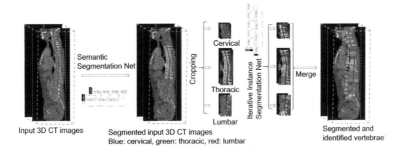

Fig. 3. A schematic view of the present approach.

Our method relies on a two-stage approach as shown in Fig. 3. The first stage aims to segment cervical, thoracic, and lumbar vertebrae from input 3D CT images. Individual vertebrae are segmented in the second stage. Moreover, vertebral centroid coordinates and their labels are also obtained. Below we first present our training dataset, followed by descriptions of the Semantic Segmentation Net and the Iterative Instance Segmentation Net.

3.1 Training Dataset

We prepared 1035 3D CT images (head: 181, chest 477, abdomen: 270, leg: 107) for training which are obtained from diverse manufacturer's equipment (e.g., GE, Siemens, Toshiba, etc.). The leg 3D CT images were prepared for the purpose of suppressing false positive in the first stage. The slice thickness ranges from 0.4 mm to 3.0 mm, and the in-plane resolution ranges from 0.34 mm to 0.97 mm. They have been selected to contain the abnormal spine curvature, metal artifacts, and the narrow FOV. Our spine model for training includes n = 25 individual vertebrae, where the regular 19 from the cervical, thoracic, and lumbar vertebrae consist irregular lumbar 6. Reference segmentations of the visible vertebrae were generated by manually correcting automatic segmentations.

3.2 Stage 1: Semantic Segmentation Net

The convolutional neural networks are widely used to solve segmentation tasks in supervised learning technique. Recent works have shown that this technique can be successfully applied to the multi-organ segmentation in 3D CT images [11]. In our method, we develop the Semantic Segmentation Net which segment cervical, thoracic, and lumbar vertebrae from 3D CT images to find the bounding boxes.

Figure 4 shows a schematic drawing of the architecture. Our architecture is based on a 3D FCN [11]. For our Semantic Segmentation Net, the convolutions

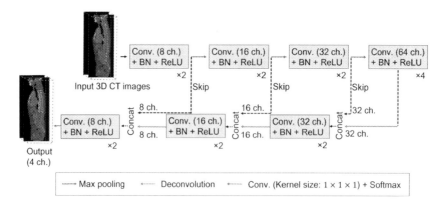

Fig. 4. Architecture of the Semantic Segmentation Net.

performed in each stage use volumetric kernels having size of $3 \times 3 \times 3$ and strides of 1 followed by batch normalization [3] and ReLU as the activate function, the max pooling uses volumetric kernels having size of $2 \times 2 \times 2$ and strides of 2, and the deconvolutions use volumetric kernels having size of $4 \times 4 \times 4$ and strides of 2.

Data Augmentation and Training. In the preprocessing steps, input 3D CT images are clipped to the $[-512.0, 1024.0]$ range and then normalized to be in the $[-1.0, 1.0]$ interval. After that, input 3D CT images are rescaled to 1.0 mm isotropic voxels. For each training iteration, we randomly crop $160 \times 160 \times 160$ voxels from the input 3D CT images and apply data augmentation. In particular, we apply an affine transformation consisting of a random rotation between -15 and $+15°$, and random scaling between -20% and $+20\%$, both sampled from uniform distributions. In addition, we apply a Gaussian noise with $\mu = 0.0$ and $\sigma = [0.0, 50.0/1536.0]$. In the training iteration, bootstrapped cross entropy loss functions [6] were optimized with the Adam optimizer [7] with a learning rate of 0.001 since the multi-class dice loss can be unstable. The idea behind bootstrapping [6] is to backpropagate cross entropy loss not from all but a subset of voxels that the posterior probabilities are less than a threshold. In our experiment, 10% of total voxels are used for the backpropagation.

3.3   Stage 2: Iterative Instance Segmentation Net

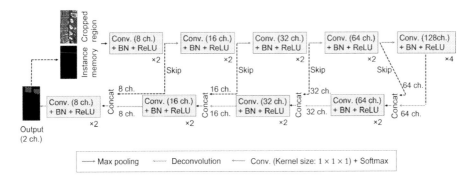

Fig. 5. Architecture of the instance segmentation net.

The goal of the second stage is segmenting, localizing, and assigning anatomical labels to each vertebra. To this end, we developed the Iterative Instance Segmentation Net inspired by Lessmann et al. [10]. The input to the Iterative Instance Segmentation Net has an auxiliary channel in addition to the 3D CT images. Given the segmented vertebra regions in the auxiliary channel, the networks output the next vertebra. The method by Lessmann et al. [10] requires lumbar 5 region as a priori knowledge, and therefore it is not able to handle general 3D

CT images. By contrast, due to using the segmentation results in the first stage, our method is able to handle general 3D CT images.

Figure 5 shows a schematic drawing of the architecture. For our Iterative Instance Segmentation Net, the convolutions performed in each stage use volumetric kernels having size of $3 \times 3 \times 3$ and strides of 1 followed by batch normalization [3] and ReLU as the activate function, the max pooling uses volumetric kernels having size of $2 \times 2 \times 2$ and strides of 2, and the deconvolutions use volumetric kernels having size of $4 \times 4 \times 4$ and strides of 2. The anatomical labels of individual vertebrae are counted starting from the boundaries of cervical and thoracic vertebrae or thoracic and lumbar vertebrae. Finally, centroids of vertebrae are calculated using the segmentation results.

Data Augmentation and Training. In the preprocessing steps, similar to the first stage, input 3D CT images are clipped to the $[-512.0, 1024.0]$ range and then normalized to be in the $[-1.0, 1.0]$ interval. After that, input 3D CT images are rescaled to 1.0 mm isotropic voxels. For each training iteration, we randomly crop the spine region from the input 3D CT images and apply data augmentation. In particular, we apply an affine transformation consisting of a random rotation between -15 and $+15°$, and random scaling between -20% and $+20\%$, both sampled from uniform distributions. In addition, we apply a Gaussian noise with $\mu = 0.0$ and $\sigma = [0.0, 50.0/1536.0]$. In the training iteration, the Dice loss of the segmented volume were optimized with the Adam optimizer [7] with a learning rate of 0.001.

4   Experimental Results

We present two sets of experimental results. The first one is on vertebral segmentation and the second one is about vertebral localization and identification. We validate our algorithm with two public datasets of 15 3D CT images with reference segmentations from the MICCAI CSI (Computational Spine Imaging) 2014 workshop challenge and 302 3D CT images of the patients with various types of pathologies introduced in [1]. There are unusual appearances in the second dataset such as abnormal spine curvature and metal artifacts. In addition, the FOV of each volume varies widely.

4.1   Segmentation Performance

We evaluated our method in terms of the segmentation accuracy with the MICCAI CSI 2014 workshop challenge. The CSI dataset consists of 15 3D CT images of healthy young adults, aged 20–34 years. The images were scanned with either a Philips iCT 256 slice CT scanner or a Siemens Sensation 64 slice CT scanner (120 kVp, with IV-contrast). The in-plane resolution ranges from 0.31 mm to 0.36 mm and the slice thickness ranges from 0.7 mm to 1.0 mm. Each volume cover thoracic and lumbar vertebrae. We evaluate the segmentation performance using Average Symmetric Surface Distance (ASSD), Hausdorff Distance (HD), and Dice score on condition that the final segmentation masks are rescaled to the resolution of the input 3D CT images. The results on the CSI dataset is summarized in Table 1. Our method achieved slightly better performance than existing methods. The examples of the segmentations and the anatomical labels obtained with our method are shown in Fig. 6. In all the 15 3D images, the Semantic Segmentation Net provided the Iterative Instance Segmentation Net with the accurate bounding boxes. Moreover, the Iterative Instance Segmentation Net segmented the vertebrae precisely and predicted all of the anatomical labels.

Table 1. Comparison of Dice scores, ASSD and HD for segmentation results.

Method	Dice score(%)	ASSD (mm)	HD (mm)
Janssens et al. [4]	95.7 %	0.37	4.32
Lessman et al. [10]	94.9 %	0.19	-
Our method	**96.6 %**	**0.10**	**2.11**

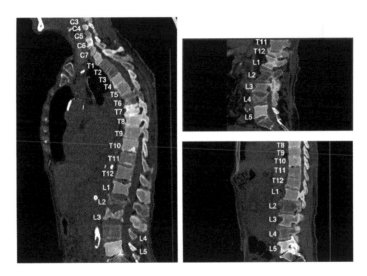

Fig. 6. Segmentation results and predicted anatomical labels obtained with the proposed method.

4.2   Identification and Localization Performance

We evaluate localization and identification performance with 302 3D CT images introduced in [1]. This dataset is challenging since it includes wide varieties of anomalies such as the abnormal spine curvature and the metal artifacts. Furthermore, the FOV of each volume is largely different. In this dataset, the reference centroid coordinates of the vertebrae and the anatomical labels were given by clinical experts. We evaluate our method with the two metrics described in [2], which are the Euclidean distance error (in mm) and identification rates (Id.Rates) defined in [1]. On calculating these metrics, the final segmentation masks are rescaled to the resolution of the input 3D CT images. Table 2 shows a comparison between our method and previous works [12,13]. The mean localization error is 8.3 mm, and the mean identification rate is 84%. Our method achieved better performance than the other existing methods.

Table 2. Comparison of localization errors in mm and identification rates.

	Method	Mean	Std	Id.rates
All	Glocker et al. [1]	12.4	11.2	70%
	Suzani et al. [12]	18.2	11.4	-
	Yang et al. [13]	9.1	**7.2**	80%
	Yang et al. [13] (+1000)	8.5	7.7	83%
	Our method	**8.3**	7.6	**84%**
Cervical	Glocker et al. [1]	7.0	4.7	80%
	Suzani et al. [12]	17.1	8.7	-
	Yang et al. [13]	6.6	3.9	83%
	Yang et al. [13] (+1000)	5.8	3.9	88%
	Our method	**5.7**	**3.8**	**89%**
Thoracic	Glocker et al. [1]	13.8	11.8	62%
	Suzani et al. [12]	17.2	11.8	-
	Yang et al. [13]	9.9	7.5	74%
	Yang et al. [13] (+1000)	9.5	8.5	78%
	Our method	**9.3**	**8.3**	**79%**
Lumbar	Glocker et al. [1]	14.3	12.3	75%
	Suzani et al. [12]	20.3	12.2	-
	Yang et al. [13]	10.9	9.1	80%
	Yang et al. [13] (+1000)	9.9	9.1	84%
	Our method	**9.8**	**9.0**	**85%**

5   Conclusion

In this paper, we propose a multi-stage framework for segmentation, localization and identification of vertebrae in 3D CT images. A novelty of this framework is to divide the three tasks into two stages. The first stage is multi-class segmentation of cervical, thoracic, and lumbar vertebrae. The second stage is iterative instance segmentation of individual vertebrae. By doing this, the method successfully works without a priori knowledge of which part of the anatomy is visible in the

3D CT images. This means that the method can be applied to a wide range of 3D CT images and applications. In the experiments using two public datasets, the method achieved the best Dice score for volume segmentation, and achieved the best mean localization error and identification rate. As far as we know, this is the first unified framework that tackles the three tasks simultaneously with the state of the art performance. We hope that the proposed method will help doctors in clinical practice.

References

1. Glocker, B., Feulner, J., Criminisi, A., Haynor, D.R., Konukoglu, E.: Automatic localization and identification of vertebrae in arbitrary field-of-view CT scans. In: Ayache, N., Delingette, H., Golland, P., Mori, K. (eds.) MICCAI 2012. LNCS, vol. 7512, pp. 590–598. Springer, Heidelberg (2012). https://doi.org/10.1007/978-3-642-33454-2_73

2. Glocker, B., Zikic, D., Konukoglu, E., Haynor, D.R., Criminisi, A.: Vertebrae localization in pathological spine CT via dense classification from sparse annotations. In: Mori, K., Sakuma, I., Sato, Y., Barillot, C., Navab, N. (eds.) MICCAI 2013. LNCS, vol. 8150, pp. 262–270. Springer, Heidelberg (2013). https://doi.org/10.1007/978-3-642-40763-5_33

3. Ioffe, S., et al.: Batch normalization: accelerating deep network training by reducing internal covariate shift. In: International Conference on Machine Learning, vol. 37, pp. 448–456 (2015)

4. Janssens, R., et al.: Fully automatic segmentation of lumbar vertebrae from ct images using cascaded 3d fully convolutional networks. In: IEEE 15th International Symposium Biomedical Imaging, pp. 893–897 (2018)

5. Jianhua, Y., et al.: A multi-center milestone study of clinical vertebral ct segmentation. Comput. Med. Imaging Graph. **49**, 16–28 (2016)

6. Keshwani, D., Kitamura, Y., Li, Y.: Computation of total kidney volume from CT images in autosomal dominant polycystic kidney disease using multi-task 3D convolutional neural networks. In: Shi, Y., Suk, H.-I., Liu, M. (eds.) MLMI 2018. LNCS, vol. 11046, pp. 380–388. Springer, Cham (2018). https://doi.org/10.1007/978-3-030-00919-9_44

7. Kingma, D.P., et al.: Adam: a method for stochastic optimization. arXiv preprint arXiv:1412.6980 (2014)

8. Klinder, T., et al.: Automated model-based vertebra detection, identification, and segmentationin CT images. Med. Image Anal. **13**, 471–482 (2009)

9. Korez, R., et al.: A framework for automated spine and vertebrae interpolation-based detection and model-based segmentation. In: IEEE 15th International Symposium Biomedical Imaging, vol. 34, pp. 1649–1662 (2015)

10. Lessmann, N., et al.: Iterative fully convolutional neural networks for automatic vertebra segmentation and identification. Med. Image Anal. **53**, 142–155 (2019)

11. Roth, H.R., et al.: An application of cascaded 3d fully convolutional networks for medical image segmentation. Comput. Med. Imaging Graph. **66**, 90–99 (2018)

12. Suzani, A., Seitel, A., Liu, Y., Fels, S., Rohling, R.N., Abolmaesumi, P.: Fast automatic vertebrae detection and localization in pathological CT scans - a deep learning approach. In: Navab, N., Hornegger, J., Wells, W.M., Frangi, A.F. (eds.) MICCAI 2015. LNCS, vol. 9351, pp. 678–686. Springer, Cham (2015). https://doi.org/10.1007/978-3-319-24574-4_81

13. Yang, D., et al.: Automatic vertebra labeling in large-scale 3D CT using deep image-to-image network with message passing and sparsity regularization. In: Niethammer, M., et al. (eds.) IPMI 2017. LNCS, vol. 10265, pp. 633–644. Springer, Cham (2017). https://doi.org/10.1007/978-3-319-59050-9_50

Discriminative Dictionary-Embedded Network for Comprehensive Vertebrae Tumor Diagnosis

Shen Zhao[1,5], Bin Chen[2], Heyou Chang[3,5], Xi Wu[4], and Shuo Li[5(✉)]

[1] School of Intelligent Systems Engineering, Sun Yat-Sen University,
Guangzhou, China
z-s-06@163.com

[2] ZheJiang University, Hangzhou, China
ttbin@hotmail.com

[3] School of Information Engineering,
Nanjing XiaoZhuang University, Nanjing, Jiangsu, China
cv_hychang@126.com

[4] Chengdu University of Information Technology, Chengdu, China
xi.wu@cuit.edu.cn

[5] Department of Medical Imaging and Medical Biophysics,
Western University, London, ON, Canada
slishuo@gmail.com

Abstract. Comprehensive vertebrae tumor diagnosis (vertebrae recognition and vertebrae tumor diagnosis from MRI images) is crucial for tumor screening and preventing further metastasis. However, this task has not yet been attempted due to challenges caused by various tumor appearance, non-tumor diseases with similar appearance, irrelevant interference information, as well as diverse MRI image field of view (FOV) and/or characteristics. We purpose a discriminative dictionary-embedded network (DECIDE) that contains an elaborated enhanced-supervision recognition network (ERN) and a discerning diagnosis network (DDN). Our ERN creatively designs projection-guided dictionary learning to leverage projections of angular point coordinates onto multiple observation axes for enhanced supervision and discriminability of different vertebrae. DDN integrates a novel label consistent dictionary learning layer into a classification network to obtain more discerning sparse codes for diagnosing performance improvement. DECIDE is trained and evaluated using a very challenging dataset consisted of 600 MRI images; the evaluation results show that DECIDE achieves high performance in both recognition (accuracy: 0.928) and diagnosis (AUC: 0.96) tasks.

Keywords: Vertebrae tumor diagnosis · Vertebrae recognition · Dictionary embedded deep learning

© Springer Nature Switzerland AG 2020
A. L. Martel et al. (Eds.): MICCAI 2020, LNCS 12266, pp. 691–701, 2020.
https://doi.org/10.1007/978-3-030-59725-2_67

1   Introduction

Comprehensive vertebrae tumor diagnosis (CVTD) means recognizing each vertebra by classifying its label and regressing its bounding box, and diagnosing whether it is invaded by tumor. CVTD is crucial because it joins recognition and diagnosis task together to enable direct diagnosis of vertebrae tumors, which are the most fatal spinal processes [1,2], from magnetic resonance imaging (MRI) images without manual processes such as vertebrae extraction. CVTD may clinically assist radiologists as an automated processor of MRI images for locating lesions, planning treatments, and preventing further metastasis [3,4].

In CVTD, both recognition and diagnosis tasks are challenging. (1) The diagnosis task may be affected by various tumor appearances, e.g., local intensity changes in approximately circular or ring-like areas, global hypointense or hyperintense depending on MRI modalities, and diffuse pepper-salt like textures (Fig. 1(a1~a3)). Furthermore, it is difficult to distinguish from other spinal diseases such as end-plate osteochondritis (Fig. 1(a4)). (2) The diagnosis task may suffer from massive irrelevant interference information in the non-vertebrae parts in the input MRI image. (3) The recognition task is troubled by unknown field of view (FOV) and corrupted MRI scans (e.g., under-sampled scans, severely deformed vertebrae) (Fig. 1(a5)), which may lead to wrong-site surgery [5], a severe medical malpractice in clinical practice [6].

Very few work have been attempted for comprehensive vertebrae tumor diagnosis in the existing literature. [7] shows a closely relative work that detects tumors from MRI patches using convolutional neural networks (CNN). This work achieves high performance, however, it requires manual vertebrae extraction. [5,8–10] singly perform vertebrae recognition, which lay solid foundation for further tumor diagnosis. [11,12] use watershed algorithm and support vector machine for diagnosing spine metastases in CT images. However, this method cannot be applied in MRI images because intensity distributions and textures in MRI images are more complicated (Fig. 1(a)). Other methods such as snakes, multi-task learning and state-space approaches can also be used in detection and/or segmentation tasks [13,14]. In all, MRI is more sensitive for evaluating spinal lesions [7,15], however, it is difficult for existing methods to distinguish vertebrae from non-vertebrae organs, and tumors from non-tumor tissues in corrupted or noisy MRI scans.

Dictionary learning (sparse coding) has the potential to obtain discriminative features from noisy images [16] for classification [17] and detection [18] tasks, however, two difficulties hinder its application in modern deep networks: (1) The dictionary and the sparse codes are generally trained in an alternate manner [19], which is difficult to be integrated into end-to-end training of CNN's. (2) The ground truth sparse codes are typically difficult to obtain. Traditional dictionary learning uses unsupervised reconstruction to obtain sparse codes, which may not be optimal for the main recognition [20] and diagnosis tasks [21].

We propose a novel discriminative dictionary-embedded network (DECIDE) for CVTD. DECIDE first recognizes each vertebra from input MRI, and then focuses on the features inside its bounding box for tumor diagnosis to eliminate

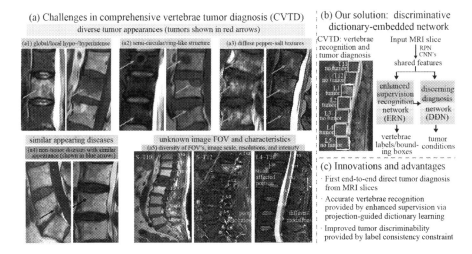

Fig. 1. Our proposed discriminative dictionary-embedded network (Fig. 1(b)) addresses the challenges of comprehensive vertebrae tumor diagnosis (Fig. 1(a), (a1~a4) show various appearances of spinal tumors and similar appearances of non-tumor diseases, (a5) shows challenge of vertebrae recognition) by embedding dictionary into CNN-based recognition/classification networks Fig. 1(c).

interference information (tumor-like patterns of non-vertebrae parts). For both recognition and diagnosis tasks, DECIDE calculates sparse codes in the forward pass, which enables end-to-end dictionary training. Furthermore, in the recognition task, the angular points of the vertebrae's bounding boxes are sparse, i.e., they only account for very few proportions of pixels in the input image. This fact is leveraged to encode each vertebra by predicting L sparse codes via embedded dictionary learning. The sparse codes are trained to approach the projections of ground truth angular points onto L observation axes (OA). For different vertebrae, since their projections on OA's of different orientations exhibit adequate discrepancy, the trained sparse codes have better distinguishability of different vertebrae. Under the projections' guidance, the ensemble of the predicted sparse codes helps to distinguish different vertebrae [22,23]. In the diagnosis task, we diagnose tumor for each recognized vertebrae using features inside its bounding box to tackle interference information. A label consistent dictionary learning layer is designed to help minimize the Mahalanobis distance of vertebrae with the same diagnosis (cancer/non-cancer), which helps distinguish tumor from similar appearing diseases.

Our Contributions are: (1) For the first time, vertebrae tumors are directly diagnosed from MRI images by an elaborated network that highlights key diagnostic information via vertebrae recognition. (2) An enhanced supervision method based on projection-guided dictionary learning is proposed for successfully recognizing sparsely distributed objects (vertebrae). (3) A label consistent strategy is introduced for improving classification (diagnosis) discriminability.

2   Methodology

Based on our pre-existing vertebrae recognition framework [5], we design our discriminative dictionary-embedded network (DECIDE, Fig. 1(b)) by introducing the dictionary learning elements. DECIDE first adopts RPN [24] to coarsely locate regions containing vertebrae. Then, two deliberate modules are designed: (1) Enhanced-supervision recognition network (ERN, Sect. 2.1, Fig. 2, in yellow background) designs a feed-forward dictionary learning layer to obtain sparse codes. These sparse codes encode the projections of each vertebra on L observation axes for enhanced supervision, which helps improve the generalized vertebrae recognition accuracy and tackle the FOV/characteristics challenges. (2) Discerning diagnosis network (DDN, Sect. 2.2, Fig. 2, in pink background) introduces a label consistent dictionary learning layer, which is embedded into a classification network to decrease the distance among sparse codes of vertebrae with the same diagnosis. This achieves accurate tumor diagnosis and alleviates the challenge caused by tumor appearances.

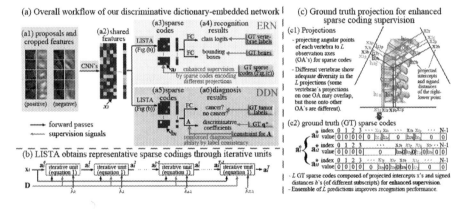

Fig. 2. Key modules of our approach. Figure 2(a) shows the overall workflow. Regional proposals and features are firstly obtained as in [24] (Fig. 2(a1∼a2)). Our main contributions are: (1) The enhanced-supervision recognition network (ERN, in yellow background) that leverages the projections of the vertebrae's angular point coordinates to various observation axes (Fig. 2(c)) for enhanced supervision. (2) The discerning diagnosis network (DDN, in pink background) that forces label consistency (i.e., vertebrae with the same diagnosis to have similar sparse codes) for enhanced diagnosis performance. (Color figure online)

2.1   Enhanced-Supervision Recognition Network (ERN)

Input of ERN. Our ERN takes the regional proposals (Fig. 2(a1)) as its input. These proposals, obtained by regional proposal network (RPN) [24], are multiscale rectangle boxes that coarsely cover vertebrae. For each proposal, its features are obtained by ROI aligning [24] and then flattened into vectors (denoted as $\mathbf{x}_i \in \mathbb{R}^M$ for the ith proposal) by cascading convolutional layers.

Acquiring Representative Sparse Codes for Recognizing Different Vertebrae. We design a dictionary learning layer to calculate the sparse code $\mathbf{a}_i$ for each $\mathbf{x}_i$. The subtlety of dictionary learning in recognition tasks is that the objects (vertebrae) are sparsely distributed, thus their positions (e.g., angular points) can be represented by $\mathbf{a}_i$ or its linear projection [18]. This draws forth the idea of enhancing the supervision of sparse code for better $\mathbf{a}_i$, which resultantly yields better object locations. Meanwhile, it is confirmed in the compressive sensing community that $\mathbf{a}_i$ is able to be recovered by $\mathbf{x}_i$ (the output of CNN's) by minimizing $\frac{1}{2}\|\mathbf{x}_i - \mathbf{D}\mathbf{a}_i\|_2^2 + \lambda\|\mathbf{a}_i\|_1$ over $\mathbf{a}_i$. Inspired by the LISTA algorithm [25], we use Eq. 1 (visually demonstrated in Fig. 2(b)) to obtain $\mathbf{a}_i$:

$$\mathbf{a}_i^t = \eta(\mathbf{a}_i^{t-1} + \beta\mathbf{D}^T(\mathbf{x}_i - \mathbf{D}\mathbf{a}_i^{t-1}); \lambda^t), \text{ where } \eta(\mathbf{r}; \lambda) = \text{sgn}(\mathbf{r})\max\{|\mathbf{r}| - \lambda, 0\} \quad (1)$$

where the sparse code of the ith proposal $\mathbf{a}_i^t$ is updated iteratively by the shrinkage function η. η is a thresholding function that processes its input $\mathbf{r}$ element by element: For each element r_j, threshold λ is subtracted from its original absolute value $|r_j|$; if $|r_j| - \lambda < 0$, this element is set to 0 in the next iteration, which reduces non-zero elements. T iterations (typically $T = 3 \sim 6$) are applied to calculate reliable $\mathbf{a}_i$. In our design, the dictionary $\mathbf{D}$ as well as all λ^t's can be trained together with the preceding CNN's in an end-to-end manner.

Designing Ground Truth Sparse Codes. The key procedure of enhancing the supervision is to design appropriate ground truth sparse codes $\mathbf{a}_i^*$. For each vertebra (corresponding to a positive proposal), its ground truth sparse codes are obtained by the intercepts and signed distances of its angular point coordinates' projections to L observation axes around the input image, i.e., the projected intercepts x's and signed distances h's with different subscripts in $\mathbf{a}_i^*$ in Fig. 2(c). The orientations of these axes are uniformly distributed (see Fig. 2(c), for clarity, the projections of only one vertebra to $L = 3$ axes are demonstrated). For each non-vertebrae region (corresponding to a negative proposal), its ground truth sparse codes are set to zero vectors.

Loss Function. After obtaining the predicted $\mathbf{a}_i$ and the ground truth $\mathbf{a}_i^*$, we design a loss function as follows: Firstly, two sibling fully connected (FC) layers are used as inverse projections; they take $\mathbf{a}_i$ as input and separately output L object class probability vectors and $4 \times L$ bounding box coordinates. Then, as in our previous work [5], message passing method is leveraged for vertebrae class probability calibration. Finally, all these calibrated class probabilities, bounding boxes, and sparse codes are supervised by the corresponding ground truths, i.e., the total loss function of the recognition task is:

$$\mathrm{L_r} = \frac{\lambda_1}{N_1}\sum_{i_1=1}^{N_1}\sum_{l=1}^{L}L_{ce}(\mathbf{u}_{i_1,l}, u_{i_1}^*) + \frac{\lambda_2}{N_2}\sum_{i_2=1}^{N_2}\sum_{l=1}^{L}L_{sl}(\mathbf{v}_{i_2,l}, \mathbf{v}_{i_2}^*) + \frac{\lambda_3}{N_3}\sum_{i_3=1}^{N_3}\sum_{l=1}^{L}L_{sl}(\mathbf{a}_{i_3,l}, \mathbf{a}_{i_3,l}^*) \quad (2)$$

where: (1) $L_{ce}(\mathbf{u}_{i_1,l}, u_{i_1}^*)$ means the cross entropy loss of the predicted class probabilities $\mathbf{u}_{i_1,l}$ produced by the lth sparse code and the ground truth label

$u_{i_1}^*$, N_1 is the total proposal number. (2) $L_{sl}(\mathbf{v}_{i_2,l}, \mathbf{v}_{i_2}^*)$ means the average of the smooth L1 loss [24] of all elements in vector $\mathbf{v}_{i_2,l} - \mathbf{v}_{i_2}^*$, i.e., the difference between the lth sparse code's prediction of the i_2th vertebra's bounding box coordinates $\mathbf{v}_{i_2,l}$ and the corresponding ground truth $\mathbf{v}_{i_2}^*$, N_2 is the positive proposal number. (3) $L_{sl}(\mathbf{a}_{i_3,l}, \mathbf{a}_{i_3,l}^*)$ means the smooth L1 loss of each predicted sparse code and its ground truth, N_3 is the total sparse code number.

Enhanced Supervision. It is shown in Eq. 2 that our ERN provides L supervision and L predictions (class probabilities, bounding boxes, and sparse codes) for each vertebra for enhanced supervision. Even if the projections of some vertebrae onto one axis overlap, those onto other axes can still show enough discrepancy because the OAs' orientations are diverse. This discrepancy helps distinguish different vertebrae. The ensemble of the L predictions determine the final recognitions, which helps lower risks of overfitting and better generalization performance [23].

Summarized Advantages. Our ERN designs an end-to-end projection guided dictionary learning layer for enhanced supervision. The ensemble of a vertebra's L predicted projections onto different OA's improves the recognition discriminability and handles the FOV/image characteristics challenge.

2.2   Discerning Diagnosis Network (DDN)

Label Consistent Dictionary Learning. Our DDN takes the features cropped by the recognized vertebrae as input, and then flattens them into vectors and solves for the sparse codes (denoted as $\mathbf{b}_i$ in diagnosis task for clarity) as mentioned above. However, since this task boils down to a classification problem, we do not have ground truth sparse codes for supervision. Thus, we impose a label consistency constraint to enhance discriminative capability by introducing the discriminative coefficient $\mathbf{q}_i \in \mathbb{R}^K$ for each vertebrae as follows:

Firstly, we uniformly assign labels to each element of $\mathbf{q}_i$; then, elements whose labels are the same with the ith proposal are assigned 1 and the others 0. In our work, $\mathbf{q}_i$ has K elements and 2 diagnosis labels (non-tumor and tumor), thus, the first $\frac{K}{2}$ elements in $\mathbf{q}_i$ are assigned label 0, and the last $\frac{K}{2}$ elements label 1; therefore, if the ith proposal has label 0, then the first $\frac{K}{2}$ elements in $\mathbf{q}_i$ are set to 1 and the others 0. $\mathbf{T}$ is the transformation matrix to predict sparse coefficients using $\mathbf{b}_i$. Thus, we have the label consistent loss term by calculating the Mahalanobis distance between the optimal sparse code $\mathbf{b}_i^*$ and predicted $\mathbf{b}_i$:

$$\|\mathbf{q}_i - \mathbf{T}\mathbf{b}_i\|_F^2 = \|\mathbf{T}\mathbf{b}_i^* - \mathbf{T}\mathbf{b}_i\|_F^2 = (\mathbf{b}_i^* - \mathbf{b}_i)^T \mathbf{T}^T \mathbf{T} (\mathbf{b}_i^* - \mathbf{b}_i) \qquad (3)$$

In Eq. 3, $\mathbf{T}^T\mathbf{T}$ can be regarded as the covariance matrix in the definition of Mahalanobis distance. Although we do not explicitly know $\mathbf{b}_i^*$ in classification tasks, this term urges $\mathbf{b}_i$ to be close to it. Furthermore, since proposals with the same diagnosis labels have same $\mathbf{q}_i$'s, this term urges the Mahalanobis distance among their $\mathbf{b}_i$'s to be small. Since Mahalanobis distance has a strong discriminative capacity between data classes, the sparse codes are mapped to a more discriminative feature space than the input space. Thus, the label consistency dictionary learning mitigates diagnosis challenge [17].

Loss Function. We formulate the label consistent loss term (Eq. 3) into the total loss:

$$L_d = \frac{\lambda_4}{N_2} \sum_{i=1}^{N_2} L_{ce}(\mathbf{c}_i, c_i^*) + \frac{\lambda_5}{N_2} \sum_{i=1}^{N_2} L_{sl}(\mathbf{q}_i, \mathbf{Tb}_i) \qquad (4)$$

where: (1) $L_{ce}(\mathbf{c}_i, c_i^*)$ is the cross entropy classification loss of the predicted cancer logits $\mathbf{c}_i$ (which is obtained by simply feeding the sparse code $\mathbf{b}_i$ into a fully connected layer) and the ground truth diagnosis label c_i^*. N_2 is the positive proposal number as mentioned above. (2) $L_{sl}(\mathbf{q}_i, \mathbf{Tb}_i)$ is the label consistent loss term. We use the smooth L1 loss in Eq. 4 instead of the L2 loss in Eq. 3 to prevent exploding gradients when training together with the CNN's.

Summarized Advantages. Our DDN purposes a label consistency constraint to prompt recognized vertebrae of the same diagnosis labels to be closer (i.e., the Mahalanobis distances among their sparse codes calculated using the features inside their bounding boxes are smaller). This tackles the interference information challenge and alleviates the tumor appearance challenge.

3   Experiments and Results

Dataset and Ground Truth Annotations. A challenging dataset consisting of 600 spinal MRI images of ~163 patients has been collected to demonstrate the effectiveness of our DECIDE for CVTD. The dataset contains arbitrary MRI images of thoracic, lumbar, and sacrum vertebrae of 6 different FOV's. For each patient, 3~4 slices covering all vertebrae are chosen from 3D scans and resized to 512×512. Our dataset contains 4600 vertebrae, 818 of them are invaded by tumors. An experienced oncologist for spinal tumors has carefully labeled all vertebrae invaded by tumors twice with a temporal interval of one month. The second annotation is blinded to his initial annotation, which is used to assess intra-operator variability. The manual labels are used to set up the ground truth for training. If the two manual annotations have different labels (one indicates tumor and the other not), the ground truth is regarded as "positive" (invaded by tumors) for training.

Evaluation Metrics. Standard five-fold cross-validation is used to evaluate our study. For the recognition performance, we follow the most recent vertebrae recognition work [5] using Image recognition accuracy (IRA), Identification rate (IDR), and mAP_{75} as evaluation metrics to respectively measure patient-wise accuracy, vertebrae-wise accuracy, and comprehensive classification-location performance. For the diagnosis performance, the ROC curve is used as the main evaluation metrics. Discussions based on this curve, such as area under curve (AUC), accuracy, precision, and recall are also conducted.

High Recognition and Diagnosis Performances. Figure 3(a) shows that in MRI slices of various image FOV/characteristics and vertebrae appearance,

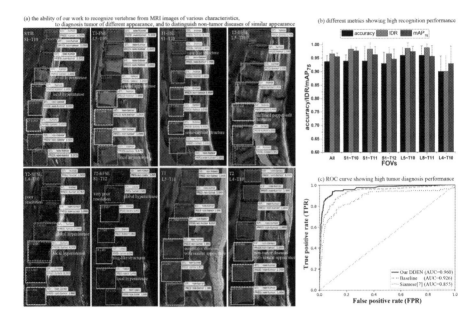

Fig. 3. Different metrics showing the effectiveness of our work. (a) Our network can accurately recognize vertebrae from images of different FOV and characteristics; it can also distinguish tumors from non-tumor diseases of various appearances. (b) Different metrics demonstrating high recognition performance. (c) The ROC curve and AUC value showing high diagnosis performance.

the recognized vertebrae bounding boxes (dashed) overlaps well with the ground truth boxes (solid) of the correct labels (colors). Furthermore, the automatic diagnosis results (PRED) are generally the same as the ground truth (GT). In Fig. 3(b), the black, red, and blue bars show high IRA (overall: 0.938, individual FOV's: >0.9), IDR (overall: 0.966, individual FOV's: >0.9), and mAP_{75} (overall: 0.956, individual FOV's: >0.92), which means that the recognition work produces very few wrongly classified, missing, or false positive recognitions. Figure 3(c) shows that our work achieves satisfactory tumor diagnosis ROC curve with an AUC of 0.96. If we regard vertebrae with predicted tumor probability greater than 0.5 as "positive" (suffering from tumors), we get a diagnostic accuracy of 0.939, precision of 0.860, and recall of 0.796. These results are comparable with those of the oncologist (second column in Table 1).

Comparison with the State-of-the-Art. Although no previous work has performed CVTD, we compare our work with those performing single recognition [5,8,24] or diagnosis [7] task. We also conduct ablation experiments using baseline methods without the dictionary learning layers. As shown in Table 1, DECIDE outperforms all compared methods in both recognition (first three rows) and diagnosis (last four rows) tasks. For recognition, IRA, IDR, and mAP_{75} all benefit from the enhanced supervision provided by embedded dic-

Table 1. Superiority of DECIDE to state-of-the-art methods and the ablation experiment without embedded dictionary.

Method	IRA	IDR	mAP_{75}	AUC	Accuracy	Precision	Recall
Our DECIDE	0.928 ± 0.021	0.956 ± 0.014	0.946 ± 0.013	0.960 ± 0.073	0.939 ± 0.102	0.860 ± 0.168	0.796 ± 0.183
Intra-observer	–	–	–	–	0.964	0.874	0.838
Ablation	0.911 ± 0.035	0.947 ± 0.047	0.925 ± 0.028	0.926 ± 0.071	0.922 ± 0.084	0.830 ± 0.177	0.731 ± 0.228
Siamese [7]	–	–	–	0.855 ± 0.121	0.884 ± 0.089	0.774 ± 0.197	0.623 ± 0.225
Hi-scene [5]	0.878 ± 0.048	0.930 ± 0.053	0.923 ± 0.039	–	–	–	–
DI2IN [8]	0.803 ± 0.149	0.904 ± 0.115	–	–	–	–	–
Faster-RCNN [24]	0.750 ± 0.138	0.869 ± 0.104	0.848 ± 0.146	–	–	–	–

tionary. By comparing with the ablation study where the supervision by the projections on the L axes is disabled (Rows 1∼3 of the third column in Table 1), the advantage of enhanced supervision is demonstrated; even for the most difficult FOV T10∼L4, the performance is only slightly lower (Fig. 3(b)). Furthermore, by comparing with [8] that uses dictionary learning as post-processing for landmark refinement, our embedded dictionary is more beneficial because it exploits the projections and introduces the ensemble of multiple predictions, which improves vertebrae recognition discriminability when their appearances are changed significantly by tumors. For diagnosis, the superiority of DECIDE to the ablation experiments (Rows 4∼7 of the third column in Table 1) shows that the label consistency strategy improves the diagnostic discriminability of sparse codes. Also, our DECIDE eliminates manual selection of patch size and resolution as in [7]; instead, the recognition network automatically deals with these issues and provides features adaptive to different vertebrae for succeeding diagnosis, i.e., the workflow of our DECIDE reinforces mutual benefits between two tasks and improves tumor diagnosis performance.

4   Conclusion

We have designed a discriminative dictionary-embedded network (DECIDE) as a novel clinical tool for comprehensive vertebrae tumor diagnosis (CVTD) from raw input MRI images. Our DECIDE integrates dictionary learning into both recognition and diagnosis networks for enhanced supervision and discriminative representations. The effectiveness of our network, as well as its advantage to the state-of-the-art, are demonstrated by extensive experiments.

References

1. Katherine, N., Theresa, A., Laurie, K.: Cancer to bone: a fatal attraction. Nat. Rev. Cancer **11**(6), 411–425 (2011)
2. Mundy, G.: Metastasis to bone: causes, consequences and therapeutic opportunities. Nat. Rev. Cancer **2**(8), 584–593 (2002)

3. O'Sullivan, G., Carty, F., Cronin, C.: Imaging of bone metastasis: an update. World J. Radiol. **7**(8), 202–211 (2015)
4. Shelly, S., Avi, B., Orit, S., Michal, M., Hayit, G., Eyal, K.: Convolutional neural networks for radiologic images: a radiologist's guide. Radiology **290**(3), 590–606 (2019)
5. Zhao, S., Wu, X., Chen, B., Li, S.: Automatic vertebrae recognition from arbitrary spine MRI images by a hierarchical self-calibration detection framework. MICCAI 2019. LNCS, vol. 11767, pp. 316–325. Springer, Cham (2019). https://doi.org/10.1007/978-3-030-32251-9_35
6. Philip, F.: Patient Safety in Surgery, 2nd edn. Springer, London (2014). https://doi.org/10.1007/978-1-4471-4369-7
7. Wang, J., Fang, Z., Lang, N., Yuan, H., Su, M., Baldi, P.: A multi-resolution approach for spinal metastasis detection using deep Siamese neural networks. Comput. Biol. Med. **84**(1), 137–146 (2014)
8. Yang, D., et al.: Automatic vertebra labeling in large-scale 3D CT using deep image-to-image network with message passing and sparsity regularization. In: Niethammer, M., et al. (eds.) IPMI 2017. LNCS, vol. 10265, pp. 633–644. Springer, Cham (2017). https://doi.org/10.1007/978-3-319-59050-9_50
9. Chen, H., et al.: Automatic localization and identification of vertebrae in spine CT via a joint learning model with deep neural networks. In: Navab, N., Hornegger, J., Wells, W., Frangi, A. (eds.) Medical Image Computing and Computer-Assisted Intervention - MICCAI 2015. Lecture Notes in Computer Science, vol. 9349, pp. 515–522. Springer, Cham (2015). https://doi.org/10.1007/978-3-319-24553-9_63
10. Lootus, M., Kadir, T., Zisserman, A.: Vertebrae detection and labeling in lumbar MR images. In: Yao, J., Klinder, T., Li, S. (eds.) Computational Methods and Clinical Applications for Spine Imaging. LNCVB, vol. 17, pp. 219–230. Springer, Cham (2014). https://doi.org/10.1007/978-3-319-07269-2_19
11. Wiese, T., Burns, J., Yao, J., Summers, R.: Computer-aided detection of sclerotic bone metastases in the spine using watershed algorithm and support vector machines. In: 2011 IEEE International Symposium on Biomedical Imaging: From Nano to Macro, pp. 152–155. IEEE, Chicago, IL, USA (2011). https://doi.org/10.1109/ISBI.2011.5872376
12. Burns, J., Yao, J., Wiese, T., Munoz, H., Jones, E., Summers, R.: Automated detection of sclerotic metastases in the thoracolumbar spine at CT. Radiology **268**(1), 69–78 (2013)
13. Zhao, S., et al.: Robust segmentation of intima-media borders with different morphologies and dynamics during the cardiac cycle. IEEE J. Biomed. Health Inf. **22**(5), 1571–1582 (2018)
14. Gao, Z., et al.: Robust estimation of carotid artery wall motion using the elasticity-based state-space approach. Med. Image Anal. **37**, 1–21 (2017)
15. Shah, L., Salzman, K.: Imaging of spinal metastatic disease. Int. J. Surg. Oncol. **2011**(1), 1–13 (2011)
16. Sun, X., Nasrabadi, N., Tran, T.: Supervised deep sparse coding networks for image classification. IEEE Trans. Image Process. **29**, 405–418 (2019)
17. Jiang, Z., Lin, Z., Davis, L.: Label consistent K-SVD: learning a discriminative dictionary for recognition. IEEE Trans. Pattern Anal. Mach. Intell. **35**(11), 2651–2664 (2013)
18. Xue, Y., Bigras, G., Hugh, J., Ray, N.: Training convolutional neural networks and compressed sensing end-to-end for microscopy cell detection. IEEE Trans. Med. Imaging **38**(11), 2632–2641 (2019)

19. Aharon, M., Elad, M., Bruckstein, A.: K-SVD: an algorithm for designing overcomplete dictionaries for sparse representation. IEEE Trans. Image Process. **54**(11), 4311–4322 (2006)
20. Zhao, S., Wu, X., Chen, B., Li, S.: Automatic spondylolisthesis grading from MRIs across modalities using faster adversarial recognition network. Med. Image Anal. **58**, 101533 (2019)
21. Adam, C., Andrew, N.: The importance of encoding versus training with sparse coding and vector quantization. In: Proceedings of the 28th International Conference on Machine Learning (ICML2011), pp. 911–928. Omnipress, Bellevue, Washington (2011)
22. Xie, H., Li, J., Xue, H.: A survey of dimensionality reduction techniques based on random projection. arXiv preprint arXiv:1706.04371 (2017)
23. Quan, Y., Xu, Y., Sun, Y., Huang, Y., Ji, H.: Sparse coding for classification via discrimination ensemble. In: Proceedings of the IEEE Conference on Computer Vision and Pattern Recognition, pp. 5839–5847. IEEE, Las Vegas (2016). https://doi.org/10.1109/CVPR.2016.629
24. Ren, S., He, K., Girshick, R., Sun, J.: Faster R-CNN: towards real-time object detection with region proposal networks. In: Advances in Neural Information Processing Systems (NIPS 2015), pp. 91–99. Springer, Montreal, Quebec, Canada (2015)
25. Gregor, K., LeCun, Y.: Learning fast approximations of sparse coding. In: Proceedings of the 27th International Conference on International Conference on Machine Learning (ICML2010), pp. 399–406. Omnipress, Madison (2010)

Multi-vertebrae Segmentation from Arbitrary Spine MR Images Under Global View

Heyou Chang[1,2,3], Shen Zhao[3], Hao Zheng[1], Yang Chen[2(✉)], and Shuo Li[3(✉)]

[1] School of Information Engineering,
Nanjing XiaoZhuang University, Nanjing, Jiangsu, China
[2] School of Computer Science and Engineering, Southeast University,
Nanjing, Jiangsu, China
chenyang.list@seu.edu.com
[3] Department of Medical Imaging and Medical Biophysics, Western University,
London, ON, Canada
slishuo@gmail.com

Abstract. Multi-vertebrae segmentation plays an important role in spine diseases diagnosis and treatment planning. Global spatial dependencies between vertebrae are essential prior information for automatic multi-vertebrae segmentation. However, due to the lack of global information, previous methods have to localize specific vertebrae regions first, then segment and recognize the vertebrae in the region, resulting in a reduction in feature reuse and increase in computation. In this paper, we propose to leverage both global spatial and label information for multi-vertebrae segmentation from arbitrary MR images in one go. Specifically, a spatial graph convolutional network (GCN) is designed to first automatically learn an adjacency matrix and construct a graph on local feature maps, then adopt stacked GCN to capture the global spatial relationships between vertebrae. A label attention network is built to predict the appearance probabilities of all vertebrae using attention mechanism to reduce the ambiguity caused by variant FOV or similar appearances of adjacent vertebrae. The proposed method is trained in an end-to-end manner and evaluated on a challenging dataset of 292 MRI scans with various fields of view, image characteristics and vertebra deformations. The experimental results show that our method achieves high performance (89.28 ± 5.21 of IDR and $85.37 \pm 4.09\%$ of mIoU) from arbitrary input images.

Keywords: Multi-vertebrae segmentation · Global information · Graph convolutional network · Attention network

1 Introduction

Automatic multi-vertebrae segmentation (*i.e.*, partitioning and labeling each of the vertebrae in an input image) from arbitrary magnetic resonance imaging (MRI) is an important step for spinal image analysis and intervention,

© Springer Nature Switzerland AG 2020
A. L. Martel et al. (Eds.): MICCAI 2020, LNCS 12266, pp. 702–711, 2020.
https://doi.org/10.1007/978-3-030-59725-2_68

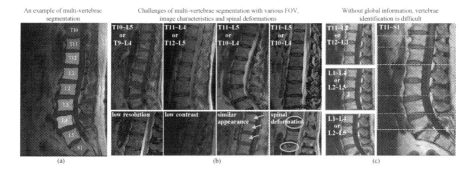

Fig. 1. (a) Each vertebra is partitioned and assigned a unique label (color coded) in multi-vertebrae segmentation. (b) Spine MRI images with various FOV, characteristics and spinal deformations make multi-vertebrae segmentation challenging. (c) Global information plays an important role in multi-vertebrae segmentation. (Color figure online)

e.g., spine diseases diagnosis, surgical treatment planning and locating spinal pathologies [1–3]. Precise segmentation is a challenging task due to high topological shape variations, different resolutions and various field of views (FOV), as shown in Fig. 1. Moreover, the similar structures and appearances in close vertebrae and various spinal deformations increase the difficulty of multi-vertebrae segmentation.

The inherent global position of vertebrae is essential prior information for multi-vertebrae segmentation, which has not been fully exploited. Compared to non-vertebra pixels, the pixels belonging to different vertebrae may have more similar appearance. The relative positions of vertebrae are also fixed. Therefore, it is necessary to exploit the relationship between vertebrae in a global view to accomplish multi-vertebrae segmentation effectively. However, most of existing methods with encoder-decoder structure (*i.e.*, FCNs [4], U-net [5]) have a limited receptive field and cannot capture longer-range relationships between vertebrae. For instance, A. Sekuboyina *et al.* [6] employed a multi-layered perception to locate lumbar vertebrae region and adopted 2D U-net to segment and annotate the lumbar vertebrae in the region. R. Janssens *et al.* [7] trained a regression 3D FCNs to find the bounding box of the lumbar region and a 3D U-net to perform segmentation and recognition. N. Lessmann *et al.* [8] first localized and recognized each vertebra in a sliding-window manner, then performed a binary segmentation (spine vs. background) neural network to segment the vertebrae. [9–12] also proposed other methods such as spine-GAN, cascade amplifier regression network and hierarchical self-calibration detection framework for vertebrae segmentation, detection and/or identification. Due to the lack of global information, these methods consist of multiple phases (vertebra localization and segmentation) with multiple networks to learn, and only work for specific vertebrae segmentation. Taken an arbitrary spine scan as input, there is no end-to-end approach that handles multi-vertebrae segmentation in one go.

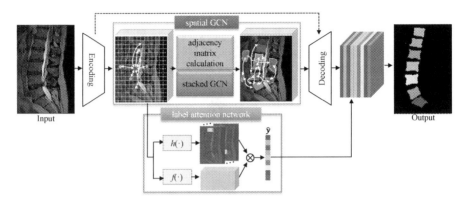

Fig. 2. An overview of the proposed method, which includes (1) a spatial GCN to learn representative features by capturing global spatial information between the vertebrae, (2) a label attention network to reduce the probability of wrong segmentation by exploiting global label information and weighting the output of the decoding network.

Graph convolutional network (GCN) [13] has shown its power for capturing global information of non-grid structure data, and been applied to various computer vision tasks, such as classification [14] and disease prediction [15]. However, these GCN based methods can not be directly applied in vertebrae segmentation because their adjacency matrices are pre-defined and constant, which is improper in segmentation task. Here, we adopt self-attention method to automatically learn the relationships between the pixels and calculate the adjacency matrix based on local feature maps.

In this paper, we propose a novel approach for multi-vertebrae segmentation (MVSeg) with spatial GCN and label attention network, both of which leverage global information to improve segmentation performance. The framework of MVSeg is illustrated in Fig. 2. A spatial GCN is introduced to capture *global spatial dependencies* between any two positions of the feature maps. The proposed spatial GCN could effectively model contextual information and generate representative features by constructing a graph on the local feature maps followed by stacked GCN. Then, a label attention network is designed to exploit *global label information* by generating a probability vector of the vertebrae using attention mechanism. The label attention network can reduce the misclassification caused by FOV variety and similarity of adjacent vertebrae by sharing feature learning with decoding network and weighting the output of the decoding network using the probability vector. By exploiting both global spatial and label information, multi-vertebrae segmentation could benefit from long-range information.

Our main contributions are summarized as follows: (1) We propose a novel method for multi-vertebrae segmentation from an arbitrary MRI by leveraging both global spatial and label information. (2) We design a spatial GCN to capture the global spatial information between the vertebrae and a label attention network to exploit the global label information. By modeling global dependencies, the proposed method could effectively reduce the probability of wrong segmentation.

2   Methodology

The proposed MVSeg mainly consists of two parts: (1) Spatial GCN (Sect. 2.1), which effectively models the inherent dependencies of vertebrae by calculating a adjacency matrix based on local feature maps, and enhances the features' representation capability by globally propagating information using stacked GCN. (2) Label attention network (Sect. 2.2), which generates a label probability vector through attention mechanism. The probability is used as a weight vector and multiplied with the output of decoding network to produce coherent predictions and reduce the probability of wrong segmentation caused by variant FOV and similar appearance of adjacent vertebrae. An encoding network and a decoding network are also included to learn input representations and recover the spatial resolution, respectively. Skip-connections between them are introduces to enable the two networks to share information.

2.1   Spatial GCN for Global Spatial Information

Spatial GCN captures global spatial relationships of pixels over local feature maps and generates representative features through four carefully designed steps: (1) feature maps reduction to save memory; (2) adjacency matrix calculation to build a graph, (3) stacked GCN to capture longer-range information; and (4) element-wise addition to fuse local and global information. The details of spatial GCN is illustrated in Fig. 3.

Given a feature map $\mathbf{X} \in \mathcal{R}^{H \times W \times D}$, we first feed it into a convolution layer with 1×1 filters to reduce the number of feature maps and generate a new feature maps $\hat{\mathbf{X}} \in \mathcal{R}^{H \times W \times d}(d < D)$. Then we reshape $\hat{\mathbf{X}}$ to $\mathcal{R}^{N \times d}$, where $N = H \times W$ is the number of pixels and each row $\hat{\mathbf{x}}_i \in \mathcal{R}^d$ is a feature vector for the i-th pixel. Since each pixel in $\hat{\mathbf{X}}$ corresponds a large patch in the original image and the patches are overlapped, the pixels are related. To calculate the relations between any two pixels, we adopt self-attention method, which outputs a map $\mathbf{S} \in \mathcal{R}^{N \times N}$ by performing a matrix multiplication between $\hat{\mathbf{X}}$ and the transpose of $\hat{\mathbf{X}}$, and applying a softmax layer:

$$\mathbf{S}_{i,j} = \frac{exp(\hat{\mathbf{x}}_i * \hat{\mathbf{x}}_j^T)}{\sum_{j=1}^{N} exp(\hat{\mathbf{x}}_i * \hat{\mathbf{x}}_j^T)} \tag{1}$$

$\mathbf{S}_{i,j}$ represents the value at location (i, j) of $\mathbf{S}$ and measures the similarity of the i-th position and j-th position. The adjacency matrix $\mathbf{A} \in \mathcal{R}^{N \times N}$ is calculated by setting $\mathbf{A}_{i,j} = 0$, if $\mathbf{S}_{i,j} < \tau$; $\mathbf{A}_{i,j} = 1$, otherwise. $\mathbf{A}_{i,j}$ indicates whether i-th and j-th pixels are adjacent or not. τ is a hyper-parameter, which controls the sparseness of $\mathbf{A}$. $\mathbf{A}$ is sparse when τ is large, and vice versa.

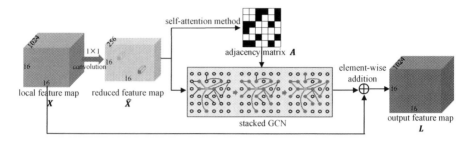

Fig. 3. The details of spatial GCN, which contains four parts: (1) feature maps reduction, (2) adjacency matrix calculation, (3) stacked GCN, and (4) element-wise addition.

Then, GCN is adopted to excavate the global spatial information over the feature map $\hat{\mathbf{X}}$, which can aggregate information from adjacent pixels and output embedding vectors of pixels based on their connections. For a one-layer GCN, the new k-dimensional feature matrix $\mathbf{L}^1 \in \mathcal{R}^{N \times k}$ is computed as

$$\mathbf{L}^1 = f(\mathbf{D}^{-1/2}\mathbf{A}\mathbf{D}^{-1/2}\hat{\mathbf{X}}\mathbf{W}_0) \tag{2}$$

where $\mathbf{D}$ is a diagonal matrix with $\mathbf{D}_{i,i} = \sum_j \mathbf{A}_{i,j}$. $\mathbf{W}_0 \in \mathcal{R}^{d \times k}$ is a learnable weight matrix, and f is an activation function. The i-th row of $\mathbf{L}^1$ represents the new feature vector of the i-th pixel, which is computed as the weighted sum of its directly adjacent pixels in $\hat{\mathbf{X}}$. GCN with one layer of convolution can only capture information about near neighbors. To incorporate larger neighborhoods information, multiple GCN layers are stacked as follows

$$\mathbf{L}^{i+1} = f(\mathbf{D}^{-1/2}\mathbf{A}\mathbf{D}^{-1/2}\mathbf{L}^i\mathbf{W}_i) \tag{3}$$

where $\mathbf{W}_i$ is the weight matrix for the i-th layer, and $\mathbf{L}^0 = \hat{\mathbf{X}}$. In the experiment, to feasibly model global information, three layers of GCN are stacked. It can be inferred that each position at $\mathbf{L}^3$ is a weighted sum of features across all positions. The activation function is set as $tanh$ at each layer.

The output of the stacked GCN is first reshaped to $H \times W \times k$ and resized to the same size of $\mathbf{X}$ through 1×1 convolution, then added with $\mathbf{X}$ to fuse the original information. It can be seen that the output of the spatial GCN captures both global and local spatial information, which could improve the features' representation ability.

Summarized Advantages. The spatial GCN 1) constructs a graph based on local feature maps by regarding each pixel as a node and computing the adjacency matrix by self-attention method, and 2) executes stacked GCN on the graph to acquire larger receptive fields and more representative features.

2.2   Label Attention Network for Global Label Information

Label attention network generates a label probability vector using attention mechanism and handles the misclassification problem caused by the morphological similarity of adjacent vertebrae and the diversity of FOV. By weighting the output of the decoding network using the label probability vector, the network could produce coherent segmentation results.

The label attention network takes the output $\mathbf{L} \in \mathcal{R}^{H \times W \times D}$ of the spatial GCN as input, and outputs a label probability vector with the dimensionality of C, where each dimension corresponds to one vertebra. In the experiment, an image contains up to nine vertebrae: T10, T11, T12, L1, L2, L3, L4, L5 and S1. Then C is set to 10 (back ground + nine vertebrae). We first design an attention estimator $h(\cdot)$ to automatically generate a label attention map $\mathbf{B} = h(\mathbf{L})$, where $\mathbf{B} \in \mathcal{R}^{H \times W \times C}$. $h(\cdot)$ consists of 3 convolution layers with 512 kernels of size 1×1, 512 kernels of size 3×3, and C kernels of size 1×1, respectively, followed by batch normalization and ReLu nonlinearity operations except the last convolution layer. Then, $\mathbf{B}$ is normalized per channel by $B_{i,j,c} = \frac{exp(B_{i,j,c})}{\sum_{i,j} exp(B_{i,j,c})}$. Intuitively, the value of $B_{i,j,c}$ should be higher if label c is related to the input image and the image region corresponding to location (i,j) is related to label c. After that, the label attention map $\mathbf{B}$ is used to calculate a weighted feature vector for each label by $\mathbf{f}_c = \sum_{i,j} B_{i,j,c} \mathbf{L}_{i,j,:}$, where $\mathbf{L}_{i,j,:} \in \mathcal{R}^D$. The weighted feature vector $\mathbf{f}_c$ is related to image regions corresponding to label c, which could improve the robustness and accuracy of classification.

To predict the appearance of each label, a linear classifier is learnt for each label by $\hat{y}_c = \mathbf{W}_c \mathbf{f}_c + \mathbf{b}_c$, where $\mathbf{W}_c$ and $\mathbf{b}_c$ are classifier parameters for label c. Considering $\hat{y}_c = \sum_{i,j} B_{i,j,c}(\mathbf{W}_c \mathbf{L}_{i,j,:} + \mathbf{b}_c)$, which can be viewed as performing a multiplication and weighted aggregation at every location of $\mathbf{L}$, the linear classifiers for all labels are modeled as a convolution layer $f(\cdot)$ with C kernels of 1×1 in the implementation. The output of $f(\cdot)$ multiplies with $\mathbf{B}$ element-wisely, and generates a label confidence vector $\hat{\mathbf{y}} = [\hat{y}_1, ..., \hat{y}_C]$ by performing sum-pooling and sigmoid nonlinearity operations. The loss function of the label attention network is defined as $\ell_{classification} = -\sum_{i=1}^{C}(\hat{y}_i * log y_i + (1 - \hat{y}_i) * log(1 - y_i))$, where y_i represents the ground label truth. $y_i = 1$ if and only if the i-th vertebra is associated with image, and 0 otherwise.

Then, the label probability vector $\hat{\mathbf{y}}$ is used as a weight vector and multiplied with the output of the decoding network $\mathbf{S} \in \mathcal{R}^{256 \times 256 \times C}$ by $\hat{\mathbf{S}}_{:,:,c} = \hat{y}_c \mathbf{S}_{:,:,c}$. The cross-entropy loss between the ground segmentation truth and $\hat{\mathbf{S}}$ is adopted as the segmentation loss $\ell_{segmentation}$. Finally, the whole network is trained using a loss function consisting of the segmentation loss and classification loss:

$$\ell = \lambda \cdot \ell_{segmentation} + (1 - \lambda) \cdot \ell_{classification} \qquad (4)$$

where λ is a parameter to balance the segmentation and classification loss.

Summarized Advantages. The label attention network 1) generates a label probability vector based on multiple linear classifiers and weighted feature vectors using attention mechanism, and 2) multiplies the segmentation output with the probability vector to improve the segmentation performance.

3   Experiments and Results

Dataset. The proposed MVSeg is evaluated on a challenging dataset, which consists of 292 arbitrary MR images with various FOV, image characteristics and vertebrae deformations. There are seven kinds of FOV in all, *i.e.*, T10~S1(63), T11~S1(105), T12~S1(64), T10~L5(19), T10~L4(12), T11~L5(15) and T11~L4 (14). The image number of each kind is listed in the brackets. It can be seen that the distributions of FOV in the dataset are quite different. The images are automatically extracted from each 3D MRI scan and resized to 256×256. The proposed approach is implemented using the Tensorflow backend and trained on Nvidia P100 GPUs with 16 GB memory for 10000 steps with a fixed learning rate of 0.0001.

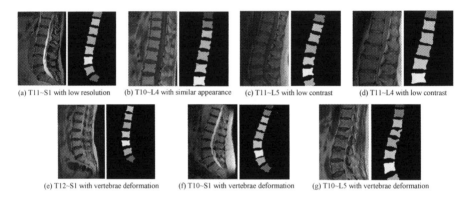

(a) T11~S1 with low resolution   (b) T10~L4 with similar appearance   (c) T11~L5 with low contrast   (d) T11~L4 with low contrast

(e) T12~S1 with vertebrae deformation   (f) T10~S1 with vertebrae deformation   (g) T10~L5 with vertebrae deformation

Fig. 4. The proposed MVSeg achieves high multi-vertebrae segmentation performance on a challenging dataset with varied FOV, different image characteristics and vertebrae deformations. Left: input image. Right: segmentation result. The best view in color.

Evaluation Metrics. Standard five-fold cross-validation is adopted for evaluation, and four metrics are used to evaluate the segmentation performance: 1) average precision (AP), which is the percentage of pixels in the total pixels that are segmented correctly in all images; 2) identification rate (IDR), which measures the accuracy of individual vertebra recognition; 3) dice coefficient (Dice), which quantifies the similarity between the segmentation and ground truth of all vertebrae in all images; and 4) mean region intersection over union (mIoU), which is the average IoU of the segmentation and ground truth mask of all vertebrae in all images.

Qualitative Demonstration. Figure 4 demonstrates that MVSeg achieves high multi-vertebrae segmentation performance. The partition of each vertebrae is highly precise and the recognition of each vertebra is correct in arbitrary images with varied FOV, image characteristics and vertebrae deformations. For example, (b), (c) and (e) have 7 vertebrae, but their FOV and image characteristics are different.

Quantitative Analysis. Table 1 gives the mean accuracy and variance of MVSeg and U-net, which is a well-known segmentation network for biomedical images. Moreover, to verify the effectiveness of the spatial GCN and the label attention network on segmentation accuracy, the performance of MVSeg without spatial GCN (MVSeg$_{/s}$) and MVSeg without label attention network (MVSeg$_{/l}$) is also listed in Table 1. It is worth noting that the parameter settings in the encoding and decoding path of all methods are the same. From Table 1, we can see that MVSeg achieves the best performance on all of the four metrics. Compared with U-net, the improvements of MVSeg on AP, IDR, Dice and mIoU are **1.47%**, **8.45%**, **8.95%** and **9.13%**, respectively. Both MVSeg$_{/s}$ and MVSeg$_{/l}$ perform better than U-net with higher mean accuracy and lower variance, which indicate the validity of the two proposed modules in multi-vertebrae segmentation.

Table 1. The mean accuracy and variance (%) of different methods for multi-vertebrae segmentation

	AP	IDR	Dice	mIoU
U-net	97.33 ± 2.08	80.83 ± 11.85	78.14 ± 10.91	76.24 ± 12.17
MVSeg$_{/l}$	98.53 ± 1.35	85.36 ± 5.25	81.93 ± 4.91	80.65 ± 5.94
MVSeg$_{/s}$	98.64 ± 0.24	88.50 ± 2.64	84.80 ± 2.97	82.58 ± 2.42
MVSeg	$\mathbf{98.80 \pm 0.49}$	$\mathbf{89.28 \pm 5.21}$	$\mathbf{87.09 \pm 4.09}$	$\mathbf{85.37 \pm 4.09}$

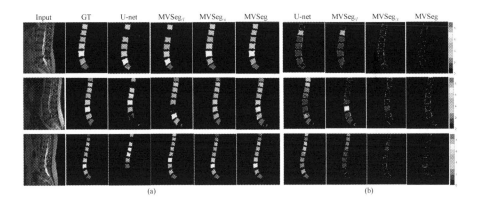

Fig. 5. Some multi-vertebrae segmentation results of different methods. (a) segmentation results, (b) error images. The best view in color.

Figure 5 presents visualization of some segmentation results of testing images. The first and second columns represent input image and ground truth (GT), respectively. The third column is the segmentation result of U-net, which has poor precision and accuracy of segmentation. The fourth and fifth columns are the results of MVSeg$_{/l}$ and MVSeg$_{/s}$, respectively. Although their performance is better than U-net, the recognitions of middle vertebrae are mixed with each other and the boundary of some vertebrae is not accurate. By fusing both global position and label information, MVSeg can segment and label the vertebrae more precisely (the last column in Fig. 5(a)). Figure 5(b) shows the error images between the outputs of different methods and the ground truth.

4   Conclusion

In this paper, we propose a novel method for multi-vertebrae segmentation from arbitrary MRI by incorporating global spatial and label information. In MVSeg, a spatial GCN is constructed to enhance the features' representation ability by constructing a graph and adopting stacked GCN to capture global position information; and a label attention network is designed to exploit global label information using attention mechanism. The method is trained in an end-to-end manner and verified on a challenging dataset with various FOV, image characteristics and vertebrae deformations. The experimental results demonstrate the effectiveness of MVSeg.

Acknowledgement. This work was supported in part by the National Natural Science Fund of China under Grant 61806098 and 61976118, in part by the State's Key Project of Research and Development Plan under Grant 2017YFA0104302, 2017YFC0109202 and 2017YFC0107900.

References

1. Cai, Y., Osman, S., Sharma, M., Landis, M., Li, S.: Multi-modality vertebra recognition in arbitrary views using 3D deformable hierarchical model. IEEE Trans. Med. Imaging **34**(8), 1676–1693 (2015). https://doi.org/10.1109/TMI.2015.2392054
2. Han, Z., Wei, B., Leung, S., Nachum, I., Laidley, D., Li, S.: Automated pathogenesis-based diagnosis of lumbar neural foraminal stenosis via deep multiscale multitask learning. Neuroinformatics **16**(3–4), 325–337 (2018)
3. Liao, H., Mesfin, A., Luo, J.: Joint vertebrae identification and localization in spinal CT images by combining short-and long-range contextual information. IEEE Trans. Med. Imaging **37**(5), 1266–1275 (2018). https://doi.org/10.1109/TMI.2018.2798293
4. Long, J., Shelhamer, E., Darrell, T.: Fully convolutional networks for semantic segmentation. In: Proceedings of the IEEE Conference on Computer Vision and Pattern Recognition, pp. 3431–3440. IEEE Press, Boston (2015). https://doi.org/10.1109/CVPR.2015.7298965
5. Ronneberger, O., Fischer, P., Brox, T.: U-Net: convolutional networks for biomedical image segmentation. In: Navab, N., Hornegger, J., Wells, W.M., Frangi, A.F. (eds.) MICCAI 2015. LNCS, vol. 9351, pp. 234–241. Springer, Cham (2015). https://doi.org/10.1007/978-3-319-24574-4_28

6. Sekuboyina, A., Valentinitsch, A., Kirschke, J.S., Menze, B.H.: A localisation-segmentation approach for multi-label annotation of lumbar vertebrae using deep nets. arXiv preprint arXiv:1703.04347 (2017)
7. Janssens, R., Zeng, G., Zheng, G.: Fully automatic segmentation of lumbar vertebrae from CT images using cascaded 3D fully convolutional networks. In: 15th IEEE International Symposium on Biomedical Imaging, pp. 893–897. IEEE Press, Washington (2018). https://doi.org/10.1109/ISBI.2018.8363715
8. Lessmann, N., van Ginneken, B., Isgum, I.: Iterative convolutional neural networks for automatic vertebra identification and segmentation in CT images. In: Medical Imaging 2018: Image Processing, vol. 10574, p. 1057408. International Society for Optics and Photonics, Huston (2018). https://doi.org/10.1117/12.2292731
9. Han, Z., Wei, B., Mercado, A., Leung, S., Li, S.: Spine-GAN: semantic segmentation of multiple spinal structures. Med. Image Anal. **50**, 23–35 (2018)
10. Pang, S., Leung, S., Ben Nachum, I., Feng, Q., Li, S.: Direct automated quantitative measurement of spine via cascade amplifier regression network. In: Frangi, A.F., Schnabel, J.A., Davatzikos, C., Alberola-López, C., Fichtinger, G. (eds.) MICCAI 2018. LNCS, vol. 11071, pp. 940–948. Springer, Cham (2018). https://doi.org/10.1007/978-3-030-00934-2_104
11. He, X., Zhang, H., Landis, M., Sharma, M., Warrington, J., Li, S.: Unsupervised boundary delineation of spinal neural foramina using a multi-feature and adaptive spectral segmentation. Med. Image Anal. **36**, 22–40 (2017)
12. Zhao, S., Wu, X., Chen, B., Li, S.: Automatic vertebrae recognition from arbitrary spine MRI images by a hierarchical self-calibration detection framework. In: Shen, D., et al. (eds.) MICCAI 2019. LNCS, vol. 11767, pp. 316–325. Springer, Cham (2019). https://doi.org/10.1007/978-3-030-32251-9_35
13. Defferrard, M., Bresson, X., Vandergheynst, P.: Convolutional neural networks on graphs with fast localized spectral filtering. In: Advances in Neural Information Processing Systems, pp. 3844–3852. Curran Associates, Barcelona (2016)
14. Zhang, M., Cui, Z., Neumann, M., Chen, Y.: An end-to-end deep learning architecture for graph classification. In: 32th Thirty-Second AAAI Conference on Artificial Intelligence, pp. 4438–4445. AAAI Press, New Orleans (2018)
15. Kazi, A., et al.: Graph convolution based attention model for personalized disease prediction. In: Shen, D., et al. (eds.) MICCAI 2019. LNCS, vol. 11767, pp. 122–130. Springer, Cham (2019). https://doi.org/10.1007/978-3-030-32251-9_14

A Convolutional Approach to Vertebrae Detection and Labelling in Whole Spine MRI

Rhydian Windsor[1(✉)], Amir Jamaludin[1], Timor Kadir[2],
and Andrew Zisserman[1]

[1] Visual Geometry Group, Department of Engineering Science,
University of Oxford, Oxford, UK
rhydian@robots.ox.ac.uk
[2] Plexalis Ltd., Thame, UK

Abstract. We propose a novel convolutional method for the detection and identification of vertebrae in whole spine MRIs. This involves using a learnt vector field to group detected vertebrae corners together into individual vertebral bodies and convolutional image-to-image translation followed by beam search to label vertebral levels in a self-consistent manner. The method can be applied without modification to lumbar, cervical and thoracic-only scans across a range of different MR sequences. The resulting system achieves 98.1% detection rate and 96.5% identification rate on a challenging clinical dataset of whole spine scans and matches or exceeds the performance of previous systems of detecting and labelling vertebrae in lumbar-only scans. Finally, we demonstrate the clinical applicability of this method, using it for automated scoliosis detection in both lumbar and whole spine MR scans.

Keywords: Vertebral bodies · Whole spine MRI · Scoliosis

1 Introduction

The objective of this paper is automated vertebrae detection and identification of vertebral levels. This is an important task for several reasons. Firstly, automated diagnosis of many spinal diseases such as disc degeneration [9,10] or spinal stenosis [13] relies on accurate localisation of vertebral structures or, in the case of pathological scoliosis, lordosis and kyphosis [8], analysing the geometry of the spinal column. Secondly, vertebral bodies can be used to infer other spinal structures of interest such as the spinal cord or ribs. Finally, vertebrae can act as points to allow registration between different scans.

There are several issues that make this task challenging. One of the most obvious is that vertebrae are highly repetitive and hence distinguishing between

Electronic supplementary material The online version of this chapter (https://doi.org/10.1007/978-3-030-59725-2_69) contains supplementary material, which is available to authorized users.

© Springer Nature Switzerland AG 2020
A. L. Martel et al. (Eds.): MICCAI 2020, LNCS 12266, pp. 712–722, 2020.
https://doi.org/10.1007/978-3-030-59725-2_69

different levels can be hard. Labelling by simply counting down from the C2 vertebra is problematic as it assumes that all vertebrae have been detected, C2 is visible, and that every patient has the same number of vertebrae which is not always true. Furthermore, for clinical use, labelling must be robust to: variations in spinal anatomy (such as collapsed vertebrae, hemivertebrae and fused vertebra); vertebrae numbers – around 11.3% of the population have one more or one less mobile vertebra [18]); different imaging parameters including MR weighting (e.g. T1, T2, STIR, TIRM and FLAIR); fields of view (e.g. lumbar, whole spine scans); scan resolution and also number/thickness of slices in the scan.

This paper proposes a new approach to this challenge, in particular in the case of 3-D sagittal *whole spine* clinical MRIs which are important for diagnosing several diseases such as ankylosing spondylitis and multiple myeloma. We make the following contributions to the tasks of *vertebral body detection* and *vertebral level labelling* in clinical MRIs: We propose a new convolutional method of detection based on localising the corners and centroids of vertebrae and then grouping them together (Sect. 2); We reformulate the labelling task as a convolution enhancement followed by a language modelling inspired sequential correction, removing the need for a recurrent network and showing robustness to variations in vertebra numbers (Sect. 3); We show that the resulting system is robust to a variety of fields of view and pathologies by evaluating on a large clinical dataset of MR lumbar scans, and also, for the first time, on whole spine scans. We achieve state-of-the-art performance at vertebra identification in both (Sect. 4); We demonstrate a clinical application of this system by using it to automatically detect cases of scoliosis (Sect. 5).

Related Work: There have been several approaches to automated detection and labelling of vertebrae in MRIs although most focus on fixed fields of view, e.g. lumbar or cervical scans only [3,9,12,13]. Previous vertebrae labelling methods tend to rely on either heuristic-based graphical models [3,11] or assuming the bottom vertebra is S1 and counting up [13]. Zhao et al. [21] perform labelling in MRIs with arbitrary fields of view, but only in the lower spine (from S1-T12 to L4-T10). Windsor and Jamaludin [19] report a method of detecting, though not labelling, vertebrae in full spine scans iteratively but also require the location of the S1 vertebra for initialization. Cai et al. [2] also perform detection and labelling by a hierarchical deformation model and even report success in a single full spine scan but do not evaluate the performance quantitatively on a whole spine dataset. Furthermore such models are slow to apply and make strong assumptions on spinal geometry such as a fixed number of vertebrae and lack of major pathology (fused/collapsed vertebrae). Greater progress has been made in whole-body CT images where the task is more straightforward; CT imaging protocols are highly standardised due to image intensities being consistent, representing X-Ray absorptions of voxel locations. Also, CT scans tend to have more 3-D information, with higher resolution in the sagittal plane than that typical of clinical MR. Approaches using graphical models and recurrent neural networks for labelling have been reported to achieve 70–89% identification rate in mixed field of view CT scans [4,5,20].

2   Detecting Vertebrae

Detection of the vertebrae proceeds in three stages, as illustrated in Fig. 1. First, corners and centroids of each vertebra are predicted in each sagittal slice. Second, each detected corner is assigned to a centroid by predicting a vector field for each corner type (e.g. top left, bottom right) which points to the corresponding centroid. The detected centroid and the four corners which point to it define the bounding quadrilateral for that vertebra in that slice. Third, these quadrilaterals are grouped across slices to define detected volumes for each vertebra.

Inference is performed here by a U-Net architecture (shown in detail in the supplementary material and also in the extended version of this paper available on arxiv with identifier 2007.02606) which ingests an MRI (one slice) and outputs 13 channels: 5 channels are the *landmark detection* channels and each is used to detect the centroid and 4 corners respectively of all vertebral bodies appearing in that image; the remaining 8 channels are the *landmark grouping* channels corresponding to the x and y coordinates of the 4 vector fields used in grouping.

In more detail, the locations of corners and centroids are identified as modes of the heatmaps in the *landmark detection* channels. For each corner detected, there exists a corresponding grouping vector field from the *landmark grouping* channels; 4 corners equals 4 vectors fields. A group is then formed from the 4 closest corners pointing to a single centroid; a group here forms a quadrilateral. This is done for each detected centroid. If two or more centroids are assigned the same corner, the centroid to which the corner's vector points closest remains, and the other centroid is discarded to stop double detection of a single vertebra. If there is no detected corner within a fixed range of a centroid, it is also discarded, making the system robust to spurious centroid detections. This is performed in each sagittal slice. Finally, the vertebral bodies detected are grouped across slices by measuring the IoU between quadrilaterals and assigning them to the same vertebra if they have an overlap of greater than 0.5.

Discussion. Our approach of detecting vertebrae as a series of points and then grouping them differs from other methods which have used region proposal networks [21] pixel/voxel-wise segmentation [23], deformable-part models [2] or simply detected vertebrae centroids [3,20]. However, this approach is receiving increasing attention in the computer vision literature, with Zhou et al. [22] showing state-of-the-art results by detecting objects as a series of keypoints. The advantage of this method is its high speed (1 s inference on GPU-enabled hardware) combined with more accurate bounding regions than using standard bounding boxes.

Training: To train the network, we use separate loss functions for the detection and grouping channels. The ground truth annotations are coordinates of the corners of the vertebral bodies (centroids are computed from these). Target

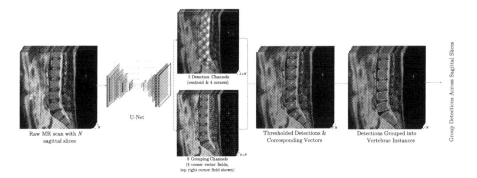

Fig. 1. The pipeline used to detect vertebrae in whole spine scans. The output detection channels are thresholded and each resulting connected volume becomes a corner or a centroid. Each corner has corresponding vector in the grouping channels at the point-of-detection and 'points' with a magnitude and direction according to that vector. Each centroid is assigned the 4 corners that point closest to it.

detection channel outputs are constructed by overlaying a Gaussian kernel on each annotated ground truth landmark in the detection channels with variance proportional to the square root of the area of the detection. These channels are trained using the weighted L1 loss; $\mathcal{L}_{detect}(Y, \hat{Y}) = \sum_{k=1}^{5} \sum_{i,j} \alpha_{ijk} |y_{ijk} - \hat{y}_{ijk}|$ where Y is the response map output by the network and $\hat{Y}$ is the target response map. y_{ijk} and $\hat{y}_{ijk}$ are the value of response and ground truth map respectively at image coordinate (i, j) in detection channel k and a_{ijk} is a weighing factor given by $\alpha_{ijk} = \frac{N_k}{N_k + P_k}$ if $\hat{y}_{ijk} \geq T$ or $\frac{P_k}{N_k + P_k}$ if $\hat{y}_{ijk} < T$, where P_k and N_k are the number of pixels respectively above and below threshold T in channel k. This weighing factor balances the loss from false positive responses in the heatmap and false negatives, speeding up training. In the experiments in this paper, $T = 0.01$ was used. *Landmark grouping* channels are trained by using the L2 loss; $\mathcal{L}_{group} = \sum_{l=1}^{4} \sum_{b} \sum_{(i,j) \in \mathcal{N}_{bl}} ||\mathbf{v}_{ij}^{l} - \mathbf{r}_{ij}^{k}||_{2}^{2}$. Here l indexes each corner type/vector field (e.g. top left, bottom right), b indexes labelled ground truth vertebral body, $\mathcal{N}_{bl}$ is a neighbourhood of pixels surrounding the lth corner of vertebral body b. $\mathbf{v}_{ij}^{l}$ is the value of the vector field corresponding to corner l at the pixel location (i, j) and $\mathbf{r}_{ij}^{b}$ is the displacement from the centroid of vertebral body b to (i, j). As suggested by [15], heavy augmentation is used during training. Scans are padded, rotated, zoomed and flipped in the coronal plane. Non-square scans are split into squares overlapping by 40% and resized to 224×224 to ensure constant sized input to the network.

3   Labelling Vertebrae

Once the vertebrae are detected, the next task is to label each vertebrae with its level (e.g. S1, L5, L4 etc.). There are two types of information to consider when labelling a vertebra; the *appearance* of the vertebrae – its intensity pattern, shape

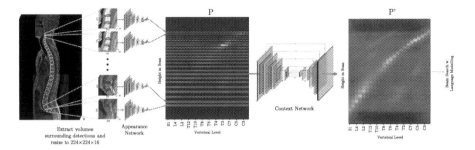

Fig. 2. Overview the labelling pipeline. Volumes surrounding each of the vertebrae detected are extracted and input to an appearance network. This is then used to construct an input image to the context network which gives the probability of a given label at a given height in the image. The resulting probability map is shown in the far right of the image, followed by a beam search to generate the final level sequence. The architectures of both networks are shown in the supplementary material.

and size – and its *context* – its position in relation to other vertebrae that have been detected in the scan. As such, we train two networks; an *appearance network* to infer the level of vertebra from its appearance alone, and a *context network* which takes as input the predictions of the appearance network along with the spatial configuration of the detections in the scan to improve the predictions. The final stage is to search for a consistent labelling of the vertebra using a sequential 'language model' that builds in ordering constraints. The labelling pipeline is outlined in Fig. 2. It should be noted that both networks are fully convolutional. This differs from the approaches outlined in [20] and [12] which use recurrent neural networks and graphical models respectively.

The labelling pipeline proceeds as follows: A volume around each detected vertebra is extracted and given as input to the appearance model which predicts a softmax probability vector over the labels. Input volumes are created by first fitting a bounding box around the detection, then expanding the box by 100% to include nearby anatomical features and resampling its size to $224 \times 224 \times 16$, where 16 is the number of sagittal slices. For the appearance network, we use a simple 3D CNN (details of this and the context network are given the supplementary material). Then, the predictions of the appearance networks are used to make a probability-height map where the probability vector at the height of each vertebra is equal to the output vector of the appearance network for that vertebra. In detail: the probability-height map, $P \in \mathbb{R}^{H \times N}$ where H is the height of the image and $N = 24$ is the number of vertebrae level classes (C2 to S1). If vertebra detection v spans from heights h_{v_1} to h_{v_2} in the scan then $P_{h'n} = p^a_{vn} \forall\ h_{v_1} \leq h' \leq h_{v_2}$ and 0 otherwise where p^a_{vn} is the probability v has level n, given by the appearance network. Next, the context network refines this probability-height map, using detections around each vertebra to update its predicted class. Finally, the output map is decoded into a logical sequence of vertebra using a language-model inspired beam search which imposes global constraints such as no repetitions. This is described in detail in Sect. 3.1.

Training: The appearance network is trained using a cross-entropy loss function. For re-calibration [7], a softmax operation with temperature $T = 10$ is applied to the logits layer. P is given to the context net which produces P', a refined version of the same map as shown on the far right of Fig. 2. The cross-entropy between the output probability map at the centroid height of each vertebra and its ground truth label is used as the loss function. When training the context network, augmentation is applied at training time by randomly removing between 0 to 4 of the highest and lowest detections from the input image and further removing each remaining vertebra's probability map from the input image with probability 0.2. This ensures the labelling pipeline is robust to missed detections and not reliant on distinct looking vertebrae at the top and bottom of the spine.

3.1 Enforcing Monotonicity and Dealing with Numerical Variations

To predict the labels for each vertebrae detected, P' shown in Fig. 2 must be decoded into a sequence of level predictions. A naive method to do this would be to take the argmax of the probability map at each vertebra centroid height and use this as a label. However, this allows implausible sequences such as repeated vertebral levels to be predicted. If the detection v at height h_v has label index l, then we know v' at $h_{v'} > h_v$ detection should have label $l' > l$ with the greatest probability of being $l + 1$ (e.g. L4 should be followed by L3). The challenge of imposing such constraints is analogous to that faced in automatic speech recognition where CTC training is followed by language modelling to predict a valid character or word sequence [6,16]. We take inspiration from this and use a beam search to obtain a valid labelling: beginning at the highest detection and searching down, sequences are generated and scored by selecting the highest probability labels for each vertebra. The k most likely sequences of levels are stored in memory at each step, where k is the beam width. Sequences with repetitions of levels are given probability 0 and those with skipped levels are penalised by multiplying the sequence probability by a penalty score. The method can also incorporate numerical variations, with ± 1 lumbar vertebrae given a small sequence probability penalty reflecting the low incidence of this.

4 Detection and Labelling Results

We evaluate the system at the task of vertebrae detection and labelling in whole spine and lumbar scans. Following [11,20], we define correct detections to be when the ground truth vertebra centroid is contained entirely within a single bounding quadrilateral. For detection, we report precision, recall, and the localisation error (LE), defined to be the mean distance of ground truth centroids from the closest detection quadrilateral centroid. For labelling, we report identification rate (IDR), the fraction of vertebrae detected and labelled correctly.

4.1 Datasets

Three datasets are used in this work: OWS, Genodisc and Zukić. OWS is a dataset of 710 sagittal whole spine scans across 196 patients from the Picture

Table 1. Performance of the pipeline on the three datasets. Our approach is compared with other methods using the same datasets and also a LSTM labelling baseline. Results are reported on a per-vertebra level. Higher is better for detection precision (Prec.), detection recall (Rec.), and level identification rate (IDR). Lower is better for localisation error (LE). We also report the percentage of vertebrae within one level of their ground truth value (IDR $\pm$ 1). Lootus [11] is tested on a subset of 291 scans from the Genodisc dataset. **Note, Windsor† [19] requires manual initialization by providing the location of the S1 vertebra, so is not directly comparable.**

Dataset	No. Scans	No. Vert	Method	Prec. (%)	Rec. (%)	IDR(%)	IDR $\pm$ 1(%)	LE (mm)
OWS (Whole Spine)	37	888	Windsor† [19]	**99.4**	**99.4**	-	-	**1.0 $\pm$ 0.9**
			Label baseline	-	-	86.9	93.4	-
			Ours	99.0	98.1	**96.5**	**97.3**	2.4 $\pm$ 1.3
Genodisc (Lumbar)	421	2947	Lootus [11]	-	-	86.9	-	3.5 $\pm$ 3.3
			Label baseline	-	-	90.1	97.4	-
			Ours	**99.7**	**99.7**	**98.4**	**99.7**	1.6 $\pm$ 1.1
Zukić (Lumbar)	17	154	Zukić [23]	98.7	92.9	-	-	**1.6 $\pm$ 0.8**
			Label baseline	-	-	87.0	94.3	-
			Ours	**99.3**	**98.7**	90.9	98.7	2.0 $\pm$ 1.5

Archiving and Communication System (PACS) of an orthopaedic centre. The dataset exhibits a wide range of pathologies such as hemivertebra, fused vertebrae, numerical variations of vertebrae and scoliosis. Scans are taken from different scanners with a range of MR parameters (T1, T2, FLAIR, TIRM and STIR). The dataset is split into training, validation and testing sets with an 60/20/20% split at the patient level. Corners of vertebrae from S1 to C2 are annotated and vertebrae levels were marked by a radiologist in one scan for each patient, with S1 being the first vertebra attached to the pelvis. In the case of 25 vertebrae between S1 and C2 instead of the normal 24, an extra lumbar vertebrae is labelled (L6). Networks were trained on the OWS training set, using an Adam optimizer with $\beta_1 = 0.9$, $\beta_2 = 0.999$ and a learning rate of 0.001. The Genodisc and Zukić datasets are used only for testing. Genodisc's test set has 421 clinical lumbar MRIs used by Lootus et al. [12]. Zukić [23] is a small dataset of 17 mostly lumbar sagittal MRIs available on the online SpineWeb platform.

4.2   Results

The results of detection and labelling on all datasets are shown in Table 1 with comparisons to other methods reported on the same datasets where available. We also compare our convolutional labelling pipeline results to a baseline recurrent approach for vertebra labelling, training a bidirectional LSTM on the appearance features extracted from each detected volume. Example predicted detection and labelling sequences across a range of pathologies are given in Fig. 3. The LSTM baseline used is detailed in the supplementary material.

Discussion: For whole spine detection on the OWS dataset, the proposed method achieves a high precision and recall of 99.0% and 98.1% respectively.

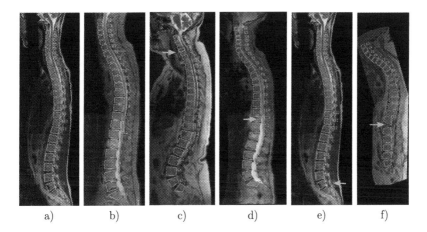

Fig. 3. Example detection and labelling of vertebrae for a range of whole spine scans. a) and b) are examples of typical spines. The other images shows examples of the system dealing with pathologies; c) shows an example of a spine with fused C3 and C4 vertebrae; d) is an example of a spine with a collapsed vertebra; e) shows a spine with an extra vertebra between S1 and C2 and f) shows an extremely kyphotic spine with a hemivertebra at the bottom of the thoracic spine.

It achieves a level identification rate of 96.5%, significantly exceeding the LSTM baseline. The few errors the system made for labelling are generally due to S2 being detected as S1, meaning all labels are out by one. In practice, this is a mistake radiologists often make as it can be difficult to tell which is the first sacral vertebra without looking at axial scans to see which bone is joint to the pelvis. This also explains why [19] achieves slightly higher precision and recall at detection than ours; it is a semi-automated algorithm given the location of S1 at initialization and thus bypasses this difficult problem of S1 recognition. On the Genodisc lumbar spine dataset, our method again significantly outperforms the baseline, and it is also outperforms the prior method of Lootus et al. [11] (98.4% compared to 86.9%). Importantly, OWS and Genodisc are scans of patients with a wide range of pathologies imaged using typical clinical MR protocols; hence strong performance here gives evidence of the clinical usefulness of this approach. We show example results for pathological spines in the supplementary material. In the supplementary material we also report results by other groups for vertebrae detection and labelling in MRIs. However, these are for different datasets to which we could not get access, with different scanning protocols, fields of view (FoV) and patient sets, and thus cannot be compared to directly.

5   Automated Scoliosis Detection

Finally, as an illustration of a potential application of the proposed approach, we explore the ability of the system to classify cases of scoliosis from sagittal MR scans. In clinical practice this is usually determined by measuring the Cobb

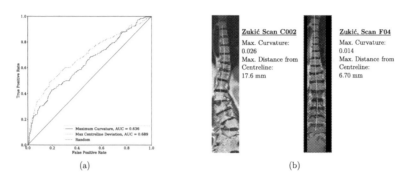

(a)   (b)

Fig. 4. Results of scoliosis classification in sagittal scans using the proposed vertebrae detection system: (a) ROC curve for simple scoliosis classifiers based on statistics of polynomial curves fitted through vertebrae centroids in Genodisc scans; (b) A qualitative comparison of curves fit through detected vertebrae in full spine scans. Curves are overlaid on coronal slices synthesised from the sagittal slices.

angle in coronal views of X-ray scans [1], however measuring scoliosis in the supine position has also been shown to be possible [8,17] and MRI can be useful for understanding disease etiology and symptoms [14]. This is a more difficult task in sagittal scans as it requires sensitive detection of the sides of vertebra with a clear decision boundary as to when a vertebra is present or not present in a slice which can be difficult in cases of partial visibility. In the entirety of the Genodisc dataset, scoliosis was reported by a radiologist in 198 of 3542 scans, across 2009 patients. By measuring statistics of a quintic polynomial fit through vertebra centroids and using them as predictive features we develop classifiers for this label. Specifically, we measure the maximum curvature of the polynomial, and the maximum deviation of the curve from a straight vertical line fit through the vertebrae, assuming that a low curvature vertebral column with little deviation from the centreline corresponds to a non-scoliotic spine. While ground truth scoliosis labels were not available for the whole spine scans, we give qualitative results, comparing the features of curves fit though vertebrae of a scoliotic and non-scoliotic scan from Zukić. Results of these experiments are shown in Fig. 4.

Discussion: The results for automated scoliosis detection from sagittal scans are promising. Using simple classifiers AUCs of 0.636–0.689 are achieved in a highly class-imbalanced problem. Of the features measured, distance from the vertical centreline of the vertebrae performed best. The system is also shown to capture scoliotic curves in full spine scans too. These experiments illustrate that the proposed method produces a strong geometric representation of the spine, which can be used in further downstream tasks.

6   Conclusion

We introduce a novel method for vertebrae detection and labelling in whole spine sagittal MRIs. It shows state-of-the-art results for vertebra identification in lumbar scans, with little performance drop in the whole spine case, and is robust to a range of spinal defects, numerical variations of vertebrae and different scanning protocols. We also demonstrate a potential diagnostic application: automated detection of scoliosis from sagittal MRIs. Future work will include integrating automated detection of different spinal pathologies and implementing the system for CT scans.

Acknowledgements. The authors would like to thank Dr. Sarim Ather for useful discussions on spinal anatomy and clinical approaches to diagnosing disease, as well as assistance labelling the data. Rhydian Windsor is supported by Cancer Research UK as part of the EPSRC CDT in Autonomous Intelligent Machines and Systems (EP/L015897/1). Amir Jamaludin is supported by EPSRC Programme Grant Seebibyte (EP/M013774/1). The Genodisc data was obtained during the EC FP7 project GENODISC (HEALTH-F2-2008-201626).

References

1. Aebi, M.: The adult scoliosis. Eur. Spine J. **14**(10), 925–948 (2005)
2. Cai, Y., Osman, S., Sharma, M., Landis, M., Li, S.: Multi-modality vertebra recognition in arbitrary views using 3D deformable hierarchical model. IEEE Trans. Med. Imaging **34**(8), 1676–1693 (2015)
3. Forsberg, D., Sjöblom, E., Sunshine, J.L.: Detection and labeling of vertebrae in MR images using deep learning with clinical annotations as training data. J. Digit. Imaging **30**(4), 406–412 (2017). https://doi.org/10.1007/s10278-017-9945-x
4. Glocker, B., Feulner, J., Criminisi, A., Haynor, D.R., Konukoglu, E.: Automatic localization and identification of vertebrae in arbitrary field-of-view CT scans. In: Ayache, N., Delingette, H., Golland, P., Mori, K. (eds.) MICCAI 2012. LNCS, vol. 7512, pp. 590–598. Springer, Heidelberg (2012). https://doi.org/10.1007/978-3-642-33454-2_73
5. Glocker, B., Zikic, D., Konukoglu, E., Haynor, D.R., Criminisi, A.: Vertebrae localization in pathological spine CT via dense classification from sparse annotations. In: Mori, K., Sakuma, I., Sato, Y., Barillot, C., Navab, N. (eds.) MICCAI 2013. LNCS, vol. 8150, pp. 262–270. Springer, Heidelberg (2013). https://doi.org/10.1007/978-3-642-40763-5_33
6. Graves, A., Fernández, S., Gomez, F., Schmidhuber, J.: Connectionist temporal classification: labelling unsegmented sequence data with recurrent neural networks. In: International Conference on Machine learning, ICML 2006 (2006)
7. Guo, C., Pleiss, G., Sun, Y., Weinberger, K.Q.: On calibration of modern neural networks. In: International Conference on Machine Learning (2017)
8. Jamaludin, A., Kadir, T., Clark, E., Zisserman, A.: Predicting spine geometry and scoliosis from DXA scans. In: MICCAI Workshop: Computational Methods and Clinical Applications in Musculoskeletal Imaging (2019)
9. Jamaludin, A., Kadir, T., Zisserman, A.: SpineNet: automated classification and evidence visualization in spinal MRIs. Med. Image Anal. **41**, 63–73 (2017)

10. Jamaludin, A., et al.: Automation of reading of radiological features from magnetic resonance images (MRIs) of the lumbar spine without human intervention is comparable with an expert radiologist. Eur. Spine J. **26**, 1374–1383 (2017)
11. Lootus, M., Kadir, T., Zisserman, A.: Vertebrae detection and labelling in lumbar MR images. In: Yao, J., Klinder, T., Li, S. (eds.) Computational Methods and Clinical Applications for Spine Imaging. LNCVB, vol. 17, pp. 219–230. Springer, Cham (2014). https://doi.org/10.1007/978-3-319-07269-2_19
12. Lootus, M., Kadir, T., Zisserman, A.: Radiological grading of spinal MRI. In: MICCAI Workshop: Computational Methods and Clinical Applications for Spine Imaging (2014)
13. Lu, J.T., et al.: Deep spine: automated lumbar vertebral segmentation, disc-level designation, and spinal stenosis grading using deep learning. In: Machine Learning for Healthcare Conference (2018)
14. Ozturk, C., Karadereler, S., Ornek, I., Enercan, M., Ganiyusufoglu, K., Hamzaoglu, A.: The role of routine magnetic resonance imaging in the preoperative evaluation of adolescent idiopathic scoliosis. Int. Orthop. **34**(4), 543–546 (2010)
15. Ronneberger, O., Fischer, P., Brox, T.: U-Net: convolutional networks for biomedical image segmentation. In: Navab, N., Hornegger, J., Wells, W.M., Frangi, A.F. (eds.) MICCAI 2015. LNCS, vol. 9351, pp. 234–241. Springer, Cham (2015). https://doi.org/10.1007/978-3-319-24574-4_28
16. Scheidl, H., Fiel, S., Sablatnig, R.: Word beam search: a connectionist temporal classification decoding algorithm. In: 2018 16th International Conference on Frontiers in Handwriting Recognition (ICFHR) (2018)
17. Taylor, H.J., Harding, I., Hutchinson, J., Nelson, I., Blom, A., Tobias, J.H., Clark, E.M.: Identifying scoliosis in population-based cohorts: development and validation of a novel method based on total-body dual-energy X-ray absorptiometric scans. Calcif. Tissue Int. **92**(6), 539–547 (2013)
18. Tins, B.J., Balain, B.: Incidence of numerical variants and transitional lumbosacral vertebrae on whole-spine MRI. Insights Imaging **7**(2), 199–203 (2016). https://doi.org/10.1007/s13244-016-0468-7
19. Windsor, R., Jamaludin, A.: The ladder algorithm: finding repetitive structures in medical images by induction. In: IEEE International Symposium on Biomedical Imaging (2020)
20. Yang, D., et al.: Automatic vertebra labeling in large-scale 3D CT using deep image-to-image network with message passing and sparsity regularization. In: Niethammer, M., et al. (eds.) IPMI 2017. LNCS, vol. 10265, pp. 633–644. Springer, Cham (2017). https://doi.org/10.1007/978-3-319-59050-9_50
21. Zhao, S., Wu, X., Chen, B., Li, S.: Automatic vertebrae recognition from arbitrary spine MRI images by a hierarchical self-calibration detection framework. In: Shen, D., et al. (eds.) MICCAI 2019. LNCS, vol. 11767, pp. 316–325. Springer, Cham (2019). https://doi.org/10.1007/978-3-030-32251-9_35
22. Zhou, X., Wang, D., Krähenbühl, P.: Objects as points. In: arXiv preprint arXiv:1904.07850 (2019)
23. Zukić, D., Vlasák, A., Egger, J., Hořínek, D., Nimsky, C., Kolb, A.: Robust detection and segmentation for diagnosis of vertebral diseases using routine MR images. Compu. Graph. Forum **33**(6), 190–204 (2014)

Keypoints Localization for Joint Vertebra Detection and Fracture Severity Quantification

Maxim Pisov[1,2]([⊠]), Vladimir Kondratenko[1], Alexey Zakharov[2,3],
Alexey Petraikin[4], Victor Gombolevskiy[4], Sergey Morozov[4],
and Mikhail Belyaev[1]

[1] Skolkovo Institute of Science and Technology, Moscow, Russia
m.pisov@skoltech.ru
[2] Kharkevich Institute for Information Transmission Problems, Moscow, Russia
[3] Moscow Institute of Physics and Technology, Moscow, Russia
[4] Research and Practical Clinical Center of Diagnostics and Telemedicine
Technologies, Department of Health Care of Moscow, Moscow, Russia

Abstract. Vertebral body compression fractures are reliable early signs of osteoporosis. Though these fractures are visible on Computed Tomography (CT) images, they are frequently missed by radiologists in clinical settings. Prior research on automatic methods of vertebral fracture classification proves its reliable quality; however, existing methods provide hard-to-interpret outputs and sometimes fail to process cases with severe abnormalities such as highly pathological vertebrae or scoliosis. We propose a new two-step algorithm to localize the vertebral column in 3D CT images and then to simultaneously detect individual vertebrae and quantify fractures in 2D. We train neural networks for both steps using a simple 6-keypoints based annotation scheme, which corresponds precisely to current medical recommendations. Our algorithm has no exclusion criteria, processes 3D CT in 2 s on a single GPU and provides an intuitive and verifiable output. The method approaches expert-level performance and demonstrates state-of-the-art results in vertebrae 3D localization (the average error is 1 mm), vertebrae 2D detection (precision is 0.99, recall is 1), and fracture identification (ROC AUC at the patient level is 0.93).

Keywords: Vertebral fractures · Object detection · Keypoints localization

1 Introduction

Osteoporotic fractures are common in older adults and resulted in more than two million Disability Adjusted Life Years in Europe [10]. The presence of vertebrae

Electronic supplementary material The online version of this chapter (https://doi.org/10.1007/978-3-030-59725-2_70) contains supplementary material, which is available to authorized users.

fractures dramatically increases the probability of subsequent fractures [13]; thus can be used as an early marker of osteoporosis. Medical imaging, such as Computed Tomography (CT), is a useful tool to identify fractures [14]. However, radiologists frequently miss fractures, especially if they are not specializing in musculoskeletal imaging; with the average error rate being higher than 50% [18]. At the same time, rapidly evolving low dose CT programs, e.g., for lung cancer, provide a solid basis for opportunistic screening of vertebral fractures.

The medical image computing community thoroughly investigated fractures detection and/or classification on vertebrae-level [1,5,24,27], whole study-level [2,26], or jointly on both levels [19], see Sect. 2 for more details. Many of these approaches require prior vertebrae detection [1,19,27], or spine segmentation [2,5,24]. Though both problems are active areas of research with prominent results, fractured vertebrae are the most complex cases for these algorithms [25], and even good average detection/segmentation accuracy may not be sufficient for accurate fracture estimation. As a result, researchers had to exclude some studies from the subsequent fracture classification due to errors in prior segmentation [27], or due to scoliosis [26].

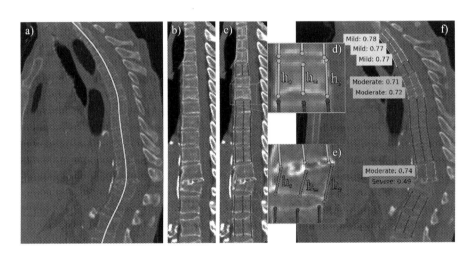

Fig. 1. Overview of the proposed model. **Step 1**: a) localizing vertebrae centers in 3D CT (a sagittal projection is shown); b) generating a new 2D image via spine 'straightening'. **Step 2**: c) identifying key-points and the corresponding heights; d–e) a closer look at some vertebrae (colors denote the fracture severity). **Finally**: f) the original image with estimated fracture severities. (Color figure online)

The second important issue is the mismatch between computer science problem statements and the radiological way to define fractures. The Genant scale [7] is a widely used medical criterion recommended by the International Osteoporosis Foundation [6]. It relies on the measurements of h_a, h_m, h_p - the anterior, middle and posterior heights of vertebral bodies (Fig. 1d, 1e):

$$G = \frac{\min\{h_a, h_m, h_p\}}{\max\{h_a, h_m, h_p\}}, \tag{1}$$

G values provide an easy to interpret continuous index, whereas existing methods are usually trained to predict a binary label extracted from radiological reports [2,26] or multiclass labels based on threshold levels for G [5,27]. A related problem is the interpretability of the methods' outputs. The only available information is the network's attention [26] or a similar score [19] somehow related to the probability of fracture presence.

Our contribution is two-fold. First, we propose a new method to identify the vertebral column in 3D CT and, as a consequence, reducing the problem to 2D by producing the corresponding mid-sagittal slice [4] to measure h_a, h_m, h_p for each vertebra (Fig. 1a, b). Our method is trained to directly solve the localization problem rather than spine segmentation and demonstrates excellent localization quality with the average error less than 1 mm. Also, it allows us to process all studies with no exceptions, including cases with severe scoliosis. Second, our method estimates six keypoints to detect each vertebra and estimate its heights h_* simultaneously (Fig. 1c–e), which results in excellent fracture classification quality with the area under ROC curve equal to 0.93. The predictions are highly interpretable as they can be validated by a doctor using a simple ruler.

2   Previous Work

The automatic classification of vertebral fractures has received much attention from the medical image analysis community. A quantitative image analysis method was proposed in [5] to classify individual vertebra. First, the spinal column is segmented by an external method detecting intervertebral intervals. Then each vertebra is split into 17 sections to extract a set of simple features such as mean density from the segmentation mask. Finally, a support vector machine classifies vertebrae based on the obtained 51 features. The system provides excellent sensitivity (98.7%) but quite low specificity (77.3%). A similar approach was used in [27] where authors calculated computer vision features such as histograms of oriented gradients from vertebra masks and achieved ROC AUC 0.88. A plain deep learning-based version of this two-step approach was proposed in [1], where classical ResNet was trained on 3-channel 2D images obtained from the prior segmentation mask by taking central sagittal, axial and coronal slices for each vertebra.

It is important to note that all the methods above rely on prior segmentation, which may result in removing some cases with severe abnormalities. Indeed, the authors of [27] reported that 11 cases out of 154 were excluded from the analysis due to incorrect prior spine segmentation largely caused by high-grade fractures.

This requirement was relaxed in several papers. In [19] the authors proposed a two-step pipeline for vertebrae detection: first, a segmentation neural network is used to generate pixel-level predictions (background, normal, fracture), then the predicted maps are aggregated. Instead of the whole spine mask, the authors used the ground-truth coordinates of vertebrae centroids to produce vertebrae-level predictions and achieved ROC AUC 0.93. A simple idea was used in [26], where the authors selected the central sagittal slices as the spine is usually located in the middle of the image. In particular, they processed only 6.9 central slices per study (on average). As a result, this approach fails to identify fractures in patients with at least moderate scoliosis, and they had to exclude 156 out of 869 subjects from the analysis, primarily due to scoliosis. Though the average prevalence of scoliosis is 8.85%, it positively correlated with age and increases from 10.95% in 60–69 to 50% in 90+ age groups [11], so this cohort can not be ignored in vertebral fractures screening. The classification method from [26] consists of a ResNet34 which processes each of the central sagittal slices separately; then the obtained scores are aggregated by a simple LSTM network.

Finally, an original approach was proposed in [2]. Though the method also relies on external spine segmentation, the mask is used to extract the spinal cord and create a new virtual sagittal slice. Next, small patches are extracted from this slice and classified by a convolutional network; finally, a recurrent neural network (RNN) is used to aggregate the predictions from each patch. Although the training database is the largest among the reviewed works (consisting of 1673 cases), the model achieves 83.9% sensitivity (with 93.8% specificity), likely due to poor study-level binary annotation extracted from the radiological reports.

3   Method

The majority of existing methods are two-step pipelines: first, the spine is localized or segmented; second, individual vertebrae are processed to identify fractures. We follow the same scheme with two major goals:

1. Replace the first part by a more task-specific alternative to avoid the exclusion of any case from the analysis due to segmentation failures.
2. Create a method capable of directly working with Genant fracture reporting.

The second, more important goal dictated the annotation protocol, which affects both steps of our method. Following [4], raters were instructed to find the mid-sagittal slice for each of the visible vertebrae and annotate six keypoints to measure anterior, middle and posterior heights (Fig. 1d, 1e).

In the first part of our pipeline we predict 3D coordinates of the middle height keypoints using a 3D fully convolutional neural network with soft-argmax activation [16]. The predicted coordinates are then used to localize the spine and to select a sagittal plane which contains all the vertebrae - an idea introduced in [3] and later applied in [2]. However, we cannot simply choose a sagittal plane on the original image, because such a plane might simply not exist, e.g. for patients with severe spinal scoliosis. For this reason the obtained 3D curve is used to 'straighten' the spine and generate a new 2D slice (Sect. 3.1). It is worth noting that though the input is a 3D CT scan, the annotation is intrinsically bidimensional, so this dimensionality reduction is essential. The key advantage of our approach is that we directly use final annotation to find the most appropriate 2D representation of the original 3D image.

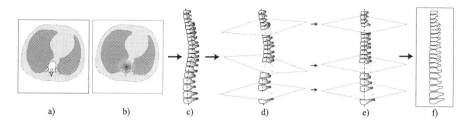

Fig. 2. The spine straightening pipeline: a) a single axial slice; b) an axial slice with the probability heatmap (red), the cross indicates the resulting point after the soft-argmax operation; c) the combined points from each slice result in a 3D curve; d) planes, orthogonal to the curve (for better visualization most planes are omitted); e) a straightened vertebral column (the planes become parallel); f) the new central sagittal plane. (Color figure online)

The second part of our network processes the obtained 2D image in order to localize each vertebra and predict positions of six keypoints. Given the fact that the number of vertebrae for a given input image is not known a priori, a natural solution is to use object detection techniques in order to make vertebra-level predictions without the need for additional postprocessing. Our major insight is that directly predicting keypoints coordinates relative to the bounding box center shows a dramatic increase in model quality (Sect. 3.2). We also propose a combined loss function to enforce good quality of G index in addition to localization and heights estimation parts.

Finally the predictions of the second network are mapped back to 3D in order to calculate the heights h_a, h_m, h_p and the G index.

3.1 Spine Straightening

We use a 3D UNet-like [17] architecture to predict a 2D probability map on each axial slice, followed by the 2D soft-argmax operation [16] to obtain spatial coordinates of the vertebral body central line. We train our network by optimizing *mean absolute error* between the predicted points and the ones smoothly interpolated from the annotation of middle height endpoints (Fig. 3a).

By combining the predictions for each axial plane, we obtain a 3D curve. We then interpolate the image onto a new 3D grid on which the obtained curve becomes a straight vertical line. The grid is constructed in such a way, so that the planes normal to the curve become parallel (Fig. 2d–e). Finally we select a new sagittal plane where all vertebrae are visible. Figure 2 shows a detailed illustration.

3.2 Vertebrae-Level Predictions

The object detection step of our method is mainly based on YOLO9000 [20] and YOLOv3 [21] with several modifications, in order to adapt it to the specifics of the given task. Similarly to YOLO9000, our architecture consists of a single Region Proposal Network (RPN), that makes predictions relative to a set of anchor boxes, and doesn't require ROI-pooling [22] and any further refinement. Additionally, YOLOv3 uses Feature Pyramid Networks (FPNs) [15], which enables it to make accurate predictions on various scales. As we know that the range of shapes a vertebra can have is quite narrow, we use a simple UNet-based [23] architecture instead. Finally, as opposed to the YOLO family, our RPN makes predictions in the original resolution, because we favor accuracy over speed.

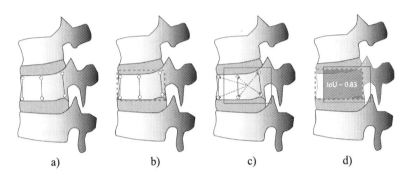

a)   b)   c)   d)

Fig. 3. Target generation steps: a) example of an annotated vertebra; b) the generated axis-aligned bounding box (green, dashed); c) the **keypoints' coordinates** relative to anchor box' center; d) the **objectness** O is 1, if IoU between the boxes is greater than 0.5 (Color figure online)

The **target generation** pipeline is shown in Fig. 3. Note, that, because our goal is to assign 6 keypoints per vertebra, we don't really need to predict bounding boxes. Instead, we generate *axis aligned bounding boxes* (Fig. 3b, d) to calculate *intersection-over-union* (IoU) at both train and test time. We encode the target keypoints coordinates by using the same scale- and shift-invariant encoding as in [22]:

$$e^x = (g^x - a^x)/a^w; \quad e^y = (g^y - a^y)/a^h, \tag{2}$$

where $(g^x, g^y), (e^x, e^y)$ are the global and encoded coordinates of a given keypoint respectively, (a^x, a^y) - is the center of an anchor box, and (a^w, a^h) are its width and height (Fig. 3). We selected the following anchor boxes' scales: 17, 23, 28, 35 mm and aspect ratios: 0.8, 1.1, 1.3, 2.

Finally, we propose the following **loss function** to train our second network:

$$L = BCE(\hat{o}, o) + \frac{1}{\sum\limits_i I[o_i = 1]} \sum\limits_i \frac{I[o_i = 1]}{G_i} \cdot MAE(\hat{e}_i, e_i), \tag{3}$$

where both sums are calculated over all vertebrae in the training batch, BCE is the standard *log-loss* between real (o) and predicted ($\hat{o}$) objectness (Fig. 3d),

MAE is the *mean absolute error* between real (e_i) and predicted $(\hat{e}_i)$ encoded keypoints' coordinates (2) for the i-th vertebra and G_i is the respective Genant score (1).

4   Data

Our dataset consists of 100 chest CT. It represents a randomly selected subset of a publicly available dataset.[1] The images have various voxel spacing ranging from $.5 \times .5 \times .8$ mm to $1 \times 1 \times .8$ mm and different numbers of visible vertebrae: from 10 to 15.

The data was annotated by 7 experts with 1 to 5 years of experience in radiology and a board-certified radiologist with 12 years of experience in the field. In total the dataset contains 1268 annotated vertebrae with 2–3 annotations per vertebra. The distribution of vertebral fractures is the following: 125 mild, 80 moderate, 17 severe deformations and 1046 normal vertebrae. Patient-wise we have a somewhat balanced distribution with 30, 16, 41 and 13 patients with none, mild, moderate and severe deformations respectively.

5   Results

5.1   Experimental Setup

In all of our experiments the only preprocessing we use is intensity normalization to zero mean and unit variance.

We trained our **spine straightening** network with the Adam [12] optimizer with default parameters ($\beta_1 = 0.9, \beta_2 = 0.999$) and a learning rate of 10^{-3}, which showed the fastest convergence rate. As the architecture operates on whole 3D images, we reduce the images' resolution to a spacing of $3 \times 3 \times 3$ mm, the predicted curve is then linearly interpolated to the original resolution. In such a setting the network reached convergence after approx. 10k batches of size 3.

Similarly we trained the **vertebrae detection** network for 4k iterations with batches of size 30. We start with a learning rate of 10^{-4} at the early stages of training and decrease it by a factor of 2 after 1k, 1.4k and 2k iterations, because gradually decreasing the learning rate enabled the model to reach better optima.

We reported results obtained using fivefold cross-validation. As we have multiple annotations per study, we also report the inter-expert variability. To obtain patient-level predictions, we use the most severe fracture among all the vertebrae, which is equivalent to taking the minimal Genant score.

5.2   Method Performance

We report the localization quality of the first step of our method in Table 1, left. To calculate these numbers, we found the closest annotated vertebrae in 3D. Also, we report 2D detection metrics for the second network, see Table 1, right.

---

[1] https://mosmed.ai/datasets/ct_lungcancer_500.

Table 1. Vertebral body centers localization and vertebrae detection metrics. Columns with white background denote the average (std) number for all vertebrae, the ones with grey background - for *Moderately* and *Severely* fractured vertebrae ($G \leq 0.74$). The ground truth G-index is not defined for false positives, so Precision is reported for all vertebrae only.

	Localization, mm		Recall		Precision
Proposed	0.97 (0.64)	1.14 (0.69)	0.997 (0.003)	1.000 (0.000)	0.993 (0.002)
Experts	1.01 (0.98)	1.17 (0.78)	0.983 (0.006)	0.973 (0.013)	0.996 (0.001)

Table 2. Binary classification metrics for various grades of fractures: at least *Mild* ($G \leq 0.8$) and at least *Moderate* ($G \leq 0.74$). All numbers are given as mean (std). Columns with white background denote vertebrae-level predictions, the ones with grey background - patient-level predictions.

Grade	G by	ROC AUC		Specificity		Sensitivity	
Mild	Proposed	.87 (.02)	.93 (.03)	.93 (.01)	.68 (.08)	.65 (.03)	.94 (.03)
	Experts	.91 (.02)	.91 (.05)	.91 (.01)	.60 (.11)	.70 (.05)	.90 (.05)
Moderate	Proposed	.94 (.02)	.93 (.03)	.98 (.01)	.86 (.05)	.75 (.05)	.84 (.05)
	Experts	.98 (.01)	.95 (.03)	.97 (.01)	.83 (.07)	.79 (.06)	.88 (.07)

To analyze the performance of vertebrae fractures classification, we report metrics for two threshold values of G following the radiological definition of severity [7], see Table 2. We assume that the most relevant problem for chest CT is the identification of at least Moderate fractures ($G \leq 0.74$) as healthy vertebrae in the thoracic spine are wedged, so normal variation can be misclassified as a Mild fracture ($0.74 < G \leq 0.8$) [14]. The obtained results are close to human-level performance on our dataset and comparable with other works. Similar values of ROC AUC were obtained at vertebra (0.88 [27], 0.93 [19]) and patient levels (0.92 [26]).

However, we cannot directly compare the performance with other works due to several factors. First, different definitions of fractured vertebrae are used across papers. Second, abdominal and pelvis CTs differ from chest CT in the number of visible vertebrae and anatomy (e.g., the above-mentioned fact concerning a higher error rate in chest CT). This fact motivates us to release our annotation and provide the community test data for further development.[2]

Figure 1 shows an example of the inference process on an image from the dataset. Due to limited space, we refer the interested reader to the supplementary materials for a broader set of examples. The overall inference takes under 2 s on Nvidia GTX 980ti, with an approximately equal time required for spine localization and all the subsequent steps (including spine straightening).

---

[2] https://github.com/neuro-ml/vertebral-fractures-severity.

6   Discussion

We proposed a new method for automatic identification of vertebrae-level fractures classification using the Genant score, which approaches, and in some cases surpasses, the inter-expert variability. Our analysis of examples on which the model performs poorly (some of which can be found in the appendix) shows that the experts' variability in these cases is also unusually high. Note that our method can be easily adapted to more common 2D X-Ray images by simply dropping the first network. Also, similarly to Mask R-CNN [9], we can extend our method to predict additional metadata for each vertebra, such as labels or segmentation masks. Particularly, it would be interesting to validate our method on the challenging localization and labeling dataset [8].

References

1. Antonio, C.B., Bautista, L.G.C., Labao, A.B., Naval, P.C.: Vertebra fracture classification from 3D CT lumbar spine segmentation masks using a convolutional neural network. In: Nguyen, N.T., Hoang, D.H., Hong, T.-P., Pham, H., Trawiński, B. (eds.) ACIIDS 2018. LNCS (LNAI), vol. 10752, pp. 449–458. Springer, Cham (2018). https://doi.org/10.1007/978-3-319-75420-8_43
2. Bar, A., Wolf, L., Amitai, O.B., Toledano, E., Elnekave, E.: Compression fractures detection on CT. In: Medical Imaging 2017: Computer-Aided Diagnosis, vol. 10134, p. 1013440. International Society for Optics and Photonics (2017)
3. Bromiley, P.A., Kariki, E.P., Adams, J.E., Cootes, T.F.: Fully automatic localisation of vertebrae in CT images using random forest regression voting. In: Yao, J., Vrtovec, T., Zheng, G., Frangi, A., Glocker, B., Li, S. (eds.) CSI 2016. LNCS, vol. 10182, pp. 51–63. Springer, Cham (2016). https://doi.org/10.1007/978-3-319-55050-3_5
4. Buckens, C.F., et al.: Intra and interobserver reliability and agreement of semiquantitative vertebral fracture assessment on chest computed tomography. PLoS ONE 8(8), e71204 (2013)
5. Burns, J.E., Yao, J., Summers, R.M.: Vertebral body compression fractures and bone density: automated detection and classification on CT images. Radiology 284(3), 788–797 (2017)
6. Genant, H.K., Bouxsein, M.L.: Vertebral Fracture Initiative: Executive Summary (2011). https://www.iofbonehealth.org/sites/default/files/PDFs/IOF_VFI-Executive_Summary-English.pdf
7. Genant, H.K., Wu, C.Y., Van Kuijk, C., Nevitt, M.C.: Vertebral fracture assessment using a semiquantitative technique. J. Bone Miner. Res. 8(9), 1137–1148 (1993)
8. Glocker, B., Zikic, D., Konukoglu, E., Haynor, D.R., Criminisi, A.: Vertebrae localization in pathological spine CT via dense classification from sparse annotations. In: Mori, K., Sakuma, I., Sato, Y., Barillot, C., Navab, N. (eds.) MICCAI 2013. LNCS, vol. 8150, pp. 262–270. Springer, Heidelberg (2013). https://doi.org/10.1007/978-3-642-40763-5_33
9. He, K., Gkioxari, G., Dollár, P., Girshick, R.: Mask R-CNN. In: Proceedings of the IEEE International Conference on Computer Vision, pp. 2961–2969 (2017)
10. Johnell, O., Kanis, J.: An estimate of the worldwide prevalence and disability associated with osteoporotic fractures. Osteoporos. Int. 17(12), 1726–1733 (2006)

11. Kebaish, K.M., Neubauer, P.R., Voros, G.D., Khoshnevisan, M.A., Skolasky, R.L.: Scoliosis in adults aged forty years and older: prevalence and relationship to age, race, and gender. Spine **36**(9), 731–736 (2011)
12. Kingma, D.P., Ba, J.: Adam: A method for stochastic optimization. arXiv preprint arXiv:1412.6980 (2014)
13. Klotzbuecher, C.M., Ross, P.D., Landsman, P.B., Abbott III, T.A., Berger, M.: Patients with prior fractures have an increased risk of future fractures: a summary of the literature and statistical synthesis. J. Bone Miner. Res. **15**(4), 721–739 (2000)
14. Lenchik, L., Rogers, L.F., Delmas, P.D., Genant, H.K.: Diagnosis of osteoporotic vertebral fractures: importance of recognition and description by radiologists. Am. J. Roentgenol. **183**(4), 949–958 (2004)
15. Lin, T.Y., Dollár, P., Girshick, R., He, K., Hariharan, B., Belongie, S.: Feature pyramid networks for object detection. In: Proceedings of the IEEE Conference on Computer Vision and Pattern Recognition, pp. 2117–2125 (2017)
16. Luvizon, D.C., Tabia, H., Picard, D.: Human pose regression by combining indirect part detection and contextual information. Comput. Graph. **85**, 15–22 (2019)
17. Milletari, F., Navab, N., Ahmadi, S.A.: V-net: fully convolutional neural networks for volumetric medical image segmentation. In: 2016 Fourth International Conference on 3D Vision (3DV), pp. 565–571. IEEE (2016)
18. Mitchell, R.M., Jewell, P., Javaid, M.K., McKean, D., Ostlere, S.J.: Reporting of vertebral fragility fractures: can radiologists help reduce the number of hip fractures? Arch. Osteoporos. **12**(1), 1–6 (2017). https://doi.org/10.1007/s11657-017-0363-y
19. Nicolaes, J., et al.: Detection of vertebral fractures in CT using 3D convolutional neural networks. arXiv preprint arXiv:1911.01816 (2019)
20. Redmon, J., Farhadi, A.: YOLO9000: better, faster, stronger. In: Proceedings of the IEEE conference on computer vision and pattern recognition, pp. 7263–7271 (2017)
21. Redmon, J., Farhadi, A.: YOLOv3: an incremental improvement. CoRR abs/1804.02767 (2018). http://arxiv.org/abs/1804.02767
22. Ren, S., He, K., Girshick, R., Sun, J.: Faster R-CNN: towards real-time object detection with region proposal networks. In: Advances in Neural Information Processing Systems, pp. 91–99 (2015)
23. Ronneberger, O., Fischer, P., Brox, T.: U-Net: convolutional networks for biomedical image segmentation. In: Navab, N., Hornegger, J., Wells, W.M., Frangi, A.F. (eds.) MICCAI 2015. LNCS, vol. 9351, pp. 234–241. Springer, Cham (2015). https://doi.org/10.1007/978-3-319-24574-4_28
24. Roth, H.R., Wang, Y., Yao, J., Lu, L., Burns, J.E., Summers, R.M.: Deep convolutional networks for automated detection of posterior-element fractures on spine CT. In: Medical Imaging 2016: Computer-Aided Diagnosis, vol. 9785, p. 97850P. International Society for Optics and Photonics (2016)
25. Sekuboyina, A., Kukačka, J., Kirschke, J.S., Menze, B.H., Valentinitsch, A.: Attention-driven deep learning for pathological spine segmentation. In: Glocker, B., Yao, J., Vrtovec, T., Frangi, A., Zheng, G. (eds.) MSKI 2017. LNCS, vol. 10734, pp. 108–119. Springer, Cham (2018). https://doi.org/10.1007/978-3-319-74113-0_10
26. Tomita, N., Cheung, Y.Y., Hassanpour, S.: Deep neural networks for automatic detection of osteoporotic vertebral fractures on ct scans. Comput. Biol. Med. **98**, 8–15 (2018)
27. Valentinitsch, A., et al.: Opportunistic osteoporosis screening in multi-detector CT images via local classification of textures. Osteoporos. Int. **30**(6), 1275–1285 (2019)

Grading Loss: A Fracture Grade-Based Metric Loss for Vertebral Fracture Detection

Malek Husseini[1,2(✉)], Anjany Sekuboyina[1,2], Maximilian Loeffler[2], Fernando Navarro[1,2], Bjoern H. Menze[1], and Jan S. Kirschke[2]

[1] Department of Computer Science,
Technical University of Munich, Munich, Germany
malek.husseini@tum.de
[2] Klinikum rechts der Isar, Technical University of Munich, Munich, Germany

Abstract. Osteoporotic vertebral fractures have a severe impact on patients' overall well-being but are severely under-diagnosed. These fractures present themselves at various levels of severity measured using the Genant's grading scale. Insufficient annotated datasets, severe data-imbalance, and minor difference in appearances between fractured and healthy vertebrae make naive classification approaches result in poor discriminatory performance. Addressing this, we propose a representation learning-inspired approach for automated vertebral fracture detection, aimed at learning latent representations efficient for fracture detection. Building on state-of-art metric losses, we present a novel *Grading Loss* for learning representations that respect Genant's fracture grading scheme. On a publicly available spine dataset, the proposed loss function achieves a fracture detection $F1$ score of 81.5%, a 10% increase over a naive classification baseline.

Keywords: Fracture detection · Metric loss · Representation learning

1 Introduction

Vertebral fractures are severely under-diagnosed. According to a 2013 study, 84% of incidental vertebral fractures were not reported in CT [2]. This is either due to the fractures being asymptomatic or to the symptoms wrongly being attributed to other factors. Osteoporotic vertebral fractures have critical consequences such as disability or increased mortality. Osteoporotic vertebral fractures cause pain and kyphosis in the short term, but are associated with an eightfold higher

M. Husseini and A. Sekuboyina—Shared first authors.

Electronic supplementary material The online version of this chapter (https://doi.org/10.1007/978-3-030-59725-2_71) contains supplementary material, which is available to authorized users.

A. L. Martel et al. (Eds.): MICCAI 2020, LNCS 12266, pp. 733–742, 2020.
https://doi.org/10.1007/978-3-030-59725-2_71

mortality in the long term [3]. Accentuating this is their high prevalence in older adult population (40% by the age of 80 years), making a missed diagnosis critical. Therefore, there is a need for an automated and reproducible detection of vertebral fractures.

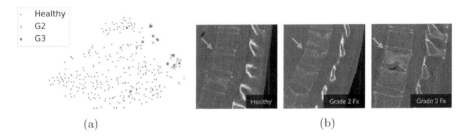

(a)   (b)

Fig. 1. Illustrating fracture grades: (a) TSNE visualisation of latent representations learnt by formulating fracture detection as a simple classification problem, resulting in poor separability. (b) An example selection of the three classes of vertebrae studied in this work, healthy, grade-2 fracture, and grade-3 fracture.

Vertebral Fracture Detection. Automatic detection of vertebral fractures is relatively unexplored. Valetinitsch et al. [7] propose the extraction of texture-based features such as histogram of gradients or local binary patterns from the trabecular of a segmented verebrae and classifying them using a random forest. From a deep learning perspective, Bar et al. [8] employ a convolutional neural network for classifying sagittal patches from the vertebral column and aggregating the classification across patches using a recurrent neural network. Along similar lines, Tomita et al. [9] work on thoraco-lumbar slices processed with a CNN and aggregated across slices using a long short-term memory (LSTM) network. However, unlike [8], the latter does not need any anatomy to be segmented to start the processing. Note that these approaches are *ad hoc* implementations of CNNs working on large data samples and provide minimal insights into the workings of the network. Recently, Nicolaes et al. [10] proposed a fully 3D approach for detecting vertebral fractures based on a voxel-level prediction regime, also providing a weak localization of the fracture. However, it being patch-based and predicting *per voxel* limits its real-time applicability.

We argue that formulating a vertebral fracture detection as a naive classification problem is sub-optimal, more so in case of limited and unbalanced data regimes. Figure 1a illustrates the TSNE representations of the latent features of one of the baselines in this work, viz. detecting vertebral fractures using a simple cross entropy loss using a convolutional neural network. Observe the resulting poor class-separation between healthy and fractured vertebrae. We attribute this to the wide variation in vertebral shapes: a healthy lumbar vertebra is 'more different' from a healthy upper-thoracic vertebra than a fractured lumbar vertebra.

Moreover, there exists a 'gradation' among vertebral fractures, further obfuscating a clear shape-based separation (cf. Fig. 1b).

Genant's Vertebral Fracture Grading. The Current gold standard in grading vertebral fractures is a semi-quantitative method developed by Genant et al. [1], according to which fractures are categorized into three grades (Grades 1, 2, and 3; cf. Fig. 1b). This is based on the height-loss a vertebra undergoes compared to its healthy counterpart. A healthy vertebra is considered to be Grade 0. Grades 2 and 3 have proven clinical consequences, while this is unknown for Grade 1; as its small height reduction results in a high inter-rater uncertainty, Grade 1 fractures are excluded in this study.

Representation Learning. In this work, we aim to incorporate the gradual shape variations, courtesy of the fracture grades, into the training process of a classifier by explicitly adjusting the latent space. Deep learning models are believed to generate useful representations as a byproduct of the task they are trying to solve. However, this is not the case in low-data regimes as shown in Fig. 1a. *Representation learning* or *metric learning* can be used to learn efficient latent representations in such scenarios. Siamese networks [15] using contrastive loss and Face-Net [14] with its triplet loss are examples of standard metric learning frameworks wherein representations of similar entities are clustered together while those of dissimilar ones are pushed apart.

Contributions. In this work, we attempt to solve vertebral fracture detection as a two-class, healthy *vs.* fractured classification problem.

- Towards accurate classification, we pre-train the neural network using fracture-grade based representation learning. For this, we propose a novel *loss function* termed *grading loss*, which encourages the learnt representations to respect the gradation in the appearance of fractures.
- Accounting for the dependence of vertebral shapes on vertebral labels, we also propose a spine-region based pre-conditioning module.
- We validate the proposed fracture detection regime on a publicly available VerSe dataset obtaining a classification $F1$ score of 82%, outperforming naive classification as well as standard representation learning approaches.

2 Methodology

Given a collection of 2D vertebral patches, the objective of our work is to classify them into two classes, fractured and healthy. As vertebral shape depends on it label and the amount of variation in this shape due to a fracture depends on the fracture grade, we hypothesize that preceding the classification stage with fracture-grade and vertebral-label dependant pre-training results in an improved class separation. Consequently, this results in improved classification performance. We model these pre-training stages with inspiration from the field of representation learning.

2.1   Grading Loss

Metric learning aims to learn better data representations by working on a notion of 'distance' in the latent space. It aims to cluster similar objects closer (by reducing the distance between them) while pushes dissimilar objects farther. This is done by optimizing loss functions, such as the contrastive [15] and triplet losses [14]. Note that these losses work on a notion of data similarity and dissimilarity. By design, they do not include a 'ranking' within this similarity or dissimilarity. For example, a 'ranking' is obvious in vertebral fractures, where a grade-2 and grade-3 vertebrae are *fractured*, but the former is more similar to a vertebra from a *healthy* class than the latter. Incorporating such 'ranking' criterion into the metric learning framework, we propose the *grading loss*.

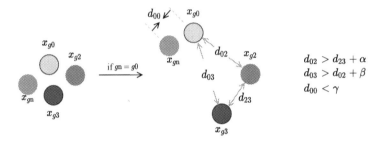

Fig. 2. Left: the arrangement of the three grades and the positive anchor during initialization. Right: $g3$ is drawn closer to $g2$ and both are pulled away from $g0$ under the constraint of $g0$ being closer to $g2$ than to $g3$. The positive anchor gn works on clustering the similar classes together in the latent space. gn belongs to the $g0$ sub-class in this figure.

Assume a 2D vertebral patch, $x \in X$, is mapped to a representation $f(x)$ by a neural network f. The Euclidean distance between the representations of two examples x^i and x^j is denoted by $d(x^i, x^j) = ||f(x^i) - f(x^j)||_2^2$. We design the *grading loss* as a quadruplet loss [11] working with quadruplets denoted by $\{x_{g0}^i, x_{g2}^j, x_{g3}^k, x_{gn}^l\}$, where x_{g0}^i denotes a healthy vertebra sample, x_{g2}^j and x_{g3}^k denote samples of grade two and three, respectively, and n $\in \{0, 2, 3\}$ can be randomly chosen as a healthy or a fractured example. Observe that the x^i, x^j, and x^k form a static triplet, i.e. they are always sampled from fixed sub-classes of fracture grades. We incorporate a grading in the embedding space as follow: a grade-3 fracture is farther away from a healthy (grade-0) vertebra than a grade-2 fracture vertebra, and grade-2 is closer to grade-3 and it is to healthy. We can formulate these requirements as:

$$d(x_{g2}, x_{g3}) + \alpha < d(x_{g2}, x_{g0}) \text{ and} \tag{1}$$

$$d(x_{g0}, x_{g2}) + \beta < d(x_{g0}, x_{g3}), \tag{2}$$

where α and β are distance thresholds. Note that the Eq. 1 uses $g2$ as a reference sample and Eq. 2 uses $g0$. Owing to the triangular inequality of distances, the restrictions on the distances from $g3$ to $g0$ and $g2$ are already satisfied by the above conditions. The above conditions can be achieved by optimizing the following loss terms:

$$\mathcal{L}_1 = \max(0, d(x_{g2}, x_{g3}) - d(x_{g2}, x_{g0}) + \alpha) \text{ and} \tag{3}$$

$$\mathcal{L}_2 = \max(0, d(x_{g0}, x_{g2}) - d(x_{g0}, x_{g3}) + \beta). \tag{4}$$

Observe that Eqs. 3 and 4 structurally represent the triplet loss. However, observe that these do not work on similarities. They form *separating* objectives between various fracture grades. Finally, a third, *clustering* objective is incorporated by virtue of x^l in the quadruplet. Recall that x_{gn} could belong to any of the three sub-classes. Based on the value of n in the sampled triplet, the clustering objective *pulls* x_{gn} closer to its match in the static triplet. We demonstrate our *grading loss* in Fig. 2 with x_{gn} belonging to $g0$ sub-class and we refer to x_{gn} by the term positive anchor. The clustering objective can be represented as:

$$\mathcal{L}_3 = \max(0, \gamma - d(x_{gn}^{\{i,j,k\}}, x_{gn}^l)) \tag{5}$$

where $n \in \{0, 2, 3\}$ and x^i and x^j are a pair of samples from the same class. Assembling the loss terms together results in the proposed objective of *grading loss*, $\mathcal{L}_G = \mathcal{L}_1 + \mathcal{L}_2 + \mathcal{L}_3$. In this work, the distance thresholds are chosen to be $\alpha > \beta > \gamma$. This is to ensure that $d(x_{g2}, x_{g0})$ is as large as possible while maintaining $d(x_{g3}, x_{g0}) > d(x_{g2}, x_{g0})$. That is, a higher separation between fractured and healthy classes is desirable compared to that between the two grades.

2.2   Conditioning Representations on Vertebral Indices

Considering the wide variation in shape from cervical to lumbar vertebrae, we claim that learning label-specific representations as a pre-training stage also improves fracture detection. Assuming the availability vertebral labels during the training process, we construct five categories of vertebra based on their shape similarity: $T1 \sim T5$, $T6 \sim T9$, $T10 \sim T12$, $L1 \sim L4$, and $L5$. Treating this as a five-class problem, we can employ any standard metric loss for learning label-specific representations. Note that our *grading loss* can also be extended for this case as there exists a 'ranking' among the classes. Another application could be in brain tumors, where grades are also present (high grade and low grade gliomas). We leave the application of our loss to such scenarios for future work.

2.3   Implementation

We perform fracture detection with the following network architecture [13], containing 5×5 filter kernels wherever applicable:

(conv32-bn-relu) $\rightarrow$ maxpool $\rightarrow$ (conv64-bn-relu) $\rightarrow$ maxpool (conv128-bn-relu)
$\rightarrow$ maxpool $\rightarrow$ (conv256-bn-relu) $\rightarrow$ maxpool $\rightarrow$ (linear256-bn-lrelu)
$\rightarrow$ (linear128-bn-lrelu) $\rightarrow$ (linear64-bn-lrelu) $\rightarrow$ linear8

where 'bn' and 'lrelu' represent batch normalization and leaky Relu layers respectively. Recall that our network consists of a pre-training stage (for representation learning) and a training stage (for fracture detection). Furthermore, the pre-training stage consists of two sub-stages: vertebral-index-based representation learning and fracture-grade-based representation learning. Once pre-trained, the network is trained by optimizing a binary cross entropy loss over the fractured and healthy classes. For this, the last linear layer (linear8) is replaced with a two node linear layer (linear2) for the two classes. The network is implemented using the Pytorch library on an Nvidia GTX 1080 gpu. All losses are optimized using the Adam optimizer with a learning rate of 0.0001. We set the hyper-parameters α, β and γ to 1.5, 1 and 0.5 respectively.

3   Experiments

In this section, we evaluate the contribution of the two main components proposed as part of our classification routine: first, the proposed grading loss' ability in learning efficient representations, and second, our complete fracture detection routine.

Dataset. Recall that the proposed approach works at a vertebra level and utilizes vertebral labels. We utilize the publicly available VerSe [4–6] dataset and its centroid annotations. As part of [1], its vertebra are annotated for fractures of three grades. We work with healthy, grade-2 and grade-3 fracture. We exclude the cervical vertebrae ($C1 \sim C7$) as vertebral fractures are extremely rare in this region. The dataset consists of 1283 vertebrae extracted from 157 scans, among which 1133 are healthy, 104 are $g2$ fractures and 46 are $g3$ fractures. The data is split into a training set containing 966 vertebrae and a test set with 312 vertebrae. The healthy:$g2$:$g3$ ratio in these sets is 851:79:36 and 282:25:10, respectively.

Data Preparation. Typically, a vertebra's mid-sagittal slice is a good indicator of a fractures. However, in cases where the vertebra presents itself in an atypical orientation, using the mid-sagittal slice is ineffective. Therefore, we utilize the vertebral centroids to extract 2D reformations of the vertebra along the mid-vertebral plane perpendicular to the vertebra's sagittal axis. Specifically, we construct a spline passing through the centroids and reformat the sagittal plane along which this spline passes. From this reformation, vertebral patches of size 112×112 pixels at 1×1 mm resolution are extracted so that additional context is provided by the vertebra above and below the vertebra-of-interest (VOI). Our network consumes these patches. Additionally, a Gaussian around the centroid is passed as an additional channel for indicating the VOI.

Table 1. Evaluating learnt representations (**representation learn → fracture train**): Performance comparison of various losses for learning fracture-specific representations. * indicates statistical insignificance (p-value = 0.44).

Setup	SN	SP	$F1$
Contrastive	54.2 ± 5.3	$\mathbf{95.6 \pm 1.5^*}$	57.4 ± 4.2
Triplet	$\mathbf{71.7 \pm 5.1}$	90.2 ± 2.0	57.3 ± 3.3
Grading	67.1 ± 10.2	$\mathbf{95.3 \pm 1.6^*}$	$\mathbf{65.2 \pm 5.1}$

3.1 Experiments and Results

We validate the proposed *grading loss* in two stages: first, it is deployed as a stand-alone representation learning loss, where the separability of the learnt representations is tested (without any fracture-oriented training), and second, it is combined with a fracture classification module as a pre-training stage along with the proposed spine region-based representation learning component. The classification performances of various setups are compared using sensitivity (SN), specificity (SP), and $F1$ scores. We report the mean scores across fifteen randomly chosen folds of the dataset. Note that the proportion of healthy:$g2$:$g3$ is preserved through all the folds.

***Grading Loss* Results in Better Representations.** In this experiment, we validate the effectiveness of the proposed *grading loss* at learning efficient representations for fracture detection. We compare our loss with the two standard metric-learning losses: contrastive and triplet losses. Specifically, the neural network is optimized on the training set using each of the metric losses. Once trained, it is used to obtained the latent representations of the training samples on which a support vector machine (SVM) with a linear kernel is learnt. The more linearly separable the representations are, better the learnt SVM performs on the test set's latent representation. Table 1 reports the classification performance of this SVM on the test set representations. Observe that the *grading loss* readily offer better 'linear' separability (∼8% increase in $F1$ score) of the fracture vs. healthy classes compared to the contrastive and triplet loss. The TSNE visualisations of these representations (cf. Fig. 3) illustrate this clustering characteristic of the *grading loss*.

Proposed Fracture Detection Regime. Our complete fracture detection pipeline consists of three stages: two pre-training stages followed by the main classification stage. The first pre-training stage includes optimizing a contrastive loss over the five regions of spine described in Sect. 2.2. Experiments justifying the choice of contrastive loss for this stage are presented in the supplement. Following this, the network goes through the second pre-training stage where our *grading loss* is minimized. Finally, the network is optimized for fracture detection using cross entropy loss. We represent the proposed pipeline as

Contrastive Loss   Triplet Loss   *Grading loss*

Fig. 3. TSNE visualisation of the representations learnt by various metric learning losses, without explicit classification-specific training. Proposed *grading loss* obtains more separability between healthy and fractured classes.

Table 2. Validating the proposed fracture detection regime (**label pre-train → representation learn → fracture train**): Comparison of the proposed training routine based on *grading loss* with naive classification as well as with other representation-learning-augmented classifications.

Label pre-train	Rep. learn	Frac. train	SN	SP	$F1$
✗	✗	✓	71.0 ± 6.4	96.4 ± 1.7	71.2 ± 4.4
✓	✗	✓	$\mathbf{72.4 \pm 7.7}$	$\mathbf{97.8 \pm 1.0}$	$\mathbf{75.9 \pm 5.5}$
✗	Contrastive	✓	73.7 ± 7.7	98.0 ± 1.1	77.6 ± 2.6
✗	Triplet	✓	73.7 ± 6.3	96.4 ± 1.4	73.6 ± 2.4
✗	**Grading**	✓	$\mathbf{76.0 \pm 5.8}$	$\mathbf{97.8 \pm 1.1}$	$\mathbf{78.6 \pm 4.4}$
✓	Contrastive	✓	74.3 ± 4.7	98.0 ± 0.6	78.2 ± 2.1
✓	Triplet	✓	75.7 ± 6.4	97.6 ± 0.8	77.5 ± 2.9
✓	**Grading**	✓	$\mathbf{76.9 \pm 5.8}$	$\mathbf{98.5 \pm 0.95}$	$\mathbf{81.5 \pm 3.8}$

label pre-train → representation learn → fracture train). Table 2 reports an ablative test of the proposed routine. We test the contribution of label-based pre-training and that of the proposed grading loss-based representation learning is evaluated. Compared with a baseline network trained end-to-end for fracture detection, pre-training with vertebral labels offers a 5% improvement in $F1$ score. On a different note, tuning the representations with fracture grades using *grading loss* also improves the classification performance, providing about 7% $F1$ score improvement over simple classification. These validate the effectiveness of the two pre-training stages incorporated in our routine. Finally, testing all the stages of the proposed pipeline, we observe that our combination results in the highest performance across the three metrics with an overall $F1$ score of 81%.

4   Conclusion

We conclude that in case of low-data regimes with severe data imbalance, augmenting classification with representation learning-based pre-training helps.

Compared to conventional metric losses which work on similarity or dissimilarity of examples, the proposed *grading loss* which incorporating a 'ranking' within the classes provides a superior performance. Going a step further, incorporating the vertebral label information using similar techniques of representation learning further improves fracture detection. The proposed fracture routine achieves an $F1$ score of 81.5%, an improvement of over 10% over naive classification baseline. In future work, we will extend this study to incorporate grade-1 fractures as well as a 3D context, thus making our approach more robust to severely deformed spines.

Acknowledgements. This work is supported by DIFUTURE, funded by the German Federal Ministry of Education and Research under (01ZZ1603[A-D]) and (01ZZ1804[A-I]).

References

1. Genant, H.K., et al.: Vertebral fracture assessment using a semiquantitative technique. J. Bone Miner. Res. **8**(9), 1137–1148 (1993)
2. Carberry, G., et al.: Unreported vertebral body compression fractures at abdominal multidetector CT. Radiology **268**(1), 120–126 (2013)
3. Cauley, J., et al.: Risk of mortality following clinical fractures. Osteoporos. Int. **11**(7), 556–561 (2000)
4. Loeffler, M., et al.: A vertebral segmentation dataset with fracture grading. Radiol. Artif. Intell. **2**(4), e190138 (2020)
5. Sekuboyina, A. et al.: VerSe: A Vertebrae Labelling and Segmentation Benchmark. arXiv eprint: 2001.09193. arXiv preprint arXiv:2001.09193 (2020)
6. Sekuboyina, A., et al.: Labelling vertebrae with 2D reformations of multidetector CT images: an adversarial approach for incorporating prior knowledge of spine anatomy. Radiol. Artif. Intell. **2**(2), e190074 (2020). https://doi.org/10.1148/ryai. 2020190074
7. Valentinitsch, A.: Opportunistic osteoporosis screening in multi-detector CT images via local classification of textures. Osteoporos. Int. **30**(6), 1275–1285 (2019). https://doi.org/10.1007/s00198-019-04910-1
8. Bar, A., et al.: Compression fractures detection on CT. In: Medical Imaging 2017: Computer-Aided Diagnosis, vol. 10134, p. 1013440. International Society for Optics and Photonics (2017)
9. Tomita, N., et al.: Deep neural networks for automatic detection of osteoporotic vertebral fractures on CT scans. Comput. Biol. Med. **98**, 8–15 (2018)
10. Nicolaes, J. et al.: Detection of vertebral fractures in CT using 3D Convolutional Neural Networks. arXiv preprint arXiv:1911.01816 (2019)
11. Chen, W., et al.: Beyond triplet loss: a deep quadruplet network for person re-identification. In: Proceedings of the IEEE Conference on Computer Vision and Pattern Recognition, pp. 403–412 (2017)
12. Husseini, M., Sekuboyina, A., Bayat, A., Menze, B.H., Loeffler, M., Kirschke, J.S.: Conditioned variational auto-encoder for detecting osteoporotic vertebral fractures. In: Cai, Y., Wang, L., Audette, M., Zheng, G., Li, S. (eds.) CSI 2019. LNCS, vol. 11963, pp. 29–38. Springer, Cham (2020). https://doi.org/10.1007/978-3-030-39752-4_3

13. Raghu, M., et al.: Transfusion: understanding transfer learning for medical imaging. Adv. Neural Inf. Process. Syst. **10**(10007/1234567890), 3342–3352 (2019)
14. Schroff, F., et al.: FaceNet: a unified embedding for face recognition and clustering. In: Proceedings of the IEEE Conference on Computer Vision and Pattern Recognition, pp. 815–823 (2015)
15. Hadsell, R., et al.: Dimensionality reduction by learning an invariant mapping. In: IEEE Computer Society Conference on Computer Vision and Pattern Recognition, vol. 2, pp. 1735–1742 (2006)
16. Finn, C., et al.: Model-agnostic meta-learning for fast adaptation of deep networks. In: Proceedings of the 34th International Conference on Machine Learning, vol. 70, pp. 1126–1135 (2017)

3D Convolutional Sequence to Sequence Model for Vertebral Compression Fractures Identification in CT

David Chettrit[1(✉)], Tomer Meir[2], Hila Lebel[3], Mila Orlovsky[3],
Ronen Gordon[3], Ayelet Akselrod-Ballin[3], and Amir Bar[3]

[1] Independent AI Consultant, Tel Aviv-Yafo, Israel
davidvcfpaper@outlook.com
[2] Weizmann Institute of Science, Rehovot, Israel
[3] Zebra Medical Vision Ltd., Shefayim, Israel

Abstract. An osteoporosis-related fracture occurs every three seconds worldwide, affecting one in three women and one in five men aged over 50. The early detection of at-risk patients facilitates effective and well-evidenced preventative interventions, reducing the incidence of major osteoporotic fractures. In this study we present an automatic system for identification of vertebral compression fractures on Computed Tomography images, which are often an undiagnosed precursor to major osteoporosis-related fractures. The system integrates a compact 3D representation of the spine, utilizing a Convolutional Neural Network (CNN) for spinal cord detection and a novel end-to-end sequence to sequence 3D architecture. We evaluate several model variants that exploit different representation and classification approaches, and present a framework combining an ensemble of models that achieves state of the art results, validated on a large data set, with a patient-level fracture identification of 0.955 Area Under the Curve (AUC). The system proposed has the potential to support osteoporosis clinical management, improve treatment pathways and to change the course of one of the most burdensome diseases of our generation.

1 Introduction

Worldwide, an osteoporotic fracture occurs on average every three seconds, with one in three women and one in five men aged over 50 experiencing an osteoporotic fracture [1,2] at significant economic cost to health and social care systems. Treatments of osteoporosis are widely available, evidence-based and cost effective [3], but a considerable diagnostic gap exists to identify patients at risk of fracture. However, in over 55% of cases, major osteoporotic fractures (MOFs) are preceded by an often asymptomatic and undiagnosed warning sign, the vertebral compression fracture (VCF) [4,5].

D. Chettrit and T. Meir—Equal contribution
A. Akselrod-Ballin and A. Bar—Equal advising.

© Springer Nature Switzerland AG 2020
A. L. Martel et al. (Eds.): MICCAI 2020, LNCS 12266, pp. 743–752, 2020.
https://doi.org/10.1007/978-3-030-59725-2_72

Despite their prognostic value, VCFs commonly go unreported on routine radiological imaging, with between 13–16% of retrospectively confirmed VCFs actually reported at the time of the initial imaging study [6,7]. Typically, VCFs are not a difficult radiological diagnosis, rather they are overlooked as 'incidental' findings relative to the primary reason for undertaking the CT examination [8].

Previous work in the domain of automatic identification of VCF utilize traditional machine learning approaches, which commonly focus on vertebra segmentation and are applied on small data sets of several hundred of studies [9,10]. For example, the work by Valentinitsch et al. [11] employed a random forest (RF) classifier based on texture features and regional vertebral Bone Mineral Density analysis to identify VCF, and was applied on approximately 200 patients.

Following recent advances in deep learning, Chen et al. [12] integrated a Convolutional Neural Network (CNN) with an initial vertebra localization using RF's. The authors in Bar et al. [13], first segment the portion of the spine included in a Computed Tomography (CT) scan and virtually construct a 2D sagittal section of the spine. Then a two-dimensional (2D) CNN classifies the sagittal patches and finally a Recurrent Neural Network (RNN) is applied on the resulting vector of probabilities. Similarly, Tomita et al. [14] proposed a 2D approach that first utilizes a CNN-based feature extraction module from 2D CT slices and then uses an RNN based feature aggregation method based on Long Short-Term Memory (LSTM) networks. RNN's and specifically LSTM's have been successfully applied in many applications [15]. Still, this type of representation is limited in exploiting the full vertebra sequence information composing the spine.

A different approach by Sekuboyina et al. [16] performed vertebra labeling based on a butterfly-shaped network architecture that combines the information from the sagittal and coronal maximal intensity projection images in addition to an adversarial energy-based auto-encoder as a discriminator for learning of the butterfly net. Our proposed system differs from the majority of studies in the field that rely on 2D/2.5D in that it uses 3D vertebral volumes. Similarly, a recent study by Nicholas et al. [17] presented a two staged VCF detection method that first predicts the class probability for every voxel using a 3D CNN, followed by aggregation of the information to VCF and patient level predictions. This study demonstrated the advantage of 3D representation, yet its major shortcoming was the use of a small dataset, of only 90 patients.

To summarize, this study proposes an automatic system for identification of VCF which integrates several novel deep learning components. Our contribution is three fold. First, we create a compact 3D representation of the spine. The representation is based on a spinal cord detection CNN, a reconstructed sagittal view, and volumetric patches along the vertebral column. The benefit of this representation is that it preserves 3D details, important for diagnosis, incorporating intensity, texture and localization features, making it effective for further classification. Second, to the best of our knowledge we are the first to combine a sequence to sequence architecture learned end-to-end for this type of application. Leveraging a sequence of 3D features based on a sequence of vertebrae

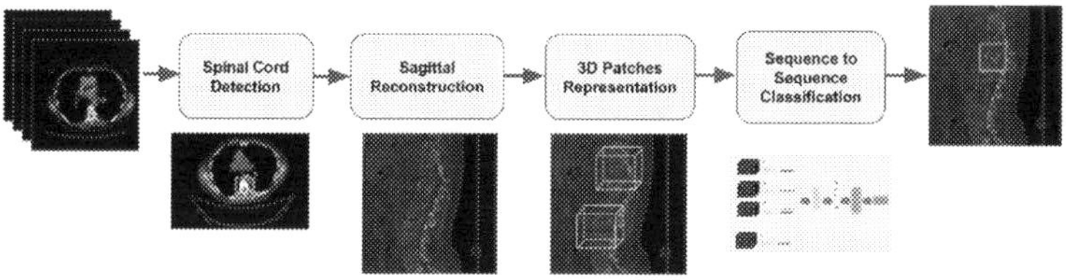

Fig. 1. System outline. Given a CT scan, the algorithm combines 3D components to output a reconstructed sagittal view detecting VCF in a bounding box.

allows incorporating local 3D features with the entire vertebra column information. Third, we report state of the art (SOTA) results, on one of the largest CT datasets in this problem domain. Finally, the system presented here can be used in clinical practice for early detection of patients suspicious for osteoporosis, enabling focused interventions and reducing the incidence, thus fracture burden on a wider scale.

2   Methods

The framework inputs chest and abdominal axial CT images and generates a compact 3D representation of the spine. This phase includes spinal cord detection based on a CNN, sagittal reconstruction, and extraction of 3D patches along the vertebral column. Then, a sequence to sequence compression fractures classification and localization component is applied. A high level description of our method is depicted in Fig. 1.

2.1   From CT to Vertebral Column Region of Interest (ROI)

A YOLO [18] detector type of CNN was utilized in order to localize the spinal cord on axial CT scans, allowing an accurate definition of the region of interest (ROI) for the pipeline. The detector was trained to identify the spinal cord location on every axial slice in the volume each 30 mm. The remaining locations were then linearly interpolated. The model was trained on 300 slices from the training set where the spinal cord location was annotated by a bounding box.

2.2   From Vertebral Column ROI to Patch Representation

Due to the large data size involved in CT imaging of the spine, a 3D compact representation of the spine VOI (volume of interest) is created as follows. First, the sagittal view is reconstructed from the axial scans of the volume. Each sagittal slice in the 3D volume is resampled to maintain equal pixel size in x and y. The resampling is done along the column of each slice column using Fourier Transforms. Then, given the spinal cord localization, the volume surrounding the spinal column is cropped and partitioned into a set of k 3D volumetric patches,

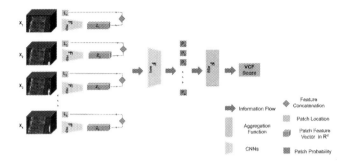

Fig. 2. Flowchart of the classification and localization architecture. It presents the flow from 3D input patches (X_i) to the series overall score (VCF score), including the patch feature extractor F_{rep}, the position information per patch (L_i), the sequence to sequence CNN F_{seq}, the fracture probabilities per patch (P_i), and the aggregation function F_{agg}.

such that the vertebra column is completely tiled with a minimal overlap. The patches are sampled, resized to (p_H, p_W, p_Z) corresponding to the size of the patches in the sagittal, coronal, and axial dimensions respectively, and normalized after a HU windowing is applied (window center 370 HU and window width 840 HU). This step produces the input ($k, p_H, p_W, p_Z, 1$) to the following classification and localization step.

2.3   Classification and Localization

A high level description of this classification and fracture localization method is depicted in Fig. 2.

Basic Setup. Let $X = (x_1, ..., x_k)$ be an input sequence of length k, where every $x_i \in \mathbb{R}^{H \times W \times Z}$ is a single 3D patch of the representation described in Sect. 2.2. Next, we use three function: F_{rep}, F_{agg} and F_{seq}, where F_{rep} is a mapping from $\mathbb{R}^{H \times W \times Z}$ to $\mathbb{R}^D$ and is typically used to map a patch into a new vector representation; F_{agg} is a mapping from $\mathbb{R}^{K \times D}$ to $\{0, 1\}$, e.g. performs aggregation of an entire sequence into a single result; and F_{seq} is a mapping from $\mathbb{R}^{K \times D}$ to $\{0, 1\}^K$, e.g, performs classification for each sequence item, considering other sequence items. F_{rep}, F_{agg} and F_{seq} are typically learned neural networks. Thus, given an input sequence X we start by applying F_{rep} to obtain an input representation per sequence item:

$$Z = (F_{rep}(x_1), ..., F_{rep}(x_k))$$

Where $Z_i \in \mathbb{R}^D$ is the representation obtained for each sequence item. We then use F_{agg} and F_{seq} to obtain an aggregated sequence result and a result per sequence item respectively. To do this, we simply need to apply F_{agg} and F_{seq}. Our loss is thus defined as follows:

$$L = L_1(Y_{agg}, F_{agg}(F_{seq}(Z))) + \lambda L_2(Y_{seq}, F_{seq}(Z))$$

Where we set L_1 and L_2 as the binary cross entropy loss. Y_{agg} and Y_{seq} are the ground truth fracture label per series and per patch, respectively. λ is here a weight to the sequence component of the loss. We train this system in an end-to-end fashion. Next, we explore possible choices for F_{rep}, F_{seq} and F_{agg}.

Patch Representation. As described, the input patch representation is typically a cropped $3D$ region of a CT image. Thus, we can obtain a natural representation of it using a $3D$ CNN. Specifically, we use a CNN of the following architecture: 3 blocks, where each blocks contains two 3D convolutions followed by 3D Max Pooling. The convolution filters is doubled for every block, and is initially starting at 8. The resulting output is used a feature vector representation per patch.

We note that a single patch is independent of the global context of the entire volume and of the specific location where it originally resides. To alleviate this difficulty, we supplement the learned CNN representation with a location descriptor. This enables learning a representation obtained by the CNN. A location feature is simply a scalar indicating the anatomical relative position of the patch center.

Sequence Items Classification. Thus far we have considered ways to map an input sequence of patches into a new sequence of representation which better captures the higher level semantics in the input sequence. Next, we describe possible ways to binary classify single sequence items. Perhaps the most natural strategy would be to convolve a *max* filter over the sequence such that every sequence item is scored according to its environment *max* value. An alternative could be to go over the sequence in a serial manner, e.g. top to bottom or vise-versa and make the decision based on the sequence representation up to now. This could be simply achieved using an LSTM layer. This approach delivered the best results.

Sequence Aggregation. Given a sequence of probabilities, we can consider multiple aggregation strategies. A natural way to aggregate a sequence into a single result is using the *max* value, e.g. a sequence is positive if a single fracture is found. The *max* value could be unreliable due to noise; thus, smoothing is applied as a preceding step. An alternative approach is by considering multiple items in the sequence, either in a certain order or not. One possible ordering is simply the sequence of patches along the spine from top to bottom or vice versa. After experimenting different approaches as described in Sect. 4, we finally opted for the *max* function.

3   Experiments

3.1   Dataset

The data includes retrospectively collected CT chest and abdomen images acquired on different types of scanners (Siemens, Philips and General Electric), with a vast representation of institutions, containing the vertebrae between T1 to L5. Patients were separated to independent training, tuning and test partitions, with 80%, 10%, and 10% of the patients, respectively. Ground truth was determined by consensus agreement of three US Board Certified radiologists. Each Radiologist was provided with full-volume sagittal CT series and identified the presence or absence of vertebral compression fracture according to the Genant scale [19].

The training set consisted of 1,832 standard CT series of patients 50 years old and higher, with 613 (33%) positive series and 1,219 (67%) negative series for VCF. Cardiac focused chest CT, lumbar spine, chest CT with lung windowing and PET-CT were excluded, since there is little value in opportunistic screening of vertebral compression fractures for this type of data.

The tuning set, used for setting the algorithm hyper-parameters consisted of 311 series, 133 (42.8%) of them were positive for VCF and 178 (57.2%) were negative. The tuning set distributed similarly to the training set and had the same exclusion criteria. Finally, the held-out validation set to test the algorithm performance was curated from the test patients partition, and consisted of 346 series, 153 (44.2%) of them considered positive. This set distributed similarly to training and tuning sets.

3.2   Implementation Details

Spinal Cord Localization. The model's validation set comprised of 200 additional axial slices unseen during training. The results were then validated by an expert. The expert reviewed the model's bounding box localization of the spinal cord and agreed with 98.99% of the cases.

Image Augmentation. Each spine patch ROI in the series sequence was preprocessed with the following additional augmentations: random horizontal and vertical flip 50% of all the images and random rotation (up to $\pm 20°$). Augmentations were chosen to align with commonly-used approaches [13,20] and limited to 50% of all cases to ensure the usual appearance of the vertebrae is preserved.

Mini-batch Training. For training, we use a learning rate of $1e-5$ and batch size of 16. The weights were randomly initiated and trained for 2000 epochs with 150 iterations per epoch and an equal proportion of the positive and negative classes per batch. Models were trained using 2 NVIDIA Tesla K80 GPUs with 12 GB memory, using Keras and Tensorflow backend. We used the Adam optimizer, saving the model with the best ROC AUC on our validation set.

3.3   Model Ensembles and Test Time Augmentation

Our final model is made of an ensemble of three models and one Test Time Augmentation (TTA) per model (flip left/right or identity). In the ensemble the models output is averaged to a single score. To find the best ensemble an exhaustive search was performed on the hyper-parameters of the model including: patch input size, HU windowing, weight initialization approaches, learning rate and batch content. In model architecture we evaluated the following: (i) a sequence model with a LSTM classifier,(ii) a sequence model with max probability, and (iii) a shallow 3D convolution model. The best ensemble was found among the candidates in the pool with respect to each model ROC AUC on the tuning set.

Table 1. Comparison of our method to previously proposed approaches in the literature. Note – indicates data is not available.

	AUC	Sen	Spec	Data (#cases)	Model type
Bar et al. [13]	–	0.839	0.938	3'701	2D
Tomita et al. [14]	0.92	0.88	0.71	1'432	2D
Valentinitsch et al. [11]	0.88	0.80	0.74	154	3D
Nicolaes et al. [17]	0.95	0.905	0.938	90	3D
Ours	**0.955**	0.822	0.951	2'489	3D

4 Results

We compared different strategies for series level classification working in the feature spaces of the 3D patches, using three classification approaches: (i) the maximum patch probability, (ii) the maximum patch probability with the patch dependant localization information, and (iii) a BiLSTM aggregation. Best results were obtained by maximizing the scores of the patch features, as shown in Fig. 3a. This classification method achieved our best performance: 0.961 AUC ROC on the tuning set.

Figure 3b shows our method patient-level fracture identification ROC curve obtained on the held-out validation set. We also compared our proposed method to previously described methods in the literature, as shown in Table 1. We note that all these results have been reported using different validation sets. Based on this table, we can see that our proposed method is on par with the best AUC ROC reported so far [17], but is trained and validated on a significantly larger and more diverse data set.

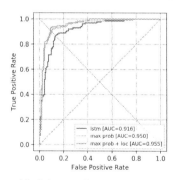 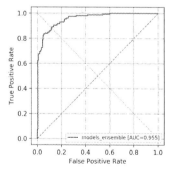

Model experiments results. ROCs computed for different aggregation approaches on the tuning set.

Results on the held-out validation set. It plots the ROC curve of our final ensemble model.

Fig. 3. Experiments and test results.

Figure 4 provides illustration of both correctly and incorrectly detected VCFs, marked by a bounding box. During development, we observed that a common sources of false positive results were Genant Mild fractures, which represent a small range of vertebral height loss between 20–25% and also obtain considerably lower agreement among clinicians [19]. But, also due to confusion with other vertebral pathologies, such as Schmorl nodes.

Regarding performance, we report an average processing time per case of 61.36 s for the entire pipeline (standard deviation 43.16 s). These measurements were made on our test set using a CPU with 16 cores.

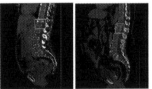

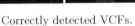
Correctly detected VCFs.   Incorrectly detected VCFs.

Fig. 4. Illustration of the results. The VCF detected by the algorithm is displayed on the reconstructed mid-sagittal slice of the pre-processed 3D input image. The first two images in (a) are the result of correctly detected VCFs, whereas the following two in (b) show incorrectly detected ones. There, our system confused L5 and Schmorl nodes with VCFs.

5   Conclusions

This study describes the development and testing of a novel end-to-end patient-level vertebral fracture identification system on CT. The system introduces a compact and effective 3D representation for VCF identification, a novel ROI extraction pipeline and leveraging a sequence to sequence architecture, obtaining SOTA results in this problem domain on a large data set. With to this architecture, location context and information from adjacent vertebrae are fused to the model. We expect the proposed system to improve VCF diagnosis in a clinical settings by pre-screening routine CT examinations and flagging suspicious cases. Future work, will focus on extending this approach to other modalities and osteoporosis applications.

References

1. Johnell, O., Kanis, J.A.: An estimate of the worldwide prevalence and disability associated with osteoporotic fractures. Osteoporos. Int. **17**, 1726–1733 (2006)
2. Kanis, J.A., et al.: Long-term risk of osteoporotic fracture in Malmö. Osteoporos. Int. **11**(8), 669–674 (2000)

3. Kanis, J.A., Borgstrom, F., Zethraeus, N., Johnell, O., Oden, A., Jönsson, B.: Intervention thresholds for osteoporosis in the UK. Bone **36**, 22–32 (2005)
4. Hasserius, R., Karlsson, M.K., Nilsson, B.E., Redlund-Johnell, I., Johnell, O., E. V. O. Study: Prevalent vertebral deformities predict increased mortality and increased fracture rate in both men and women: a 10-year population-based study of 598 individuals from the Swedish cohort in the european vertebral osteoporosis study. Osteoporos. Int. **14**, 61–68 (2003)
5. Edwards, B.J., Bunta, A.D., Simonelli, C., Bolander, M., Fitzpatrick, L.A.: Prior fractures are common in patients with subsequent hip fractures. Clin. Orthop. Relat. Res. **461**, 226–230 (2007)
6. Carberry, G.A., Pooler, B.D., Binkley, N., Lauder, T.B., Bruce, R.J., Pickhardt, P.J.: Unreported vertebral body compression fractures at abdominal multidetector CT. Radiology **268**, 120–126 (2013)
7. Williams, A.L., Al-Busaidi, A., Sparrow, P.J., Adams, J.E., Whitehouse, R.W.: Under-reporting of osteoporotic vertebral fractures on computed tomography. Eur. J. Radiol. **69**, 179–183 (2009)
8. Bartalena, T., et al.: Incidental vertebral compression fractures in imaging studies: lessons not learned by radiologists. World J. Radiol. **2**, 399–404 (2010)
9. Jakubicek, R., Chmelik, J., Jan, J., Ourednicek, P., Lambert, L., Gavelli, G.: Learning-based vertebra localization and labeling in 3D CT data of possibly incomplete and pathological spines. Comput. Methods Programs Biomed. **183**, 105081 (2020)
10. Ramos, J.S., Watanabe, C.Y.V., Nogueira-Barbosa, M.H., Traina, A.J.M.: BGrowth: an efficient approach for the segmentation of vertebral compression fractures in magnetic resonance imaging. In: Proceedings of the 34th ACM/SIGAPP Symposium on Applied Computing - SAC 2019, pp. 220–227. ACM Press, New York (2019)
11. Valentinitsch, A., et al.: Opportunistic osteoporosis screening in multi-detector CT images via local classification of textures. Osteoporos. Int. **30**, 1275–1285 (2019)
12. Chen, H., et al.: Automatic localization and identification of vertebrae in spine CT via a joint learning model with deep neural networks. In: Navab, N., Hornegger, J., Wells, W.M., Frangi, A.F. (eds.) MICCAI 2015. LNCS, vol. 9349, pp. 515–522. Springer, Cham (2015). https://doi.org/10.1007/978-3-319-24553-9_63
13. Bar, A., Wolf, L., Amitai, O.B., Toledano, E., Elnekave, E.: Compression fractures detection on CT. In: Armato III, S.G., Petrick, N.A. (eds.) Medical Imaging 2017: Computer-Aided Diagnosis, vol. 10134, pp. 1036–1043. International Society for Optics and Photonics, SPIE (2017)
14. Tomita, N., Cheung, Y.Y., Hassanpour, S.: Deep neural networks for automatic detection of osteoporotic vertebral fractures on CT scans. Comput. Biol. Med. **98**, 8–15 (2018)
15. Sherstinsky, A.: Fundamentals of recurrent neural network (RNN) and long short-term memory (LSTM) network. Physica D **404**, 132306 (2020)
16. Sekuboyina, A., et al.: Btrfly net: vertebrae labelling with energy-based adversarial learning of local spine prior. In: Frangi, A.F., Schnabel, J.A., Davatzikos, C., Alberola-López, C., Fichtinger, G. (eds.) MICCAI 2018. LNCS, vol. 11073, pp. 649–657. Springer, Cham (2018). https://doi.org/10.1007/978-3-030-00937-3_74
17. Nicolaes, J., et al.: Detection of vertebral fractures in CT using 3D convolutional neural networks. In: Cai, Y., Wang, L., Audette, M., Zheng, G., Li, S. (eds.) CSI 2019. LNCS, vol. 11963, pp. 3–14. Springer, Cham (2020). https://doi.org/10.1007/978-3-030-39752-4_1

18. Redmon, J., Farhadi, A.: YOLO9000: better, faster, stronger. In: 2017 IEEE Conference on Computer Vision and Pattern Recognition (CVPR), pp. 6517–6525. IEEE, July 2017
19. Genant, H.K., Wu, C.Y., van Kuijk, C., Nevitt, M.C.: Vertebral fracture assessment using a semiquantitative technique. J. Bone Miner. Res. **8**, 1137–1148 (1993)
20. Peris, A., Casacuberta, F.: A neural, interactive-predictive system for multimodal sequence to sequence tasks. In: Proceedings of the 57th Annual Meeting of the Association for Computational Linguistics: System Demonstrations, Stroudsburg, PA, USA, pp. 81–86. Association for Computational Linguistics (2019)

SIMBA: Specific Identity Markers for Bone Age Assessment

Cristina González[1]([envelope]) [iD], María Escobar[1] [iD], Laura Daza[1], Felipe Torres[1],
Gustavo Triana[2], and Pablo Arbeláez[1] [iD]

[1] Center for Research and Formation in Artificial Intelligence,
Universidad de los Andes, Bogotá, Colombia
{ci.gonzalez10,mc.escobar11,a.daza10,f.torres11,
pa.arbelaez}@uniandes.edu.co
[2] Fundación Santa Fe de Bogotá, Bogotá, Colombia

Abstract. Bone Age Assessment (BAA) is a task performed by radiologists to diagnose abnormal growth in a child. In manual approaches, radiologists take into account different *identity markers* when calculating bone age, i.e., chronological age and gender. However, the current automated Bone Age Assessment methods do not completely exploit the information present in the patient's metadata. With this lack of available methods as motivation, we present SIMBA: Specific Identity Markers for Bone Age Assessment. SIMBA is a novel approach for the task of BAA based on the use of identity markers. For this purpose, we build upon the state-of-the-art model, fusing the information present in the identity markers with the visual features created from the original hand radiograph. We then use this robust representation to estimate the patient's relative bone age: the difference between chronological age and bone age. We validate SIMBA on the Radiological Hand Pose Estimation dataset and find that it outperforms previous state-of-the-art methods. SIMBA sets a trend of a new wave of Computer-aided Diagnosis methods that incorporate all of the data that is available regarding a patient. To promote further research in this area and ensure reproducibility we will provide the source code as well as the pre-trained models of SIMBA.

Keywords: Bone age assessment · Computer-aided diagnosis ·
Identity markers · Relative bone age

1 Introduction

The height of a child is one of the best indicators for general health and overall well-being. Early diagnosis of abnormal growth in children is relevant not only for predicting the final adult height but also for detecting potential endocrine disorders. Prior studies have found that early recognition of abnormal growth

C. González and M. Escobar—Both authors contributed equally to this work.

A. L. Martel et al. (Eds.): MICCAI 2020, LNCS 12266, pp. 753–763, 2020.
https://doi.org/10.1007/978-3-030-59725-2_73

in children is necessary for timely treatment of pathological conditions such as precocious puberty [9, 16]. Physicians evaluate the growth rate of a child through Bone Age Assessment (BAA), a measurement of a child's skeletal development in months that varies according to *identity markers*, such as the child's chronological age, gender, and ethnicity.

Currently, the way in which radiologists establish a child's bone age is by comparing the child's hand radiograph against atlases for bone age measurement. These atlases contain standard reference images portraying male and female bone development from birth to an estimate of the last years of bone development for each gender. In the Greulich and Pyle (G & P) [7] atlas, the physician finds the reference image that presents the most similarities with the hand radiograph of the patient and uses it as a guideline to determine bone age. Tanner and Whitehouse's (TW2) [19] method is based on identifying anatomical Regions of Interest (RoIs) on the epiphysis of the hand and wrist, assigning a score to each RoI and combining them to calculate an overall bone age. In both approaches, the radiologist considers the patient's specific identity markers, particularly the gender and chronological age.

Figure 1 shows an example of the influence that *identity markers* have on bone age. The three hand radiographs present in Fig. 1 belong to children with virtually the same chronological age. However, it is visible that the ossification patterns present in each of the hand radiographs vary significantly. First, gender is an important identity marker to take into account. Comparing Fig. 1b (a female patient) and Fig. 1c (a male patient), it is possible to observe that the bone structures in the region surrounded by the red box are more developed for the female than for the male. This finding is supported by the fact that skeletal development is faster in females than in males.

Nonetheless, when comparing two hand radiographs of patients of the same gender with the same chronological age, as it happens between Fig. 1a and Fig. 1b, the expected result would be that the bone patterns did not vary much. Because most patients have a regular growth pattern, physicians use the chronological age as a starting point and compute the difference in skeletal development. This *relative bone age* between the patient's chronological age and the patient's bone age is the information that the radiologists use to diagnose growth disorders. However, Fig. 1b belongs to a patient with regular growth, having a relative bone age of +1 month, while Fig. 1a belongs to a patient with accelerated growth, hence the relative bone age of −38 months.

Since physicians have to take into account different factors before determining bone age, BAA is highly dependent on the radiologist's expertise level. Automated methods for BAA have been proposed as an alternative for manual approaches in order to reduce the variability among radiologists. The only commercial automated method is BoneXpert [20], a private software based on edge detection and active appearance models [2] currently used in clinical settings. However, the algorithm was developed using patients from a single cohort. Therefore there is no guarantee that it can generalize the BAA for children with different *identity markers* from those of the patients with whom the model was

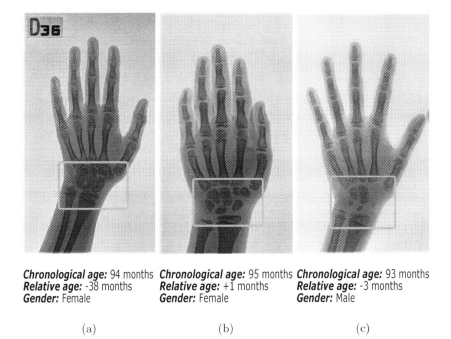

Chronological age: 94 months **Chronological age:** 95 months **Chronological age:** 93 months
Relative age: -38 months **Relative age:** +1 months **Relative age:** -3 months
Gender: Female **Gender:** Female **Gender:** Male

(a)   (b)   (c)

Fig. 1. Hand radiograph of three patients with virtually the same chronological age. Despite the small variations in chronological age, the anatomical structures, surrounded by the red box, have a very diverse appearance due to differences in the ossification patterns. These difference are related to the relative bone age of the patient with respect to the chronological age and their gender. (Color figure online)

trained. Like BoneXpert, the first wave of automated BAA methods [15, 22] and digital atlases [4–6] were private. Limiting the comparison that could be done among methods and the disposition that researchers had in choosing BAA as a relevant problem to tackle.

To motivate the development of more general automatic BAA methods, the Radiological Society of North America (RSNA) organized a challenge in 2017 with a dataset containing patients from different hospitals [8]. The winners of this challenge, 16 bit [1], developed a method that uses global information of the hand radiograph image and handcrafted embedding for gender. In our previous work [3], we created the Radiological Hand Pose Estimation (RHPE) dataset, which includes information of bone age, gender, anatomical RoIs, and chronological age for a cohort with a different ethnicity than the previously available dataset. There have been several automated methods for BAA that focus on a global approach [8,13,17,21], like what physicians do in G & P, and other approaches that exploit the local information [1,3,10,11,14], in a way inspired by TW2, to predict the bone age of the child.

In this paper, we present Specific Identity Markers for Bone age Assessment (SIMBA). Figure 2 shows an overview of our method. Motivated by the way physicians estimate bone age in children, SIMBA builds upon the state-of-the-art method, BoNet [3], and incorporates patient-specific identity markers, *i.e.* chronological age and gender, to perform BAA. State-of-the-art methods that use gender information for BAA introduce this input to the model both, directly as an additional input [1,11], and indirectly training an additional model [8]. Experimentally, we demonstrate that our way of incorporating gender information is more effective for BAA, as our model significantly outperforms the state-of-the-art methods in the RHPE dataset. We extract high-level features from the hand radiograph implicitly guided by an attention heatmap over the anatomical RoIs, following the idea suggested by [3]. Our model then uses learnable independent multipliers for each identity marker and combines them with the image features to generate the prediction. We also introduce relative bone age as a new way of approaching the problem of BAA in a fashion similar to physicians. We evaluate SIMBA on the open source RHPE dataset, outperforming the current state-of-the-art.

Our main contributions can be summarized as follows:

1. We propose a novel way to incorporate identity markers into BAA methods. We demonstrate that using a patient's gender and chronological age as prior for the model is relevant for better BAA.
2. We demonstrate that addressing the problem of BAA by estimating the relative bone age with the prior of chronological age is relevant for better BAA.

In order to ensure the reproducibility of our results and promote further research on BAA, we provide the pre-trained models, the source code for SIMBA and the additional metadata corresponding to the chronological age for the RHPE dataset.[1]

2   SIMBA

Our method is inspired by how radiologists take advantage of all the available information for each patient when computing BAA. Thus, SIMBA not only considers visual information from the hand radiograph and local information from anatomical RoIs, but it also leverages identity markers, particularly, chronological age and gender. Additionally, we propose a novel paradigm of predicting the relative bone age of a patient as the deviation from the chronological age. Figure 2 depicts our approach for incorporating a patient's specific identity markers to predict relative bone age. In the following sections, we explain the different components of SIMBA in more detail.

2.1   Specific Identity Markers

Similarly to the state-of-the-art method BoNet, given an image I and a heatmap H our architecture incorporates the Inception-V3 [18] (I_{V3}) network to extract visual features, $\mathbb{I}$.

$$\mathbb{I} = I_{V3}(I, H)$$

---

[1] https://github.com/BCV-Uniandes/SIMBA.

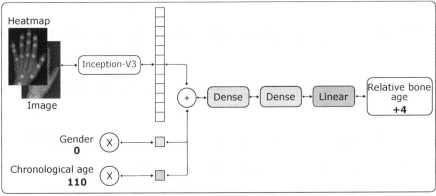

Fig. 2. Overview of our method. Our model takes as input the hand radiograph and the attention heatmap generated from the anatomical RoIs. We calculate visual features from these two inputs using an Inception-V3 architecture. We multiply the identity markers with two learnable and independent multipliers and concatenate this information with the visual features extracted. Finally, we process all this information, the visual features and the identity markers, jointly to predict the bone age deviation from the chronological age. Best viewed in color. (Color figure online)

Among a patient's identity markers, the most relevant ones for BAA are gender and chronological age, due to skeletal development varying with gender and being correlated to chronological age. We incorporate these identity markers directly by processing them jointly with the previously extracted visual features. Instead of using handcrafted embeddings for incrementing dimensionality [3], we learn multipliers m_g and m_c to balance the importance of each of the inputs from which our model makes the final prediction, regardless of their size. Therefore, our model learns weighted representations for gender $\mathbb{G}$ and chronological age $\mathbb{C}$, according to their relevance to the final prediction. These representations given a gender g and a chronological age c are defined as follows:

$$\mathbb{G} = m_g \cdot g$$

$$\mathbb{C} = m_c \cdot c$$

Thus, the joint final representation $\mathbb{J}$, which our model uses to estimate the child's bone age, corresponds to the concatenation of the visual features extracted from the image and the heatmap, along with the weighted representations of the gender and chronological age.

$$\mathbb{J} = [\mathbb{I}; \mathbb{G}; \mathbb{C}]$$

2.2 Relative Bone Age

We propose a paradigm shift in terms of the formulation of the BAA task. We define a new task for BAA equivalent to the one previously formulated for this problem. Based on the priors it receives, our model is optimized to predict the difference between the chronological age c and the bone age b of the patient, defined as the relative bone age r_b. In other words, our model learns to take as input a chronological age and outputs a residual bone age.

$$r_b = c - b$$

For this purpose, our model learns two intermediate representations from linear layers followed by $ReLU$ non-linear activation function ($Dense$). From these layers, the model learns to generate a joint representation $\widehat{\mathbb{J}}$ of the visual information and the specific identity markers of the patient. Finally, SIMBA predicts the relative bone age with a fully-connected layer.

$$Dense(x) = ReLU(W(x) + b)$$

$$\widehat{\mathbb{J}} = Dense(Dense(x))$$

$$r_b = W\widehat{\mathbb{J}} + b$$

Implementation Details: we train our method on an NVIDIA TITAN-X Pascal GPU for 150 epochs with an initial learning rate of 0.001, 17 images per batch and use an Adam [12] optimizer with the standard parameters. Additionally, we use a dynamic learning rate scheduler to reduce the learning rate when reaching a plateau with a patience of 2 epochs, a reducing factor of 0.8, and a cooldown of 5 epochs.

3   Experiments

3.1   Experimental Setup

For our experimental validation, we use the RHPE dataset with the original data splits. We perform an ablation study to determine the individual contribution of each module. For the ablation experiments, we train our method on the training set and select the model that performs better on the validation set. Additionally, for the comparison of our method with respect to the state-of-the-art, we train our best model using the data from both the training and validation set and evaluate on the official RHPE test server.

In accordance with our new formulation of the task, we aim at estimating the deviation of the bone age from the chronological age of the patient in months for each given image in the dataset. To evaluate our experimental results, we rely on the Mean Absolute Distance (MAD) previously used in the RSNA 2017 Pediatric Bone Age Challenge [8].

Table 1. Comparison of our method against the state-of-the-art methods, as reported on [3], on the RHPE test set. Our method SIMBA significantly outperforms the state-of-the-art-methods.

Method	MAD
16 bit [1]	8.57
BoNet [3]	7.60
SIMBA (Ours)	**5.47**

3.2   Experimental Validation

Comparison with the State-of-the-Art. We compare the results of our method, SIMBA, with respect to the state-of-the-art methods in this dataset. Table 1 shows results in the test set of the RHPE dataset for the methods as reported by [3], 16 bit and BoNet.

The results reported in Table 1 demonstrate that SIMBA significantly outperforms the state-of-the-art method in the RHPE dataset test set. Since the other state-of-the-art methods do not include the chronological age information as prior for the model, our results empirically demonstrate that including identity markers, specifically the gender and chronological age of the patient, is important to improve performance in BAA. Additionally, we demonstrate that when we replace the task of estimating bone age with estimating its deviation with respect to chronological age, it is evident in performance improvements for BAA.

Relative Age Bias Analysis. To gain further insight, we calculated the correlation between the relative age and the MAD metric for all the patients in the validation split. As shown in Fig. 3, we estimated the correlation coefficient to measure the linear relationship between these variables. The correlation coefficient is 0.016, and thus, we can state that there is no strong linear dependence. Furthermore, we performed a linear regression on the data and found that the slope of the line is 0.097, which is consistent with an approximately uniform distribution of MAD with respect to relative age. Based on this analysis, we can conclude that SIMBA is not biased towards relative age and learns to predict a residual bone age based entirely on the visual input and the guidance of the identity markers.

Ablation Study. We designed an ablation study of our method for its different components. For all our experiments we train our model incrementally starting from the baseline. We use the source code publicly available for BoNet. In this way, to build the final model, we add our modifications in the following order: gender multiplier, chronological age, and relative bone age, according to the components included in each ablation experiment. Table 2 shows the results of

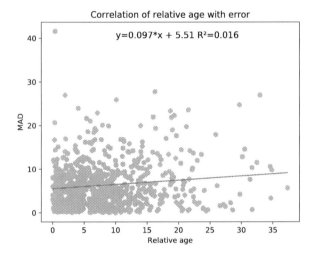

Fig. 3. Correlation between relative age and MAD for the validation set of RHPE. There is no strong linear dependence (the pink line has a tendency to be horizontal), therefore SIMBA is not biased towards relative age. Best viewed in color. (Color figure online)

our final model, the ablation experiments, and the baseline in the validation split of the RHPE dataset.

Relative Bone Age Ablation. If we train our model without establishing the final task as the estimation of the deviation of the patient's bone age with respect to their chronological age, the error of our model increases by 0.16 months in the validation set of the RHPE dataset. These results demonstrate that the task that we propose allows our model to be able to exploit the input information, that is, the image and the identity markers more adequately.

Table 2. Ablation experiments for our method on the RHPE validation set. We report the results of our final method, ablation experiments for: relative bone age, chronological age and gender multiplier, and the baseline of our experimental setup.

Method	Identity markers		Relative bone age	MAD
	Gender	Chronological age		
BoNet [3]				7.48
SIMBA	✓	✓		6.50
	✓		✓	8.72
		✓	✓	7.33
	✓	✓	✓	**6.34**

Chronological Age Ablation. By eliminating the chronological age as a prior for our model, the MAD increases by 2.38 months. This result empirically supports that the decrease in error associated with the introduction of the task of the estimation of relative bone age is determined by having the prior of chronological age. We consider the above to be intuitive if we understand that, by not including the chronological age, the model must learn to estimate not only the bone age but also the chronological age of the patient to finally estimate the deviation between them. However, this does not represent the real case since the radiologist usually knows the patient's chronological age.

Gender Multiplier Ablation. If we change our gender multiplier for the hand-crafted embedding used in the state-of-the-art methods for BAA, the MAD increases by 0.99 months. This result shows that our multiplier exploits gender information more efficiently. Additionally, the effectiveness of the other contributions of our model are highly related to the joint processing of the identity markers of the patient.

4   Conclusions

In this work, we present a new paradigm for the task of BAA by estimating the deviation between the bone age and the chronological age of a patient. To the best of our knowledge, SIMBA is the first method for this task that leverages information from specific identity markers, particularly the gender and the chronological age of a patient. Our model outperforms the state-of-the-art method in the test set of the benchmark RHPE dataset. The previous state-of-the-art methods do not consider chronological age as an identity marker and directly estimate the child's bone age. We demonstrate experimentally that including prior information related to specific identity markers of the patient into the network, inspired by the way radiologists do it in their medical practice, results in a more accurate Bone Age Assessment.

Acknowledgments. This project was partially funded by the Colombian Ministry of Science, Technology and Innovation under the Colciencias grant: 841-2017 code 120477758362.

References

1. Cicero, M., Bilbily, A.: Machine Learning and the Future of Radiology: How We Won the 2017 RSNA ML Challenge (2017). https://16bit.ai/blog/ml-and-future-of-radiology
2. Cootes, T.F., Edwards, G.J., Taylor, C.J.: Active appearance models. In: Burkhardt, H., Neumann, B. (eds.) ECCV 1998. LNCS, vol. 1407, pp. 484–498. Springer, Heidelberg (1998). https://doi.org/10.1007/BFb0054760

3. Escobar, M., González, C., Torres, F., Daza, L., Triana, G., Arbeláez, P.: Hand pose estimation for pediatric bone age assessment. In: Shen, D., et al. (eds.) MICCAI 2019. LNCS, vol. 11769, pp. 531–539. Springer, Cham (2019). https://doi.org/10.1007/978-3-030-32226-7_59

4. Gaskin, C.M., Kahn, M.M.S.L., Bertozzi, J.C., Bunch, P.M.: Skeletal Development of the Hand and Wrist: A Radiographic Atlas and Digital Bone Age Companion. Oxford University Press, Oxford (2011). https://doi.org/10.1093/med/9780199782055.001.0001

5. Gertych, A., Zhang, A., Sayre, J., Pospiech-Kurkowska, S., Huang, H.: Bone age assessment of children using a digital hand atlas. Comput. Med. Imaging Graph. **31**(4–5), 322–331 (2007). https://doi.org/10.1016/j.compmedimag.2007.02.012

6. Gilsanz, V., Ratib, O.: Hand Bone Age: A Digital Atlas of Skeletal Maturity. Springer, Heidelberg (2005). https://doi.org/10.1007/978-3-642-23762-1

7. Greulich, W.W., Pyle, S.I., Todd, T.W.: Radiographic Atlas of Skeletal Development of the Hand and Wrist, vol. 2. Stanford University Press, Stanford (1959)

8. Halabi, S.S., et al.: The rsna pediatric bone age machine learning challenge. Radiology **290**(2), 498–503 (2019). https://doi.org/10.1148/radiol.2018180736

9. Haymond, M., Kappelgaard, A.M., Czernichow, P., Biller, B.M., Takano, K., Kiess, W.: Participants in the global advisory panel meeting on the effects of growth hormone: early recognition of growth abnormalities permitting early intervention. Acta Paediatr. **102**(8), 787–796 (2013). https://doi.org/10.1111/apa.12266

10. Iglovikov, V.I., Rakhlin, A., Kalinin, A.A., Shvets, A.A.: Paediatric bone age assessment using deep convolutional neural networks. In: Stoyanov, D., et al. (eds.) DLMIA/ML-CDS 2018. LNCS, vol. 11045, pp. 300–308. Springer, Cham (2018). https://doi.org/10.1007/978-3-030-00889-5_34

11. Ji, Y., Chen, H., Lin, D., Wu, X., Lin, D.: PRSNet: part relation and selection network for bone age assessment. In: Shen, D., et al. (eds.) MICCAI 2019. LNCS, vol. 11769, pp. 413–421. Springer, Cham (2019). https://doi.org/10.1007/978-3-030-32226-7_46

12. Kingma, D.P., Ba, J.: Adam: a method for stochastic optimization. arXiv preprint arXiv:1412.6980 (2014)

13. Larson, D.B., Chen, M.C., Lungren, M.P., Halabi, S.S., Stence, N.V., Langlotz, C.P.: Performance of a deep-learning neural network model in assessing skeletal maturity on pediatric hand radiographs. Radiology **287**(1), 313–322 (2018). https://doi.org/10.1148/radiol.2017170236

14. Liu, C., Xie, H., Liu, Y., Zha, Z., Lin, F., Zhang, Y.: Extract bone parts without human prior: end-to-end convolutional neural network for pediatric bone age assessment. In: Shen, D., et al. (eds.) MICCAI 2019. LNCS, vol. 11769, pp. 667–675. Springer, Cham (2019). https://doi.org/10.1007/978-3-030-32226-7_74

15. Liu, J., Qi, J., Liu, Z., Ning, Q., Luo, X.: Automatic bone age assessment based on intelligent algorithms and comparison with TW3 method. Comput. Med. Imaging Graph. **32**(8), 678–684 (2008). https://doi.org/10.1016/j.compmedimag.2008.08.005

16. Oerter Klein, K.: Precocious puberty: who has it? who should be treated? J. Clin. Endocrinol. Metabol. **84**(2), 411–414 (1999). https://doi.org/10.1210/jcem.84.2.5533

17. Pan, X., Zhao, Y., Chen, H., Wei, D., Zhao, C., Wei, Z.: Fully automated bone age assessment on large-scale hand x-ray dataset. Int. J. Biomed. Imaging (2020). https://doi.org/10.1155/2020/8460493

18. Szegedy, C., Vanhoucke, V., Ioffe, S., Shlens, J., Wojna, Z.: Rethinking the inception architecture for computer vision. In: 2016 IEEE Conference on Computer Vision and Pattern Recognition, CVPR 2016, Las Vegas, NV, USA, June 27–30, 2016, pp. 2818–2826 (2016). https://doi.org/10.1109/CVPR.2016.308
19. Tanner, J., Whitehouse, R., Marshall, W., Carter, B.: Prediction of adult height from height, bone age, and occurrence of menarche, at ages 4 to 16 with allowance for midparent height. Arch. Dis. Child. **50**(1), 14–26 (1975)
20. Thodberg, H., Kreiborg, S., Juul, A., Pedersen, K.: The bonexpert method for automated determination of skeletal maturity. IEEE Trans. Med. Imaging **28**(1), 52–66 (2009). https://doi.org/10.1109/tmi.2008.926067
21. Torres, F., González, C., Escobar, M., Daza, L., Triana, G., Arbeláez, P.: An empirical study on global bone age assessment. In: 15th International Symposium on Medical Information Processing and Analysis. International Society for Optics and Photonics (2020). https://doi.org/10.1117/12.2542431
22. Tsao, S., Gertych, A., Zhang, A., Liu, B.J., Huang, H.K.: Automated bone age assessment of older children using the radius. In: Andriole, K.P., Siddiqui, K.M. (eds.) Medical Imaging 2008: PACS and Imaging Informatics. SPIE, March 2008. https://doi.org/10.1117/12.770018

Doctor Imitator: A Graph-Based Bone Age Assessment Framework Using Hand Radiographs

Jintai Chen[1], Bohan Yu[1], Biwen Lei[1], Ruiwei Feng[1], Danny Z. Chen[2], and Jian Wu[1(✉)]

[1] College of Computer Science and Technology, Zhejiang University, Hangzhou, China
wujian2000@zju.edu.cn
[2] Department of Computer Science and Engineering, University of Notre Dame, Notre Dame, IN 46556, USA

Abstract. Bone age assessment is challenging in clinical practice due to the complicated bone age assessment process. Current automatic bone age assessment methods were designed with rare consideration of the diagnostic logistics and thus may yield certain uninterpretable hidden states and outputs. Consequently, doctors can find it hard to cooperate with such models harmoniously because it is difficult to check the correctness of the model predictions. In this work, we propose a new graph-based deep learning framework for bone age assessment with hand radiographs, called Doctor Imitator (DI). The architecture of DI is designed to learn the diagnostic logistics of doctors using the scoring methods (e.g., the Tanner-Whitehouse method) for bone age assessment. Specifically, the convolutions of DI capture the local features of the anatomical regions of interest (ROIs) on hand radiographs and predict the ROI scores by our proposed Anatomy-based Group Convolution, summing up for bone age prediction. Besides, we develop a novel Dual Graph-based Attention module to compute patient-specific attention for ROI features and context attention for ROI scores. As far as we know, DI is the first automatic bone age assessment framework following the scoring methods without fully supervised hand radiographs. Experiments on hand radiographs with only bone age supervision verify that DI can achieve excellent performance with sparse parameters and provide more interpretability.

Keywords: Graph-based convolution · Bone age · Interpretability

1 Introduction

Bone age differs from the chronological age and often varies with the gender and ethnicity [11]. Bone age assessment (BAA) is typically used to estimate the skeletal maturity and diagnose the growth problem of children. There are two

---

J. Chen, B. Yu and B. Lei—These authors are co-first author.

© Springer Nature Switzerland AG 2020
A. L. Martel et al. (Eds.): MICCAI 2020, LNCS 12266, pp. 764–774, 2020.
https://doi.org/10.1007/978-3-030-59725-2_74

widely employed methods for BAA using hand radiographs in clinical practice: the Greulich-Pyle (GP) method [14] and the scoring methods (e.g., the Tanner-Whitehouse method [13]). In the GP method, bone age is estimated by referring the entire hand radiographs to the atlas. In the scoring methods, doctors analyze and score the region of interests (e.g., the joints) of the hands individually, and use the weighted sum of these scores to estimate the bone age. It was verified that the scoring methods were more accurate and reliable than the GP method [11].

Recently, various automatic BAA methods were proposed [1,2,10,19]. Similar to the processing of the GP method, most of these models predicted bone ages by capturing the features of the entire hand, which benefited from the capability of deep learning models. Larson et al. [9] built a classifier to predict bone ages, using ResNet50 [4] as the backbone. Iglovikov et al. [5] trained several end-to-end regression models and predicted bone ages using the model ensemble strategy. In [15], the Gaussian process regression was employed to increase the sensibility to the hand pose. Wang et al. [17] followed the structure of Faster-RCNN to predict bone ages. Besides, attention methods were utilized in [2,10] to highlight the essential parts of the hands. A relation computing module was introduced in [6] to better deal with the important parts of the hands. A new annotation for the Central Positions of Anatomical ROIs (CPAR) to the Radiological Society of North America BAA (RSNA-BAA) dataset was published [1], which provided the ground truth central positions of the anatomical region of interests (ROIs). Using the CPAR annotations, BoNet [1] promoted the performance of the challenge champion methods. Although these methods could obtain good performance in BAA, they used features of the entire hands and incurred poor interpretability. Following the scoring methods, TW-AI [20] predicted bone ages based on the scores of ROIs, and thus promoted interpretability. However, training TW-AI required full supervision of the ground truth ROI scores, which was uneconomical and hard to popularize.

Motivated by the ROI-dependent classification method with weak supervision in [7], in this paper we propose a novel deep learning framework, called Doctor Imitator (DI), for predicting ROI scores and bone ages using hand radiographs with only bone age supervision. DI is designed by imitating the diagnostic logistics of doctors and the processing of the scoring methods, and obtains excellent and interpretable results with extremely low model complexity. Specifically, we propose an Anatomy-based Group Convolution (AG-Conv) to predict ROI scores using the local features of ROIs and sum up the ROI scores for the bone age prediction. In clinical practice, an experienced doctor may assign the ROI scores with the consideration of some patient-specific characteristics of the bones. Motivated by this, we develop a novel Dual Graph-based Attention Module (DGAM) to assist ROI score prediction, which consists of two new graph-based convolution (GConv) blocks. These two GConv blocks compute a patient-specific attention map for the ROI features and a context attention map for ROI scores, respectively. Different from the previous graph-based convolution (GConv) methods, our new GConv constructs two graphs on one radiograph and updates the features of nodes according to these two graphs simultaneously. Experiments on the

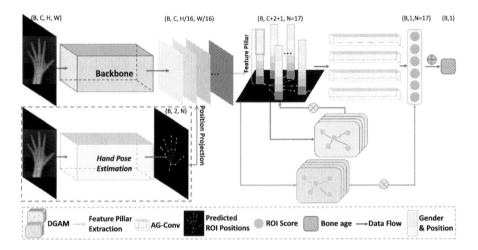

Fig. 1. Illustrating our two-stage Doctor Imitator framework. The first stage (*Hand Pose Estimation*), framed in the blue box, predicts the central positions of ROIs. In the second stage, the local features of ROIs are extracted to predict ROI scores, which are summed up for bone age prediction. A Dual Graph-based Attention Module (DGAM) computes two attention maps to help ROI score prediction. The feature sizes are marked above the feature maps. (Color figure online)

RSNA-BAA dataset and a private dataset verify that our DI framework achieves good performance on bone age prediction and ROI score prediction with only bone age supervision.

There are three main contributions in this work. **(A)** We propose a novel deep learning model to predict bone ages and ROI scores, following the diagnostic logistics of doctors and the scoring methods. **(B)** We introduce a novel dual graph-based attention module to compute patient-specific attention and context attention, updating the node features with two graphs simultaneously. **(C)** Experiments show that our DI framework can predict ROI scores with only bone age supervision, and thus increase the interpretability of the model.

2 Doctor Imitator Architecture

When estimating bone ages using the scoring methods, doctors often analyze the local characteristics of anatomical ROIs (e.g., some joints), and assign a score to every ROI according to the criteria. When assigning scores, experienced doctors also consider some common characteristics of the patient's bones. Finally, the ROI scores are summed up for bone age prediction. Imitating this process, we design a two-stage Doctor Imitator (DI) framework for bone age prediction, as illustrated in Fig. 1. In the first stage, a *Hand Pose Estimation* model is trained to predict the central positions of ROIs. Since the first stage is not the focus of this work, we use just the *Hand Pose Estimation* model of BoNet [1] by re-implementing it. In the second stage, we train a model to predict ROI scores and bone ages. We take the modified SSN [12] as the backbone, called Improved SSN (ImSSN), to extract the features of hand radiographs. Then we extract the feature pillars (the local features) of the predicted ROIs and perform an Anatomy-based Group Convolution (AG-Conv) module on the ROI feature pillars to predict the ROI scores. During the score prediction, a novel Dual Graph-based Attention Module (DGAM) is utilized to compute patient-specific attention for ROI features and context attention for ROI scores. Finally, we sum up the weighted ROI scores (the weights are provided by context attention) as the predicted bone age. We train the second stage of DI using L_1 loss. In our DI framework, there are 17 ROIs (see Fig. 2), since we train the first stage (*Hand Pose Estimation*) using the CPAR ground truth data [1].

In what follows, we describe the backbone and feature pillar extraction in Sect. 2.1, the AG-Conv module for ROI score prediction in Sect. 2.2, and the DGAM for attention computing in Sect. 2.3.

2.1 Backbone and Feature Pillar Extraction

In clinical practice, doctors assign ROI scores based on ROI local features. We use SSN [12] with some modification as the backbone, called improved SSN (ImSSN), which is a network similar to U-Net and extracts multi-scaled features. We add a 3×3 average pooling layer on the top-most layer of SSN, and preserve only the $8\times$ and $16\times$ down-sampled branches and output a $16\times$ down-sampled feature map.

We use ImSSN to process a hand radiograph and obtain a feature map $F_m \in \mathbb{R}^{C \times H/16 \times W/16}$, where (H, W) is the size of an input hand radiograph, and C is the number of feature channels. To obtain the local features of the ROIs, we project the predicted ROI central positions onto the feature map F_m, and take the features on the projected positions as the local features of ROIs. This position projection can be formulated by:

$$(i, j) = (\lfloor \frac{I}{s} \rfloor, \lfloor \frac{J}{s} \rfloor) \tag{1}$$

where (I, J) indicates the original central position of an ROI, (i, j) is the corresponding position on the feature map, and s is the down-sampling rate of ImSSN ($s = 16$ in this work). As a convolution captures features by fusing the features around, the feature pillars on the projected positions can represent the features of different ROIs. We extract the feature pillars as:

$$F_p = F_m[:, i, j] \tag{2}$$

where $F_p \in \mathbb{R}^{C \times 1 \times 1}$ is the feature pillar of an ROI. Besides, since the gender and ROI central positions are helpful to ROI score prediction, we add a binary value representing the gender and the original ROI central position to the corresponding feature pillar, thus $F_p \in \mathbb{R}^{(C+2+1) \times 1 \times 1}$, as illustrated in Fig. 1.

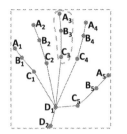 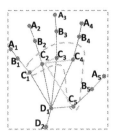

Fig. 2. Illustrating the ROI central positions and the receptive fields on different graphs (see Sect. 2.3). The receptive field of B_3 on graph $\mathcal{G}_1$ is in the green curve region shown in the left sub-figure. The receptive field of C_2 on graph $\mathcal{G}_2$ is shown in the red curve region in the right sub-figure. (Color figure online)

2.2 Anatomy-Based Group Convolution Module

In the scoring methods (e.g., the TW methods), the ROIs with similar anatomy usually use similar scoring criteria. In Fig. 2, the ROIs marked with the same letters (e.g, A_1, A_2, A_3, A_4, A_5) are anatomically similar, and the ROIs can be divided into four *anatomy groups*: A, B, C, and D. Following this, we propose an Anatomy-based Group Convolution (AG-Conv) module with four convolution blocks corresponding to the four *anatomy groups*. For the ROIs in an *anatomy group*, one convolution block is used to process the feature pillars and predict the scores. These convolution blocks can be implemented by one-by-one convolutions with batch normalization and ReLU activation, and thus the AG-Conv module shall be very light.

2.3 Dual Graph-Based Attention Module

When assigning scores to ROIs, an experienced doctor not only focuses on the local characteristics of bones but also pays attention to some patient-specific characteristics. Besides, the ROI scores are weighted and summed up for bone age prediction in the scoring methods. Following this process, we propose a Dual Graph-based Attention Module (DGAM) with two graph-based Convolution blocks to compute patient-specific attention maps for ROI feature pillars and a context attention map for ROI scores, which are different from the self-attention style in the known graph-based attention methods (e.g., GAT [16]).

Graph Construction. Different from the previous graph-based methods, we construct two undirected graphs $\mathcal{G}_1 = (\mathcal{V}, \mathcal{E}_1)$ and $\mathcal{G}_2 = (\mathcal{V}, \mathcal{E}_2)$ on one hand radiograph with N ROIs. The node set $\mathcal{V} = \{v_i \in \mathbb{R}^f\}$ includes all the ROIs presented with the corresponding feature pillars, for $i = 1, 2, \ldots, N$ and the feature dimension[1] $f = C + 3$. Graphs $\mathcal{G}_1, \mathcal{G}_2$ have the same node set $\mathcal{V}$, but their edge sets $\mathcal{E}_1$ and $\mathcal{E}_2$ are different. The edge set $\mathcal{E}_1$ contains the natural

---

[1] For simplicity, the last two dimensions (1×1) of the feature pillar $F_p \in \mathbb{R}^{(C+3) \times 1 \times 1}$ are discarded.

connections of joints (ROIs), and $\mathcal{E}_2$ contains the full connections among the ROIs in the same *anatomy group*. For example, in Fig. 2, the nodes connected with B_3 are A_3 and C_3 in graph $\mathcal{G}_1$, while in graph $\mathcal{G}_2$, nodes B_1, B_2, B_3, B_4, B_5 (in the same *anatomy group*) are connected to one another. Besides, self-connections are also available to all the nodes in both $\mathcal{G}_1$ and $\mathcal{G}_2$. Since the relation among the ROIs in an *anatomy group* and the relation among the naturally connected ROIs (joints) are different, constructing two graphs is helpful to model these relations.

GConv Blocks. DGAM consists of two GConv blocks (see Fig. 1), one for patient-specific attention computing, called Patient-specific Attention Block (PAB), and the other for context attention computing, called Context Attention Block (CAB). Both blocks are implemented in a spatial graph-based convolution manner [18] to make the model light. To feed the nodes (ROIs) presented by feature pillars to a GConv, we reformat the ROI feature pillars on a radiograph into a matrix $X_{(N \times f)}$, where f is the feature dimension and N is the number of ROIs (nodes). Given the graphs $\mathcal{G}_1, \mathcal{G}_2$, the GConv operation can be defined by:

$$X_{(N \times f_{i+1})} = \frac{1}{2} \sum_{j=\{1,2\}} L_{(N \times N)}^{(j)} X_{(N \times f_i)} W_{(f_i \times f_{i+1})}^{(j)} \tag{3}$$

where j indexes the graphs $\mathcal{G}_j$ ($j \in \{1, 2\}$). $W^{(j)}$ is a learnable weight matrix for node feature updating and can be implemented by a one-dimensional convolution. The subscripts of the matrices indicate the matrix sizes. f_i, f_{i+1} indicate the feature dimensions before and after feature updating, respectively. $L^{(j)} = [D^{(j)}]^{-\frac{1}{2}} (A^{(j)} + I)[D^{(j)}]^{\frac{1}{2}}$ is the normalized Signless Laplacian matrix on $\mathcal{G}_j$, and $A^{(j)}, D^{(j)}$ are the first order adjacency matrix and the degree matrix, respectively. By Eq. (3), the node features are updated by aggregating the features from the neighboring nodes based on the two graphs. Both PAB and CAB are implemented by sequentially stacking the GConvs (as in Eq. (3)), with different output channel dimensions. Specifically, PAB outputs a feature map in the same size as the input node features as $Att^X \in \mathbb{R}^{N \times f}$, and CAB outputs a feature map $Att^S \in \mathbb{R}^{N \times 1}$.

Attention Computing. The patient-specific attention presents the importance of the feature channels. Thus, we apply the node-wise average to compute the patient-specific attention map $\overline{Att}_{(N \times f)}^X$ by:

$$\overline{att}^X = \frac{1}{N} \sum_{n=1}^{N} att_n^X \tag{4}$$

where $[att_1^X, att_2^X, \ldots, att_N^X] = Att_{(N \times f)}^X$ and $\overline{Att}_{(N \times f)}^X = [\overline{att}^X, \overline{att}^X, \ldots, \overline{att}^X]$. att_n^X and $\overline{att}^X$ are feature vectors of size f. In the scoring methods, the weights of ROIs are the same for every patient of the same gender. Hence, in training, we use the exponential moving average (EMA) to compute the general context attention map conditioned by the gender, with the updating parameter $\theta = 0.01$, as specified in Eq. (5). In testing, we just used the attention maps learned in training stage.

$$\overline{Att}_g^S \leftarrow (1-\theta)\overline{Att}_g^S + \theta Att_{g|B}^S \tag{5}$$

where $\overline{Att}_g^S$ indicates the general context attention map of the gender g while $Att_{g|B}^S$ indicates the average of the predicted context attention of gender g in the batch B. $\overline{Att}_g^S$ is initialized as $Att_{g|B=1}^S$. Then the attention maps are applied to the feature pillars $X_{(N \times f)}$ and the ROI scores $S_{(N \times 1)}$ of one radiograph by:

$$\begin{cases} X_{(N \times f)}^* = \overline{Att}_{(N \times f)}^X \odot X_{(N \times f)} \\ S_{(N \times 1)}^* = \overline{Att}_{(N \times 1)}^S \odot S_{(N \times 1)} \end{cases} \tag{6}$$

where $\odot$ denotes the Hadamard production, and $X_{(N \times f)}^*$ and $S_{(N \times 1)}^*$ are the feature pillars and ROI scores updated by the attention maps.

3   Experiments

Dataset. We evaluate Doctor Imitator (DI) on the RSNA-BAA dataset [1,3] and RHPE dateset [1]. RSNA-BAA contains 12,611 hand radiographs in the training set, 1,425 radiographs in the validation set, and 200 radiographs in the test set. The ground truth bone ages are from 0 to 18 years. RHPE is composed of 5,492 hand radiographs in the training set, 715 radiographs in the validation set, and 80 radiographs in the test set. The ground truth bone ages of RHPE vary from 0 to 20 years. Before processing by DI, all the radiographs are resized to 512×512. Besides, we use the CPAR ground truth in [1] to guide the ROI central position prediction by the *Hand Pose Estimation* models [1]. Similar to the previous work, we report the Mean Absolute Difference (MAD) between the predicted bone ages and the corresponding ground truth bone ages.

Experimental Setup. We implement DI by PyTorch 1.3. We train the second stage of DI with 200 epochs. The batch size is 48. The initial learning rate is 10^{-3}, and is reduced by 10× after 60 epochs and is reduced by 10× again after 120 epochs. The optimizer is Adam [8]. We employ random flip, rotation ($-5° \sim 5°$), and the Gaussian blur Operation for data augmentation.

Performance and Complexity Comparisons. We compare DI with the state-of-the-art BoNet [1] on the MAD and model complexity. As the ground truth annotations of the RHPE test set are not available and the evaluation server [1] is not stable, we compare DI and BoNet on the validation set of RHPE. The performances of BoNet reported on the RHPE validation set are obtained by running the open source codes[1]. As shown in Table 1, our DI outperforms BoNet on the RSNA and RHPE with various percentages (10%, 20%, 40%, 100%) of

---

[1] https://github.com/BCV-Uniandes/Bonet

Table 1. Comparison on the performances and the model complexity with the state-of-the-art BoNet [1]. The percentages in the parentheses indicate what percentage of the training samples are used in training. "RH" indicates "RHPE" and "RS" indicates "RSNA". "*" marks the results of BoNet obtained from the open source codes we run.

Model	10%		20%		40%		100%		Model size	FLOPs	fps
	RH	RS	RH	RS	RH	RS	RH	RS			
BoNet [1]	12.65*	6.78	11.73*	6.18	10.14*	5.31	8.19*	4.37	123.2 M	17.8	39
DI (ours)	**12.28**	**6.42**	**11.36**	**6.04**	**10.12**	**5.26**	**8.15**	**4.30**	**9.8 M**	**13.1**	**98.2**

Table 2. Ablation study on the proposed modules of DI. "PA" denotes the patient-specific attention and "CA" denotes the context attention.

Exp	ImSSN	AG-Conv	RG-Conv	PA	CA	MAD
1	✓					5.38
2	✓	✓				4.86
3	✓		✓			4.99
4	✓	✓		✓		4.57
5	✓	✓			✓	4.51
6	✓	✓		✓	✓	4.30

the training samples available. Comparing to the public implemented version of BoNet [1], the bone age prediction part of DI is over 12× smaller in model size than the bone age prediction part of BoNet. Also, evaluated on the number of floating-point multiplication-adds (FLOPs) on our GPU, DI outperforms BoNet by a clear margin. In inference, DI is ∼3× faster than BoNet, handling 98.2 bone age radiograph frames per second (fps). Experiments show that DI is very efficient, attains good performance, and has low model complexity.

Doctor Imitating Improvement Study. DI is designed by imitating the diagnostic logistics of doctors. The experimental results on RSNA-BAA dataset in Table 2 verify that imitating doctors obtains improvements. To show the capability of AG-Conv, we compare it with the Random Group Convolution (RG-Conv). In performing RG-Conv, we group the ROIs into four groups randomly and process the ROIs in the groups with the same convolution module. One can see that it is helpful by using the same convolutions on the anatomically similar ROIs. Comparing Exp.4, Exp.5, and Exp.6 in Table 2, it is evident that both of the patient-specific attention and context attention are helpful.

Interpretability and Standard ROI Score Prediction. To evaluate the interpretability of DI, we test DI on a private dataset that includes 38 hand radiographs with 11 ROI scores annotated following the Chinese-CHN method. DI is first trained on the RSNA-BAA dataset and then fine-turned on the private dataset. As shown in Fig. 3, one can see that the predicted ROI scores highly coincide with the ground truth ROI scores. In practice, the predicted ROI scores

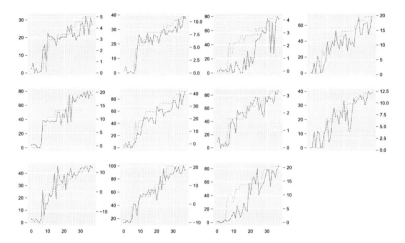

Fig. 3. Illustrating the ROI scores (blue points) and the ground truth ROI scores (yellow points). The x-axis indexes the patients, while the left y-axis shows the ground truth ROI scores and the right y-axis shows the predicted ROI scores. The points are connected to better show the consistence of the ground truth scores and the predicted scores. (Color figure online)

can be translated to the standard scores of the scoring methods by a mapping function (similar to the inverse mapping from ROI scores to bone ages in the scoring methods, which is used in many scoring methods).

4 Conclusions

In this paper, we proposed an automatic bone age assessment model, Doctor Imitator (DI), by imitating the diagnostic logistics of doctors using the scoring methods. An Anatomy-based Group Convolution was proposed to predict the ROI scores by processing the local features of ROIs. Besides, a novel Dual Graph-based Attention Module was introduced to compute the patient-specific attention and context attention for ROI score prediction. As far as we know, DI is the first BAA framework following the processing of the scoring methods with only bone age supervision. Compared with the state-of-the-art, DI achieves good performance with low model complexity and excellent interpretability.

Acknowledgements. The research of Real Doctor AI Research Centre was partially supported by the Zhejiang University Education Foundation under grants No. K18-511120-004, No. K17-511120-017, and No. K17-518051-021, the National Natural Science Foundation of China under grant No. 61672453, the National key R&D program sub project "large scale cross-modality medical knowledge management" under grant No. 2018AAA0102100, the Zhejiang public welfare technology research project under grant No. LGF20F020013, the National Key R&D Program Project of "Software Testing Evaluation Method Research and its Database Development on Artificial Intelligence Medical Information System" under the Fifth Electronics Research Institute of the Ministry of Industry and Information Technology (No. 2019YFC0118802), and The National Key R&D Program Project of "Full Life Cycle Detection Platform and Application Demonstration of Medical Artificial Intelligence Product" under the National Institutes for Food and Drug Control (No. 2019YFB1404802), and the Key Laboratory of Medical Neurobiology of Zhejiang Province. The research of D.Z. Chen was partially supported by NSF Grant CCF-1617735.

References

1. Escobar, M., González, C., Torres, F., Daza, L., Triana, G., Arbeláez, P.: Hand pose estimation for pediatric bone age assessment. In: Shen, D., et al. (eds.) MICCAI 2019. LNCS, vol. 11769, pp. 531–539. Springer, Cham (2019). https://doi.org/10.1007/978-3-030-32226-7_59

2. Gasmallah, M., Zulkernine, F., Rivest, F., Mousavi, P., Sedghi, A.: Fully end-to-end super-resolved bone age estimation. In: Meurs, M.-J., Rudzicz, F. (eds.) Canadian AI 2019. LNCS (LNAI), vol. 11489, pp. 498–504. Springer, Cham (2019). https://doi.org/10.1007/978-3-030-18305-9_51

3. Halabi, S.S., et al.: The RSNA pediatric bone age machine learning challenge. Radiology **290**(2), 498–503 (2019)

4. He, K., et al.: Deep residual learning for image recognition. In: CVPR (2016)

5. Iglovikov, V.I., Rakhlin, A., Kalinin, A.A., Shvets, A.A.: Paediatric bone age assessment using deep convolutional neural networks. In: Stoyanov, D., et al. (eds.) DLMIA/ML-CDS 2018. LNCS, vol. 11045, pp. 300–308. Springer, Cham (2018). https://doi.org/10.1007/978-3-030-00889-5_34

6. Ji, Y., Chen, H., Lin, D., Wu, X., Lin, D.: PRSNet: part relation and selection network for bone age assessment. In: Shen, D., et al. (eds.) MICCAI 2019. LNCS, vol. 11769, pp. 413–421. Springer, Cham (2019). https://doi.org/10.1007/978-3-030-32226-7_46

7. Kazi, A., et al.: Automatic classification of proximal femur fractures based on attention models. In: International Workshop on Machine Learning in Medical Imaging (2017)

8. Kingma, D.P., Ba, J.: Adam: a method for stochastic optimization. ArXiv preprint arXiv:1412.6980 (2014)

9. Larson, D.B., et al.: Performance of a deep-learning neural network model in assessing skeletal maturity on pediatric hand radiographs. Radiology (2018)

10. Liu, C., Xie, H., Liu, Y., Zha, Z., Lin, F., Zhang, Y.: Extract bone parts without human prior: end-to-end convolutional neural network for pediatric bone age assessment. In: Shen, D., et al. (eds.) MICCAI 2019. LNCS, vol. 11769, pp. 667–675. Springer, Cham (2019). https://doi.org/10.1007/978-3-030-32226-7_74

11. Mughal, A.M., et al.: Bone age assessment methods: a critical review. Pak. J. Med. Sci. **30**(1), 211–215 (2014)

12. RuiWei, F., et al.: SSN: a stair-shape network for real-time polyp segmentation in colonoscopy images. In: ISBI (2020)

13. Tanner, J., Whitehouse, R., Marshall, W., Carter, B.: Prediction of adult height from height, bone age, and occurrence of menarche, at ages 4 to 16 with allowance for midparent height. Arch. Dis. Child. **50**(1), 14–26 (1975)

14. Todd, T., Greulich, W., Pyle, S.: Radiographic atlas of skeletal development of hand and wrist (1950)

15. Van Steenkiste, T., et al.: Automated assessment of bone age using deep learning and Gaussian process regression. In: EMBC (2018)
16. Veličković, P., et al.: Graph attention networks. In: ICLR (2017)
17. Wang, S., et al.: Bone age assessment using convolutional neural networks. In: International Conference on Artificial Intelligence and Big Data (2018)
18. Wu, Z., et al.: A comprehensive survey on graph neural networks. ArXiv preprint arXiv:1901.00596 (2019)
19. Zhang, X., et al.: A deep framework for bone age assessment based on finger joint localization. ArXiv preprint arXiv:1905.13124 (2019)
20. Zhou, X., et al.: Diagnostic performance of artificial neural network-based TW3 skeletal maturity assessment. In: ESPE (2018)

Inferring the 3D Standing Spine Posture from 2D Radiographs

Amirhossein Bayat[1,2]($\boxtimes$), Anjany Sekuboyina[1,2], Johannes C. Paetzold[1],
Christian Payer[3], Darko Stern[3], Martin Urschler[4], Jan S. Kirschke[2],
and Bjoern H. Menze[1]

[1] Department of Informatics, Technical University of Munich, Munich, Germany
amir.bayat@tum.de
[2] Department of Neuroradiology, Klinikum rechts der Isar, Munich, Germany
[3] Institute of Computer Graphics and Vision,
Graz University of Technology, Graz, Austria
[4] School of Computer Science, University of Auckland, Auckland, New Zealand

Abstract. The treatment of degenerative spinal disorders requires an understanding of the individual spinal anatomy and curvature in 3D. An upright spinal pose (i.e. standing) under natural weight bearing is crucial for such bio-mechanical analysis. 3D volumetric imaging modalities (e.g. CT and MRI) are performed in patients lying down. On the other hand, radiographs are captured in an upright pose, but result in 2D projections. This work aims to integrate the two realms, i.e. it combines the upright spinal curvature from radiographs with the 3D vertebral shape from CT imaging for synthesizing an upright 3D model of spine, loaded naturally. Specifically, we propose a novel neural network architecture working vertebra-wise, termed *TransVert*, which takes orthogonal 2D radiographs and infers the spine's 3D posture. We validate our architecture on digitally reconstructed radiographs, achieving a 3D reconstruction Dice of 95.52%, indicating an almost perfect 2D-to-3D domain translation. Deploying our model on clinical radiographs, we successfully synthesise full-3D, upright, patient-specific spine models for the first time.

Keywords: 3D reconstruction · Fully convolutional neworks · Spine posture · Digitally reconstructed radiographs

1 Introduction

A biomechanical study of spine and its load analysis in upright standing position is an active research topic, especially in cases of spine disorders [1]. Most

---

A. Bayat and A. Sekuboyina—Equal contribution.
J. S. Kirschke and B. H. Menze—Joint supervising authors.

Electronic supplementary material The online version of this chapter (https://doi.org/10.1007/978-3-030-59725-2_75) contains supplementary material, which is available to authorized users.

A. L. Martel et al. (Eds.): MICCAI 2020, LNCS 12266, pp. 775–784, 2020.
https://doi.org/10.1007/978-3-030-59725-2_75

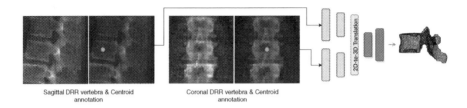

Sagittal DRR vertebra & Centroid
annotation

Coronal DRR vertebra & Centroid
annotation

Fig. 1. Overview of 2D image to 3D shape translation. The network inputs are 2D orthogonal view vertebrae patches and the centroid indicating the vertebra of interest.

common approaches for load estimation on the spine either use a general computational model of the spine for all patients or acquire subject-specific models from magnetic resonance imaging (MRI) or computed tomography (CT) [3]. While these typical 3D image acquisition schemes capture rich 3D anatomical information, they require the patient to be in a *prone* or *supine* position (lying on one's chest or back), for imaging the spine. But, analysis of the spine's shape and vertebral arrangement needs to be done in a physiologically upright standing position under weight bearing, making 2D plain radiographs a *de facto* choice. A combination of both these worlds is of clinical interest to fully assess the biomechanical situation, i.e. to capture patient-specific complex pathological spinal arrangement in a standing position and with 3D information [2,3,8].

In literature, numerous registration-based methods have been proposed for relating 2D radiographs with 3D CT or MR images. In [8], the authors propose a rough manual registration of 3D data to 2D sagittal radiographs for the lumbar vertebrae. For the same purpose, in [4], manual annotations of the vertebral bodies are used as guideline for measuring the vertebral orientations in upright standing position. These methods are time and manual-labour-intensive and thus prone to error. Moreover, both these works use only the sagittal radiographs for vertebra positioning, while ignoring the coronal reformation which is a strong indicator of the spine's natural curvature. Aiming at this objective, [9] introduced an automatic 3D–2D spine registration algorithm, where the authors propose a multi-stage optimization-based registration method by introducing a metric for comparing a CT projection with a radiograph. However, this metric is hand-crafted, parameter-heavy, and is not learning-based, thus limiting its generalizability. In [10], the 3D shape of the spine is reconstructed using a biplanar X-ray device called 'EOS'. Hindering its applicability is the high device cost and the lack of its presence in a clinical routine. Recently the problem of reconstructing 3D shapes given 2D images have been explored using deep learning approaches. An approach closest to ours was proposed by Ying et al. [12], where they introduce a deep neural network to synthesize 3D CT images given orthogonal radiographs using adversarial networks. However, this model is highly memory intensive and fails to synthesise smaller anatomies like vertebrae in 3D. Moreover, it has been evaluated only on digitally reconstructed radiographs (DRR), and its clinical applicability remains to be validated.

Motivation. The problem of 3D reconstruction of a spine in an anatomically upright position from 2D radiograph images relies on retrieving information from radiographs, which are 2D projections of a 3D object. Spine's sagittal reformation captures crucial information in the form of the vertebral body's and process' shape and its orientation around the sagittal (left-right) axis. However, its orientation around the cranio-caudal and anterior-posterior axes is obfuscated (cf. Fig. 1). This information is available when combining sagittal with coronal reformations (or lateral with a.p. radiographs). Motivated by this, we propose a fully-supervised, computationally efficient, and registration-free approach combining sagittal and coronal 2D images to synthesise the vertebra's 3D shape model. Specifically:

- We introduce a novel fully convolutional network (FCN) architecture for fusing orthogonal radiographs to generate 3D shapes.
- We identify an approach for training the network on synthetically generated radiographs from CT, being supervised by the CT's 3D vertebral masks.
- Validating our approach, we achieve dice score of **95.52%** on digitally reconstructed radiographs. We also successfully reconstruct 3D, patient-specific spine models on real clinical radiographs.

2   Methods

Generating 3D shapes from 2D information is an ill-posed problem. For solving this, we utilize information from two orthogonal radiographs and an annotation on the vertebra of interest while relying on the shape prior learnt by the network.

2.1   TransVert: Translating 2D Information to 3D Shapes

The network performing 2D-to-3D synthesis needs to address the following requirements: First, it needs to appropriately combine information in the sagittal and coronal projections to recover 3D information. Second, recovering 3D shapes from 2D projections is inherently an ill-posed problem, requiring incorporation of prior knowledge. Lastly, the size of certain vertebra (towards the scan's periphery) is larger in radiographs compared to their true size due to the cone-beam of gamma-ray source. This effect should be negated when reconstructing the 3D model, i.e. the mapping should not be purely image-based. We address these requirements by proposing the *TransVert* architecture.

Overview. TransVert takes four 2D inputs, the sagittal and coronal vertebral image patches and their corresponding annotation images indicating the vertebra-of-interest (VOI). Denoting the 2D vertebral sagittal and coronal reformations by x_s and x_c, and their corresponding VOI annotation by y_s and y_c, we desire the vertebra's full-body 3D shape, $\mathbf{y}$, as a discrete voxel-map:

$$\mathbf{y} = G(x_s, x_c, y_s, y_c), \tag{1}$$

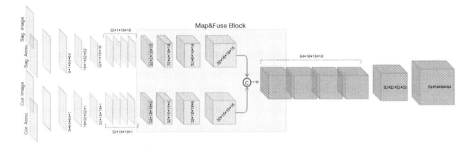

Fig. 2. Architecture of *TransVert*. Our model is composed of sagittal and coronal 2D encoders (self-attention module in red), a 'map&fuse' block, and a 3D decoder. (Color figure online)

where G denotes the mapping performed by TransVert. In our case, the VOI-annotation image is obtained by placing a **disc of radius 1** around the vertebral centroid. In Sect. 3, we analyze denser annotation choices (vertebral body and full vertebral masks). Ideally, training the TransVert mapping requires radiograph images and their corresponding 'real world' 3D spine models. However, this correspondence does not exist and is, in fact, the problem we intend to solve. Thus, TransVert is trained on sagittal and coronal digitally reconstructed radiographs (DRR) constructed from CT images. It is supervised by the corresponding CT images' voxel-level, vertebral segmentation masks. As DRRs are similar in appearance to real radiographs, a DRR-trained TransVert architecture paired with a robust training regime, can be readily deployed on clinical radiographs.

Architecture. TransVert consists of three blocks: a 2D sagittal encoder, a 2D coronal encoder, and a 3D decoder. The three blocks are combined by a 'map&fuse' block. Refer to Fig. 2 for a detailed illustration. The map&fuse block is responsible for *mapping* 2D representations of each the sagittal and coronal views into intermediate 3D latent representations followed by *fusing* them into a single 3D representation by channel-wise concatenation. This representation is then decoded into a viable 3D voxelized representation by the decoder. Note that the intermediate 3D representation is constructed from orthogonal views. Therefore, map&fuse block consists of anisotropic convolutions, with an anisotropy along the dimensions that need to be expanded. For example: the anterior-posterior dimension needs to be expanded for a coronal view. Consequently, the convolutional strides and padding directions are orthogonal for each of the view. At the network encoders' input, the vertebral images and VOI-annotations are combined using a self-attention layer. It was empirically observed that the attention mechanism yielded a better performance than a naive fusion by concatenating them as multiple channels.

Loss. Using solely a regression loss leads to converging to a local optimum where a mean (or median) shape is predicted, especially in the highly varying regions of the vertebra such as the vertebral processes. This is rectified by augmenting the loss with an adversarial component which checks the validity of a prediction at a global level. Therefore, TransVert is trained in a fully supervised manner by optimizing a combination of an ℓ_1 distance-based regression loss and an adversarial loss based on the least-squared GAN (LSGAN [5]). Formally, the TransVert and the Discriminator combination is trained by minimizing the following losses:

$$\mathcal{L}_G = \alpha_G \|\mathbf{y} - G(x_s, x_c, y_s, y_c)\|_1 + \alpha_D \left(D(G(x_s, x_c, y_s, y_c)) - 1 \right)^2 \text{ and} \quad (2)$$

$$\mathcal{L}_D = (D(\mathbf{y}) - 1)^2 + D(G(x_s, x_c, y_s, y_c))^2, \quad (3)$$

where D represents the discriminator network and G represents the TranVert. α_G and α_D are weights of loss terms and fixed to $\alpha_G = 10$ and $\alpha_D = 0.1$. Note that $\mathbf{y}$ is binary valued containing $\{0, i\}$, where $i \in \{8, 9, \ldots 24\}$ denotes the vertebral index from T1 to L5. Forcing the network to predict the vertebral index implicitly incorporates an additional prior relating the shape to the vertebral index. Details about the discriminator architecture and the adversarial training regime are provided in the supplemental material. The network is implemented with Pytorch framework on a Quadro P6000 GPU. It is trained till convergence using an Adam optimizer with initial learning rate is 0.0001.

3   Results

In this section, we describe the creation of DRRs, present an ablative study quantitatively analyzing the contribution of various architectural components, compare various VOI-annotation types, and finally deploy TransVert on real clinical radiographs.

3.1   Data

Recall that TransVert works with two data modalities: it is trained on DRRs extracted from CT images while being supervised by their corresponding 3D segmentation mask, and it is deployed on clinical radiographs.

CT Data. We work with two datasets: a publicly available dataset for lung nodule detection with 800 chest CT scans [13], and an in-house dataset with 154 CT scans. In all, we work with ~12 k vertebrae split 5:1 forming the training and validation set, reporting 3-fold cross validated results. Note that very few lumbar vertebrae are visible in [13] as it is lung-centred.

Data Preparation: The CT scans are segmented using [11] and the generated masks are validated by an experienced neuro-radiologist in order to consider only accurate ones for the study. These vertebral masks are used for supervision.

Generation of the corresponding digitally reconstructed radiographs (DRR) is performed using a ray-casting approach [7], wherein a line is drawn from the radiation source (focal point) to every single pixel on the digitally reconstructed radiographs (DRR) image and the integral of the CT intensities over this line are calculated. Parameters for this generation include the radiation source-to-detector (=180 cm in this work) and the source-to-object distance (=150 cm here). Post the generation of the sagittal and coronal digitally reconstructed radiographs (DRR), patches of size 64 × 64 are extracted around each vertebral centroid, constituting the image input to TransVert. The second input, viz. the VOI-annotation, can be extracted from the projected segmentation mask.

Clinical Radiographs. We clinically validate TransVert on real long standing radiographs in corresponding lateral and anterior-posterior (a.p.) projections obtained in 30 patients. Acquisition parameters such as source-to-detector and source-to-object distances were similar to those used for DRR generation. Vertebral centroids needed for the VOI-annotations were automatically generated on both views using [6].

Data Normalization. TransVert is trained on DRRs and tested on clinical radiographs. These data modalities have different intensity ranges, requiring normalization. We observe that z-score normalization works well, i.e. $\mathcal{I} = (\mathcal{I}-\mu_{\mathcal{I}})/\sigma_{\mathcal{I}}$, where $\mu_{\mathcal{I}}$ and $\sigma_{\mathcal{I}}$ are the mean and standard deviation of the image $\mathcal{I}$.

3.2 Experiments

We perform three sets of experiments validating our proposed approach, aimed at analysing the architectural aspects of TransVert, the data fed into it, and finally its applicability in a clinical setting. Note that a quantitative comparison with the ground truth can be performed only in experiments dealing with DRRs and CT images. Performance evaluation of various settings is compared by computing Dice coefficient and Hausdorff Distance between the predicted 3D vertebral mask and its ground truth from the CT mask.

Analysing TransVert's Architecture. The proposed architecture for TransVert consists of the following architectural choices: fusion of sagittal and coronal views, anisotropic convolutions in the map&fuse block, a self-attention layer combining the image and the VOI-annotation, and finally, an adversarial component on the loss function. An ablative study over these components is reported in Table 1. First, *do we need two views?*. For this, we evaluate the performance of a model that tries to reconstruct 3D shape from only a sagittal image. Next, *do we need anisotropic convolutions?*. For this, we compare two versions of map&fuse: one with a simple outer product for combining the orthogonal views (Naive View-Fusion) and one with the proposed anisotropic convolutions (TransVert). Observe that a simple fusion of views already outperforms a 'sagittal only' reconstruction. Also, anisotropic convolutions outperform

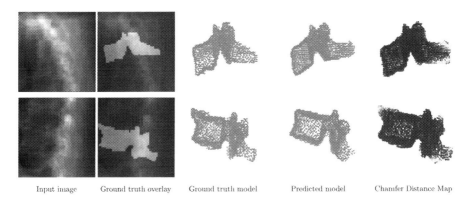

Input image   Ground truth overlay   Ground truth model   Predicted model   Chamfer Distance Map

Fig. 3. Shape modelling with TransVert on DRRs: First column indicates the image input. Second and third columns visualise the ground truth (GT) vertebral mask and the fourth visualises the predicted 3D shape model. Last column shows an overlayed Chamfer distance map between point clouds of GT and prediction.

Table 1. Architectural ablative study: The performance progressively improves with addition of each component. (Vertebral centroids are the VOI-annotations here.

Setup	Dice (%)	Hausdorff (mm)
Sagittal only	88.40	7.43
Naive View-Fusion (Outer Product)	92.59	6.45
TransVert	94.75	5.75
TransVert + Self Attn	95.31	5.27
TransVert + SelfAttn + Adv.	95.52	5.11

fusion of views using outer-products. This can be attributed to the 2D-to-3D learning component involved in the latter. Lastly, *do we need the bells & whistles on top of TransVert?* Observe that incorporating the self-attention layer in the encoders and an adversarial training regime progressively improved performance, resulting in a Dice of 95.5% and a Hausdorff Distance of 5.11 mm. Figure 3 illustrates the 3D shape models reconstructed using the proposed architecture. Extracting a point cloud (with 2048 points) from these shapes, we also illustrate a point-wise Chamfer distance map. Observe that a vertebra's posterior region (vertebral process) is hardly visible in the image inputs. Despite this, TransVert is capable of recovering the process, albeit with a certain disagreement between the prediction and ground truth.

Analysing VOI-Annotation Type. Recall that alongside the image input, TransVert requires an auxiliary input indicating the vertebra of interest. We argue that a vertebral centroid suffices. In this study we show that our choice of vertebral centroid performs at a level comparable to a far denser full-vertebra annotation as reported in Table 2. We compare our centoids-to-vertebra (C2V) setup to two other, denser annotations: one where the vertebral body is annotated in the DRR (B2V) and one where the full vertebral body is annotated (V2V). As baseline, we include a setup without any VOI-annotation as an auxiliary input. Note that including the annotation input offers approximately 20% improvement in the mean Dice coefficient. Observe that a most dense V2V annotations and our C2V annotations perform comparably with only <1% difference. Therefore, C2V is an obvious choice owing to the ease of marking centroids, more so because of existing automated labelling approaches.

Table 2. VOI annotation study: Performance drop from a denser (V2V) to a sparser annotation (C2V) is minor, while annotation effort decreases manifold.

Input	Dice (%)	Hausdorff (mm)
No annotation	76.44	14.74
V2V	96.24	4.18
B2V	95.67	4.95
C2V	95.31	5.27

2D-to-3D Translation in Clinical Radiographs. TransVert works with individual vertebral images and their centroids. A 3D model of the spine can be constructed by stacking the predicted 3D vertebrae models at their corresponding 3D centroid locations. Vertebra's position along the axial and coronal axes is obtained from the sagittal reformation and its sagittal position from the coronal reformation. Figure 4 illustrates the results of this process. The top row visualises a 3D spine reconstruction based on 2D DRRs and compares it with the ground truth. More importantly, the bottom row depicts a successful deployment of TransVert in reconstructing the 3D, patient-specific posture of upright standing spine. Note that no 3D ground truth spine model exists for these cases. We visualise the 2D overlay of the segmentation on the radiographs, and the sagittal and coronal view of its 3D shape model, the former overlaid on the radiograph too. Observe that the 3D model's posture matches with that of the radiographs.

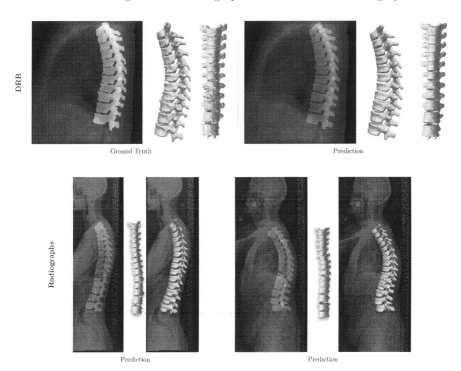

Fig. 4. Full 3D spine models: (Top row) Comparison of a DRR-based spine model reconstruction with its CT ground truth mask. (Bottom row) 3D patient-specific spine models constructed from real clinical radiographs. (Best viewed by zooming in.)

4   Conclusion

We propose TransVert, a novel architecture trained to infer a full-3D spine model from 2D sagittal and coronal radiographs and sparse centroid annotations. We identify an approach to train TransVert on DRRs in a fully-supervised manner. Along with an ablative study on TransVert's architectural components, we show a successful use case of deploying it on a real-world clinical radiograph.

References

1. Dreischarf, M., et al.: Estimation of loads on human lumbar spine: a review of in vivo and computational model studies. J. Biomech. **49**(6), 833–845 (2016)
2. El Ouaaid, Z., et al.: Effect of changes in orientation and position of external loads on trunk muscle activity and kinematics in upright standing. J. Electromyogr. Kinesiol. **24**(3), 387–393 (2014)
3. Akhavanfar, M.H., et al.: Obesity and spinal loads; a combined MR imaging and subject-specific modeling investigation. J. Biomech. **70**, 102–112 (2018)
4. Bauer, S., et al.: Effects of individual spine curvatures-a comparative study with the help of computer modelling. Biomed. Eng./Biomedizinische Technik **57**, 132–135 (2012)

5. Mao, X., et al.: Least squares generative adversarial networks. In: IEEE ICCV, pp. 2794–2802 (2017)
6. Bayat, A., Sekuboyina, A., Hofmann, F., Husseini, M.E., Kirschke, J.S., Menze, B.H.: Vertebral labelling in radiographs: learning a coordinate corrector to enforce spinal shape. In: Cai, Y., Wang, L., Audette, M., Zheng, G., Li, S. (eds.) CSI 2019. LNCS, vol. 11963, pp. 39–46. Springer, Cham (2020). https://doi.org/10.1007/978-3-030-39752-4_4
7. Staub, D., et al.: A digitally reconstructed radiograph algorithm calculated from first principles. Med. Phys. **40**(1), 011902 (2013)
8. Eskandari, A.H., et al.: Subject-specific 2D/3D image registration and kinematics-driven musculoskeletal model of the spine. J. Biomech. **57**, 18–26 (2017)
9. Ketcha, M.D., et al.: Multi-stage 3D–2D registration for correction of anatomical deformation in image-guided spine surgery. Phys. Med. Biol. **62**(11), 4604 (2017)
10. Humbert, L., et al.: 3D reconstruction of the spine from biplanar X-rays using parametric models based on transversal and longitudinal inferences. Med. Eng. Phys. **31**(6), 681–687 (2009)
11. Sekuboyina, A., et al.: VerSe: a vertebrae labelling and segmentation benchmark. arXiv preprint arXiv:2001.09193 (2020)
12. Ying, X., et al.: X2CT-GAN: reconstructing CT from biplanar X-rays with generative adversarial networks. In: IEEE CVPR (2019)
13. Armato III, S.G.: The lung image database consortium (LIDC) and image database resource initiative (IDRI): a completed reference database of lung nodules on CT scans. Med. Phys. **38**(2), 915–931 (2011)

Generative Modelling of 3D In-Silico Spongiosa with Controllable Micro-structural Parameters

Emmanuel Iarussi[1,2], Felix Thomsen[1,3(✉)], and Claudio Delrieux[1,3]

[1] Consejo Nacional de Investigaciones Científicas y Técnicas, Buenos Aires, Argentina
[2] Facultad Regional Buenos Aires, Universidad Tecnológica Nacional, Buenos Aires, Argentina
[3] Universidad Nacional del Sur, Bahía Blanca, Argentina
felix.thomsen@uns.edu.ar

Abstract. Research in vertebral bone micro-structure generally requires costly procedures to obtain physical scans of real bone with a specific pathology under study, since no methods are available yet to generate realistic bone structures *in-silico*. Here we propose to apply recent advances in generative adversarial networks (GANs) to develop such a method. We adapted style-transfer techniques, which have been largely used in other contexts, in order to transfer style between image pairs while preserving its informational content. In a first step, we trained a volumetric generative model in a progressive manner using a Wasserstein objective and gradient penalty (PWGAN-GP) to create patches of realistic bone structure *in-silico*. The training set contained 7660 purely spongeous bone samples from twelve human vertebrae (T12 or L1) with isotropic resolution of 164 μm and scanned with a high resolution peripheral quantitative CT (Scanco XCT). After training, we generated new samples with tailored micro-structure properties by optimizing a vector z in the learned latent space. To solve this optimization problem, we formulated a differentiable goal function that leads to valid samples while compromising the appearance (content) with target 3D properties (style). Properties of the learned latent space effectively matched the data distribution. Furthermore, we were able to simulate the resulting bone structure after deterioration or treatment effects of osteoporosis therapies based only on expected changes of micro-structural parameters. Our method allows to generate a virtually infinite number of patches of realistic bone micro-structure, and thereby likely serves for the development of bone-biomarkers and to simulate bone therapies in advance.

Keywords: Bone micro-structure · Progressive generative adversarial network · Structural morphing · Style-transfer · XCT

Electronic supplementary material The online version of this chapter (https://doi.org/10.1007/978-3-030-59725-2_76) contains supplementary material, which is available to authorized users.

© Springer Nature Switzerland AG 2020
A. L. Martel et al. (Eds.): MICCAI 2020, LNCS 12266, pp. 785–794, 2020.
https://doi.org/10.1007/978-3-030-59725-2_76

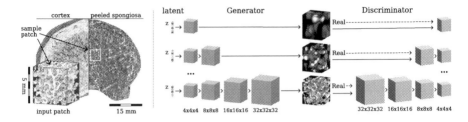

Fig. 1. Left: 7660 purely spongeous HR-pQCT patches of 32^3 voxels were sampled from twelve human vertebrae phantoms. Right: Two 3D Convolutional Neural Networks, Generator and Discriminator, were progressively trained to mimic the vertebrae volume distribution at incremental resolutions.

1 Introduction

The development of new methods for characterizing the micro-structure of spongeous bone is an active research field in constant progress. For the development and analysis of specific structural parameters (e.g. bone volume ratio, trabecular separation or plate-to-rod ratio), often very simple *in-silico* bone models are used [11,19] containing only few rods and plates intersecting each other, since they allow easiest control of the desired output. These very simple models reflect rather poorly the real structure of bone. A first attempt to generate more realistic bone has recently been presented, evolving a 3D structure from separated 2D slices, generated with a technique in Fourier domain [14]. The authors reported high accordance between generated and real samples regarding trabecular thickness, however further parameters have not been evaluated. In this work we developed a direct method to create 3D micro-structural bone samples *in-silico* that 1) contain realistic structures, and 2) allow to steer the properties to simulate changes of micro-structural parameters from deterioration or medical bone treatment.

In order to achieve these goals, we first trained a generative volumetric convolutional neural network (CNN) on isotropic patches of $32 \times 32 \times 32$ voxels (box of 5 mm diameter) to generate *in-silico* HR-pQCT patches. Working in 3D severely increases the complexity of the generative models with respect to images (2D) or videos (2D+t). Therefore we adapted a WGAN-GP architecture [6] to work with 3D data and trained it by progressively growing the resolution up to the target shape [8], accordingly we call this architecture *progressive* WGAN-GP (PWGAN-GP). The latent vector z corresponds to a random point in a 32-dimensional hypersphere, which defines entirely the generated volume that the discriminator network is intended to differentiate from a real sample.

A key factor of our approach is to provide control over the generated samples, to be able to navigate in the learned latent space and to customize the morphological properties of the output volumes. Instead of shaping the latent space by means of a conditional GAN [10], conditioned on target micro-structural parameters, we took advantage of generative style transfer techniques [3] largely used in photo manipulation applications and other types of art-work. In contrast to conditional GANs, this two-step approach, GAN and style transfer, does not require to retrain the generative model if the user desires to incorporate new control properties, i.e. a new set of micro-structural parameters. The style transfer mechanism allows to steer the properties and to simulate changes of micro-structural parameters, for instance from deterioration or medical bone treatment. This is achieved by mapping the latent vector of an original sample to a vector of new micro-structural parameters (style) but still with a similar micro-structure (content), hence by solving an optimization problem with two conditions. Thus, we show that not only neural networks but also style transfer strategies can be employed to generate and control structural properties in 3D bone CT scans. To the best of our knowledge, our work is the first attempt to address micro-structural customization of 3D volumes formulating it as a style transfer problem and also the first generative adversarial method to synthesize bone micro-structure [16]. Code, data and trained models will be released at http://github.com/emmanueliarussi/generative3DSpongiosa.

2   Methods

In this chapter we describe the examined generative models and the neural style transfer function. We adapted a 2D state of the art generative network model [8] and trained it with HR-pQCT volumetric scans of human vertebrae (Fig. 1). Once trained, we assessed the suitability of generated volumes qualitatively via 3D renderings, and quantitatively with standard metrics of trabecular bone. Next, we generated samples with tailored micro-structural properties using a style transfer perspective over the learned latent space. Synthetic samples contained new micro-structural properties but preserved the general content of the original sample.

2.1   Examined Deep Generative Models

Generative adversarial networks (GANs) [5] are trained to stochastically generate samples close to a distribution represented by the training set. Despite the high success of these network architectures, training is still a very unstable process. Therefore, several alternatives have emerged to deal with training issues.

In particular, we based our framework on Wasserstein GANs [1], consisting of a generator network $G : z \mapsto \tilde{x}$, and a discriminator network (also called *critic* in the context of WGANs) $D : x \mapsto D(x) \in \mathbb{R}$ which were simultaneously trained to try to fool each other: while G learned to generate fake samples $\tilde{x}$ from an unknown distribution or noise z, the critic D learned meanwhile to distinguish fake from real samples. Formally, the training objective function optimizes:

$$\min_{G} \max_{D \in \mathcal{D}} \mathop{\mathbb{E}}_{x \sim \mathbb{P}_r} [D(x)] - \mathop{\mathbb{E}}_{\tilde{x} \sim \mathbb{P}_g} [D(\tilde{x})], \tag{1}$$

where $\mathcal{D}$ is the set of 1-Lipschitz functions, $\mathbb{P}_r$ the data distribution and $\mathbb{P}_g$ the model distribution defined by $\tilde{x} = G(z), z \sim p(z)$. Since enforcing the Lipschitz constraint is not trivial, we applied a gradient penalty mechanism (PG) [6] at every iteration with a critic parameter update (WGAN-PG).

We employed three additional non-progressive GANs to compare performance. 1) We trained the exact same architecture as described before but with an adversarial loss [5] and without extra regularization terms, gradient penalty and critic drift. Instead, we added a sigmoid layer at the final discriminator step in order to output labels in range $[0, 1]$. 2) As a more advanced alternative we applied the full framework with gradient penalty and critic drift but without progressive training (WGAN-GP). 3) We additionally tested an alternative WGAN enforcing Lipschitz constraints by means of weight clipping. However, we omitted the third network from the detailed analysis since it suffered from mode collapse: all generated samples resulted extremely similar and did not resemble any realistic bone structure. All hyperparameters (total epochs, training rate, etc.) were kept fixed among all methods.

2.2   Generating Samples with Custom Properties

We framed the problem of generating samples with tailored micro-structural properties by formulating an optimization problem over the latent space of our generative model. In the spirit of style-transfer techniques, we defined the *content* x_t as the target volume we want to stick to. The target volume could be a sample from our training data set, or also be produced by the generator network, in that case $x_t = G(z_t)$. A key contribution of our approach is to redefine the notion of *style* for 3D scans of human vertebrae. If x_t provides the overall volume constraint, the style is given by a set of micro-structural properties w_t, our generated sample has to satisfy. Formally, we set up an optimization problem over z' minimizing the sum of two conditions:

$$\min_{z'} \mu \| x_t - G(z') \|_2^2 + \| w_t - P(G(z')) \|_2^2, \qquad (2)$$

with $\mu \geq 0$ and $P(\cdot)$ the algorithm to compute a vector of differentiable micro-structural properties as given by w_t, Sect. 3.3. The objective function is differentiable and can be minimized using gradient descent. Notice that the term accounting for the content is hard to minimize and that the global minimum may be not unique. In practice, we set μ to a small value (e^{-4}) and optimized Eq. 2 with the Limited Memory Broyden–Fletcher–Goldfarb–Shanno algorithm (L-BFGS). As we show in the Results section, this optimization allows us to navigate the latent space learned by our network, retrieving plausible synthetic samples with the desired micro-structural properties.

3   Experimental Setup

In this section we describe the sampling procedure, choice of hyperparameters and the applied statistics.

3.1   Training Dataset

Twelve human vertebrae (T12 and L1) were embedded into epoxy resin without damaging any trabeculae to become cylindrical vertebrae phantoms. These phantoms were scanned on a high-resolution peripheral QCT (HR-pQCT) with isotropic resolution of 82 μm, 59.4 kVp and 900 μAs (XtremeCT I, Scanco Medical AG, Brüttisellen, Switzerland) and automatically calibrated to density values. The spongiosa has been peeled from the cortex with a semi-automatic procedure [18] and down-sampled to an isotropic resolution of 164 μm, thereby increasing the signal-to-noise ratio and obtaining a lower voxel number per patch but still keeping most structural information (see Fig. 1).

We defined then 7660 purely spongeous patches on the entire set of vertebra phantoms with isotropic size of $32 \times 32 \times 32$ voxels (box of diameter 5 mm) and regular offset 8 voxels in all directions (1.3 mm). Patches were normalized from $[-350, 1100]$ mg/cm^3 to $[-1, 1]$ (only 0.001% or 1764 from 152 million voxels were thereby clamped), and denormalized again in production mode to compute structural parameters correctly. Data have been augmented to 122,560 patches by employing all possible 16 axis-aligned rotations and reflections without imposing a misalignment of the vertical axis. This procedure avoids any interpolation artifact and respects the preferential structural and load orientation of bone.

3.2   Parameter Settings

We progressively trained our network in four stages with isotropic samples of 4^3, 8^3, 16^3 and 32^3 voxels. Each training stage consisted on five training epochs followed by five blending epochs to fade smoothly in the new layers. We used a batch size of 16 in all stages and the ADAM optimizer [9] with $\beta_1 = 0$, $\beta_2 = 0.99$ and learning rate $= 0.001$ for the generator and critic networks. We set the number of critic iterations per generator iteration to 3 and used gradient penalty with $\lambda = 10$. In order to further regularize the discriminator, we penalized outputs drifting away from 0 by adding the average of the critic output squared to its loss with weight $\epsilon_{drift} = 0.001$. Instead of batch normalization we used pixel normalization [8] after each 3D convolutional layer in the generator network. Additionally, we updated the generator weights using Exponential Moving Average (EMA) [20]. In total, our generator and discriminator networks contained more than 200,000 trainable parameters each.

3.3   Quantitative and Qualitative Evaluation

We implemented a set of commonly applied bone micro-structural parameters in Python: Bone mineral density (BMD), standard deviation of density values (BMD.SD), bone volume ratio (BV/TV), tissue mineral density (TMD), mean intercept length (MIL), parallel plate model dependent trabecular separation (Tb.Sp) and thickness (Tb.Th) [17]. The vector $P(G(z'))$ as defined in Eq. 2 contained BMD, BV/TV, TMD and BMD.SD, that were specifically implemented as differentiable PyTorch functions, similar techniques have been used elsewhere [7].

Table 1. Means, $\pm$ standard deviations of micro-structural parameters and p-values of Tukey's range test's ($\alpha = 0.05$). Bold values indicate no statistical difference.

Parameter	Real	GAN	WGAN-GP	PWGAN-GP
BMD[mg/cm^3]	123.58 ±36.21	122.85 ±31.01 (**99.54%**)	120.07 ±33.04 (**19.01%**)	124.76 ±34.45 (**95.04%**)
BMD.SD[mg/cm^3]	125.97 ±22.96	118.85 ±17.62 (<0.01%)	131.38 ±21.70 (<0.01%)	123.63 ±21.26 (**21.28%**)
TMD[mg/cm^3]	341.94 ±30.90	334.18 ±22.34 (<0.01%)	348.52 ±26.04 (<0.01%)	338.40 ±26.43 (**7.48%**)
BV/TV[%]	18.90 ±7.11	18.04 ±5.45 (**5.82%**)	18.67 ±6.54 (**95.12%**)	18.77 ±7.42 (**99.44%**)
MIL[mm]	1.18 ±0.42	1.13 ±0.39 (**16.33%**)	1.17 ±0.37 (**99.06%**)	1.19 ±0.40 (**68.20%**)
Tb.Sp[mm]	0.98 ±0.43	0.95 ±0.39 (**37.81%**)	0.97 ±0.37 (**98.45%**)	0.99 ±0.41 (**76.95%**)
Tb.Th[μm]	197.87 ±32.62	189.65 ±44.79 (0.03%)	199.21 ±47.77 (**96.17%**)	199.52 ±36.78 (**70.02%**)

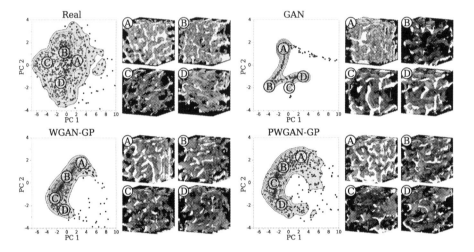

Fig. 2. Distributions of principal components of 700 randomly selected patches, and each four 3D renderings of representative samples of real and generated sets of three different architectures.

Since micro-structural parameters were not independent from each other, either for physical (e.g. BMD cannot be higher than TMD), or for physiological reasons (e.g. correlation between BMD and BV/TV), we reduced the parameters for simplicity to two (linearly independent) principal components (PC1 and PC2) that explained 99% of the variation on real patches. We considered 700 real and generated patches for all resulting metrics. Quantitative analyses of the networks were conducted by computing means and standard deviations of all considered micro-structural parameters, computed with a fixed threshold of 225 mg/cm^3. We employed for all methods Tukey's range tests with $\alpha = 0.05$ between real and generated patches.

4   Results

Figure 2 shows the distribution of real and generated patches. The upper left plot indicates the target distribution, noticeably GAN differs most and PWGAN-GP

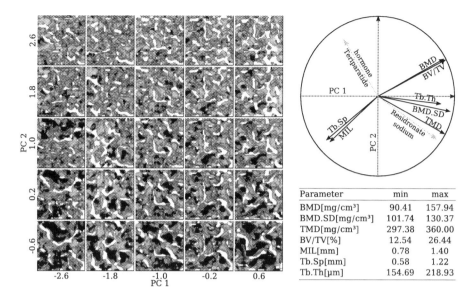

Parameter	min	max
BMD[mg/cm³]	90.41	157.94
BMD.SD[mg/cm³]	101.74	130.37
TMD[mg/cm³]	297.38	360.00
BV/TV[%]	12.54	26.44
MIL[mm]	0.78	1.40
Tb.Sp[mm]	0.58	1.22
Tb.Th[µm]	154.69	218.93

Fig. 3. Generated samples with varying micro-structural parameters but fixed content. Right: Direction of change of structural parameters and two specific drugs to strengthen bone (Fig. 4), range of measured parameters.

least from the real distribution. Besides that all neural networks still underpopulated certain areas in parameter space, e.g. PWGAN-GP at $(PC1, PC2) = (2, -1)$, single generated patches were visually difficult to distinguish from real patches. Table 1 shows statistics of micro-structural parameters. Only PWGAN-GP (right column) was indistinguishable on all considered statistics, while WGAN-GP was statistically different from real data on parameters BMD.SD and TMD, and GAN additionally on Tb.Th. Figure 3 shows the application of the neural style transfer method on 25 bone samples with different micro-structural parameters. The patches are variations of a single patch which was directly generated by the GAN with principal components close to the one in the center, thereby keeping bone micro structure as fixed as possible by simultaneously fitting the individual target structural parameters. The circle on the top right shows the change of principal components for each micro-structural parameter. Also treatment effects of drugs for osteoporosis therapy can be expressed in such a way, as shown for Residronate and Teriparatide. Therefore, we used reported treatment effects of the examined structural parameters [4].

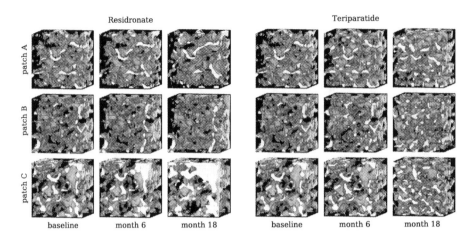

Fig. 4. Simulated bone structure under Residronate and Teriparatide treatment after 6 and 18 months. While Residronate (left) is known to calcify existing bone only, Teripartide (right) is able to form new bone.

5   Discussion and Conclusion

We implemented a method to generate realistic *in-silico* patches of bone micro-structure with defined micro-structural parameters. Qualitative analyses showed high similarity with real bone. Micro-structural parameters were indistinguishable between generated and real patches, which is particularly important since the loss-function of the neural network was not explicitly considering these statistics, hence further micro-structural characteristics might be modeled correctly as well.

We still see potential to improve our method, i.e. to closer align the distributions of generated and real patches (Fig. 2). Latent vectors that are an explicit combination of style-generating and structural variables might be worth of consideration. Furthermore, network architectures applied for similar problems like the generation of blood vessels [15], lung nodules [13], liver lesions [2] or other applications [16] seem to be promising alternatives to consider. Our method generated only patches of 5 mm (or 32^3 voxels). We tried also to train for patches with 10 mm (or 64^3) thereby using a larger number of epochs and latent dimensions, but application on larger scale became unstable according to the distribution criterion. An alternative might be a generative model to produce continuous bone structures which required however a more complex logic to maintain a smooth continuum of bone structural parameters.

We foresee our method being useful in a number of applications. For instance, the generation of bone structures for developing and testing of new micro-structural parameters that enhance our current understanding of bone stability. Also, the simulation of micro-structural changes by bone-forming drugs or immobility- or age related bone-reduction is still an open issue. Since micro-structural properties of specific treatments for osteoporosis are generally known, the content-preserving method to vary micro-structural parameters can be used to simulate bone micro-structure after a specific treatment. Such simulations are shown in Fig. 4 based on reported treatment effects of the bisphosphonate Residronate and the parathyroid hormone Teriparatide using reported effects on BMD, BV/TV and TMD [4]. The proposed method could potentially be refined by incorporating differentiable formulas of additional structural parameters, such as Tb.Sp, Tb.Th, the anisotropy and elongation indices. The simulated micro-structure might also serve to compute minimum bone failure load with the finite-element-method or the mean-intercept-length tensor [12], which both require exact models of the bone.

Acknowledgements. We thank C.-C. Glüer for providing the phantoms. This study was supported by Agencia Nacional de Promoción Científica y Tecnológica, Argentina (PICT 2017-1731), PID UTN 2018 (SIUTNBA0005139), PID UTN 2019 (SIUTNBA0005534), and NVIDIA GPU hardware grant that supported this research with the donation of two Titan Xp graphic cards.

References

1. Arjovsky, M., Chintala, S., Bottou, L.: Wasserstein GAN. arXiv preprint arXiv:1701.07875 (2017)
2. Frid-Adar, M., Diamant, I., Klang, E., Amitai, M., Goldberger, J., Greenspan, H.: GAN-based synthetic medical image augmentation for increased CNN performance in liver lesion classification. Neurocomputing **321**, 321–331 (2018). https://doi.org/10.1016/j.neucom.2018.09.013
3. Gatys, L.A., Ecker, A.S., Bethge, M.: A neural algorithm of artistic style. arXiv preprint arXiv:1508.06576 (2015)
4. Glüer, C.C., et al.: Comparative effects of teriparatide and risedronate in glucocorticoid-induced osteoporosis in men: 18-month results of the EuroGIOPs trial. J. Bone Miner. Res. **28**, 1355–1368 (2013). https://doi.org/10.1002/jbmr.1870
5. Goodfellow, I., Bengio, Y., Courville, A.: Deep Learning. MIT Press, Cambridge (2016)
6. Gulrajani, I., Ahmed, F., Arjovsky, M., Dumoulin, V., Courville, A.C.: Improved training of Wasserstein GANs. In: Advances In Neural Information Processing Systems, pp. 5767–5777 (2017)
7. Karam, C., Sugimoto, K., Hirakawa, K.: Fast convolutional distance transform. IEEE Signal Process. Lett. **26**(6), 853–857 (2019). https://doi.org/10.1109/LSP.2019.2910466
8. Karras, T., Aila, T., Laine, S., Lehtinen, J.: Progressive growing of GANs for improved quality, stability, and variation. arXiv preprint arXiv:1710.10196 (2017)
9. Kingma, D.P., Ba, J.: Adam: A method for stochastic optimization. In: International Conference on Learning Representations, pp. 1–15 (2015)
10. Mirza, M., Osindero, S.: Conditional generative adversarial nets. arXiv preprint arXiv:1411.1784 (2014)
11. Moreno, R., Borga, M., Smedby, Ö.: Evaluation of the plate-rod model assumption of trabecular bone. In: IEEE International Symposium on Biomedical Imaging, pp. 470–473 (2012). https://doi.org/10.1109/ISBI.2012.6235586

12. Moreno, R., Borga, M., Smedby, Ö.: Techniques for computing fabric tensors: a review. In: Westin, C.-F., Vilanova, A., Burgeth, B. (eds.) Visualization and Processing of Tensors and Higher Order Descriptors for Multi-Valued Data. MV, pp. 271–292. Springer, Heidelberg (2014). https://doi.org/10.1007/978-3-642-54301-2_12

13. Onishi, Y., et al.: Automated pulmonary nodule classification in computed tomography images using a deep convolutional neural network trained by generative adversarial networks. BioMed Res. Int. **2019**, 9 p. (2019). https://doi.org/10.1155/2019/6051939

14. Peña-Solórzano, C.A., et al.: Development of a simple numerical model for trabecular bone structures. Med. Phys. **46**(4), 1766–1776 (2019). https://doi.org/10.1002/mp.13435

15. Russ, T., et al.: Synthesis of CT images from digital body phantoms using Cycle-GAN. Int. J. Comput. Assist. Radiol. Surg. **14**(10), 1741–1750 (2019). https://doi.org/10.1007/s11548-019-02042-9

16. Sorin, V., Barash, Y., Konen, E., Klang, E.: Creating artificial images for radiology applications using generative adversarial networks (GANs)-a systematic review. Acad. Radiol. **27**(8), 1175–1185 (2020). https://doi.org/10.1016/j.acra.2019.12.024

17. Thomsen, F.: Medical 3D image processing applied to computed tomography and magnetic resonance imaging. Ph.D. Thesis, Universidad Nacional del Sur, Bahía Blanca, Argentina (2017). https://doi.org/10.13140/RG.2.2.10998.80966

18. Thomsen, F., Peña, J., Delrieux, C., Glüer, C.C.: Structural insight v3: a standalone program for micro structural analysis of computed tomography volumes. In: Congreso Argentino de Informática y Salud (2016). https://doi.org/10.13140/RG.2.2.29351.14245

19. Thomsen, F., et al.: A new algorithm for estimating the rod volume fraction and the trabecular thickness from in vivo computed tomography. Med. Phys. **43**(12), 6598–6607 (2016). https://doi.org/10.1118/1.4967479

20. Yazıcı, Y., Foo, C.S., Winkler, S., Yap, K.H., Piliouras, G., Chandrasekhar, V.: The unusual effectiveness of averaging in GAN training. arXiv preprint arXiv:1806.04498 (2018)

GAN-Based Realistic Bone Ultrasound Image and Label Synthesis for Improved Segmentation

Ahmed Z. Alsinan[1]([✉]), Charles Rule[2]([✉]), Michael Vives[3]([✉]),
Vishal M. Patel[4]([✉]), and Ilker Hacihaliloglu[5,6]([✉])

[1] Department of Electrical and Computer Engineering, Rutgers University,
Piscataway, NJ, USA
ahmed.alsinan@rutgers.edu
[2] Department of Computer Science, Rutgers University, Piscataway, NJ, USA
cer157@scarletmail.rutgers.edu
[3] Department of Orthopedics, Rutgers New Jersey Medical School,
Newark, NJ, USA
vivesmj@njms.rutgers.edu
[4] Department of Electrical and Computer Engineering, Johns Hopkins University,
Baltimore, MD, USA
vpatel36@jhu.edu
[5] Department of Biomedical Engineering, Rutgers University, Piscataway, NJ, USA
ilker.hac@soe.rutgers.edu
[6] Department of Radiology, Rutgers University Robert Wood Johnson Medical
School, New Brunswick, NJ, USA

Abstract. To provide a safe alternative, for intra-operative fluoroscopy,
ultrasound (US) has been investigated as an alternative safe imaging
modality for various computer assisted orthopedic surgery (CAOS) procedures. However, low signal to noise ratio, imaging artifacts and bone
surfaces appearing several millimeters (mm) in thickness have hindered
the wide spread application of US in CAOS. In order to provide a solution
for these problems, research has focused on the development of accurate,
robust and real-time bone segmentation methods. Most recently methods based on deep learning have shown very promising results. However,
scarcity of bone US data introduces significant challenges when training
deep learning models. In this work, we propose a computational method,
based on a novel generative adversarial network (GAN) architecture, to
(1) produce synthetic B-mode US images and (2) their corresponding segmented bone surface masks in real-time. We show how a duality concept
can be implemented for such tasks. Armed by two convolutional blocks,
referred to as self-projection and self-attention blocks, our proposed GAN
model synthesizes realistic B-mode bone US image and segmented bone
masks. Quantitative and qualitative evaluation studies are performed on
1235 scans collected from 27 subjects using two different US machines
to show comparison results of our model against state-of-the-art GANs
for the task of bone surface segmentation using U-net.

This work was supported in part by 2017 North American Spine Society Young Investigator grant.

A. L. Martel et al. (Eds.): MICCAI 2020, LNCS 12266, pp. 795–804, 2020.
https://doi.org/10.1007/978-3-030-59725-2_77

Keywords: Orthopedic surgery · Segmentation · Ultrasound · Bone · Generative adversarial network · Deep learning

1   Introduction

Segmentation of bone surfaces from intra-operative US data is an important step for US-guided CAOS procedures. Due to the success of deep learning methods in medical image analysis, recent research has focused on the use of convolutional neural networks (CNNs) for accurate, robust, and real-time segmentation of bone surfaces [1,14]. However, scarcity of data size, due to a lack of standardized data and patient privacy concerns, is a major challenge in applying deep learning methods in the medical imaging field. This is specifically a challenge due to the fact that US is not a standard imaging modality in CAOS and US-guided CAOS procedures are not common. Another limiting factor is the manual data collection procedure: sub-optimal orientation of the US transducer with respect to the imaged bone anatomy will result in the acquisition of low quality bone scans [4].

Increasing the size of existing datasets through data augmentation in order to improve models' performance is extensively investigated by various researchers [9]. Earlier work has focused on the introduction of hand crafted image transformations such as random rotations, translations, nonlinear deformations. However, such augmentation methods are limited in their ability to mimic real variations and are highly sensitive to the parameter choice [17]. While transfer learning methods [11], that first train on large datasets then fine-tune on smaller datasets achieve state-of-the-art results on natural image datasets, these methods often do not suit medical image data and offer relatively little benefit to performance [11]. This is especially very problematic for bone US data since its very limited compared to larger medical data such as chest X-ray images. This gap in performance is due to the difference between medical images' features and natural images' features. Furthermore, medical images are often 3D, and there is no streamlined way to transfer 2D feature knowledge into 3D feature knowledge. One approach to overcome this problem is by using unsupervised feature extractors that have only been trained on medical images, however, this requires the target network architecture to be similar to the feature extractors' source architecture, which is uncommon. Image generation methods have recently become a popular solution for the challenge of creating large amounts of training data for deep learning [13]. Generative Adversarial Networks (GANs) have been used in diverse contexts such as unsupervised representation learning [10], image-to-image translation [5] and unsupervised domain adaptation of multi-modal medical imaging data [6]. This groundwork of successful research demonstrates GANs' potential for augmenting small datasets of medical images.

In this work, we propose a computational method, based on a GAN architecture specifically designed to (1) produce synthetic B-mode bone US images and (2) generate their corresponding segmented bone surfaces which can be used as labels. Based on [8,16], we show that a duality concept can be adopted for such tasks when implemented by two convolutional blocks, referred to as self-projection and self-attention blocks. We have conducted quantitative and qualitative evaluation studies on 1235 scans collected from 27 subjects using two different US machines. Furthermore, we show comparison results of our model against state-of-the-art GANs presented in [5,10] for the task of generating B-mode bone US images. We also evaluate bone surface segmentation accuracy using synthesized B-mode bone US images generated by the networks investigated when tested on Ronneberger's et al. [12] U-net architecture. Our work is the first report for generating simultaneous B-mode bone US data and corresponding segmentation labels which we believe to be a novel contribution in the field of US-guided CAOS.

2   Proposed Method

2.1   Network Architecture

Our architecture is based on the common GAN layout utilizing two co-existing neural networks; a generator G that generate synthetic samples and a discriminator D which attempts to discriminate between these generated synthetic samples and real ones [3]. The generator network transforms some pure random noise vectors z (typically a Gaussian) sampled from a prior distribution $p_z(z)$ into new samples such that $\mathbf{x} = G(\mathbf{z})$. The generated image x_g is expected to resemble the real images x_r. On the other hand, the discriminator D has both: (1) real samples with distribution $p_r(x)$ as well as (2) generated samples with distribution $p_g(x)$ and its output $y_s = D(\mathbf{x})$. The gradient information is back-propagated from the discriminator to the generator and hence, the generator optimizes its parameters to generate better images. Gradient-based methods have been proposed to train such a GAN as saddle point optimization problem. However, an imbalance between the training of the generator and the discriminator might occur if the Jensen–Shannon (JS) divergence was used [15] and the discriminator will more likely be too strong, which makes the generator weakly-trained. Moreover, the problem of mode collapse would arise when the distribution $p_g(x)$ learned by the generator was based on limited modes of the real samples distribution $p_r(x)$. This results in weak and limited generations of images. The training of our proposed GAN follows the typical optimization problem such that the discriminator D is trying to maximize and the generator G is trying to minimize the following objective function $\mathcal{L}(D, G)$:

$$\min_{G} \max_{D} \mathcal{L}(D, G) = \underset{x_r \sim p_{r(x)}}{E} \left[\log D(x, y)\right] + \underset{z \sim p_{z(z)}}{E} \left[\log(1 - D(x, z))\right];$$

In our generator architecture design the encoder maps the input image into a low-dimensional latent space, and the decoder maps the latent representation into the original space. It is trained to generate both US images and their corresponding segmentation images. We adopt the duality concept presented by [8] with our generator G and discriminator D both incorporating dual information into account. Therefore, our proposed GAN architecture generates segmentation masks/label in addition to the synthesized B-mode US images. This is achieved by modifying the GAN architecture to use two-channel images. In vivo real B-mode US data was assigned to the first channel and expert bone segmentation was assigned to the second channel. Based on [7,16], we also employ a self-projection and self-attention blocks into the GAN model as shown in Fig. 1. Our input is processed through convolutional blocks, with each block consisting of several convolutional layers. Our projection blocks, denoted as P, we add a 1×1 convolution to the projected input that is fed-forward through a 1×1 convolution, a 3×3 convolution, and another 1×1 convolution with each convolution operation followed by batch normalization and rectified linear unit (ReLU) activation. We also use a stride of 2 convolutions to upsample the feature maps. On the other hand, our self-attention block, denoted as A, consists of a 1×1 convolution (followed by batch normalization and Leaky ReLU activation) that is (1) multiplied by a transposed 1×1 convoluted replica resulting in an attention map and (2) multiplied by the attention map to generate self-attention feature maps. The self-attention approach helps modeling wider range image regions. With self-attention features, the generator can associate fine details at every location and associate them with similar portions of the image. In addition, the discriminator can now enforce complicated geometric constraints relative to the overall image [16]. The architecture of the generator can be summarized as:

- **encoder:** $A32$ $P32$ - $A64$ $P64$ - $A128$ $P128$ - $A256$ $P256$ - $A512$ $P512$
- **decoder:** $A512$ $P512$ - $A256$ $P256$ - $A128$ $P128$ - $A64$ $P64$ - $A32$ $P32$

In our discriminator model a two-input $N \times N$ PatchGAN-like discriminator [5] was used to classify $N \times N$ patches of the input image as real or synthetic. Our discriminator architecture consists of five convolutional blocks, with a final convolution is applied to the last layer to map the 1-dimensional output before applying a Sigmoid function. Batch normalization operations were followed by 0.2-slope leaky ReLU. An Adam solver with a 0.0002 learning rate was used and the structure of the discriminator can be expressed as follows:

- **discriminator:** $A32$ $P32$ - $A64$ $P64$ - $A128$ $P128$ - $A256$ $P256$ - $A512$ $P512$

3   Experimental Results

3.1   Data Acquisition

To conduct our experiments that particularly target the problem of data limitation in the US-guided CAOS field, we have collected 1235 in vivo B-mode US images categorized into four groups of bone structures: radius, femur, spine and tibia. Data were collected upon obtaining the approval of the institutional review board (IRB). Depth settings and image resolutions varied between 3–8 cm, and 0.12–0.19 mm, respectively. All the collected scans were scaled to a standardized size of 256×256 and manually segmented by an expert ultrasonographer. Two imaging devices were used to collect data:

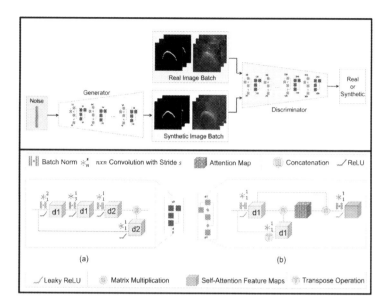

Fig. 1. Top: an overview of our proposed GAN architecture with its self-projection and attention blocks based generator and patchGAN-like discriminator. Bottom: our proposed (a) self-projection block and (b) self-attention block.

1. Sonix-Touch US machine (Analogic Corporation, Peabody, MA, USA) with a 2D C5-2/60 curvilinear probe and L14-5 linear probe. Using this device we have collected 1000 scans from 23 subjects. 400 scans from the Sonix touch, using random split, were used for training the GANS, 300 scans were used for training the U-net, and 300 scans were used for testing. We repeated this process 3 times and during random split same patient data was not included in the training and testing data.
2. Clarius C3 hand-held wireless ultrasound probe (Clarius Mobile Health Corporation, BC, Canada). Using this device we have collected 235 scans from 4 subjects. All Clarius data was used for testing.

   We conducted our experiments using the Keras framework and Tensorflow as backend with an Intel Xeon CPU at 3.00 GHz and an Nvidia Titan-X GPU with 8 GB of memory. Our GAN converged in about 2 h during the training process. Testing on average took 35 ms. For our experiments, the proposed network and those presented in [5, 10] were implemented as per the recommendations by their respective authors. For consistency, we used an Adam solver with learning rate of 0.0002, an exponential decay rate for the first and second moment estimates of $\beta_1 = 0.5$ and $\beta_2 = 0.999$, with a mini-batch SGD for all models considered.

3.2 Quantitative Results

Quantitative evaluation of our proposed GAN architecture was performed against three methods [2,5,10]. In order to show that the synthesized US images are useful for improving the performance of a supervised segmentation network we use the well known U-net architecture described in [12]. We would like to mention that the U-net architecture used in this work is not the main contribution of this work, since the synthesized images can be used in conjunction with other CNN-based network architectures [1,14]. If a GAN architecture captures the target distribution correctly it should generate a new set of training images (synthesized images) that should be indistinguishable from the in vivo real B-mode US data. Therefore, a U-net trained on either of these datasets, assuming they have the same size, should produce similar results. To evaluate this we have performed the following studies: (1) train U-net using limited in vivo real B-mode US data and test using in vivo real B-mode US data, (2) train U-net using limited in vivo real B-mode US data together with synthesized B-mode US data and test using in vivo real B-mode US data, (3) train U-net using synthesized B-mode US data and test U-net using in vivo real B-mode US data, (4) train U-net using real in vivo B-mode US data and test U-net using synthesized B-mode US data. Bone segmentation results are evaluated by calculating Dice, Rand error (Rand), the structural similarity index (SSIM), Hamming Loss, intersection over union (IoU) and average Euclidean distance (AED) [1].

Table 1 shows the performance of bone surface segmentation when U-net [12] is trained on various combinations of in vivo real B-mode US data and synthesized B-mode US data. We observe that adding synthesized images to the real in vivo B-mode US images improves the accuracy over the corresponding real-only counterpart. Overall our method outperforms previous state-of-the-art GAN architectures. In particular, it achieves 7%/7% and 5%/4% improvement for both data sets (Sonix/Clarius), in IoU value, over the GAN architectures proposed in [5,10] respectively. A paired t-test, for IoU, Dice and AED results at a %5 significance level, between our proposed network and the networks in [5,10] achieved p-values less than 0.05 indicating that the improvements of our method are statistically significant. Quantitative results presented in Tables 2 and 3 show that our proposed GAN architecture captures the target distribution better compared to the methods in [5,10] achieving improved results for IoU, Dice, Rand and AED evaluation metrics. Results in Table 2 were obtained when U-net [12] was trained using 600 synthetic B-mode US data generated using the proposed and two other architectures [5,10]. Testing was performed using 300 in vivo real B-mode US data obtained from Sonix Touch and 235 in vivo real B-mode US data obtained from Clarius probe. In Table 3 results were obtained when U-net [12] was trained using 535 in vivo real B-mode US data obtained from SonixTouch and Clarius probe. Testing was performed using 600 synthetic B-mode US data generated using the proposed method and two other GAN architectures [5,10].

Table 1. Quantitative results for bone surface segmentation using U-net [12]. Testing was done using 300 in vivo real B-mode US data obtained from Sonix Touch for Dataset I. For Dataset II testing was performed using all the 235 scans collected from Clarius C3 US probe. Notation note: number of in vivo real B-mode US images/number of synthetic B-mode US images used for training- GAN method used.

Method	IoU%	Dice	Rand	SSIM	Hamming	AED
Dataset I - Sonix-Touch US						
300/000 - N/A	0.7703	0.8642	0.9264	0.1106	0.2280	0.9386
300/300 - Radford et al. [10]	0.8391	0.9036	0.8522	0.3588	0.1608	0.8146
300/300 - Isola et al. [5]	0.8516	0.9117	0.8477	0.5401	0.1483	0.5687
300/300 - Ours	**0.8977**	**0.9400**	**0.7899**	**0.7038**	**0.1022**	**0.2985**
300/600 - Radford et al. [10]	0.8621	0.9183	0.7084	0.5540	0.3826	0.1378
300/600 - Arjovsky et al. [2]	0.8827	0.9255	0.6876	0.6038	0.1244	0.3220
300/600 - Isola et al. [5]	0.8943	0.9395	0.6657	0.7021	0.1035	0.2896
300/600 - Ours	**0.9309**	**0.9580**	**0.6125**	**0.7586**	**0.0690**	**0.1596**
Dataset II - Clarius C3 US						
300/000 - N/A	0.7594	0.8564	0.9350	0.1086	0.2405	0.7821
300/300 - Radford et al. [10]	0.8128	0.8869	0.8678	0.2750	0.1871	0.7536
300/300 - Isola et al. [5]	0.8322	0.9126	0.8463	0.3483	0.1593	0.8211
300/300 - Ours	**0.8753**	**0.9193**	**0.8381**	**0.5861**	**0.1278**	**0.1970**
300/600 - Radford et al. [10]	0.8458	0.9128	0.8483	0.4822	0.1486	0.6217
300/600 - Arjovsky et al. [2]	0.8531	0.9196	0.8104	0.5480	0.1311	0.4853
300/600 - Isola et al. [5]	0.8646	0.9214	0.7903	0.5728	0.1275	0.3482
300/600 - Ours	**0.9225**	**0.9536**	**0.7636**	**0.7408**	**0.0774**	**0.1583**

Table 2. Quantitative results for bone surface segmentation. Results were obtained when U-net [12] was trained using 600 synthetic B-mode US data generated using the proposed method and [5,10]. Testing was performed using 300 in vivo real B-mode US data (Sonix Touch) and 235 in vivo real B-mode US data (Clarius probe). Notation note: method used-blocks type.

Method	IoU%	Dice	Rand	Hamming	AED
Radford et al. [10]	0.8471	0.9158	0.8483	0.1783	0.7133
Isola et al. [5]	0.8625	0.9115	0.8284	0.1183	0.4540
Ours-none	0.6952	0.8068	0.9845	0.1967	0.9347
Ours-self-projection only	0.8356	0.9023	0.8615	0.1883	0.8053
Ours-self-attention only	0.8502	0.9104	0.8816	0.1668	0.5063
Ours-self-projection & self-attention	**0.9054**	**0.9766**	**0.8169**	**0.1208**	**0.1852**

Table 3. Quantitative results for bone surface segmentation. Results were obtained when U-net [12] was trained using 535 in vivo real B-mode US data obtained from Sonix Touch and Clarius probe. Testing was performed using 600 synthetic B-mode US data generated using the proposed method and two other GAN architectures [5,10]. Notation note: number of synthetic B-mode images used for testing - method used.

Method	IoU%	Dice	Rand	Hamming	AED
600-B-mode-Radford et al. [10]	0.8726	0.9158	0.8464	0.1405	0.4610
600-B-mode-Isola et al. [5]	0.8933	0.9304	0.7629	0.1108	0.2814
600-B-mode-Ours	**0.9357**	**0.9640**	**0.7195**	**0.0496**	**0.1952**

3.3   Qualitative Results

Qualitative results of our proposed GAN model are shown in Fig. 2. In each row of Fig. 2, we demonstrate one example of in vivo real B-mode US image (four examples in total). Columns are labeled alphabetically where we show in (a)-right: real in vivo B-mode US images and in (a)-left: their corresponding bone surface segmentations obtained by an expert. Figure 2 columns (b) through (d) demonstrate synthetic B-mode US images (right) and their corresponding synthetic bone surface segmentations as generated by [5,10] and our proposed model, respectively. Investigating the results we can infer that our proposed method results in fewer artifacts compared to the state-of-the-art [5,10].

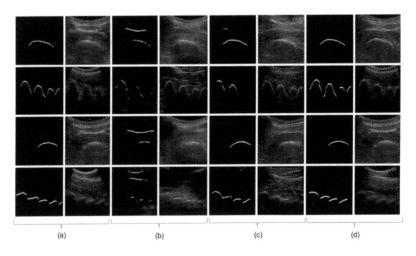

Fig. 2. Four examples of B-mode US images and their corresponding bone segmentation mask images are displayed in four rows. Columns are labeled alphabetically where we show in (a)-right: real in vivo B-mode US images and in (a)-left: their corresponding bone surface segmentations mask as obtained by an expert. Columns (b) through (d) demonstrate synthetic B-mode US images (right) and their corresponding synthetic bone surface segmentations as generated by [5,10] and our proposed model.

4   Discussion and Conclusion

In this paper, a novel GAN model for real-time and accurate B-mode bone US image generation is proposed. Our model has been implemented using two main components: (1) a generator that produces synthesized B-mode US as well as bone surface images and (2) a PatchGAN-like discriminator [5] that was used to classify $N \times N$ patches of the input images as real or synthetic. We have employed two integral components of building the generator and discriminator: a self-projection and self-attention blocks. With self-attention features the generator can associate fine details at every location and associate them with similar portions of the image. The main benefit of our self-attention blocks is that they leverage complementary features in distant portions of the image rather than local regions of xed shape especially for images with complex structural patterns, e.g. US B-mode images. The relationship between near and far pixels is learned, which allows the model to focus on separated structurally relevant features. Since the task is to replicate the relationship between the US B-mode and segmentation images, our model's ability to span a larger region in the image to create features gives it an advantage over the classic GAN model, which is limited by its filter size. In a classic GAN model, the relationship between the segment and US features is likely to be diluted across local features, while in a self-attention model the relationship is preserved by these larger feature regions. Additionally, the self-attention discriminator used checks for consistency in features in distant areas, which enforces accurate reproduction of geometric patterns in the B-mode US images and leads to higher-quality augmented data. On the other hand, self-projection blocks allow semantic information to be more efficiently passed forward in the network while progressively increasing feature map sizes, compared to simple convolutions. They allow us to have more comprehensive feature maps. Furthermore, self-projection blocks are also convolutional blocks, and therefore are computationally less expensive to train and infer on. To the best of our knowledge, this was not previously investigated for generating B-mode bone US images. Based on the quantitative results presented, we can conclude that having a self-attention mechanism can significantly improve the results for the image synthesis task at hand. Our future work will involve more extensive clinical validation of the proposed GAN model.

References

1. Alsinan, A.Z., Patel, V.M., Hacihaliloglu, I.: Automatic segmentation of bone surfaces from ultrasound using a filter layer guided CNN. Int. J. Comput. Assist. Radiol. Surg. **14**(5), 775–783 (2019)
2. Arjovsky, M., Chintala, S., Bottou, L.: Wasserstein generative adversarial networks. In: Precup, D., Teh, Y.W. (eds.) Proceedings of the 34th International Conference on Machine Learning. Proceedings of Machine Learning Research, PMLR, International Convention Centre, Sydney, Australia, 06–11 August 2017, vol. 70, pp. 214–223 (2017)
3. Goodfellow, I., et al.: Generative adversarial nets. Advances in Neural Information Processing Systems, pp. 2672–2680 (2014)

4. Hacihaliloglu, I., Guy, P., Hodgson, A.J., Abugharbieh, R.: Volume-specific parameter optimization of 3D local phase features for improved extraction of bone surfaces in ultrasound. Int. J. Med. Robot. Comput. Assist. Surg. **10**(4), 461–473 (2014)

5. Isola, P., Zhu, J.Y., Zhou, T., Efros, A.A.: Image-to-image translation with conditional adversarial networks. In: 2017 IEEE Conference on Computer Vision and Pattern Recognition (CVPR), pp. 5967–5976. IEEE (2017)

6. Kamnitsas, K., et al.: Unsupervised domain adaptation in brain lesion segmentation with adversarial networks. In: Niethammer, M., et al. (eds.) IPMI 2017. LNCS, vol. 10265, pp. 597–609. Springer, Cham (2016). https://doi.org/10.1007/978-3-319-59050-9_47

7. Laina, I., Rupprecht, C., Belagiannis, V., Tombari, F., Navab, N.: Deeper depth prediction with fully convolutional residual networks. In: 2016 Fourth International Conference on 3D Vision (3DV), pp. 239–248. IEEE (2016)

8. Neff, T., Payer, C., Štern, D., Urschler, M.: Generative adversarial networks to synthetically augment data for deep learning based image segmentation, May 2018 (2018). https://doi.org/10.3217/978-3-85125-603-1-07

9. Payer, C., Štern, D., Bischof, H., Urschler, M.: Regressing heatmaps for multiple landmark localization using CNNs. In: Ourselin, S., Joskowicz, L., Sabuncu, M.R., Unal, G., Wells, W. (eds.) MICCAI 2016. LNCS, vol. 9901, pp. 230–238. Springer, Cham (2016). https://doi.org/10.1007/978-3-319-46723-8_27

10. Radford, A., Metz, L., Chintala, S.: Unsupervised representation learning with deep convolutional generative adversarial networks. In: 4th International Conference on Learning Representations, ICLR 2016, San Juan, Puerto Rico, 2–4 May 2016, Conference Track Proceedings (2016). http://arxiv.org/abs/1511.06434

11. Raghu, M., Zhang, C., Kleinberg, J., Bengio, S.: Transfusion: understanding transfer learning for medical imaging. In: Advances in Neural Information Processing Systems, pp. 3342–3352 (2019)

12. Ronneberger, O., Fischer, P., Brox, T.: U-Net: convolutional networks for biomedical image segmentation. In: Navab, N., Hornegger, J., Wells, W.M., Frangi, A.F. (eds.) MICCAI 2015. LNCS, vol. 9351, pp. 234–241. Springer, Cham (2015). https://doi.org/10.1007/978-3-319-24574-4_28

13. Shin, H.-C., et al.: Medical image synthesis for data augmentation and anonymization using generative adversarial networks. In: Gooya, A., Goksel, O., Oguz, I., Burgos, N. (eds.) SASHIMI 2018. LNCS, vol. 11037, pp. 1–11. Springer, Cham (2018). https://doi.org/10.1007/978-3-030-00536-8_1

14. Villa, M., Dardenne, G., Nasan, M., Letissier, H., Hamitouche, C., Stindel, E.: FCN-based approach for the automatic segmentation of bone surfaces in ultrasound images. Int. J. Comput. Assist. Radiol. Surg. **13**(11), 1707–1716 (2018)

15. Yadav, A.K., Shah, S., Xu, Z., Jacobs, D.W., Goldstein, T.: Stabilizing adversarial nets with prediction methods. In: 6th International Conference on Learning Representations, ICLR 2018, Vancouver, BC, Canada, 30 April–3 May 2018, Conference Track Proceedings. OpenReview.net (2018). https://openreview.net/forum?id=Skj8Kag0Z

16. Zhang, H., Goodfellow, I., Metaxas, D., Odena, A.: Self-attention generative adversarial networks (2018)

17. Zhao, A., Balakrishnan, G., Durand, F., Guttag, J.V., Dalca, A.V.: Data augmentation using learned transformations for one-shot medical image segmentation. In: Proceedings of the IEEE Conference on Computer Vision and Pattern Recognition, pp. 8543–8553 (2019)

Robust Bone Shadow Segmentation from 2D Ultrasound Through Task Decomposition

Puyang Wang[1(✉)], Michael Vives[2], Vishal M. Patel[1], and Ilker Hacihaliloglu[3]

[1] Department of Electrical and Computer Engineering, Johns Hopkins University,
Baltimore, MD, USA
pwang47@jhu.edu
[2] Department of Orthopedics, Rutgers New Jersey Medical School,
New Brunswick, NJ, USA
[3] Department of Biomedical Engineering, Rutgers University, Piscataway, USA

Abstract. Acoustic bone shadow information in ultrasound (US) is important during imaging bones in US-guided orthopedic procedures. In this work, an end to end deep learning-based method is proposed to segment the bone shadow region from US data. In particular, we decompose the bone shadow segmentation task into two subtasks, coarse bone shadow enhancement (BSE) and horizontal bone interval mask (HBIM) estimation. Outputs from two subtasks are processed by a masking operation to generate the final bone shadow segmentation. To better leverage the mutual information in different tasks, our model features a shared encoder as deep feature extractor for both subtasks and two multi-scale pyramid pooling decoders. Additionally, we propose a conditional shape discriminator to regularize the shape of the output segmentation map. The proposed method is validated on 814 in vivo US scans obtained from knee, femur, distal radius and tibia bones. Validation against expert annotation achieved statistically significant improvements in segmentation of bone shadow regions compared to the state-of-the-art method.

1 Introduction

In order to provide a safe alternative to intra-operative fluoroscopy, ultrasound (US)has been investigated as an alternative intra-operative imaging modality in various orthopedic procedures [4]. US provides real-time, safe, and 2D/3D imaging. However, low signal-to-noise (SNR) ratio, limited field of view, and various imaging artifacts have hindered the wide spread use of US in computer assisted orthopedic surgery (CAOS) applications. Furthermore, regions corresponding to bone boundaries appear several millimeters (mm) in thickness due to the width of the US beam further complicating the interpretation of the collected US data. In order to alleviate some of these difficulties, various groups have proposed bone segmentation or enhancement methods [4].

In the context of bone imaging, using US, bone boundaries have the highest intensity in the image followed by a region with low intensity values denoted

© Springer Nature Switzerland AG 2020
A. L. Martel et al. (Eds.): MICCAI 2020, LNCS 12266, pp. 805–814, 2020.
https://doi.org/10.1007/978-3-030-59725-2_78

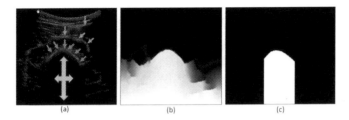

Fig. 1. (a) B-mode US image of in vivo femur. Thick yellow arrows point to the bone shadow region. Red arrows point to the bone surface response. Green arrows point to soft tissue interface resembling bone response. (b) Bone shadow enhanced image obtained using [3]. (c) Gold standard bone shadow obtained by expert manual segmentation. In both (a) and (c) regions corresponding to soft tissue are displayed with black color coding, regions corresponding to bone shadow are displayed with gray/white color coding. (Color figure online)

as the shadow region. Shadow region is the result of a high acoustic impedance mismatch between the soft tissue and the bone boundary resulting in most of the US signal being reflected back to the transducer surface. In order to improve the accuracy and robustness of bone segmentation, several groups have incorporated bone shadow information into their framework [4]. Bone shadow information can also be used in order to guide the orthopedic surgeon to a standardized diagnostic viewing plane with minimal artifacts. Most recently, bone shadow information was also incoporated into deep learning-based bone segmentation methods [9,10]. In [10], the authors have proposed a simultaneous bone enhancement, classification and segmentation framework based on deep learning. The bone enhancement stage [10] uses bone shadow image features extracted using the method proposed in [3]. The bone shadow enhancement method, proposed in [3], is based on the construction of a signal transmission map from the local phase bone image features. Although the method improves general appearance of the bone shadow region, it produces suboptimal bone shadow enhancement results and can not run in real-time (Fig. 1).

In this work, our goal is to improve the bone shadow segmentation by proposing a deep learning-based method which yields better performance over other methods. The motivation and contribution of the proposed method are as follows:

- Because of US imaging principle and anatomy of bone structures, bone shadows share some common shape profiles. In Fig. 1(c), the gold standard bone shadow will ideally have sharp horizontal cut-off for non-bone area and certain bone surfaces on top. Thus we propose an adversarial network to implicitly impose the shape regularization.
- Expert manual annotations of medical images are expensive and time consuming. We leverage the bone shadow image features extracted using the method proposed in [3] and use it as surrogate ground truth to not only

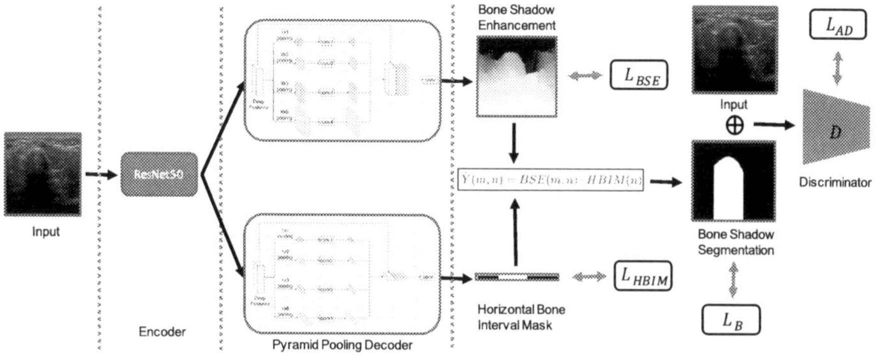

Fig. 2. An overview of the proposed multi-task learning-based method for bone shadow segmentation from US images.

provide additional supervision on intermediate results, but also enable the semi-supervised learning for US bone shadow segmentation.

- By using only left and right boundary of bone, one can create a horizontal bone interval mask and apply it on bone shadow enhanced image, Fig. 1(b), to output bone shadow segmentation results that are close to ground truth. We propose a subnetwork that estimate the bone regions horizontally by only learning from manually annotated bone landmarks which has lower annotation cost than full segmentation. This could lead to larger scale dataset for training.

2 Proposed Method

In the proposed method, two subnetworks are first trained separately to produce a coarse bone shadow enhancement (BSE) and horizontal bone interval mask (HBIM). After obtaining both coarse BSE and HBIM, a masking operation is used to generate the final bone shadow. As a result, we provide a joint trainable end-to-end deep learning model for robust bone shadow segmentation. The proposed CNN model consists of one shared encoder and two independent multi-scale decoders for coarse BSE and HBIM estimation. To further regularize the shape of the output bone shadow, we introduce a conditional shape discriminator which can guide the training of bone shadow segmentation network by adding the adversarial loss on the shape information. Figure 2 provides an overview of our framework.

2.1 Conditional Shape Discriminator

Unlike other semantic segmentation tasks, bone shadow segmentation is different in many ways. One major difference of the output segmentation map is the

general shape. The type of bones (knee, fibia, femur, etc.), view planes (longitudinal and transverse), and most importantly the orientation of the US transducer
with respect to the imaged bone anatomy would affect the bone shadow shape
individually.

To ensure specific shape on the estimated bone shadows by a CNN, a conditional shape discriminator D is added in the training stage and designed following
a conditional Generative Adversarial Network (cGAN) framework [6]. It takes
both the input image and its corresponding bone shadow segmentation (segmentation from proposed network or ground truth) to identify if the segmentation is
ground truth on the basis of binary images. From the perspective of segmentation
network, it regularizes N output $\hat{Y}$ using the binary cross entropy loss:

$$L_{AD} = -\frac{1}{N} \sum_{i=1}^{N} [\log(1 - D(X_i, \hat{Y}_i)) + \log(D(X_i, Y_i))], \tag{1}$$

where X_i is input image and Y_i is the corresponding ground truth. Because,
for binary segmentation task, the output segmentation is binary which varies in
different shapes, this adversarial loss can effectively enforce the output segmentation map to follow a reasonable shape even with different types of bones and
view planes.

2.2   Coarse Bone Shadow Enhancement

One of the main challenges in deep learning-based medical image analysis is the
generalization ability of the trained model due to the lack of large amounts of
manually annotated data. However, recent studies have shown that, by training
the model through semi supervised learning on automatic annotated or weakly
labelled data, the model gains better generalization ability and improves the
overall performance even for different imaging modalities [2].

In this work, we propose to use Bone Shadow Enhancement (BSE) method,
proposed in [3], to filter the US image and generate a coarse estimation of the
bone shadow regions. BSE image signal at position (x, y) is computed by modeling the interaction of the US signal within the tissue as scattering and attenuation information using:

$$BSE(x,y) = [(CM_{LP}(x,y) - \rho)/[max(US_A(x,y), \epsilon)]^{\delta}] + \rho, \tag{2}$$

where $CM_{LP}(x, y)$ is the confidence map image obtained by modeling the propagation of US signal inside the tissue taking into account bone features present
in local phase bone image $LP(x, y)$ [3]. $US_A(x, y)$ maximizes the visibility of
high intensity bone features inside a local region and satisfies the constraint
that the mean intensity of the local region is less than the echogenicity of the
tissue confining the bone [3]. Tissue attenuation coefficient is represented by δ.
ρ is a constant related to tissue echogenicity confining the bone surface, and ϵ is
a small constant used to avoid the division by zero [3].

2.3 Horizontal Bone Interval Mask

As shown in Fig. 1(b) and (c), the previously defined BSE image can be regarded as a coarse estimation of bone shadow regions. While the sharp boundary of the bone surface is usually well preserved, it can also have high confidence shadows leaking into non-bone regions horizontally. To solve this shadow leakage problem, image processing technique that can remove shadows corresponding to non-bone structure while keeping the bone shadow needs to be applied on the BSE image. From the observation that the shadow leakage usually happens below the bone surface and expands horizontally, a Horizontal Bone Interval Mask (HBIM) is proposed to mask out the non-bone shadows. Given a US image $X(m, n)$ of size $N \times M$, its corresponding BSE image $BSE(m, n)$ and the manually segmented bone shadow $Y(m, n)$, HBIM is defined as follows:

$$HBIM(n) = \begin{cases} 1, & \text{if } \exists\, m,\ Y(m, n) > 0 \\ 0, & \text{otherwise.} \end{cases} \tag{3}$$

HBIM can be seen as a vector in which 1 indicates the presence of bone surface along corresponding vertical line in US image. Thus we can derive the final fine bone shadow segmentation $\hat{Y}$ using HBIM as follows,

$$\hat{Y}(m, n) = BSE(m, n) \cdot HBIM(n). \tag{4}$$

As a result, one is able to calculate a high quality bone shadow segmentation using only the input US image and the horizontal location information of the bone in the US image. Moreover, as will be shown later, this leads to a much more robust and predictable bone shadow segmentation than a simple end-to-end training scheme.

2.4 Network Structure

The proposed framework features three tasks: (1) Coarse BSE estimation, (2) HBIM estimation, and (3) final bone shadow segmentation. Noticeably, with three different tasks, our proposed framework is a multi-task learning (MTL) model.

We view the first two tasks as intermediate tasks that are highly correlated with the final task. In the proposed method, we use a ResNet50 [5] pretrained on ImageNet [1] as the shared encoder to take the advantage of very deep neural network. The first convolutional layer is modified to take a single channel input. While the ResNet50 encoder is shared across all tasks for deep feature extraction, each of the intermediate tasks has its own decoder. As noted in U-Net [8], the key part of precise pixel-wise prediction for biomedical image segmentation task is to make good use of the multi-scale features. In our network, we adopt the decoder that was first proposed in [11]. For HBIM estimation, the desired output is a one-dimensional row vector. In order to achieve that, we changed the pyramid pooling to $(1, 1), (1, 2), (1, 3), (1, 6)$ and add another average pooling layer between the input deep feature and concatenation to align the feature size.

Finally, we complete the final bone shadow segmentation by using the estimated HBIM and BSE of previous two tasks following Eq. 4. Given the proposed MTL model for these three tasks, it turns out helpful to extract comprehensive image features by sharing a shared encoder and then branching out for task-specific losses for each task. To further maximize the synergy across all the tasks, we propose a combined loss function containing four task-specific losses: $L = L_{BSE} + L_{HBIM} + L_B + \lambda L_{AD}$, where L_{HBIM} and L_B are binary cross entropy loss of estimated HBIM and bone shadow, and L_{BSE} is the L_1 loss of estimated BSE. L_{AD} represents the adversarial loss (loss from the discriminator D) as defined in Eq. 1 with weight λ. As for the structure of the discriminator D, we follow the structure that was proposed in [7].

3   Dataset and Training

After obtaining the institutional review board (IRB) approval, a total of 814 different US images, from 20 healthy volunteers, were collected using SonixTouch US machine (Analogic Corporation, Peabody, MA, USA). The scanned anatomical bone surfaces include knee, femur, radius, and tibia. All bone shadows of the collected data were manually annotated by an expert ultrasonographer in the preprocessing stage. The BSE images were obtained using the filter parameters defined in [3] and the HBIMs were obtained using Eq. 3 with bone shadow annotations. The datasets were randomly separated on the subject level into training and testing sets by an 60%/40% split (573/241 in images level). Any subject with data included in the training set were excluded from the testing set. During preprocessing, the images were resampled into 0.15 mm isotropic resolution, and resized to 256×256.

The coarse BSE and HBIM estimation tasks are trained first with a batch size of 32 for 100 epochs in which only L_{BSE} and L_{HBIM} are used to train the network. The base network is optimized by the Adam optimizer with a learning rate of 10^{-4}. A joint training using all four losses is applied afterwards with a batch size of 32 for 50 epochs with $\lambda = 0.1$. During testing, the image can be forwarded though the network for all tasks by one shot. The experiments are performed on a Linux workstation equipped with an Intel 3.50 GHz CPU and a 12 GB NVidia Titan Xp GPU using the PyTorch framework. The average running time of our model for single testing image is around 0.03 s which makes real-time application possible.

4   Experimental Results

4.1   Bone Shadow Segmentation

We compare the performance of our method with that of the following four methods: Unet [8], PSPnet [11], PSPGAN and PSPnet-MTL. PSPGAN denotes the method that combines the proposed conditional shape discriminator in Sect. 2

and PSPnet. PSPnet-MTL is the multi-task version of PSPnet without conditional shape discriminator. The comparison between PSPGAN, PSPnet-MTL and PSPGAN-MTL is for the purpose of ablation study. For all the compared methods, parameters are set as suggested in their corresponding papers and trained using the same training dataset as used to train our network.

Table 1. Bone shadow segmentation and bone surface localization comparison of methods on various metrics. The proposed PSPGAN-MTL achieves statistically significant improvements using two-tailed t test with p values < 0.05.

	Bone shadow segmentation			Bone surface localization			
	Dice	mIoU(%)	pAcc.(%)	AED	Recall	Precision	F-score
U-net [8]	0.890 ± 0.068	80.97	87.86	2.11 ± 1.05	0.625	0.616	0.620
PSPnet [11]	0.911 ± 0.062	85.69	92.76	1.36 ± 1.41	0.730	0.825	0.775
PSPGAN	0.927 ±0.056	86.98	92.83	1.49 ±1.69	0.727	**0.826**	0.774
PSPnet-MTL	0.956 ± 0.052	92.08	96.47	0.25 ± 0.19	0.894	0.748	0.918
PSPGAN-MTL	**0.962 ± 0.046**	**92.97**	**96.63**	**0.19 ± 0.13**	**0.907**	0.775	**0.934**

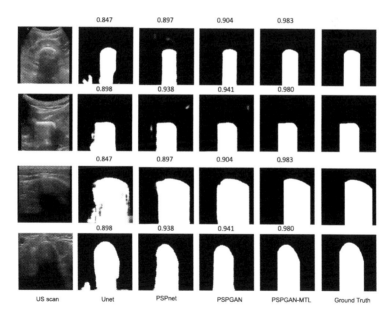

Fig. 3. Bone shadow segmentation results for in vivo tibia, distal radius, knee and femur. Dice coefficients computed against the ground truth are shown on top of the each result.

The Dice coefficient, mean Intersection over Union (mIoU) and pixel-wise accuracy (Acc.) are used to measure the segmentation performance of different methods. Average results of all test scans are shown in Table 1. As can be

Fig. 4. From left to right: in vivo US scan of spine, estimated BSE, estimated HBIM, PSPGAN-MTL, PSPGAN.

seen from this table, in all three metrics, our method provides the best performance compared to the other methods. Going directly from PSPGAN to PSPGAN-MTL provides implicit data augmentation and bone shadow prior for the tasks with limited data, thus results in a much more robust and accurate bone shadow segmentation. By adding proposed conditional shape discriminator, both PSPGAN and PSPGAN-MTL can outperform their counterparts, PSPnet, PSPnet-MTL. These experiments clearly show the significance of each component of proposed method, integrating coarse BSE estimation and HBIM for bone shadow segmentation and conditional shape discriminator.

Apart from the quantitative comparison of Dice, mean IoU and pixel accuracy, we also compared our method PSPGAN-MTL with others qualitatively by visual inspection. The segmentation results corresponding to different methods and the intermediate outputs of PSPGAN-MTL are shown in Fig. 3. The more shape alike PSPGAN result shows the effect of the proposed conditional shape discriminator comparing to PSPnet.

For the final experiment of bone shadow segmentation, we compare two methods: PSPGAN-MTL and PSPGAN, in term of their ability to correctly segment spine (multiple bones) which is not present in the dataset. From the results shown in Fig. 4, it is clear that with the help of the proposed multi-task bone shadow segmentation, PSPGAN-MTL suffers no mis-segmentation and provides a more complete segmentation compared with PSPGAN.

4.2   Bone Surface Localization

One main application of bone shadow segmentation is bone surface localization from bone shadow in which accurate and robust localization is important for the improved guidance in US-based CAOS procedures. In this experiment, we applied raycasting method to perform bone surface localization from bone shadows.

The Average Euclidean Distance (AED) results (mean+std) in Table 1 show that the proposed PSPGAN-MTL outperforms the other methods on test scans by a large margin. Note that the bone surface localization experiment was carried out using previous bone shadow segmentation results for all methods. Therefore, the networks are not trained specifically on the bone surface localization task. A further paired t-test between PSPGAN-MTL and PSPGAN at a 5%

significance level with p-value of 0.0009 clearly indicates that the improvements of our method are statistically significant.

5 Conclusion

In this paper, we proposed an end-to-end deep learning framework that enabled robust and accurate bone shadow segmentation for bone ultrasound examination. The main novelty lies in (1) the introduction of conditional shape discriminator to shape specific image segmentation problem, (2) the design of two subtasks, coarse bone shadow enhancement and horizontal bone interval mask to improve the performance of each task and (3) the integration of the highly-related homogeneous tasks into a single unified bone shadow segmentation network. Formulating the network with a single powerful encoder based on Resnet50 and two pyramid pooling decoders, the proposed network brings strong synergy across all tasks when extracting shared deep features. Future work will include more extensive validation and extension to 3D data for processing volumetric US scans.

Acknowledgment. This work was supported in part by 2017 North American Spine Society Young Investigator grant.

References

1. Deng, J., Dong, W., Socher, R., Li, L.J., Li, K., Fei-Fei, L.: ImageNet: a large-scale hierarchical image database. In: 2009 IEEE Conference on Computer Vision and Pattern Recognition, pp. 248–255. IEEE (2009)
2. Ganaye, P.-A., Sdika, M., Benoit-Cattin, H.: Semi-supervised learning for segmentation under semantic constraint. In: Frangi, A.F., Schnabel, J.A., Davatzikos, C., Alberola-López, C., Fichtinger, G. (eds.) MICCAI 2018. LNCS, vol. 11072, pp. 595–602. Springer, Cham (2018). https://doi.org/10.1007/978-3-030-00931-1_68
3. Hacihaliloglu, I.: Enhancement of bone shadow region using local phase-based ultrasound transmission maps. Int. J. Comput. Assist. Radiol. Surg. **12**(6), 951–960 (2017). https://doi.org/10.1007/s11548-017-1556-y
4. Hacihaliloglu, I.: Ultrasound imaging and segmentation of bone surfaces: a review. Technology **5**(02), 74–80 (2017)
5. He, K., Zhang, X., Ren, S., Sun, J.: Deep residual learning for image recognition. In: Proceedings of the IEEE Conference on Computer Vision and Pattern Recognition, pp. 770–778 (2016)
6. Mirza, M., Osindero, S.: Conditional generative adversarial nets. arXiv preprint arXiv:1411.1784 (2014)
7. Radford, A., Metz, L., Chintala, S.: Unsupervised representation learning with deep convolutional generative adversarial networks. arXiv preprint arXiv:1511.06434 (2015)
8. Ronneberger, O., Fischer, P., Brox, T.: U-Net: convolutional networks for biomedical image segmentation. In: Navab, N., Hornegger, J., Wells, W.M., Frangi, A.F. (eds.) MICCAI 2015. LNCS, vol. 9351, pp. 234–241. Springer, Cham (2015). https://doi.org/10.1007/978-3-319-24574-4_28

9. Villa, M., Dardenne, G., Nasan, M., Letissier, H., Hamitouche, C., Stindel, E.: FCN-based approach for the automatic segmentation of bone surfaces in ultrasound images. Int. J. Comput. Assist. Radiol. Surg. **13**(11), 1707–1716 (2018)

10. Wang, P., Patel, V.M., Hacihaliloglu, I.: Simultaneous segmentation and classification of bone surfaces from ultrasound using a multi-feature guided CNN. In: Frangi, A.F., Schnabel, J.A., Davatzikos, C., Alberola-López, C., Fichtinger, G. (eds.) MICCAI 2018. LNCS, vol. 11073, pp. 134–142. Springer, Cham (2018). https://doi.org/10.1007/978-3-030-00937-3_16

11. Zhao, H., Shi, J., Qi, X., Wang, X., Jia, J.: Pyramid scene parsing network. In: IEEE Conference on Computer Vision and Pattern Recognition (CVPR), pp. 2881–2890 (2017)

Correction to: The Case of Missed Cancers: Applying AI as a Radiologist's Safety Net

Michal Chorev⬤, Yoel Shoshan, Ayelet Akselrod-Ballin,
Adam Spiro, Shaked Naor, Alon Hazan, Vesna Barros,
Iuliana Weinstein, Esma Herzel, Varda Shalev, Michal Guindy,
and Michal Rosen-Zvi

Correction to:
Chapter "The Case of Missed Cancers: Applying AI
as a Radiologist's Safety Net" in: A. L. Martel et al. (Eds.):
Medical Image Computing and Computer Assisted Intervention
– MICCAI 2020, **LNCS 12266,**
https://doi.org/10.1007/978-3-030-59725-2_22

The original version of this chapter was revised. Dr. Ayelet Akselrod-Ballin con-tributed to the development of the conference paper and was therefore added to the list of coauthors.

The updated version of this chapter can be found at
https://doi.org/10.1007/978-3-030-59725-2_22

© Springer Nature Switzerland AG 2021
A. L. Martel et al. (Eds.): MICCAI 2020, LNCS 12266, p. C1, 2021.
https://doi.org/10.1007/978-3-030-59725-2_79

Author Index

Printed in the United States
By Bookmasters